谨以此书祝贺

陈公宁教授75寿辰！

陈公宁文集

解析函数插值与矩量问题

李仲来 / 主编

北京师范大学出版集团
BEIJING NORMAL UNIVERSITY PUBLISHING GROUP
北京师范大学出版社

2013 · 北京

图书在版编目(CIP)数据

解析函数插值与矩量问题——陈公宁文集/陈公宁著，李仲来主编．—北京：北京师范大学出版社，2013.12
（北京师范大学数学家文库）
ISBN 978-7-303-16762-3

Ⅰ.①解…　Ⅱ.①陈…　②李…　Ⅲ.①解析函数—文集②插值—文集③矩量问题—文集　Ⅳ.①O174.55-53②O174.42-53③O174.43-53

中国版本图书馆 CIP 数据核字（2013）第 172762 号

营销中心电话　010-58802181 58805532
北师大出版社高等教育分社网　http://gaojiao.bnup.com
电子信箱　gaojiao@bnupg.com

出版发行：北京师范大学出版社 www.bnup.com
北京新街口外大街 19 号
邮政编码：100875
印　刷：北京京师印务有限公司
经　销：全国新华书店
开　本：155 mm × 235 mm
印　张：34
插　页：4
字　数：523 千字
版　次：2013 年 12 月第 1 版
印　次：2013 年 12 月第 1 次印刷
定　价：85.00 元

策划编辑：岳昌庆　　责任编辑：岳昌庆
美术编辑：毛　佳　　装帧设计：毛　佳
责任校对：李　菡　　责任印制：孙文凯

▲ 1978年8月回长乐探亲时与父母及三位弟弟、弟妹家人合影。

▲ 1980年8月访美前夕与家人合影。左起：陈公宁、田树芸；后排：女儿陈仲。

▲ 1981年9月在UCSB的South Hall前留影。陈公宁在该楼5502室工作。

1985 年 10 月在西安交通大学全国泛函分析学术会议上做学术报告。

1991 年 11 月访问莫斯科大学时在其广场上留影。

1993 年 2 月在北京师范大学英东楼 215 室讨论班做报告。

▶ 1993 年 5 月 与樊畿先生（左）在北京友谊宾馆下榻处合影。

▶ 1994 年 5 月 摄于美国密执安大学音乐厅前。

▶ 1999 年 6 月与两位博士合影。左起：胡永建、陈公宁、赵斌。

2003 年 6 月主持张晓南（右侧立者）博士论文答辩会。

2010 年秋，与大哥陈公宪合影于扬州个园。

2012 年 7 月与女儿陈仲一家人合影。前排左起：郑欣、陈公宁、陈仲、胡毅；后排左起：胡冰玥、胡冰野。

序 1

本书为我(或与合作者)在 1974 年至 2012 年间发表的 80 篇数学论文的选集,共分 4 个部分,大体上反映了我学习矩阵与算子理论及其应用(特别在解析函数插值与矩量问题的应用)4 个阶段的过程与成果.这些是我们辛勤耕耘的收获,得之不易.重要的是,在这孤寂研究过程中,享受到难以言表的数学美感与精神上的愉悦——“人生如此自可乐,岂必局束为人鞿”.

1939 年十月初六(农历)我出生在福建省长乐县(后改市)城关.父亲为银行职员,也搞过工程技术,母亲为妇产科医生.在我童年印象中,父母对我们 5 位兄弟教育宽严有度,孩子们得以自由地发展.小学 4 年级一次全校性算术竞赛中我得冠军,颇有点成就感.初中 2 年级时喜欢上平面几何,时常与同学在沙地上演算题,彼此考问考答.(看来,几何直觉是增进对数学理解力,提高数学兴趣的很有效的途径.)我对数学的进一步兴趣发生在 1953 年考入福州一中之后,时任数学老师林景贤在省内颇有名气.我担任班上数学课代表,与林老师有较多的接触.至今,我还很佩服他的独特授课方式.(但总觉得那时中学数学内容较狭窄,与大学高等数学之间实难沟通.经历了半个世纪回头看来,这个问题似乎依然如故,且有变大的趋势.)

1956年高中毕业,起初想报考工科或医科,子承父母业,但思量再三,最后还是报考了北京师范大学数学系.从此与数学、与北京师范大学数学系结缘,立志将数学研究与教育作为自己的终身职业.1962年研究生班毕业,留校当助教.之后,曾在本系与物理系本科讲授微积分基础课程,还参加一个讨论班,但数学研究乏善可陈.1966年“文化大革命”开始,教学与研究全部停顿.政治运动,大串联,教育革命小分队下厂与劳动锻炼(去临汾干校,挖战备防空洞,烧砖等)接二连三.我从“三十不立”,流年似水,临近“四十而惑”,岁月蹉跎,心中很惶恐不安.等到“大学还是要办的”最高指示后,与在系的工农兵学员一块重新学习数学业务,开门办学,下乡下厂,顺便解决一些实际问题,包括在京郊昌平水利局的近半年的设计水库库区与坝体的工作.

10年“文化大革命”后,大学逐渐地恢复正常秩序.20世纪七八十年代国家开始改革开放,1980年9月我有幸受公派赴美国加州大学圣塔巴巴拉分校(UC,Santa Barbara)留学2年,并拜识该校著名数学家樊繼(Ky Fan)教授.可谓:“运命唯所遇”.樊师先帮助我研究矩阵理论,亲自批阅我的论文;第二年收我为弟子,指导我专攻Hilbert空间算子理论与非交换复分析.在他的提携下,我有了明确而适意的学术研究方向,并很快地取得一些基础性的论文成果.个中的辛苦与努力只有自己知晓.平日间樊师还在私人言谈中传授为学之道,从他身上我也学到许多做人做事的哲理,受益终身.1982年9月回国后5年期间,继续前两年的研究工作,也有一些成果.本书中前两部分的论文就是这两段时期的研究成果:第一部分是关于一般矩阵的;第二部分关于全纯映射与算子函数的.

1989年左右,我们开始转而重视当时很热门的特殊矩阵理论与应用.我们瞄准捷克布拉格与以色列学派的一系列工作,从中得到启示与动力.当时,最感兴趣的有:Bezout矩阵,Vandermonde矩阵,Loewner矩阵,Hankel矩阵,Toeplitz矩阵与Pick矩阵等;研究内容包括:这些矩阵的结构特征与性质,核,求逆,分解以及彼此之间的关联.这些研究及其块矩阵形式推广为我们20世纪90年代之后有理函数插值与解析函数插值以及矩量问题的研究打下扎实的根基.反过来,它们也对特殊矩阵类研究提供新的源泉.这个因果关系,我们是始料未及的,冥冥之中如有神助.

1992 年我们得到有理插值的完全解.用的是自创的所谓 Hankel 向量方法:将该插值求解转换为寻找相关 Hankel 向量的生成函数,这里,Loewner 与 Hankel 矩阵的一种关系起着关键作用.该项工作主要由赵斌与张惠品独立完成.此时,我正在莫斯科大学访问,全力研究奇异 Hankel 矩阵.1994 年在访问美国密执安大学的一个冬日黄昏,在系资料室里"鬼使神差"地省悟到我们先前的 Hankel 向量方法完全适用于 Navanlinna-Pick 插值的求解,并很有可能扩展到更广的矩阵值解析函数插值问题的情形.当下便收集有关方向的大量经典文献资料,并记下许多宝贵的书目.回国后,随即向研究小组报告了这个成果,并写成论文发表.此后近 20 年间,我们基本上围绕着矩阵值解析函数类的插值与相关的矩量问题,开展比较全面的探究,得到自创的 Hankel 块向量方法/Toeplitz 块向量方法在前述问题求解中的应用成果(李小青,胡永建分别完成基础工作部分).其中有些居国际上领先地位,为同行、专家肯定、引证与采用.我们的工作目标简而言之是:对某些区域内或边界上的给定插值数据,在特定的矩阵值解析函数类内寻求有解的条件,有唯一解的条件以及(有无穷多解时)解的参数化表达式.我们的技巧将多个重点情形等价地归纳为单个重点情形;前者为插值,后者本质上是矩量问题.

2003 年送走最后一位博士生后,我先后应聘去临沂、郑州与许昌基层高校工作 6 年多,主要精力用在指导一部分青年教师科学研究.但我与北京师范大学数学科学学院胡永建(还有中国农业大学的杨正宏)的合作研究一直持续着.这期间也取得一些可喜的成果,内容包括:边界插值问题,多项式零点定位问题,矩阵值耗散函数的极小 w-熵的内插式与块 Pick 矩阵的极大行列式填充,块广义 Pick 矩阵迹不变性,以及 Carathéodory 矩阵系数问题与 NP 型插值在非退化情形下极端耗散函数解的刻画等.前述特殊矩阵理论与插值以及矩量问题的部分成果分别组成本书的后两个部分,也是本书的重点.

我深深感念已故樊畿教授多年来的帮助与指引,老师的恩情山高海深.记得 1989 年 5 月樊师应邀访问我校时,曾有几次私人长谈.谈及并要我关注苏联(现乌克兰)大师 M. G. Krein(1982 年 Wolf 奖得主)领导的学派工作："这个方向的算子论与经典复分析关系密切，也是算子论中很深的一部分". 他还告诉我，"这些工作既有理论上丰富内涵，又有系统理论与控制的背景与重要作用；以纯粹数学的眼光看来，也是主流的分析，且近年来这方向研究非常活跃."此行他还带来一箱为我精心挑选的这方向的参考书(它们业已成为我们基本文献与必读书目). 这些教导切合我们当时的状况，至今还影响着我们研究工作的发展.

我这些年以来，有幸结识许多年青学者. 他们是我的学生、朋友或先或后合作者，在帮助本书有关内容表述与发展方面起着重要作用. 特别地我想对胡永建、杨正宏、赵斌、张惠品、李小青、贾利新、郝辉、吴化璋、张晓南、秦建国、牛少彰、许松、梁景伟、赵生富等表达我的谢意.

我还对多年来一直培养我，给予热情鼓励与提携的众多师长们致以衷心的感谢.

北京师范大学数学科学学院领导，特别是李仲来教授在这部文集的出版过程给予我很多帮助；北京师范大学出版社为本文集的出版做了许多工作，我一并向他们表示谢意.

最后，我愿趁此机会向我的亲人表示感谢，感谢他们一贯地给予我的呵护与支持.

陈公宁

2012 年 11 月写于

北京师范大学乐育楼区

序 2

我非常高兴得知陈公宁老师的著作选集将由北京师范大学出版社出版. 陈老师邀请我为此书写一序言,我感觉备受推崇和荣耀.

我是 1993 年到北京师范大学数学系师从陈公宁老师读硕士研究生的. 当时陈老师和他的学生张惠品、赵斌等刚刚完成了一项关于一般有理插值问题在数值情况的研究. 这个一般有理插值问题就是:给定一组数据 $y_i, C_{ik}, k=0,1,\cdots,\tau_i-1; i=1,2,\cdots,\theta$,总共包含 $\sum_{i=1}^{\theta}\tau_i$ 对数据,这里,$y_i\neq y_j$ 对 $i\neq j$,目的是找出所有有理数值函数 Φ 使得

$$\frac{1}{k!}\Phi^{(k)}(y_i)=C_{ik}, \forall i,k.$$

这个问题是首先由 Antoulas, Ball, Kang, Williems 和 Anderson 用广义 Loewner 矩阵的办法解决的. 后来又有人提出多项式生成矩阵的办法. 陈老师、张惠品和赵斌发现了一种崭新的计算特殊多项式生成矩阵的办法. 令人惊讶的是,他们发现对于每一个有理数值插值问题,都对应一个唯一的向量,称之为 Hankel 向量. 这个向量非常重要,

因为一般有理插值问题的复杂度和相应的多项式生成矩阵被它完全地决定. 他们的这项成果发表在 Linear Algebra and Its Applications(《线性代数及其应用》),1996 年,244 卷. 这篇开创性的工作为以后这个研究领域的深入发展奠定了基础.

在陈老师的指导下,在我的硕士学位论文中,我将上面提到的陈老师、张惠品和赵斌的工作推广到矩阵值解析函数插值问题的情况,特别是我们用块 Hankel 向量的办法给出了矩阵值 Nevanlinna-Pick 插值问题和 Stieltjes 矩阵幂矩量问题之间的一个一一对应. 这不仅使我们能够给出 Nevanlina-Pick 插值问题的解的存在性一个判定法则,而且基于幂矩量问题的解,我们可以给出 Nevanlinna-Pick 插值问题的解. 反过来也是这样的. 这推广了 Krein 和 Nudel'man 1970 年的著作 The Markov Moment Problem and Extremal Problems(《马尔可夫矩量问题和极值问题》)中关于数值插值的一个定理到矩阵插值的情况. 虽然 Nevanlinna-Pick 矩阵值插值及算子值插值问题都在我们之前解决了,但是我们所提出的使用 Hankel 向量的办法却是新颖和独特的. 与此同时,我的师弟贾利新、胡永建等也作出了相关的优秀的工作. 后来在美国留学期间,我的研究方向转移到解析数论. 陈老师和我的师弟胡永建等又继续在这一插值领域作出很多杰出的文章. 对此我不是很了解的,愿读者能在陈老师的著作中去涉猎.

下面从我个人在国外的一些经历谈一谈陈老师工作的国际影响力. 2010 年,我有幸在普林斯顿高等研究院工作一年,接触到世界著名的解析数论学家 Enrico Bombieri. 他于 1974 年因大筛法方面的工作获得菲尔兹奖. 我问他可否给我写一封申请美国永久居民的推荐信,他很爽快地答应了. 我想作为同行,他可能只会写我在解析数论方向的工作. 可是让我吃惊的是,在他的信中他竟然用一小段专门地讲了我在陈老师指导下做的硕士学位论文,这表现了他对解析插值这一领域的兴趣. 2011 年,我在美国的明尼苏达大学数学系工作面试的时候,见到著名数学家 Andrew Odlyzko. 他在解析数论、计算复杂度与密码等领域中都有过非常杰出的成就. 在面试中,他表现出对我的硕士学位论文的很大兴趣,并问及我关于这一领域的最新进展. 这更让我对陈老师曾经带我进入到这一前沿领域非常感激;对于他自己一人在国内带出一个赶超国际水平的团队也是非常佩服的.

在准备写这一序言的过程中，我也了解到陈老师早年在矩阵算子理论和多复变函数方面的研究. 在他 1984 年发表在 Journal of Mathematical Analysis and Applications 杂志第 98 卷的文章"Iteration for Holomorphic Maps of the Open Unit Ball and the Generalized Upper Half-plane of C^n"(《C^n 的单位开球和广义上半平面的全纯映照的迭代》)中，他将 J. Wolff 的一个经典的关于单复变函数全纯迭代定理推广到了多复变的情况. 在他的 1990 年发表在中国《数学年刊》第 11 卷的文章《von Neumann, Heinz 和樊畿定理的推广》中，他将 von Neumann, Heinz 和樊畿有关希尔伯特空间上有界线性算子的结果推广到更一般的两种情况.

愿读者在陈老师的这一著作选集中品尝到他的数学工作的广度和深度，也愿读者能很快地抓住其中的精髓，进入到矩阵算子理论的前沿.

李小青
美国纽约州立大学布法罗分校副教授
2012 年 12 月 25 日

目　录

四、解析函数插值和矩量问题以及有关分析问题

附录

Contents

Ⅲ. Special Matrices and Their Applications, Rational Interpolation Problem

Ⅳ. Interpolations for Analytic Functions and Moment Problems, and Some Related Questions in Analysis

Appendix

Appendix

一、矩阵的若干一般结果

I.

General Results for Matrices

北京师范大学学报(自然科学版)
1984,(1):1－8

矩阵的正则性的若干条件①

On Some Regularity Conditions for Matrices

设 $\mathbf{C}^{n,n}(\mathbf{R}^{n,n})$ 表示 $n\times n$ 复(实)矩阵的空间;$\mathbf{C}^n(\mathbf{R}^n)$ 是 n 维复(实)的行向量空间,$\boldsymbol{e}_1,\boldsymbol{e}_2,\cdots,\boldsymbol{e}_n$ 是 $\mathbf{R}^n$ 的典型基. $\mathbf{C}^n$ 上范数 Φ(只要求弱齐性,即 $\Phi(\alpha\boldsymbol{x})=\alpha\Phi(\boldsymbol{x})$ 对一切数量 $a\geqslant 0$ 成立)是单调的,如果对任意 $\mathbf{C}^n$ 内向量 $\boldsymbol{x}=(x_1,x_2,\cdots,x_n)$ 与 $\boldsymbol{y}=(y_1,y_2,\cdots,y_n)$,$|x_i|\leqslant|y_i|(i=1,2,\cdots,n)$ 蕴涵 $\Phi(\boldsymbol{x})\leqslant\Phi(\boldsymbol{y})$. 已经知道单调范数 Φ 是绝对的与严格齐性的,即分别有 $\Phi(\boldsymbol{x})=\Phi(|\boldsymbol{x}|)$ 对任意 $\boldsymbol{x}=(x_1,x_2,\cdots,x_n)\in\mathbf{C}_n$ 成立,这里 $|\boldsymbol{x}|=(|x_1|,|x_2|,\cdots,|x_n|)$,以及 $\Phi(\lambda\boldsymbol{x})=|\lambda|\Phi(\boldsymbol{x})$ 对任意数量 λ 成立. 同时,由公式

$$\Psi(\boldsymbol{y})=\sup_{\Phi(\boldsymbol{x})=1}\left|\sum_{i=1}^{n}x_iy_i\right| \tag{1}$$

定义的 Φ 在 $\mathbf{C}^n$ 上的共轭范数 Ψ,当 Φ 单调时,也是单调的,并且强的 Hölder 不等式

$$\sum_{i=1}^{n}|x_iy_i|\leqslant\Phi(\boldsymbol{x})\Psi(\boldsymbol{y}) \tag{2}$$

对任 $\boldsymbol{x},\boldsymbol{y}\in\mathbf{C}^n$ 成立[4,5].

我们称 $\mathbf{A}=(a_{ij})\in\mathbf{R}^{n,n}$ 是非奇异的 M-矩阵,如果 $a_{ij}\leqslant 0$ 对一切 $i\neq j$,且有 $\boldsymbol{x}=(x_1,x_2,\cdots,x_n)>\mathbf{0}$(即各 $x_i>0$)使得 $\mathbf{A}\boldsymbol{x}^{\mathrm{T}}>\mathbf{0}$. 为方便计,命 $\mathbf{N}=\{1,2,\cdots,n\}$;用 $\mu(\mathbf{A})=(m_{ij})$ 表示 $\mathbf{A}=(a_{ij})\in\mathbf{C}^{n,n}$ 的比较矩阵,其元素定义为

$$m_{ij}=\begin{cases}|a_{ij}|, & i=j,\\ -|a_{ij}|, & i\neq j.\end{cases}$$

众所周知,$\mu(\mathbf{A})$ 是非奇异的 M-矩阵蕴涵 $\mathbf{A}$ 是非奇异的[3]. 又用 $(\mathbf{A})_i$

① 收稿日期:1983-01-20.

表示行向量$(|a_{i1}|,|a_{i2}|,\cdots,|a_{i,i-1}|,0,|a_{i,i+1}|,\cdots,|a_{in}|)$. 今后总假定$\Phi((\mathbf{A})_i)>0(i=1,2,\cdots,n)$，因为如果不然，$\mathbf{A}$ 有一行的非对角元素均为0，此时问题容易归纳到 $n-1$ 阶的情形.

在[2]中，Fiedler 与 Pták 讨论了一些矩阵正则性条件，他们基本上是针对 $l_p(p\geqslant 1)$ 范数的. 本文应用单调范数与矩阵有向图的知识将上述条件推广到 $\mathbf{C}^n$ 上的一般单调范数，并建立相应的谱集估计.

定理 1　设 $\mathbf{A}=(a_{ij})\in\mathbf{C}^{n,n}$，$\Phi$ 与 Ψ 是 $\mathbf{C}^n$ 上的一对共轭范数，满足 $\Psi(\boldsymbol{e}_i)\leqslant 1(i\in\mathbf{N})$. 如果

$$\sigma_1=\sum_{i=1}^{n}\frac{\Phi((\mathbf{A})_i)\Psi(\boldsymbol{e}_i)}{|a_{ii}|+\Phi((\mathbf{A})_i)}<\min_{i\in\mathbf{N}}\{\Psi(\boldsymbol{e}_i)\},\tag{3}$$

那么 $\mathbf{A}$ 是非奇异的.

证　定义　$x_i=\dfrac{\Phi((\mathbf{A})_i)\Psi(\boldsymbol{e}_i)}{|a_{ii}|+\Phi((\mathbf{A})_i)}$，$i=1,2,\cdots,n$. 由于约定 $\Phi((\mathbf{A})_i)>0$ 对一切 $i\in\mathbf{N}$ 成立，因此各 $x_i>0$. 下面证明 $\mu(\mathbf{A})\boldsymbol{x}^{\mathrm{T}}>\mathbf{0}$，这里 $\boldsymbol{x}^{\mathrm{T}}=(x_1,x_2,\cdots,x_n)^{\mathrm{T}}$. 由强 Hölder 不等式，对各 $i\in\mathbf{N}$ 有

$$\begin{aligned}
&\sum_{j\neq i}|a_{ij}|x_j\leqslant\Phi((\mathbf{A})_i)\Psi(x_1,x_2,\cdots,x_{i-1},0,x_{i+1},\cdots,x_n)\\
&=\Phi((\mathbf{A})_i)\Psi\Big(\sum_{j\neq i}x_j\boldsymbol{e}_j\Big)\leqslant\Phi((\mathbf{A})_i)\sum_{j\neq i}x_j\Psi(\boldsymbol{e}_j)\leqslant\Phi((\mathbf{A})_i)\sum_{j\neq i}x_j\\
&=\Phi((\mathbf{A})_i)\left[\sigma_1-\frac{\Phi((\mathbf{A})_i)\Psi(\boldsymbol{e}_i)}{|a_{ii}|+\Phi((\mathbf{A})_i)}\right]\\
&<\Phi((\mathbf{A})_i)\left[\Psi(\boldsymbol{e}_i)-\frac{\Phi((\mathbf{A})_i)\Psi(\boldsymbol{e}_i)}{|a_{ii}|+\Phi((\mathbf{A})_i)}\right]=|a_{ii}|x_i.
\end{aligned}\tag{4}$$

因此 $\mu(\mathbf{A})\boldsymbol{x}^{\mathrm{T}}>\mathbf{0}$，因而 $\mathbf{A}$ 是非奇异矩阵. 证毕. ■

定理 2　设 $\mathbf{A}$，Φ 与 Ψ 同定理 1，又命 $S=\{i\in\mathbf{N}:|a_{ii}|\leqslant r_i(A)\equiv\sum_{j\neq i}|a_{ij}|\}$. 如果 S 是空集或者

$$\sigma_2=\sum_{i\in S}\frac{\Phi((\mathbf{A})_i)\Psi(\boldsymbol{e}_i)}{|a_{ii}|+\Phi((\mathbf{A})_i)}+\beta\sum_{i\in\mathrm{N}\backslash S}\frac{r_i(\mathbf{A})}{|a_{ii}|}<\min_{i\in\mathrm{N}}\{\Psi(\boldsymbol{e}_i)\},\tag{5}$$

这里 $\beta=\max\limits_{i\in S}\dfrac{\Phi((\mathbf{A})_i)\Psi(\boldsymbol{e}_i)}{|a_{ii}|+\Phi((\mathbf{A})_i)}$，那么 $\mathbf{A}$ 是非奇异矩阵.

证　当 S 是空集时，$|a_{ii}|>r_i(\mathbf{A})(i=1,2,\cdots,n)$，即 $\mathbf{A}$ 是严格对角占优的，因而 $\mathbf{A}$ 是非奇异的. 现在假设 S 不空，此时有正数 ε 使得下列二式均成立：

$$\sigma'_2 = \sum_{i\in S}\frac{\Phi((\mathbf{A})_i)\Psi(\boldsymbol{e}_i)}{|a_{ii}|+\Phi((\mathbf{A})_i)} + (\beta+\varepsilon)\sum_{i\in \mathbf{N}\backslash S}\frac{r_i(\mathbf{A})}{|a_{ii}|} < \min_{i\in\mathbf{N}}\{\Psi(\boldsymbol{e}_i)\},$$

$$\beta+\varepsilon \leqslant \beta\frac{|a_{ii}|}{r_i(\mathbf{A})}\ \text{对所有}\ i\in\mathbf{N}\backslash S.$$

定义 n 个实数 $\xi_i(i=1,2,\cdots,n)$ 如下:

$$\xi_i = \begin{cases}\dfrac{\Phi((\mathbf{A})_i)\Psi(\boldsymbol{e}_i)}{|a_{ii}|+\Phi((\mathbf{A})_i)}, & i\in S,\\ (\beta+\varepsilon)\dfrac{r_i(\mathbf{A})}{|a_{ii}|}, & i\in\mathbf{N}\backslash S.\end{cases} \tag{6}$$

我们的约定 $\Phi((\mathbf{A})_i)>0$ 蕴涵 $r_i(\mathbf{A})>0(i=1,2,\cdots,n)$,因此,$\boldsymbol{\xi}=(\xi_1,\xi_2,\cdots,\xi_n)>\mathbf{0}$. 当 $i\in S$ 时,类似于(4)式的推导,我们有

$$\sum_{j\neq i}|a_{ij}|\xi_j \leqslant \Phi((\mathbf{A})_i)\left[\sigma'_2 - \frac{\Phi((\mathbf{A})_i)\Psi(\boldsymbol{e}_i)}{|a_{ii}|+\Phi((\mathbf{A})_i)}\right]$$

$$<\Phi((\mathbf{A})_i)\left[\Psi(\boldsymbol{e}_i) - \frac{\Phi((\mathbf{A})_i)\Psi(\boldsymbol{e}_i)}{|a_{ii}|+\Phi((\mathbf{A})_i)}\right] = |a_{ii}|\xi_i.$$

当 $i\in\mathbf{N}\backslash S$ 时有

$$\sum_{j\neq i}|a_{ij}|\xi_j \leqslant r_i(\mathbf{A})\max_{j\in\mathbf{N}}\xi_j.$$

但对任意 $i\in\mathbf{N}\backslash S$,由于 $\beta+\varepsilon\leqslant\beta\dfrac{|a_{ii}|}{r_i(\mathbf{A})}$,有

$$\xi_j = (\beta+\varepsilon)\frac{r_i(\mathbf{A})}{|a_{ii}|}\leqslant\beta,$$

从而,$\beta=\max\limits_{j\in\mathbf{N}}\xi_j$. 由此推出当 $i\in\mathbf{N}\backslash S$ 时,

$$\sum_{j\neq i}|a_{ij}|\xi_j \leqslant r_i(\mathbf{A})\beta \leqslant r_i(\mathbf{A})\frac{\beta+\varepsilon}{|a_{ii}|}\frac{|a_{ii}|\beta}{\beta+\varepsilon}$$

$$=\frac{\beta}{\beta+\varepsilon}|a_{ii}|\xi_i < |a_{ii}|\xi_i.$$

综合上述两种情况,有 $\mu(\mathbf{A})\boldsymbol{\xi}^{\mathrm{T}}>\mathbf{0}$,于是 $\mu(\mathbf{A})$ 是非奇异 M-矩阵,因而 $\mathbf{A}$ 是非奇异矩阵. 证毕. ■

显然地,当 $S=\mathbf{N}$ 时,定理 2 退化为定理 1.

定理 3 设 $\mathbf{A}$,Φ 与 Ψ 同定理 1,又设 $\{k_i\}$ 是满足 $|a_{ii}|>k_i\Phi((\mathbf{A})_i)$ 的 n 个任意正数$(i=1,2,\cdots,n)$;$W=\left\{i\in\mathbf{N}:\dfrac{r_i(\mathbf{A})}{\Phi((\mathbf{A})_i)}>k_i\right\}$;$m=\max\limits_{j\in W}[(1+k_i)^{-1}\Psi(\boldsymbol{e}_i)]$ 以及

$$J_0 = \left\{i\in\mathbf{N}:\frac{\Psi(\boldsymbol{e}_i)}{1+k_i}<\frac{mr_i(\mathbf{A})}{k_i\Phi((\mathbf{A})_i)}\right\}. \tag{7}$$

如果 W 是空集或者

$$\sigma_3 = \sum_{i\in J_0}\frac{\Psi(\boldsymbol{e}_i)}{1+k_i} + m\sum_{i\in \mathbf{N}\backslash J_0}\frac{r_i(\mathbf{A})}{k_i\Phi((\mathbf{A})_i)} \leqslant \min_{i\in\mathbf{N}}\{\Psi(\boldsymbol{e}_i)\}, \tag{8}$$

那么 $\mathbf{A}$ 是非奇异矩阵.

证　如果 W 是空集，则对一切 $i\in\mathbf{N}$，$\frac{r_i(\mathbf{A})}{\Phi((\mathbf{A})_i)}\leqslant k_i$，于是 $|a_{ii}|>k_i\Phi((\mathbf{A})_i)\geqslant r_i(\mathbf{A})(i=1,2,\cdots,n)$，因此 $\mathbf{A}$ 是非奇异的. 如果 W 不空，那么对每个 $i\in W$ 有

$$\frac{\Psi(\boldsymbol{e}_i)}{1+k_i}\leqslant m < \frac{mr_i(\mathbf{A})}{k_i\Phi((\mathbf{A})_i)},$$

此蕴涵 $W\subset J_0$. 定义 n 个实数 ξ_i 如下($i\in\mathbf{N}$)：

$$\xi_i = \begin{cases}\dfrac{\Psi(\boldsymbol{e}_i)}{1+k_i}, & i\in J_0,\\[2ex] \dfrac{mr_i(\mathbf{A})}{k_i\Phi((\mathbf{A})_i)}, & i\in\mathbf{N}\backslash J_0.\end{cases} \tag{9}$$

因为 $k_i>0$ 与 $m>0$，所以 $\boldsymbol{\xi}=(\xi_1,\xi_2,\cdots,\xi_n)>\mathbf{0}$. 现在如果 $i\in J_0$ 由强 Hölder 不等式得

$$\begin{aligned}\sum_{j\neq i}|a_{ii}|\xi_j &\leqslant \Phi((\mathbf{A})_i)\sum_{j\neq i}\xi_j = \Phi((\mathbf{A})_i)\left[\sigma_3-\frac{\Psi(\boldsymbol{e}_i)}{1+k_i}\right]\\ &\leqslant \Phi((\mathbf{A})_i)\left[\Psi(\boldsymbol{e}_i)-\frac{\Psi(\boldsymbol{e}_i)}{1+k_i}\right]= k_i\Phi((\mathbf{A})_i)\xi_i<|a_{ii}|\xi_i.\end{aligned} \tag{10}$$

如果 $i\in\mathbf{N}\backslash J_0$，那么更有 $i\in\mathbf{N}\backslash W$，于是$\frac{r_i(\mathbf{A})}{\Phi((\mathbf{A})_i)}\leqslant k_i$，并且，

$$\sum_{j\neq i}|a_{ii}|\xi_j\leqslant r_i(\mathbf{A})\max_{j\in\mathbf{N}}\xi_j. \tag{11}$$

我们接着证明$\max\limits_{j\in\mathbf{N}}\xi_j=m$. 事实上，显然有 $m=\max\limits_{j\in W}\xi_j$，因为 $W\subset J_0$. 假如 $j\in\mathbf{N}\backslash J_0$，则 $j\in\mathbf{N}\backslash W$，于是有 $r_j(\mathbf{A})/\Phi((\mathbf{A})_j)\leqslant k_j$，从而 $\xi_j=\frac{mr_j(\mathbf{A})}{k_j\Phi((\mathbf{A})_j)}\leqslant m$；假如 $j\in J_0\backslash W$，则$\frac{r_j(\mathbf{A})}{\Phi((\mathbf{A})_j)}\leqslant k_j$，于是 $\xi_j=\frac{\Psi(\boldsymbol{e}_j)}{1+k_j}<\frac{mr_j(\mathbf{A})}{k_j\Phi((\mathbf{A})_j)}\leqslant m$. 因此，$m=\max\limits_{j\in\mathbf{N}}\xi_j$. 继续(11)式得出对所有 $i\in\mathbf{N}\backslash J_0$，

$$\sum_{j\neq i}|a_{ij}|\xi_j\leqslant\frac{r_i(\mathbf{A})}{k_i\Phi((\mathbf{A})_i)}k_i\Phi((\mathbf{A})_i)m<m\frac{r_i(\mathbf{A})}{k_i\Phi((\mathbf{A})_i)}|a_{ii}|=|a_{ii}|\xi_i.$$

连同(10)式推得 $\mu(\mathbf{A})\boldsymbol{\xi}^{\mathrm{T}}>\mathbf{0}$. 因此定理成立. ■

定理 3 有如下的变种，它的证明类似于定理 3，我们略去.

定理 4 设 $\mathbf{A},\Phi$ 与 Ψ 定理 1,又设 $\{k_i\}$ 是满足 $|a_{ii}|>k_i\Phi((\mathbf{A})_i)$ 的 n 个任意正数 $(i=1,2,\cdots,n)$;$W=\left\{i\in\mathbf{N}:\dfrac{r_i(\mathbf{A})}{\Phi((\mathbf{A})_i)}>k_i\right\}$;$m'=\max\limits_{i\in W}[(1+k_i)^{-1}]$ 以及

$$J=\left\{i\in\mathbf{N}:\frac{\Psi(\mathbf{e}_i)}{1+k_i}<\frac{m'r_i(\mathbf{A})}{k_i\Phi((\mathbf{A})_i)}\right\}. \tag{12}$$

如果 W 是空集或者

$$\sigma_4=\sum_{i\in J}\frac{\Psi(\mathbf{e}_i)}{1+k_i}+m'\sum_{i\in\mathbf{N}\setminus J}\frac{r_i(\mathbf{A})}{k_i\Phi((\mathbf{A})_i)}\leqslant\min_{i\in\mathbf{N}}\{\Psi(\mathbf{e}_i)\}, \tag{13}$$

那么 $\mathbf{A}$ 是非奇异矩阵.

从定理 4 直接得一个推论,它与[1]的定理 1.5 有密切关系.

推论 1 设 $\mathbf{A},\Phi$ 与 Ψ 同定理 1,又设存在正数 α 满足 $|a_{ii}|>\alpha\Phi((\mathbf{A})_i)(i=1,2,\cdots,n)$. 如果

$$\sum_{i\in\mathbf{N}}\frac{r_i(\mathbf{A})}{k_i\Phi((\mathbf{A})_i)}\leqslant\alpha(1+\alpha)\min_{i\in\mathbf{N}}\{\Psi(\mathbf{e}_i)\}, \tag{14}$$

那么 $\mathbf{A}$ 是非奇异矩阵.

证 利用定理 4 的记号,令 $k_i=\alpha(i=1,2,\cdots,n)$ 即可. ■

推论 2 设 $\mathbf{A},\Phi$ 与 Ψ 同定理 1,又设 $\{k_i\}$ 是满足 $|a_{ii}||a_{jj}|>k_ik_j\Phi((\mathbf{A})_i)\Phi((\mathbf{A})_j)$ $(i,j=1,2,\cdots,n;i\neq j)$ 的 n 个任意正数;$W=\left\{i\in\mathbf{N}:\dfrac{r_i(\mathbf{A})}{\Phi((\mathbf{A})_i)}>k_i\right\}$;$m'=\max\limits_{j\in W}[(1+k_i)^{-1}\Psi(\mathbf{e}_i)]$ 以及

$$J_0=\left\{i\in\mathbf{N}:\frac{\Psi(\mathbf{e}_i)}{1+k_i}<\frac{mr_i(\mathbf{A})}{k_i\Phi((\mathbf{A})_i)}\right\}.$$

如果 W 是空集或者

$$\sigma'_3=\sum_{i\in J_0}\frac{\Psi(\mathbf{e}_i)}{1+k_i}+m\sum_{i\in\mathbf{N}\setminus J_0}\frac{r_i(\mathbf{A})}{k_i\Phi((\mathbf{A})_i)}\leqslant\min_{j\in\mathbf{N}}\{\Psi(\mathbf{e}_i)\}, \tag{15}$$

那么 $\mathbf{A}$ 是非奇异矩阵.

证 如果 W 是空集,那么 $|a_{ii}||a_{jj}|>k_ik_j\Phi((\mathbf{A})_i)\Phi((\mathbf{A})_j)\geqslant r_i(\mathbf{A})r_j(\mathbf{A})(i,j=1,2,\cdots,n;i\neq j)$. 由[2]中定理 6.9 推得 $\mu(\mathbf{A})$ 是非奇异 M-矩阵,因而 $\mathbf{A}$ 是非奇异的. 如 W 不空,定义 ξ_i 如(9)式,用类似于定理 3 的推证我们有

$$\sum_{k\neq i}|a_{ik}|\xi_k\leqslant k_i\Phi((\mathbf{A})_i)\xi_i\quad(\text{对 } i\in J_0),$$

$$\sum_{p\neq i}|a_{ip}|\xi_p\leqslant k_i\Phi((\mathbf{A})_i)\xi_i\quad(\text{对 } i\in\mathbf{N}\setminus J_0).$$

联合上述两式推出

$$\left[\sum_{k\neq i}|a_{ik}|\xi_k\right]\left[\sum_{p\neq j}|a_{jp}|\xi_p\right]\leqslant k_ik_j\Phi((\mathbf{A})_i)\Phi((\mathbf{A})_j)\xi_i\xi_j$$
$$<|a_{ii}||a_{jj}||\xi_i\xi_j\quad(i,j=1,2,\cdots,n;i\neq j).$$

再次应用[2]中定理 6.9 推出矩阵 $\boldsymbol{D}^{-1}\mu(\mathbf{A})\boldsymbol{D}$ 是非奇异 M-矩阵，这里 $\boldsymbol{D}=\mathrm{diag}(\xi_1,\xi_2,\cdots,\xi_n)$，于是 $\mathbf{A}$ 是非奇异矩阵. 证毕. ■

接着，我们应用矩阵有向图的一些结果进一步地推广定理 3 及其推论. 设 $\Gamma(\mathbf{A})$表示 $\mathbf{A}\in\mathbf{C}^{n,n}$的有向图，它的一个回路 ν 是 $\Gamma(\mathbf{A})$顶点 $i_1,i_2,\cdots,i_p,i_{p+1}=i_1$ 的序列，这里规定 $p\geqslant 2,i_1,i_2,\cdots,i_p$ 是各不相同的，并且 $a_{i_1i_2}\neq 0,\cdots,a_{i_pi_1}\neq 0$. 将 $\Gamma(\mathbf{A})$的所有回路组成的集合记为 $\mathscr{G}(\mathbf{A})$. 最近，Brualdi[6]得出如下结果：

定理(Brualdi) 设 $\mathbf{A}=(a_{ij})\in\mathbf{C}^{n,n}$的所有主对角元素 $a_{ii}\neq 0$. 则$\mu(\mathbf{A})$是非奇异 M -矩阵当且仅当存在 n 个正数 $x_1,x_2,\cdots,x_n$ 使得

$$\prod_{i\in\nu}|a_{ii}|>\prod_{i\in\nu}\left[\frac{1}{x_i}\sum_{j\neq i}|a_{ij}|x_j\right]\tag{16}$$

对 $\mathscr{G}(\mathbf{A})$中所有回路 ν 成立.

推论 3 设 $\mathbf{A}$,Φ 与 Ψ 同定理 1 但所有 $a_{ii}\neq 0$，又设$\{k_i\}$是满足$\prod_{i\in\nu}|a_{ii}|>\prod_{i\in\nu}[k_i\Phi((\mathbf{A})_i)]$的 n 个任意正数，这里 ν 是 $\Gamma(\mathbf{A})$的任意回路；$W=\left\{i\in\mathbf{N}:\dfrac{r_i(\mathbf{A})}{\Phi((\mathbf{A})_i)}>k_i\right\}$；$m=\max\limits_{i\in W}[(1+k_i)^{-1}\Psi(\boldsymbol{e}_i)]$以及

$$J_0=\left\{i\in\mathbf{N}:\frac{\Psi(\boldsymbol{e}_i)}{1+k_i}<\frac{mr_i(\mathbf{A})}{k_i\Phi((\mathbf{A})_i)}\right\}.$$

如果 W 是空集或者

$$\sigma''_3=\sum_{i\in J_0}\frac{\Psi(\boldsymbol{e}_i)}{1+k_i}+m\sum_{i\in\mathbf{N}\backslash J_0}\frac{r_i(\mathbf{A})}{k_i\Phi((\mathbf{A})_i)}\leqslant\min_{i\in\mathbf{N}}\{\Psi(\boldsymbol{e}_i)\},\tag{17}$$

那么 $\mathbf{A}$ 是非奇异矩阵.

证 由 Brualdi 定理与类似于推论 2 的证明直接证出最后结论. 这里略去证明细节. ■

注 设 H_1,H_2 与 H_3 分别是满足定理 3，推论 2 与推论 3 的$\{k_i\}_{i=1}^{i=n}$的集合. 则容易看出，$H_1\subset H_2\subset H_3$，并且 $H_1\neq H_2$，(当 $\mathscr{G}(\mathbf{A})$有一回路含有三个不同顶点时)$H_2\neq H_3$. 因此，一般地说，推论 3 优于推论 2，而后者优于定理 3.

另外，当$\mathbf{A}\in\mathbf{C}^{n,n}$不可约时，则推论3条件中"设$\{k_i\}$是满足$\prod_{i\in\nu}|a_{ii}|>\prod_{i\in\nu}[k_i\Phi((\mathbf{A})_i)]$…… 任意回路"可以改换为"设$\{k^i\}_{i=1}^{i=n}$是满足$\prod_{i\in\nu}|a_{ii}|\geqslant\prod_{i\in\nu}[k_i\Phi((\mathbf{A})_i)](\nu\in\mathscr{G}(\mathbf{A}))$但至少对某个$\nu\in\mathscr{G}(\mathbf{A})$严格不等式成立"，推论3仍保持有效. 有关理由可见[6].

十分自然地如将前面诸结论用于矩阵$\mathbf{A}-\lambda\boldsymbol{I}$，这里$\boldsymbol{I}$是$n$阶单位矩阵，我们将得到矩阵$\mathbf{A}$的谱集一些估计.

定理1′ 设$\mathbf{A}$，Φ与Ψ同定理1. 则$\mathbf{A}$的谱集包含在复平面上如下集合内：

$$\left\{z\in\mathbf{C}:\sum_{i\in\mathrm{N}}\frac{\Phi((\mathbf{A})_i)\Psi(\boldsymbol{e}_i)}{|a_{ii}-z|+\Phi((\mathbf{A})_i)}\geqslant\min_{i\in\mathrm{N}}\{\Psi(\boldsymbol{e}_i)\}\right\}.$$

证 设$\lambda\in\mathbf{C}$不属于上述集合. 则有

$$\sum_{i\in\mathrm{N}}\frac{\Phi((\mathbf{A})_i)\Psi(\boldsymbol{e}_i)}{|a_{ii}-\lambda|+\Phi((\mathbf{A})_i)}>\min_{i\in\mathrm{N}}\{\Psi(\boldsymbol{e}_i)\}.$$

因为按定义$\Phi((\mathbf{A})_i)=\Phi((\mathbf{A}-\lambda\boldsymbol{I})_i)$，$i\in\mathbf{N}$，将定理1用于矩阵$\mathbf{A}-\lambda\boldsymbol{I}$推出$\mathbf{A}-\lambda\boldsymbol{I}$是非奇异的，即$\lambda$不是$\mathbf{A}$的特征值. 定理证毕. ■

类似地，对应于定理2，定理3及其推论分别有矩阵$\mathbf{A}$谱集的如下估计.

定理2′ 设$\mathbf{A}$，Φ与Ψ同定理1，又设集合$S=\{i\in\mathbf{N}:|a_{ii}|\leqslant r_i(\mathbf{A})\equiv\sum_{j\neq i}|a_{ij}|\}$. 如果$S$不空，那么$\mathbf{A}$的谱集包含在复平面上如下两个集合的并集内：

$$P=\bigcup_{i\in S}\{z\in\mathbf{C}:|a_{ii}-z|>r_i(\mathbf{A})\},$$

$$Q=\left\{z\in\mathbf{C}:\frac{\Phi((\mathbf{A})_i)\Psi(\boldsymbol{e}_i)}{|a_{ii}-z|+\Phi((\mathbf{A})_i)}+\sum_{i\in\mathrm{N}\backslash S}\frac{r_i(\mathbf{A})}{|a_{ii}-z|}\geqslant\min_{i\in\mathrm{N}}\{\Psi(\boldsymbol{e}_i)\}\right\}.$$

证 假定λ是$\mathbf{A}$的特征值，并且$\lambda\notin P$. 则对每一个$i\in S$有

$$|a_{ii}-\lambda|\leqslant r_i(\mathbf{A}).$$

因为$\mathbf{A}-\lambda\boldsymbol{I}$是奇异矩阵，根据定理2我们有

$$\sum_{i\in S}\frac{\Phi((\mathbf{A})_i)\Psi(\boldsymbol{e}_i)}{|a_{ii}-\lambda|+\Phi((\mathbf{A})_i)}+\beta\sum_{i\in\mathrm{N}\backslash S}\frac{r_i(\mathbf{A})}{|a_{ii}-\lambda|}\geqslant\min_{i\in\mathrm{N}}\{\Psi(\boldsymbol{e}_i)\},$$

但是$\beta=\max\limits_{i\in S}\dfrac{\Phi((\mathbf{A})_i)\Psi(\boldsymbol{e}_i)}{|a_{ii}-\lambda|+\Phi((\mathbf{A})_i)}\leqslant1$，所以，$\lambda\in Q$. 证毕.

定理3′ 设$\mathbf{A}$，Φ与Ψ同定理1，又设$\{k_i\}$是n个正数；$W=$

$\left\{i\in\mathbf{N}:\dfrac{r_i(\mathbf{A})}{\Phi((\mathbf{A})_i)}>k_i\right\}$;(如果 W 不空)$m=\max\limits_{i\in W}[(1+k_i)^{-1}\Psi(\boldsymbol{e}_i)]$以及

$$J_0=\left\{i\in\mathbf{N}:\frac{\Psi(\boldsymbol{e}_i)}{1+k_i}<\frac{mr_i(\mathbf{A})}{k_i\Phi((\mathbf{A})_i)}\right\}.$$

如果 W 是空集或者(8)式成立,那么 $\mathbf{A}$ 的谱集包含在复平面上如下 n 个圆盘的并集内:

$$\boldsymbol{D}_i=\{z\in\mathbf{C}:|a_{ii}-z|\leqslant k_i\Phi((\mathbf{A})_i)\}\quad(i=1,2,\cdots,n).$$

对应于定理 4 与推论 1 也有类似的谱集估计,这里不再赘述.

推论 2′　设 $\mathbf{A},\Phi$ 与 Ψ 同定理 1,又设 $\{k_i\},W,m$ 以及 J_0 同推论 2. 如果 W 是空集或者(15)式成立,那么 $\mathbf{A}$ 的谱集包含在复平面上如下卵形的并集内:

$$\boldsymbol{D}_{ij}=\{z\in\mathbf{C}:|a_{ii}-z||a_{jj}-z|\leqslant k_ik_j\Phi((\mathbf{A})_i)\Phi((\mathbf{A})_j)\}$$
$$(i,j=1,2,\cdots,n;i\neq j).$$

推论 3′　设 $\mathbf{A},\Phi$ 与 Ψ 同推论 3,又设 $\{k_i\},W,m$ 与 J_0 也同推论 3. 如果 W 是空集或者(17)式成立,那么 $\mathbf{A}$ 的谱集包含在复平面上如下区域的并集内:

$$\boldsymbol{D}_\nu=\left\{z:\mathbf{C}:\prod_{i\in\nu}|a_{ii}-z|\leqslant\prod_{i\in\nu}[k_i\Phi((\mathbf{A})_i)]\right\}\quad(\nu\in\mathscr{G}(\mathbf{A})).$$

注　为比较前三个结果,我们假定 $\mathbf{A}\in\mathbf{C}^{n,n}$ 满足 $a_{ii}\neq0(i=1,2,\cdots,n)$,且在此三个结果中,取$\{k_i\}$是一样的. 此时,条件(8)(15)与(17)也是相同的. 不难看到

$$\bigcup_{\nu\in\mathscr{G}(\mathbf{A})}\boldsymbol{D}_\nu\subset\bigcup_{i\neq j}\boldsymbol{D}_{ij}\subset\bigcup_{i\in\mathbf{N}}\boldsymbol{D}_i.$$

因此,一般地说,推论 3′的谱集估计优于推论 2′,而后者又优于定理 3′.

下列两个有趣结果可看为推论 3′的直接推论.

推论 4　设 $\mathbf{A}\in\mathbf{R}^{n,n}$ 是主对角元素 a_{ii} 皆为 0 的非负不可约矩阵,又设 g 是 $\mathscr{G}(\mathbf{A})$ 中最短回路中不同的顶点数目. 沿用推论 3′的记号,并将 n 个正数$\{k_i\Phi((\mathbf{A})_i)\}$按大小顺序排列为

$$k_{i_1}\Phi((\mathbf{A})_{i_1})\leqslant\cdots\leqslant k_{i_n}\Phi((\mathbf{A})_{i_n}).$$

如果 W 是空集或者(17)式成立,那么 $\mathbf{A}$ 的谱半径 $\rho(\mathbf{A})$ 满足

$$\rho(\mathbf{A})\leqslant\sqrt[g]{\prod_{j=n-g+1}^{n}[k_{i_j}\Phi((\mathbf{A})_{i_j})]}.\tag{18}$$

证　Perron-Frobenius 理论的一个定理表明,在我们的假设下,$\rho(\mathbf{A})$

是 $\mathbf{A}$ 的正特征值. 显然地,$\mathbf{A}-\rho(\mathbf{A})\boldsymbol{I}$ 的所有主对角元素不为 0. 如果(18)式不算,那么

$$\rho(\mathbf{A}) > \sqrt[g]{\prod_{j=n-g+1}^{n}[k_{i_j}\Phi((\mathbf{A})_{i_j})]}.$$

设 ν 是 $\mathscr{G}(\mathbf{A})$ 中任一回路,其不同的顶点数目为 p. 显然,$p\geqslant g$,且有

$$\sqrt[g]{\prod_{j=n-g+1}^{n}[k_{i_j}\Phi((\mathbf{A})_{i_j})]} \geqslant \sqrt[p]{\prod_{j=n-p+1}^{n}[k_{i_j}\Phi((\mathbf{A})_{i_j})]} \geqslant \sqrt[p]{\prod_{i\in\nu}[k_i\Phi((\mathbf{A})_i)]}.$$

因此,当 $i\in\nu$ 时,

$$|a_{ii}-\rho(\mathbf{A})| = \rho(\mathbf{A}) > \sqrt[p]{\prod_{j\in\nu}[k_j\Phi((\mathbf{A})_j)]}.$$

应用推论 3′到 $\mathbf{A}-\rho(\mathbf{A})\boldsymbol{I}$,我们推出 $\mathbf{A}-\rho(\mathbf{A})\boldsymbol{I}$ 是非奇异的,此与 $\rho(\mathbf{A})$ 是 $\mathbf{A}$ 的特征值矛盾. 证毕. ■

推论 5 设 $\mathbf{A}$ 与 g 同推论 4. 如果正数 α 使得(14)式成立,那么 $\mathbf{A}$ 的谱半径 $\rho(\mathbf{A})$满足

$$\rho(\mathbf{A}) \leqslant \alpha\sqrt[g]{\prod_{j=n-g+1}^{n}[\Phi((\mathbf{A})_{i_j})]} \tag{19}$$

这里 $\Phi((\mathbf{A})_{i_1})\leqslant\Phi((\mathbf{A})_{i_2})\leqslant\cdots\leqslant\Phi((\mathbf{A})_{i_n})$.

证 只要注意到推论 1 矩阵 $\mathbf{A}\in\mathbf{C}^{n,n}$ 增加各主对角元素 $a_{ii}\neq0$ 的条件;"α 满足 $|a_{ii}|>\alpha\Phi((\mathbf{A}_i)(i=1,2,\cdots,n)$"改为"$\alpha$ 满足 $\prod_{i\in\nu}|a_{ii}|>\prod_{i\in\nu}[\alpha\Phi((\mathbf{A})_i)]$,其中 ν 是 $\mathscr{G}(\mathbf{A})$ 中任一回路"之后命题仍然正确. ■

参考文献

[1]K Fan and A J Hoffman. Nat. Bur. Stand. Appl. Math. Ser., 1954,39:117—130.

[2]M Fiedler and V Pták. Czech. Math. Journal, 1962,87:382—400.

[3]G -N Chen. Linear Algebra Appl. 1983,55:125—132.

[4]F L Bauer, J Stoer and C Witzgall. Numer. Math., 1961,3:257—264.

[5]P Nowosad and R Tovar. Linear Algebra Appl., 1980, 31:179—197.

[6]R A Brualdi. Linear and Multilinear Algebra, 1982,11:143—165.

ON SOME REGULARITY CONDITIONS FOR MATRICES

Abstract A norm F on $\mathbf{C}^n$, the n-dimensional complex coordinate space, is called monotonic if for every $\boldsymbol{x}=(x_1, x_2, \cdots, x_n)$ and $\boldsymbol{y}=(y_1, y_2, \cdots, y_n)$ in $\mathbf{C}^n$, $|x_i| \leqslant |y_i|$ for $i=1,2,\cdots,n$ implies $F(\boldsymbol{x}) \leqslant F(\boldsymbol{y})$. In the present note, the author uses monotonic norms on $\mathbf{C}^n$ to investigate some regularity conditions for matrices, and sufficient conditions for the comparison matrix of a matrix to be nonsingular M-matrix. These results are further used to obtain connected estimates for spectra of matrices.

数学学报

1992,35(5):620－622

对"关于半正定 Hermite 矩阵乘积迹的一个不等式"一文的注记①

Remarks on "A Trace Inequality on Product of Semipositive Definite Hermite Matrices"

摘要 本文指出文献[1]中结果及其一种推广均可视为矩阵论中两个已知定理的简单推论.

设 $\boldsymbol{A}$ 为 n 阶复矩阵,用 $\mathrm{tr}(\boldsymbol{A})$ 记 $\boldsymbol{A}$ 的迹;用 $\lambda_1(\boldsymbol{A}),\lambda_2(\boldsymbol{A}),\cdots,\lambda_n(\boldsymbol{A})$ 与 $\sigma_1(\boldsymbol{A}),\sigma_2(\boldsymbol{A}),\cdots,\sigma_n(\boldsymbol{A})$ 分别表示 $\boldsymbol{A}$ 的 n 个特征值与奇异值,且约定它们已排序为 $\mathrm{Re}\,\lambda_1(\boldsymbol{A})\geqslant \mathrm{Re}\,\lambda_2(\boldsymbol{A})\geqslant\cdots\geqslant \mathrm{Re}\,\lambda_n(\boldsymbol{A})$ 与 $\sigma_1(\boldsymbol{A})\geqslant\sigma_2(\boldsymbol{A})\geqslant\cdots\geqslant\sigma_n(\boldsymbol{A})$. 显然,$\sigma_i(\boldsymbol{A})=\lambda_i^{1/2}(\boldsymbol{A}^*\boldsymbol{A}),i=1,2,\cdots,n$.

陈道琦[1]证明了 Bellman 的一个不等式的如下形式的推广结果.

定理 1 设 $\boldsymbol{A}_1,\boldsymbol{A}_2,\cdots,\boldsymbol{A}_m$ 皆为 $\boldsymbol{n}$ 阶 Hermite 半正定矩阵. 则有

$$|\,\mathrm{tr}(\boldsymbol{A}_1\boldsymbol{A}_2\cdots\boldsymbol{A}_m)\,|\leqslant\Big[\sum_{k=1}^{m}\mathrm{tr}(\boldsymbol{A}_k^m)\Big]^{1/m}\leqslant\frac{1}{m}\sum_{k=1}^{m}\mathrm{tr}(\boldsymbol{A}_k^m). \tag{1}$$

显然,上述定理为下面定理 2 的特殊情形. 在本文我们直接证明定理 2,从而顺便地给出定理 1 的十分简单的证明.

定理 2 设 $\boldsymbol{A}_1,\boldsymbol{A}_2,\cdots,\boldsymbol{A}_m$ 皆为 n 阶复矩阵. 则

$$\begin{aligned}\mathrm{Re}\,\mathrm{tr}(\boldsymbol{A}_1\boldsymbol{A}_2\cdots\boldsymbol{A}_m)&\leqslant|\,\mathrm{tr}(\boldsymbol{A}_1\boldsymbol{A}_2\cdots\boldsymbol{A}_m)\,|\leqslant\Big[\prod_{k=1}^{m}\mathrm{tr}(\boldsymbol{A}_k^*\boldsymbol{A}_k)^{m/2}\Big]^{1/m}\\&\leqslant\frac{1}{m}\sum_{k=1}^{m}\mathrm{tr}(\boldsymbol{A}_k^*\boldsymbol{A}_k)^{m/2}.\end{aligned} \tag{2}$$

---

① 收稿日期:1991-03-11

本文为国家教委博士点基金资助的课题.

为证明定理 2,我们需要下列两个已知的结论. 当 $\boldsymbol{x}=(x_1,x_2,\cdots,x_n)$ 与 $\boldsymbol{y}=(y_1,y_2,\cdots,y_n)$ 为实向量时,$\boldsymbol{y}$ 弱控制 $\boldsymbol{x}$,记为 $\boldsymbol{x}\prec_w\boldsymbol{y}$,假如将 $\boldsymbol{x}$ 与 $\boldsymbol{y}$ 分量按递减顺序排列为 $x_{[1]}\geqslant x_{[2]}\geqslant\cdots\geqslant x_{[n]}$ 与 $y_{[1]}\geqslant y_{[2]}\geqslant\cdots\geqslant y_{[n]}$,则有

$$\sum_{i=1}^{j}x_{[i]}\leqslant\sum_{i=1}^{j}y_{[i]},\quad j=1,2,\cdots,n.$$

引理 1 (Fan [3]) 若 $\boldsymbol{A}$ 为 n 阶复矩阵,其对角元素为 $a_1,a_2,\cdots,a_n$,则有

$$\begin{aligned}(\operatorname{Re}a_1,\operatorname{Re}a_2,\cdots,\operatorname{Re}a_n)&\prec_w(|a_1|,|a_2|,\cdots,|a_n|)\\&\prec_w(\sigma_1(\boldsymbol{A}),\sigma_2(\boldsymbol{A}),\cdots,\sigma_n(\boldsymbol{A})).\end{aligned}\tag{3}$$

引理 2 ([4,p. 250]) 若 $\boldsymbol{A}_1,\boldsymbol{A}_2,\cdots,\boldsymbol{A}_m$ 皆为 n 阶复矩阵,则有

$$\begin{aligned}\sum_{i=1}^{j}\sigma_i(\boldsymbol{A}_1\boldsymbol{A}_2\cdots\boldsymbol{A}_m)&\leqslant\sum_{i=1}^{j}\sigma_i(\boldsymbol{A}_1)\,\sigma_i(\boldsymbol{A}_2)\,\cdots\,\sigma_i(\boldsymbol{A}_m)\\&\leqslant\frac{1}{m}\sum_{i=1}^{j}\sum_{k=1}^{m}\sigma_i^m(\boldsymbol{A}_k).\end{aligned}\tag{4}$$

$$j=1,2,\cdots,n.$$

定理 2 的证明 令 $\boldsymbol{B}=\boldsymbol{A}_1\boldsymbol{A}_2\cdots\boldsymbol{A}_m=(b_{ij})$. 应用引理 1,2 与多因子的 Hölder 不等式,我们有

$$\begin{aligned}\operatorname{Re}\operatorname{tr}(\boldsymbol{B})&\leqslant|\operatorname{tr}(\boldsymbol{B})|\leqslant\sum_{i=1}^{n}|b_{ii}|\leqslant\sum_{i=1}^{n}\sigma_i(\boldsymbol{B})\\&\leqslant\sum_{i=1}^{n}\sigma_i(\boldsymbol{A}_1)\sigma_i(\boldsymbol{A}_2)\cdots\sigma_i(\boldsymbol{A}_m)\leqslant\prod_{k=1}^{m}\Big[\sum_{i=1}^{n}\sigma_i^m(\boldsymbol{A}_k)\Big]^{1/m}\\&=\prod_{k=1}^{m}\Big[\sum_{i=1}^{n}\lambda_i^{m/2}(\boldsymbol{A}_k^*\boldsymbol{A}_k)\Big]^{1/m}=\prod_{k=1}^{m}[\operatorname{tr}(\boldsymbol{A}_k^*\boldsymbol{A}_k)^{m/2}]^{1/m}\\&\leqslant\frac{1}{m}\sum_{k=1}^{m}\operatorname{tr}(\boldsymbol{A}_k^*\boldsymbol{A}_k)^{m/2},\end{aligned}$$

这里在最后估计中用到算术平均与几何平均不等式. 定理证毕. ■

下面推论 1 的取 $m=2$ 的情形即为 Bellman 不等式的原始形式.

推论 1 设 $\boldsymbol{A}_1,\boldsymbol{A}_2,\cdots,\boldsymbol{A}_m$ 皆为 n 阶实对称半正定矩阵. 则有

$$\operatorname{tr}(\boldsymbol{A}_1\boldsymbol{A}_2\cdots\boldsymbol{A}_m)\leqslant\prod_{k=1}^{m}[\operatorname{tr}(\boldsymbol{A}_k^m)]^{1/m}\leqslant\frac{1}{m}\sum_{k=1}^{m}\operatorname{tr}(\boldsymbol{A}_k^m).\tag{5}$$

当 $\boldsymbol{A}_1,\boldsymbol{A}_2,\cdots,\boldsymbol{A}_m$ 为 n 阶 Hermite 半正定矩阵时,由前知道,

$$|\operatorname{tr}(\boldsymbol{A}_1\boldsymbol{A}_2\cdots\boldsymbol{A}_m)|\leqslant\sum_{i=1}^{n}\lambda_i(\boldsymbol{A}_1)\lambda_i(\boldsymbol{A}_2)\cdots\lambda_i(\boldsymbol{A}_m),$$

接着应用多因子的 Hölder 不等式的各种推广，例如应用 Jensen 不等式(见[2. p. 29])可得新的不等式.

推论 2 设 $\boldsymbol{A}_1, \boldsymbol{A}_2, \cdots, \boldsymbol{A}_m$ 皆为 n 阶 Hermite 半正定矩阵，则有

$$|\operatorname{tr}(\boldsymbol{A}_1 \boldsymbol{A}_2 \cdots \boldsymbol{A}_m)| \leqslant \prod_{k=1}^{m} \operatorname{tr}(\boldsymbol{A}_k) \leqslant \left[\frac{1}{m}\sum_{k=1}^{m} \operatorname{tr}(\boldsymbol{A}_k)\right]^m. \tag{6}$$

推论 3 (Ault 不等式，见[4, p. 250])设 $\boldsymbol{A}_1, \boldsymbol{A}_2, \cdots, \boldsymbol{A}_m$ 为 n 阶复矩阵，记 $\boldsymbol{B} = \boldsymbol{A}_1\boldsymbol{A}_2\cdots\boldsymbol{A}_m$. 则

$$\operatorname{tr}\left(\frac{\boldsymbol{B}+\boldsymbol{B}^*}{2}\right) \leqslant \operatorname{tr}(\boldsymbol{B}^*\boldsymbol{B})^{1/2} \leqslant \frac{1}{m}\operatorname{tr}\left[\sum_{k=1}^{m} (\boldsymbol{A}_k^*\boldsymbol{A}_k)^{m/2}\right]. \tag{7}$$

证 首先有

$$\operatorname{tr}\left(\frac{\boldsymbol{B}+\boldsymbol{B}^*}{2}\right) = \frac{1}{2}\operatorname{tr}(\boldsymbol{B}) + \frac{1}{2}\operatorname{tr}(\boldsymbol{B}^*) = \operatorname{Re}\operatorname{tr}(\boldsymbol{B})$$

$$\leqslant \sum_{i=1}^{n} \sigma_i(\boldsymbol{B}) = \sum_{i=1}^{n} \lambda_i^{1/2}(\boldsymbol{B}^*\boldsymbol{B}) = \operatorname{tr}(\boldsymbol{B}^*\boldsymbol{B})^{1/2},$$

再按定理 2 的证明便知(7)式成立. 证毕. ■

参考文献

[1] 陈道琦. 关于半正定 Hermite 矩阵乘积迹的一个不等式. 数学学报，1988，31(4)：565－569.

[2] G. H. 哈代，J. E. 李特伍德，G. 波利亚. 不等式. 越民义，译，科学出版社，1965.

[3] Fan Ky. Maximum properties and inequalities for the eigenvalues of completely continuous operators. Proc. Nat. Acad. Sci. U. S. A., 1951, 37: 760－766.

[4] A W Marshall and I Olkin. Inequalities: Theory of Majorization and Its Applications. Academic Press, New York, 1979.

Linear Algebra and Its Applications
1983,49:137－151

奇异矩阵的一些性质①

On Some Properties of Singular Matrices

Abstarct Some results of Ostrowski in [5] are generalized to the case of monotonic norms.

§ 1. Introduction. Preliminaries

Let $\mathbf{C}^{n,n}(n>1)$ be the space of $n\times n$ complex matrices. For any $\boldsymbol{A}\in\mathbf{C}^{n,n}$, let $(\boldsymbol{A})_i$ denote the i th row of $\boldsymbol{A}$, $\boldsymbol{A}^{\mathrm{T}}$ the transpose matrix to $\boldsymbol{A}$, and $\boldsymbol{A}(i_1,i_2,\cdots,i_m)$ the principal submatrix of $\boldsymbol{A}$ formed by rows and columns with indices $i_1,i_2,\cdots,i_m$.

We recall that a matrix $\boldsymbol{A}\in\mathbf{C}^{n,n}$ is said to be *reducible* if there exists a permutation matrix $\boldsymbol{P}$ such that

$$\boldsymbol{A}=\boldsymbol{P}^{\mathrm{T}}\begin{bmatrix}\boldsymbol{A}_{11} & \boldsymbol{O}\\ \boldsymbol{A}_{21} & \boldsymbol{A}_{22}\end{bmatrix}\boldsymbol{P} \tag{1}$$

where $\boldsymbol{A}_{11}$ and $\boldsymbol{A}_{22}$ are square matrices; if $\boldsymbol{A}$ is not reducible it is called *irreducible*.

An *M-matrix* is real square matrix with nonpositive off-diagonal entries and having all principal minors positive (see (4.4) in [3]). A singular *M*-matrix is, by definition, a singular matrix in the closure of the set of *M*-matrices (see (5.2) in [3]). For convenience we will also refer to the matrices in this closure as *M*-matrices. The original *M*-matrices are then the nonsingular ones in this set (cf. Theorems

① Received 12 November 1981; revised 1 June 1982. Submitted by Hans Schneider.

(4.3) (5.1) and (5.5) in [3]).

Given any $\mathbf{A}=(a_{ij})\in\mathbf{C}^{n,n}$, we define its *comparison matrix* $M(\mathbf{A})=(m_{ij})$ *by*

$$m_{ii}=|a_{ii}|,\ m_{ij}=-|a_{ij}|\ (i\neq j;\ i,j=1,2,\cdots,n).$$

A norm Φ on $\mathbf{C}^n$, the n-dimensional complex coordinate space, is called *monotonic* [*strictly monotonic*] if for every $\boldsymbol{x}=(x_1,x_2,\cdots,x_n)$ and $\boldsymbol{y}=(y_1,y_2,\cdots,y_n)$ in $\mathbf{C}^n$, $|x_i|\leqslant|y_i|$ for $i=1,2,\cdots,n$ (with strict inequality holding for at least one index) implies

$$\Phi(\boldsymbol{x})\leqslant\Phi(\boldsymbol{y})\ [\Phi(\boldsymbol{x})<\Phi(\boldsymbol{y})].$$

Let Φ be any monotonic norm on $\mathbf{C}^n$, and let Ψ be its polar norm (then Ψ is also monotonic). It is well known [1] that Ψ is absolute and strictly homogeneous, i. e., $\Phi(\boldsymbol{x})=\Phi(|\boldsymbol{x}|)$ for all $\boldsymbol{x}\in\mathbf{C}^n$ and $\Phi(b\boldsymbol{x})=|b|\Phi(\boldsymbol{x})$ for every scalar b, respectively, and that the *strong Hölder inequality* holds:

$$\sum_{i=1}^{n}|x_iy_i|\leqslant\Phi(\boldsymbol{x})\Psi(\boldsymbol{y})\tag{2}$$

for any $\boldsymbol{x},\ \boldsymbol{y}\in\mathbf{C}^n$. We will call $(\boldsymbol{x},\ \boldsymbol{y})$ a *dual pair* if equality holds in (2).

Introduce the notation that $\mathbf{A}_\beta=(|a_{ij}|^\beta)-\mathrm{diag}(|a_{11}|^\beta,|a_{22}|^\beta,\cdots,|a_{nn}|^\beta)$ for $0<\beta<1$ if $\mathbf{A}=(a_{ij})\in\mathbf{C}^{n,n}$.

Nowosad and Tovar [4] proved that every $\mathbf{A}\in\mathbf{C}^{n,n}$ satisfying

$$f_i(\mathbf{A})\overset{\mathrm{df}}{\equiv}\alpha(n)\Phi((\mathbf{A}_\beta)_i)\Psi((\mathbf{A}_{1-\beta})_i^{\mathrm{T}})<|a_{ii}|$$
$$(i=1,2,\cdots,n;\ 0<\beta<1)\tag{3}$$

is nonsingular, where $\alpha(n)$ is the constant associated with Φ, and that if $\Phi_1(\boldsymbol{B})=\Phi_1(\boldsymbol{B}^{\mathrm{T}})$ for every $\boldsymbol{B}\in\mathbf{C}^{n,n}$, where $\Phi_1(\boldsymbol{B})=\Phi(\Phi((\boldsymbol{B})_1),\cdots,\Phi((\boldsymbol{B})_n))$, then $\alpha(n)=1$. Examples of such norms are the l_p norms $(p\geqslant1)$.

In [5], Ostrowski stated and proved some elegant theorems based on the l_p norms. Following Ostrowski [5], in the present paper we will provide some characterizations of the singular matrix $\mathbf{A}$ in $\mathbf{C}^{n,n}$ under the condition

$$f_i(\mathbf{A})\leqslant|a_{ii}|\quad(i=1,2,\cdots,n)\tag{4}$$

and some connected estimates for the moduli of eigenvalues of matrices.

§ 2. Singular Matrices with the Condition(4)

We may assume in our discussion that none of the $\Phi((A_\beta)_i)$, $\Psi((A_{1-\beta})_i^{\mathrm{T}})$ vanishes. For, if $\Phi((A_\beta)_i)=0$ [or $\Psi((A_{1-\beta})_i^{\mathrm{T}})=0$], all off-diagonal entries in the ith row [or ith column] are 0 and the problem is essentially reduced to the $(n-1)$-square case.

We begin with

Lemma 1 *Suppose that $\mathbf{A}\in \mathbf{C}^{n,n}$ is irreducible and singular, and that $M(\mathbf{A})$ is an M-matrix. Then the rank of $\mathbf{A}$ is $n-1$, and the equations*

$$\sum_{j=1}^{n} a_{ij}x_j = 0 \quad (i=1,2,\cdots,n) \tag{5}$$

have a nontrivial solution $\boldsymbol{x}=(x_1,x_2,\cdots,x_n)$, where none of the x_i vanishes.

Proof We see easily that $M(\mathbf{A})$ then is an irreducible and singular M-matrix. Indeed, (5) implies

$$|a_{ii}||x_i| - \sum_{j\neq i}|a_{ij}||x_j| \leqslant 0 \quad (i=1,2,\cdots,n),$$

and therefore if $M(\mathbf{A})$ were nonsingular, then, by 11° in Theorem (4. 3) in [3], it would follow that $|x_i|\leqslant 0$ $(i=1,2,\cdots,n)$, a contradiction. According to the Theorem 5. 7 in [3], we have that each principal submatrix of $M(\mathbf{A})$ of order $<n$ is a nonsingular M matrix. Thus, each principal submatrix of $\mathbf{A}$ of order $<n$ is also nonsingular, so that rank $(\mathbf{A})=n-1$. Since $\mathbf{A}$ is singular, the system (5) can be solved so that not all x_i vanish. If we had for instance $x_n=0$, then from the $n-1$ first of these equations it would follow that a principal submatrix of $\mathbf{A}$ of order $n-1$ is singular, which contradicts the above result. ■

Assume that Φ is a strictly monotonic norm on $\mathbf{C}^n$ such that $\Phi_1(\boldsymbol{B})=\Phi_1(\boldsymbol{B}^{\mathrm{T}})$ for every $\boldsymbol{B}\in \mathbf{C}^{n,n}$. Then the relations (4) become

$$g_i(\mathbf{A}) \overset{\mathrm{df}}{\equiv} \Phi((\mathbf{A}_\beta)_i)\Psi((\mathbf{A}_{1-\beta})_i^{\mathrm{T}}) \leqslant |a_{ii}| \quad (i=1,2,\cdots,n). \tag{4'}$$

Theorem 1 *Let $\mathbf{A}\in \mathbf{C}^{n,n}$, and suppose that Φ is a strictly monotonic norm on $\mathbf{C}^n$ such that $\Phi_1(\boldsymbol{B})=\Phi_1(\boldsymbol{B}^{\mathrm{T}})$ for every $\mathbf{B}\in \mathbf{C}^{n,n}$, and that*

Ψ is the polar norm of Φ. If $\mathbf{A}$ is singular under the condition (4′), then we have

$$\Phi((\mathbf{A}_\beta)_i)\Psi((\mathbf{A}_{1-\beta})_i^{\mathrm{T}}) = |a_{ii}| \quad (i=1,2,\cdots,n; x_i \neq 0), \tag{6}$$

$$|a_{ii}||x_i| = \sum_{j\neq i}|a_{ji}||x_j| \quad (i=1,2,\cdots,n), \tag{7}$$

and

$$|a_{ii}||x_i| = \Phi((\mathbf{A}_\beta^{\mathrm{T}}\mathbf{X})_i)\Psi((\mathbf{A}_{1-\beta})_i^{\mathrm{T}}) \quad (i=1,2,\cdots,n), \tag{8}$$

where $\mathbf{X}=\mathrm{diag}(x_1,x_2,\cdots,x_n)$, $\boldsymbol{x}=(x_1,x_2,\cdots,x_n)$ is a nontrivial solution of the system

$$\sum_{j=1}^{n} a_{ji}x_j = 0 \quad (i=1,2,\cdots,n). \tag{5′}$$

Proof Since $\mathbf{A}\in\mathbf{C}^{n,n}$ is singular, the system (5′) has a nontrivial solution $\boldsymbol{x}=(x_1,x_2,\cdots,x_n)$, and therefore

$$|a_{ii}||x_i| \leqslant \sum_{j\neq i}|a_{ji}||x_j| = \sum_{j\neq i}|a_{ji}|^{\beta}|x_j||a_{ji}|^{1-\beta} \quad (i=1,2,\cdots,n). \tag{9}$$

It follows from (2) and (9) that

$$|a_{ii}||x_i| \leqslant \Phi((\mathbf{A}_\beta^{\mathrm{T}}\mathbf{X})_i)\Psi((\mathbf{A}_{1-\beta})_i^{\mathrm{T}}) \quad (i=1,2,\cdots,n), \tag{10}$$

where $\mathbf{X}=\mathrm{diag}(x_1,x_2,\cdots,x_n)$. Using (4′) we get

$$\Phi((\mathbf{A}_\beta)_i)\Psi((\mathbf{A}_{1-\beta})_i^{\mathrm{T}})|x_i| \leqslant \Phi((\mathbf{A}_\beta^{\mathrm{T}}\mathbf{X})_i)\Psi((\mathbf{A}_{1-\beta})_i^{\mathrm{T}}) \quad (i=1,2,\cdots,n)$$

and, dividing this by $\Psi((\mathbf{A}_{1-\beta})_i^{\mathrm{T}})$,

$$\Phi((\mathbf{X}\mathbf{A}_\beta)_i) \leqslant \Phi((\mathbf{A}_\beta^{\mathrm{T}}\mathbf{X})_i) \quad (i=1,2,\cdots,n). \tag{11}$$

According to the hypothesis about the norm Φ, we have from (11)

$$\Phi_1(\mathbf{X}\mathbf{A}_\beta) \leqslant \Phi_1(\mathbf{A}_\beta^{\mathrm{T}}\mathbf{X}). \tag{12}$$

But here we have equality as $\mathbf{A}_\beta^{\mathrm{T}}\mathbf{X} = (\mathbf{X}\mathbf{A}_\beta)^{\mathrm{T}}$, and thus also equality in (11) (10), and also in (9), and finally, for each i with $x_i\neq 0$, in (4′). We obtain therefore (6) (7), and (8). The Theorem 1 is now proved. ■

Theorem 2 *Let $\mathbf{A}\in\mathbf{C}^{n,n}$, and suppose that Φ is a monotonic norm on $\mathbf{C}^n$, and Ψ is its polar norm. If $\mathbf{A}$ is irreducible and singular under the condition (4), then we have that*

$$f_i(\mathbf{A}) = |a_{ii}| \quad (i=1,2,\cdots,n), \tag{13}$$

$$|a_{ii}||y_i| = \sum_{j\neq i} |a_{ij}||y_i| \quad (i = 1,2,\cdots,n), \tag{14}$$

where $\mathbf{y}=(y_1, y_2, \cdots, y_n)$ *is a solution of the system* (5) *with all* $y_i \neq 0$.

Proof If $\mathbf{A}$ is irreducible and singular under (4), then $M(\mathbf{A})$ is an irreducible and singular M-matrix. By the Lemma 1 there exists a $\mathbf{y}=(y_1, y_2, \cdots, y_n)$ with all $y_i \neq 0$ such that

$$\sum_{j=1}^{n} a_{ij} y_j = 0 \quad (i = 1,2,\cdots,n),$$

and therefore

$$|a_{ii}||y_i| \leqslant \sum_{j=1}^{n} |a_{ij}||y_j| \quad (i = 1,2,\cdots,n). \tag{15}$$

From this we have, for an arbitrary $d_i > 0$,

$$|a_{ii}||y_i| d_i \leqslant \sum_{j\neq i} |a_{ij}||y_j| d_i \quad (i = 1,2,\cdots,n),$$

which implies by (4)

$$d_i |y_i| \alpha(n)\Phi((\mathbf{A}_\beta)_i)\Psi((\mathbf{A}_{1-\beta})_i^{\mathrm{T}}) \leqslant \sum_{j\neq i} |a_{ij}||y_j| d_i \quad (i = 1,2,\cdots,n). \tag{16}$$

Summing (16) over $i=1,2,\cdots,n$, and using the homogeneity of Φ and Ψ, we obtain

$$\alpha(n)\sum_i \Phi((\mathbf{DA}_\beta)_i)\Psi(y_i(\mathbf{A}_{1-\beta})_i^{\mathrm{T}}) \leqslant \sum_i \sum_{j\neq i} |a_{ij}||y_j| d_i,$$

where $\mathbf{D}=\mathrm{diag}(d_1, d_2, \cdots, d_n)$. Now, following [4], we have by (2)

$$\begin{aligned}\sum_i \sum_{j\neq i} |a_{ij}||y_j| d_i &= \sum_j \sum_{i\neq j} |a_{ij}||y_j| d_i \\ &= \sum_i \sum_{j\neq i} |a_{ji}|^\beta d_j |a_{ji}|^{1-\beta} |y_i| \\ &\leqslant \sum_i \Phi((\mathbf{A}_\beta^{\mathrm{T}}\mathbf{D})_i)\Psi(y_i(\mathbf{A}_{1-\beta})_i^{\mathrm{T}})\end{aligned} \tag{17}$$

and thus

$$\alpha(n)\sum_i \Phi((\mathbf{DA}_\beta)_i)\Psi(y_i(\mathbf{A}_{1-\beta})_i^{\mathrm{T}}) \leqslant \sum_i \Phi((\mathbf{A}_\beta^{\mathrm{T}}\mathbf{D})_i)\Psi(y_i(\mathbf{A}_{1-\beta})_i^{\mathrm{T}}). \tag{18}$$

Applying to $\mathbf{A}$ Lemma 2.2 in [4], we have for some positive diagonal matrix $\mathbf{D}=\mathrm{diag}(d_1, d_2, \cdots, d_n)$

$$\Phi((\mathbf{A}_\beta^{\mathrm{T}}\mathbf{D})_i) = \alpha(\mathbf{A}_\beta)\Phi((\mathbf{DA}_\beta)_i) \leqslant \alpha(n)\Phi((\mathbf{DA}_\beta)_i) \quad (i = 1,2,\cdots,n),$$

where $\alpha(\boldsymbol{A}_\beta)$ is a constant depending on $\boldsymbol{A}_\beta$ and Φ. We obtain therefore

$$\alpha(n)\sum_i \Phi((\boldsymbol{D}\boldsymbol{A}_\beta)_i)\Psi(y_i(\boldsymbol{A}_{1-\beta})_i^{\mathrm{T}}) = \sum_i \Phi((\boldsymbol{A}_\beta^{\mathrm{T}}\boldsymbol{D})_i)\Psi(y_i(\boldsymbol{A}_{1-\beta})_i^{\mathrm{T}}).$$

Hence (13) and (14) hold. The Theorem 2 is proved. ■

Theorem 3 *Let* $\boldsymbol{A}\in \mathbf{C}^{n,n}$ *and let* Φ *and* Ψ *be as in Theorem* 1. *If* $\boldsymbol{A}$ *is singular under the condition* (4′), *and of rank* $n-k$ $(1\leqslant k\leqslant n-1)$, *then* $\boldsymbol{A}$ *can brought, by an* $n\times n$ *permutation matrix* $\boldsymbol{P}$, *into the form*

$$\boldsymbol{PAP}^{\mathrm{T}} = \left[\begin{array}{ccc:c} \boldsymbol{B}_1 & & \boldsymbol{O} & \\ & \ddots & & \boldsymbol{O} \\ \boldsymbol{O} & & \boldsymbol{B}_k & \\ \hdashline & \boldsymbol{U} & & \boldsymbol{V} \end{array}\right], \tag{19}$$

where each square submatrix $\boldsymbol{B}_i$ *is irreducible and singular (eventually* $\boldsymbol{B}_1=\boldsymbol{A}$ *if* $\boldsymbol{A}$ *is irreducible);* $\boldsymbol{V}$ *is a nonsingular and square submutrix of* $\boldsymbol{A}$. *For each* i *for which* a_{ii} *lies in* $\boldsymbol{B}_1+\cdots+\boldsymbol{B}_k$, *the relation* (4′) *holds with equality. If, further,* Ψ *is also strictly monotonic, then* $\boldsymbol{U}=\boldsymbol{O}$ *in* (19).

Proof We first observe that if $\boldsymbol{A}$ is irreducible and singular under (4′), then $M(\boldsymbol{A})$ is an irreducible and singular M-matrix. From Lemma 1 it follows that $\mathrm{rank}(\boldsymbol{A})=n-1$, whence $\boldsymbol{A}=\boldsymbol{B}_1$, and that the system (5) has a nontritial solution $\boldsymbol{x}=(x_1,x_2,\cdots,x_n)$ with all $x_i\neq 0$. We have by Theorem 1 that all relations (4′) hold with equality.

Next, if $\boldsymbol{A}$ is reducible and singular under (4′), we can consider the reduced normal form of $\boldsymbol{A}$ (see [7, p. 46]):

$$\boldsymbol{A}=(a_{ij})=\boldsymbol{Q}^{\mathrm{T}}\begin{bmatrix} \boldsymbol{A}_{11} & \boldsymbol{A}_{12} & \cdots & \boldsymbol{A}_{1s} \\ & \boldsymbol{A}_{22} & \cdots & \boldsymbol{A}_{2s} \\ & & \ddots & \vdots \\ \boldsymbol{O} & & & \boldsymbol{A}_{ss}\end{bmatrix}\boldsymbol{Q}, \tag{20}$$

where $\boldsymbol{Q}$ is an $n\times n$ permutation matrix, and each $\boldsymbol{A}_{ii}$ is square, irreducible, and of order n_i. We have, of course, $n_1+n_2+\cdots+n_s=n$. Obviously, some of the $\boldsymbol{A}_{ii}$ are singular, as $\boldsymbol{A}$ is singular, and the number of such submatrices in (20) is exactly equal to k, $k\leqslant s$, by Lemma 1. We may assume, without loss of generality, that they are $\boldsymbol{A}_{11},\cdots,\boldsymbol{A}_{kk}$, and $\boldsymbol{A}_{11}=\boldsymbol{A}(1,2,\cdots,n_1)$.

Put now

$$\boldsymbol{A}'=(a'_{ij})=\boldsymbol{Q}^{\mathrm{T}}\begin{bmatrix}\boldsymbol{A}_{11} & \boldsymbol{O} & \cdots & \boldsymbol{O}\\ & \boldsymbol{A}_{22} & \cdots & \boldsymbol{A}_{2s}\\ & & \ddots & \vdots\\ \boldsymbol{O} & & & \boldsymbol{A}_{ss}\end{bmatrix}\boldsymbol{Q}. \tag{21}$$

It is clear that $|a'_{ij}|\leqslant|a_{ij}|(i,j=1,2,\cdots,n)$, and thus

$$\Phi((\boldsymbol{A}'_{\beta})_i)\leqslant\Phi((\boldsymbol{A}_{\beta})_i)\quad(i=1,2,\cdots,n), \tag{22}$$

$$\Psi((\boldsymbol{A}'_{1-\beta})_i^{\mathrm{T}})\leqslant\Psi((\boldsymbol{A}_{1-\beta})_i^{\mathrm{T}})\quad(i=1,2,\cdots,n). \tag{23}$$

It follows from this that for $i=1,2,\cdots,n$

$$\Phi((\boldsymbol{A}'_{\beta})_i)\Psi((\boldsymbol{A}'_{1-\beta})_i^{\mathrm{T}})\leqslant\Phi((\boldsymbol{A}_{\beta})_i)\Psi((\boldsymbol{A}_{1-\beta})_i^{\mathrm{T}})\leqslant|a_{ii}|. \tag{24}$$

According to the generalized version of Ky Fan's theorem, due to Carlson and Varga [2, Cor. 2 of Theorem 6], we have, in particular, for n suitably chosen positive numbers $x_1,x_2,\cdots,x_n$,

$$\frac{1}{x_i}\sum_{\substack{j=1\\ j\neq i}}^{n_1}|a_{ij}|x_j\leqslant\Phi((\boldsymbol{A}'_{\beta})_i)\Psi((\boldsymbol{A}'_{1-\beta})_i^{\mathrm{T}})\quad(i=1,2,\cdots,n_1). \tag{25}$$

By (24) and (25) we obtain that

$$\frac{1}{x_i}\sum_{\substack{j=1\\ j\neq i}}^{n_1}|a_{ij}|x_j\leqslant|a_{ii}|\quad(i=1,2,\cdots,n_1). \tag{26}$$

Since $\boldsymbol{A}_{11}$ is irreducible and singular, all the relations (26) hold with equality. We have therefore equality in (24) for $i=1,2,\cdots,n_1$, and also equality for $i=1,2,\cdots,n_1$ in (22) and (23). From the hypothesis about Φ it follows easily that $\boldsymbol{A}_{12}=\boldsymbol{O},\cdots,\boldsymbol{A}_{1s}=\boldsymbol{O}$ in (20). Similarly, we have that $\boldsymbol{A}_{ij}=\boldsymbol{O}$, $i\neq j$, $i=2,3,\cdots,k$, $j=1,2,\cdots,s$, and that (24) holds with equality for each i contained in the set of the indices of the matrices $\boldsymbol{A}_{22},\cdots,\boldsymbol{A}_{kk}$. Thus we infer, in general, that there exists a permutation matrix $\boldsymbol{P}$ such that $\boldsymbol{PAP}^{\mathrm{T}}$ has the form of (19).

Similarly, if Ψ is also strictly monotonic, then we have that either $\boldsymbol{A}=\boldsymbol{B}_1$, or $\boldsymbol{A}_{ji}=\boldsymbol{O}$, $j\neq i$, for each i for which $\boldsymbol{A}_{ii}$ is singular in (20). This means that $\boldsymbol{A}$ has the form of (19) with $\boldsymbol{U}=\boldsymbol{O}$. *This completes the proof of Theorem* 3. ■

Remark Consider G-functions $(b_1, b_2, \cdots, b_n)$ and $(c_1, c_2, \cdots, c_n)$ (cf. [2, p. 98]) such that $b_i(\mathbf{A})$ and $c_i(\mathbf{A})$ depend only on the moduli of the off diagonal entries of the ith row and the ith column of $\mathbf{A}$, $i = 1, 2, \cdots, n$, respectively, and that $b_i(\mathbf{A})$ and $c_i(\mathbf{A})$ are strictly monotonic and monotonic, $i = 1, 2, \cdots, n$, respectively. We see, by using arguments similar to those used for Theorem 3, that the corresponding conclusion stated in the preceding theorem holds also if $\mathbf{A} \in \mathbf{C}^{n,n}$ is singular under the condition

$$\{b_i(\mathbf{A})\}^{\beta}\{c_i(\mathbf{A})\}^{1-\beta} \leqslant |a_{ii}| \quad (i = 1, 2, \cdots, n;\ 0 < \beta < 1). \tag{27}$$

Theorem 4 *Let* $\mathbf{A} \in \mathbf{C}^{n,n}$, *and let* Φ *and* Ψ *be as in Theorem* 1. *If* $\mathbf{A}$ *is irreducible and singular under the condition* (4′), *then we have for* n *suitable numbers* δ_i *of modulus* 1

$$a_{ij} = -\varepsilon_i \delta_i \bar{\delta}_j |a_{ij}| \qquad (i \neq j;\ i, j = 1, 2, \cdots, n), \tag{28}$$

where $\varepsilon_j = a_{jj}/|a_{jj}|$.

Proof It is similar to the proof of the Theorem 2 in [5], and is omitted. ■

Theorem 5 *Let* $\mathbf{A} \in \mathbf{C}^{n,n}$, *and let* Φ *and* Ψ *be as in the Theorem* 2. *If* $\mathbf{A}$ *is irreducible and singular under the condition* (4), *then we have for* n *suitable numbers* σ_i *of modulus* 1

$$a_{ij} = -\varepsilon_j \delta_j \bar{\delta}_i |a_{ij}| \qquad (i \neq j;\ i, j = 1, 2, \cdots, n), \tag{29}$$

where $\varepsilon_j = a_{jj}/|a_{jj}|$.

Proof It is also similar to the proof of the Theorem 2 in [5], and therefore is omitted. ■

From the relations (28) and (29) it follows directly that

$$\det(\mathbf{A}) = \det(M(\mathbf{A}))\prod_{i=1}^{n}\varepsilon_i, \tag{30}$$

where $\left|\prod_{i=1}^{n}\varepsilon_i\right| = 1$. Hence, Theorem 4 or Theorem 5 reduces the discussion of conditions for the singularity of $\mathbf{A}$ to the discussion of the case where all a_{ij} are real and

$$a_{ii} = |a_{ii}|, \quad a_{ij} = -|a_{ij}| \quad (i \neq j). \tag{31}$$

Note that under the assumptions of Theorem 4 or Theorem 5, $M(\boldsymbol{A})$ is an irreducible and singular M-matrix.

In further discussion, if $\boldsymbol{A}$ is irreducible and Φ is a monotonic norm such that $\Phi_1(\boldsymbol{B}) = \Phi_1(\boldsymbol{B}^{\mathrm{T}})$ for every $\boldsymbol{B} \in \mathbf{C}^{n,n}$, then the relations (4′) become

$$\Phi((\boldsymbol{A}_\beta)_i)\Psi((\boldsymbol{A}_{1-\beta})_i^{\mathrm{T}}) = | a_{ii} | \quad (i = 1,2,\cdots,n), \tag{32}$$

and we assume that none of the $\Phi((\boldsymbol{A}_\beta)_i)$, $\Psi((\boldsymbol{A}_{1-\beta})_i^{\mathrm{T}})$ vanishes.

Introduce n^2 numbers ε_{ij} where

$$\begin{aligned} \varepsilon_{ii} &= 0; \\ \varepsilon_{ij} &= 1 \quad \text{if} \quad a_{ij} \neq 0 \quad (i \neq j), \\ \varepsilon_{ij} &= 0 \quad \text{if} \quad a_{ij} = 0 \quad (i \neq j). \end{aligned} \tag{33}$$

Let now

$$m_i = \max_{j \neq i} | a_{ji} | \quad (i = 1,2,\cdots,n), \tag{34}$$

$$\beta_i = | a_{ii} | / m_i \quad (i = 1,2,\cdots,n), \tag{35}$$

$$\tau = \sqrt{\beta_1^2 + \beta_2^2 + \cdots + \beta_n^2}, \quad c_i = \tau/\beta_i \quad (i = 1,2,\cdots, n). \tag{36}$$

Theorem 6 *Let $\mathbf{A} \in \mathbf{C}^{n,n}$, and let Φ and Ψ be as in Theorem 1. Suppose that $\Psi(\boldsymbol{x})$ depends only on the $\max_i |x_i|$ for every $\boldsymbol{x} \in \mathbf{C}^n$, and that all relations (32) are valid. Then $\mathbf{A}$ irreducible and singular implies that*

$$| a_{ij} | = \varepsilon_{ij} m_j \quad (i \neq j;\ i,j = 1,2,\cdots, n), \tag{37}$$

and that $\tau\{\Psi(\boldsymbol{e})\}^{-1}$ is an eigenvalue of the operator $F: \mathbf{R}^n - \{\mathbf{0}\} \to \mathbf{R}^n$ ($\mathbf{R}^n$ denotes the n-dimensional real coordinate space), where $F = (F_1, F_2, \cdots, F_n)$ is defined by

$$F_i(\boldsymbol{d}) = \Phi((\boldsymbol{ED})_i) \quad (i = 1,2,\cdots,n), \tag{38}$$

in which $\boldsymbol{e} = (1,1,\cdots, 1) \in \mathbf{R}^n$, $\boldsymbol{E} = (\varepsilon_{ij} c_i) \in \mathbf{C}^{n,n}$, $\boldsymbol{d} = (d_1, d_2, \cdots, d_n)$, and $\mathbf{D} = \operatorname{diag}(d_1, d_2, \cdots, d_n)$; and to $\tau\{\Psi(\boldsymbol{e})\}^{-1}$ corresponds an eigenvector with positive components.

Proof From the hypothesis of this theorem we have equality in (10), and thus it follows that each $((\boldsymbol{A}_\beta^{\mathrm{T}}\boldsymbol{X})_i, (\boldsymbol{A}_{1-\beta})_i^{\mathrm{T}})$ is a dual pair for $i = 1,2,\cdots,n$, that is,

$$\sum_{j\neq i} | a_{ji} |^{\beta} | x_j || a_{ji} |^{1-\beta} = \Phi((\mathbf{A}_{\beta}^{\mathrm{T}}\mathbf{X})_i)\Psi(\,(\mathbf{A}_{1-\beta})_i^{\mathrm{T}}) \quad (i = 1,2,\cdots,n). \tag{39}$$

Since $\Psi(\boldsymbol{x})$ depends only on max $i\ |x_i|$ for every $\boldsymbol{x}\in\mathbf{C}^n$, we have that

$$\Psi(\,(\mathbf{A}_{1-\beta})_i^{\mathrm{T}}) = \Psi(\max_{j\neq i} | a_{ji} |^{1-\beta}\boldsymbol{e}) = m_i^{1-\beta}\Psi(\boldsymbol{e}) \quad (i = 1,2,\cdots,n) \tag{40}$$

where $\boldsymbol{e}=(1,\cdots,1)\in\mathbf{R}^n$. Introducing this into (39), we see easily

$$\sum_{j\neq i} | a_{ji} |^{\beta} | x_j || a_{ji} |^{1-\beta} = \max_{j\neq i} | a_{ji} |^{1-\beta}\sum_{j\neq i} | a_{ji} |^{\beta} | x_j |$$
$$= m_i^{1-\beta}\Psi(\boldsymbol{e})\Phi((\mathbf{A}_{\beta}^{\mathrm{T}}\mathbf{X})_i),$$

which implies that

$$| a_{ji} | = m_i\varepsilon_{ji} \quad (j\neq i;\ i,j = 1,2,\cdots,n). \tag{41}$$

and

$$\{\Psi(\boldsymbol{e})\}^{-1}\sum_{j\neq i}\varepsilon_{ji} | x_j | = \Phi(\varepsilon_{1i}x_1,\cdots,\varepsilon_{i-1i}x_{i-1},0,\varepsilon_{i+1i}x_{i+1},\cdots,\varepsilon_{ni}x_n)$$
$$(i = 1,2,\cdots,n). \tag{42}$$

We now have from (32) (40) and (41)

$$| a_{ii} | = m_i^{1-\beta}\Psi(\boldsymbol{e})\Phi(\varepsilon_{i1}m_1^{\beta},\varepsilon_{i2}m_2^{\beta},\cdots,\varepsilon_{in}m_n^{\beta}) \quad (i = 1,2,\cdots,n). \tag{43}$$

Put

$$h_i = m_i^{\beta} \quad (i = 1,2,\cdots,n). \tag{44}$$

Then the equations (43) become by (35) and (36)

$$\beta_i\{\Psi(\boldsymbol{e})\}^{-1}h_i = \Phi(\varepsilon_{i1}h_1,\varepsilon_{i2}h_2,\cdots,\varepsilon_{in}h_n) \quad (i = 1,2,\cdots,n)$$

or

$$\tau\{\Psi(\boldsymbol{e})\}^{-1}h_i = \Phi(\varepsilon_{i1}c_ih_1,\varepsilon_{i2}c_ih_2,\cdots,\varepsilon_{in}c_ih_n) \quad (i = 1,2,\cdots,n). \tag{45}$$

Let $F=(F_1,F_2,\cdots,F_n)$ be the operator defined by (38). From (45) we have

$$\tau\{\Psi(\boldsymbol{e})\}^{-1}(h_1,h_2,\cdots,h_n) = (\Phi((\boldsymbol{EH})_1),\Phi((\boldsymbol{EH})_2),\cdots,\Phi((\boldsymbol{EH})_n))$$
$$= F(h_1,h_2,\cdots,h_n), \tag{46}$$

where $\boldsymbol{E}=(\varepsilon_{ij}c_i)\in\mathbf{C}^{n,n}$, $\boldsymbol{H}=\mathrm{diag}(h_1,h_2,\cdots,h_n)$. This means that the operator possesses a positive eigenvalue $\tau\{\Psi(\boldsymbol{e})\}^{-1}$, to which corresponds an eigenvector $(h_1,h_2,\cdots,h_n)\in\mathbf{R}^n$ with all $h_i>0$. The Theorem

6 is proved. ■

Remark From Lemma 1 in [6], it follows

$$0 < c_\Phi \sum_{j=1}^{n} | y_j | \leqslant \Phi(\boldsymbol{y}) \leqslant f_\Phi \sum_{j=1}^{n} | y_j | \tag{47}$$

for any $\boldsymbol{y} \in \mathbf{C}^n$, $\boldsymbol{y} \neq \mathbf{0}$, and any monotonic norm Φ on $\mathbf{C}^n$, where c_Φ, f_Φ depend only on Φ. By (42) it is seen easily that

$$c_\Phi \leqslant \{\Psi(\boldsymbol{e})\}^{-1} \leqslant f_\Phi. \tag{48}$$

Suppose that τ' is the maximum eigenvalue of the matrix $(\varepsilon_{ij} c_i)$. To τ' corresponds an eigenvector $(h_1, h_2, \cdots, h_n)$ given as in (44) by Theorem 3 of [5]. Then we have from (45) that

$$c_\Phi \tau' \leqslant \tau\{\Psi(\boldsymbol{e})\}^{-1} \leqslant f_\Phi \tau'. \tag{49}$$

In particular, if $c_\Phi = f_\Phi$, we have, by (48) (49) and (43), $\tau = \tau'$ and

$$| a_{ii} | = m_i^{1-\beta} \sum_{j=1}^{n} \varepsilon_{ij} m_j^\beta \quad (i = 1,2,\cdots,n). \tag{50}$$

Thus the Theorem 6 then reduces to the Theorem 3 of Ostrowski [5], and we see that the conditions (37) and (46) are not only necessary but also sufficient for an irreducible matrix **A** satisfying (32) to be singular under the assumptions of Theorem 6.

§ 3. Some Connected Estimates for the Moduli of Eigenvalues of Matrices

Suppose that $\mathbf{A} \in \mathbf{C}^{n,n}$ has λ as the eigenvalue with geometric multiplicity k $(1 \leqslant k \leqslant n-1)$ and having the maximal modulus ρ. From the condition (3) it follows easily that

$$\rho \leqslant \max_i [\, | a_{ii} | + f_i(\mathbf{A})]. \tag{51}$$

Theorem 7 *Let $\mathbf{A} \in \mathbf{C}^{n,n}$, and let Φ and Ψ be as in the Theorem 1 [note then that $f_i(\mathbf{A}) = g_i(\mathbf{A})$, $i = 1, 2, \cdots, n$]. If we have equality in (51), then ρ is the maximal eigenvalue of the matrix $(|a_{ij}|) \in \mathbf{C}^{n,n}$ and **A** can be brought into the form of (19) where each square submatrix $\boldsymbol{B}_i$ is irreducible (eventually $\mathbf{A} = \boldsymbol{B}_1$) and has λ as eigenvalue. For each i for which a_{ii} lies in the matrix $\boldsymbol{B}_1 \dot{+} \boldsymbol{B}_2 \dot{+} \cdots \dot{+} \boldsymbol{B}_k$, we have*

$$\rho = | a_{ii} | + g_i(\mathbf{A}). \tag{52}$$

If Ψ is also strictly monotonic, then $\mathbf{U}=\mathbf{O}$ in (19). If, further, $\mathbf{A}$ is irreducible (in this case $\mathbf{A}=\mathbf{B}_1$), we have

$$\mathbf{A}=\varepsilon \mathbf{D}(|a_{ij}|)\mathbf{D}^{-1}, \tag{53}$$

where $\mathbf{D}$ is a diagonal matrix in $\mathbf{C}^{n,n}$ with the diagonal entries of modulus 1, and $|\varepsilon|=1$. In particular, the quotients a_{ii}/λ are real and nonnegative.

Proof Let τ be the maximal eigenvalue of the nonnegative matrix $(|a_{ii}|)$. Put $\lambda=\varepsilon\rho$, $|\varepsilon|=1$, and $a_{ii}=\varepsilon_i|a_{ii}|$, $|\varepsilon_i|=1$, $i=1,2,\cdots,n$. Then we have

$$\rho\leqslant\tau\leqslant\max_i[|a_{ii}|+g_i(\mathbf{A})]. \tag{54}$$

If equality holds in (51), then we have also equality in (54) and $\rho=\tau$. But then it follows that

$$|a_{ii}-\lambda|\geqslant\rho-|a_{ii}|\geqslant f_i(\mathbf{A})=g_i(\mathbf{A})\quad(i=1,2,\cdots,n), \tag{55}$$

and Theorem 3 can be applied to the matrix $\mathbf{A}-\lambda\mathbf{I}$, which is of rank $n-k$. We see that $\mathbf{A}$ has k square and irreducible submatrices $\mathbf{B}_1,\mathbf{B}_2,\cdots,\mathbf{B}_k$ (eventually $\mathbf{A}=\mathbf{B}_1$ if $\mathbf{A}$ is irreducible) each of which has λ as its eigenvalue, and that

$$|a_{ii}-\lambda|=g_i(\mathbf{A})=\rho-|a_{ii}| \tag{56}$$

for each i for which a_{ii} lies in $\mathbf{B}_1\dotplus\mathbf{B}_2\dotplus\cdots\dotplus\mathbf{B}_k$.

If Ψ is strictly monotonic, too, then we have $\mathbf{U}=\mathbf{O}$ in (19). If, further, $\mathbf{A}$ is irreducible, applying Theorem 3 again to $\mathbf{A}-\lambda\mathbf{I}$, we have that

$$|a_{ii}-\lambda|=g_i(\mathbf{A})=\rho-|a_{ii}| \tag{57}$$

for $i=1,2,\cdots,n$. But then it follows from (57) that $|\varepsilon_i|a_{ii}|-\varepsilon\rho|=\rho-|a_{ii}|$, whence $\varepsilon_i=\varepsilon$. Applying Theorem 4 to the irreducible matrix $\mathbf{A}-\lambda\mathbf{I}$, since $a_{ii}-\lambda=-\varepsilon|a_{ii}-\lambda|$ by (57), we have

$$a_{ij}=\varepsilon\,\delta_i\bar{\delta}_j\,|a_{ij}|\qquad(i,j=1,2,\cdots,n), \tag{58}$$

which is equivalent to (53). In particular, as $(|a_{ij}|)$ is an irreducible and nonnegative matrix, by a theorem of Frobenius we have $\rho>0$, $\lambda\neq0$, and therefore

$$\frac{a_{ii}}{\lambda}-1=-\frac{\varepsilon}{\lambda}|a_{ii}-\lambda|=-\frac{\rho-|a_{ii}|}{\rho},$$

$$\frac{a_{ii}}{\lambda}=\frac{|a_{ii}|}{\rho}\geqslant 0.$$

This theorem is now proved. ■

Theorem 8 *Let $\mathbf{A}\in\mathbf{C}^{n,n}$, and let Φ and Ψ be as in the Theorem 2. Suppose that $\mathbf{A}$ is irreducible and that we have equality in (51). Then we have that ρ is the maximal eigenvalue of the matrix $(|a_{ij}|)$,*

$$\rho=|a_{ii}|+f_i(\mathbf{A})\qquad(i=1,2,\cdots,n),\tag{59}$$

and

$$\mathbf{A}=\varepsilon\,\mathbf{D}(|a_{ij}|)\mathbf{D}^{-1},$$

where ε, $\mathbf{D}$ are as in Theorem 7. In particular, the quotients a_{ii}/λ are real and nonnegative.

Proof It is similar to the proof of the preceding theorem, and is therefore omitted. ■

References

[1] F L Bauer, J Stoer and C Witzgall. Absolute and monotonic norms. Numer. Math. 1961,3:257—264.

[2] D H Carlson and R S Varga. Minimal G -functions. Linear Algebra Appl., 1973, 6:97—117.

[3] M Fiedler and V Pták. On matrices with non-positive off-diagonal and positive principal minors. Czech. Math. J., 1962, 12:382—400.

[4] P Nowosad and R Tovar. Spectral inequalities and G-functions. Linear Algebra Appl., 1980, 31:179—197.

[5] A M Ostrowski. On some inequalities in the theory of matrices. Scripta Math., 1962, 26:201—222.

[6] A M Ostrowski. On some metrical properties of operator matrices and matrices partitioned into blocks. J. Math. Anal. Appl., 1961, 2:161—209.

[7] R S Varga. Matrix Iterative Analysis. Prentice-Hall, Englewood Cliffs, N J, 1962.

Linear Algebra and Its Applications
1983, 55:169—180

关联 G-函数的对角占优的一些推广①

Some Generalizations of Diagonal Dominance Associated with G-functions

Abstract The concept of a G-function introduced by Nowosad and Hoffman is used to characterize classes of complex square matrices, resulting from various degrees of diagonal dominance associated with G-functions. Their relationship to the set of M-matrices is established.

§ 1. Introduction

As is well known, an $n\times n$ complex matrix $\boldsymbol{A}=(a_{ij})$ is called diagonally dominant if

$$|a_{ii}|\geqslant r_i(\boldsymbol{A})=\sum_{j\neq i}|a_{ij}| \qquad (i=1,2,\cdots,n). \tag{1.1}$$

This suggests the following extension making use of the concept of a G-function (see Section 2). An $n\times n$ complex matrix $\boldsymbol{A}=(a_{ij})$ is called G-diagonally dominant if there exists a G-function $\boldsymbol{f}=(f_1,f_2,\cdots,f_n)$ such that

$$|a_{ii}|\geqslant f_i(\boldsymbol{A}) \qquad (i=1,2,\cdots,n). \tag{1.2}$$

Denote by DG the set of such $n\times n$ complex matrices. Our objective in this paper is to characterize and interrelate some subclasses of DG, re-

① Received 28 June 1982; revised 3 November 1982.
Submitted by Richard A. Brualdi.

sulting from various degrees of diagonal dominance associated with G-functions. With the restriction that all G-functions used are monotonic, the major classes considered are the ones of G-diagonally dominant matrices, matrices of G-generalized positive type, and G-semistrictly diagonally dominant matrices, denoted by RDG (LDG), RPG (LPG), and RMG (LMG, MG), respectively. Their relationship to the sets of M-matrices and nonsingular M-matrices is established.

This paper is divided into five sections. In Section 2, we explain the necessary notation and definitions. Two of our main results, Theorems 3.3 and 3.4, appear in Section 3. Section 4 is devoted to investing properties of the matrices of G-generalized positive type; Theorem 4.1 is related very closely to Theorem 1 of [12]. Finally, in Section 5 we discuss some results on three classes of G-semistrictly diagonally dominant matrices, generalizing some theorems by Beauwens [1] and Varga [12].

§ 2. Notation and Definitions

In all that follows, $\mathbf{C}^{n,n}$ ($\mathbf{R}^{n,n}$) denotes the collection of all $n\times n$ complex (real) matrices; $\mathbf{C}_\pi^{n,n}$ denotes the subset of $\mathbf{C}^{n,n}$ with all diagonal entries nonzero. Denote by $\mathbf{R}^n$ the collection of all real column n-vectors $\boldsymbol{x}=(x_1,x_2,\cdots,x_n)^{\mathrm{T}}$. If $\boldsymbol{x}=(x_1,x_2,\cdots,x_n)^{\mathrm{T}}\in\mathbf{R}^n$, then $\boldsymbol{x}>\mathbf{0}$ ($\boldsymbol{x}\geqslant\mathbf{0}$) implies that $x_i>0$ ($x_i\geqslant 0$) for all $1\leqslant i\leqslant n$. Following Fiedler and Ptak [7], denote by $\mathbf{Z}^{n,n}$ the subset of $\mathbf{R}^{n,n}$ with nonpositive off-diagonal entries.

The concept of a G-function introduced by Nowosad [9] and Hoffman [8] will play an important role throughout the paper. Let $\mathscr{P}_n, n\geqslant 2$, be the set of all n-tuples of real-valued functions $\boldsymbol{f}=(f_1,f_2,\cdots,f_n)$ such that each $f_i(\mathbf{A})$ is nonnegative for any $\mathbf{A}\in\mathbf{C}^{n,n}$, and depends only on the moduli of the off-diagonal entries of the matrices. Recall (cf. [4, p. 98]) that $\boldsymbol{f}=(f_1,f_2,\cdots,f_n)\in\mathscr{P}_n$ is a G-function if each $\mathbf{A}=(a_{ij})\in\mathbf{C}^{n,n}$ satisfying

$$|a_{ii}|>f_i(\mathbf{A})\quad(i=1,2,\cdots,n)\tag{2.1}$$

is nonsingular. We denote by $\mathcal{G}_n$ the set of G-functions in $\mathcal{P}_n$. A G-function $\boldsymbol{f}=(f_1, f_2, \cdots, f_n)\in\mathcal{G}_n$ is said to be *monotonic* [respectively, *strictly row monotonic*] if, for any $\boldsymbol{A}$, $\boldsymbol{B}\in\mathbf{C}^{n,n}$,

$$|a_{ij}|\leqslant|b_{ij}| \quad (i,j=1,2,\cdots,n;\ i\neq j) \tag{2.2}$$

[respectively, with strict inequality for at least one pair (i_0, j_0), $j_0\neq i_0$] implies

$$f_i(\boldsymbol{A})\leqslant f_i(\boldsymbol{B}) \quad (i=1,2,\cdots,n) \tag{2.3}$$

[respectively, with strict inequality for $i=i_0$]. The set of strictly row monotonic G-functions in $\mathcal{G}_n$ will be denoted by $\mathcal{G}_n^r$ (Cf. [4] and [13].) As examples, if $\boldsymbol{x}=(x_1, x_2, \cdots, x_n)^{\mathrm{T}}>\boldsymbol{0}$, $\boldsymbol{y}=(y_1, y_2, \cdots, y_n)^{\mathrm{T}}>\boldsymbol{0}$ and

$$r_i^x(\boldsymbol{A})=x_i^{-1}\sum_{j\neq i}|a_{ij}|x_j \quad (i=1,2,\cdots,n), \tag{2.4}$$

$$c_i^y(\boldsymbol{A})=r_i^y(\boldsymbol{A}^{\mathrm{T}}) \quad (i=1,2,\cdots,n), \tag{2.5}$$

then both $\boldsymbol{r}^x=(r_1^x, r_2^x, \cdots, r_n^x)$ and $\boldsymbol{c}^y=(c_1^y, c_2^y, \cdots, c_n^y)$ are in $\mathcal{G}_n^r$. Similarly, a G-function $\boldsymbol{f}=(f_1, f_2, \cdots, f_n)\in\mathcal{G}_n$ is said to be strictly row-column monotonic if, for any $\boldsymbol{A}$, $\boldsymbol{B}$ from $\mathbf{C}^{n,n}$ satisfying (2.2) and having one index i_0 such that in (2.2) strict inequality holds for some (i_0, j_0) with $j_0\neq i_0$ or for some (t_0, i_0) with $t_0\neq i_0$, we have that (2.3) holds with strict inequality for $i=i_0$ unless $f_{i_0}(\boldsymbol{A})=f_{i_0}(\boldsymbol{B})=0$. The set of such G-functions in $\mathcal{G}_n$ will be denoted by $\mathcal{G}_n^s$. (Cf. [13].)

It is well known that, for any $\boldsymbol{x}>\boldsymbol{0}$ and $\boldsymbol{y}>\boldsymbol{0}$, both $(\{r_1^x c_1^y\}^{1/2}, \{r_2^x c_2^y\}^{1/2}, \cdots, \{r_n^x c_n^y\}^{1/2})$ and $\frac{1}{2}\boldsymbol{r}^x+\frac{1}{2}\boldsymbol{c}^y=(\frac{1}{2}r_1^x+\frac{1}{2}c_1^y, \frac{1}{2}r_2^x+\frac{1}{2}c_2^y, \cdots, \frac{1}{2}r_n^x+\frac{1}{2}c_n^y)$ *are in* $\mathcal{G}_n^s$.

A matrix $\boldsymbol{A}=(a_{ij})\in\mathbf{Z}^{n,n}$ is called a (possibly singular) *M-matrix* if $\boldsymbol{A}+\lambda\boldsymbol{I}_n$ is nonsingular for any scalar $\lambda>0$. For irreducible $\boldsymbol{A}\in\mathbf{Z}^{n,n}$, $\boldsymbol{A}$ is an M-matrix if $\boldsymbol{A}+\mathrm{diag}(d_1, d_2, \cdots, d_n)$ is nonsingular whenever $\boldsymbol{d}=(d_1, d_2, \cdots, d_n)^{\mathrm{T}}\geqslant\boldsymbol{0}$ with $\boldsymbol{d}\neq\boldsymbol{0}$. (See, for example, [7] and [4].) We shall denote by P_0 and P the sets of all M-matrices and all nonsingular M-matrices in $\mathbf{Z}^{n,n}$ respectively.

Given any $\mathbf{A}=(a_{ij})\in \mathbf{C}^{n,n}$, we denote by $\mu(\mathbf{A})=(a_{ij})\in \mathbf{Z}^{n,n}$ the *comparison matrix* defined by

$$a_{ij}=\begin{cases} |a_{ij}|, & \text{for all} \quad i=j, \\ -|a_{ij}|, & \text{for all} \quad i\neq j. \end{cases}$$

It is seen that if $\mathbf{A}\in \mathrm{DG}$ then $\mu(\mathbf{A})\in \mathrm{DG}$ and P_0.

If $\mathbf{A}\in \mathbf{C}^{n,n}$ is reducible, there exists a permutation matrix $\mathbf{Q}\in \mathbf{R}^{n,n}$ such that

$$\mathbf{A}=\mathbf{Q}^{\mathrm{T}}\begin{bmatrix} \mathbf{A}_{11} & \mathbf{A}_{12} & \cdots & \mathbf{A}_{1m} \\ & \mathbf{A}_{22} & \cdots & \mathbf{A}_{2m} \\ & & \ddots & \vdots \\ \mathbf{O} & & & \mathbf{A}_{mm} \end{bmatrix}\mathbf{Q} \tag{2. 6}$$

where $2\leqslant m\leqslant n$, and each square submatrix $\mathbf{A}_{tt}$ is either irreducible or a 1×1 null matrix (cf. [11, p. 46]). Denote by $S_t(\mathbf{A})$ the set of indices of $\mathbf{A}_{tt}$, $1\leqslant t\leqslant m$. Continuing, for irreducible $\mathbf{A}\in \mathbf{C}^{n,n}$, we write simply $\mathbf{A}=[\mathbf{A}_{11}]$.

§ 3. *G*-Diagonal Dominance with Monotonic *G*-functions

In this section, the following subclasses of DG are investigated:

$$\mathrm{RDG}=\{\mathbf{A}\in \mathbf{C}^{n,n}: \exists\, \boldsymbol{f}\in \mathscr{G}_n^r \text{ such that (1. 2) holds}\}\ ; \tag{3. 1}$$

$$\mathrm{LDG}=\{\mathbf{A}\in \mathbf{C}^{n,n}: \exists\, \boldsymbol{f}\in \mathscr{G}_n^s \text{ such that (1. 2) holds}\}\ . \tag{3. 2}$$

We first collect some observations.

Lemma 3. 1 For any $\mathbf{A}\in \mathbf{C}^{n,n}$, the following are equivalent:

(a) $\mu(\mathbf{A})\in P$.

(b) *There exists an* $\boldsymbol{f}=(f_1,f_2,\cdots,f_n)\in \mathscr{G}_n$ *such that* (2. 1) *holds.*

(c) *There exists an* $\boldsymbol{f}=(f_1,f_2,\cdots,f_n)\in \mathscr{G}_n$ *such that*

$$|a_{ii}|\,|a_{jj}|>f_i(\mathbf{A})f_j(\mathbf{A}) \qquad \textit{for all} \quad i\neq j.$$

Proof The step from (a) to (b) is obvious by 2° of Theorem (4. 3) of [7]. The implication (b) to (c) is trivial. The fact that (c)→(a) follows immediately from Corollary 4. 1 of [8] and the definition of *M*-matrices. ■

Lemma 3.2 *Let* $\mathbf{A}\in\mathbf{C}^{n,n}$ *be of the form* (2.6).

(a) *If* $\mathbf{A}\in\mathrm{RDG}$ *with* $\boldsymbol{f}=(f_1,f_2,\cdots,f_n)\in\mathscr{G}_n^r$, *then* $\mathbf{A}_{tt}$ *singular implies* $\mathbf{A}_{tj}=\mathbf{O}$ *for all* $j\neq t$, *and* $|a_{ii}|=f_i(\mathbf{A})$ *whenever* $i\in S_t(\mathbf{A})$.

(b) *If* $\mathbf{A}\in\mathrm{LDG}\cap\mathbf{C}_\pi^{n,n}$ *with* $\boldsymbol{f}=(f_1,f_2,\cdots,f_n)\in\mathscr{G}_n^s$, *then* $\mathbf{A}_{tt}$ *singular implies* $\mathbf{A}_{tj}=\mathbf{A}_{jt}=\mathbf{O}$ *for all* $j\neq i$, *and* $|a_{ii}|=f_i(\mathbf{A})$ *whenever* $i\in S_t(\mathbf{A})$.

Proof Observe that if $\mathbf{A}$ is irreducible ($\mathbf{A}=[\mathbf{A}_{11}]$), then both (a) and (b) obviously hold by a theorem of Ky Fan [6]. Thus we may assume that $\mathbf{A}$ is reducible, and is of the form (2.6) with $2\leqslant m\leqslant n$. Put $\mathbf{A}_R=\boldsymbol{Q}^{\mathrm{T}}\operatorname{diag}(\mathbf{A}_{11},\mathbf{A}_{12},\cdots,\mathbf{A}_{mm})\boldsymbol{Q}$, where the $\mathbf{A}_{ii}$ and $\boldsymbol{Q}$ are as in (2.6). Clearly, $|(\mathbf{A})_{ij}|\geqslant|(\mathbf{A}_R)_{ij}|$ $(i,j=1,2,\cdots,n)$, where $(\mathbf{A})_{ij}$ denotes the (i,j) entry of $\mathbf{A}$, and therefore

$$f_i(\mathbf{A})\geqslant f_i(\mathbf{A}_R)\qquad(i=1,2,\cdots,n)\tag{3.3}$$

for any monotonic G-function $\boldsymbol{f}=(f_1,f_2,\cdots,f_n)$ by definition. According to Corollary 2 of Theorem 6 of [4], we can choose an $\boldsymbol{x}=(x_1,x_2,\cdots,x_n)^{\mathrm{T}}>\mathbf{0}$, depending on $\mathbf{A}_R$, such that

$$f_i(\mathbf{A}_R)\geqslant\hat{r}_i^x(\mathbf{A}_R)\qquad(i=1,2,\cdots,n),\tag{3.4}$$

where

$$\hat{r}_i^x(\mathbf{A}_R)=x_i^{-1}\sum_{\substack{j\neq i\\ j\in\langle i\rangle}}|a_{ij}|x_j\geqslant 0,$$

$\langle i\rangle$ denoting the $S_k(\mathbf{A}_R)$ containing i. [When $\langle i\rangle=\{i\}$, take $\hat{r}_i^x(\mathbf{A}_R)=0$.]

(a): Suppose that $\mathbf{A}\in\mathrm{RDG}$ with $\boldsymbol{f}\in\mathscr{G}_n^r$, and that $\mathbf{A}_{tt}$ is singular. It follows from (1.2) (3.3) and (3.4) that

$$|a_{ii}|\geqslant f_i(\mathbf{A})\geqslant f_i(\mathbf{A}_R)\geqslant\hat{r}_i^x(\mathbf{A}_R)\qquad(i=1,2,\cdots,n).\tag{3.5}$$

As $\mathbf{A}_{tt}$ is either irreducible or a 1×1 null matrix, $\mathbf{A}_{tt}$ singular implies, by (3.5),

$$|a_{ii}|=\hat{r}_i^x(\mathbf{A}_R)\quad \textit{for all}\quad i\in S_t(\mathbf{A}_R),$$

so that

$$|a_{ii}|=f_i(\mathbf{A})=f_i(\mathbf{A}_R)\quad\text{whenever}\quad i\in S_t(\mathbf{A}_R).\tag{3.6}$$

Since $f\in\mathscr{G}_n^r$ and $S_t(\mathbf{A}_R)=S_t(\mathbf{A})$, the conclusion follows immediately from (3.6).

(b): Suppose that $\mathbf{A}\in \mathrm{LDG}\cap \mathbf{C}_{\pi}^{n,n}$ with $f\in \mathscr{G}_n^s$, and that $\mathbf{A}_{tt}$ is singular. $\mathbf{A}$ similar argument applied to prove (a) above shows that (3.6) holds again in this case. Since all $a_{ii}\neq 0$ and $f\in \mathscr{G}_n^s$, we have by definition that $\mathbf{A}_{tj}=\mathbf{A}_{jt}=\mathbf{O}$ *for all* j≠t, *proving* (*b*). ■

We remark that Lemma 3.2(*b*) *can be considered as a generalization of Theorem* 3 *of* [5].

Two of our main results are the following theorems.

Theorem 3.3 *Let* $\mathbf{A}\in\mathbf{C}^{n,n}$ *be of the form* (2.6). *Then the following are equivalent*:

(a) $\mu(\mathbf{A})\in P_0$, *and* $\mu(\mathbf{A}_{tt})$ *singular implies* $\mathbf{A}_{tj}=\mathbf{O}$ *whenever* $j\neq t$.

(b) *There exists an* $\boldsymbol{x}=(x_1,x_2,\cdots,x_n)^{\mathrm{T}}>\mathbf{0}$ *such that* $\mu(A)\boldsymbol{x}\geqslant\mathbf{0}$.

(c) $\mathbf{A}\in$ RDG.

If $\mathbf{A}\in\mathbf{C}_{\pi}^{n,n}$, *each of the following is equivalent to the first three properties of this theorem*:

(d) *There exists an* $\boldsymbol{x}=(x_1,x_2,\cdots,x_n)^{\mathrm{T}}>\mathbf{0}$ *such that* $|a_{ii}||a_{jj}|\geqslant r_i^x(\mathbf{A})r_j^x(\mathbf{A})$ *whenever* $i\neq j$.

(e) *There exists an* $\boldsymbol{f}=(f_1,f_2,\cdots,f_n)\in\mathscr{G}_n^r$ *such that* $|a_{ii}||a_{jj}|\geqslant f_i(\mathbf{A})f_i(\mathbf{A})$ *whenever* $i\neq j$.

Proof (a) ↔ (b): This is known; see Theorem 1 of [2].

(b)→ (c),(d) and (d) → (e) are trivial.

(c)→ (a): Clearly, (c) implies $\mu(\mathbf{A})\in$ RDG, so that $\mu(\mathbf{A})\in P_0$. The second part of (a) follows at once from Lemma 3.2(a), since for $j\neq i$, $\mu(\mathbf{A}_{ij})=\mathbf{O}$ implies $\mathbf{A}_{ij}=\mathbf{O}$.

(*e*)→(*a*): *Observe that* $\mu(\mathbf{A})\in P_0$ by Lemma 3.1 and by definition. As before, we may assume that $\mathbf{A}$ is reducible, and is of the form (2.6) with $2\leqslant m\leqslant n$. $\mathbf{A}$ similar argument to that used in the proof of Lemma 3.2 shows that both (3.3) and (3.4) hold for some $\boldsymbol{x}=(x_1,x_2,\cdots,x_n)^{\mathrm{T}}>\mathbf{0}$, and therefore

$$|a_{ii}||a_{jj}|\geqslant f_i(\mathbf{A})f_j(\mathbf{A})\geqslant f_i(\mathbf{A}_R)f_j(\mathbf{A}_R)$$
$$\geqslant \hat{r}_i^x(\mathbf{A}_R)\hat{r}_j^x(\mathbf{A}_R)\quad \text{for all}\quad i\neq j. \tag{3.7}$$

Suppose now that $\mu(\mathbf{A}_{tt})$ is singular. Since all $a_{ii}\neq 0$, we have that

$\mu(\mathbf{A}_{tt})$ is irreducible with order $\geqslant 2$, so that $\hat{r}_i^x(\mathbf{A}_R)>0$ for all $i\in S_t(\mathbf{A})$. If some $\mathbf{A}_{ts}$, were a nonzero submatrix of $\mathbf{A}$, where $t<s\leqslant m$, then we would have $f_{i_0}(\mathbf{A})>f_{i_0}(\mathbf{A}_R)$ for some fixed $i_0\in S_t(\mathbf{A})$, as $\boldsymbol{f}\in\mathscr{G}_n^r$. It follows from (3.3) and (3.7) that

$$|a_{i_0i_0}||a_{jj}|>\hat{r}_{i_0}^x(\mathbf{A}_R)\hat{r}_j^x(\mathbf{A}_R) \quad \text{whenever} \quad j\in S_t(\mathbf{A}), \quad j\neq i_0.$$

Together with (3.7), this in turn implies that $\mathbf{X}_t^{-1}\mu(\mathbf{A}_{tt})\mathbf{X}_t$ [and also $\mu(\mathbf{A}_{tt})$] is nonsingular by a theorem of Brauer [3], where $\mathbf{X}_t=\operatorname{diag}(x_{i_1},x_{i_2},\cdots,x_{i_p})$ and $\{i_1,i_2,\cdots,i_p\}=S_t(\mathbf{A})$. This contradiction completes the proof of the theorem. ■

Theorem 3.4 *Let $\mathbf{A}\in\mathbf{C}_\pi^{n,n}$ be of the form (2.6). Then the following are equivalent:*

(a) *$\mu(\mathbf{A})\in P_0$ and $\mu(\mathbf{A}_{tt})$ singular implies $\mathbf{A}_{tj}=\mathbf{A}_{jt}=\mathbf{O}$ whenever $j\neq t$.*

(b) *There exists a positive diagonal matrix $\mathbf{D}\in\mathbf{R}^{n,n}$ such that $\mu(\mathbf{A})\mathbf{D}+\mathbf{D}\mu(\mathbf{A}^{\mathrm{T}})$ is positive semidefinite.*

(c) $\mathbf{A}\in$ LDG.

(d) *There exists an $\boldsymbol{f}=(f_1,f_2,\cdots,f_n)\in\mathscr{G}_n^s$ such that $|a_{ii}||a_{jj}|\geqslant f_i(\mathbf{A})f_j(\mathbf{A})$ whenever $i\neq j$.*

Proof (a)↔ (b) is known; see Theorem 1 of [2].

(c)→(a): Obviously, $\mu(\mathbf{A})\in P_0$. The second part of (a) follows immediately from Lemma 3.2(b).

(c) → (d) is trivial.

(a) → (c): Since $\mu(\mathbf{A})\in P_0$, $\mu(\mathbf{A}^{\mathrm{T}})\in P_0$, by definition. Moreover, if $\mu(\mathbf{A})$ satisfies the second part of (a), then so does $\mu(\mathbf{A}^{\mathrm{T}})$. Thus, in particular, both $\mu(\mathbf{A})$ and $\mu(\mathbf{A}^{\mathrm{T}})$ satisfy (b) of Theorem 3.3, so that $\mu(\mathbf{A})\boldsymbol{x}\geqslant\mathbf{0}$ and $\mu(\mathbf{A}^{\mathrm{T}})\boldsymbol{y}\geqslant\mathbf{0}$ for some pair of vectors $\boldsymbol{x}>\mathbf{0}$ and $\boldsymbol{y}>\mathbf{0}$. It follows that

$$|a_{ii}|\geqslant\frac{1}{2}r_i^x(\mathbf{A})+\frac{1}{2}c_i^y(\mathbf{A}) \quad (i=1,2,\cdots,n), \tag{3.8}$$

proving (c).

(d)→(a): The proof is similar to that of (e)→(a) of Theorem 3.3, and therefore is omitted. The theorem is proved. ■

§ 4. G-generalized Positive Type with Monotonic G-functions

Let $\mathbf{N}=\{1,2,\cdots,n\}$, and write $J(\mathbf{A},\boldsymbol{f})=\{i\in\mathbf{N}:|a_{ii}|>f_i(\mathrm{A})\}$ for any $\mathbf{A}\in\mathbf{C}^{n,n}$, and for $\boldsymbol{f}=(f_1,f_2,\cdots,f_n)\in\mathscr{P}_n$. It is clear that $J(\mathbf{A},\boldsymbol{f})=J(\mu(\mathbf{A}),\boldsymbol{f})$. In this section, we consider two subclasses of RDG and LDG:

RPG: The class of matrices $\mathbf{A}\in\mathbf{C}^{n,n}$ with the following properties. For each $\mathbf{A}\in$ RPG there exists an $\boldsymbol{f}=(f_1,f_2,\cdots,f_n)\in\mathscr{G}_n^r$ such that:

(i) (1.2) holds with

$$J(\mathbf{A},\boldsymbol{f})\text{ nonempty},\tag{4.1}$$

(ii) for each $i_0\notin J(\mathbf{A},\boldsymbol{f})$ there exist indices $i_1,i_2,\cdots,i_t$ in $\mathbf{N}$ for which

$$a_{i_j i_{j+1}}\neq 0\quad (j=0,1,\cdots,t-1)\quad\text{and}\quad i_t\in J(\mathbf{A},\boldsymbol{f}).\tag{4.2}$$

LPG: The class of matrices $\mathbf{A}\in\mathbf{C}^{n,n}$ with the following properties. For $\mathbf{A}\in$ LPG there exists an $\boldsymbol{f}=(f_1,f_2,\cdots,f_n)\in\mathscr{G}_n^s$ such that:

(i) (4.1) holds, and

(ii) for each $i_0\neq J(\mathbf{A},\boldsymbol{f})$ there exist indices $i_1,i_2,\cdots,i_t$ in $\mathbf{N}$ for which

$$|a_{i_j i_{j+1}}|+|a_{i_{j+1} i_j}|>0\quad (j=0,1,\cdots,t-1)\quad\text{and}\quad i_t\in J(\mathbf{A},\boldsymbol{f}).\tag{4.3}$$

A matrix $\mathbf{A}\in\mathbf{C}^{n,n}$ is said to be of *generalized positive type* (cf. [12]) if $\mathbf{A}$ is in RPG with respect to $f_i(\mathbf{A})=r_i^x(\mathbf{A})$ $(i=1,2,\cdots,n)$ for an $\boldsymbol{x}=(x_1,x_2,\cdots,x_n)^{\mathrm{T}}>\mathbf{0}$.

Theorem 4.1 *For any* $\mathbf{A}\in\mathbf{C}^{n,n}$ *the following are equivalent*:

(a) $\mu(\mathbf{A})\in P$.

(b) $\mu(\mathbf{A})$ *is of generalized positive type*.

(c) $\mathbf{A}\in$ RPG.

If $\mathbf{A}\in\mathbf{C}_\pi^{n,n}$, *the following is equivalent to the first three properties of this theorem*:

(d) $\mathbf{A}\in$ LPG.

Proof (a)↔(b) is well known; see [12].

(b)→(c) is trivial.

(b)→(d): If $\mu(\boldsymbol{A})$ satisfies (b), then so does $\mu(\boldsymbol{A}^{\mathrm{T}})$, and therefore by definition there exist $\boldsymbol{x}>\boldsymbol{0}$ and $\boldsymbol{y}>\boldsymbol{0}$ such that $\mu(\boldsymbol{A})\boldsymbol{x}\geqslant\boldsymbol{0}$ with $J(\mu(\boldsymbol{A}), \boldsymbol{r}^{x})$ nonempty and $\mu(\boldsymbol{A}^{\mathrm{T}})\boldsymbol{y}\geqslant\boldsymbol{0}$ with $J(\mu(\boldsymbol{A}^{\mathrm{T}}), \boldsymbol{c}^{y})$ nonempty. Thus, we have that (3.8) holds, and that $J(\boldsymbol{A}, \frac{1}{2}\boldsymbol{r}^{x}+\frac{1}{2}\boldsymbol{c}^{y})$ is nonempty, since $J(\boldsymbol{A}, \frac{1}{2}\boldsymbol{r}^{x}+\frac{1}{2}\boldsymbol{c}^{y})\supset J(\mu(\boldsymbol{A}), \boldsymbol{r}^{x})$. Next, if $i_0\notin J(\boldsymbol{A}, \frac{1}{2}\boldsymbol{r}^{x}+\frac{1}{2}\boldsymbol{c}^{y})$, then $i_0\notin J(\mu(\boldsymbol{A}), \boldsymbol{r}^{x})$, so that by definition there is some chain from i_0 to $i_t\in J(\mu(\boldsymbol{A}), \boldsymbol{r}^{x})$. Thus, i_t is also in $J(\boldsymbol{A}, \frac{1}{2}\boldsymbol{r}^{x}+\frac{1}{2}\boldsymbol{c}^{y})$, proving (d).

(c)→(a): Assume $\boldsymbol{A}\in$ RPG with $\boldsymbol{f}=(f_1, f_2, \cdots, f_n)\in\mathscr{G}_n^r$. Then, obviously, $\mu(\boldsymbol{A})\in$ RPG, so that $\mu(\boldsymbol{A})\in P_0$. As before, we may assume that $\boldsymbol{A}$ is reducible, and is of the form (2.6) with $2\leqslant m\leqslant n$. Note that $\mu(\boldsymbol{A})\in$ RPG implies $\mu(\boldsymbol{A})\in$ RDG. A similar argument used in the proof of Lemma 3.2(a) shows that $\mu(\boldsymbol{A})$ is nonsingular. Indeed, if $\mu(\boldsymbol{A})$ were a singular M-matrix, then there would exist a fixed integer t, $1\leqslant t\leqslant m$, such that $\mu(\boldsymbol{A}_{tt})$ is singular, so that, by Lemma 3.2(a), $\boldsymbol{A}_{tj}=\boldsymbol{O}$ *whenever* j≠t, *and*

$$|a_{ii}|=f_i(\boldsymbol{A}) \quad \text{for all} \quad i\in S_t(\boldsymbol{A}).$$

These mean that there is no chain from $\mathrm{i}\in \mathrm{S_t}(\boldsymbol{A})$ to i_t, satisfying (4.2), such that $i_r\in J(\boldsymbol{A}, \boldsymbol{f})$, in contradiction to (c). Hence $\mu(\boldsymbol{A})\in P$.

(d)→(a): This proof is not dissimilar to the one of (c)→(a) above, using Lemma 3.2(b), and therefore is omitted. ■

It is not devoid of interest to consider two results which may be helpful in determining whether a matrix $\boldsymbol{A}\in\mathbf{R}^{n,n}$ satisfies the conditions (4.2) and (4.3). They expand a result of Vandergraft [10].

Proposition 4.2 *Let* $\boldsymbol{H}=\{(x_1, x_2, \cdots, x_n)^{\mathrm{T}}\in\mathbf{R}^n: x_i\geqslant 0$ *for all* $i\in\mathbf{N}\}$, *and let* $\boldsymbol{A}\in\mathbf{R}^{n,n}$. *Suppose that there exists an* $\boldsymbol{f}=(f_1, f_2, \cdots, f_n)\in$

$\mathscr{P}_n$ *such that*

$$|a_{ii}| \geqslant f_i(\mathbf{A}) \quad (i = 1,2,\cdots,n) \tag{4.4}$$

with $J(\mathbf{A}, f)$ *nonempty. Then the following are equivalent:*

(a) *For each* $i_0 \notin J(\mathbf{A}, f)$ *there exists a chain in* $\mathbf{N}$ *from* i_0 *to* $i_t \in J(\mathbf{A}, f)$ *satisfying* (4.2).

(b) *The vector* $\boldsymbol{u} \in \mathbf{R}^n$ *defined by*

$$\boldsymbol{u} = (|a_{11}| - f_1(\mathbf{A}), \cdots, |a_{nn}| - f_n(\mathbf{A}))^{\mathrm{T}} \tag{4.5}$$

does not lie in any face of $\mathbf{H}$ *which is invariant under* $\mathbf{A}$, *that is*, $\boldsymbol{u}$ *does not lie in any proper subspace of* $\mathbf{R}^n$, *spanned by unit coordinate vectors, which is invariant under* $\mathbf{A}$.

Proposition 4.3 *Let* $\mathbf{A} \in \mathbf{R}^{n,n}$. *Suppose that there exists an* $f = (f_1, f_2, \cdots, f_n) \in \mathscr{P}_n$ *such that* $\boldsymbol{u} \neq \mathbf{0}$, *where* $\boldsymbol{u}$ *is defined by* (4.5), *and the set* $\bar{J}(\mathbf{A}, f) = \{i \in \mathbf{N}: |a_{ii}| \neq f_i(\mathbf{A})\}$ *is nonempty. Then the follu-wing are equivalent:*

(a) *For each* $i_0 \notin \bar{J}(\mathbf{A}, f)$, *there exists a chain from* i_0 *to* $i_r \in \bar{J}(\mathbf{A}, f)$, *satisfying* (4.2) (*respectively, satisfying* (4.3)).

(b) $\boldsymbol{u}$ *does not lie in any proper subspace of* $\mathbf{R}^n$, *spanned by unit coordinate vectors, which is invariant under* $\mathbf{A}$ (*respectively*, $|\mathbf{A}| + |\mathbf{A}^{\mathrm{T}}|$, *where* $|\mathbf{A}| = (|a_{ij}|)$).

The proofs of Propositions 4.2 and 4.3 are almost the same as that of Theorem 4.1 of [10], and therefore are omitted.

§ 5. *G*-semistrict Diagonal Dominance with Monotonic *G*-functions

Let $\mathbf{A} \in \mathbf{C}^{n,n}$, and let q be the permutation which applies the sequence $(p_1, p_2, \cdots, p_n)$ onto $(1, 2, \cdots, n)$. Define the matrix $\mathbf{A}_t \in \mathbf{C}^{n,n}$ for any $t \in \mathbf{N}$ by

$$(\mathbf{A}_t)_{ij} = \begin{cases} 0, & if \quad q(i) > t \quad \text{or} \quad q(j) > t, \quad i \neq j, \\ a_{ij}, & \text{otherwise.} \end{cases} \tag{5.1}$$

It is obvious that $\mathbf{A}_t$ depends only on $\mathbf{A}$, t, and q, and that $\mathbf{A}_n = \mathbf{A}$,

$$|(\mathbf{A}_t)_{ij}| \leqslant |(\mathbf{A}_{t+1})_{ij}| \quad \text{for all} \quad i,j \in \mathbf{N} \text{ and } 1 \leqslant t \leqslant n-1. \tag{5.2}$$

In this section, we consider three subclasses of the class of G-semistrictly diagonally dominant matrices, denoted by RMG, LMG, and MG.

RMG: The class of matrices $\mathbf{A} \in \mathbf{C}^{n,n}$ with the following properties. For each $\mathbf{A} \in$ RMG there exists an $\boldsymbol{f}=(f_1, f_2, \cdots, f_n) \in \mathscr{G}_n^r$ and a permutation q which applies $(p_1, p_2, \cdots, p_n)$ onto $(1, 2, \cdots, n)$ such that (1.2) holds and

$$|a_{p_t p_t}| > f_{p_t}(\mathbf{A}_t) \qquad (t = 1, 2, \cdots, n). \tag{5.3}$$

Obviously, if $\mathbf{A} \in$ RMG with $\boldsymbol{f}=(f_1, f_2, \cdots, f_n)=(r_1, r_2, \cdots, r_n)$, and if q is the identity permutation, then $\mathbf{A}$ is just a so-called *lower semistrictly diagonally dominant* matrix (cf. [1]).

Similarly, we can define the class LMG (MG), provided "$\boldsymbol{f}=(f_1, f_2, \cdots, f_n) \in \mathscr{G}_n^r$" in the definition of RMG is replaced by "$\boldsymbol{f} \in \mathscr{G}_n^s$" (by "monotonic $\boldsymbol{f} \in \mathscr{G}_n$").

The following result is a generalization of Theorem 2.1 of [1]:

Theorem 5.1 *A matrix* $\mathbf{A} \in \mathbf{C}^{n,n}$ *is* G *-irreducibly diagonally dominant with* $\boldsymbol{f}=(f_1, f_2, \cdots, f_n) \in \mathscr{G}_n^r$ *(that is,* $\mathbf{A}$ *is irreducible, and* (1.2) *holds for* $\boldsymbol{f}=(f_1, f_2, \cdots, f_n) \in \mathscr{G}_n^r$ *with equality in at most* $n-1$ *cases) if and only if it is irreducible and is in* RMG *with the same* $\boldsymbol{f} \in \mathscr{G}_n$.

Proof The proof is quite similar to that of Beauwens [1]. Assume irreducible $\mathbf{A}$ is in RMG with some $\boldsymbol{f}=(f_1, f_2, \cdots, f_n) \in \mathscr{G}_n^r$. Let q be the permutation given in the definition of RMG. Then by the condition (5.3) with $t=n$

$$|a_{p_n p_n}| > f_{p_n}(\mathbf{A}_n) = f_{p_n}(\mathbf{A}), \tag{5.4}$$

so that the sufficient condition holds.

Conversely, assume that $\mathbf{A}$ is G-irreducibly diagonally dominant with some $\boldsymbol{f} \in \mathscr{G}_n^r$, and let Q_{p_n} be a node of its directed graph such that (1.2) is strict for $i=p_n$. Let $Q_{p_{n-1}} \neq Q_{p_n}$ be any node such that there exists a path from $Q_{p_{n-1}}$ to Q_{p_n}, let $Q_{p_{n-2}}$ be any node distinct from Q_{p_n} and $Q_{p_{n-1}}$ such that there exists a path from $Q_{p_{n-2}}$ to Q_{p_n} or $Q_{p_{n-1}}$, and so on.

Since $\mathbf{A}$ is irreducible, this procedure will exhaust all the nodes of its directed graph. Letting q be the permutation which applies $(p_1, p_2, \cdots, p_n)$ onto $(1, 2, \cdots, n)$, and $\mathbf{A}_t$, $t \in \mathbf{N}$, be the corresponding matrix (5.1), we observe that, for each t $(1 \leqslant t \leqslant n-1)$, $\mathbf{A}$ has a nonzero entry $a_{p_t p_j}$ with $t < j$. On the other hand, $(\mathbf{A}_t)_{p_t p_j} = 0$ for $t < j$ by (5.1), so that

$$| a_{p_t p_t} | \geqslant f_{p_t}(\mathbf{A}) > f_{p_t}(\mathbf{A}_t) \quad \text{for} \quad t = 1, 2, \cdots, n-1.$$

(5.4) is satisfied by construction. Thus, $\mathbf{A}$ is in RMG with the same $f \in \mathscr{G}_n^r$. This completes the proof of the theorem. ■

Theorem 5.2 *For any* $\mathbf{A} \in \mathbf{C}_\pi^{n,n}$, *the following are equivalent*:

(a) $\mu(\mathbf{A}) \in P$.

(b) *There exists an* $\boldsymbol{x} = (x_1, x_2, \cdots, x_n)^{\mathrm{T}} > \mathbf{0}$ *such that* $\mu(\mathbf{AX})$ *is lower semistrictly diagonally dominant*, *where* $\mathbf{X} = \operatorname{diag}(x_1, x_2, \cdots, x_n)$.

(c) $\mathbf{A} \in$ RMG.

(d) $\mathbf{A} \in$ LMG.

(e) $\mathbf{A} \in$ MG.

Proof (a)↔(b) is known; see [12].

(b)→(c), (c)→(e), and (d)→ (e) are all trivial.

(b)→(d): If $\mathbf{A}$ satisfies (b), then so does $\mathbf{A}^{\mathrm{T}}$, so that there exist $\boldsymbol{x} > \mathbf{0}$ and $\boldsymbol{y} > \mathbf{0}$ such that $\mu(\mathbf{A})\boldsymbol{x} + \mu(\mathbf{A}^{\mathrm{T}})\boldsymbol{y} \geqslant \mathbf{0}$, and

$$| a_{ii} | > \frac{1}{2} x_i^{-1} \sum_{j < i} | a_{ij} | x_j + \frac{1}{2} y_i^{-1} \sum_{j < i} | a_{ji} | y_j \qquad \text{for all} \quad i \in \mathbf{N}.$$

It follows that (1.2) and (5.3) hold for $\boldsymbol{f} = \frac{1}{2}\boldsymbol{r}^x + \frac{1}{2}\boldsymbol{c}^y \in \mathscr{G}_n^s$ and for the identity permutation. Thus $\mathbf{A}$ is in LMG.

(e)→(a): Assume $\mathbf{A} \in$ MG with monotonic $\boldsymbol{f} \in \mathscr{G}_n$, and the permutation q. A similar argument to that used in the proof of Theorem 3 of [13] shows that

$$\det \mu(\mathbf{A}) \geqslant \prod_{t=1}^{n} \{ | a_{p_t p_t} | - f_{p_t}(\mathbf{A}_t) \}.$$

Since $\mu(\mathbf{A}) \in P_0$, the last inequality implies $\mu(\mathbf{A}) \in P$ by (5.3). This completes the proof of Theorem 5.2. ■

With Theorems 4.1 and 5.2, we have

Corollary 5.3 *As subsets of* $\mathbf{Z}^{n,n} \cap \mathbf{C}_{\pi}^{n,n}$ *we have*

$$\mathrm{RPG}=\mathrm{LPG}=\mathrm{RMG}=\mathrm{LMG}=\mathrm{MG}=P.$$

The author would like to thank the referee for his helpful comments on property (e) of Thememe 3.3, and (d) of Theorem 3.4.

References

[1] R Beauwens. Semistrict diagonal dominance. SIAM J. Numer. Anal., 1976, 13:109—112.

[2] A Berman, R S Varga and R C Ward. 𝔄𝔏𝔓𝔖: Matrices with nonpositive off-diagonal entries. Linear Algebra Appl., 1978, 21:233—244.

[3] A Brauer. Limits for the characteristic roots of a matrix IV. Duke Math. J., 1952, 19:75—91.

[4] D H Carlson and R S Varga. Minimal *G*-functions. Linear Algebra Appl., 1973, 6:97—117.

[5] G-N Chen. On some properties of singular matrices. Linear Algebra Appl., 1983, 49:137—151.

[6] Ky Fan. Note on circular disks containing the eigenvalues of a matrix. Duke Math. J., 1958, 25:441—445.

[7] M. Fiedler and V Pták. On matrices with nonpositive off-diagonal elements and positive principal minors. Czech. Math., 1962, 12:382—400.

[8] A J Hoffman. Generalizations of Geršhgorin's theorem: *G*-generating families. Lecture Notes, Univ. of California at Santa Barbara, 1969: 46.

[9] P Nowosad, On the functional $(x^{-1}, \boldsymbol{A}x)$ and some of its applications. An. Acad. Brasil Ciênc., 1965, 37:163—165.

[10] J S Vandergraft. A note on irreducibility for linear operators on partially ordered finite dimensional vector spaces, Linear Algebra Appl., 1976, 13:139—146.

[11] R S Varga, Matrix Iterative Analysis. Prentice-Hall, Englewood Cliffs, N J, 1962.

[12] R S Varga. On recurring theorems on diagonal dominance. Linear Algebra Appl., 1976, 13:1—9.

[13] B-Y Wang and G-N Chen. On bounds of determinants and regularity theorems associated with *G*-functions (in Chinese). J. Beijing Normal Univ., 1979, 3:41—49.

Linear Algebra and Its Applications
1983, 55:125—132

关于矩阵秩下界的注记①

Note on Lower Bounds for the Rank of a Matrix

Abstract　The following results are proved: Let $\mathbf{A}=(a_{ij})$ be an $n\times n$ complex matrix, $n\geqslant 2$, and let k be a fixed integer, $1\leqslant k\leqslant n-1$. (1) If there exists a monotonic G-function $\boldsymbol{f}=(f_1,f_2,\cdots,f_n)$ such that for every subset S of $\{1,2,\cdots,n\}$ consisting of $k+1$ elements we have

$$\prod_{i\in S} f_i(\mathbf{A}) < \prod_{i\in S} |a_{ii}|,$$

then the rank of $\mathbf{A}$ is $\geqslant n-k+1$. (2) If $\mathbf{A}$ is irreducible and if there exists a G-function $\boldsymbol{f}=(f_1,f_2,\cdots,f_n)$ such that for every subset S of $\{1,2,\cdots,n\}$ consisting of $k+1$ elements we have

$$\prod_{i\in S} f_i(\mathbf{A}) \leqslant \prod_{i\in S} |a_{ii}|,$$

then the rank of $\mathbf{A}$ is $\geqslant n-k+1$ if $k\geqslant 2, n\geqslant 3$; it is $\geqslant n-1$ if $k=1$.

§ 1. Definitions

For a positive integer n, let $\mathbf{N}$ be the set of indices $1,2,\cdots,n$; let $\mathbf{R}^n$ and $\mathbf{C}^{n,n}$ be the n-dimensional real vector space and the set of all $n\times n$ complex matrices, respectively. Denote by $\mathbf{A}[i_1,i_2,\cdots,i_m]$ the principal submatrix of $\mathbf{A}\in\mathbf{C}^{n,n}$ formed by rows and columns with indices $i_1,i_2,\cdots,$

① Received 2 October 1981; revised 29 October 1982.
Submitted by David H. Carlson.

i_m, $1\leqslant i_1 < i_2 < \cdots < i_m \leqslant n$. Introduce for a matrix $\mathbf{A}\in\mathbf{C}^{n,n}$ and for an $\boldsymbol{x}\in\mathbf{R}^n$ with $\boldsymbol{x}>\mathbf{0}$ (i. e., $x_i>0$ for all $i\in\mathbf{N}$) the expressions

$$r_{i,s} = r_{i,s}(\mathbf{A}) = \Big(\sum_{j\neq i} |a_{ij}|^s\Big)^{1/s} \qquad \text{for all} \quad i\in\mathbf{N},$$

$$c_{i,s} = c_{i,s}(\mathbf{A}) = \Big(\sum_{j\neq i} |a_{ji}|^s\Big)^{1/s} \qquad \text{for all} \quad i\in\mathbf{N}, \tag{1}$$

$$r_i^x(\mathbf{A}) = \frac{1}{x_i}\sum_{j\neq i} |a_{ij}|\, x_j \qquad \text{for all} \quad i\in\mathbf{N}.$$

Let $f_i=f_i(\mathbf{A})$, $i\in\mathbf{N}$, be a nonnegative function which depends only on the moduli of the $n(n-1)$ off-diagonal entries of the matrices in $\mathbf{C}^{n,n}$. We say [1,5] that $\boldsymbol{f}=(f_1,f_2,\cdots,f_n)$ is a *G-function* if every $\mathbf{A}\in\mathbf{C}^{n,n}$ satisfying

$$f_i(\mathbf{A}) < |a_{ii}| \qquad \text{for all} \quad i\in\mathbf{N} \tag{2}$$

is nonsingular. A G-function $\boldsymbol{f}=(f_1,f_2,\cdots,f_n)$ is called *monotonic* if, for any $\mathbf{A}=(a_{ij})$ and $\boldsymbol{B}=(b_{ij})\in\mathbf{C}^{n,n}$, $|a_{ij}|\leqslant|b_{ij}|$ for all $i\neq j$ $(i,j\in\mathbf{N})$ implies

$$f_i(\mathbf{A}) \leqslant f_i(\boldsymbol{B}) \qquad \text{for all} \quad i\in\mathbf{N}.$$

§ 2. Background

A. Ostrowski [8] has proved the following theorem:

Theorem (a) *Let* $\mathbf{A} = (a_{ij})\in\mathbf{C}^{n,n}$, $n\geqslant 2$. *Suppose*

$$r^{\alpha}_{i,\alpha p}c^{1-\alpha}_{i,(1-\alpha)q}r^{\alpha}_{j,\alpha p}c^{1-\alpha}_{j,(1-\alpha)q} < |a_{ii}||a_{jj}|$$
$$\textit{for all} \quad i\neq j \quad (i,j\in\mathbf{N}), \tag{3}$$

where $0\leqslant\alpha\leqslant 1$, p, $q\geqslant 1$, $1/p+1/q=1$. *Then* $det(\mathbf{A})\neq 0$, *and thus the rank of* $\mathbf{A}$, rank($\mathbf{A}$), *is* n.

The following theorem, generalizing Theorem (a) with $\alpha p=1$, was obtained by Shemesh in [2]:

Theorem (b) *Let* $\mathbf{A}=(a_{ij})\in\mathbf{C}^{n,n}$, $n\geqslant 2$, *and let* k *be a fixed integer*, $1\leqslant k\leqslant n-1$. *If for any real* α *with* $0\leqslant\alpha\leqslant 1$ *and for every subset* S *of* $\mathbf{N}$ *consisting of* $k+1$ *elements we have*

$$\prod_{i\in S} r^{\alpha}_{i,1}c^{1-\alpha}_{i,1} < \prod_{i\in S}|a_{ii}|, \tag{4}$$

then rank($\mathbf{A}$)$\geqslant n-k+1$.

In [4], we have given the following result:

Theorem (c) *For any* $\mathbf{A}\in\mathbf{C}^{n,n}$, $n\geq 2$, *the following properties are equivalent*:

(1) *There exists a G-function* $\boldsymbol{f}=(f_1,f_2,\cdots,f_n)$ *such that the relations* (2) *hold.*

(2) *There exists a G-function* $\boldsymbol{f}=(f_1,f_2,\cdots,f_n)$ *such that*

$$f_i(\mathbf{A})f_j(\mathbf{A})<|a_{ii}a_{jj}| \quad \text{for all} \quad i\neq j \quad (i,j\in\mathbf{N}). \tag{5}$$

(3) $M(\mathbf{A})$ *is a nonsingular M-matrix, where* $M(\mathbf{A})=(m_{ij})$ *denotes the* $n\times n$ *real matrix defined by* (*see* [7])

$$m_{ij}=\begin{cases}|a_{ij}|, & if \quad i=j,\\ -|a_{ij}|, & if \quad i\neq j.\end{cases}$$

Remark Property (1) of Theorem (c) implies that $\mathbf{A}$ *is nonsingular by the definition of G-function, and therefore property* (2) *or* (3) *of Theorem* (*c*) *also is sufficient for* $\mathbf{A}$ *to be nonsingular; this result contains as a particular case Theorem* (*a*) *quoted above.*

§ 3. Results

The purpose of this note is to prove the following theorems.

Theorem 1 *Let* $\mathbf{A}=(a_{ij})\in\mathbf{C}^{n,n}$, $n\geq 2$, *and let k be a fixed integer with* $1\leq k\leq n-1$. *If for a monotonic G-function* $\boldsymbol{f}=(f_1,f_2,\cdots,f_n)$ *and for every subset S of* $\mathbf{N}$ *consisting of* $k+1$ *elements we have*

$$\prod_{i\in S}f_i(\mathbf{A})<\prod_{i\in S}|a_{ii}| \tag{6}$$

then $\operatorname{rank}(\mathbf{A})\geq n-k+1$.

Theorem 2 *Let* $\mathbf{A}=(a_{ij})\in\mathbf{C}^{n,n}$, $n\geq 2$, *be irreducible, and let k be a fixed integer,* $0\leq k\leq n-1$. *If for a G-function* $\boldsymbol{f}=(f_1,f_2,\cdots,f_n)$ *and for every subset S of* $\mathbf{N}$ *consisting of* $k+1$ *elements we have*

$$\prod_{i\in S}f_i(\mathbf{A})\leq\prod_{i\in S}|a_{ii}|, \tag{7}$$

then $\operatorname{rank}(\mathbf{A})\geq n-k+l$ *if* $2\leq k\leq n-1$; $\operatorname{rank}(\mathbf{A})\geq n-1$ *if* $k=0,1$.

Proof of Theorem 1 The relations (6) imply that none of the a_{ii} vanishes, and that there exist only $p\leq k$ indices i for which $f_i(\mathbf{A})\geq|a_{ii}|$. *If* $p=0$, then the relations (2) hold, and thus rank $(\mathbf{A})=n\geq$

$n-k+1$. We may assume, without loss of generality, that for $p\geqslant 1$, $f_i(\mathbf{A})\geqslant |a_{ii}|$, $i=1,2,\cdots,p$.

If $1\leqslant p<k$, then $\mathbf{A}$ can be written in the form

$$\mathbf{A}=\begin{bmatrix}\mathbf{A}_{11} & \mathbf{A}_{12}\\ \mathbf{A}_{21} & \mathbf{A}_{22}\end{bmatrix}, \tag{8}$$

where $\mathbf{A}_{22}=\mathbf{A}[p+1,\cdots,n]$. According to our assumption,

$$f_i(\mathbf{A})<|a_{ii}|,\quad i=p+1,p+2,\cdots,n. \tag{9}$$

Suppose that $\mathbf{A}$ is irreducible. Then, by Ky Fan's theorem [6], there exists an $\boldsymbol{x}=(x_1,x_2,\cdots,x_n)^{\mathrm{T}}\in\mathbf{R}^n$ with $\boldsymbol{x}>\mathbf{0}$ (depending on $\mathbf{A}$) such that

$$r_i^x(\mathbf{A})\leqslant f_i(\mathbf{A})\quad \text{for all } i\in\mathbf{N}, \tag{10}$$

whence

$$\frac{1}{x_i}\sum_{\substack{j=p+1\\ j\neq i}}^{n}|a_{ij}|x_j\leqslant r_i^x(\mathbf{A})<|a_{ii}|,\quad i=p+1,p+2,\cdots,n. \tag{11}$$

This signifies that $\mathbf{D}^{-1}\mathbf{A}_{22}\mathbf{D}$ is strictly diagonally dominant, where $\mathbf{D}=\mathrm{diag}(x_{p+1},\cdots,x_n)$ is a positive diagonal matrix, so that $\det(\mathbf{A}_{22})\neq 0$. Thus $\mathrm{rank}(\mathbf{A})\geqslant\mathrm{rank}(\mathbf{A}_{22})=n-p\geqslant n-k+1$.

For reducible $\mathbf{A}$, the result follows easily from the generalized Ky Fan's theorem due to Carlson and Varga in [1].

If $p=k$ we have by (9)

$$f_i(\mathbf{A})<|a_{ii}|,\quad i=k+1,k+2,\cdots,n, \tag{9'}$$

and [by (6)]

$$f_k(\mathbf{A})f_j(\mathbf{A})<|a_{kk}a_{jj}|,\quad j=k+1,k+2,\cdots,n,$$

so that

$$f_i(\mathbf{A})f_j(\mathbf{A})<|a_{ii}a_{jj}|,\quad i,j=k,k+1,\cdots,n,\ i\neq j. \tag{12}$$

Put now, corresponding to $\mathbf{A}$,

$$\mathbf{A}'=(a'_{ij})=\left(\begin{array}{ccc|c} a_{11} & & \mathbf{O} & \\ & \ddots & & \mathbf{O}\\ \mathbf{O} & & a_{k-1k-1} & \\ \hline & \mathbf{O} & & \mathbf{B}\end{array}\right)$$

where $\mathbf{B}=\mathbf{A}[k,\cdots,n]$. It is obvious that $a'_{ii}=a_{ii}$ for all $i\in\mathbf{N}$, and that $|a'_{ij}|\leqslant|a_{ij}|$ for all $i\neq j$, and thus

$$f_i(\mathbf{A}') \leqslant f_i(\mathbf{A}) \qquad \text{for all} \quad i \in \mathbf{N}. \tag{13}$$

Suppose that the normal form of $\mathbf{A}'$ (cf. [9, p. 46]) is given by

$$\mathbf{A}' = \mathbf{P}^{\mathrm{T}} \left[\begin{array}{ccc|ccc} a_{11} & & \mathbf{O} & & & \\ & \ddots & & & \mathbf{O} & \\ \mathbf{O} & & a_{k-1k-1} & & & \\ \hline & & & A_{11} & \cdots & A_{1m} \\ & \mathbf{O} & & & \ddots & \vdots \\ & & & \mathbf{O} & & A_{mm} \end{array}\right] \mathbf{P}, \tag{14}$$

where $\mathbf{P}$ is a permutation matrix, and each $\mathbf{A}_{ii}$, $1 \leqslant i \leqslant m$, is square and irreducible (if $\mathbf{B}$ is irreducible, then let $\mathbf{P}=\mathbf{I}_n$ and $\mathbf{A}_{11}=\mathbf{B}$). Obviously, $\operatorname{rank}(\mathbf{A}') = k-1+\operatorname{rank}(\mathbf{B})$. From the generalized Ky Fan's theorem it follows that there exists an $\boldsymbol{x}=(x_1, x_2, \cdots, x_n) \in \mathbf{R}^n$ with $\boldsymbol{x} > \mathbf{0}$ (depending on $\mathbf{A}'$) for which

$$f_i(\mathbf{A}') \geqslant \hat{r}_i^x(\mathbf{A}') = 0, \quad i = 1, 2, \cdots, k-1,$$

$$f_i(\mathbf{A}') \geqslant \hat{r}_i^x(\mathbf{A}') \overset{\mathrm{df}}{=} \frac{1}{x_i} \sum_{\substack{j \in \langle i \rangle \\ j \neq i}} |a_{ij}| x_j, \quad i = k, \cdots, n, \tag{15}$$

where $\langle i \rangle = \{j \in \mathbf{N}: a_{jj}$ and a_{ii} lie in the same $\mathbf{A}_{ll}$ of (14)$\}$, and $r_i^x(\mathbf{A}')=0$ if $\langle i \rangle = \{i\}$, $i \in \mathbf{N}$.

By (12) (13) and (15) we have that

$$\hat{r}_i^x(\mathbf{A}')\hat{r}_j^x(\mathbf{A}') < |a'_{ii}a'_{jj}| = |a_{ii}a_{jj}| \qquad \text{for all} \quad i \neq j,\ i, j \in \mathbf{N}.$$

Note that $\hat{\boldsymbol{r}}=(\hat{r}_1^x, \hat{r}_2^x, \cdots, \hat{r}_n^x)$ is a G-function (see [1]). It follows from the remark on Theorem (c) that $\mathbf{A}'$ is nonsingular and consequently $\operatorname{rank}(\mathbf{B}) = \operatorname{rank}(\mathbf{A}') - (k-1) = n-k+1$. Thus $\operatorname{rank}(\mathbf{A}) \geqslant \operatorname{rank}(\mathbf{B}) = n-k+1$, which completes the proof of Theorem 1. ■

Remark Since the G-functions given in Theorems (a) and (b) are monotonic, Theorem 1 is, as is easily seen, a generalization of Theorems (a) and (b).

To show Theorem 2, first we introduce the following

Lemma 1 *Let* $\mathbf{A} \in \mathbf{C}^{n,n}$, $n \geqslant 2$, *be irreducible. If for a* G*-function* $\boldsymbol{f}=(f_1, f_2, \cdots, f_n)$ *we have*

$$f_i(\mathbf{A}) \leqslant |a_{ii}| \qquad \textit{for all} \quad i \in \mathbf{N} \tag{16}$$

or

$$f_i(\mathbf{A})f_j(\mathbf{A}) \leqslant |a_{ii}a_{jj}| \quad \text{for all} \quad i \neq j,\ i,j \in \mathbf{N}, \tag{17}$$

then $\operatorname{rank}(\mathbf{A}) \geqslant n-1$.

Proof It is seen that $M(\mathbf{A})$ is an M-matrix in both cases, because by (16) or (17), $M(\mathbf{A})+\operatorname{diag}(d_1,d_2,\cdots,d_n)$ is a nonsingular M-matrix whenever $\boldsymbol{d}=(d_1,d_2,\cdots,d_n)^{\mathrm{T}}>\mathbf{0}$. Note that then $M(\mathbf{A})$ is irreducible. According to Theorem 5.7 in [3], we get that all principal submatrices of order $n-1$ of $M(\mathbf{A})$ are nonsingular M-matrices. Thus $\operatorname{rank}(\mathbf{A}) \geqslant n-1$. ■

Lemma 2 *Let* $\mathbf{A} \in \mathbf{C}^{n,n}$, $n \geqslant 3$, *be irreducible and let* k *be an integer with* $2 \leqslant k \leqslant n-1$. *If for an* $\boldsymbol{x}=(x_1,x_2,\cdots,x_n)^{\mathrm{T}} \in \mathbf{R}^n$ *with* $\boldsymbol{x}>\mathbf{0}$ *and for every subset* S *of* $\mathbf{N}$ *consisting of* $k+1$ *elements we have*

$$\prod_{i \in S} r_i^x(A) \leqslant \prod_{i \in S} |a_{ii}|, \tag{18}$$

then $\operatorname{rank}(\mathbf{A}) \geqslant n-k+1$.

Proof Obviously the relations (18) imply that none of the a_{ii}, $r_i^x(\mathbf{A})$ vanish, as $\mathbf{A}$ is irreducible, and that there exist only $p \leqslant k$ indices i such that $r_i^x(\mathbf{A})>|a_{ii}|$. If $p=0$, then for all $i \in \mathbf{N}$, $r_i^x(\mathbf{A}) \leqslant |a_{ii}|$. From Lemma 1 it follows that $\operatorname{rank}(\mathbf{A}) \geqslant n-1 \geqslant n-k+1$. For the case $p \geqslant 1$, without loss of generality, we may assume that

$$r_1^x(\mathbf{A})>|a_{11}|.$$

Let $\mathbf{A}_{22}$ denote the principal submatrix of $\mathbf{A}$ indexed by rows and columns $2,3,\cdots,n$. Now $\mathbf{A}_{22}$ satisfies the hypotheses of Theorem 1 for $n-1$ and $k-1$, so that

$$\operatorname{rank}(\mathbf{A}) \geqslant \operatorname{rank}(\mathbf{A}_{22}) \geqslant (n-1)-(k-1)+1=n-k+1. \quad ■$$

Proof of Theorem 2 The proof is an immediate consequence of Lemma 1, Lemma 2 and Ky Fan's theorem. ■

From Theorem 1 we have directly the following

Corollary *Let* $\mathbf{A} \in \mathbf{C}^{n,n}$, $n \geqslant 2$, *and let* k *be an integer*, $1 \leqslant k < n-1$. *Then for any monotonic* G*-function* $\boldsymbol{f}=(f_1,f_2,\cdots,f_n)$ *each eigenvalue of* $\mathbf{A}$ *with geometric multiplicity* $\geqslant k$ *lies in at least one of the* $\binom{n}{k+1}$ *sets of the form*

$$\left\{\lambda:\prod_{i\in S}|\lambda-a_{ii}|\leqslant\prod_{i\in S}f_i(\mathbf{A})\right\}$$

in the complex plane, where S is a subset of $\mathbf{N}$ *having* $k+1$ *elements.*

Proof If λ does not belong to any set of the form (24) for some monotonic G-function f, then

$$\prod_{i\in S}|\lambda-a_{ii}|>\prod_{i\in S}f_i(\mathbf{A})=\prod_{i\in S}f_i(\lambda\mathbf{I}-\mathbf{A})$$

for every subset S of $\mathbf{N}$ having $k+1$ elements. From Theorem 1 it follows that $\mathrm{rank}(\lambda\mathbf{I}-\mathbf{A})\geqslant n-k+1$, that is, λ is not an eigenvalue of $\mathbf{A}$ with geometric multiplicity $\geqslant k$. ■

The author would like to thank Professor David Carlson for his recommendation of this note, and the referee for his helpful comments, which could shorten considerably the proof of Lemma 2.

References

[1] D H Carlson and R S Varga. Minimal G-functions. Linear Algebra Appl., 1973, 6:97－117.

[2] Dan Shemesh. On a theorem of A. Ostrowski. Linear and Multilinear Algebra, 1977, 5:139－143.

[3] M Fiedler and V Pták. On matrices with non-positive off-diagonal elements and positive principal minors. Czechoslovak Math. J., 1962, 12:382－400.

[4] G-N Chen. Some generalizations of diagonal dominance associated with G-functions. Linear Algebra Appl., 1983, 55:169－180.

[5] A J Hoffman. Generalizations of Geršgorin's theorem: G-generating families. lecture notes, Univ. of California at Santa Barbara, Aug. 1969.

[6] Ky Fan. Note on circular disks containing the eigenvalue of a matrix. Duke Math. J., 1958, 25:441－445.

[7] A Ostrowski. Über die Determinanten mit überwiegender Hauptdiagonale. Comment. Math. Helv., 1937, 10:69－96.

[8] A Ostrowski. On some conditions for the non-vanishing of determinants. Proc. Amer. Math. Soc., 1961, 12:268－273.

[9] R S Varga. Matrix Iterative Analysis. Prentice-Hall, Englewood Cliffs, N J, 1962.

陈公宁文集

二、全纯映射与算子解析函数若干结果

Ⅱ. Holomorphic Maps and Analytic Functions of Operators

北京师范大学学报(自然科学版)
1985,(4):14－22

关于 Hilbert 空间上算子解析函数的若干结果①

Some Results on Analytic Functions for Operators of Hilbert Spaces

摘要 设 $\boldsymbol{H}$ 是复 Hibert 空间. 对 $\boldsymbol{H}$ 上算子 A 与复解析函数 f 在某些通常假定下,$f(A)$ 表示由 Riesz-Dunford 积分定义的 $\boldsymbol{H}$ 上算子,遵循 Ky Fan 的近期工作,一些算子解析函数的补充结果被得到. 这些结果之中包含有关于上半平面 $\boldsymbol{\Omega}$ 解析函数 f,$f(\boldsymbol{\Omega}) \subset \boldsymbol{\Omega}$ 对应的算子解析函数 $f(A)$ 的迭代定理,以及关于自单位开圆 $\boldsymbol{\Delta}$ 到右半平面 $\boldsymbol{\Pi}$ 或自 $\boldsymbol{\Pi}$ 到 $\boldsymbol{\Delta}$ 内解析函数 f 对应的算子解析函数 $f(A)$ 的特征性质.

§1. 引言与预备知识

假定 $\boldsymbol{H}$ 表示某一复 Hilbert 空间,我们用 $\mathscr{L}(\boldsymbol{H})$ 表示 $\boldsymbol{H}$ 上全体算子(有界线性变换)组成的(具有算子范数的)Banach 空间. 当 $A \in \mathscr{L}(\boldsymbol{H})$,Re A 与 Im A 分别表示 A 的实部与虚部:$\mathrm{Re}\,A = (A + A^*)/2$,$\mathrm{Im}\,A = (A - A^*)/(2\mathrm{i})$,这里 $\mathrm{i} = \sqrt{-1}$. 显然地,Re A 与 Im A 都是 $\boldsymbol{H}$ 上的 Hermite 算子. 当 $A, B \in \mathscr{L}(\boldsymbol{H})$,且 A 与 B 都是 Hermite 算子时,如果 $B-A$ 是正算子(即有 $\langle (B-A)\rangle x, x \geqslant 0$ 对所有 $x \in \boldsymbol{H}$ 成立),我们记为 $A \leqslant B$ 或 $B \geqslant A$. 同时,记号 $A < B$ 或 $B > A$ 表示 $B - A$ 是可逆的正算子. 此外,用 $\boldsymbol{\Delta}$,$\boldsymbol{\Pi}$ 与 $\boldsymbol{\Omega}$ 分别表示复平面 $\mathbf{C}$ 上的单位开圆、右半平面与上半平面:

$$\boldsymbol{\Delta} = \{z \in \mathbf{C}: |z| < 1\}, \boldsymbol{\Pi} = \{x \in \mathbf{C}: \mathrm{Re}\, z > 0\}$$

$$\boldsymbol{\Omega} = \{z \in \mathbf{C}: \mathrm{Im}\, z > 0\},$$

① 收稿日期:1985-01-07.

而 $\overline{\boldsymbol{\Delta}},\overline{\boldsymbol{\Pi}}$ 与 $\overline{\boldsymbol{\Omega}}$ 表示它们的闭包. 同时,

$$\boldsymbol{\Delta}(\boldsymbol{H})=\{A\in\mathscr{L}(\boldsymbol{H}):\|A\|<1\},\boldsymbol{\Pi}(\boldsymbol{H})=\{A\in\mathscr{L}(\boldsymbol{H}):\mathrm{Re}\,A>0\},$$
$$\boldsymbol{\Omega}(\boldsymbol{H})=\{A\in\mathscr{L}(\boldsymbol{H}):\mathrm{Im}\,A>0\},$$

$\overline{\boldsymbol{\Delta}}(\boldsymbol{H}),\overline{\boldsymbol{\Pi}}(\boldsymbol{H})$ 与 $\overline{\boldsymbol{\Omega}}(\boldsymbol{H})$ 表示它们的闭包.

现令 $A\in\mathscr{L}(\boldsymbol{H})$,$\sigma(A)$ 是 A 的谱集. 对于在 $\sigma(A)$ 的某一开邻域 $\boldsymbol{\Gamma}$ 上解析的复函数 f,$f(A)$ 将表示由 Riesz-Dunford 积分(见[3],p. 568)定义的 $\boldsymbol{H}$ 上的算子:

$$f(A)=\frac{1}{2\pi\mathrm{i}}\int_{\mathscr{L}}f(z)(zI-A)^{-1}\mathrm{d}z.$$

上式中,$\mathscr{L}$ 是一个正向、简单、封闭与可求长的曲线,其内部区域 $\mathscr{D}$ 包含 $\sigma(A)$ 且 $\mathscr{L}\cup\mathscr{D}\subset\boldsymbol{\Gamma}$;$I$ 表示 $\mathscr{L}(\boldsymbol{H})$ 内的单位算子. 一个著名的 Von Neumann 定理指出[8]:假设 f 在单位闭圆 $\overline{\boldsymbol{\Delta}}$ 的某个邻域上解析,且 $f(\overline{\boldsymbol{\Delta}})\subset\overline{\boldsymbol{\Delta}}$. 如果 $A\in\overline{\boldsymbol{\Delta}}(\boldsymbol{H})$,即 A 是 $\boldsymbol{H}$ 上的压缩算子:$\|A\|\leqslant 1$,那么 $f(A)\in\overline{\boldsymbol{\Delta}}(\boldsymbol{H})$. 与此定理密切相关且等效的有如下 Ky Fan 定理[5].

定理(Ky Fan) 假设 f 与 g 在 $\boldsymbol{\Delta}$ 上解析,且 $f(\boldsymbol{\Delta})\subset\boldsymbol{\Delta}$,$g(\boldsymbol{\Delta})\subset\boldsymbol{\Pi}$. 如果 $A\in\boldsymbol{\Delta}(\boldsymbol{H})$,即 A 是 $\boldsymbol{H}$ 上的真正压缩算子:$\|A\|<1$,那么 $f(A)\in\boldsymbol{\Delta}(\boldsymbol{H})$ 与 $g(A)\in\boldsymbol{\Pi}(\boldsymbol{H})$.

类似地,假设 g 与 h 在 $\boldsymbol{\Pi}$ 上解析,且 $g(\boldsymbol{\Pi})\subset\boldsymbol{\Pi}$ 与 $h(\boldsymbol{\Pi})\subset\boldsymbol{\Delta}$. 如果 $A\in\boldsymbol{\Pi}(\boldsymbol{H})$,那么 $g(A)\in\boldsymbol{\Pi}(\boldsymbol{H})$ 与 $h(A)\in\boldsymbol{\Delta}(\boldsymbol{H})$.

对于算子解析函数的迭代,Ky Fan[6] 得到两个有意义的结果,其中第一个是单复变函数论中一个古典的 J. Wolff 定理的推广. 我们引用它们作为本文的出发点.

定理 Ⅰ 设 f 在 $\boldsymbol{\Delta}$ 上解析,且 $f(\boldsymbol{\Delta})\subset\boldsymbol{\Delta}$,又设 $f^{[n]}$ 表示 f 的第 n 次迭代. 则存在 $w\in\overline{\boldsymbol{\Delta}}$,使得对所有 $A\in\boldsymbol{\Delta}(\boldsymbol{H})$,我们有

$$(1-|w|^2)I\leqslant[I-\overline{w}f(A)][I-f(A)^*f(A)]^{-1}[I-wf(A)^*]$$
$$\leqslant(I-\overline{w}A)(I-A^*A)^{-1}(I-wA^*),\tag{1}$$

$$(1-|w|^2)I\leqslant[I-\overline{w}f(A)][(1+|w|^2)I-\overline{w}f(A)-wf(A)^*]^{-1}[I-wf(A)^*]\leqslant(I-\overline{w}A)[(1+|w|^2)I-\overline{w}A-wA^*]^{-1}(I-wA^*),\tag{2}$$

$$\left\|f^{[n]}(A)-\frac{w}{d(w,A)+|w|^2}I\right\|\leqslant\left\|A-\frac{w}{d(w,A)+|w|^2}I\right\|$$

$$= \frac{\{d(w,A)[d(w,A)+|w|^2-1]\}^{1/2}}{d(w,A)+|w|^2}, \tag{3}$$

$$\| [f^{[n]}(A)-wI][I-\overline{w}f^{[n]}(A)]^{-1} \| \leqslant \| (A-wI)(I-\overline{w}A)^{-1} \|$$
$$= \left[\frac{d(w,A)+|w|^2-1}{d(w,A)}\right]^{1/2}, \tag{4}$$

其中,$n=1,2,\cdots$,并且

$$d(w,A) = \| (I-\overline{w}A)(I-A^*A)^{-1}(I-wA^*) \|. \tag{5}$$

此外,只要 $|w|<1$,一定有 $f(w)=w$.

定理 Ⅱ 设 g 在 $\boldsymbol{\Pi}$ 上解析,且 $g(\boldsymbol{\Pi})\subset\boldsymbol{\Pi}$,又设 $g^{[n]}$ 表示 g 的第 n 次迭代. 则或者有

$$\operatorname{Re} g(B) \geqslant \operatorname{Re} B \tag{6}$$

对所有 $B\in\boldsymbol{\Pi}(\boldsymbol{H})$ 成立,或者存在 $w\in\overline{\boldsymbol{\Pi}}$,使得对所有 $B\in\boldsymbol{\Pi}(\boldsymbol{H})$ 我们有

$$(4\operatorname{Re} w)I \leqslant [g(B)+\overline{w}I][\operatorname{Re} g(B)]^{-1}[g(B)^*+wI]$$
$$\leqslant (B+\overline{w}I)(\operatorname{Re} B)^{-1}(B^*+wI), \tag{7}$$

$$\left\| g^{[n]}(B) - \left[\frac{\delta(w,B)}{2}-\overline{w}\right]I \right\| \leqslant \left\| B - \left[\frac{\delta(w,B)}{2}-\overline{w}\right]I \right\|$$
$$= \left\{\frac{[\delta(w,B)]^2}{4} - \delta(w,B)\operatorname{Re} w\right\}^{1/2}, \tag{8}$$

$$\| [g^{[n]}(B)-wI][g^n(B)+\overline{w}I]^{-1} \| \leqslant \| (B-wI)(B+\overline{w}I)^{-1} \|$$
$$= \left[1-\frac{4\operatorname{Re} w}{\delta(w,B)}\right]^{1/2}, \tag{9}$$

其中,$n=1,2,\cdots$,并且

$$\delta(w,B) = \| (B+\overline{w}I)(\operatorname{Re} B)^{-1}(B^*+wI) \|. \tag{10}$$

此外,只要 $\operatorname{Re} w>0$,一定有 $g(w)=w$.

循着 Ky Fan 的工作[4~6],本文旨在给出有关 Hilbert 空间上算子解析函数的若干补充结果. §2 主要讨论由 $\boldsymbol{\Omega}(\boldsymbol{H})$ 到 $\boldsymbol{\Omega}(\boldsymbol{H})$ 内的算子解析函数的迭代,其中定理 1 与作者近期工作[1] 密切相关. §3 集中讨论定理 1 的退化情形,即 $\boldsymbol{H}=\mathbf{C}$ 的情形. 最后,在 §4 中,我们讨论由 $\boldsymbol{\Pi}(\boldsymbol{H})$ 到 $\boldsymbol{\Delta}(\boldsymbol{H})$ 内与由 $\boldsymbol{\Delta}(\boldsymbol{H})$ 到 $\boldsymbol{\Pi}(\boldsymbol{H})$ 内的算子解析函数的特征,主要结果是定理 3 与定理 4.

§2. 算子解析函数 $k:\boldsymbol{\Omega}(\boldsymbol{H})\to\boldsymbol{\Omega}(\boldsymbol{H})$ 的迭代

首先,讨论两个有关的引理.

引理 1　设 k 在 $\boldsymbol{\Omega}$ 上解析，且 $k(\boldsymbol{\Omega})\subset\boldsymbol{\Omega}$. 则对任意 $A\in\boldsymbol{\Omega}(\boldsymbol{H})$，一定有 $k(A)\in\boldsymbol{\Omega}(\boldsymbol{H})$.

证　定义 $\tilde{k}(z)=-\mathrm{i}k(\mathrm{i}z)$，则 $\tilde{k}$ 在 $\boldsymbol{\Pi}$ 上解析，且 $\tilde{k}(\boldsymbol{\Pi})\subset\boldsymbol{\Pi}$. 现设 $A\in\boldsymbol{\Omega}(\boldsymbol{H})$，并令 $B=-\mathrm{i}A$，则 $\mathrm{Re}\,B=\dfrac{A-A^*}{2\mathrm{i}}=\mathrm{Im}\,A>0$. 因此，由引言部分知道，$\mathrm{Re}\,\tilde{k}(B)>0$. 但是，

$$\mathrm{Im}\,k(A)=\frac{k(A)-k(A)^*}{2\mathrm{i}}=\frac{\mathrm{i}\,\tilde{k}(B)+\mathrm{i}\,\tilde{k}(B)^*}{2\mathrm{i}}=\mathrm{Re}\,\tilde{k}(B),$$

因此 $\mathrm{Im}\,k(A)>0$. 证毕. ■

引理 2　设 $A\in\boldsymbol{\Omega}(\boldsymbol{H})$，$\tau\in\overline{\boldsymbol{\Omega}}$，且实数 $\rho>0$，则不等式

$$\left\|(A-\bar{\tau}I)(\mathrm{Im}\,A)^{-1}(A^*-\tau I)\right\|\leqslant\rho \tag{11}$$

等价于

$$\left\|A-\left(\bar{\tau}+\frac{\mathrm{i}\rho}{2}\right)\right\|\leqslant\left(\frac{\rho^2}{4}-\rho\,\mathrm{Im}\,\tau\right)^{1/2}, \tag{12}$$

并且，(11) 式中"="成立当且仅当(12) 式中"="成立. 当 $\tau\in\boldsymbol{\Omega}$ 时，(11) 与(12) 两式都等价于

$$\left\|(A-\tau I)(A-\bar{\tau}I)^{-1}\right\|\leqslant\left(1-\frac{4\mathrm{Im}\,\tau}{\rho}\right)^{1/2}, \tag{13}$$

并且，"="在(11)(12) 与(13) 式中同时出现.

证　令 $B=-\mathrm{i}A$ 与 $w=-\mathrm{i}\tau$，则 $B\in\boldsymbol{\Pi}(\boldsymbol{H})$ 与 $\mathrm{Re}\,w\geqslant 0$，$\mathrm{Im}\,A=\mathrm{Re}\,B$，$\mathrm{Im}\,\tau=\mathrm{Re}\,w$. 这样一来，引理 2 便是〔6〕中引理 2 的一个直接推论. ■

下面是本节的主要结果.

定理 1　设 k 在 $\boldsymbol{\Omega}$ 上解析，且 $k(\boldsymbol{\Omega})\subset\boldsymbol{\Omega}$. 则或者有

$$\mathrm{Im}\,k(A)\geqslant\mathrm{Im}\,A \tag{14}$$

对所有 $A\in\boldsymbol{\Omega}(\boldsymbol{H})$ 成立，或者存在复数 $\tau\in\overline{\boldsymbol{\Omega}}$，使得对所有 $A\in\boldsymbol{\Omega}(\boldsymbol{H})$ 我们有

$$\begin{aligned}(4\mathrm{Im}\,\tau)I&\leqslant[k(A)-\bar{\tau}I][\mathrm{Im}\,k(A)]^{-1}[k(A)^*-\tau I]\\&\leqslant(A-\bar{\tau}I)(\mathrm{Im}\,A)^{-1}(A^*-\tau I),\end{aligned} \tag{15}$$

$$\begin{aligned}\left\|k^{[n]}(A)-\left[\bar{\tau}+\frac{\mathrm{i}\rho(\tau,A)}{2}\right]I\right\|&\leqslant\left\|A-\left[\bar{\tau}+\frac{\mathrm{i}\rho(\tau,A)}{2}\right]I\right\|\\&=\left\{\frac{[\rho(\tau,A)]^2}{4}-\rho(\tau,A)\mathrm{Im}\,\tau\right\}^{1/2},\end{aligned} \tag{16}$$

$$\left\|[k^{[n]}(A)-\tau I][k^{[n]}(A)-\bar{\tau}I]^{-1}\right\|\leqslant\left\|(A-\tau I)(A-\bar{\tau}I)^{-1}\right\|$$

$$=\left[1-\frac{4\mathrm{Im}\,\tau}{\rho(\tau,A)}\right]^{1/2}, \tag{17}$$

$$\rho(\tau,k^{[n]}(A))\leqslant\rho(\tau,A), \tag{18}$$

其中，$n=1,2,3,\cdots$ 并且

$$\rho(\tau,A)=\left\|(A-\bar{\tau}I)(\mathrm{Im}\,A)^{-1}(A^{*}-\tau I)\right\|. \tag{19}$$

此外，只要 $\mathrm{Im}\,\tau>0$，一定有 $k(\tau)=\tau$.

证 如同引理 1 的证明，设 $\tilde{k}(z)=-\mathrm{i}k(\mathrm{i}z)$，则 $\tilde{k}$ 在 $\boldsymbol{\Pi}$ 上解析，且 $\tilde{k}(\boldsymbol{\Pi})\subset\boldsymbol{\Pi}$. 根据定理 Ⅱ，我们或者有

$$\mathrm{Re}\,\tilde{k}(B)\geqslant\mathrm{Re}\,B \tag{20}$$

对所有 $B\in\boldsymbol{\Pi}(\boldsymbol{H})$ 成立，或者存在 $\omega\in\overline{\boldsymbol{\Pi}}$，使得对所有 $B\in\boldsymbol{\Pi}(\boldsymbol{H})$ 有相应的(7) ~ (10) 式成立. 特别地，当 $\mathrm{Re}\,w>0$ 时，一定有 $\tilde{k}(w)=w$. 现令 $\tau=\mathrm{i}w$，则当 $\mathrm{Im}\,\tau>0$ 时，$k(\tau)=\tau$. 设 $A=\mathrm{i}B$，则(20) 式等价于 $\mathrm{Re}(-\mathrm{i}k(\mathrm{i}B))\geqslant\mathrm{Re}\,B$，此即

$$\mathrm{Im}\,k(A)\geqslant\mathrm{Re}(-\mathrm{i}A)=\mathrm{Im}\,A$$

对所有 $A\in\boldsymbol{\Omega}(\boldsymbol{H})$ 成立. 同样地，(7) 式变为

$$\begin{aligned}(4\mathrm{Im}\,\tau)I&\leqslant[-\mathrm{i}k(\mathrm{i}B)+\mathrm{i}\bar{\tau}I][\mathrm{Re}(-\mathrm{i}k(\mathrm{i}B))]^{-1}[\mathrm{i}k(\mathrm{i}B)^{*}-\mathrm{i}\tau I]\\&=[k(A)-\bar{\tau}I][\mathrm{Im}\,k(A)^{-1}][k(A)^{*}-\tau I]\\&\leqslant(-\mathrm{i}A+\mathrm{i}\bar{\tau}I)[\mathrm{Re}(-\mathrm{i}A)]^{-1}(\mathrm{i}A^{*}-\mathrm{i}\tau I)\\&=(A-\bar{\tau}I)(\mathrm{Im}\,A)^{-1}(A^{*}-\tau I),\end{aligned}$$

因此，(15) 式得证. 因为 $k^{[n]}(\boldsymbol{\Omega})\subset\boldsymbol{\Omega}$，由引理 1 知道，$k^{[n]}(A)\in\boldsymbol{Q}(\boldsymbol{H})$，$n=1,2,3,\cdots$，对所有 $A\in\boldsymbol{\Omega}(\boldsymbol{H})$ 成立. 注意到(15) 式对任意的这样算子 A 皆成立与 $k^{[n]}(A)=k(k^{[n-1]}(A))$ 的事实，我们有

$$\begin{aligned}0&\leqslant[k^{[n]}(A)-\bar{\tau}I][\mathrm{Im}\,k^{[n]}(A)]^{-1}[k^{[n]}(A)-\tau I]\\&\leqslant[k^{[n-1]}(A)-\bar{\tau}I][\mathrm{Im}\,k^{[n-1]}(A)]^{-1}[k^{[n-1]}(A)-\tau I]\\&\leqslant\cdots\leqslant(A-\bar{\tau}I)(\mathrm{Im}\,A)^{-1}(A-\tau I),\end{aligned}$$

此蕴涵

$$\begin{aligned}&\left\|[k^{[n]}(A)-\bar{\tau}I][\mathrm{Im}\,k^{[n]}(A)]^{-1}[k^{[n]}(A)-\tau I]\right\|\\&\quad=\left\|(A-\bar{\tau}I)(\mathrm{Im}\,A)^{-1}(A-\tau I)\right\|\end{aligned}$$

(因为上式两端 $\|\quad\|$ 内的算子是正算子)，此即是(18) 式. 取 $\rho=\left\|(A-\bar{\tau}I)(\mathrm{Im}\,A)^{-1}(A-\tau I)\right\|$，应用引理 2，则(16) 与(17) 式(当 $\mathrm{Im}\,\tau>0$ 时) 得

证. 余下只需验明(17) 式对 Im $\tau=0$ 情形也成立. 但这是显而易见的，因为这时$\bar{\tau}=\tau$，(17) 式中三个数皆等于 1. 定理 1 证毕. ■

附注 1　如果定理 1 中 k 还满足条件：$k(z)\neq z, z\in\boldsymbol{\Omega}$(即 k 在$\boldsymbol{\Omega}$ 内没有不动点)，此时由定理 1 知道 Im $\tau=0$ 一定成立，即 τ 是实数. 这样，(17) 式变为平凡情形，(15) 式中左端项变为零算子. 同时，(16) 式变为

$$\left\| k^{[n]}(A)-\left[\tau+\frac{\mathrm{i}\rho(\tau,A)}{2}\right]I\right\| \leqslant \left\| A-\left[\tau+\frac{\mathrm{i}\rho(\tau,A)}{2}\right]I\right\|$$

$$=\frac{\rho(\tau,A)}{2},(n=1,2,\cdots).$$

这个式子的几何意义是：$k^{[n]}$ 将 $\mathscr{L}(\boldsymbol{H})$ 内以$\left[\tau+\frac{\mathrm{i}\rho(\tau,A)}{2}\right]I$ 为球心，以 $\frac{\rho(\tau,A)}{2}$ 为半径的球的边界点 A 映射到此闭球内，并且，如果 B 也是此球的边界点，即有

$$\left\| B-\left[\tau+\frac{\mathrm{i}\rho(\tau,A)}{2}\right]I\right\| =\frac{\rho(\tau,A)}{2},$$

那么由引理 2 知道，$\rho(\tau,A)=\rho(\tau,B)$. 因此，我们进一步断言，$k^{[n]}$ 将上述球的所有边界点映射到该闭球内. 此外，此球与 $\mathscr{L}(\boldsymbol{H})$ 的超平面 $\partial\boldsymbol{\Omega}(\boldsymbol{H})$ 相切于 τI.

值得指出的是，定理 1 以及附注 1 在 Hibert 空间的广义上半平面上的全纯自映射中也有十分相似的结果(见〔2〕定理 2).

§3. 退化情形：$\boldsymbol{H}=\mathbf{C}$ 的讨论

当 $\boldsymbol{H}$ 退化为 $\mathbf{C}$ 时，定理 1 与附注 1 中的有关结论仍然有效，而且尚有一些进一步的结果. 此时，(14) ～ (19) 诸式分别变为十分简单形式：

$$\operatorname{Im} k(z)\geqslant \operatorname{Im} z, \tag{14'}$$

$$4\operatorname{Im}\tau\leqslant |k(z)-\bar{\tau}|^2/\operatorname{Im} k(z), \tag{15'}$$

$$\left|k^{[n]}(z)-\left[\bar{\tau}+\frac{\mathrm{i}\rho(\tau,z)}{2}\right]\right|\leqslant\left|z-\left[\bar{\tau}+\frac{\mathrm{i}\rho(\tau,z)}{2}\right]\right|$$

$$=\left\{\frac{[\rho(\tau,z)]^2}{4}-\rho(\tau,z)\operatorname{Im}\tau\right\}^{1/2}, \tag{16'}$$

$$|[k^{[n]}(z)-\tau]/[k^{[n]}(z)-\bar{\tau}]|\leqslant|(z-\tau)/(z-\bar{\tau})|$$

$$=\left[1-\frac{4\operatorname{Im}\tau}{\rho(\tau,z)}\right]^{1/2}, \tag{17'}$$

$$\rho(\tau, k^{[n]}(z)) \leqslant \rho(\tau, z), \tag{18'}$$

其中，$n = 1, 2, \cdots$，并且

$$\rho(\tau, z) = |z - \bar{\tau}|^2 / \operatorname{Im} z. \tag{19'}$$

因此，如果 k 在 $\boldsymbol{\Omega}$ 上解析且 $k(\boldsymbol{\Omega}) \subset \boldsymbol{\Omega}$，那么或者有(14′) 式对所有 $z \in \boldsymbol{\Omega}$ 成立，或者存在复数 τ，满足 $\operatorname{Im} \tau \geqslant 0$，使得(15′) ~ (18′) 诸式对所有正整数 n 与所有 $z \in \boldsymbol{\Omega}$ 成立，其中 $\rho(\tau, z)$ 由(19′) 式定义. 此外，当 $\operatorname{Im} \tau > 0$ 时，$k(\tau) = \tau$.

从几何上看，前一种可能性(即(14′) 式对所有 $z \in \boldsymbol{\Omega}$ 成立) 十分直观. 对后一种可能我们分两种情形讨论.

情形 Ⅰ ($\operatorname{Im} \tau > 0$ 情形). 此时由引理 2 知道，(16′)(17′) 与(18′) 彼此等价. 显然地，三者皆等价于

$$d(\tau, k^{[n]}(z)) \leqslant d(\tau, z), \tag{21}$$

这里，当 $u \in \boldsymbol{\Omega}$ 时，$d(\tau, u) = \tanh^{-1} \left| \dfrac{z - u}{\bar{z} - u} \right|$，它与 $\alpha(z, u) = \dfrac{|u|}{2\operatorname{Im} z}$ 给出 $\boldsymbol{\Omega}$ 上所谓 CRF(Carathéodory-Reiffen-Finsler) 距离[7]. 因此，当 $\operatorname{Im} \tau > 0$ 时，如果存在 $z_0 \in \boldsymbol{\Omega}$，使得 $\operatorname{Im} k(z_0) < \operatorname{Im} z_0$，那么存在 k 的在 $\boldsymbol{\Omega}$ 上的不动点 τ，使得 $k^{[n]}$ 不增大 z 与 τ 之间的 CRF 距离，这里 z 可以是 $\boldsymbol{\Omega}$ 上任意一点，n 是任意的正整数.

情形 Ⅱ ($\operatorname{Im} \tau = 0$，即 τ 是实数情形). 此时，(15′) 与(17′) 变为平凡情形，而(16′) 与(18′) 两式变为

$$\left| k^{[n]}(z) - \left(\tau - \frac{\mathrm{i}\,|z - \tau|^2}{2\operatorname{Im} z} \right) \right| \leqslant \left| z - \left(\tau - \frac{\mathrm{i}\,|z - \tau|^2}{2\operatorname{Im} z} \right) \right| = \frac{|z - \tau|^2}{2\operatorname{Im} z}, \tag{16''}$$

$$\frac{|k^{[n]}(z) - \tau|^2}{\operatorname{Im} k^{[n]}(z)} \leqslant \frac{|z - \tau|^2}{\operatorname{Im} z}. \tag{18''}$$

由引理 2 看出，上述两式是彼此等价的. 同时它们也具有明显的几何意义. 对给定的实数 τ 与给定的 $z \in \boldsymbol{\Omega}$，用 $\mathrm{D}_{z\tau}$ 表示复平面上以点 $\left(\tau, \dfrac{|z - \tau|^2}{2\operatorname{Im} z} \right)$ 为圆心，以 $\dfrac{|z - \tau|^2}{2\operatorname{Im} z}$ 为半径的闭圆. 则(16″) 式告诉我们，z 在 $\mathrm{D}_{z\tau}$ 的边界 $\partial \mathrm{D}_{z\tau}$ 上，$\mathrm{D}_{z\tau}$ 的内部在 $\boldsymbol{\Omega}$ 上，且 $\mathrm{D}_{z\tau}$ 与实轴相切于点$(\tau, 0)$，此外，$k^{[n]}(z) \in \mathrm{D}_{z\tau}$ 对一切正整数 n 成立.

如果 k 在 $\boldsymbol{\Omega}$ 上没有不动点，那么由定理 1 知道 $\operatorname{Im} \tau = 0$. 对于这样的

解析函数,我们有

定理 2 设 k 在 $\boldsymbol{\Omega}$ 上解析,$k(\boldsymbol{\Omega})\subset\boldsymbol{\Omega}$,且 $k(z)\neq z$,$z\in\boldsymbol{\Omega}$. 则或者对所有 $z\in\boldsymbol{\Omega}$,$\operatorname{Im}k(z)\geqslant\operatorname{Im}z$ 成立,或者存在唯一的实数 τ,使得对所有正整数 n 与所有 $z\in\boldsymbol{\Omega}$,我们有

$$k^{[n]}(\mathrm{D}_{z\tau})\subset\mathrm{D}_{z\tau}\subset\overline{\boldsymbol{\Omega}}. \tag{22}$$

证 由前面分析知道,如果 $\operatorname{Im}k(z)<\operatorname{Im}z$ 对某一个 $z\in\boldsymbol{\Omega}$ 成立,因为 k 在 ω 上没有不动点,那么存在实数 τ,使得对任意 $z\in\boldsymbol{\Omega}$,我们有 $z\in\partial\mathrm{D}_{z\tau}$,且 $k^{[n]}(z)\in\mathrm{D}_{z\tau}$ 对任意正整数 n 与任意 $z\in\boldsymbol{\Omega}$ 成立. 类似于附注 1 的讨论,我们得到

$$k^{[n]}(\partial\mathrm{D}_{z\tau})\subset\mathrm{D}_{z\tau}\subset\overline{\boldsymbol{\Omega}}.$$

接着我们证明 $k^{[n]}(\mathrm{D}_{z\tau}-\partial\mathrm{D}_{z\tau})\subset\mathrm{D}_{z\tau}$. 事实上,如果 u 位于 $\mathrm{D}_{z\tau}$ 的内部,那么闭圆 $\mathrm{D}_{z\tau}$ 以 u 为其边界点,且此闭圆与实轴也切于点 $(\tau,0)$. 此时,显然地 $\mathrm{D}_{u\tau}\subset\mathrm{D}_{z\tau}$,因此,$k^{[n]}(u)\in\mathrm{D}_{u\tau}\subset\mathrm{D}_{z\tau}$,于是(22)式得证. 余下只需证明实数 τ 的唯一性. 假若存在另一实数 $\tau'(\neq\tau)$,使得 $k(\mathrm{D}_{z\tau'})\subset\mathrm{D}_{z\tau'}$ 对所有 $z\in\boldsymbol{\Omega}$ 成立. 这时可选取 z 与 $z'\in\boldsymbol{\Omega}$,使得 $\mathrm{D}_{z\tau}$ 与 $\mathrm{D}_{z'\tau'}$ 相外切于 $w\in\boldsymbol{\Omega}$,按前面推证,$k(w)\in\mathrm{D}_{z\tau}$ 与 $k(w)\in\mathrm{D}_{z'\tau'}$,因此,$k(w)=w$,但此与 k 在 $\boldsymbol{\Omega}$ 上没有不动点的假定抵触,定理 2 证毕. ■

作为定理 2 的一个简单应用,我们得到定理 1 的一个补充结果.

推论 设 k 在 $\boldsymbol{\Omega}$ 上解析,$k(\boldsymbol{\Omega})\subset\boldsymbol{\Omega}$,且 $k(z)\neq z$ 对所有 $z\in\boldsymbol{\Omega}$ 成立. 则定理 1 中的复数 τ 是唯一的.

证 类似于〔4〕中定理的证明,我们看出复数 τ 只依赖于函数 k 本身,而与 $\mathscr{L}(\boldsymbol{H})$ 无关. 因此,当 k 在 $\boldsymbol{\Omega}$ 上无不动点时,τ 是实数,这时 τ 的唯一性便是定理 2 的一个直接结果. 推论证毕. ■

附注 2 对定理 Ⅰ 与定理 Ⅱ 也有类似于上述推论的结果.

§4. 算子解析函数 $h:\boldsymbol{\Pi}(\boldsymbol{H})\to\boldsymbol{\Delta}(\boldsymbol{H})$ 与 $p:\boldsymbol{\Delta}(\boldsymbol{H})\to\boldsymbol{\Pi}(\boldsymbol{H})$

考虑两个复函数 ϕ 与 ψ:

$$\phi(z)=\frac{1+z}{1-z},\quad z\neq1, \tag{23}$$

$$\psi(z)=\frac{z-1}{z+1},\quad z\neq-1.$$

则 ϕ 与 ψ 互逆,且 $\phi(\boldsymbol{\Delta})=\boldsymbol{H}$,$\psi(\boldsymbol{\Pi})=\boldsymbol{\Delta}$. 现在我们研究本段标题所表明的

两类算子解析函数的性质作为定理 Ⅰ 与定理 Ⅱ 的补充. 同时,这里的证明技巧也是基于〔6〕中有关方法的.

定理 3 设 h 在 $\boldsymbol{\Pi}$ 上解析,且 $h(\boldsymbol{\Pi})\subset\boldsymbol{\Delta}$. 则或者有

$$[I-h(A)^*]^{-1}[I-h(A)h(A)^*][I-h(A)]^{-1}\geqslant \mathrm{Re}\,A \tag{24}$$

对所有 $A\in\boldsymbol{\Pi}(\boldsymbol{H})$ 成立,或者存在复数 $\xi\in\overline{\boldsymbol{\Pi}}$,使得对所有 $A\in\boldsymbol{\Pi}(\boldsymbol{H})$ 我们有

$$(4\mathrm{Re}\,\xi)I\leqslant[(\bar{\xi}+1)I-(\bar{\xi}-1)h(A)][I-h(A)^*h(A)]^{-1}[(\xi+1)I-(\xi-1)h(A)^*]\leqslant(A+\bar{\xi}I)(\mathrm{Re}\,A)^{-1}(A^*+\xi I), \tag{25}$$

$$\left\| h(A)-\frac{(\xi-1)(\bar{\xi}+1)}{|\xi-1|^2+e(\xi,A)}I\right\| \leqslant \left\| \psi(A)-\frac{(\xi-1)(\bar{\xi}+1)}{|\xi-1|^2+e(\xi,A)}I\right\| =\frac{\{e(\xi,A)[e(\xi,A)-4\mathrm{Re}\,\xi]\}^{1/2}}{|\xi-1|^2+e(\xi,A)}, \tag{26}$$

$$\left\| \left[h(A)-\frac{\xi-1}{\xi+1}I\right]\left[I-\frac{\bar{\xi}-1}{\xi+1}h(A)\right]^{-1}\right\| \leqslant \left\|(A-\xi I)(A+\bar{\xi}I)^{-1}\right\| =\left[\frac{e(\xi,A)-4\mathrm{Re}\,\xi}{e(\xi,A)}\right]^{1/2}, \tag{27}$$

其中,

$$e(\xi,A)=\left\|(A+\bar{\xi}I)(\mathrm{Re}\,A)^{-1}(A^*+\xi I)\right\|. \tag{28}$$

此外,只要 $\mathrm{Re}\,\xi>0$,一定有 $h(\xi)=\psi(\xi)$.

证 令 $f=h\circ\phi$,则 f 在 $\boldsymbol{\Delta}$ 上解析,且 $f(\boldsymbol{\Delta})\subset\boldsymbol{\Delta}$. 根据定理 Ⅰ,存在复数 w, $|w|\leqslant 1$, 满足该定理的有关结论. 特别地当 $|w|<1$ 时, $f(w)=w$. 现令 $\xi=\phi(w)$,且对任意给定的 $A\in\boldsymbol{\Pi}(\boldsymbol{H})$,令 $B=\psi(A)$,则 $\|B\|<1, A=\phi(B)$,因而

$$f(B)=(h\circ\phi\circ\psi)(A)=h(A). \tag{29}$$

应用关系式

$$B=\psi(A)=(A-I)(A+I)^{-1},$$

我们求得

$$I-B=2(A+I)^{-1}, \tag{30}$$

$$I-B^*B=\psi(A^*+I)^{-1}\mathrm{Re}\,A(A+I)^{-1}. \tag{31}$$

现区分两种互斥情况讨论: $w=1$ 与 $w\neq 1$. 当 $w=1$ 时,(1) 式给出

$$0\leqslant[I-f(B)][I-f(B)^*f(B)]^{-1}[I-f(B)^*]$$
$$\leqslant(I-B)(I-B^*B)^{-1}(I-B^*).$$

应用(29) ~ (31) 式,上述最后不等式可改写为

$$0 \leqslant [I-h(A)][I-h(A)^{*}h(A)]^{-1}[I-h(A)^{*}]$$

$$\leqslant 2(A+I)^{-1}\cdot\frac{1}{4}(A+I)(\mathrm{Re}\,A)^{-1}(A^{*}+I)\cdot 2(A^{*}+I)^{-1}=(\mathrm{Re}\,A)^{-1}. \tag{32}$$

因为上式两端都是可逆的正算子,所以(32) 式等价于(24) 式.

当 $w\neq 1$ 时,$\xi=\phi(w)\neq\infty$,且 $\mathrm{Re}\,\xi\geqslant 0$. 由前面分析知道,只要 $|w|<1$,一定有 $f(w)=w$,因此由于 $f(w)=(h\circ\phi)(w)=h(\xi)$ 以及

$$1-|w|^{2}=\frac{4\mathrm{Re}\,\xi}{|1+\xi|^{2}}, \tag{33}$$

我们得出:只要 $\mathrm{Re}\,\xi>0$,便有 $h(\xi)=w=\psi(\xi)$. 经过简单计算我们有

$$I-\overline{w}f(B)=I-\frac{\bar{\xi}-1}{\bar{\xi}+1}h(A), \tag{34}$$

$$I-\overline{w}B=I=\frac{\bar{\xi}-1}{\bar{\xi}+1}(A-I)(A+I)^{-1}=\frac{2}{\bar{\xi}+1}(A+\bar{\xi}I)(A+I)^{-1}, \tag{35}$$

$$\left\|[\psi(A)-wI][I-\overline{w}\psi(A)^{-1}]\right\|=\left\|(A-\xi I)(A+\bar{\xi}I)^{-1}\right\|, \tag{36}$$

$$\left\|(I-\overline{w}B)(I-B^{*}B)^{-1}(I-wB^{*})\right\|=\frac{e(\xi,A)}{|1+\xi|^{2}}, \tag{37}$$

这里,$e(\xi,A)$ 由(28) 式定义. 借助于(33) 式与上述诸式,由(1) 式直接推出(25) 式. 同样地,取定 $n=1$,由(3) 与(4) 式经过简单计算可得(26) 与(27) 两式. 定理 3 证毕. ■

定理 4　设 p 在 $\boldsymbol{\Delta}$ 上解析,且 $p(\boldsymbol{\Delta})\subset\boldsymbol{\Pi}$. 则或者有

$$\mathrm{Re}\,p(A)\geqslant(I-A^{*})^{-1}(I-A^{*}A)(I-A)^{-1} \tag{38}$$

对所有 $A\in\boldsymbol{\Delta}(\boldsymbol{H})$ 成立,或者存在复数 η, $|\eta|\leqslant 1$,使得对所有 $A\in\boldsymbol{\Delta}(\boldsymbol{H})$ 我们有

$$4(1-|\eta|^{2})I\leqslant[(1-\overline{\eta})p(A)+(1+\overline{\eta})I][\mathrm{Re}\,p(A)]^{-1}[(1-\eta)p(A)^{*}$$
$$+(1-\eta)I]\leqslant 4(I-\overline{\eta}A)(I-A^{*}A)(I-\eta A^{*}), \tag{39}$$

$$\left\|p(A)-\frac{2\varepsilon(\eta,A)-(1+\eta)(1-\overline{\eta})}{|1-\eta|^{2}}I\right\|$$
$$\leqslant\left\|\phi(A)-\frac{2\varepsilon(\eta,A)-(1+\eta)(1-\overline{\eta})}{|1-\eta|^{2}}I\right\|$$
$$=\frac{2}{|1-\eta|^{2}}\{[e(\eta,A)]^{2}-(1-|\eta|^{2})\varepsilon(\eta,A)\}^{1/2}, \tag{40}$$

$$\left\| \left[p(A)-\frac{1+\eta}{1-\eta}I\right]\left[p(A)+\frac{1+\bar{\eta}}{1-\bar{\eta}}I\right]^{-1} \right\|$$

$$\leqslant \left\| (A-\eta I)(I-\eta A)^{-1} \right\| = \left[1-\frac{1-|\eta|^2}{\varepsilon(\eta,A)}\right]^{1/2}, \qquad (41)$$

其中,

$$\varepsilon(\eta,A) = \left\| (I-\bar{\eta}A)(I-A^*A)^{-1}(I-\eta A^*) \right\|. \qquad (42)$$

此外,只要 $|\eta|<1$,一定有 $p(\eta)=\phi(\eta)$.

证 令 $g=p\circ\psi$,则 g 在 $\boldsymbol{\Pi}$ 上解析,且 $g(\boldsymbol{\Pi})\subset\boldsymbol{\Pi}$. 根据定理 Ⅱ,或者对所有 $B\in\boldsymbol{\Pi}(\boldsymbol{H})$,

$$\mathrm{Re}\, g(B)\geqslant \mathrm{Re}\, B,$$

或者存在 ω,$\mathrm{Re}\,\omega\geqslant 0$,使得(7) ~ (10) 诸式对所有 $B\in\boldsymbol{\Pi}(\boldsymbol{H})$ 成立. 此外,只要 $\mathrm{Re}\,\omega>0$,定有 $g(\omega)=\omega$. 令 $\eta=\phi(\omega)$,此等价于:只要 $|\eta|<1$,定有 $p(\eta)=\phi(\eta)$. 现取定 $A\in\boldsymbol{\Delta}(\boldsymbol{H})$,并令 $B=\phi(A)$,则 $B\in\boldsymbol{\Pi}(\boldsymbol{H})$,且 $A=\psi(B)$. 应用关系式 $B=(I+A)(I-A)^{-1}$,经计算得出

$$B+\bar{\omega}I=(1+\bar{\omega})(I-\bar{\omega}A)(I-A)^{-1},$$

$$\mathrm{Re}\, B=(I-A^*)^{-1}(I-A^*A)(I-A)^{-1},$$

$$\mathrm{Re}\,\omega=\frac{1-|\eta|^2}{|1-\eta|^2},\ |1+\omega|^2=\frac{4}{|1-\eta|^2},$$

$$(B-\omega I)(B+\bar{\omega}I)=\frac{1+\omega}{1+\bar{\omega}}(A+\eta I)(I-\bar{\eta}A)^{-1},$$

$$\delta(\omega,A)=4\varepsilon(\eta,A)/|1-\eta|^2.$$

应用 $g(B)=(p\circ\psi)(B)=p(A)$,以及上述诸式,连同定理 Ⅱ 中的(6) ~ (10) 式直接推出本定理(38) ~ (42) 诸式. 定理 4 证毕. ■

致谢

感谢 Ky Fan 教授在本文准备过程中的支持与帮助,其中包括提供他的文章〔6〕的手稿.

参考文献

[1]G-N Chen. J. Math. Anal. Appl. ,1984:98,305-313.

[2]G-N Chen. J. Math Anal. Appl. ,1985,112(1):274-278.

[3]N Dunford,J T Schwartz. Linear Operators,Part I:General Theory. New

York-London-Sydney-Toronto: Wiley-Interscience, 1958.

[4]Ky Fan. Math. Z., 1982, 179: 293 — 298.

[5]Ky Fan. Math. Z. 1978, 160: 275 — 290.

[6]Ky Fan. Linear Multilinear Algebra, 1983, 12: 295 — 304.

[7]L A Harris. Advances in Holomorphy. J A Barroso ed., North-Holland Publishing Comp., 1979: 345 — 406.

[8]Von Neumann. J Math. Nachr., 1951, 4: 258 — 281.

Some Results on Analytic Functions for Operators of Hilbert Spaces

Abstract Let $\boldsymbol{H}$ be a complex Hilbert Space. For an operator A on $\boldsymbol{H}$ and a complex analytic function f, under some conditions, denote by $f(A)$ the operator on $\boldsymbol{H}$ defined by Riesz-Dunford integral. Following the recent work by Ky Fan, some results on analytic functions of operators are obtained. Among these are the theorem on iteration of analytic function $f(A)$ of operator when f is a complex function analytic on the open upper half-plane $\boldsymbol{\Omega}$ such that $f(\boldsymbol{\Omega}) \subset \boldsymbol{\Omega}$, and properties for analytic function $f(A)$ when f is either a complex function analytic on the open right half-plane $\boldsymbol{\Pi}$ such that $f(\boldsymbol{\Pi}) \subset \boldsymbol{\Delta}$, the open unit disk, or that on $\boldsymbol{\Delta}$ such that $f(\boldsymbol{\Delta}) \subset \boldsymbol{\Pi}$.

北京师范大学学报(自然科学版)
1987,(1):13—17

Banach 代数中对谱半径的 Schwarz 引理①

Schwarz's Lemma for the Spectral Radius in Complex Banach Algebras

摘要 本文旨在研究复 Banach 代数中 Schwarz 引理的一种类似形式的某些结果,在这一形式中范数替换为谱半径. 作为它的应用,我们得到包括 Harnack 双重不等式与 Carathéodory 不等式的谱半径变种的一些结论.

关键词 Banach 代数,Schwarz 引理,谱半径,Harnack 双重不等式,Carathéodory 不等式

众所周知,复分析中著名的 Schwarz 引理有多种的推广与变异形式,其中最基本的是如下结果[1]:

定理 令 X 与 Y 是复 Banach 空间,且 X_0 是 X 的单位开球. 如果 $h:X_0\to Y$ 是满足 $\|h(x)\|\leqslant 1(x\in X_0)$ 与 $h(0)=0$ 的全纯函数,那么

$$\|h(x)\|\leqslant\|x\|\ (x\in X_0).$$

这里当 X 与 Y 是复 Banach 空间,Ω 是 X 的开子集时,我们称 $h:\Omega\to Y$ 是全纯的,假如在 Ω 的各点 x 处,h 的 Fréchet 导数 $h'(x)$ 存在并为自 X 到 Y 内的有界复线性映射([2],p. 248). 用 $\mathrm{Hol}(\Omega,Y)$ 表示定义在 Ω 上而值域在 Y 内的所有全纯函数的集合.

本文旨在讨论复 Banach 代数中对谱半径的上述定理(Schwarz 引理)的变异形式以及它的推论. 此后总假定字母 U 表示复域 $\mathscr{C}$ 上的有单位元 $e\neq 0$ 的 Banach 代数. 若 $x\in\mathscr{U}$,用 $\sigma(x)$ 与 $|x|_\sigma$ 分别记 x 的谱与谱半径.

---

① 收稿日期:1986-04-13.

令$\mathscr{U}_\Delta=\{x\in U:\sigma(x)\subseteq\Delta\}$，这里$\Delta=\{\lambda\in\mathscr{C}:|\lambda|<1\}$. 则$U_\Delta$是$U$内的开集(〔2〕, p. 243). 对于$U$内元素的谱半径，在下文中常要引用如下两个事实：第一，如果$x,y\in U$且$xy=yx$，那么$|x^n|_\sigma=|x|_\sigma^n$($n$为正整数)，$|xy|_\sigma\leqslant|x|_\sigma|y|_\sigma$与$|x+y|_\sigma\leqslant|x|_\sigma+|y|_\sigma$；第二，如果$f\in \mathrm{Hol}(\Delta,U)$，那么函数$|\lambda|\to|f(\lambda)|_\sigma$是$\Delta$上的次调和函数，因而有其相应的极大原理(见〔3〕, p. 10 与〔4〕).

现设$f\in \mathrm{Hol}(\Delta,\mathscr{C})$或$\mathrm{Hol}(\Delta,U)$，又设$x\in U_\Delta$，定义函数$\hat{f}$如下：

$$\hat{f}(x)=\frac{1}{2\pi\mathrm{i}}\int_\Gamma f(\lambda)(\lambda e-x)^{-1}\mathrm{d}\lambda,\tag{1}$$

这里，Γ是Δ内围绕$\sigma(x)$的任意周线. 可以证明，这时$\hat{f}$不依赖于这样Γ的选择，并且，$\hat{f}\in \mathrm{Hol}(U_\Delta,U)$[5,2]. 我们从如下两个引理开始.

引理 Ⅰ 设$f\in \mathrm{Hol}(\Delta,U)$满足$|f(\lambda)|_\sigma\leqslant 1(\lambda\in\Delta)$，且对某$n\geqslant 1$，$f(0)=f'(0)=\cdots=f^{(n-1)}(0)=0$. 则或者$|f(\lambda)|_\sigma<|\lambda|^n(\lambda\in\Delta-\{0\})$，或者$|f(\lambda)|_\sigma\equiv|\lambda|^n(\lambda\in\Delta)$.

证 基本上类似于〔6〕中定理1的证明. 令$g(\lambda)=f(\lambda)/\lambda^n(\lambda\in\Delta-\{0\})$. 按题设$g\in \mathrm{Hol}(\Delta,\mathscr{C})$. 应用次调和函数$\lambda\to|g(\lambda)|_\sigma(\lambda\in\Delta)$的极大原理，由不等式

$$|g(\lambda)|_\sigma=\frac{1}{|\lambda|^n}|f(\lambda)|_\sigma\leqslant|\lambda|^{-n}\quad(\lambda\in\Delta-\{0\})$$

容易推得$|g(\lambda)|_\sigma\leqslant 1(\lambda\in\Delta)$，并且，除了$|g(\lambda)|_\sigma\equiv 1(\lambda\in\Delta)$情形外，必定有$|g(\lambda)|_\sigma<1(\lambda\in\Delta-\{0\})$. 因此本引理成立. ■

引理 Ⅱ[5] 设Λ是$\mathscr{C}$内开集，且$f\in \mathrm{Hol}(\Lambda,U)$. 如果$x\in U_\Lambda$且与$f$可换(即对所有$\lambda\in\Lambda$，$xf(\lambda)=f(\lambda)x$)，那么有

$$\sigma(\hat{f}(x))\subseteq\sigma(f(\sigma(x))).$$

我们的基本结果是如下定理，它已在〔5〕中出现，这里我们采用稍为不同的证法.

定理 Ⅰ 设$f\in \mathrm{Hol}(\Delta,U)$满足$|f(\lambda)|_\sigma\leqslant 1(\lambda\in\Delta)$，且对某$n\geqslant 1$，$f(0)=f'(0)=\cdots=f^{(n-1)}(0)=0$. 如果$x\in U$有谱半径$|x|_\sigma<1$且与$f$可换，那么

$$|\hat{f}(x)|_\sigma=|x^n|_\sigma,\tag{2}$$

这里，$\hat{f}$由(1)式定义.

证 设$x\in U$有$|x|_\sigma<1$. 取实数t使得$|x|_\sigma<t<1$. 因为$\hat{f}\in \mathrm{Hol}(U_\Delta,U)$，且对所有$\lambda\in\Delta$，$\lambda t^{-1}x\in U_\Delta$，故$G(\lambda)=\hat{f}(\lambda t^{-1}x)$定义一个

自 Δ 到 U 内的全纯函数,即 $G \in \mathrm{Hol}(\Delta,U)$. 按条件 $f(0)=f'(0)=\cdots=f^{(n-1)}(0)=0$ 以及全纯函数的链规则容易算出,$G(0)=\hat{f}(0)=0$, $G'(0)=\hat{f}'(0)t^{-1}x=0,\cdots,G^{(n-1)}(0)=0$. 按题设,对所有 $\lambda\in\Delta$, $|f(\lambda)|_\sigma\leqslant 1$,且 $\lambda t^{-1}x$ 与 f 可换,根据引理Ⅱ容易推出,$|G(\lambda)|_\sigma=|\hat{f}(\lambda t^{-1}x)|_\sigma\leqslant 1$. 再利用引理Ⅰ得出,$|G(\lambda)|_\sigma=|\hat{f}(\lambda t^{-1}x)|_\sigma\leqslant|\lambda|^n(\lambda\in\Delta)$,亦即,对所有满足条件 $|x|_\sigma<t<1$ 的实数 t 都有 $|\hat{f}(x)|_\sigma\leqslant t^n$. 因此,$|\hat{f}(x)|_\sigma\leqslant|x|_\sigma^n=|x^n|_\sigma$. 证毕. ■

特殊地,当 $f\in\mathrm{Hol}(\Delta,\mathscr{C})$ 时,$x\in U$ 与 f 的可换性自然成立,因此有

推论Ⅰ 设 $f\in\mathrm{Hol}(\Delta,\mathscr{C})$ 满足 $|f(\lambda)|\leqslant 1(\lambda\in\Delta)$,且对某 $n\geqslant 1$, $f(0)=f'(0)=\cdots=f^{(n-1)}(0)=0$. 如果 $x\in U_\Delta$,那么 $|\hat{f}(x)|_\sigma\leqslant|x^n|_\sigma$.

如下推论与定理Ⅰ的 $n=1$ 情形相仿,但假设条件似更接近于 Schwarz 引理的"谱半径形式".

推论Ⅱ 设 $f\in\mathrm{Hol}(\Delta,U)$ 满足 $|f(\lambda)|_\sigma\leqslant 1(\lambda\in\Delta)$, $|f(0)|_\sigma=0$,且 $f(0)f(\lambda)=f(\lambda)f(0)(\lambda\in\Delta)$. 如果 $x\in U_\Delta$ 与 f 可换,那么

$$|\hat{f}(x)|_\sigma\leqslant|x|_\sigma.$$

证 容易看出由 $g(\lambda)=f(\lambda)-f(0)(\lambda\in\Delta)$ 定义的函数 $g\in\mathrm{Hol}(\Delta,U)$,且 $g(0)=0$. 由谱半径性质有(因为 $f(\lambda)$ 与 $f(0)$ 可换)

$$|g(\lambda)|_\sigma\leqslant|f(\lambda)|_\sigma+|f(0)|_\sigma=|f(\lambda)|_\sigma\leqslant 1\quad(\lambda\in\Delta),$$

对 g 与 $n=1$ 引用定理Ⅰ,得出 $|\hat{g}(x)|_\sigma\leqslant|x|_\sigma$. 另一方面,从关于(1)式运算的基本代数法则知道 $\hat{g}(x)=\hat{f}(x)-f(0)$,且因 $f(\lambda)$ 与 $f(0)$, x 可换($\lambda\in\Delta$),有 $\hat{f}(x)f(0)=f(0)\hat{f}(x)$,于是 $\hat{g}(x)f(0)=f(0)\hat{g}(x)$. 再次应用谱半径性质可得

$$|\hat{f}(x)|_\sigma\leqslant|\hat{g}(x)|_\sigma+|f(0)|_\sigma=|\hat{g}(x)|_\sigma\leqslant|x|_\sigma.\quad \text{证毕.}\ ■$$

注 一般地说,当以 $|f(0)|_\sigma=0$(即 $f(0)\in U$ 是拓扑幂零的)代替定理Ⅰ中 $f(0)=0$(即 $\|f(0)\|=0$)假定时,上述推论Ⅱ中的附加条件 $f(0)f(\lambda)=f(\lambda)f(0)(\lambda\in\Delta)$ 是必需的,即使对 $x=\lambda e$(λ 是模小于 1 的复数)的特殊情形. 其反例可见〔6〕p. 1011. 有趣的是,当 U 限制为有限维 Hilbert 空间 H 上的所有有界线性算子组成的 Banach 代数时,记它为 $\mathscr{B}(H)$,我们有如下与 H 维数有关的结论,其中 $f(0)$ 与 $f(\lambda)$ 可换性虽可以去掉,但结果比前一推论减弱.

推论Ⅲ 设 $H\neq\{0\}$ 是 m 维复 Hilbert 空间,又设 $f\in\mathrm{Hol}(\Delta,$

$\mathscr{B}(H)$) 满足 $|f(\lambda)|_\sigma \leqslant 1(\lambda \in \Delta)$ 与 $|f(0)|_\sigma = 0$. 如果 $T \in \mathscr{B}(H)$ 满足 $|T|_\sigma < 1$ 且与 f 可换,那么

$$|\hat{f}(T)|_\sigma \leqslant |T|_\sigma^{1/m}.$$

证　令 $g(\lambda) = f(\lambda)^m/\lambda(\lambda \in \Delta - \{0\})$. 则有

$$|g(\lambda)|_\sigma = |f(\lambda)|_\sigma^m / |\lambda| \leqslant |\lambda|^{-1} \quad (\lambda \in \Delta - \{0\}). \tag{3}$$

由于 $\dim H = m < \infty$,且 $f(0) \in \mathscr{B}(H)$ 是拓扑幂零的,故必定有 $f(0)^m = 0$,因而按前面定义的 $g \in \mathrm{Hol}(\Delta, \mathscr{B}(H))$. 这时,$\lambda \mapsto |g(\lambda)|_\sigma$ 是次调和的,因此由(3) 式推得 $|g(\lambda)|_\sigma \leqslant 1(\lambda \in \Delta)$. 应用(1) 式运算代数法则到 $f(\lambda)^m = \lambda g(\lambda)$ 可得(因 T 与 f,因而也与 g 可换)$\hat{f}(T)^m = T\hat{g}(T)$,且 $T\hat{g}(T) = \hat{g}(T)T$. 再次引用谱半径性质以及引理 Ⅱ,最后推出

$$|\hat{f}(T)^m|_\sigma = |\hat{f}(T)|_\sigma^m \leqslant |T|_\sigma |\hat{g}(T)|_\sigma \leqslant |T|_\sigma \sup_{\lambda \in \sigma(T)} |g(\lambda)|_\sigma$$
$$\leqslant |T|_\sigma \sup_{\lambda \in \Delta} |g(\lambda)|_\sigma \leqslant |T|_\sigma.$$

证毕. ■

下述定理 Ⅱ 与定理 Ⅰ 的 $n = 1$ 情形紧密相关,并且,前者可视为后者的推广.

定理 Ⅱ　设 $f \in \mathrm{Hol}(\Delta, U)$ 满足 $|f(\lambda)|_\sigma \leqslant 1(\lambda \in \Delta)$ 与 $f(0) = \alpha e$(α 是复数). 如果 $x \in U_\Delta$ 与 f 可换,那么

$$|\hat{f}(x)|_\sigma \leqslant (|\alpha| + |x|_\sigma)/(1 - |\alpha||x|_\sigma). \tag{4}$$

证　先考虑 $|f(\lambda)|_\sigma \not\equiv 1(\lambda \in \Delta)$ 的情形. 此时由次调和函数的极大原理得 $|f(\lambda)|_\sigma < 1(\lambda \in \Delta)$,于是,$|\alpha| < 1$,且 $|\bar{\alpha}f(\lambda)|_\sigma < 1(\lambda \in \Delta)$. 因此,由 Möbius 变换 $g(\lambda) = [f(\lambda) - \alpha e][e - \bar{\alpha}f(\lambda)]^{-1}(\lambda \in \Delta)$ 确定的函数 $g \in \mathrm{Hol}(\Delta, U)$,且 $g(0) = 0$. 按〔3〕定理 1.6.4,又因 $\sigma(f(\lambda)) \subseteq \Delta$,有

$$|g(\lambda)|_\sigma = \max_{\xi \in \sigma(f(\lambda))} \left| \frac{\xi - \alpha}{1 - \bar{\alpha}\xi} \right| \leqslant 1.$$

应用定理 Ⅰ,对题设中的 $x \in U_\Delta$,有 $|\hat{g}(x)|_\sigma \leqslant |x|$. 另一方面,按(1) 式运算 基本代数法则,

$$\hat{g}(x) = [\hat{f}(x) - \alpha e][e - \bar{\alpha}\hat{f}(x)]^{-1}.$$

由上式可推出 $\hat{f}(x)\hat{g}(x) = \hat{g}(x)\hat{f}(x)$,于是从谱半径性质有

$$|\hat{f}(x)|_\sigma - |\alpha| \leqslant |\hat{f}(x) - \alpha e|_\sigma \leqslant |\hat{g}(x)|_\sigma |e - \bar{\alpha}\hat{f}(x)|_\sigma$$
$$\leqslant |\hat{g}(x)|_\sigma [1 + |\alpha||\hat{f}(x)|_\sigma] \leqslant |x|_\sigma [1 + |\alpha||f(x)|_\sigma].$$

整理一下便得(4) 式. 倘若 $|f(\lambda)|_\sigma \equiv 1(\lambda \in \Delta)$,即有 $|\alpha| = 1$,这时从引理 Ⅱ,$|\hat{f}(x)|_\sigma \leqslant \sup_{\lambda \in \sigma(x)} |f(\lambda)|_\sigma = 1$,则知(4) 式也成立. 定理 Ⅱ 证毕. ■

特殊地，当 $f \in \mathrm{Hol}(\Delta, \mathscr{C})$ 时，由谱映射定理，对 $x \in U_\Delta$ 有 $\sigma(f(x)) = f(\sigma(x))$(〔2〕,p. 224). 此时，定理 Ⅱ 有如下的改进.

推论 Ⅳ 设 $f \in \mathrm{Hol}(\Delta, \mathscr{C})$ 满足 $|f(\lambda)| \leqslant 1 (\lambda \in \Delta)$. 则当 $x \in U_\Delta$ 时，有

$$|\hat{f}(x)|_\sigma \leqslant (|f(0)| + |x|_\sigma)/(1 + |f(0)||x|_\sigma).$$

证 如前，令 $g(\lambda) = [f(\lambda) - \alpha][1 - \bar{\alpha} f(\lambda)]^{-1} (\lambda \in \Delta)$，这里，$\alpha = f(0)$，只要 $|f(\lambda)| \not\equiv 1 (\lambda \in \Delta)$，由最大模原理知道，$|f(\lambda)| < 1$. 由谱映射定理以及前一定理所证可得

$$|\hat{g}(x)|_\sigma = \sup_{\lambda \in \sigma(x)} |g(\lambda)| = \sup_{\lambda \in \sigma(x)} \left| \frac{f(\lambda) - \alpha}{1 - \bar{\alpha} f(\lambda)} \right| \leqslant |x|_\sigma.$$

由此推出，$|f(\lambda)| \leqslant (|x|_\sigma + |\alpha|)/(1 + |\alpha||x|_\sigma)$，$(\lambda \in \sigma(x))$，因此，

$$|\hat{f}(x)|_\sigma = \sup_{\lambda \in \sigma(x)} |f(\lambda)| \leqslant (|x|_\sigma + |\alpha|)/(1 + |\alpha||x|_\sigma).$$

倘若 $|f(\lambda)| \equiv 1 (\lambda \in \Delta)$，则有 $|\alpha| = |f(0)| = 1$，这时上述不等式显然也成立. 证毕. ■

下面是 Harnack 双边不等式的谱半径变异形式.

推论 Ⅴ 设 $g \in \mathrm{Hol}(\Delta, \mathscr{C})$ 满足 $\mathrm{Re}\, g(\lambda) > 0 (\lambda \in \Delta)$ 与 $g(0) = 1$，且对某 $n \geqslant 2$，$g'(0) = g''(0) = \cdots = g^{(n-1)}(0) = 0$. 则对 $x \in U_\Delta$ 我们有

$$(1 - |x^n|_\sigma)/(1 + |x^n|_\sigma) \leqslant |\hat{g}(x)|_\sigma \leqslant (1 + |x^n|_\sigma)/(1 - |x^n|_\sigma).$$

证 定义 $f(\lambda) = (g(\lambda) - 1)/(g(\lambda) + 1)$. 则 $f \in \mathrm{Hol}(\Delta, \mathscr{C})$，且 $f(0) = f'(0) = \cdots = f^{(n-1)}(0) = 0$. 当 $x \in U_\Delta$ 时，有 $\hat{f}(x) = [\hat{g}(x) - e][\hat{g}(x) + e]^{-1}$，或者 $\hat{g}(x) - e = \hat{f}(x)[\hat{g}(x) + e]$. 于是，由于 $\mathrm{Hol}(\Delta, \mathscr{C})$ 是可换族，故 $\hat{f}(x)\hat{g}(x) = \hat{g}(x)\hat{f}(x)$(〔2〕,p. 244)，$|\hat{g}(x) - e|_\sigma \leqslant |\hat{f}(x)|_\sigma [|\hat{g}(x)|_\sigma + 1] \leqslant |x^n|_\sigma [|\hat{g}(x)|_\sigma + 1]$. 因此，我们有要证的第二个不等式：

$$|\hat{g}(x)|_\sigma \leqslant (1 + |x^n|_\sigma)/(1 - |x^n|_\sigma).$$

另外，由 $h(\lambda) = 1/g(\lambda)$ 定义的函数 $h \in \mathrm{Hol}(\Delta, \mathscr{C})$ 也满足 $h(0) = 1$，$h'(0) = \cdots = h^{(n-1)}(0) = 0 (n \geqslant 2)$. 这时，按运算法则 $\hat{h}(x) = \hat{g}(x)^{-1}$. 利用前一部分证明的结果，我们有

$$|\hat{g}(x)|_\sigma^{-1} \leqslant |\hat{g}(x)^{-1}|_\sigma = |\hat{h}(x)|_\sigma \leqslant (1 + |x^n|_\sigma)/(1 - |x^n|_\sigma),$$

于是，本推论结果中的第一个不等式也成立. ■

作为前一推论的一个应用，我们有如下的 Carathéodory 不等式的谱

半径变异形式.

推论 Ⅵ　设 $h \in \mathrm{Hol}(\Delta,\mathscr{C})$ 对某 $n \geqslant 1$ 满足 $h(0) = h'(0) = \cdots = h^{(n-1)}(0) = 0$,且 $\mathrm{Re}\, h(\lambda) < 1(\lambda \in \Delta)$. 则对 $x \in U_\Delta$ 我们有

$$|\hat{h}(x)|_\sigma \leqslant 2|x^n|_\sigma/(1-|x^n|_\sigma).$$

证　定义函数 $g: g(\lambda) = 1 - h(\lambda)$. 这时,$g \in \mathrm{Hol}(\Delta,\mathscr{C})$ 满足前一推论诸条件,故从 $\hat{g}(x) = e - \hat{h}(x)$ 以及前一推论的证明可得

$$|\hat{h}(x)|_\sigma = |\hat{g}(x) - e|_\sigma \leqslant |x^n|_\sigma |\hat{g}(x) + e|_\sigma$$

$$= |x^n|_\sigma |2e - \hat{h}(x)|_\sigma \leqslant |x^n|_\sigma [2 + |\hat{h}(x)|_\sigma],$$

于是,最后导出要证的不等式.　■

在〔7〕中,Fan Ky 给出前两个推论关于范数的类似结果,但在那里 U 只限于是特殊的 Banach 代数 $\mathscr{B}(H)$,其中 H 是复 Hilbert 空间.

参考文献

[1] Hille H and Phillips R S. Functional Analysis and Semigroups. Math. Soc. Colloq. Pub. 31, Providence 1957.

[2] Rudin W. Functional Analysis. McGraw-Hill, New York, 1973.

[3] Rickart C E. General Theory of Banach Algebras. D. Van Nostrand Company Inc. , 1960.

[4] Vesentini E. Boll. Un. Mat. Ital. , 1968, 1: 427.

[5] Chen Gongning. Chinese Annals of Math, 1990, 11A(3): 314－320.

[6] Globernik J. Rev. Rovm. Math. Pures et Appl. , XIX, Bucaresi, 1974, 8: 1 009.

[7] Fan Ky. Math. Z. , 1978, 160: 275.

SCHWARZ'S LEMMA FOR THE SPECTRAL RADIUS IN COMPLEX BANACH ALGEBRAS

Abstract　Some results, analogous to the Schwarz's Lemma in complex Banach algebras are obtained, where the norm is replaced with the spectral radius. As applications, the versions of Harnack's double inequality, and of Caretheódory's inequality are established among others.

Keywords　Banach algebra; Schwarz's lemma; spectral radius; Harnack's double inequality; Carathéodory's inequality

数学年刊

1990,11A(3):314－320

关于 von Neumann-Heinz 定理与樊畿定理的推广①

The Generalizations of Theorems of von Neumann-Heinz, and Ky Fan

摘要 应用复 Banach 代数中极大原理与谱映射定理，本文得到 von Neumann-Heinz 定理与樊畿定理的两种推广结果及其若干推论.

§1. 引言

在本文中，A 总是表示具有单位元 e 的复 Banach 代数；$B(H)$ 表示非零复 Hilbert 空间 H 上的有界线性算子组成的 Banach 代数. 设 Ω 是复平面 **C** 上的一个区域，$\mathrm{Hol}(\Omega,A)$ 是定义在 Ω 上的 A- 值全纯函数的集合. 当 $x\in A_\Omega=\{y\in A:\sigma(y)\subset\Omega\}$ 这里，$\sigma(y)$ 表示 y 的谱，且 $k\in\mathrm{Hol}(\Omega,A)$ 时，$\tilde{k}(x)$ 将表示由下列积分确定的 A 内元素：

$$\tilde{k}(x)=\frac{1}{2\pi\mathrm{i}}\int_\Gamma k(\lambda)(\lambda e-x)^{-1}\mathrm{d}\lambda,\tag{1}$$

其中，Γ 是 Ω 内围绕 $\sigma(x)$ 的围道. 有关上述运算微积的讨论可见[8]. 特别地，有如下结论.

引理 1 如果 $f_1,f_2\in\mathrm{Hol}(\Omega,A)$，$f(\lambda)=f_1(\lambda)f_2(\lambda)$，$\lambda\in\Omega$，并且 $x\in A_\Omega$，那么当且仅当

$$\int_\Gamma[f_1(x)-f_1(\lambda)](\lambda e-x)^{-1}[f_2(\lambda)x-xf_2(\lambda)](\lambda e-x)^{-1}\mathrm{d}\lambda=0$$

---

① 收稿日期：1987-03-30.

本文摘要曾在 1986 年全国泛函分析第四次学术会议上宣读.

对 Ω 内围绕 $\sigma(x)$ 的围道 Γ 成立时，$f(x)=f_1(x)f_2(x)$.

上述结论表明，如果 $f_1(\lambda)=f_1(x)$，即 $f_1(\lambda)$ 为 A 的常数元素，或者如果 $f_2(\lambda)x=xf_2(\lambda)$ 对所有 $\lambda\in\Omega$ 成立，那么有 $f(x)=f_1(x)f_2(x)$.

假如 $A=B(H)$ 与 $f\in\mathrm{Hol}(\Omega,\mathbf{C})$，(1) 式内的积分便是通常的 Riesz-Dunford 积分，即当 $T\in B(H)$，$\sigma(T)\subset\Omega$，且 $f\in\mathrm{Hol}(\Omega,\mathbf{C})$ 时，(1) 式变为

$$\tilde{f}(T)=\frac{1}{2\pi\mathrm{i}}\int_{\Gamma}f(\lambda)(\lambda I-T)^{-1}\mathrm{d}\lambda,\tag{2}$$

这里，I 为 H 上的单位算子. 对于这样定义的 $\tilde{f}:B(H)_{\Omega}\to B(H)$，$B(H)_{\Omega}=\{S\in B(H):\sigma(S)\subset\Omega\}$，有如下两个基本定理(见[10]，[3，定理 1]).

定理 1(von Neumann-Heinz) 如果 $T\in B(H)$ 是压缩，即 $\|T\|\leqslant 1$，$\Omega\supset\overline{\Delta}=\{\lambda\in C,\ |\lambda|\leqslant 1\}$，$u$〔或 v〕$\in H(\Omega,\mathbf{C})$ 满足 $u(\lambda)\leqslant 1$〔或 $\mathrm{Re}\,v(\lambda)\geqslant 0$〕，$\lambda\in\overline{\Delta}$，那么有 $\|\tilde{u}(T)\|\leqslant 1$[或 $\mathrm{Re}\,\tilde{v}(T)\geqslant 0$].

在上述定理中，$\mathrm{Re}\,\tilde{v}(T)$ 表示 $\tilde{v}(T)\in B(H)$ 的实部，一般地，如果 $T\in B(H)$，那么

$$\mathrm{Re}\,T=\frac{1}{2}(T+T^*).$$

并且，如果 T_1 与 T_2 是 $B(H)$ 的两个 Hermite 算子，那么符号 $T_1\geqslant T_2$ 表示 T_1-T_2 是正的，即 $((T_1-T_2)w,w)\geqslant 0$ 对所有 $w\in H$ 成立；$T_1>T_2$ 表示 T_1-T_2 是正的与可逆的.

定理 2(樊畿) 如果 $T\in B(H)$ 是真正压缩，即 $\|T\|<1$，$\Delta=\{\lambda\in\mathbf{C}:\ |\lambda|<1\}$，$h$[或 k]$\in\mathrm{Hol}(\Delta,\mathbf{C})$ 满足 $\|h(\lambda)\|<1$[或 $\mathrm{Re}\,k(\lambda)>0$]，$\lambda\in\Delta$，那么有 $\|\tilde{h}(T)\|<1$[或 $\mathrm{Re}\,\tilde{k}(T)>0$].

在[2]中，我们给出上述两个定理的简化证明. 本文应用复 Banach 代数的极大原理与谱映射定理给出它们的两种推广以及若干有关推论.

在后面要引用的另一结果是(见[13，引理 6.2 与定理 6.3])

引理 2 设 E 是 H 上正规算子 V 的单位谱分解，$\Omega\supset\sigma(V)$，并且，$f\in\mathrm{Hol}(\Omega,B(H))$ 使得 $f(\lambda)V=Vf(\lambda)$，$\lambda\in\Omega$. 则有

$$\tilde{f}(V)=\int_{\sigma(V)}f(\lambda)\mathrm{d}E(\lambda),\tag{3}$$

$$\|\tilde{f}(V)\|\leqslant\max\{\|f(\lambda)\|:\lambda\in\sigma(V)\}.\tag{4}$$

§2. 基本定理的第一种推广

在本节中假定 A 有 Hermite 与连续的对合 $*$. U 与 U_e 分别表示 A 的酉元素的集合与 U 的主分量. 当 $x \in A$ 时, $|x|_\sigma$ 表示 x 的谱半径, $p(x)=|x^*x|_\sigma^{1/2}$. 此时, $|x|_\sigma \leqslant p(x)$, 并且, p 是 A 上的一个伪范数, 它在 A 上连续(有关这一方面的内容可参阅[9,6]). 当 Q 是 A 的子集时, $\overline{\mathrm{co}}Q$ 表示 Q 的闭凸包.

特别地, 当 $A=B(H)$ 时, 取对合 $*$ 为 H 上算子的 Hilbert 空间共轭, 显然, 它是 Hermite 与连续的, 且有 $|T|_\sigma \leqslant p(T)=\|T\|$ 对所有 $T\in B(H)$ 成立. 我们首先证明一个引理, 它含有极大原理的一种较弱形式[2,定理1].

引理 3 设 A 有 Hermite 与连续的对合 $*$, $\Omega \supset r\bar{\Delta}$, 这里, $0<r\leqslant 1$. 令 $A_0=\{x\in A: p(x)<1\}$ 与 $A_1=\{x\in A: p(x)\leqslant 1\}$. 如果 $k\in \mathrm{Hol}(\Omega, A)$, 那么有

$$\tilde{k}(rx)=\frac{1}{2\pi}\int_0^{2\pi}\tilde{k}(r\psi_x(\mathrm{e}^{it}))\mathrm{d}t \quad (x\in A_0), \tag{5}$$

$$\tilde{k}(rA_1)\subset \overline{co}\,\tilde{k}(rU_e), \tag{6}$$

$$\|\tilde{k}(rx)\| \leqslant \sup\{\|\tilde{k}(ry)\|: y\in U_e\} \quad (x\in A_1), \tag{7}$$

这里, $\tilde{k}$ 由(1) 式确定, ψ_x 是如下的特征函数:

$$\psi_x(\lambda)=(e-xx^*)^{-1/2}(\lambda e+x)(e+\lambda x^*)^{-1}(e-x^*x)^{1/2} \quad (x\in A_0, |\lambda|\leqslant 1). \tag{8}$$

证 由前所述, p 是 A 上的连续的伪范数, 因而 A_0 是 A 的开集[6,p.12]. 此外, 由于 $|x|_\sigma \leqslant p(x)$, 因此, 只要 $x\in A_1$, 便有 $rx\in A_\Omega$. 当 $k\in \mathrm{Hol}(\Omega,A)$ 时, $\tilde{k}: A_\Omega \to A$ 是全纯的. 事实上, 对任意 $x\in A_\Omega$, 因 A_Ω 是 A 的开集[11,p.243], 所以当 $h\in A$ 具有充分小的范数时, $\sigma(x+h)\subset\Omega$. 现设 Γ 是 Ω 内围绕 $\sigma(x)$ 与 $\sigma(x+h)$ 的围道, 并设当 $\lambda\in\Gamma$ 时,

$$\|(\lambda e-x)^{-1}\|<M, \quad \|h\|<1/(2M).$$

此时

$$\left\|\tilde{k}(x+h)-\tilde{k}(x)-\frac{1}{2\pi i}\int_\Gamma k(\lambda)(\lambda e-x)^{-1}h(\lambda e-x)^{-1}\mathrm{d}\lambda\right\|$$

$$=\left\|\frac{1}{2\pi i}\int_\Gamma k(\lambda)[(\lambda e-x-h)^{-1}-(\lambda e-x)^{-1}-(\lambda e-x)^{-1}h(\lambda e-x)^{-1}\mathrm{d}\lambda]\right\|$$

$$\leqslant \frac{L}{2\pi}\max_{\lambda\in\Gamma}\|k(\lambda)\|\ \|(\lambda e-x-h)^{-1}-(\lambda e-x)^{-1}-(\lambda e-x)^{-1}h(\lambda e-x)^{-1}\|$$

$$\leqslant \frac{L}{2\pi}\max_{\lambda\in\Gamma}\|k(\lambda)\|\ \|(e-(\lambda e-x)^{-1}h)^{-1}-(\lambda e-x)^{-1}h-e\|\ \|(\lambda e-x)^{-1}\|$$

$$\leqslant \frac{L}{2\pi}\max_{\lambda\in\Gamma}\|k(\lambda)\|\cdot 2\|(\lambda e-x)^{-1}\|^{3}\|h\|^{2}$$

$$\leqslant \frac{L}{\pi}\max_{\lambda\in\Gamma}\|k(\lambda)\|M^{3}\|h\|^{2},$$

这里，L 表示 Γ 的长度，上式表明 $\tilde{k}$ 在 x 处的 Fréchet 导数存在，因而 $\tilde{k}$ 在 A_Ω 上全纯.

此外，$\Omega\supset r\overline{\Delta}$ 蕴涵，$A_\Omega\supset r\,\overline{\mathrm{co}}U_e$，这是因为如果 $x\in r\,\overline{\mathrm{co}}U_e$，那么 $|x|_\sigma\leqslant p(x)=r$，因而 $\sigma(x)\subset r\overline{\Delta}\subset\Omega$. 应用[2，定理 1]便得(6)与(7)式，而(5)式可用向量值全纯函数的平均值性质[7]与当 $|\lambda|=1$ 与 $x\in A_0$ 时便有 $\psi_x(\lambda)\in U_e$ 的事实[6]直接从 $\tilde{k}(r\psi_x)|_{\lambda=0}$ 的表达式推出. 证毕. ■

现在应用引理 3 导出第一节中两个基本定理的一种推广结果. 为简单起见，我们采用记号 $T\cup S$，假如 $TS=ST$，这里，$T,S\in B(H)$（或 A).

定理 3 设 $\Omega\supset\overline{\Delta}$，$f$[或 g] $\in \mathrm{Hol}(\Omega,B(H))$ 满足 $\|f(\lambda)\|\leqslant 1$[或 $\mathrm{Re}\,g(\lambda)\geqslant 0$]；$\forall\lambda\in\overline{\Delta}$，并且，$T\in B(H)$ 是 H 上的压缩，$T\cup f(\lambda)$ 与 $T^*\cup f(\lambda)$[或 $T\cup g(\lambda)$ 与 $T^*\cup g(\lambda)$]对所有 $\lambda\in\Omega$ 成立. 则有

$$\|\tilde{f}(T)\|\leqslant 1\quad[\text{或 }\mathrm{Re}\,\tilde{g}(T)\geqslant 0].$$

证 显然，对 H 上的酉算子 V，有 $\Omega\supset\overline{\Delta}\supset\sigma(V)$. 今令 $T_j=d_jT$，这里，$0<d_j<1$ 且 $\lim\limits_{j\to\infty}d_j=1$. 按题设，$f(\lambda)\cup\psi_{T_j}(\lambda')$ 对所有 j 与 $\lambda,\lambda'\in\overline{\Delta}$ 成立. 应用当 $|\lambda|=1$ 时 $\psi_{T_j}(\lambda)$ 为 H 上酉算子的事实，引理 3 中(5)式(取 $r=1$)与引理 2，得出

$$\|\tilde{f}(T_j)\|\leqslant\sup\{\|\tilde{f}(\psi_{T_j}(\lambda))\|:|\lambda|=1\}\leqslant\sup\{\|f(\lambda)\|:|\lambda|=1\}\leqslant 1.$$

另一方面，由于 $\|T_j\|<1$，故 $T_j\in B(H)_\Omega=\{S\in B(H):\sigma(S)\subset\Omega\}$，应用 $\tilde{f}$ 在 $B(H)_\Omega$ 上的连续性，即得

$$\|\tilde{f}(T)\|\leqslant 1.$$

类似地，由(5)式得出

$$\mathrm{Re}\,\tilde{g}(T_j)=\frac{1}{2\pi}\int_0^{2\pi}\mathrm{Re}\,\tilde{g}(\psi_{T_j}(\mathrm{e}^{it}))\mathrm{d}t,$$

但对任意的 $t\in[0,2\pi]$，由引理 3 知道

$$\mathrm{Re}\,\tilde{g}(\psi_{T_j}(e^{it}))=\int_{\sigma(\psi_{T_j}(e^{it}))}\mathrm{Re}\,g(\lambda)\mathrm{d}E(\lambda).$$

上式中,$\mathrm{Re}\,g(\lambda)$ 与 $\psi_{T_j}(e^{it})$ 可交换,因为 $g(\lambda)\cup T_j, T_j^*$. 再根据[13,引理6.1]得出,$\mathrm{Re}\,\tilde{g}(\psi_{T_j}(e^{it}))\geqslant 0$, 于是 $\mathrm{Re}\,\tilde{g}(T_j)\geqslant 0$. 应用连续性便得 $\mathrm{Re}\,\tilde{g}(T)\geqslant 0$. 证毕. ■

注 当 f[或 g]$\in \mathrm{Hol}(\Omega,\mathbf{C})$ 时,定理 3 中的条件等同于前面的 von Neumann-Heinz 定理的假定条件,(因为这时 $T\cup f(\lambda)$ 与 $T^*\cup f(\lambda)$ 自然成立,)因此定理 3 确实是定理 1 的推广.

定理 4 假定 f [或 g] $\in \mathrm{Hol}(\Delta,B(H))$ 满足 $\|f(\lambda)\|<1$[或 $\mathrm{Re}\,g(\lambda)>0$],$\forall\lambda\in\Delta$,且 $T\in B(H)$ 是真正压缩,即 $\|T\|<1$,它满足 $T\cup f(\lambda)$ 与 $T^*\cup f(\lambda)$[或 $T\cup g(\lambda)$ 与 $T^*\cup g(\lambda)$],$\forall\lambda\in\Delta$,则有

$$\|\tilde{f}(T)\|<1\quad[\text{或}\ \mathrm{Re}\,\tilde{g}(T)>0].$$

证 如果 $\|T\|<1$,那么存在 r 满足 $0<r<1$ 与真正压缩 $\hat{T}\in B(H)$ 使得 $T=r\hat{T}$. 此时,$\Delta\supset r\overline{\Delta}$ 与 $\sigma(\hat{T})\subset\Delta$. 类似于定理 3 的证明,应用引理 3 与引理 2(取 $\Omega=\Delta$),我们推出

$$\begin{aligned}\|\tilde{f}(T)\|&=\|\tilde{f}(r\hat{T})\|\leqslant\max\{\|\tilde{f}(r\psi_{\hat{T}}(\lambda))\|:|\lambda|=1\}\\&\leqslant\max\{\|f(\lambda)\|:|\lambda|=r\}<1.\end{aligned}$$

为证明本定理的另一半结果,令 $f(\lambda)=(g(\lambda)-I)(g(\lambda)+I)^{-1}$. 因为 $g\in\mathrm{Hol}(\Delta,B(H))$ 与 $\mathrm{Re}\,g(\lambda)>0$,$\forall\lambda\in\Delta$,所以有 $f\in\mathrm{Hol}(\Delta,B(H))$ 与 $\|f(\lambda)\|<1$,这后一结论可由 $B(H)$ 内 Cayley 变换性质得到. 由于 $T\cup g(\lambda)$,$\forall\lambda\in\Delta$,故从引理 1(取 $\Omega=\Delta$,$A=B(H)$)看出,$\tilde{f}(T)=(\tilde{g}(T)-I)(\tilde{g}(T)+I)^{-1}$,此时容易验证,$\tilde{g}(T)=(\tilde{f}(T)+I)(I-\tilde{f}(T))^{-1}$ 与 $\mathrm{Re}\,\tilde{g}(T)=(I-\tilde{f}(T)^*)^{-1}(I-\tilde{f}(T)^*\tilde{f}(T))(I-\tilde{f}(T))^{-1}$. 因此,当且仅当 $I-\tilde{f}(T)^*\tilde{f}(T)>0$,即 $\|\tilde{f}(T)\|<1$ 时,$\mathrm{Re}\,\tilde{g}(T)>0$. 但由前一部分证明知道 $\|\tilde{f}(T)\|<1$,因而 $\mathrm{Re}\,\tilde{g}(T)>0$ 成立. ■

注 容易看出,当 f 与 g 属于 $\mathrm{Hol}(\Delta,\mathbf{C})$ 时,定理 4 变为樊畿定理,因而前者是后者的一种推广.

推论 1 设 $T_1,T_2,\cdots,T_n\in B(H)$ 满足 $|T_j|_\sigma<1$ 与 $T_i\cup T_j$,$i,j=1,2,\cdots,n$,$f\in\mathrm{Hol}(\Delta,B(H))$ 满足 $\|f(\lambda)\|<1$,$\forall\lambda\in\Delta$,且 $T_j\cup f(\lambda)$ 与 $T_j^*\cup f(\lambda)$,$\forall\lambda\in\Delta$,$j=1,2,\cdots,n$. 令 Φ 是由$\{T_1,T_2,\cdots,T_n\}$生成的半群(在复合运算下). 则存在与范数 $\|\cdot\|$ 等价的 H 上的 Hilbert 范数 $\|\cdot\|_*$,使得对每个 $T\in\Phi$,都有 $\|\tilde{f}(T)\|_*<1$.

证　根据[12]中定理2.1与3.1，存在与范数 $\|\cdot\|$ 等价的 H 上的 Hilbert 范数 $\|\cdot\|_*$，使得只要 $T\in\Phi$，便有 $\|T\|_*<1$. 这样，本推论的结果可由定理4与 $T\cup f(\lambda)$，$T^*\cup f(\lambda)$，$\forall\lambda\in\Delta$，$T\in\Phi$ 的事实直接得出. ■

下面的推论给出著名的 Schwarz 引理的算子形式.

推论2(对算子的 Schwarz 引理)　设 $f\in \mathrm{Hol}(\Delta,B(H))$ 满足 $\|f(\lambda)\|<1$，$\forall\lambda\in\Delta$，且 $T\in B(H)$ 是真正压缩，它满足 $T\cup f(\lambda)$ 与 $T^*\cup f(\lambda)$，$\forall\lambda\in\Delta$. 如果对某整数 $n\geqslant 1$，$f(0)=f'(0)=\cdots=f^{(n-1)}(0)=0$，那么有

$$\tilde{f}(T)\tilde{f}(T)^*\leqslant T^nT^{*n},\tag{9}$$

$$\|\tilde{f}(T)\|\leqslant\|T^n\|.\tag{10}$$

证　应用定理4与[3]中推论2的推证办法即可得到本推论. ■

推论3(最大模原理)　假设 $f\in\mathrm{Hol}(\Delta,B(H))$. 对满足 $0\leqslant r<1$ 的实数 r，令

$$m(r)=\max\{\|f(\lambda)\|:|\lambda|=r\}.\tag{11}$$

则有

$$m(r)=\max\{\|\tilde{f}(T)\|:\|T\|\leqslant r\},\tag{12}$$

其中 max 取遍 H 上范数 $\leqslant r$ 且满足 $T\cup f(\lambda)$ 与 $T^*\cup f(\lambda)$，$\forall\lambda\in\Delta$ 的所有算子 T.

证　除去 $m(r)=0$ 与 $r=0$ 的平凡情形，可设 $0<r<1$ 与 $m(r)>0$. 定义函数 g：$g(\lambda)=f(r\lambda)/m(r)$. 则 g 在 $|\lambda|<1/r$ 内全纯，且由(11)式，只要 $|\lambda|=1$，便有 $\|g(\lambda)\|\leqslant 1$. 按算子值函数的最大模原理可推出，对所有 $|\lambda|<1$，$\|g(\lambda)\|\leqslant 1$. 因此，对 $B(H)$ 中任意真正压缩 T，只要 $T\cup g(\lambda)$ 与 $T^*\cup g(\lambda)$，$\forall\lambda\in\Delta$，(此由 $f(\lambda)$ 与 T,T^* 可换性推出，) 按定理4的证明可得到 $\|\tilde{g}(T)\|\leqslant 1$. 这表示 $\|\tilde{f}(T)\|\leqslant m(r)$ 对所有范数 $<r$ 且满足条件 $T\cup f(\lambda)$ 与 $T^*\cup f(\lambda)$ $(\lambda\in\Delta)$ 的任意 $T\in B$ 成立，于是，由 $\tilde{f}$ 的连续性推出，$\max\{\|\tilde{f}(T)\|:\|T\|\leqslant r\}\leqslant m(r)$，这里 max 的取值范围同于(12)式. 另一方面，相反的不等式成立可从取 $T=\lambda_0 I$ 直接看出，这里，λ_0 是满足条件 $|\lambda_0|=r$ 与 $\|f(\lambda_0)\|=m(r)$ 的复数. 证毕. ■

§3. 基本定理的谱半径推广形式

由[10,12]知道，当我们局限于考虑复值全纯函数时，则定理1,2与推论3的谱半径形式变种(即诸命题中范数为谱半径所代替)仍然成立. 这里的关键在于对由(2)式定义的$\widetilde{f}(T)$，谱映射定理成立. 然而当我们考虑由(1)式定义的$\widetilde{k}(x)$时，如果$k \in \mathrm{Hol}(\Omega, A)$，那么$\sigma(\widetilde{k}(x)) = k(\sigma(x))$一般失去意义. 有幸的是，我们可以推出如下形式的谱映射定理(见下面定理5). 应用它容易得到基本定理的谱半径推广结果.

在本节中，只假定A是具有单位元e的复Banach代数(不必有对合).

定理5(谱映射定理)　设Ω是$\mathbf{C}$内区域，$f \in \mathrm{Hol}(\Omega, A)$，且$x \in A_\Omega$满足$x \cup f(\lambda)$，$\forall \lambda \in \Omega$. 则有

$$\sigma(\widetilde{f}(x)) \subset \bigcup_{\lambda \in \sigma(x)} \sigma(f(\lambda)). \tag{13}$$

假若f还满足$f(\lambda) \cup f(\mu)$，$\forall \lambda, \mu \in \Omega$，则有

$$\{\lambda \in \mathbf{C} : \lambda e \in f(\sigma(x))\} \subset \sigma(\widetilde{f}(x)) \subset \bigcup_{\lambda \in \sigma(x)} \sigma(f(\lambda)). \tag{14}$$

证　(13)式即是[8，定理2]. 这里给出较为简单的证明. 令$\mu \in \sigma(\widetilde{f}(x))$但$\mu \notin \bigcup_{\lambda \in \sigma(x)} \sigma(f(\lambda))$. 这时由$h(\lambda) = (f(\lambda) - \mu e)^{-1}$定义的$A$-值函数$h$将在含有$\sigma(x)$的$\Omega$的某个开集内全纯. 由$f(\lambda) - \mu e$与$x$的可换性与引理1得到$\widetilde{h}(x)(\widetilde{f}(x) - \mu e) = (\widetilde{f}(x) - \mu e)\widetilde{h}(x) = e$，因而$\mu \notin \sigma(\widetilde{f}(x))$，导出矛盾. 因此，(13)式成立. 为证明(14)式，设存在复数β，使得$\beta e \in f(\sigma(x))$，于是，对某个$\lambda_0 \in \sigma(x)$，$f(\lambda_0) = \beta e$. 今定义A-值函数g，它与f有相同的定义域，并且，$f(\lambda) - f(\lambda_0) = g(\lambda)(\lambda - \lambda_0)$. 这时，$g \in \mathrm{Hol}(\Omega, A)$，且$g(\lambda) \cup x$(因为$f(\lambda) \cup x$)，$\forall \lambda \in \Omega$. 再用引理1得出，$\widetilde{f}(x) - \beta e = \widetilde{g}(x)(x - \lambda_0 e) = (x - \lambda_0 e)\widetilde{g}(x)$. 因此，如果$\widetilde{f}(x) - \beta e$在$A$内有逆$y$，即$\widetilde{g}(x)(x - \lambda_0 e)y = y\widetilde{g}(x)(x - \lambda_0 e) = e$，那么$y\widetilde{g}(x)$将是$x - \lambda_0 e$在$A$内的逆，因为由假设保证$\widetilde{g}(x)$与$x, y$均可换. 这表明$\beta \in \sigma(\widetilde{f}(x))$. 证毕. ■

注　假如$f \in \mathrm{Hol}(\Omega, \mathbf{C})$，则显然有$\bigcup_{\lambda \in \sigma(x)} \sigma(f(\lambda)) = f(\sigma(x))$与$\{\lambda \in \mathbf{C} : \lambda e \in f(\sigma(x))\} = f(\sigma(x))$，且这时$f(\lambda)$与$f(\mu)$之间的可换性自动成立. 因此，在此时定理5退化为通常的复Banach代数中的谱映射定理[11, p.244].

应用刚才证明过的定理5可以得到一些有价值的推论，其中包括基

本定理的关于谱半径的推广形式(见后面推论 5 与推论 6).下面的推论与[14] 中谱半径极大原理以及[1] 中一个问题密切相关.

推论 4 设 $f\in \mathrm{Hol}(\Delta,A)$.对满足 $0<r<1$ 的实数 r,令 $m(r)$ 如同(11) 式,并令

$$M(r)=\sup\{|f(\lambda)|_\sigma:|\lambda|=r\}. \tag{15}$$

则有

$$\sup\{|\tilde{f}(x)|_\sigma:|x|_\sigma\leqslant r\}\leqslant M(r)\leqslant m(r), \tag{16}$$

这里,sup 取遍所有谱半径 $\leqslant r$,且与 f 可换的 $x\in A$.如果 A 还是一个 B^*-代数,且对每个 $\lambda\in\Delta$,$f(\lambda)$ 是 A 内正规元素,那么有

$$\max\{|\tilde{f}(x)|_\sigma:|x|_\sigma\leqslant r\}=M(r)=m(r), \tag{17}$$

这里,max 的取值范围同于(16) 式中 sup 的取值范围.

证 可以看到,(16) 式可直接从(13) 式以及 $|f(\lambda)|_\sigma\leqslant\|f(\lambda)\|$ 的事实推出.假如 A 是 B^*-代数,且 $f(\lambda)$ 在 A 内是正规的,那么,对任意 $\lambda\in\Delta$,$|f(\lambda)|_\sigma=\|f(\lambda)\|$,因而 $M(r)=m(r)$,$0<r<1$.另一方面,存在复数 β,使得 $|\beta|=r$ 与 $|f(\beta)|_\sigma=M(r)$,因而 $\beta e\in A$ 满足 $|\beta e|_\sigma=r$ 与 $|\tilde{f}(\beta e)|_\sigma=|f(\beta)|_\sigma=M(r)$.因此,(17) 式得证. ■

推论 5 设 $\Omega\supset\overline{\Delta}$,$f\in\mathrm{Hol}(\Omega,A)$ 满足 $|f(\lambda)|_\sigma\leqslant 1$,$\forall\lambda\in\overline{\Delta}$.则只要 $x\in A$ 有 $|x|_\sigma\leqslant 1$,且与 f 可换,便有 $|\tilde{f}(x)|_\sigma\leqslant 1$.

推论 6 设 $f\in\mathrm{Hol}(\Delta,A)$ 满足 $|f(\lambda)|_\sigma<1$,$\forall\lambda\in\Delta$,则只要 $x\in A$ 有 $|x|_\sigma<1$,且与 f 可换,便有 $|\tilde{f}(x)|_\sigma<1$.

前述两个推论可直接由定理 5 以及 $|f(\lambda)|_\sigma$ 是 Ω(或 Δ) 上的次调和函数,因而上半连续,它在紧集 $\sigma(x)$ 上达到上确界的事实[14,5] 得到.

推论 7(对谱半径的 Schwarz 引理) 设 $f\in\mathrm{Hol}(\Delta,A)$ 满足 $|f(\lambda)|_\sigma\leqslant 1$,$\forall\lambda\in\Delta$,且 $f(0)=f'(0)=\cdots=f^{(n-1)}(0)=0$,这里 $n\geqslant 1$.则有

$$|\tilde{f}(x)|_\sigma\leqslant|x^n|_\sigma, \tag{18}$$

这里 $x\in A$ 满足 $|x|_\sigma<1$,且 $x\cup f(\lambda)$,$\forall\lambda\in\Delta$.

证 定义 h 为 $h(\lambda)=f(\lambda)/\lambda^n$.由题设条件,$h\in\mathrm{Hol}(\Delta,A)$.对任意 ε,$0<\varepsilon<1$,选取 $\delta>0$ 使得 $|\lambda^n|\geqslant 1-\varepsilon$ 只要 $1-\delta<|\lambda|\leqslant 1$.这时

$$1\geqslant|f(\lambda)|_\sigma=|\lambda^n||h(\lambda)|_\sigma\geqslant(1-\varepsilon)|h(\lambda)|_\sigma$$

对 $1-\delta<|\lambda|<1$ 成立.应用[14] 中一个结果,我们有 $|h(\lambda)|_\sigma\leqslant 1/(1-\varepsilon)$,$|\lambda|<1$,于是,$|h(\lambda)|_\sigma\leqslant 1$ 对 $|\lambda|<1$ 成立.因此,对任意 $x\in A$ 满

足 $|x|_\sigma<1$，且与 h 可换，有 $|\tilde{h}(x)|_\sigma\leqslant 1$（按定理 5）. 再应用 x^n 与 f（因而与 h）可换的事实，得到[11, p. 289]

$$|\tilde{f}(x)|_\sigma=|\tilde{h}(x)x^n|_\sigma\leqslant|\tilde{h}(x)|_\sigma|x^n|_\sigma\leqslant|x^n|_\sigma$$

对所有 A 中满足 $|x|_\sigma<1$ 与 $x\cup f(\lambda)(\lambda\in\Delta)$ 的 x 均成立. 证毕. ■

注 当 $n=1$ 与 $x=\lambda e$（λ 是复数）时上述推论即为[4]中对于谱半径的 Schwarz 引理的原始形式.

致谢

对樊瓛教授在本文准备过程中给予的帮助与指导，作者谨表谢意.

参考文献

[1] Brown A and Douglas R G. On maximum theorems for analytic operator functions. Acta Sci. Math. Szeged, 1966. 26: 325－327.

[2] Chen Gongning. On simplifled proofs of theorems of von Neumann, Heinz, and Ky Fan, and their extended versions, in Nonlinear and convex analysis, Marcel Dekker, 1987, Ch. 2, 15－21.

[3] Fan Ky. Analytic functions of a proper contraction. Math. Z., 1978, 160: 275－290.

[4] Globernik J. Schwarz's lemma for the spectral radius. Rev. Roum. Math. Pures et Appl. Tome XIX, 1974, 8: 1 009—1 012.

[5] Halmos R R. A Hilbert Space Problem Book. Princeton: Van Nostrand, 1967.

[6] Harris L A. Banach algebras with involution and Möbius transformation. J Funct. Anal., 1972, 11: 1－16.

[7] Hille E and Phillps R S. Functional Analysis and Semi-Groups. Amer. Math. Soc., Providence R. I, 1957.

[8] Kantorovitz S. Operational calculus in Banach algebras for algebra-valued functions. Trans. Amer Math, Soc., 1964, 110: 519－537.

[9] Pták V. On the spectral radius in Banach algebras with involution. Bull. London Math. Soc., 1970, 2: 327－334.

[10] Riesz F and Sz. Nagy B. Functional Analysis. New York, Ungar, 1960.

[11] Rudin W. Functional Analysis. McGraw-Hill, Inc. New York, 1973.

[12] Shin M H and Tan K K. Analytic functions of topological proper contractions. Math. Z., 1984, 187: 317－323.

[13] Tao Z G. Analytic operator functions, J. Math Anal. Appl. 1984, 103: 293－320.

[14] Vesentini E. On the subharmonicity of the spectral radius. Boll. Un. Mat. Ital., 1968, 1: 427－429.

Journal of Mathematical Analysis and Applications
1984,98(2):305－313

对 $\mathbf{C}^n$ 的单位开球与广义上半平面的全纯映射的迭代①

Iteration for Holomorphic Maps of the Open Unit Ball and the Generalized Upper Half-Plane of $\mathbf{C}^n$

§ 1. Introduction

Recently, Ky Fan [2,3] extended a classical theorem of Wolff [5] on iteration of holomorphic functions in complex analysis to the theory of holomorphic functions of operators. It is our purpose here to generalize Wolff's theorem to holomorphic maps of the open unit ball B of $\mathbf{C}^n$, and to that of the generalized upper half-plane E of $\mathbf{C}^n$. The writing of the present paper was inspired by the results of [2,3]. Our Theorems 1 and 2 below are analogous to the Theorems 1 and 2 in [3] in the setting of functional calculus.

A result we shall need is the following theorem of Earle and Hamilton [1]:

Theorem (Earle-Hamilton). *If D is a non-empty bounded connected open subset of a complex Banach space, then any holomorphic map g from D strictly inside D (i. e., there exists $\varepsilon > 0$ such that $\| g(x) - y \| > \varepsilon$ for all $x \in D$, $y \notin D$) has a unique fixed point in D.*

Here by holomorphic we mean a map of D into a Banach space having a Fréchet derivative at each point of D.

① 收稿日期:1982-08-07. Submitted by Ky Fan.

§ 2. Iteration for Holomorphic Self-Maps of B

In all that follows, $\mathbf{C}^n$ will denote the n-dimensional complex coordinate space. We take the standard Hilbert space structure on $\mathbf{C}^n$. Let B be the open unit ball in $\mathbf{C}^n$, i. e., $B=\{z\in\mathbf{C}^n:\|z\|<1\}$. Fix $b\in B$. Let P_b be the orthogonal projection of $\mathbf{C}^n$ to the subspace spanned by b, and let $Q_b=I-P_b$. Here I stands for the identity map of $\mathbf{C}^n$. Consider the Möbius transformation

$$\varphi_b(z)=\frac{b-P_bz-(1-\|b\|^2)^{1/2}Q_bz}{1-\langle z,b\rangle} \tag{1}$$

where $\langle\,,\rangle$ denotes the inner product of $\mathbf{C}^n$. If $\Omega=\{z\in\mathbf{C}^n:\langle z,b\rangle\neq 1\}$ (clearly $\Omega\supset\overline{B}$, the closure of B in $\mathbf{C}^n$, since $\|b\|<1$), then $\varphi_b:\Omega\to\mathbf{C}^n$ is holomorphic. Furthermore, it is well known[4, Theorem 2. 2. 2] that φ_b is a biholomorphic map of B onto itself with $\varphi_b(0)=b$ and $\varphi_b^{-1}=\varphi_b$. In the case $\|b\|=1$, it is easy to verify $\|\varphi_b(z)\|=1$ for all z in B (see also identity (6) below).

The following lemma is analogous to Lemma 1 in [3], and is fundamental in the proof of Theorem 1.

Lemma 1 *Let b be a fixed point of $\overline{B}$, $z\in B$ and $d>0$. Then inequality*

$$\frac{|1-\langle z,b\rangle|^2}{1-\|z\|^2}\leqslant d \tag{2}$$

holds if and only if

$$T_b(z,d)\leqslant\frac{d(d+\|b\|^2-1)}{(d+\|b\|^2)^2}, \tag{3}$$

where

$$T_b(z,d)=\left\|P_bz-\frac{b}{d+\|b\|^2}\right\|^2+\frac{d}{d+\|b\|^2}\|Q_bz\|^2. \tag{4}$$

Equality occurs in (2) if and only if it occurs in (3).

In the case $\|b\|<1$, both (2) and (3) are equivalent to

$$\|\varphi_b(z)\|^2\leqslant\frac{d+\|b\|^2-1}{d}; \tag{5}$$

and equalities hold simultaneously in (2)(3) and (5).

Proof Since $|\langle z,b\rangle| = |\langle P_b z, b\rangle| = \|P_b z\| \|b\|$ and $\|z\|^2 = \|P_b z\|^2 + \|Q_b z\|^2$ for any $z \in B$, inequality (2) is equivalent to

$$1 - \langle P_b z, b\rangle - \langle b, P_b z\rangle + \|P_b z\|^2 \|b\|^2 \leqslant d(1 - \|P_b z\|^2 - \|Q_B z\|^2)$$

or

$$T_b(z,d) \leqslant \frac{d-1}{d+\|b\|^2} + \frac{\|b\|^2}{(d+\|b\|^2)^2} = \frac{d(d+\|b\|^2-1)}{(d+\|b\|^2)^2}$$

Hence (2)and(3) are equivalent to one another.

In the case $\|b\| < 1$, the equivalence of (5) and (2) (and therefore also (3)) is an immediate consequence of the identity

$$1 - \|\varphi_b(z)\|^2 = \frac{(1-\|b\|^2)(1-\|z\|^2)}{|1-\langle z,b\rangle|^2} \quad (b \in \overline{B}, z \in B). \tag{6}$$

The above argument remains valid if each inequality sign $\leqslant$ is replaced by a strict inequality sigh $<$. Consequently, equalities in (2) and (3)(and also in (5) when $\|b\| < 1$) hold simultaneously. This completes the proof. ■

Theorem 1 *Let $f_1, f_2, \cdots, f_n$ be holomorphic functions of n complex variables with domain B, and let $F = (f_1, f_2, \cdots, f_n)$ be a holomorphic map of B into B. Let $F^{[m]}$ denote the mth iterate of F(i. e., $F^{[1]} = F, F^{[m]} = F \circ F^{[m-1]}$ for $m \geqslant 2$). Then there exists a w in $\overline{B}$ such that*

$$T_w(F^{[m]}(z), d(w,z)) \leqslant T_w(z, d(w,z)) = \frac{d(w,z)\{d(w,z)+\|w\|^2-1\}}{\{d(w,z)+\|w\|^2\}^2}, \tag{7}$$

$$\|\varphi_w(F^{[m]}(z))\| \leqslant \|\varphi_w(z)\| \doteq \left[\frac{d(w,z)+\|w\|^2-1}{d(w,z)}\right]^{1/2}, \tag{8}$$

and

$$d(w, F^{[m]}(z)) \leqslant d(w,z) \tag{9}$$

hold for any $z \in B$ and $m = 1, 2, \cdots$, where $T_b(z,d)$ is defined by (4) *and*

$$d(w,z) = \frac{|1-\langle z,w\rangle|^2}{1-\|z\|^2}. \tag{10}$$

Furthermore, $F(w) = w$ unless $\|w\| = 1$.

Proof Take a sequence $\{a_k\}$ such that $0 < a_k < 1$ and $\lim\limits_{k\to\infty} a_k = 1$.

Let $F_k = a_k F$. As for all $z \in B, y \notin B$

$$\| F_k(z) - y \| \geqslant \| y \| - a_k \| F(z) \| > 1 - a_k,$$

Earle - Hamilton's theorem implies that there exists a $z^{(k)}$ such that $F_k(z^{(k)}) = z^{(k)}$, whence $\| z^{(k)} \| < a_k < 1$. By using a subsequence if necessary, we may assume that $\{z^{(k)}\}$ tends to a limit as $k \to \infty$. Let $w = \lim_{k\to\infty} z^{(k)}$. Then $\| w \| \leqslant 1$. Clearly, we have $F(w) = w$ unless $\| w \| = 1$. Note that the convergence $\lim_{k\to\infty} F_k(z) = F(z)$ is uniform in B, since $\| F_k(z) - F(z) \| < 1 - a_k$ for all z in B.

As $F_k(z^{(k)}) = z^{(k)}$ and $F_k(B) \subset B$, we conclude from the version of Schwarz lemma [4, p. 163] that

$$\| \varphi_{z^{(k)}}(F_k(z)) \| \leqslant \| \varphi_{z^{(k)}}(z) \| \quad (z \in B).$$

By the identity (6) with $b = z^{(k)}$, this implies

$$\frac{| 1 - \langle F_k(z), z^{(k)} \rangle |^2}{1 - \| F_k(z) \|^2} \leqslant \frac{| 1 - \langle z, z^{(k)} \rangle |^2}{1 - \| z \|^2} \quad (z \in B).$$

When $k \to \infty$, we obtain the case $m = 1$ of (9) for any z in B, i. e. ,

$$d(w, F(z)) \leqslant d(w, z). \tag{11}$$

Denote the mth iterate of F by $F^{[m]}$, which is also a holomorphic map of B into B. Since $F^{[m]}(z) = F(F^{[m-1]}(z))$ and (11) is proved for any z in B, we have

$$d(w, F^{[m]}(z)) \leqslant d(w, F^{[m-1]}(z)) \leqslant \cdots \leqslant d(w, z)$$

for any z in B and $m = 1, 2, \cdots$, so that we obtain (9). Then, by an application of Lemma 1, (7) and the case of $\| w \| < 1$ of (8) follow from (9) and (10). When $\| w \| = 1$, (8) remains true and the inequality therein becomes equality according to the identity (6). This completes the proof of Theorem 1. ■

Remark 1 Using (10) and the identity

$$\begin{aligned} &1 - 2\mathrm{Re}\langle z, w \rangle + \| w \|^2 (1 - \| Q_w z \|^2) \\ &= | 1 - \langle z, w \rangle |^2 + \| w \|^2 (1 - \| z \|^2), \end{aligned} \tag{12}$$

we can give another expression for the quantity on the right side of (7)

$$\frac{d(w,z)(d(w,z) + \| w \|^2 - 1)}{(d(w,z) + \| w \|^2)^2}$$

$$= \| \varphi_w(z) \|^2 \frac{|1-\langle z,w\rangle|^4}{\{1-2\mathrm{Re}\langle z,w\rangle + \|w\|^2(1-\|Q_w z\|^2)^2\}}. \tag{13}$$

In Theorem 1, if F is fixed-point-free in B, then we must have $\|w\| = 1$, and (7)(combined with (4)(10)(12)and(13))gives

$$\|P_w(F^{[m]}(z)) - c(w,z)\|^2 + r(w,z)\|Q_w(F^{[m]}(z))\|^2$$
$$\leqslant \|P_w z - c(w,z)\|^2 + r(w,z)\|Q_w z\|^2 = r^2(w,z),$$

for any z in B and $m = 1,2,\cdots$, where

$$c(w,z) = \frac{(1-\|z\|^2)w}{|1-\langle z,w\rangle|^2 + 1 - \|z\|^2},$$

$$r(w,z) = \frac{|1-\langle z,w\rangle|^2}{|1-\langle z,w\rangle|^2 + 1 - \|z\|^2}.$$

Then, $\|c(w,z)\| + r(w,z) = 1$ for all z in B and all w in $\mathbf{C}^n$ with $\|w\| = 1$. In the simplest case when $n = 1$, it is seen that $P_w = I$ and $Q_W = 0$. If F is fixed point-free in the open unit disk of the complex plane, by identity (12) we have

$$c(w,z) = \frac{(1-\|z\|^2)w}{2-2\mathrm{Re}\,(\overline{w}z)},$$

$$r(w,z) = \frac{|1-\overline{w}z|^2}{2-\mathrm{Re}(\overline{w}z)}.$$

Hence in this case Theorem 1 reduces to the classical Wolff's theorem.

Remark 2 In the preceding theorem, if $\|w\| = 1$, then

$$\lim_{s\uparrow 1}\langle F^{[m]}(sw),w\rangle = 1 \quad (m = 1,2,\cdots). \tag{14}$$

Indeed, when $\|w\| = 1$, (9) with $z = sw\,(0 < s < 1)$ and (10) give

$$|1-\langle F^{[m]}(sw),w\rangle|^2 \leqslant \frac{(1-s)^2}{1-s^2}(1-\|F^{[m]}(sw)\|^2) \leqslant \frac{1-s}{1+s}.$$

In the case of $n = 1$, since $\langle F(sw),w\rangle = w^{-1}F(sw)$, (14) with $m = 1$ becomes

$$\lim_{s\uparrow 1} F(sw) = w,$$

which is the same as a result pointed out by T. Ando (see [3, Remark 3]).

§ 3. Iteration for Holomorphic Self-Maps of E

Let E denote the generalized upper half-plane of $\mathbf{C}^n$, i. e., the set of all $z \in \mathbf{C}^n$ with IM z > 0, where IM z $= \mathrm{Im}\, z_1 - \|z'\|^2$, $z = (z_1, z')$, $z' = (z_2, z_3, \cdots, z_n)$ and $\|z'\|^2 = |z_2|^2 + |z_3|^2 + \cdots + |z_n|^2$. Then, E

is an unbounded convex domain in $\mathbf{C}^n$, and $\overline{E} = \{z \in \mathbf{C}^n : \mathrm{IM}\ z \geqslant 0\}$. Consider the Cayley transformation Φ defined by

$$u = \Phi(z) = \mathrm{i}\,\frac{e_1 + z}{1 - z_1} \tag{15}$$

where $e_1 = (1, 0')$. This sends z in $\mathbf{C}^n$ with $z_1 \neq 1$ to u in $\mathbf{C}^n$, and implies

$$\mathrm{IM}\ u = \frac{1 - \|z\|^2}{|1 - z_1|^2}, \tag{16}$$

and

$$z = \Psi(u) = \frac{2u}{\mathrm{i} + u_1} - e_1. \tag{17}$$

Hence Φ is a biholomorphic map of B onto E (see [4, p. 31]).

In this section, we shall consider iterations of holomorphic self-maps of E. To prove Theorem 2 below we shall need the following

Lemma 2 *Let $v = (v_1, v')$ be a fixed vector of $\overline{E}$, $u = (u_1, u') \in E$ and $h > 0$. Then inequality*

$$\frac{|\mathrm{i}(u_1 - \overline{v}_1) + 2\langle u', v'\rangle|^2}{\mathrm{IM}\ u} \leqslant h \tag{18}$$

holds if and only if

$$R_v(u, h, h + 4\|v'\|^2) \leqslant \frac{h - 4\mathrm{IM}\ v}{4(h + \|v'\|^2)}, \tag{19}$$

where

$$R_v(u, h, t) = h^{-1}\|P_{v'}u' + 2\mathrm{i}t^{-1}(u_1 - \overline{v}_1)v'\|^2 + t^{-1}\|Q_{v'}u'\|^2 + t^{-2}\left|(u_1 - \overline{v}_1) - \frac{1}{2}\mathrm{i}t\right|^2 \tag{20}$$

for $u \in E$ and $t > 0$. Equality occurs in (18) *if and only if it occurs in* (19).

Proof Note that $\|u'\|^2 = \|P_{v'}u'\|^2 + \|Q_{v'}u'\|^2$ and $|\langle u', v'\rangle| = \|P_{v'}u'\|\,\|v'\|$. By multiplying out everything we see that (18) is equivalent to

$$(h + 4\|v'\|^2)\|P_{v'}u'\|^2 + 2\mathrm{i}(u_1 - \overline{v}_1)\langle v', P_{v'}u'\rangle - 2\mathrm{i}(\overline{u}_1 - v_1)\langle P_{v'}u', v'\rangle + h\|Q_{v'}u'\|^2 + |u_1|^2 - u_1\left(v_1 - \frac{1}{2}\mathrm{i}h\right) - \overline{u}_1\left(\overline{v}_1 + \frac{1}{2}\mathrm{i}h\right) + |v_1|^2 \leqslant 0.$$

By completing the square, the last inequality can be rewritten as (19).

The above argument remains true if every inequality sign $\leqslant$ is replaced by a strict inequality sign $<$. Hence, equalities in (18) and (19) hold simultaneously. This proves the lemma. ■

Theorem 2 *Let $g_1, g_2, \cdots, g_n$ be holomorphic functions of n complex variables with domain E. Let $G = (g_1, g_2, \cdots, g_n)$ be a holomorphic map of E into itself. Then either*

$$\mathrm{IM}\, G(u) \geqslant \mathrm{IM}\, u \tag{21}$$

holds for any $u \in E$; or there exists a $v = (v_1, v')$ in $\overline{E}$ such that

$$R_v(G^{[m]}(u), h, h + 4\|v'\|^2) \leqslant R_v(u, h, h + 4\|v'\|^2) = \frac{h - 4\mathrm{IM}\, v}{4(h + \|v'\|^2)} \tag{22}$$

and

$$h(v, G^{[m]}(u)) \leqslant h(v, u) \tag{23}$$

hold for any u in E and $m = 1, 2, \cdots$, where $R_v(u, h, t)$ is defined by (20) and

$$h = h(v, u) = \frac{|\,\mathrm{i}(u_1 - \overline{v}_1) + 2\langle u', v'\rangle\,|^2}{\mathrm{IM}\, u} \tag{24}$$

Furthermore, $G(v) = v$ unless $\mathrm{IM}\, v = 0$.

Proof Consider the Cayley transformation Φ and its holomorphic inverse Ψ defined by (15) and (17), respectively. Then $\Phi(B) = E$ and $\Psi(E) = B$. It is seen that $F = \Psi \circ G \circ \Phi$ is a holomorphic map such that $F(B) \subset B$, so by Theorem 1 there exists a w in $\overline{B}$ having the properties stated in Theorem 1. Put $\Phi(w) = v$. Since $F(z) = \Psi(G(u))$ for $z \in B$, we have by (17)

$$F(z) = \frac{2G(u)}{\mathrm{i} + g_1(u)} - e_1. \tag{25}$$

Thus

$$1 - \langle F(z), e_1\rangle = \frac{2\mathrm{i}}{\mathrm{i} + g_1(u)} \tag{26}$$

and, in particular,

$$1 - z_1 = \frac{2\mathrm{i}}{\mathrm{i} + u_1}. \tag{27}$$

It is convenient to distinguish two cases.

Case 1 $w = e_1$. By (9) with $m = 1$, we then obtain

$$\frac{|1-\langle F(z), e_1\rangle|^2}{1-\|F(z)\|^2} \leqslant \frac{|1-z_1|^2}{1-\|z\|^2} \quad (z \in B).$$

From (16)(26) and (27) it follows that the last inequality can be rewritten as

$$(\mathrm{IM}\ G(u))^{-1} \leqslant (\mathrm{IM}\ u)^{-1} \quad (u \in E),$$

which is equivalent to (21).

Case 2 $w \neq e_1$. Then, by (15) and (16), $v = \Phi(w)$ is finite, and IM $v > 0$ if and only if $\|w\| < 1$; IM $v = 0$ if $\|w\| = 1$ but $w \neq e_1$. Since $F(w) = w$ unless $\|w\| = 1$ by Theorem 1, we have $G(v) = v$ unless IM $v = 0$. Simple computations show that for $z \in B$

$$|1-\langle F(z), w\rangle|^2 = \frac{4\,|\,\mathrm{i}(g_1(u)-\overline{v}_1)+2\langle G(u)', v'\rangle|^2}{|\mathrm{i}+g_1(u)|^2\,|\mathrm{i}+v_1|^2} > 0 \quad (28)$$

and

$$|1-\langle z, w\rangle|^2 = \frac{4\,|\,\mathrm{i}(u_1-\overline{v}_1)+2\langle u', v'\rangle|^2}{|\mathrm{i}+u_1|^2\,|\mathrm{i}+v_1|^2} > 0, \quad (29)$$

where $G(u)' = (g_2(u), g_3(u), \cdots, g_n(u))$. Thus by (16)(26)(27)(28) and (29), (9) with $m = 1$ becomes

$$h(v, G(u)) \leqslant h(v, u) \quad (u \in E), \quad (30)$$

where $h(v,u)$ is defined by (24). It follows from Lemma 2 that the case of $m = 1$ of (22) holds for any $u \in E$. Suppose as before that $G^{[m]}$ is the mth iterate of G. Obviously, $G^{[m]}(E) \subset E$. Since (30) is proved for any $u \in E$, and $G^{[m]}(u) = G(G^{[m-1]}(u))$, we have

$$h(v, G^{[m]}(u)) \leqslant h(v, G^{[m-1]}(u)) \leqslant \cdots \leqslant h(v, u) \quad (u \in E).$$

Then, (22) follows easily from Lemma 2. Theorem 2 is proved. ■

Remark 3 In the simplest case when $n = 1$, (21)(22) and (23) in the preceding theorem become respectively

$$\mathrm{Im}\ g_1(u) \geqslant \mathrm{Im}\ u_1,$$

$$\left|g_1^{[m]}(u) - \left(\overline{v}_1 + \frac{1}{2}\mathrm{i}h_1\right)\right|^2 \leqslant \left|u_1 - \left(\overline{v}_1 + \frac{1}{2}\mathrm{i}h_1\right)\right|^2 = \frac{1}{4}h_1^2 - h_1\,\mathrm{Im}\ v_1,$$

and

$$h_1(v_1, g_1^{[m]}(u)) \leqslant h_1(v_1, u_1).$$

Here $h_1 = h_1(v_1, u_1) = |u_1 - \overline{v}_1|^2 / \mathrm{Im}\, u_1$.

Acknowledgment

I wish to express my appreciation to Professor Ky Fan for his direction in the writing of this paper.

References

[1] C J Earle and R S Hamilton. A fixed point theorem for holomorphic mappings. in Global Analysis. Proceedings, Symposium Pure Mathematics Vol. XVI, Berkeley, Calif. ,1968, Amer. Math. Soc. ,Providence, R. I. ,1970:61－65.

[2] Ky Fan. Iteration of analytic functions of operators. Math Z. ,1982,179:293－298.

[3] Ky Fan. Iteration of analytic functions of operators, Ⅱ. Linear Multilinear Algebra,1983,12(4):295－304.

[4] W Rudin. Function Theory in the Unit Ball of $\mathbf{C}^n$. Springer-Verlag, New York, 1980.

[5] J Wolff. Sur une généralisation d'un théorème de Schwartz. C. R. Acad. Sci. Paris,1926,182:918－920;1926,183:500－502.

Journal of Mathematical Analysis and Applications
1985,112(1):274－278

同一 Hilbert 空间之间全纯映射的迭代[①]

Iteration of Holomorphic Maps in Hilbert Spaces

Abstract Previous work (Gong-Ning Chen, J. Math. Anal. Appl. 98(1984), 305～313) on iteration of holomorphic maps of $\mathbf{C}^n$ is continued. The purpose of this note is to extend results given in the above mentioned reference to the case of complex Hilbert spaces. Other comments are appended.

1. A theorem of Wolff [9](see also Burckel [1])asserts that if $f:\Delta\to\Delta$, the open unit disc in the complex plane, is analytic and fixed point-free, then there exists $u\in\partial\Delta$ such that every closed disc in Δ which is tangent to $\partial\Delta$ at u is mapped into itself by every iterate of f. Recent papers by Ky Fan [4,5], and by the author [2]show that Wolff's theorem can be extended to analytic functions of operators, and to holomorphic maps of $\mathbf{C}^n$, respectively. The present note is a continuation of [2], and deals with iteration of holomorphic maps in Hilbert spaces. The purpose here is to generalize results givin in [2] to the case of complex Hilbert spaces of any dimension. As in the previous work [2], this writing was also inspired by the elegent results of [4,5].

Throughout this note, X and Y will be complex Hilbert spaces of any dimension, and B and B_1 the open unit ball of X and Y, respectively.

---

① Submitted by Ky Fan.

A map F defined on an open connected set D in X with values in Y is said to be holomorphic if for each $x \in D$ its Fréchet derivative at x exists as a bounded complex-linear map of X into Y. By a biholomorphic map we mean a holomorphic map with a holomorphic inverse.

Fix $b \in B$, and let P_b be the orthogonal projection of X to the subspace spanned by b and set $Q_b = I - P_b$. Here I stands for the identity map on X. Denote by Φ_b the Möbius transformation

$$\Phi_b = \frac{b - P_b x - (1 - \| b \|^2)^{1/2} Q_b x}{1 - \langle x, b \rangle} \quad (x \in \overline{B}), \tag{1}$$

where $\langle , \rangle$ denotes the inner product of X. It is well known [6,8] that the identity

$$1 - \| \Phi_b(x) \|^2 = \frac{(1 - \| b \|^2)(1 - \| x \|^2)}{| 1 - \langle x, b \rangle |^2} \tag{2}$$

holds for all $x \in \overline{B}$, and that Φ_b is a holomorphic map of B ($\overline{B}$) onto B ($\overline{B}$).

2. In this section, we extend Theorem 1 in [2] to the case of iteration for holomorphic self-maps of B of complex Hilbert spaces. Other comments are appended. We begin with a version of Lemma 1 in [2] by replacing $\mathbf{C}^n$ by X (with the same proof).

Lemma 1 *Let X be a complex Hilbert space and let b be a fixed point of $\overline{B}$, $x \in B$, and $d > 0$. Then inequality*

$$\frac{| 1 - \langle x, b \rangle |^2}{1 - \| x \|^2} \leqslant d \tag{3}$$

holds if and only if

$$T_b(x,d) \equiv \left\| P_b x - \frac{b}{d + \| b \|^2} \right\|^2 + \frac{d}{d + \| b \|^2} \| Q_b x \|^2 \leqslant \frac{d(d + \| b \|^2 - 1)}{(d + \| b \|^2)^2}. \tag{4}$$

Equality occurs in (3) *if and only if it occurs in* (4). *When $b \in B$, both* (3) *and* (4) *are equivalent to*

$$\| \Phi_b(x) \|^2 \leqslant \frac{d + \| b \|^2 - 1}{d}, \tag{5}$$

and equalities hold simultaneously in (3) (4) *and* (5).

Theorem 1 *Let F be a holomorphic self-map of B, and $F^{[m]}$ denote the mth iterate of F. Then there exists a $w \in \overline{B}$ such that*

$$d(w,F^{[m]}(x)) \leqslant d(w,x), \tag{6}$$

$$T_w(F^{[m]}(x),d(w,x)) \leqslant T_w(x,d(w,x)) = \frac{d(w,x)\{d(w,x)+\|w\|^2-1\}}{\{d(w,x)+\|w\|^2\}^2} \tag{7}$$

and

$$\|\Phi_w(F^{[m]}(x))\|^2 \leqslant \|\Phi_w(x)\|^2 = \frac{d(w,x)+\|w\|^2-1}{d(w,x)} \tag{8}$$

hold for any $x \in B$, and $m = 1,2,\cdots$, where

$$d(w,x) = \frac{|1-\langle x,w\rangle|^2}{1-\|x\|^2}, \tag{9}$$

and $T_w(y,d(w,x))$ is defined as in (4) for $y \in B$. If, further, $F: B \to B$ is continuous in the weak topology of X, then we have $F(w) = w$ unless $\|w\| = 1$.

Proof Take $0 < t_k < 1$ with $\lim\limits_{k\to\infty} t_k = 1$, and set $F_k = t_k F$. As was stated in the proof of Theorem 1 in [2], by an application of a theorem of Earle and Hamilton [3] we can choose a sequence $\{x_k\}$ in B such that $F_k(x_k) = x_k$ for all k. Since $\overline{B}$ is weakly compact in X, we may assume by replacing a suitable subsequence that $\{x_k\}$ converges weakly to w, i. e., $\lim\limits_{k\to\infty}\langle y,x_k\rangle = \langle y,w\rangle$ for any $y \in X$. Then $\|w\| \leqslant \varliminf\limits_{k\to\infty}\|x_k\| \leqslant 1$, so that $w \in \overline{B}$. From the Schwarz-Pick lemma it follows that

$$\|\Phi_{x_k}(F_k(x))\| \leqslant \|\Phi_{x_k}(x)\| \quad \text{for any } x \in B.$$

In view of the identity (2), this is equivalent to

$$\frac{|1-t_k\langle F(x),x_k\rangle|^2}{1-t_k^2\|F(x)\|^2} \leqslant \frac{|1-\langle x,x_k\rangle|^2}{1-\|x\|^2}.$$

By letting $k \to \infty$, we derive that the case $m = 1$ of (6) holds for any $x \in B$.

If, further, F is weakly continuous in B, i. e., $F(x_k) \to^w F(x_0)$ whenever $x_k \to^w x_0$ and $x_k's, x_0 \in B$, then we have $F(w) = w$ unless $\|w\| = 1$. The rest of this proof follows now from Lemma 1 by a similar argument to that used in Theorem 1 in [2]. This completes the proof. ■

For finite dimensional X, the condition here on F coincides with that of Theorem 1 in [2], since the weak continuity of F is then a consequence of the condition of holomorphy. Thus, Theorem 1 here is a generalization of Theorem 1 in [2].

It is incidentally seen from the proof of Theorem 1 that if F is a holomorphic map of B into B, and is weakly continuous in $\overline{B}$, then F has a fixed point in $\overline{B}$. In particular, if F is a biholomorphic map of B onto B, it is well known [6,8] that $F = U\Phi_b$, where U is unitary and $b = F^{-1}(0)$; therefore F is weakly continuous in $\overline{B}$ since Φ_b is. This leads to the following.

Corollary 1 (Hayden and Suffridge [6]). *Any biholomorphic map of B onto B fixes a point in $\overline{B}$.*

Like Wolff's theorem, quoted above, Theorem 1 has an appealing geometric interpretation. For any $x \in B$, let $E(w,x)$ denote the set of all $z \in X$ that satisfy

$$T_w(z, d(w,x)) \leqslant \frac{d(w,x)\{d(w,x) + \| w \|^2 - 1\}}{\{d(w,x) + \| w \|^2\}^2},$$

where $d(w,x)$ is as in (9). Then $E(w,x)$ is the closed ellipsoid with center at $w/(d(w,x) + \| w \|^2)$. Let $\partial E(w,x)$ be the boundary of $E(w,x)$. We have:

Corollary 2 *Suppose that $F: B \to B$ is holomorphic. Then there exists $w \in \overline{B}$ such that $F^{[m]}(E(w,x)) \subset E(w,x) \subset \overline{B}$ for all $x \in B$ and $m = 1,2,\cdots$. If $\| w \| = 1$, $E(w,x)$ is tangent to ∂B at w for all $x \in B$.*

Proof A little manipulation shows that $E(w,x) \subset \overline{B}$ for any $x \in B$, and if, in particular, $\| w \| = 1$, then $E(w,x)$ is tangent to ∂B at w. Then, (7) means that $x \in \partial E(w,x)$ for any $x \in B$, and $F^{[m]}(x) \in E(w,x)$ for all $x \in B$ and $m = 1,2,\cdots$. If $y \in \partial E(w,x)$, we have by Lemma 1 that $d(w,y) = d(w,x)$, whence $F^{[m]}(y) \in E(w,y) = E(w,x)$. If y is in the interior of $E(w,x)$, then we also have $F^{[m]}(y) \in E(w,x)$. for any m, for, if not, we would have that $F^{[m]}(y) \notin E(w,x)$ for some m. This implies that there exists u belonging to $\partial E(w,x) \cap \partial E(w,y)$, so that $d(w,x) = d(w,u) = d(w,y)$, a contradiction. The corollary is proven. ■

As for (6) and (8) with the case of $\| w \| < 1$, both of them are equivalent to the fact that

$$\rho(w, F^{[m]}(x)) \leqslant \rho(w,x)$$

holds for all $x \in B$ and $m = 1,2,\cdots$, where $\rho(x,y) = \tanh^{-1} \| \Phi_x(y) \|$ is called the CRF(Carathéodory - Reiffen - Finsler)metric on B(see[3,7]).

3. The rest of this note is devoted to a generalization of Theorem 2 in [2]. Let e be a unit vector in X, and let $\mathrm{IM}\ u = \mathrm{Im}\langle u,e\rangle - \| Q_e u \|^2$ for any $u \in X$. Then the corresponding upper half-plane of X is defined by

$$E = \{u \in X: \mathrm{IM}\ u > 0\}.$$

It is well known [7] that E is an unbounded convex domain in X, and $\overline{E} = \{u \in X: \mathrm{IM}\ u \geqslant 0\}$. As a consequence of the preceding theorem, we have:

Theorem 2 *Let G be a holomorphic self-map of E, and let $G^{[m]}$ be the mth iterate of G. Then either*

$$\mathrm{IM}\ G^{[m]}(u) \geqslant \mathrm{IM}\ u \tag{10}$$

holds for any $u \in E$ and $m = 1,2,\cdots$, or there exists a $v \in \overline{E}$ such that

$$h(v, G^{[m]}(u)) \leqslant h(v,u) \tag{11}$$

holds for any $u \in E$ and $m = 1,2,\cdots$, where

$$h(v,u) = \frac{| \mathrm{i}\{\langle u,e\rangle - \langle e,v\rangle\} + 2\langle Q_e u, v\rangle |^2}{\mathrm{IM}\ u}. \tag{12}$$

If, further, G: E → E is continuous in the weak topology of X, then we have $G(v) = v$ unless $IM\ v = 0$.

Proof Let ψ be the Cayley transformation defined by

$$\psi(x) = \frac{\mathrm{i}(x+e)}{1 - \langle x,e\rangle}$$

for any $x \in X$ with $\langle x,e\rangle \neq 1$. It is well known [7] that ψ is a biholomorphic map of B onto E. Thus $F = \psi^{-1} \circ G \circ \psi$ is a holomorphic map of B into B. This proof is then completed in a manner similar to that of Theorem 2 in [2], and therefore is omitted. ■

Acknowledgment

I wish to thank Professor Ky Fan for his helpful comments during the preparation of this short note.

References

[1] R B Burckel. Iterating analytic self-maps of discs. Amer. Math. Monthly, 1981, 88: 396 — 407.

[2] G-N Chen. Iteration for holomorphic maps of the open unit ball and the generalized upper half-plane of $\mathbf{C}^n$. J. Math. Anal, Appl., 1984, 98: 305—313.

[3] C J Earle and R S Hamilton. A fixed point theorem for holomorphic mappings. in Global Analysis. Proceedings, Symposium Pure Math., Vol. XVI. Berkeley, Calif., 1968, Amer. Math. Soc., Providence, R. I., 1970; 61 — 65.

[4] Ky Fan. Iteration of analytic functions of operators, Math. Z. 1982, 179: 293 — 298.

[5] Ky Fan. Iteration of analytic functions of operators, Ⅱ. Linear and Multilinear Algebra, 12, 1983, 13: 295 — 304.

[6] T L Hayden and T J Suffridge. Biholomorphic maps in Hilbert space have a fixed point. Pacific J. Math, 1971, 38: 419 — 422.

[7] L A Harris. Bounded symmetric homogeneous domains in infinite dimensional spaces. in Lecture Notes in Math. Vol. 364, A Dold, Heidelberg and B Eckmann, Zürich, Eds, Springer Verlag, Berlin/New York/Heidelberg, 1071.13 10.

[8] W Rudin. Function Theory in the Unit Ball of $\mathbf{C}^n$, Springer-Verlag, New York, 1980.

[9] J Wolff. Sur une généralisation d'un théorème de Schwartz. C. R. Acad. Sci. Paris, 1926, 182: 918 — 920; 1926, 183: 500 — 502.

B.-L. Lin & S. Simons, Nonlinear and Convex Analysis
1987:15－21

Von Neumann, Heinz, 与 Ky Fan 定理及其推广结论的简化证明①

On Simplified Proofs of Theorems of von Neumann, Heinz, and Ky Fan, and Their Extended Versions

Abstract This note introduces the use of a weak maximum principle in the simplified discussion of Theorems of Von Neumann, Heinz, and Ky Fan, and their extended versions.

§ 1. Introduction and Notation

Let H be a Hilbert space over the complex field $\mathbf{C}$, $H \neq \{0\}$, and denote by $B(H)$ the Banach algebra of all bounded linear operators T on H with respect to the operator norm $\|\cdot\|$, by $\widetilde{U}$ the set of all unitary operators of $B(H)$. If $T \in B(H)$, we denote by $\sigma(T)$ the spectrum of T. For a complex holomorphic function $u(\lambda)$ on a domain Ω containing $\sigma(T)$ in $\mathbf{C}$, $u(T)$ will denote an operator of $B(H)$ defined by the Riesz-Dunford integral [1, p. 568]

$$u(T) = (2\pi i)^{-1}\int_{\Gamma} u(\lambda)(\lambda I - T)^{-1}\mathrm{d}\lambda, \tag{1}$$

where I stands for the identity of $B(H)$ and Γ is any contour that surrounds $\sigma(T)$ in Ω. In [7] (see also [9, p. 437]), von Neumann proved the following

Theorem I *If T is a contraction of $B(H)$ (i. e., $\|T\| \leqslant 1$) and*

① Dedicated to Professor Ky Fan on the occasion of his retirement in June, 1985.

if u is a complex holomorphic function on a domain Ω containing the closed unit disk $\overline{\Delta}=\{\lambda : |\lambda| \leqslant 1\}$ such that $|u(\lambda)| \leqslant 1$ for all λ in $\overline{\Delta}$, then $\|u(T)\| \leqslant 1$.

In his original proof, he showed, by a result of Schur, that the general case of functions $u(\lambda)$ in question can be reduced to the particular case of Möbius transformations of the form

$$\lambda \to (\lambda+a)/(1+\bar{a}\lambda),\ |a|<1,$$

for which the theorem is demonstrated by an immediate calculation. It is well known that there exist other elegant proofs of Theorem I, most of which can be found in [1, p. 933], [2] for references. Among these are the proofs due to Heinz [4] and Ky Fan [2] by deriving Theorem I from their theorems which are stated here as Theorems Ⅱ and Ⅲ, respectively.

Theorem Ⅱ *If $\|T\| \leqslant 1$ and if v is a complex holomorphic function on a domain Ω containing $\overline{\Delta}$ such that Re $v(\lambda) \geqslant 0$ for all λ in $\overline{\Delta}$, then Re $v(T) \geqslant 0$.*

Here Re T is the real part of T in $B(H)$, i. e., Re $T=\frac{1}{2}(T+T^*)$. For two self-adjoint operators T_1 and T_2 in $B(H)$ we write $T_1 \geqslant T_2$ if $((T_1-T_2)w, w) \geqslant 0$ for all w in H; write $T_1 > T_2$ if $T_1 \geqslant T_2$ and T_1-T_2 is invertible in H.

Theorem Ⅲ *If $\|T\|<1$ and if h and k are complex holomorphic on the open unit disk $\Delta=\{\lambda: |\lambda|<1\}$ such that $|h(\lambda)|<1$ and* Re $k(\lambda)>0$ *for all λ in Δ, then $\|h(T)\|<1$ and* Re $k(T)>0$.

Besides the preceding methods, there exists a simple proof proposed by Harris [3] by considering Theorem Ⅰ as an immediate consequence of a maximum principle and the spectral theorem for unitary operators on a Hilbert space. The novelty of this approach, in contrast with the original von Neumann's proof, consists in that it refers the problem to make an estimate of the norm of $u(T)$ for a general $\|T\| \leqslant 1$ to the one of $u(V)$, where V is in $\widetilde{U}$, for which Theorem Ⅰ is valid.

In this note, we investigate a weak maximum principle (Theorem

Ⅰ in Sect. 2) and its corollary and use them to give the simplified proofs of Theorems Ⅰ, Ⅱ, Ⅲ and their extended versions of Tao [11]. All the proofs here are surprisingly simple and are almost uniform.

§ 2. A Weaker Form of The Maximum Principle

Let A denote a Banach algebra over **C** having an identity e and an involution $*$. An element x in A is said to be self-adjoint if $x = x^*$. An element x in A is said to be unitary if $xx^* = x^* x = e$. Denote by U the set of all unitary elements of A, by U_e the identity component of U. If $x \in A$, we denote by $\sigma(x)$ the spectrum of x and by $|x|_\sigma$ the spectral radius of x. An involution for A is said to be hermitian if self-adjoint elements have real spectra, and is said to be continuous if $\| x^* \| \leqslant Mx$ for all x in A and for some constant M. Define $p(x) = |x^* x|_\sigma^{1/2}$ for x in A. Let D be an open subset of A. An A-valued function $g: D \to A$ is said to be holomorphic on D if the Fréchet derivative of g at each x in D exists as a bounded complex linear self-mapping of A [10, p. 248].

It is known [3] that holomorphic functions are continuous and that the chain rule holds. For a subset S of A, let $\overline{\text{co}}\, S$ denote the closed convex hull of S, and Co(S) the (closed) holomorphic convex hull of S (see [3, p. 11]).

The following basic theorem contains a weakened form of the maximum principle of Harris [3, Theorem 8], however, it is quite enough for our purposes.

Theorem 1 *Put* $A_1 = \{x \in A: |x|_\sigma \leqslant 1, p(x) \leqslant 1\}$ *and let* D *be an open subset of* A *containing* $r\,\overline{\text{co}}\, U_e$ *for some* r, $0 < r \leqslant 1$. *If* $h: D \to A$ *is a holomorphic function, then*

$$A_1 \subset \overline{\text{co}}\, U_e, \tag{2}$$

$$h(rA_1) \subset \overline{\text{co}}\, h(rU_e), \tag{3}$$

$$\| h(rx) \| \leqslant \sup\{ \| h(ry) \| : y \in U_e\} \leqslant \infty \text{ for all } x \text{ in } A_1, \tag{4}$$

and

$p(h(rx)) \leqslant \sup\{p(h(ry)): y \in U_e\} \leqslant \infty$ for all x in A_1 whenever the involution for A is hermitian and continuous. (5)

Moreover, $h|_{rA_1}$ is completely determined by its values on rU_e.

Proof Observe that both (2) and (3) of Theorem 1 with $r=1$ can be deduced from Theorem 8 of [3] together with the fact that $\mathrm{Co}(S) \subset \overline{\mathrm{co}}\, S$ for each subset S of A. We, however, prefer to sketch a direct proof as follows. Clearly we may assume that $|x|_\sigma < 1$ and $p(x) < 1$. To prove (3), consider first the case of $r=1$. Then, $D \supset \overline{\mathrm{co}}\, U_e$. Let $\ell \in A^*$, the conjugate space of A, and define $F(\lambda) = \exp(\ell(h(f(\lambda))))$ for $|\lambda| \leqslant 1$. (First take as h the identity function on A to obtain (2).) Here f is the characteristic function defined by

$$f(\lambda) = (e - xx^*)^{-1/2}(\lambda e + x)(e + \lambda x^*)^{-1}(e - x^* x)^{1/2}, \tag{6}$$

which is holomorphic in a neighborhood of $\overline{\Delta}$, $f(0) = x$, and maps the unit circle in $\mathbf{C}$ into U_e [3, p. 2]. By the chain rule and the maximum modulus principle,

$$|F(0)| \leqslant \sup_{|\lambda|=1} |F(\lambda)|,$$

and therefore

$$\mathrm{Re}\, \ell(h(x)) \leqslant \sup_{|\lambda|=1} \mathrm{Re}\, \ell(h(f(\lambda))) \leqslant \sup \mathrm{Re}\, \ell(h(U_e)).$$

It then follows that $h(x) \in \overline{\mathrm{co}}\, h(U_e)$, for, if not, by a separation theorem [1, p. 417] there exist constants c and t, $t > 0$, and $\ell_1 \in A^*$ such that

$$\sup \mathrm{Re}\, \ell_1(\overline{\mathrm{co}}\, h(U_e)) \leqslant c - t < c \leqslant \mathrm{Re}\, \ell_1(h(x)),$$

a contradiction, so that (3) holds when $r=1$. In the general case of $0 < r \leqslant 1$, the function k defined by $k(y) = h(ry)$ is holomorphic on $r^{-1}D$ into A. (3) now follows directly from the result just proved and from the convexity of balls. Obviously, (4) is a simple consequence of (3) whether $\overline{\mathrm{co}}\, U_e$ is bounded in A or not. If the involution for A is hermitian and continuous, we have ([3], [8]) that p is an algebra pseudonorm on A, and is continuous on A, whence (5) is valid by (3). For the last part of the theorem, as before, it will be enough to verify the result in the case $r=1$. But this follows directly from the mean value property for vector-valued holomorphic functions [5, p. 99]:

$$h(x) = h(f(0)) = (2\pi)^{-1}\int_0^{2\pi} h(f(e^{i\theta}))d\theta \tag{7}$$

for each x in A with $|x|_\sigma < 1$ and $p(x) < 1$, where f is as in (6). That completes the proof of the theorem. ■

Remark We note that there exists and alternative proof of Theorem 1 by deriving it from the formula (7).

Corollary 1 *Suppose that A has hermitian and continuous involution. Let Ω be a domain containing $r\overline{\Delta}$ for some r, $0 < r \leqslant 1$, and define A_1 as in Theorem 1 and $A_\Omega = \{x \in A : \sigma(x) \subset \Omega\}$. If k is a complex holomorphic function on Ω, then (3) ~ (5) of Theorem 1 are valid with D replaced by A_Ω and $h(x)$ by the function $k(x)$ defined by*

$$k(x) = (2\pi i)^{-1}\int_\Gamma k(\lambda)(\lambda e - x)^{-1}d\lambda, \tag{8}$$

where Γ is any contour that surrounds $\sigma(x)$ in Ω.

Proof It is well known [10]that A_Ω is open in A and that the function k defined by(8) is a continuously differentiable mapping of A_Ω into A, so that $k: A_\Omega \to A$ is holomorphic by definition. In virtue of the properties of p given above, we have $A_\Omega \supset r\,\overline{\text{co}}\,U_e$ whenever $\Omega \supset r\,\overline{\Delta}$. Corollary 1 now follows from Theorem 1. ■

§ 3. Simplified Discussion of Theorems Ⅰ, Ⅱ, Ⅲ, and Their Extended Versions

In this section we shall introduce the use of the results of the preceding section in the simplified proofs of Theorems Ⅰ, Ⅱ, Ⅲ, and their extended versions of Tao [11]. A result (see [6, p. 109] and [10, p. 309])of fundamental importance for us is the following

Lemma Ⅰ *Let E be the spectral resolution of the identity for a normal operator V in B(H), and suppose that g is a complex holomorphic function on a domain Ω containing $\sigma(rV)$ in* **C** *for some r, $0 < r \leqslant 1$. Then*

$$g(rV) = \int_{\sigma(V)} g(rt)dE(t), \tag{9}$$

where the integral exists with respect to the norm topology in $B(H)$, *and*

$$\| g(rV) \| = \sup\{ | g(r\lambda) | : \lambda \in \sigma(V) \}. \tag{10}$$

We note that the involution on $B(H)$, the Hilbert space adjoint, is clearly hermitian and continuous, and that $| T |_\sigma \leqslant p(T) = \| T \|$, so that $B(H)_1 = \{ T \in B(H) : \| T \| \leqslant 1 \}$.

Proof of Theorem Ⅰ Observe that $\Omega \supset \overline{\Delta} \supset \sigma(V)$ for each V in $\widetilde{U}$ since $\sigma(V)$ lies on the unit circle. By Corollary 1 with $r = 1$ and by (10),

$$\| u(T) \| \leqslant \sup\{ \| u(V) \| : V \in \widetilde{U} \} \leqslant \sup\{ | u(\lambda) | : | \lambda | = 1 \} \leqslant 1,$$

as desired. ■

Proof of Theorem Ⅱ By (9), with $r = 1$ and since $v(V)^* = \overline{v}(V)$ for each V in $\widetilde{U}$, we have

$$\operatorname{Re} v(V) = \int_0^{2\pi} \operatorname{Re} v(e^{i\theta}) dE(\theta),$$

and therefore

$$(\operatorname{Re} v(V)w, w) = \int_0^{2\pi} \operatorname{Re} v(e^{i\theta}) dE_{w,w}(\theta)$$

for all w in H, where $E_{w,w}(\theta) = (E(\theta)w, w)$. It is well known [10] that each $E_{w,w}$ is a positive measure on $\sigma(V)$ whose total variation is $\| w \|^2$. Since $\operatorname{Re} v(\lambda) \geqslant 0$ is continuous on the compact set $\sigma(V)$ in $\mathbf{C}$, $(\operatorname{Re} v(V)w, w) \geqslant 0$ for each w in H, i. e., $\operatorname{Re} v(V) \geqslant 0$ by definition. The result now follows directly from (3) of Corollary 1. ■

Proof of Theorem Ⅲ If T is a proper contraction of $B(H)$, there exist real $r, 0 < r < 1$, and a contraction $\hat{T}$ of $B(H)$ such that $T = r\hat{T}$. Thus, as before, $\| h(T) \| \leqslant \sup\{ | h(r\lambda) | : | \lambda | = 1 \} < 1$. By applying a similar argument to that used in the preceding theorem, we can obtain $\operatorname{Re} k(T) > 0$ for each $\| T \| < 1$. ■

It is natural to expect that the proofs of the extended versions of Theorems Ⅰ, Ⅱ, and Ⅲ, given by Tao [11], should be also greatly simplified by the same approach mentioned above. These are indeed the case. For notations, let Ω be a domain in $\mathbf{C}$ and denote by $N_H(\Omega)$ the set of all $B(H)$-valued holomorphic functions $\hat{g}$ on Ω such that $\hat{g}(\beta)\hat{g}(\lambda) = \hat{g}(\lambda)\hat{g}(\beta)$ and $\hat{g}(\lambda)$ is normal in $B(H)$ for all λ, β in Ω. Recall [11] that $\hat{g}(T)$ denotes a $B(H)$-valued function, with domain $B(H)_\Omega = \{T \in B(H): \sigma(T) \subset \Omega\}$, that arise from a $B(H)$-valued function $\hat{g}$ on Ω by the formula

$$\hat{g}(T) = (2\pi i)^{-1}\int_\Gamma \hat{g}(\lambda)(\lambda I - T)^{-1}\mathrm{d}\lambda, \tag{11}$$

where Γ is as in (1).

Theorem Ⅳ *Suppose that $\hat{u}$, $\hat{v}$ are in $N_H(\Omega)$ and $\hat{h}$ is in $N_H(\Delta)$, each of which commutes with T in $B(H)$. Then Theorems* Ⅰ, Ⅱ, *and* Ⅲ *hold if u, v, and h therein are replaced by $\hat{u}$, $\hat{v}$ and $\hat{h}$, respectively.*

Proof consider only the extended result with respect to Theorem Ⅰ. In a similar way one can verify the others. By a slight extension of Theorem 10. 38 in [10], we can assert that the function $\hat{g}(T)$ defined by (11) is a holomorphic mapping of $B(H)_\Omega$ into $B(H)$. Instead of Lemma I we use an analogous lemma, that is Lemma 6. 2 of [11]. Thus, as before, we are able to conclude that the extended result is valid. ■

we conclude with the remark that each of Theorems 3. 3, 3. 4(2) ~ (4), and 3. 5 of [11] may be shown to be valid by means of the same methods, and is merely an immediate consequence of Theorem 1 and Corollary 1 together with either Lemma I or Lemma 6. 2 of [11].

References

[1] N Dunford and J T Schwartz. Linear Operators. Interscience, New York, pt. Ⅰ, 1958; pt. Ⅱ, 1963.

[2] Ky Fan. Analysis functions of a proper contraction. Math. z., 1978, 160: 275—290.

[3] L Harris. Banach algebras with involution and Möbius transformations. J. Functional Analysis, 1972, 11: 1—16.

[4] E Heinz. Ein v. Neumannscher Satz über beschränkte Operatoren im Hilbertschen Raum. Nachr. Akad. Wiss. Göttingen Math.-Phys. Kl. Ⅱ, 1952, 5—6.

[5] E Hille and R S Phillips. Functional Analysis and Semigroups. Rv. Ed.,

Amer. Math. Soc. ,Providence,R I. 1957.

[6]B Sz-Nagy and C Foias. Harmonic Analysis of Operators on a Hilbert Space. North-Holland,Amsterdam,1970.

[7]J Von Neumann. Eine Spektraltheorie für allgemeine Operatoren eines unitären Raumes. Math. Nachr. ,1951,4:258－281.

[8]V Pták. On the spectral radius in Banach algebras with involution. Bull. London Math. Soc. ,1970,2:327－334.

[9]F Riesz and B Sz-Nagy. Functional Analysis. Fredrick Ungar,New York, 1955.

[10] W Rudin. Functional Analysis. McGraw-Hill,New York,1973.

[11]Z-G Tao. Analytic operator functions, J. Math. Anal. Appl. , 1984,103:293－320.

三、特殊矩阵及其应用，有理插值问题

Ⅲ. Special Matrices and Their Applications，Rational Interpolation Problem

北京师范大学学报(自然科学版)
1990,(4):14-18

多项式零点定位基本定理的简化证明①

Short Proofs for Classical Theorems on Root Localization of Polynomials

摘要 基于 Djaferis 与 Mitter, Wimmer 的模论方法,给出关于 Routh-Hurwitz 问题与 Schur-Cohn 问题的基本定理的简化证明.

关键词 多项式惯性;模论方法;Bezout 矩阵

§1. 引言与预备知识

近年来有许多文章[1~8]及其引用的文献讨论如下 2 种分别称为 Lyapunov 方程与 Stein 方程的线性矩阵方程:

$$\boldsymbol{AX}-\boldsymbol{XB}=\boldsymbol{C}, \tag{1}$$

$$\boldsymbol{X}-\boldsymbol{AXB}=\boldsymbol{C}, \tag{2}$$

式中,$\boldsymbol{A}\in M_m(F)$(即代数闭域 F 上 m 阶矩阵的空间),$\boldsymbol{B}\in M_n(F)$,$\boldsymbol{C}\in M_{m,n}(F)$均为给定矩阵,$\boldsymbol{X}\in M_{m,n}(F)$待求.这 2 种方程在矩阵或多项式稳定性与惯性理论中起着重要作用.我们说矩阵 $\mathbf{A}\in M_m(F)$是稳定的,假如它的谱 $\sigma(\mathbf{A})\subset\mathbf{C}_{左}=\{\lambda\in\mathbf{C}:\mathrm{Re}\,\lambda<0\}$.$\mathbf{A}$(关于虚轴)的惯性为 $\mathrm{In}\,\mathbf{A}=\{\pi(\mathbf{A}),\nu(\mathbf{A}),\delta(\mathbf{A})\}$,这里,$\pi(\mathbf{A})$,$\nu(\mathbf{A})$与 $\delta(\mathbf{A})$分别表示位于虚轴右半平面 $\mathbf{C}_{右}$,$\mathbf{C}_{左}$与虚轴上 $\mathbf{A}$ 的特征值个数(计入重数).因此,$\mathbf{A}$ 稳定当且仅当$\nu(\mathbf{A})=m$.复系数的多项式

$$f(x)=f_0+f_1x+\cdots+f_{n-1}x^{n-1}+f_nx^n,f_n\neq 0 \tag{3}$$

① 收稿日期:1990-05-20.

(关于虚轴)的惯性,记为 In(f),指的是 f 的友阵

$$C_f=\begin{bmatrix}0 & 1 & 0 & \cdots & 0\\ 0 & 0 & 1 & \cdots & 0\\ \vdots & \vdots & \vdots & & \vdots\\ 0 & 0 & 0 & \cdots & 1\\ -f_0/f_n & -f_1/f_n & -f_2/f_n & \cdots & -f_{n-1}/f_n\end{bmatrix}\in M_n(F) \quad (4)$$

的惯性:$\mathrm{In}(f)=\mathrm{In}(\boldsymbol{C}_f)$,即位于 $\mathbf{C}_{右}$,$\mathbf{C}_{左}$与虚轴上 f 零点个数(计入重数)的三重数 $\mathrm{In}f=\{\pi(f),\nu(f),\delta(f)\}$. 类似地可定义 $\boldsymbol{A}\in M_m(F)$(或$f(x)$)关于实轴与关于单位圆周的惯性,记为 $\mathrm{In}'(\boldsymbol{A})$(或 $\mathrm{In}'(f)$)与 $\mathrm{In}^\circ(\boldsymbol{A})$(或 $\mathrm{In}^\circ(f)$). 如令 $\tilde{f}(x)=f(-\mathrm{i}x)$,$\mathrm{i}^2=-1$,则有

$$\mathrm{In}(-\mathrm{i}\boldsymbol{A})=\mathrm{In}'(\boldsymbol{A}),\quad \mathrm{In}(\boldsymbol{A})=\mathrm{In}'(\mathrm{i}\boldsymbol{A}),\quad \mathrm{In}(f)=\mathrm{In}'(\tilde{f}).$$

本文旨在利用求解方程(1)与(2)的模论方法[3,8]简化证明多项式惯性理论中最基本的定理及其推广. 为此先引入一些预备知识.

矩阵 $\boldsymbol{M}=(m_{ij})_{i,j=0}^{l-1}\in M_l(F)$的生成函数为

$$\sum_{i,j=0}^{l-1} m_{ij}x^iy^j. \quad (5)$$

给定多项式对 $f(x)=\sum\limits_{j=0}^{l} f_jx^j$ ($f_l\neq 0$)与 $g(x)=\sum\limits_{j=0}^{l} g_jx^j$,它们的 Bezout 矩阵为 l 阶矩阵 Bez(f,g),假如此矩阵的生成函数为

$$\frac{f(x)g(y)-g(x)f(y)}{x-y}. \quad (6)$$

按此可证,$\mathrm{Bez}(f,g)=-\mathrm{Bez}(g,f)$,$\mathrm{Bez}(f,g)=\mathrm{Bez}(f,g)^{\mathrm{T}}$,

$$\mathrm{Bez}(f,g)=S(f)g(C_f), \quad (7)$$

式中,$S(f)$为 $f(x)$的对称化子:

$$S(f)=\begin{bmatrix}f_1 & f_2 & \cdots & f_l\\ f_2 & \cdots & f_l & 0\\ \vdots & \therefore & & \vdots\\ f_l & 0 & \cdots & 0\end{bmatrix}\in M_l(F). \quad (8)$$

命题 1(Sylvester **惯性律**) 设 $\boldsymbol{A}$ 与 $\tilde{\boldsymbol{A}}$ 为 n 阶 Hermite 矩阵,$\boldsymbol{C}$ 为非奇异矩阵. 如果 $\boldsymbol{A}=\boldsymbol{C}^*\tilde{\boldsymbol{A}}\boldsymbol{C}$,那么有 $\mathrm{In}(\boldsymbol{A})=\mathrm{In}(\tilde{\boldsymbol{A}})$.

命题 2[4] 设 $f(x)=\sum\limits_{j=0}^{l} f_jx^j(f_l=1)$与 $\bar{f}(x)=\sum\limits_{j=0}^{l}\bar{f}_jx^j$ 的最大公

因式为 $d(x)$：$f(x)=f_0(x)d(x)$，$\bar{f}(x)=\bar{f}_0(x)d(x)$. 则 $d(x)$ 为实多项式，且有

$$\mathrm{In}(\mathrm{i}^{-1}\mathrm{Bez}(f,\bar{f}))=\mathrm{In}(\mathrm{i}^{-1}\mathrm{Bez}(f_0,\bar{f}_0))+(0,0,\deg d),$$

式中，$\deg d$ 为 $d(x)$ 的次数，$\det \mathrm{Bez}(f_0,\bar{f}_0)\neq 0$.

命题 3[8] 设方程(1)中 $\boldsymbol{A}$ 与 $\boldsymbol{B}$ 无公共特征值，且 $\boldsymbol{C}=\boldsymbol{LR}$，$\boldsymbol{L}\in M_{m,t}(F)$，$\boldsymbol{R}\in M_{t,n}(F)$，$r=\max(m,n)$，$S_r(\boldsymbol{A},\boldsymbol{L})=[\boldsymbol{L}\quad \boldsymbol{AL}\quad \cdots\quad \boldsymbol{A}^{r-1}\boldsymbol{L}]\in M_{m,rt}(F)$，

$$O_r(\boldsymbol{R},\boldsymbol{B})=\begin{bmatrix}\boldsymbol{R}\\ \boldsymbol{RB}\\ \vdots\\ \boldsymbol{RB}^{r-1}\end{bmatrix}\in M_{rt,n}(F).$$

若 rank $\boldsymbol{C}=1$ 与 $t=1$，则方程(1)有唯一解 $\boldsymbol{X}$：

$$\boldsymbol{X}=-b(\boldsymbol{A})^{-1}S_m(\boldsymbol{A},\boldsymbol{L})\mathrm{Bez}(a,b)O_n(\boldsymbol{B},\boldsymbol{R})a(\boldsymbol{B})^{-1},\tag{9}$$

其中，$a(x)=\det(x\boldsymbol{I}-\boldsymbol{A})$，$b(y)=\det(y\boldsymbol{I}-\boldsymbol{B})$.

命题 4[5] 设 $\boldsymbol{A}\in M_m(F)$ 与 $\boldsymbol{L}\in M_{m,t}(F)$ 组成的矩阵对 $(\boldsymbol{A},\boldsymbol{L})$ 可控制，即有 rank $S_m(\boldsymbol{A},\boldsymbol{L})=m$. 若 $\boldsymbol{X}=\boldsymbol{X}^*$ 满足 $\boldsymbol{AX}+\boldsymbol{XA}^*=\boldsymbol{LL}^*$，则

$$\mathrm{In}(\boldsymbol{A})=\mathrm{In}(\boldsymbol{X}),\quad \delta(\boldsymbol{A})=\delta(\boldsymbol{X})=0.$$

对于方程(2)，命题 3 与 4 的相应结果也成立[5,8].

§2. Routh-Hurwitz 与 Schur-Cohn 问题基本定理的推广形式的简化证明

定理 1(Hermite) 设 $a(x)=\sum_{j=0}^{l}a_jx^j\ (a_l=1)$，$\bar{a}(x)=\sum_{j=0}^{l}\bar{a}_jx^j$，$\boldsymbol{B}=\mathrm{i}^{-1}\mathrm{Bez}(a,\bar{a})$，则有

$$\mathrm{In}'(a)=\mathrm{In}(\boldsymbol{B})+(q(a),q(a),-2q(a)),\tag{10}$$

这里 $q(a)$ 表示 $a(x)$ 复共轭零点对的个数(计入重根). 若 $\det \boldsymbol{B}\neq 0$，则(10)式中 $q(a)=0$.

证 先讨论 $\det \boldsymbol{B}\neq 0$ 的情形. 这时因 a 与 $\bar{a}$ 无公共零点，$\boldsymbol{C}_a$ 与 $\boldsymbol{C}_a^*$ 的特征多项式分别为 $a(x)=\det(x\boldsymbol{I}-\boldsymbol{C}_a)$ 与 $\bar{a}(x)=\det(x\boldsymbol{I}-\boldsymbol{C}_a^*)$，按命题 3，

$$\boldsymbol{C}_a^*\boldsymbol{X}-\boldsymbol{XC}_a=\boldsymbol{e}_1\boldsymbol{e}_1^*,\quad \boldsymbol{e}_1=(1,0,\cdots,0)^{\mathrm{T}}\in\mathbf{C}^n\tag{11}$$

有唯一解 $\boldsymbol{X}$：

$$\boldsymbol{X}=-a(\boldsymbol{C}_a^*)^{-1}S_l(\boldsymbol{C}_a^*,\boldsymbol{e}_1)\mathrm{Bez}(\bar{a},a)O_l(\boldsymbol{C}_a,\boldsymbol{e}_1^*)\bar{a}(\boldsymbol{C}_a)^{-1},$$

但是，$S_l(\boldsymbol{C}_a^*,\boldsymbol{e}_1)=[\boldsymbol{e}_1\quad \boldsymbol{C}_a^*\boldsymbol{e}_1\quad \cdots\quad \boldsymbol{C}_a^{*\,l-1}\boldsymbol{e}_1]=I_l$，$O_l(\boldsymbol{C}_a,\boldsymbol{e}_1^*)=I_l$，因而上式变为

$$\boldsymbol{X}=a(\boldsymbol{C}_a^*)^{-1}\mathrm{Bez}(a,\bar{a})\bar{a}(\boldsymbol{C}_a)^{-1}.$$

代入(11)并经变换后得到

$$\boldsymbol{C}_a^*\mathrm{Bez}(a,\bar{a})-\mathrm{Bez}(a,\bar{a})\boldsymbol{C}_a=a(\boldsymbol{C}_a^*)\boldsymbol{e}_1\boldsymbol{e}_1^*\bar{a}(\boldsymbol{C}_a).$$

显然有 $\boldsymbol{B}=\boldsymbol{B}^*$，$(-\mathrm{i}\boldsymbol{C}_a^*,a(\boldsymbol{C}_a^*)\boldsymbol{e}_1)$可控制，依据命题 4，我们从

$$(-\mathrm{i}\boldsymbol{C}_a)^*\boldsymbol{B}+\boldsymbol{B}(-\mathrm{i}\boldsymbol{C}_a)=a(\boldsymbol{C}_a^*)\boldsymbol{e}_1\boldsymbol{e}_1^*\bar{a}(\boldsymbol{C}_a)$$

可推出 $\mathrm{In}(-\mathrm{i}\boldsymbol{C}_a)=\mathrm{In}(-\mathrm{i}\boldsymbol{C}_a)^*=\mathrm{In}(\boldsymbol{B})$，$\delta(-\mathrm{i}\boldsymbol{C}_a)=\delta(\boldsymbol{B})=0$，因而

$$\mathrm{In}'(\boldsymbol{C}_a)=\mathrm{In}(-\mathrm{i}\boldsymbol{C}_a)=\mathrm{In}'(a)=\mathrm{In}(\boldsymbol{B}).$$

现在考虑 $\det\boldsymbol{B}=0$ 的情形. 设 $d(x)$为 $a(x)$与 $\bar{a}(x)$的最大公因式：$a(x)=a_0(x)d(x)$，$\bar{a}(x)=\bar{a}_0(x)d(x)$，并记 $\boldsymbol{B}_0=\mathrm{i}^{-1}\mathrm{Bez}(a_0,\bar{a}_0)$. 由命题 2，$\mathrm{In}(\boldsymbol{B})=\mathrm{In}(\boldsymbol{B}_0)+(0,0,\deg d)$，其中 $\det\boldsymbol{B}_0\neq 0$. 但按刚才证明，$\mathrm{In}'(a_0)=\mathrm{In}(\boldsymbol{B}_0)$，因而 $\delta(\boldsymbol{B})=\deg d$. 另一方面，由于 $a_0(x)$无 $a(x)$的实根与复共轭根对 $\lambda,\bar{\lambda}$(否则容易验证 $a_0(x)$与 $\bar{a}_0(x)$有公共零点 $\bar{\lambda}$)，故有 $\deg d=\delta'(a)+2q(a)$，因而 $\delta'(a)=\delta(\boldsymbol{B})-2q(a)$，并且，$\pi'(a)=\pi'(a_0)+q(a)=\pi(\boldsymbol{B})+q(a)$，$\nu'(a)=\nu'(a_0)+q(a)=\nu(\boldsymbol{B})+q(a)$. 定理证毕. ■

定理 2(Hermite-Fujiwara)　设 $a(x)=\sum_{j=0}^{l}a_jx^j\ (a_l=1)$，$\hat{a}(x)=\bar{a}(-x)$，$\widetilde{\boldsymbol{B}}=\mathrm{Bez}(a,\hat{a})\boldsymbol{D}$ 为 $a(x)$的 Hermite-Fujiwara 矩阵，这里 $\boldsymbol{D}=\mathrm{diag}[1,(-1),1,\cdots,(-1)^{l-1}]$. 则有

$$\mathrm{In}(a)=\mathrm{In}(-\widetilde{\boldsymbol{B}})+(q_1(a),q_1(a),-2q_1(a)),\tag{12}$$

这里 $q_1(a)$表示 $a(x)$关于虚轴对称零点对的个数(计入重根).

证　将前一定理用于多项式对 $a_1(x)\equiv a(-\mathrm{i}x)$与 $\bar{a}_1(x)=\bar{a}(\mathrm{i}x)$，取 $\boldsymbol{B}=\mathrm{i}^{-1}\mathrm{Bez}(a_1,\bar{a}_1)$，便得

$$\mathrm{In}(a)=\mathrm{In}'(a_1)=\mathrm{In}(\boldsymbol{B})+(q_1(a),q_1(a),-2q_1(a)),\tag{13}$$

这里 $q_1(a)$为 $a_1(x)$复共轭零点对的个数. 但另一方面可以直接验证，对 $\boldsymbol{F}=\mathrm{diag}[\mathrm{i}^{l-1},\mathrm{i}^{l-2},\cdots,\mathrm{i},1]$，$\boldsymbol{F}^*\boldsymbol{B}\boldsymbol{F}=-\widetilde{\boldsymbol{B}}$. 按命题 1，$\mathrm{In}(\boldsymbol{B})=\mathrm{In}(-\widetilde{\boldsymbol{B}})$，因而由(13)式得(12). 证毕. ■

下面结论是一个关于实多项式的扰动定理，应用它可以简化著名的 Liénard-Chipart 稳定性准则的证明. 我们采用前面的办法可得到这个扰动定理的简单证明，它比原先结果更一般.

定理 3(Lancaster-Tismenetsky)[7] 设 $a(x)=\sum_{j=0}^{l}a_jx^j\ (a_l=1)$ 与 $b(x)=\sum_{j=0}^{l-1}b_jx^j$ 均为实多项式,$\tilde{a}(x)=a(x)+\mathrm{i}b(x)$,$\boldsymbol{B}=\mathrm{Bez}(a,b)$,则有

$$\delta'(\tilde{a})\leqslant\delta'(a),$$

$$\mathrm{In}'(\tilde{a})=\mathrm{In}(-\boldsymbol{B})+(q(\tilde{a}),q(\tilde{a}),-2q(\tilde{a})). \tag{14}$$

若还有 $\det\boldsymbol{B}\neq0$,则 $\delta'(\tilde{a})=0$,

$$\pi'(a)\leqslant\pi'(\tilde{a})=\pi(-\boldsymbol{B}),\nu'(a)\leqslant\nu'(\tilde{a})=\nu(-\boldsymbol{B}),$$

且当 $\delta'(a)=0$ 时,$\mathrm{In}'(a)=\mathrm{In}'(\tilde{a})$.

证 按照题设,$\tilde{a}(x)$ 为 l 次首一多项式,$\bar{\tilde{a}}(x)=a(x)-\mathrm{i}b(x)$,$\mathrm{i}^{-1}\mathrm{Bez}(\tilde{a},\bar{\tilde{a}})=-2\mathrm{Bez}(a,b)$. $\delta'(\tilde{a})\leqslant\delta'(a)$是显然的,因为 $\tilde{a}(x)$的实根必为$a(x)$的实根. 应用定理 1 即得(14)式.

当 $\det\boldsymbol{B}\neq0$ 时,从 $\tilde{a}(x)$无实根与复共轭根(否则 a 与 b 有公共零点)即得 $\delta'(\tilde{a})=q(\tilde{a})=0$. 余下结论可由[5]中定理 13.1.3 与命题 13.1.2 直接导出. 证毕. ■

本段最后一个定理可视为多项式惯性理论中 Schur-Cohn 问题最基本结果的推广.

定理 4(Schur-Cohn) 设 $a(x)$与 $\bar{a}(x)$同定理 1,$\hat{a}(x)=x^l\bar{a}(x^{-1})$($\bar{a}$的倒易多项式),$\breve{\boldsymbol{B}}=\boldsymbol{J}\ \mathrm{Bez}(a,\hat{a})$为 $a(x)$的 Schur-Cohn 矩阵,这里 $\boldsymbol{J}=(\delta_{i,l-1-j})_{i,j=0}^{l-1}$为 l 阶翻转矩阵. 则有

$$\mathrm{In}^{\circ}(a)=\mathrm{In}(\breve{\boldsymbol{B}})+(q_2(a),q_2(a),-2q_2(a)), \tag{15}$$

其中,$q_2(a)$表示 $a(x)$满足 $\bar{\lambda}=\mu^{-1}$的零点对 λ,μ 的个数. 若 $\mathrm{Bez}(a,\hat{a})$非奇异,则 $q_2(a)=0$.

证 先考虑 $\det\mathrm{Bez}(a,\hat{a})\neq0$ 的情形. 用类似于定理 1 的证法,由方程 $\boldsymbol{X}-\boldsymbol{C}_a^*\boldsymbol{X}\boldsymbol{C}_a=\boldsymbol{e}_1\boldsymbol{e}_1^*$ 可得唯一解 $\boldsymbol{X}=\hat{a}(\boldsymbol{C}_a^*)^{-1}\breve{\boldsymbol{B}}\hat{\bar{a}}(\boldsymbol{C}_a)^{-1}$,此蕴涵 $\mathrm{In}(\boldsymbol{X})=\mathrm{In}(\breve{\boldsymbol{B}})$. 但矩阵对$(\boldsymbol{C}_a^*,\boldsymbol{e}_1)$可控制,$\breve{\boldsymbol{B}}$为 Hermite 矩阵,应用类似于命题 4 的结果可得

$$\mathrm{In}(\boldsymbol{X})=\mathrm{In}(\breve{\boldsymbol{B}})=\mathrm{In}^{\circ}(\boldsymbol{C}_a^*)=\mathrm{In}^{\circ}(\boldsymbol{C}_a)=\mathrm{In}^{\circ}(a).$$

当 $\mathrm{Bez}(a,\hat{a})$奇异时,$a(x)$与$\hat{a}(x)$有公共零点. 记 $d(x)$为 a 与$\hat{a}$的最大公因式:$a(x)=a_0(x)d(x)$,$\hat{a}(x)=\hat{a}_0(x)d(x)$,则容易看出,$a_0(x)$不含

$a(x)$的互为共轭倒数的零点，因而$a(x)$的模为1或互为共轭倒数的零点都在$d(x)$的零点集合内. 同时$d(x)=\hat{\bar{d}}(x)$，即对称多项式.

根据 Schur-Cohn 矩阵$\check{B}$相合于与某个l次首一多项式$b(x)$相对应的 Hermite 矩阵 $i^{-1}\,\mathrm{Bez}(b,\bar{b})$的著名事实以及$q_2(a)=q(b)$的结果[4]，由定理1与命题1即得(15)式. 证毕. ■

参考文献

[1] Chen Ch T. A generalization of the inertia theorem. SIAM J Appl Math, 1973,25:158.

[2] Datta B N. Applications of Hankel matrices of Markov parameters to the solutions of the Routh-Hurwitz and the Schur-Cohn problems. J Math Anal Appl, 1978,68:276.

[3] Djaferis T E, Mitter S K. Algebraic methods for the study of some linear matrix equations. Linear Algebra Appl, 1982,44:125.

[4] Heinig G, Rost K. Algebraic methods for Toeplitz-like matrices and operators. Berlin: Akademie-Verlag,1984.

[5] Lancaster P, Tismenetsky M. The theory of matrices with applications. 2nd ed. New York: Academic Press, 1985.

[6] Lancaster P, et al. Factored forms for solutions of $AX-XB=C$ and $X-AXB=C$ in companion matrices. Linear Algebra Appl, 1984,62:19.

[7] Lancaster P, Tismenetsky M. Inertia characteristics of self-adjoint matrix polynomials. Linear Algebra Appl, 1983, 52/53:479.

[8] Wimmer H K. Linear matrix equations: The module theoretic approach. Linear Algebra Appl, 1989, 120:149.

SHORT PROOFS FOR CLASSICAL THEOREMS ON ROOT LOCALIZATION OF POLYNOMIALS

Abstract Short proofs for classical theorems on Routh-Hurwitz and Schur-Cohn problems are given, based on the module theoretic approach of Djaferis and Mitter, and Wimmer.

Keywords inertia of polynomial; module theoretic approach; Bezout matrix

北京师范大学学报(自然科学版)
1991,27(2):152－157

关于矩阵惯性的若干基本定理①

Some Basic Theorems on the Inertia for Matrices

摘要 应用数学归纳法给出主要惯性定理及其推广的初等证明,并讨论有关的推论.

关键词 矩阵的惯性;主要惯性定理;稳定矩阵

§1. 预备知识

用 $\mathbf{C}^{n\times n}$ 表示 n 阶复矩阵的全体. 给定 $\boldsymbol{A}\in\mathbf{C}^{n\times n}$,它的惯性定义为三元整数组 $\mathrm{In}\,\boldsymbol{A}=(\pi(\boldsymbol{A}),\nu(\boldsymbol{A}),\delta(\boldsymbol{A}))$,这里,$\pi(\boldsymbol{A})$,$\nu(\boldsymbol{A})$ 与 $\delta(\boldsymbol{A})$ 分别表示具有正实部、负实数与零实部的 $\boldsymbol{A}$ 的特征值个数(计入重数). 显然,$\pi(\boldsymbol{A})+\nu(\boldsymbol{A})+\delta(\boldsymbol{A})=\boldsymbol{A}$ 的阶数. 设 $\boldsymbol{A},\boldsymbol{B}\in\mathbf{C}^{n\times n}$,用 $\mathrm{In}\,\boldsymbol{A}\leqslant\mathrm{In}\,\boldsymbol{B}$ 表示

$$\pi(\boldsymbol{A})\leqslant\pi(\boldsymbol{B}),\quad \nu(\boldsymbol{A})\leqslant\nu(\boldsymbol{B}).$$

矩阵惯性理论为矩阵稳定性理论的直接推广. 它有如下 2 个最古老的经典定理:

定理 1[1] 若 $\boldsymbol{Q}>\boldsymbol{O}$(Hermite 正定)且 $\boldsymbol{H}$ 为与 $\boldsymbol{Q}$ 同阶的 Hermite 矩阵,则 $\mathrm{In}(\boldsymbol{QH})=\mathrm{In}\,\boldsymbol{H}$.

定理 2[2] 设 $\boldsymbol{A}\in\mathbf{C}^{n\times n}$. 则当且仅当存在 Hermite 矩阵 $\boldsymbol{H}\in\mathbf{C}^{n\times n}$ 使得 $\boldsymbol{H}>\boldsymbol{O}$ 与 $\mathrm{Re}(\boldsymbol{AH})=(\boldsymbol{AH}+\boldsymbol{HA}^*)/2>\boldsymbol{O}$ 时,$\boldsymbol{A}$ 为正稳定矩阵,即有

$$\mathrm{In}\,\boldsymbol{A}=(n,0,0).$$

说明一下,定理 1 等价于熟知的 Sylvester 惯性律,后者断言:对任意

① 收稿日期:1990-11-15.

国家教委博士点基金资助项目.

Hermite 矩阵 $\boldsymbol{H}\in\mathbf{C}^{n\times n}$ 与可逆矩阵 $\boldsymbol{S}\in\mathbf{C}^{n\times n}$，$\mathrm{In}(\boldsymbol{S}^*\boldsymbol{HS})=\mathrm{In}\,\boldsymbol{H}$. 事实上，从

$$\mathrm{In}(\boldsymbol{S}^*\boldsymbol{HS})=\mathrm{In}(\boldsymbol{SS}^*\boldsymbol{HSS}^{-1})=\mathrm{In}(\boldsymbol{SS}^*\boldsymbol{H})=\mathrm{In}\,\boldsymbol{H}$$

看出，定理 1 蕴涵 Sylvester 惯性律. 反之，若 $\boldsymbol{Q}>\boldsymbol{O}$，则对任意 $\boldsymbol{H}=\boldsymbol{H}^*$，$\mathrm{In}(\boldsymbol{QH})=\mathrm{In}(\boldsymbol{Q}^{-1/2}\boldsymbol{QHQ}^{-1/2})=\mathrm{In}(\boldsymbol{Q}^{1/2}\boldsymbol{HQ}^{-1/2})=\mathrm{In}\,\boldsymbol{H}$，因而惯性律蕴涵定理 1.

定理 2 的最重要的推广是下列的主要惯性定理[3~5]及其推广[6].

定理 3（主要惯性定理） 设 $\boldsymbol{A}\in\mathbf{C}^{n\times n}$. 则当且仅当存在 Hermite 矩阵 $\boldsymbol{H}\in\mathbf{C}^{n\times n}$ 使得 $\mathrm{Re}(\boldsymbol{AH})>\boldsymbol{O}$ 时，$\delta(\boldsymbol{A})=0$. 若 $\boldsymbol{H}$ 为 Hermite 矩阵使得 $\mathrm{Re}(\boldsymbol{AH})>\boldsymbol{O}$，则 $\mathrm{In}\,\boldsymbol{A}=\mathrm{In}\,\boldsymbol{H}$.

定理 4（主要惯性定理的推广） 设 $\boldsymbol{A}\in\mathbf{C}^{n\times n}$ 且 $\mathscr{S}$ 为 $\mathbf{C}^n$（$\mathbf{C}^n$ 视为有标准内积的酉空间）的子空间. 则当且仅当存在 Hermite 矩阵 $\boldsymbol{H}$ 使得

$$\mathrm{Re}(\boldsymbol{AH})\geqslant\boldsymbol{O}, \tag{1}$$

$$\mathrm{Ker}(\mathrm{Re}(\boldsymbol{AH}))=\mathrm{Ker}\,\boldsymbol{H}=\mathscr{S} \tag{2}$$

时，

$$\boldsymbol{A}\mathscr{S}^{\perp}\subset\mathscr{S}^{\perp}, \tag{3}$$

$$\delta(\boldsymbol{A}\,|\,\mathscr{S}^{\perp})=0. \tag{4}$$

此外，若(1)与(2)成立，则有

$$\mathrm{In}\,\boldsymbol{H}=\mathrm{In}(\boldsymbol{A}\,|\,\mathscr{S}^{\perp})+(0,0,\mathrm{dimKer}\,\boldsymbol{H})\leqslant\mathrm{In}\,\boldsymbol{A}. \tag{5}$$

定理中约定 $\mathrm{In}(\boldsymbol{A}\,|\,\{\boldsymbol{0}\})=(0,0,0)$. 我们指出，如取 $\mathscr{S}=\{\boldsymbol{0}\}\subset\mathbf{C}^n$，(3)与(4)这时变为 $\delta(\boldsymbol{A})=0$，(1)与(2)变为有可逆 Hermite 矩阵 $\boldsymbol{H}$ 使得 $\mathrm{Re}(\boldsymbol{AH})>\boldsymbol{O}$，而(5)变为 $\mathrm{In}\,\boldsymbol{H}=\mathrm{In}\,\boldsymbol{A}$. 因此，定理 4 确为定理 3 的一种推广.

本文旨在用数学归纳法与很少的预备知识简单地证明定理 3 与 4. 接着考虑它们的应用，其中包括对一些著名的结果的简化证明.

在证明定理 3,4 之前，先引入若干约定与基本表达式. 在下文中 $\boldsymbol{H}$ 与 $\boldsymbol{K}$ 总表示 Hermite 矩阵. 如果 $\mathrm{Re}(\boldsymbol{AH})>\boldsymbol{O}(\geqslant\boldsymbol{O})$，$\boldsymbol{B}=\boldsymbol{PAP}^{-1}$ 与 $\boldsymbol{K}=\boldsymbol{PHP}^*$，那么 $\mathrm{Re}(\boldsymbol{BK})=\boldsymbol{P}\mathrm{Re}(\boldsymbol{AH})\boldsymbol{P}^*>\boldsymbol{O}(\geqslant\boldsymbol{O})$，并且，$\mathrm{In}\,\boldsymbol{A}=\mathrm{In}\,\boldsymbol{B}$，$\mathrm{In}\,\boldsymbol{H}=\mathrm{In}\,\boldsymbol{K}$ 与 $\mathrm{In}(\boldsymbol{AH})=\mathrm{In}(\boldsymbol{BK})$. 因此，在后面证明中不失普遍性可假定 $\boldsymbol{A}$ 或 $\boldsymbol{H}$ 有 Jordan 法式或其变形的形式. 此外，若 $\mathrm{Re}\,\boldsymbol{A}>\boldsymbol{O}$，则有 $\mathrm{In}\,\boldsymbol{A}=(n,0,0)$，因而 $\boldsymbol{A}$ 非奇异，于是 $\mathrm{Re}(\boldsymbol{AH})>O$ 蕴涵 $\boldsymbol{A}$ 与 $\boldsymbol{H}$ 均为非奇异的. 如果 $\boldsymbol{A}$ 与 $\boldsymbol{H}$ 有形式

$$\boldsymbol{A}=\begin{pmatrix}\lambda & \boldsymbol{\alpha}^* \\ \boldsymbol{0} & \boldsymbol{B}\end{pmatrix},\ \boldsymbol{H}=\begin{pmatrix}h & \boldsymbol{\beta}^* \\ \boldsymbol{\beta} & \boldsymbol{F}\end{pmatrix}, \tag{6}$$

这里，$\lambda\in\sigma(\boldsymbol{A})$（$\boldsymbol{A}$ 的谱），h 为实数，$\boldsymbol{F}=\boldsymbol{F}^*\in\mathbf{C}^{(n-1)\times(n-1)}$，$\boldsymbol{\alpha},\boldsymbol{\beta}\in\mathbf{C}^{n-1}$，那么

$$2\mathrm{Re}(\boldsymbol{AH})=\begin{bmatrix}h(\lambda+\bar{\lambda})+\boldsymbol{\alpha}^*\boldsymbol{\beta}+\boldsymbol{\beta}^*\boldsymbol{\alpha} & \boldsymbol{\beta}^*(\lambda\boldsymbol{I}+\boldsymbol{B}^*)+\boldsymbol{\alpha}^*\boldsymbol{F} \\ (\bar{\lambda}\boldsymbol{I}+\boldsymbol{B})\boldsymbol{\beta}+\boldsymbol{F}\boldsymbol{\alpha} & \boldsymbol{BF}+\boldsymbol{FB}^*\end{bmatrix}=\boldsymbol{I}_n \tag{7}$$

等价于方程组

$$h(\lambda+\bar{\lambda})=1-\boldsymbol{\alpha}^*\boldsymbol{\beta}-\boldsymbol{\beta}^*\boldsymbol{\alpha},\quad \boldsymbol{\beta}^*(\lambda\boldsymbol{I}+\boldsymbol{B}^*)=-\boldsymbol{\alpha}^*\boldsymbol{F},\quad \boldsymbol{BF}+\boldsymbol{FB}^*=\boldsymbol{I}_{n-1}. \tag{8}$$

§2. 定理 3,4 的归纳法证明

定理 3 的证明 先证若有 $\boldsymbol{H}$ 使得 $\mathrm{Re}(\boldsymbol{AH})>\boldsymbol{O}$，则 $\delta(\boldsymbol{A})=0$. 设 $\lambda\in\sigma(\boldsymbol{A}^*)$ 与 $\boldsymbol{A}^*\boldsymbol{x}=\lambda\boldsymbol{x}\,(\boldsymbol{x}\neq\boldsymbol{0})$，这时有 $\boldsymbol{x}^*\boldsymbol{A}=\bar{\lambda}\boldsymbol{x}^*$. 分别用 $\boldsymbol{x}^*$ 与 $\boldsymbol{x}$ 左乘与右乘 $\mathrm{Re}(\boldsymbol{AH})>\boldsymbol{O}$ 便得 $(\lambda+\bar{\lambda})\boldsymbol{x}^*\boldsymbol{Hx}>0$，因而 $\lambda+\bar{\lambda}\neq0$，$\forall\lambda\in\sigma(A^*)$，此表明 $\delta(\boldsymbol{A})=\delta(\boldsymbol{A}^*)=0$. 其次证明：若 $\delta(\boldsymbol{A})=0$，则有 $\boldsymbol{H}$ 使得 $2\mathrm{Re}(\boldsymbol{AH})=\boldsymbol{I}$. 对 $\boldsymbol{A}$ 的阶数 n 应用归纳法. 当 $n=1$ 时，命题为平凡的. 现假定命题对所有满足 $1\leqslant k<n\,(n\geqslant2)$ 的 k 均成立. 考虑 n 阶矩阵的情形. 设 $\boldsymbol{A}\in\mathbf{C}^{n\times n}$ 有Jordan法式形式且有 s 个 Jordan 块. 讨论如下 2 种可能情形.

情形Ⅰ $s>1$. 这时，$\boldsymbol{A}$ 可写为 $\boldsymbol{A}=\boldsymbol{A}_1\oplus\boldsymbol{A}_2$，其中，$\boldsymbol{A}_1\in\mathbf{C}^{q\times q}$，$\boldsymbol{A}_2\in\mathbf{C}^{(n-q)\times(n-q)}$，$1\leqslant q<n$. 将归纳假定用于 $\boldsymbol{A}_1$ 与 $\boldsymbol{A}_2$（注意 $\delta(\boldsymbol{A}_1)=\delta(\boldsymbol{A}_2)=0$），我们有 Hermite 矩阵 $\boldsymbol{H}_1\in\mathbf{C}^{q\times q}$ 与 $\boldsymbol{H}_2\in\mathbf{C}^{(n-q)\times(n-q)}$ 使得 $2\mathrm{Re}(\boldsymbol{A}_1\boldsymbol{H}_1)=\boldsymbol{I}_q$ 与 $2\mathrm{Re}(\boldsymbol{A}_2\boldsymbol{H}_2)=\boldsymbol{I}_{n-q}$. 令 $\boldsymbol{H}=\boldsymbol{H}_1\oplus\boldsymbol{H}_2$，则有 $2\mathrm{Re}(\boldsymbol{AH})=\boldsymbol{I}$.

情形Ⅱ $s=1$. 这时，$\boldsymbol{A}$ 可改写为(6)中形式，但 $\boldsymbol{\alpha}=(1,0,\cdots,0)^*\in\mathbf{C}^{n-1}$ 与 $\boldsymbol{B}=\lambda\boldsymbol{I}_{n-1}+\boldsymbol{U}_{n-1}$，这里 $\boldsymbol{U}_{n-1}=(\delta_{i+1,j})\in\mathbf{C}^{(n-1)\times(n-1)}$. 假定 $\boldsymbol{H}=\boldsymbol{H}_{h,\beta,F}$ 有如(6)中形式. 我们要证明存在 $\boldsymbol{h},\boldsymbol{\beta}$ 与 $\boldsymbol{F}$ 使得 $2\mathrm{Re}(\boldsymbol{AH}_{h,\beta,F})=\boldsymbol{I}$. 显然，此等价于方程(8)关于 $\boldsymbol{h},\boldsymbol{\beta}$ 与 $\boldsymbol{F}$ 的可解性. 因 $\delta(\boldsymbol{B})=0$，按归纳假定，(8)中第 3 个方程有解 $\boldsymbol{F}=\boldsymbol{F}^*\in\mathbf{C}^{(n-1)\times(n-1)}$. 另一方面，由于 $\sigma(\boldsymbol{A})=\sigma(\boldsymbol{B})=\{\lambda\}$ 与 $\delta(\boldsymbol{A})=\delta(\boldsymbol{B})=0$ 蕴涵 $\lambda+\bar{\lambda}\neq0$ 与 $-\lambda\notin\sigma(\boldsymbol{B}^*)$，故(8)中前 2 个方程有解：

$$\boldsymbol{\beta}^*=-\boldsymbol{\alpha}^*\boldsymbol{F}(\lambda\boldsymbol{I}+\boldsymbol{B}^*)^{-1},\quad \boldsymbol{h}=(\lambda+\bar{\lambda})^{-1}(1-\boldsymbol{\alpha}^*\boldsymbol{\beta}-\boldsymbol{\beta}^*\boldsymbol{\alpha}).$$

最后证明：若 $\boldsymbol{H}$ 使得 $2\mathrm{Re}(\boldsymbol{AH})=\boldsymbol{I}$，则 $\mathrm{In}\ \boldsymbol{A}=\mathrm{In}\ \boldsymbol{H}$. 也对 $\boldsymbol{A}$ 的阶数 n 应用归纳法. 对 $n=1$，命题显然成立. 现假定本命题对 $n-1$ 成立. 考虑 n

阶情形. 不妨设 $\boldsymbol{A}$ 与 $\boldsymbol{H}$ 有(6)中的形式. 这时,$\boldsymbol{F}=\boldsymbol{F}^*$ 与(7)推出 $2\mathrm{Re}(\boldsymbol{BF})=\boldsymbol{I}_{n-1}$,因而 $\boldsymbol{F}$ 非奇异,且按归纳假定,$\mathrm{In}\ \boldsymbol{B}=\mathrm{In}\ \boldsymbol{F}$. 记$[\boldsymbol{H}/\boldsymbol{F}]=\boldsymbol{h}-\boldsymbol{\beta}^*\boldsymbol{F}^{-1}\boldsymbol{\beta}$ 为 $\boldsymbol{F}$ 在 $\boldsymbol{H}$ 内的 Schur 余量. 直接验证知道,

$$\boldsymbol{MHM}^*=[\boldsymbol{H}/\boldsymbol{F}]\oplus\boldsymbol{F},\quad \boldsymbol{M}=\begin{pmatrix}1 & -\boldsymbol{\beta}^*\boldsymbol{F}^{-1}\\ \boldsymbol{0} & \boldsymbol{I}_{n-1}\end{pmatrix},$$

因而 $\mathrm{In}\ \boldsymbol{H}=\mathrm{In}(\boldsymbol{MHM}^*)=\mathrm{In}[\boldsymbol{H}/\boldsymbol{F}]+\mathrm{In}\ \boldsymbol{F}$. 另一方面,由于 $\boldsymbol{F}^{-1}\boldsymbol{B}+\boldsymbol{B}^*\boldsymbol{F}^{-1}=\boldsymbol{F}^{-2}$从方程组(8)经过简单计算可得

$$(\lambda+\bar{\lambda})[\boldsymbol{H}/\boldsymbol{F}]=1+\boldsymbol{\beta}^*\boldsymbol{F}^{-2}\boldsymbol{\beta}>0.$$

因此,$\mathrm{In}\ \boldsymbol{A}=\mathrm{In}[\lambda]+\mathrm{In}\ \boldsymbol{B}=\mathrm{In}[\boldsymbol{H}/\boldsymbol{F}]+\mathrm{In}\ \boldsymbol{F}=\mathrm{In}\ \boldsymbol{H}$. 定理证毕. ■

在下面的定理 4 证明中,我们采用[6]中的有关约定与记号. 特别对给定 $\boldsymbol{B}\in\mathbf{C}^{n\times n}$ 与 $\mathbf{C}^n$ 的子空间 $\mathscr{S}$,$r=\dim\mathscr{S}^{\perp}$,总可以假设 $\boldsymbol{B}$ 已经过合适的酉相似变换使得 $\mathscr{S}^{\perp}$ 与 $\mathscr{S}$分别由 $\mathbf{C}^n$ 的前 r 个与后 $n-r$ 个单位坐标向量生成. 这时,$\boldsymbol{B}$ 可改写为所谓关于 $\mathscr{S}$的形式:

$$\boldsymbol{B}=\begin{pmatrix}\boldsymbol{B}_{11} & \boldsymbol{B}_{12}\\ \boldsymbol{B}_{21} & \boldsymbol{B}_{22}\end{pmatrix},\quad \boldsymbol{B}_{11}\in\mathbf{C}^{r\times r}. \tag{9}$$

倘若 $\mathscr{S}=\mathrm{Ker}\ \boldsymbol{B}$,则上式变为

$$\boldsymbol{B}=\begin{pmatrix}\boldsymbol{B}_{11} & \boldsymbol{O}\\ \boldsymbol{B}_{21} & \boldsymbol{O}\end{pmatrix}. \tag{10}$$

这里,$\begin{pmatrix}\boldsymbol{B}_{11}\\ \boldsymbol{B}_{21}\end{pmatrix}\in\mathbf{C}^{n\times r}$列满秩;若还有 $\boldsymbol{B}=\boldsymbol{B}^*$,则有 $\boldsymbol{B}_{21}=\boldsymbol{O}$ 与 $\boldsymbol{B}_{11}$ 非奇异,于是 $\boldsymbol{B}=\boldsymbol{B}_{11}\oplus\boldsymbol{O}$.

定理 4 的证明　当 $\mathscr{S}=\{\boldsymbol{0}\}\subset\mathbf{C}^n$时,由前面分析知道本定理退化为定理 3. 现对阶数 n 应用归纳法. 当 $n=1$ 时,只需考虑 $\mathscr{S}=\mathbf{C}$ 的情形,但此时定理结论显然成立. 假定本定理对所有满足 $1\leqslant k<n(n\geqslant 2)$的 k 均成立. 考虑 n 阶的情形. 如前只要讨论 $\mathscr{S}\subset\mathbf{C}^n$ 不为$\{\boldsymbol{0}\}$的情形. 设 $\dim\mathscr{S}=n-r(<n)$且 $A\in\mathbf{C}^{n\times n}$有关于 $\mathscr{S}$的形式(9). 这时,条件(3)与(4)等价于 $\boldsymbol{A}_{21}=\boldsymbol{O}$ 与 $\delta(\boldsymbol{A}_{11})=0$. 按归纳假定(用 $\boldsymbol{A}_{11}$代替 $\boldsymbol{A}$,用 $\mathbf{C}^r$代替 $\mathscr{S}$),存在 $\boldsymbol{H}_{11}=\boldsymbol{H}_{11}^*$使得 $\mathrm{Re}(\boldsymbol{A}_{11}\boldsymbol{H}_{11})\geqslant\boldsymbol{O}$ 与 $\mathrm{Ker}(\mathrm{Re}(\boldsymbol{A}_{11}\boldsymbol{H}_{11}))=\mathrm{Ker}\boldsymbol{H}_{11}=\mathbf{C}^r$,此表明有非奇异 Hermite矩阵 $\boldsymbol{H}_{11}$使得 $\mathrm{Re}(\boldsymbol{A}_{11}\boldsymbol{H}_{11})>\boldsymbol{O}$. 令 $\boldsymbol{H}=\boldsymbol{H}_{11}\oplus\boldsymbol{O}\in\mathbf{C}^{n\times n}$,则 $\boldsymbol{H}$ 显然满足(1)与(2). 反之,若对某 Hermite 矩阵 $\boldsymbol{H}$,(1)与(2)成立,令 $\boldsymbol{H}=\boldsymbol{H}_{11}\oplus$

$\boldsymbol{O}$，则 $\boldsymbol{H}_{11}\in\mathbf{C}^{r\times r}$ 非奇异，$r=\dim\mathscr{S}^{\perp}$，因而从

$$2\mathrm{Re}(\boldsymbol{AH})=\begin{bmatrix}2\mathrm{Re}\ (\boldsymbol{A}_{11}\boldsymbol{H}_{11}) & \boldsymbol{H}_{11}\boldsymbol{A}_{21}^{*}\\ \boldsymbol{A}_{21}\boldsymbol{H}_{11} & \boldsymbol{O}\end{bmatrix}\geqslant\boldsymbol{O}\tag{11}$$

推出 $\boldsymbol{A}_{21}\boldsymbol{H}_{11}=\boldsymbol{O}$，因而 $\boldsymbol{A}_{21}=\boldsymbol{O}$，此等价于(3). 另一方面，(11)也蕴涵 $\mathrm{Re}(\boldsymbol{A}_{11}\boldsymbol{H}_{11})\geqslant\boldsymbol{O}$. 条件(2)与 $\boldsymbol{H}_{11}=\boldsymbol{H}_{11}^{*}$ 非奇异，这时等价于 $\mathrm{Ker}(\mathrm{Re}(\boldsymbol{A}_{11}\boldsymbol{H}_{11}))=\mathrm{Ker}\boldsymbol{H}_{11}=\{\boldsymbol{0}\}\subset\mathbf{C}^{r}$，应用归纳假定可得 $\delta(\boldsymbol{A}_{11})=\delta(\boldsymbol{A}\,|\,\mathscr{S}^{\perp})=0$，因而(4)也成立.

最后，若(1)与(2)对 $\boldsymbol{H}\in\mathbf{C}^{n\times n}$ 与 $\mathbf{C}^{n}$ 的某个非零的子空间 $\mathscr{S}$ 成立，$\dim\mathscr{S}=n-r=d(<n)$，这时按前面的证明，$\boldsymbol{A}_{21}=\boldsymbol{O}$，因而 $\mathrm{In}\,\boldsymbol{A}=\mathrm{In}\,\boldsymbol{A}_{11}+\mathrm{In}\,\boldsymbol{A}_{22}$；同时有 $\delta(\boldsymbol{A}_{11})=0$. 应用归纳假定便得 $\mathrm{In}\,\boldsymbol{H}_{11}=\mathrm{In}(\boldsymbol{A}_{11}\,|\,\mathbf{C}^{r})=\mathrm{In}\,\boldsymbol{A}_{11}$. 因此，$\mathrm{In}\,\boldsymbol{H}=\mathrm{In}\,\boldsymbol{H}_{11}+(0,0,d)=\mathrm{In}\,\boldsymbol{A}_{11}+(0,0,d)\leqslant\mathrm{In}\,\boldsymbol{A}_{11}+\mathrm{In}\,\boldsymbol{A}_{22}=\mathrm{In}\,\boldsymbol{A}$，由于 $\mathrm{In}\,\boldsymbol{A}_{11}=\mathrm{In}(\boldsymbol{A}\,|\,\mathscr{S}^{\perp})$，(5)式得证. 定理 4 证毕. ■

§3. 有关推论

由于在定理 3,4 证明中只涉及少量的预备知识，因而作为应用可以从它们(特别从定理 4)出发推证一些矩阵惯性的结果.

推论 1[7,8] 设 $\boldsymbol{A}\in\mathbf{C}^{n\times n}$. 则下列诸命题等价：

(a) $\mathrm{Re}\,\boldsymbol{A}>\boldsymbol{O}$.

(b)对每个 Hermite 矩阵 $\boldsymbol{H}\in\mathbf{C}^{n\times n}$，$\mathrm{In}(\boldsymbol{AH})=\mathrm{In}\,\boldsymbol{H}$.

(c)存在实对角矩阵 $\boldsymbol{R}$ 与可逆矩阵 $\boldsymbol{P}$ 使得

$$\boldsymbol{PAP}^{*}=\boldsymbol{I}+\mathrm{i}\boldsymbol{R},\tag{12}$$

亦即 $\boldsymbol{A}$ 相合于矩阵 $\boldsymbol{I}+\mathrm{i}\boldsymbol{R}$.

证 (a)⇒(b)：不妨设 $\boldsymbol{O}\neq\boldsymbol{H}=\boldsymbol{H}_{11}\oplus\boldsymbol{O}$，$\boldsymbol{H}_{11}\in\mathbf{C}^{r\times r}$ 非奇异，且 $\boldsymbol{A}$ 有分块形式如(9). 这时有 $\mathrm{In}(\boldsymbol{AH})=\mathrm{In}(\boldsymbol{A}_{11}\boldsymbol{H}_{11})+(0,0,n-r)$，因为

$$\boldsymbol{AH}=\begin{pmatrix}\boldsymbol{A}_{11}\boldsymbol{H}_{11} & \boldsymbol{O}\\ \boldsymbol{A}_{21}\boldsymbol{H}_{11} & \boldsymbol{O}\end{pmatrix}.$$

若令 $\boldsymbol{H}^{+}=\boldsymbol{H}_{11}^{-1}\oplus\boldsymbol{O}$，则 $\mathrm{Re}(\boldsymbol{AHH}^{+})=\mathrm{Re}\,\boldsymbol{A}_{11}\oplus\boldsymbol{O}\geqslant\boldsymbol{O}$. 另一方面，$\mathrm{Re}\,\boldsymbol{A}>\boldsymbol{O}$ 蕴涵 $\mathrm{Re}\,\boldsymbol{A}_{11}>\boldsymbol{O}$，因而有

$$\mathrm{Ker}(\mathrm{Re}(\boldsymbol{AHH}^{+}))=\mathrm{Ker}\,\boldsymbol{H}^{+}\ (=\mathrm{Ker}\,\boldsymbol{H}).$$

依据定理 4 我们有 $\mathrm{In}\,\boldsymbol{H}=\mathrm{In}\,\boldsymbol{H}^{+}=\mathrm{In}((\boldsymbol{AH})\,|\,(\mathrm{Ker}\,\boldsymbol{H})^{\perp})+(0,0,n-r)=$

$\mathrm{In}(\boldsymbol{A}_{11}\boldsymbol{H}_{11})+(0,0,n-r)=\mathrm{In}(\boldsymbol{AH})$.

(b)⇒(c)：因 $\mathrm{Re}\,\boldsymbol{A}$ 为 Hermite 矩阵，所以有酉矩阵 $\boldsymbol{U}$ 使得 $\boldsymbol{U}\mathrm{Re}\,\boldsymbol{AU}^*=\boldsymbol{D}=\mathrm{diag}[d_1,d_2,\cdots,d_n]$（各 d_j 为实数），因而

$$\boldsymbol{UAU}^*=\boldsymbol{D}+\mathrm{i}\boldsymbol{Q}, \tag{13}$$

式中 $\boldsymbol{Q}\in\mathbf{C}^{n\times n}$ 为 Hermite 矩阵. 现证明所有 $d_j>0$，即 $\boldsymbol{D}>\boldsymbol{O}$. 倘若有一个 $d_j\leqslant 0$，不妨设 $d_1\leqslant 0$，选取 $\boldsymbol{H}=\mathrm{diag}[1,0,\cdots,0]\in\mathbf{C}^{n\times n}$，则按(b)得

$$\mathrm{In}(\boldsymbol{UAU}^*\boldsymbol{H})=\mathrm{In}(\boldsymbol{AU}^*\boldsymbol{HU})=\mathrm{In}(\boldsymbol{U}^*\boldsymbol{HU})=\mathrm{In}\,\boldsymbol{H}=(1,0,n-1).$$

但从 $\boldsymbol{UAU}^*\boldsymbol{H}=\boldsymbol{DH}+\mathrm{i}\boldsymbol{QH}=\mathrm{diag}[d_1,0,\cdots,0]+\mathrm{i}[\boldsymbol{Q}^{(1)}\ \boldsymbol{0}\ \cdots\ \boldsymbol{0}]$

（式中 $\boldsymbol{Q}^{(1)}$ 表示 $\boldsymbol{Q}$ 的第一列）看出，$\pi(\boldsymbol{UAU}^*\boldsymbol{H})=0$，得出矛盾. 因此，$\boldsymbol{D}>\boldsymbol{O}$. 最后，(13)蕴涵(12)是显见的，因为(13)式可变为 $\boldsymbol{D}^{-1/2}\boldsymbol{UAU}^*\boldsymbol{D}^{-1/2}=\boldsymbol{I}+\mathrm{i}\boldsymbol{D}^{-1/2}\boldsymbol{QD}^{-1/2}$，如酉矩阵 $\boldsymbol{V}$ 使得 $\boldsymbol{VD}^{-1/2}\boldsymbol{QD}^{-1/2}\boldsymbol{V}^*$ 为实对角矩阵 $\boldsymbol{R}$，则(12)中 $\boldsymbol{P}$ 可选取为 $\boldsymbol{VD}^{-1/2}\boldsymbol{U}$.

(c)⇒(a)：(12)式推出 $\boldsymbol{P}\mathrm{Re}\,\boldsymbol{AP}^*=\boldsymbol{I}$，因而 $\mathrm{Re}\,\boldsymbol{A}=\boldsymbol{P}^{-1}(\boldsymbol{P}^*)^{-1}>\boldsymbol{O}$. ■

推论 2[6] 设 $\boldsymbol{A}\in\mathbf{C}^{n\times n}$ 且 $\boldsymbol{H}\in\mathbf{C}^{n\times n}$ 为 Hermite 矩阵. 若 $\mathrm{Re}(\boldsymbol{AH})\geqslant\boldsymbol{O}$ 与 $\delta(\boldsymbol{A}|(\mathrm{Ker}\,\boldsymbol{H})^{\perp})=0$，则有 $\mathrm{In}\,\boldsymbol{H}=\mathrm{In}(\boldsymbol{A}|(\mathrm{Ker}\,\boldsymbol{H})^{\perp})+(0,0,d)\leqslant\mathrm{In}\,\boldsymbol{A}$，这里 $d=\dim\mathrm{Ker}\,\boldsymbol{H}$. 特殊地，若 $\mathrm{Re}(\boldsymbol{AH})\geqslant\boldsymbol{O}$，$\boldsymbol{H}$ 非奇异与 $\delta(\boldsymbol{A})=0$，则有

$$\mathrm{In}\,\boldsymbol{H}=\mathrm{In}\,\boldsymbol{A}.$$

证 从定理 4 证明看出，$\mathrm{Re}(\boldsymbol{AH})\geqslant O$ 蕴涵 $\boldsymbol{A}(\mathrm{Ker}\,\boldsymbol{H})^{\perp}\subset(\mathrm{Ker}\,\boldsymbol{H})^{\perp}$，于是按定理 4 中(1)(2)与(3)(4)的等价性便得要证的前一结果. 最后结论为前者取 $\mathrm{Ker}\,\boldsymbol{H}$ 为 $\mathbf{C}^n$ 的零子空间的特殊情形. 证毕. ■

应用推论 2 最后结果与如下事实：若 $\mathrm{Re}\,\boldsymbol{A}\geqslant\boldsymbol{O}$，则对任意 $\boldsymbol{H}=\boldsymbol{H}^*$，$\mathrm{In}(\boldsymbol{AH})\leqslant\mathrm{In}(\boldsymbol{H})$（这是推论 1 与特征值作为矩阵元素函数的连续性结果的直接推论），可得以下 2 个有用的推论.

推论 3[6] 设 $\boldsymbol{A},\boldsymbol{H}\in\mathbf{C}^{n\times n}$.

(a)若 $\mathrm{Re}(\boldsymbol{AH})\geqslant\boldsymbol{O}$，则 $\mathrm{In}(\boldsymbol{A}|(\mathrm{Ker}\,\boldsymbol{H})^{\perp})+(0,0,\dim\mathrm{Ker}\,\boldsymbol{H})\leqslant\mathrm{In}\,\boldsymbol{H}$，特殊地，如 $\boldsymbol{H}$ 为非奇异的，则有 $\mathrm{In}\,\boldsymbol{A}\leqslant\mathrm{In}\,\boldsymbol{H}$.

(b)若 $\mathrm{Re}(\boldsymbol{AH})\geqslant\boldsymbol{O}$，$\boldsymbol{H}$ 非奇异与 $\delta(\boldsymbol{A})=0$，则有 $\mathrm{In}\,\boldsymbol{A}=\mathrm{In}\,\boldsymbol{H}$.

推论 4[4] 设 $\boldsymbol{A}\in\mathbf{C}^{n\times n}$，则下列诸命题等价：

(a) $\mathrm{Re}\,\boldsymbol{A}\geqslant\boldsymbol{O}$.

(b) $\boldsymbol{A}$ 为正 $\boldsymbol{H}$-半稳定的，即对任意 $\boldsymbol{H}\geqslant\boldsymbol{O}$，有 $\nu(\boldsymbol{AH})=0$.

(c)存在实对角矩阵 $\boldsymbol{D}\geqslant\boldsymbol{O}$，Hermite 矩阵 $\boldsymbol{Q}$ 与可逆矩阵 $\boldsymbol{P}$ 使得

$$PAP^* = D + \mathrm{i}Q.$$

推论 4 的证明类似于推论 1，但要用到前述事实：Re $A \geqslant O$ 与 $H = H^*$ 蕴涵 In$(AH) \leqslant$ In H.

我们称 $A \in C^{n\times n}$ 为 Sylvester 型的，假如对任意 n 阶可逆矩阵 S，

$$\mathrm{In}(SAS^*) = \mathrm{In}\ A.$$

推论 5[8] 设 $A = A^* \in C^{n\times n}$，$B \in C^{n\times n}$ 为 Sylvester 型的，且 $\pi(AB) = n$. 则有 In A = In B.

证 因 $\pi(AB) = n$，按定理 2 有 $G > O$ 使得 Re$(ABG) > O$，因而 A 与 B 皆为可逆的，且 Re$(GB^*A) > O$. 应用推论 1 便得 In(GB^*) = In(A^{-1}) = In A. 另一方面，因 $G > O$ 与 B 为 Sylvester 型的，我们有

$$GB^* = G^{1/2}G^{1/2}B^*G^{1/2}G^{-1/2},$$

于是，In(GB^*) = In$(G^{1/2}B^*G^{1/2})$ = In B = In A. 证毕. ■

我们指出，Newman[9] 在 $B \in C^{n\times n}$ 为 Hermite 矩阵条件下证明了上述推论. 他的结果显然为推论 5 的特殊情形.

最后顺便说明一下，前述定理的证明对于相对于单位圆周的矩阵惯性的有关结果也适用或基本适用.

参考文献

[1] Sylvester J J. A demonstration of the theorem that every homogoneous quadratic polynomial is reducible by real orthogonal substitutions to the form of a sum of positive and negative squares. Phil Mag, 1852(4): 142; Math Papers I, Cambridge, 1904. 378－381.

[2] Lyapunov A. Problème Général de la stabilité du mouvement. Kharkov: Commun Soc Math, 1892,1893.

[3] Taussky O. A generalization of a theorem by Lyapunov. J Soc Indust Appl Math, 1961, 9: 640.

[4] Ostrowski A, Schneider H. Some theorems on the inertia of general matrices. J Math Anal Appl, 1962, 4:72.

[5] Daleckii Ju L, Krein M G. Stability of solutions of differential equations in Banach space. In: Amer Math Soc Translations Vol 43. Providence, 1974.

[6] Carlson D, Schneider H. Inertia theorems for matrices: The semidefinite case. J Math Anal Appl, 1963, 6:430.

[7] Wielandt H. On the eigenvalues of $A+B$ and AB. Natl Bur Standards Rept, 1367, 1951.

[8] Cain B E. Inertia theory. Lett Linear Algebra, 1980, 30: 211.

[9] Newman M. Lyapunov revisited. Linear Algebra Appl, 1983, 52—53: 525.

SOME BASIC THEOREMS ON THE INERTIA FOR MATRICES

Abstract The elementary proofs of main inertia theorem and its generalization are presented by induction. Related corollaries are considered as applications.

Keywords inertia of matrices; main inertia theorem; stable matrix

北京师范大学学报(自然科学版)
1992,28(3):292－294

奇异 Hankel 矩阵[①]

Singular Hankel Matrices

关键词 Hankel 矩阵;拟直分解;H-多项式;相容多项式;最小度

本文研究有限奇异 Hankel 矩阵的相容多项式与最小度.按定义,非零多项式 $f(z)=\sum_{j=0}^{n} f_j z^j$ 为 n 阶 Hankel 矩阵 $\boldsymbol{H}=(h_{i+j})_{i,j=0}^{n-1}$ 的相容多项式,假如 $\hat{H}[f]_n=\mathbf{0}$,这里 $[f]_n=(f_0,f_1,\cdots,f_n)^{\mathrm{T}}\in\mathbf{C}^{n+1}$ 与 $\hat{\boldsymbol{H}}=(h_{i+j})_{i,j=0}^{n-2,n}$;每个 n 阶 Hankel 矩阵 $\boldsymbol{H}$ 均可由某个有理函数 $g/f(\deg g\leqslant\deg f)$ 生成:$\boldsymbol{H}=\boldsymbol{H}_n(g/f)$,如果有 $\deg f=q$,但不存在满足 $\deg g\leqslant\deg f<q$ 的生成函数 g/f,那么称 $\boldsymbol{H}$ 有最小度 q(见[1]).n 阶 Hankel 矩阵 $\boldsymbol{H}$ 为真正的,假如其 rank $\boldsymbol{H}$ 阶前主子阵非奇异;它为退化的,假如其第 1 行与第 1 列元素均为零.Fiedler 指出[2],奇异 Hankel 矩阵可唯一地表示为真正 Hankel 矩阵 $\boldsymbol{H}_p$ 与退化 Hankel 矩阵 $\boldsymbol{H}_d$ 的拟直和,即有 $\boldsymbol{H}=\boldsymbol{H}_p+\boldsymbol{H}_d$,rank $\boldsymbol{H}=$ rank $\boldsymbol{H}_p+$ rank $\boldsymbol{H}_d$.对奇异的 Hankel 矩阵 $\boldsymbol{H}$,由它的真正部分 $\boldsymbol{H}_p$ 可以定义所谓 $\boldsymbol{H}$-多项式 $a(z)=z^m-\sum_{j=0}^{m-1}a_j z^j$,这里 $m=$ rank $\boldsymbol{H}_p$,并且 $a_0,\cdots,a_{m-1}$ 满足

$$h_{i+m}=a_0h_i+a_1h_{i+1}+\cdots+a_{m-1}h_{i+m-1},\ i=0,1,\cdots,2n-m-2.$$

下面定理给出奇异真正 Hankel 矩阵的特征.

定理 1 设 $\boldsymbol{H}=(h_{i+j})_{i,j=0}^{n-1}$ 为奇异 Hankel 矩阵,其秩为 m,其 $\boldsymbol{H}$-多项式为 $a(z)=z^m-\sum_{j=0}^{m-1}a_j z^j=\prod_{j=1}^{s}(z-t_j)^{n_j}$ (t_j 彼此不同且 $n_j\geqslant 1$).则以下

① 收稿日期:1992-03-15.
国家教委博士点基金资助项目,本文与张惠品合作.

8 个命题等价：

(1)$\boldsymbol{H}$ 为真正的；

(2)Ker $\hat{\boldsymbol{H}}=\{[f]_n:\deg f\leqslant n$ 且 $a(z)\mid f(z)\}$.

(3)$\boldsymbol{H}$ 有 n 次首一相容多项式 f；

(4)若 $\mathbf{0}\neq[f]_n\in$ Ker $\hat{\boldsymbol{H}}$，则有多项式 g 满足 $\deg g<\deg f=k$ 且 $\boldsymbol{H}=\boldsymbol{H}_n(g/f)$. 这时 g 由公式 $g(z)=(1,z,\cdots,z^{k-1})S(f)(h_0,h_1,\cdots,h_{k-1})^{\mathrm{T}}$ 唯一确定，其中 $S(f)$ 表示 f 的对称化子；

(5)$\boldsymbol{H}$ 可表示为 $\boldsymbol{H}_n(p/a)$，这里多项式 $p(z)=(1,z,\cdots,z^{m-1})S(a)(h_0,h_1,\cdots,h_{m-1})^{\mathrm{T}}$，$a$ 与 p 互素；

(6)$\boldsymbol{H}$ 可分解为 $\boldsymbol{H}=\sum\limits_{j=1}^{s}P_{nn_j}(t_j)\boldsymbol{S}_jP_{nn_j}^{\mathrm{T}}(t_j)$，这里 $P_{nk}(t)=[\binom{i}{j}t^{i-j}]_{i,j=0}^{n-1,k-1}$ 且 $\boldsymbol{S}_j=p(\boldsymbol{J}_{n_j}(t_j))\prod\limits_{i\neq j}(\boldsymbol{J}_{n_j}(t_j)-t_iI_{n_j})^{-n_i}\boldsymbol{R}_j$，式中，$p$ 同(5)，$\boldsymbol{J}_m(t)$ 为对角元素是 t 的 m 阶 Jordan 矩阵，而 $\boldsymbol{R}_j$ 为 n_j 阶翻转矩阵；

(7)$\boldsymbol{H}$ 可分解为(6)中的形式，但诸 $\boldsymbol{S}_j$ 为 n_j 阶非奇异上三角 Hankel 矩阵；

(8)$\boldsymbol{H}$ 有最小度 m.

对于退化 Hankel 矩阵情形我们有

定理 2　设 $\boldsymbol{H}=(h_{i+j})_{i,j=0}^{n-1}\neq\boldsymbol{O}$ 为奇异 Hankel 矩阵，有秩 r. 则以下 7 个命题等价：

(1)$\boldsymbol{H}$ 为退化的且有形式 diag$[\boldsymbol{O},\boldsymbol{Z}]$，其中 $\boldsymbol{Z}$ 为 r 阶非奇异下三角 Hankel 矩阵；

(2)Ker $\hat{\boldsymbol{H}}=\{[f]_n:\deg f\leqslant n-r\}$；

(3)$\boldsymbol{H}$ 有 $n-r$ 次的相容多项式；

(4)$\boldsymbol{H}$ 有相容多项式 $a(z)=1$；

(5)$\boldsymbol{H}$ 的 $\boldsymbol{H}$-多项式为$a(z)=1$；

(6)存在 r 次首一多项式 $b(z)$ 使得 $\boldsymbol{H}=\boldsymbol{H}_n(h_{2n-r-1}/(bz^{2n-2r}))$；

(7)$\boldsymbol{H}$ 有最小度 $2n-r$ 且 $\boldsymbol{H}$ 为 $\boldsymbol{H}$-不可分解的，即 $\boldsymbol{H}$ 不能表示为非零 Hankel 矩阵的拟直和.

对于一般奇异非真正 Hankel 矩阵，我们有

定理 3　设 $\boldsymbol{H}=(h_{i+j})_{i,j=0}^{n-1}\neq\boldsymbol{O}$ 为秩 r 的奇异的 Hankel 矩阵，其 $\boldsymbol{H}$-多项式同定理 1(但当 $m=0$ 时取 $a(z)=1$). 则以下 6 个命题等价：

(1)$\boldsymbol{H}$ 为非真正的 Hankel 矩阵;

(2)Ker $\hat{\boldsymbol{H}}=\{[f]_n:\deg f\leqslant n-r+m$ 且 $a(z)\mid f(z)\}$;

(3)如 f 为 $\boldsymbol{H}$ 的相容多项式,则 $\deg f\leqslant n-r+m$ 且 $a(z)\mid f(z)$;

(4)H 有唯一的拟直分解:$\boldsymbol{H}=\sum_{j=1}^{s}P_{nn_j}(t_j)\boldsymbol{S}_jP_{nn_j}^{\mathrm{T}}(t_j)+\boldsymbol{H}_d$,式中,$P_{nn_j}(t_j)$ 与 $\boldsymbol{S}_j$ 同定理 1 的命题(6);$\boldsymbol{H}_d$ 为秩 $r-m$ 的退化 Hankel 矩阵;所有和项 $\boldsymbol{H}$-不可分解;

(5)存在一对互素多项式 p 与 b,$\deg p<m$ 与 $\deg b=2n-r$,以及常数 c 使得 $\boldsymbol{H}$ 有拟直分解:$\boldsymbol{H}=\boldsymbol{H}_n(p/a)+\boldsymbol{H}_n(c/b)$,这里,$\boldsymbol{H}_d=\boldsymbol{H}_n(c/b)\neq\boldsymbol{O}$;

(6)$\boldsymbol{H}$ 有最小度 $2n-r$.

综上所述,最后我们得到

定理 4 设 $\boldsymbol{H}$ 为 n 阶奇异 Hankel 矩阵,其秩为 r,其 $\boldsymbol{H}$-多项式 $a(z)$ 的次数为 m. 则 Ker $\hat{\boldsymbol{H}}=\{[f]_n:\deg f\leqslant n-r+m$ 且 $a(z)\mid f(z)\}=$ Ker $\hat{\boldsymbol{H}}_p\cap$ Ker $\hat{\boldsymbol{H}}_d$. 此外,若 $\boldsymbol{H}$ 为真正的,则 $\boldsymbol{H}$ 有最小度 m,否则有最小度 $2n-r$.

参考文献

[1] Heinig G, Jungnickel U. Hankel matrices generated by the Markov parameters of rational functions. Linear Algebra Appl, 1986, 76: 121.

[2] Fiedler M. Quasidirect decomposition of Hankel and Toeplitz matrices. Linear Algebra Appl, 1984, 61: 155.

SINGULAR HANKEL MATRICES

Abstract Finite Hankel matrices are characterized in terms of the compatible polynomials and the minimal degree. Let $\boldsymbol{H}$ of order n be a singular Hankel matrix with rank $\boldsymbol{H}=r>0$ and the $\boldsymbol{H}$-polynomial $a(z)$ of degree m. Then $\boldsymbol{H}$ is proper (resp. degenerate) if $\boldsymbol{H}$ is compatible with some monic polynomial of degree n (resp. with polynomial 1); or iff the family of all polynomials compatible with $\boldsymbol{H}$ is the set $\{f\neq 0$: $\deg f\leqslant n$ and $a(z)\mid f(z)\}$ (resp. $\{f\neq 0$: $\deg f\leqslant n-r\}$). Moreover, $\boldsymbol{H}$ has the minimal degree equalled m if $\mathbf{H}$ is proper, and $2n-r$ otherwise.

Keywords Hankel matrix; quasidirect decomposition; $\boldsymbol{H}$-polynomial; compatible polynomial; the minimal degree

北京师范大学学报(自然科学版)
1992,28(4):438－445

Bezout 和 Hankel 矩阵(Ⅱ)——非奇异情形①

Bezout and Hankel Matrices(Ⅱ)——The Non-singular Case

摘要　简明地证明了非奇异 Hankel 矩阵与 Bezout 矩阵的特征定理,并讨论这两类矩阵的求逆问题.

关键词　Bezout 矩阵;Hankel 矩阵;相容多项式;Vandermonde 矩阵;Bezout 矩阵求逆

§1. 预备知识

本文为文献〔1〕的继续. 除了沿用该文的记号外,用 $P_n(\mathbf{C})$〔$P_n^0(\mathbf{C})$〕表示所有次数为 n〔至多为 n〕的复多项式集合. 称多项式 f 与 g 在 $P_n^0(\mathbf{C})$ 内完全互素,假如 $\max\{\deg f,\deg g\}=n$ 且 f 与 g 互素. 多项式 $f(\lambda)=\sum_{j=0}^{n} f_j\lambda^j$ 的 n 阶广义友阵 $\boldsymbol{C}_n(f)$ 定义为

$$\boldsymbol{C}_n(f)=\begin{bmatrix} 0 & f_n & & & \\ & \ddots & \ddots & & \\ & & \ddots & & \\ & & & \ddots & f_n \\ -f_0 & -f_1 & \cdots & \cdots & f_{n-1} \end{bmatrix}.$$

①　收稿日期:1992-04-13.
国家教委博士点基金资助项目. 本文与张惠品合作.

显然，若 $\deg f=n$，则 $\boldsymbol{C}_n(f)=f_n\boldsymbol{C}_f$，这里 $\boldsymbol{C}_f$ 为 f 的（第一）友阵.

本文主要应用文献〔1〕与广义 Vandermonde 矩阵的合同变换将 Bezout 与 Hankel 矩阵块对角化的最近结果〔2~4〕，汇总并简明地研究非奇异 Bezout 与 Hankel 矩阵的特征及其求逆问题. 其中一部分结果，特别是 Bezout 矩阵的求逆问题的结果是新的；多数已知结果赋予新的证明.

设 $f\in P_n(\mathbf{C})$ 为首一多项式且有因式分解：

$$f(\lambda)=\sum_{j=0}^{n}f_j\lambda^j=\prod_{j=1}^{s}(\lambda-t_j)^{n_j},f_n=1,t_j\text{ 彼此不同},n_j\geqslant 1. \quad (1)$$

对 $k\geqslant l>0$，记 $P_{kl}(t)=\left[\binom{i}{j}t^{i-j}\right]_{i=0,j=0}^{k-1,l-1}$，则称下列 n 阶矩阵为(1)中多项式 f 的广义 Vandermonde 矩阵：

$$\mathbf{V}(f)=[P_{nn_1}(t_1),P_{nn_2}(t_2),\cdots,P_{nn_s}(t_s)]. \quad (2)$$

已知 $\mathbf{V}(f)$ 为非奇异的〔4〕，特别地，当 $s=n$ 时，$\mathbf{V}(f)$ 退化为通常的 Vandermonde 矩阵.

定理Ⅰ〔4〕　设 f 如(1)所示，$g\in P_n^0(\mathbf{C})$. 则

$$\mathbf{V}^{\mathrm{T}}(f)\boldsymbol{B}_n(f,g)\mathbf{V}(f)=\operatorname{diag}[\boldsymbol{R}_i\hat{f}_i(\boldsymbol{J}_i)g(\boldsymbol{J}_i)]_{i=1}^{s}, \quad (3)$$

其中，$\hat{f}_i(\lambda)=f(\lambda)/(\lambda-t_i)^{n_i}$；$\boldsymbol{R}_i$ 为 n_i 阶翻转矩阵；$\boldsymbol{J}_i$ 表示关于 t_i 的 n_i 阶 Jordan 块. 特别地，当 f 有 n 个单根 $t_1,t_2,\cdots,t_n$ 时，(3)式变为

$$\mathbf{V}^{\mathrm{T}}(f)\boldsymbol{B}_n(f,g)\mathbf{V}(f)=\operatorname{diag}[f'(t_i)g(t_i)]_{i=1}^{n}. \quad (3')$$

定理Ⅱ〔4〕　设 $f,g,\hat{f}_i,\boldsymbol{R}_i$ 与 $\boldsymbol{J}_i$ 同定理Ⅰ. 则由 g/f 的 Markov 参数构成的 n 阶 Hankel 矩阵 $\boldsymbol{H}_n(g/f)$ 满足

$$\boldsymbol{H}_n(g/f)=\mathbf{V}(f)\operatorname{diag}[g(\boldsymbol{J}_i)\hat{f}_i(\boldsymbol{J}_i)^{-1}\boldsymbol{R}_i]_{i=1}^{s}\mathbf{V}^{\mathrm{T}}(f). \quad (4)$$

特别地，当 f 有 n 个单根 $t_1,t_2,\cdots,t_n$ 时，(4)式变为

$$\boldsymbol{H}_n(g/f)=\mathbf{V}(f)\operatorname{diag}[g(t_i)/f'(t_i)]_{i=1}^{n}\mathbf{V}^{\mathrm{T}}(f). \quad (4')$$

为将 Bezout 或 Hankel 矩阵对角化，需要应用下列引理〔5,6〕.

引理Ⅰ　设 $f,g\in P_n^0(\mathbf{C})$ 完全互素，则存在（无穷多个）复数 λ 使得 $f+\lambda g\in P_n(\mathbf{C})$ 有 n 个单根.

§2. 非奇异 Hankel 矩阵的特征与求逆问题

本节的基本结果是如下定理 1. 它汇总了非奇异 Hankel 矩阵的基本特征〔5,7,8〕.

定理 1　设 $\boldsymbol{H}$ 为 n 阶非零矩阵. 则下列各条彼此等价：

(1)$\boldsymbol{H}$ 为非奇异 Hankel 矩阵：$\boldsymbol{H}=(h_{i+j})_{i,j=0}^{n-1}$.

(2)$\boldsymbol{H}=(h_{i+j})_{i,j=0}^{n-1}$ 为 Hankel 矩阵，且方程

$$\boldsymbol{Hx}=(h_n,h_{n+1},\cdots,h_{2n-2},0)^{\mathrm{T}},\tag{5}$$

$$\boldsymbol{Hy}=(0,0,\cdots,0,1)^{\mathrm{T}},\tag{6}$$

分别有解 $\boldsymbol{x}$ 与 $\boldsymbol{y}$.

(3)$\boldsymbol{H}=(h_{i+j})_{i,j=0}^{n-1}$ 为 Hankel 矩阵，且有完全互素的多项式 $a(\lambda)=\sum_{j=0}^{n}a_j\lambda^j$ 与 $b(\lambda)=\sum_{j=0}^{n}b_j\lambda^j$ 满足

$$\begin{bmatrix} a_0 & a_1 & \cdots & a_{n-1} & a_n & & \\ & \ddots & \ddots & & \vdots & \ddots & \\ & & a_0 & a_1 & a_2 & \cdots & a_n \\ b_0 & b_1 & \cdots & b_{n-1} & b_n & & \\ & \ddots & \ddots & & \vdots & \ddots & \\ & & b_0 & b_1 & b_2 & \cdots & b_n \end{bmatrix}\begin{bmatrix} h_0 \\ h_1 \\ \vdots \\ \vdots \\ \vdots \\ h_{2n-2} \end{bmatrix}=\boldsymbol{0}.\tag{7}$$

(4)$\boldsymbol{H}=(h_{i+j})_{i,j=0}^{n-1}$ 为 Hankel 矩阵，且有完全互素的多项式 $a,b\in P_n^0(\mathbf{C})$ 使得下列缠绕关系成立

$$\boldsymbol{HC}_n^{\mathrm{T}}(a)=\boldsymbol{C}_n(a)\boldsymbol{H},\ \boldsymbol{HC}_n^{\mathrm{T}}(b)=\boldsymbol{C}_n(b)\boldsymbol{H}.\tag{8}$$

(5)$\boldsymbol{H}$ 为 n 阶 Hankel 矩阵，它与一对完全互素的多项式 a 与 b 相容.

(6)存在多项式 $a,b,p,q\in P_n^0(\mathbf{C})$，使得 $\max\{\deg a,\deg b\}=n$ 与 $aq+bp=1$，且若复数 α 与 β 使得 $\alpha a+\beta b$ 为 n 次首一多项式，则有

$$\boldsymbol{H}=\boldsymbol{H}_n((\alpha p-\beta q)/(\alpha a+\beta b)).\tag{9}$$

(7)存在完全互素多项式 $a,b\in P_n^0(\mathbf{C})$，且若复数 α 与 β 使得 n 次首一多项式 $\alpha a(\lambda)+\beta b(\lambda)=\prod_{j=1}^{s}(\lambda-t_j)^{n_j}$（$t_j$ 彼此不同，$n_j\geqslant1$），复数 δ 与 γ 满足 $\alpha\delta-\beta\gamma\neq0$，则有

$$\boldsymbol{H}=\sigma\boldsymbol{V}(\alpha a+\beta b)\boldsymbol{M}^{-1}\boldsymbol{V}^{\mathrm{T}}(\alpha a+\beta b),\tag{10}$$

其中，$\sigma\neq0$ 为常数，$\boldsymbol{M}=\operatorname{diag}[\boldsymbol{M}_1,\boldsymbol{M}_2,\cdots,\boldsymbol{M}_s]$ 为非奇异块对角矩阵. 而 $\boldsymbol{M}_i=\boldsymbol{R}_i[\gamma a(\boldsymbol{J}_i)+\delta b(\boldsymbol{J}_i)]\prod_{j\neq i}(\boldsymbol{J}_i-t_j\boldsymbol{I}_i)^{n_j}$ 为下三角 n_i 阶的 Hankel 矩阵，$\boldsymbol{R}_i,\boldsymbol{J}_i$ 含义同(3)式，$\boldsymbol{I}_i$ 为 n_i 阶单位矩阵.

(8)存在完全互素多项式 $a,b\in P_n^0(\mathbf{C})$，且若复数 $\alpha,\beta,\delta,\gamma$ 使得 $\alpha a+\beta b$ 有 n 个单根 $t_1,t_2,\cdots,t_n$，$\alpha\delta-\beta\gamma\neq0$，则有 $\boldsymbol{H}=\sigma\boldsymbol{V}(\alpha a+\beta b)\boldsymbol{M}^{-1}\boldsymbol{V}^{\mathrm{T}}(\alpha a+

βb)，其中，$\sigma\neq 0$ 为常数，$\boldsymbol{M}=\mathrm{diag}(m_1,m_2,\cdots,m_n)$非奇异，

$$m_i=(\gamma a(t_i)+\delta b(t_i))\prod_{j\neq i}(t_i-t_j),\ i=1,2,\cdots,n.$$

(9)存在完全互素多项式 $a,b\in P_n^0(\mathbf{C})$，且存在复数 $\alpha,\beta,\delta,\gamma$ 使得(8)中有关结果成立.

(10)存在 n 个不同复数 $t_1,t_2,\cdots,t_n$ 使得 $\boldsymbol{H}=\boldsymbol{V}(t)\boldsymbol{C}\boldsymbol{V}^{\mathrm{T}}(t)$，其中 $\boldsymbol{C}$ 为非奇异对角矩阵，$\boldsymbol{V}(t)$为与 $t_1,t_2,\cdots,t_n$ 对应的 Vandermonde 矩阵.

证 (1)⇒(2)，(3)⟺(5)，(7)⇒(8)与(9)⇒(10)显然.(1)⟺(5)：见〔1〕中定理 3.(8)⇒(9)：见引理Ⅰ.(10)⇒(1)：只要对 $\boldsymbol{H}$ 的元素进行简单计算，便知 $\boldsymbol{H}$ 为 n 阶非奇异 Hankel 矩阵.余下要证：(2)⇒(5)⇒(6)⇒(7).

(2)⇒(5)：设方程(5)与方程(6)的解分别为 $\boldsymbol{x}=(a_0,a_1,\cdots,a_{n-1})^{\mathrm{T}}$ 与 $\boldsymbol{y}=(b_0,b_1,\cdots,b_{n-1})^{\mathrm{T}}$.记 $a(\lambda)=\lambda^n-\sum_{j=0}^{n-1}a_j\lambda^j$ 与 $b(\lambda)=\sum_{j=0}^{n-1}b_j\lambda^j$.则显然有 $\hat{\boldsymbol{H}}[a]_n=\mathbf{0}$ 与 $\hat{\boldsymbol{H}}[b]_n=\mathbf{0}$，即 a 与 b 为 $\boldsymbol{H}$ 的相容多项式.下面证明 a 与 b 完全互素.为此只需证明：$aq+bp=1$，其中

$$p(\lambda)=(1,\lambda,\cdots,\lambda^{n-1})S(a)(h_0,h_1,\cdots,h_{n-1})^{\mathrm{T}},$$

$$q(\lambda)=-(1,\lambda,\cdots,\lambda^{n-2})S(b)(h_0,h_1,\cdots,h_{n-2})^{\mathrm{T}}=\sum_{j=0}^{n-2}q_j\lambda^j. \qquad (11)$$

事实上，假定 $aq+bp=ar+s$，$r(\lambda)=\sum_{j=0}^{n-2}r_j\lambda^j$ 与 $s(\lambda)=\sum_{j=0}^{n-1}s_j\lambda^j$.按〔1〕中定理 3，$\boldsymbol{H}=\boldsymbol{H}_n(p/a)$，于是 p/a 有 Laurent 展开 $\sum_{j=1}^{\infty}h_{j-1}\lambda^{-j}$.现设 s/a 有 Laurent 展开 $\sum_{j=1}^{\infty}\alpha_{j-1}\lambda^{-j}$.比较恒等式 $bp/a+q=r+s/a$ 两侧 λ^k($-n\leqslant k\leqslant n-2$)的系数便得出：$\boldsymbol{H}(b_0,b_1,\cdots,b_{n-1})^{\mathrm{T}}=(\alpha_0,\alpha_1,\cdots,\alpha_{n-1})^{\mathrm{T}}$ 与 $S(b)(h_0,h_1,\cdots,h_{n-2})^{\mathrm{T}}=(r_0-q_0,\cdots,r_{n-2}-q_{n-2})^{\mathrm{T}}$，因而从式(11)与式(6)推得 $r_i=\alpha_i=0(0\leqslant i\leqslant n-2)$，$\alpha_{n-1}=1$.于是，$r(\lambda)=0$ 与 $s/a=1/\lambda^n+\alpha_n/\lambda^{n+1}+\cdots$.但 $a(\lambda)$为首一 n 次多项式，前式蕴涵 $s(\lambda)=1$，从而 $aq+bp=1$.

(5)⇒(6)：设(5)成立.因 a 与 b 的线性组合仍为 $\boldsymbol{H}$ 的相容多项式，所以不妨可设 $a,b\in P_n(\mathbf{C})$.按〔1〕定理 3，存在 $p,q\in P_{n-1}^0(\mathbf{C})$ 使得 $\boldsymbol{H}=\boldsymbol{H}_n(p/a)=\boldsymbol{H}_n(q/b)$，因而 $\boldsymbol{H}_n((bp-aq)/(ab))=\boldsymbol{O}$，其中 $\deg(ab)=2n$ 与 $\deg(bp-aq)\leqslant 2n-1$.这个事实表明：$bp-aq$ 为常数 c，且 $c\neq 0$.(否则因

a 与 b 完全互素,必有 b 整除 q,因而 $q=0$ 与 $\boldsymbol{H}=\boldsymbol{O}$,与 $\boldsymbol{H}\neq\boldsymbol{O}$ 的假定条件矛盾.)现记 $\tilde{p}=p/\sqrt{c}$, $\tilde{q}=-q/\sqrt{c}$, $\tilde{a}=a/\sqrt{c}$ 与 $\tilde{b}=b/\sqrt{c}$. 则 $\tilde{a}$ 与 $\tilde{b}$ 仍为完全互素,并且,$\tilde{a}\tilde{q}+\tilde{b}\tilde{p}=1$. 如果复数 α 与 β 使得 $\alpha\tilde{a}+\beta\tilde{b}\in P_n(\mathbf{C})$ 为首一的,那么

$$\tilde{p}/\tilde{a}-(\alpha\tilde{p}-\beta\tilde{q})/(\alpha\tilde{a}+\beta\tilde{b})=\beta/(\tilde{a}(\alpha\tilde{a}+\beta\tilde{b})).$$

但 $\deg(\tilde{a}(\alpha\tilde{a}+\beta\tilde{b}))=2n$,上式右侧对应的 Hankel 矩阵 $\boldsymbol{H}_n(\beta/(\tilde{a}(\alpha\tilde{a}+\beta\tilde{b})))=\boldsymbol{O}$,于是

$$\boldsymbol{H}=\boldsymbol{H}_n(p/a)=\boldsymbol{H}_n(\tilde{p}/\tilde{a})=\boldsymbol{H}_n((\alpha\tilde{p}-\beta\tilde{q})/(\alpha\tilde{a}+\beta\tilde{b})).$$

(6)⇒(7):由(6)的第 1 部分条件知 a 与 b 完全互素,且有 $p,q\in P_n^0(\mathbf{C})$满足 $aq+bp=1$. 现若复数 $\alpha,\beta,\delta,\gamma$ 满足(7)中条件,则有

$$(\alpha a+\beta b)(\delta q-\gamma p)+(\gamma a+\delta b)(\alpha p-\beta q)=\alpha\delta-\gamma\beta\neq 0.$$

因为$(\alpha a+\beta b)(\boldsymbol{J}_i)=\boldsymbol{O}$,上式推出

$$\boldsymbol{O}\neq(\alpha\delta-\beta\gamma)\boldsymbol{I}_i=[\gamma a(\boldsymbol{J}_i)+\delta b(\boldsymbol{J}_i)][\alpha p(\boldsymbol{J}_i)-\beta q(\boldsymbol{J}_i)],$$

因而

$$\alpha p(\boldsymbol{J}_i)-\beta q(\boldsymbol{J}_i)=(\alpha\delta-\beta\gamma)[\gamma a(\boldsymbol{J}_i)+\delta b(\boldsymbol{J}_i)]^{-1}.$$

将定理Ⅱ应用于式(9)中矩阵 $\boldsymbol{H}$ 直接得式(10),其中 $\sigma=\alpha\delta-\beta\gamma\neq 0$,诸 $\boldsymbol{M}_i$ 显然为下三角 Hankel 矩阵,这是因为 $\boldsymbol{R}_i\boldsymbol{M}_i=(\gamma a+\delta b)(\boldsymbol{J}_i)\prod_{j\neq i}(\boldsymbol{J}_i-t_j\boldsymbol{I}_i)^{n_j}$ 为 n_i 阶的 Toeplitz 矩阵,$i=1,2,\cdots,s$. ■

作为定理 1 的应用,我们有如下两推论.

推论 1[5,9] 设 $f,g\in P_n^0(\mathbf{C})$为完全互素的. 则 $\boldsymbol{H}\in\mathbf{C}^{n\times n}$是与 f,g 相容的 Hankel 矩阵的充分与必要条件为

$$\boldsymbol{H}\boldsymbol{C}_n^{\mathrm{T}}(f)=\boldsymbol{C}_n(f)\boldsymbol{H},\ \boldsymbol{H}\boldsymbol{C}_n^{\mathrm{T}}(g)=\boldsymbol{C}_n(g)\boldsymbol{H}.$$

我们指出,当前推论中条件成立时,$\boldsymbol{H}=\boldsymbol{O}$ 或者 $\boldsymbol{H}$ 非奇异.

以下推论 2 提供 Hankel 矩阵求逆的行之有效的方法.

推论 2[7,8,10] 设 Hankel 矩阵 $\boldsymbol{H}=(h_{i+j})_{i,j=0}^{n-1}$使得方程(5)与方程(6)有解 $\boldsymbol{x}$ 与 $\boldsymbol{y}$,并设 $a(\lambda)=\lambda^n-(1,\lambda,\cdots,\lambda^{n-1})\boldsymbol{x}$ 与 $b(\lambda)=(1,\lambda,\cdots,\lambda^{n-1})\boldsymbol{y}$. 则 $\boldsymbol{H}$ 可逆且有

$$\boldsymbol{H}^{-1}=\boldsymbol{B}_n(a,b). \tag{12}$$

证 按定理 1 中(2)⇒(5)的证明,这时 a 与 b 完全互素,且有 $p,q\in P_{n-1}^0(\mathbf{C})$满足 $aq+bp=1$ 使得 $\boldsymbol{H}=\boldsymbol{H}_n(p/a)$非奇异. 根据 Cayley-Hamilton 定理,$\boldsymbol{I}=a(\boldsymbol{C}_a)q(\boldsymbol{C}_a)+b(\boldsymbol{C}_a)p(\boldsymbol{C}_a)=b(\boldsymbol{C}_a)p(\boldsymbol{C}_a)$. 因此

$$\boldsymbol{H}^{-1}=\boldsymbol{H}_n^{-1}(p/a)=(p(\boldsymbol{C}_a)S(a)^{-1})^{-1}=S(a)b(\boldsymbol{C}_a)=\boldsymbol{B}_n(a,b).$$ ■

§3. 非奇异 Bezout 矩阵的特征与求逆问题

以下定理是本节的主要结论，它刻画了非奇异 Bezout 矩阵.

定理 2 设 $\boldsymbol{B}$ 为 n 阶非零矩阵，$f,g\in P_n^0(\mathbf{C})$ 完全互素. 则下列各条彼此等价：

(1)$\boldsymbol{B}=\rho\boldsymbol{B}_n(f,g)$ 为非奇异的 Bezout 矩阵，这里 $\rho\neq0$ 为复数.

(2)$\boldsymbol{B}\boldsymbol{C}_n(f)=\boldsymbol{C}_n^{\mathrm{T}}(f)\boldsymbol{B},\boldsymbol{B}\boldsymbol{C}_n(g)=\boldsymbol{C}_n^{\mathrm{T}}(g)\boldsymbol{B}$.

(3)若复数 $\alpha,\beta,\delta,\gamma$ 满足 $\alpha\delta-\beta\gamma\neq0$ 且使得 $\alpha f+\beta g$ 为 n 次多项式：$\alpha f(\lambda)+\beta g(\lambda)=\prod\limits_{j=1}^{s}(\lambda-t_j)^{n_j}$，$t_j$ 不同，$n_j\geqslant1$，则有

$$\boldsymbol{V}^{\mathrm{T}}(\alpha f+\beta g)\boldsymbol{B}\boldsymbol{V}(\alpha f+\beta g)=\sigma\mathrm{diag}[\boldsymbol{C}_1,\boldsymbol{C}_2,\cdots,\boldsymbol{C}_s],\tag{13}$$

其中 σ 为非零复数，$\boldsymbol{C}_i=\boldsymbol{R}_i(\gamma f+\delta g)(\boldsymbol{J}_i)\prod\limits_{j\neq i}(\boldsymbol{J}_i-t_j\boldsymbol{I}_i)^{n_j}$ 是非奇异下三角 Hankel 矩阵，这里 $\boldsymbol{R}_i,\boldsymbol{J}_i$ 与 $\boldsymbol{I}_i$ 的含义同式(10)中有关 $\boldsymbol{M}$ 的记号.

(4)若复数 $\alpha,\beta,\delta,\gamma$ 满足 $\alpha\delta-\beta\gamma\neq0$ 且使得 $\alpha f+\beta g$ 为首一，有 n 个单根 $t_1,t_2,\cdots,t_n$，则

$$\boldsymbol{V}^{\mathrm{T}}(\alpha f+\beta g)\boldsymbol{B}\boldsymbol{V}(\alpha f+\beta g)=\sigma\mathrm{diag}[c_1,c_2,\cdots,c_n],\tag{14}$$

其中，σ 为非零复数，$c_i=(\gamma f(t_i)+\delta g(t_i))\prod\limits_{j\neq i}(t_i-t_j)\neq0,i=1,2,\cdots,n$.

(5)存在复数 $\alpha,\beta,\delta,\gamma$ 使得 $\alpha\delta-\beta\gamma\neq0$，多项式 $\alpha f+\beta g$ 为首一，有 n 个单根 $t_1,t_2,\cdots,t_n$，且式(14)成立.

证 由文献〔1〕中定理 1 易得(1)⇒(2). (3)⇒(4)是显然的，而按引理Ⅰ，(4)⇒(5)也显然. 余下只需证明：(2)⇒(1)⇒(3)；(5)⇒(1).

(2)⇒(1)：设(2)且 $f,g\in P_n^0(\mathbf{C})$ 完全互素. 因 $\boldsymbol{C}_n(f+g)=\boldsymbol{C}_n(f)+\boldsymbol{C}_n(g)$ 与 $\boldsymbol{C}_n(\alpha f)=\alpha\boldsymbol{C}_n(f)$，$\forall\alpha\in\mathbf{C}$，我们不妨可设 $f,g\in P_n(\mathbf{C})$. 这时文献〔1〕中定理 1 保证存在非零多项式 $f_1,g_1\in P_{n-1}^0(\mathbf{C})$ 使得 $\boldsymbol{B}=\boldsymbol{B}_n(f,f_1)=\boldsymbol{B}_n(g,g_1)$. 但按一个已知事实[8]，此时有非零复数 β,γ 使得 $f-\beta g=\gamma f_1$. 现令 $\rho=-\beta/\gamma(\neq0)$. 则 $\boldsymbol{B}=\boldsymbol{B}_n(f,f/\gamma+\rho g)=\rho\boldsymbol{B}_n(f,g)$，它是非奇异的 Bezout 矩阵.

(1)⇒(3)：令 $\sigma=\rho/(\alpha\delta-\beta\gamma)(\neq0)$. 应用定理Ⅰ我们有

$$\begin{aligned}&\boldsymbol{V}^{\mathrm{T}}(\alpha f+\beta g)\boldsymbol{B}\boldsymbol{V}(\alpha f+\beta g)=\boldsymbol{V}^{\mathrm{T}}(\alpha f+\beta g)\rho\boldsymbol{B}_n(f,g)\boldsymbol{V}(\alpha f+\beta g)\\&=\sigma\boldsymbol{V}^{\mathrm{T}}(\alpha f+\beta g)\boldsymbol{B}_n(\alpha f+\beta g,\gamma f+\delta g)\boldsymbol{V}(\alpha f+\beta g)=\sigma\mathrm{diag}[\boldsymbol{C}_1,\boldsymbol{C}_2,\cdots,\boldsymbol{C}_s],\end{aligned}$$

其中，$\boldsymbol{C}_i=\boldsymbol{R}_i(\gamma f+\delta g)(\boldsymbol{J}_i)\prod_{j\neq i}(\boldsymbol{J}_i-t_j\boldsymbol{I}_i)^{n_j}$ 为非奇异的 n_i 阶 Hankel 矩阵，这是因为 $\boldsymbol{B}$ 非奇异且 $\boldsymbol{R}_i\boldsymbol{C}_i$ 为 n_i 阶 Toeplitz 矩阵，$i=1,2,\cdots,s$.

(5)⇒(1)：令 $\rho=\alpha\delta-\beta\gamma(\neq 0)$. 由定理Ⅰ，

$$\boldsymbol{V}^{\mathrm{T}}(\alpha f+\beta g)\boldsymbol{B}_n(\alpha f+\beta g,\gamma f+\delta g)\boldsymbol{V}(\alpha f+\beta g)=\mathrm{diag}[c_1,c_2,\cdots,c_n],$$

其中诸 $c_i\neq 0$，因而从 $\boldsymbol{V}(\alpha f+\beta g)$ 非奇异与式(14)便得 $\boldsymbol{B}=\rho\boldsymbol{B}_n(f,g)$，它显然是非奇异的 Bezout 矩阵. ■

作为前一定理的直接推论，我们有

推论 3　设 $f,g\in P_n^0(\mathbf{C})$ 完全互素. 则有 n 个不同复数 $t_1,t_2,\cdots,t_n$ 使得 $\boldsymbol{V}^{\mathrm{T}}\boldsymbol{B}_n(f,g)\boldsymbol{V}$ 为对角矩阵，这里，$\boldsymbol{V}$ 为与 $t_1,t_2,\cdots,t_n$ 相对应的 Vandermonde 矩阵.

联系定理 1，定理 2 与推论 2，可得如下经典结论[1]及其补充[5].

推论 4　非奇异 Bezout(Hankel)矩阵之逆是 Hankel (Bezout)矩阵，且它们有相同的相容多项式.

推论 5　设 $f,g\in P_n^0(\mathbf{C})$ 完全互素，且有 $p,q\in P_n^0(\mathbf{C})$ 满足 $\deg p\leqslant\deg f$ 与 $fq+gp=1$. 则 $\boldsymbol{B}=\boldsymbol{B}_n(f,g)$ 可逆，并且，存在复数 $\rho\neq 0$，使得只要 $\alpha f+\beta g\in P_n(\mathbf{C})(\alpha,\beta\in\mathbf{C})$，就有

$$\boldsymbol{B}^{-1}=\rho\boldsymbol{H}_n((\alpha p-\beta q)/(\alpha f+\beta g)).$$

证　从推论 4 与定理 1 即可证得. ■

下面研究具有重要应用价值的 Bezout 矩阵求逆问题. 据我们所知，如下有关结果都是崭新的.

定理 3　设 $f(\lambda)=\lambda^n-\sum_{j=0}^{n-1}f_j\lambda^j\,(f_0\neq 0)$ 与 $g(\lambda)=\sum_{j=0}^{n}g_j\lambda^j$ 完全互素. 则 $\boldsymbol{B}=\boldsymbol{B}_n(f,g)$ 可逆，$\boldsymbol{B}^{-1}$ 为 Hankel 矩阵 $\boldsymbol{H}=(h_{i+j})_{i,j=0}^{n-1}$，其元素满足方程：

$$\boldsymbol{B}(h_n,h_{n+1},\cdots,h_{2n-2},\xi)^{\mathrm{T}}=(f_0,f_1,\cdots,f_{n-1})^{\mathrm{T}},\tag{15}$$

$$\boldsymbol{B}(\eta,h_0,h_1,\cdots,h_{n-2})^{\mathrm{T}}=-(f_1,f_2,\cdots,f_{n-1},-1)^{\mathrm{T}}/f_0,\tag{16}$$

其中，$\xi=h_{n-1}f_0+h_nf_1+\cdots+h_{2n-2}f_{n-1}$ 与 $\eta=(h_{n-1}-h_{n-2}f_{n-1}-\cdots-h_0f_1)/f_0$. 此外，若 $\boldsymbol{B}$ 的元素 $b_{0,n-1}\neq 0$，则 $\boldsymbol{H}$ 的元素 $h_0,h_1,\cdots,h_{n-2}$ 也满足方程：

$$\boldsymbol{B}(\nu,h_0,h_1,\cdots,h_{n-2})^{\mathrm{T}}=-(b_{1,n-1},b_{2,n-2},\cdots,b_{n-1,n-1},0)^{\mathrm{T}}/b_{0,n-1},\tag{17}$$

其中，ν 为类似于上述 ξ 与 η 的合适的复数.

证 由于 f 是 Hankel 矩阵 $\boldsymbol{H}=\boldsymbol{B}^{-1}$ 的相容多项式，故有 $\boldsymbol{H}(f_0,f_1,\cdots,f_{n-1})^{\mathrm{T}}=(h_n,h_{n+1},\cdots,h_{2n-2},\xi)^{\mathrm{T}}$，这里，$\xi=h_{n-1}f_0+\cdots+h_{2n-2}f_{n-1}$，此等价于方程(15). 现设 $\hat{f}(\lambda)=\lambda^n f(\lambda^{-1})$ 与 $\hat{g}(\lambda)=\lambda^n g(\lambda^{-1})$，$\boldsymbol{P}$ 为 n 阶翻转矩阵. 则有 $\boldsymbol{PB}_n(f,g)\boldsymbol{P}=-\boldsymbol{B}_n(\hat{f},\hat{g})$，因而 $\boldsymbol{B}_n(\hat{f},\hat{g})^{-1}=-\boldsymbol{PHP}$. 引用推论 4，这表明 $-\hat{f}/f_0$ 为 Hankel 矩阵 $\boldsymbol{PHP}$ 的首一相容多项式. 用刚才的推证便得方程(16). 特别地，令 $p(\lambda)=(1,\lambda,\cdots,\lambda^{n-1})\boldsymbol{B}(0,0,\cdots,0,1)^{\mathrm{T}}$. 容易看出，$p$ 也是 $\boldsymbol{B}$ 的相容多项式，且 $b_{0,n-1}\neq 0$ 等价于 $p(0)\neq 0$，因而 $\hat{p}$ 为 $\boldsymbol{PBP}$ 与 $\boldsymbol{PHP}$ 的 n 次相容多项式. 用前面的推证便得方程(17). ■

假如 Bezout 矩阵没有明显地与它的相容多项式联系在一起，则应用以下定理求逆更为方便.

定理 4 若 $\boldsymbol{B}$ 为 n 阶 Bezout 矩阵，则 $\boldsymbol{B}$ 可逆当且仅当方程

$$\boldsymbol{Bx}=\boldsymbol{e}_0=(1,0,\cdots,0)^{\mathrm{T}}, \tag{18}$$

$$\boldsymbol{By}=\boldsymbol{e}_{n-1}=(0,0,\cdots,0,1)^{\mathrm{T}} \tag{19}$$

有解 $\boldsymbol{x}=(x_0,x_1,\cdots,x_{n-1})^{\mathrm{T}}$ 与 $\boldsymbol{y}=(y_0,x_n,\cdots,x_{2n-2})^{\mathrm{T}}$. 此时，$y_0=x_{n-1}$，$\boldsymbol{B}^{-1}=(x_{i+j})_{i,j=0}^{n-1}$.

证 必要性是显然的. 现证条件的充分性. 方程(19)有解表明 $\boldsymbol{B}$ 的最后一列(行)不为零向量，但按 Bezout 矩阵的定义，这蕴涵 $\boldsymbol{B}$ 有 n 次相容多项式 f. 引用文献〔1〕定理 1，有 $\boldsymbol{BC}_f=\boldsymbol{C}_f^{\mathrm{T}}\boldsymbol{B}$，从而 $\boldsymbol{BC}_f^k=(\boldsymbol{C}_f^k)^{\mathrm{T}}\boldsymbol{B}$，$k=0,1,\cdots$. 这时若有 $\boldsymbol{u}=(u_0,u_1,\cdots,u_{n-1})^{\mathrm{T}}$ 满足 $\boldsymbol{u}^{\mathrm{T}}\boldsymbol{B}=\boldsymbol{0}^{\mathrm{T}}$，则 $u_k=\boldsymbol{u}^{\mathrm{T}}\boldsymbol{e}_k=\boldsymbol{u}^{\mathrm{T}}(\boldsymbol{C}_f^k)^{\mathrm{T}}\boldsymbol{e}_0=\boldsymbol{u}^{\mathrm{T}}(\boldsymbol{C}_f^k)^{\mathrm{T}}\boldsymbol{Bx}=\boldsymbol{u}^{\mathrm{T}}\boldsymbol{BC}_f^k\boldsymbol{x}=0$，$k=0,1,\cdots,n-1$，因而 $\boldsymbol{u}=\boldsymbol{0}$. 这表明 $\boldsymbol{B}$ 为可逆的. 最后，若式(18)与式(19)有解 $\boldsymbol{x}$ 与 $\boldsymbol{y}$，则 $x_{n-1}=\boldsymbol{e}_{n-1}^{\mathrm{T}}\boldsymbol{x}=\boldsymbol{y}^{\mathrm{T}}\boldsymbol{Bx}=\boldsymbol{y}^{\mathrm{T}}\boldsymbol{e}_0=y_0$，且从式(18)与式(19)立即推得 Hankel 矩阵

$$\boldsymbol{B}^{-1}=(x_{i+j})_{i,j=0}^{n-1}.$$ ■

推论 6 设 $\boldsymbol{B}$ 为 n 阶 Bezout 矩阵，其最后一列不为零向量. 则 $\boldsymbol{B}$ 可逆当且仅当方程(18)有解 $\boldsymbol{x}$. 此时，若方程(18)与方程(19)的解 $\boldsymbol{x}$ 与 $\boldsymbol{y}$ 如定理 4 所示，则 $x_{n-1}=y_0$，且 $\boldsymbol{B}^{-1}=(x_{i+j})_{i,j=0}^{n-1}$.

推论 7 设 $\boldsymbol{B}$ 为 n 阶 Bezout 矩阵，其第 1 列不为零向量. 则 $\boldsymbol{B}$ 可逆当且仅当方程(19)有解 $\boldsymbol{y}$. 此时，若方程(18)与方程(19)的解 $\boldsymbol{x}$ 与 $\boldsymbol{y}$ 如定理 4 所示，则 $x_{n-1}=y_0$，且 $\boldsymbol{B}^{-1}=(x_{i+j})_{i,j=0}^{n-1}$.

推论 8 设 $\boldsymbol{B}$ 为 n 阶 Bezout 矩阵，且方程(18)与方程(19)分别有解 $\boldsymbol{x}$ 与 $\boldsymbol{y}$，则多项式 $a(\lambda)=\lambda^n-(1,\lambda,\cdots,\lambda^n)\boldsymbol{BUy}$ 与 $b(\lambda)=1-(\lambda,\lambda^2,\cdots,$

$\lambda^n)\boldsymbol{B}\boldsymbol{U}^{\mathrm{T}}\boldsymbol{x}$皆为 $\boldsymbol{B}$ 的相容多项式，这里 $\boldsymbol{U}$ 表示 n 阶移位矩阵$(\delta_{i,j-1})_{i,j=0}^{n-1}$.

值得指出的是，n 阶非奇异 Bezout(Hankel)矩阵一定有 n 次相容多项式，因而它满足文献[1]中定理 1(定理 3)所罗列的众多性质. 下面仅举 1 个例子说明一下这些性质的应用：让我们重新证明下述的多项式惯性理论中著名的 Schur-Cohn 定理[12].

定理 5　设 $a(\lambda)=\sum_{j=0}^{n}a_j\lambda^j(a_n=1)$ 与$\hat{a}(\lambda)=\lambda^n\bar{a}(\lambda^{-1})$使得 $\boldsymbol{B}=\boldsymbol{B}_n(a,\hat{a})$非奇异，且令$\breve{\boldsymbol{B}}=\boldsymbol{P}\boldsymbol{B}$，这里 $\boldsymbol{P}$ 为 n 阶翻转矩阵. 则有 $\mathrm{In}^0(a)=\mathrm{In}(\breve{\boldsymbol{B}})$，这里，$\mathrm{In}^0(a)$与 $\mathrm{In}(\breve{\boldsymbol{B}})$分别表示多项式 a 关于单位圆周的惯性与矩阵$\breve{\boldsymbol{B}}$关于虚轴的惯性.

证　容易看出，$\breve{\boldsymbol{B}}=\breve{\boldsymbol{B}}^*=\boldsymbol{P}\boldsymbol{B}$ 为可逆的. 按文献[1]中定理 1，若令 $w(\lambda)=\sum_{j=0}^{n-1}w_j\lambda^j=\hat{a}(\lambda)-\bar{a}_0a(\lambda)\ (\neq 0)$，$\boldsymbol{u}=(a_1,a_2,\cdots,a_{n-1},1)^{\mathrm{T}}$ 与$[w]_n=(w_0,w_1,\cdots,w_{n-1})^{\mathrm{T}}$，则有 $\boldsymbol{B}-\boldsymbol{U}\boldsymbol{B}\boldsymbol{C}_a=\boldsymbol{u}[w]_n^{\mathrm{T}}$，因而 $\boldsymbol{P}\boldsymbol{B}-\boldsymbol{U}^{\mathrm{T}}\boldsymbol{P}\boldsymbol{B}\boldsymbol{C}_a=\breve{\boldsymbol{B}}-\boldsymbol{U}^{\mathrm{T}}\breve{\boldsymbol{B}}\boldsymbol{C}_a=\boldsymbol{P}\boldsymbol{u}[w]_n^{\mathrm{T}}$，其中 $\boldsymbol{U}$ 为 n 阶移位矩阵. 因此，$\breve{\boldsymbol{B}}-\boldsymbol{C}_a^*\breve{\boldsymbol{B}}\boldsymbol{C}_a=(-\boldsymbol{C}_a^*+\boldsymbol{U}^{\mathrm{T}})\breve{\boldsymbol{B}}\boldsymbol{C}_a+\boldsymbol{P}\boldsymbol{u}[w]^{\mathrm{T}}=[\boldsymbol{P}\boldsymbol{U}-a_0(\bar{a}_0,\bar{a}_1,\cdots,\bar{a}_{n-1})^{\mathrm{T}}][w]_n^{\mathrm{T}}=[\bar{w}]_n[w]_n^{\mathrm{T}}\geqslant\boldsymbol{O}$，从而按 Stein 稳定性定理的推广[12]，我们有 $\mathrm{In}^0(\boldsymbol{C}_a^*)=\mathrm{In}^0(a)=\mathrm{In}(\breve{\boldsymbol{B}})$.

■

参考文献

[1] 杨淑蓉，张惠品，陈公宁. Bezout 和 Hankel 矩阵. 北京师范大学学报(自然科学版)，1991，27(增刊 2)：20.

[2] Fiedler M，Ptak V. Bezoutians and intertwining matrices. Linear Algebra Appl，1987，86：43.

[3] Sansigre G，Alvarez M. On Bezoutian reduction with Vandermonde matrix. Linear Algebra Appl，1989，121：401.

[4] Chen G N，Yang Z H. Bezoutian representation via Vandermonde matrices. J BJ Nor Univ (Nat sci)，1991，27(3)：307.

[5] Fiedler M. Hankel and Löwner matrices. Linear Algebra Appl，1984，58：75.

[6] Fiedler M. Special matrices and its applications in numerical mathematics. Boston：Martinus Nijhoff Publ，1986.

[7] Heinig G，Jungnickel U. Hankel matrices generated by the Markov parame-

ters of rational functions. Linear Algebra Appl, 1986, 76: 121.

[8] Heinig G, Rost K. Algebraic methods for Toeplitz-like matrices and operators. In: Operator theory: Advance and applications, Vol. 13. Basel: Birkhäuser Verlag, 1984.

[9] Fiedler M. Polynomials and Hankel matrices. Linear Algebra Appl, 1985, 66: 235.

[10] Trench W F. An algorithm for the inversion of finite Toeplitz matrices. J SIAM, 1964, 12: 515.

[11] Lander F J. The Bezoutian and inversion of Hankel and Toeplitz matrices (in Russion). Mat Issled, 1974, 9(32): 69.

[12] 陈公宁.矩阵理论与应用.北京:高等教育出版社,1990.

BEZOUT AND HANKEL MATRICES(Ⅱ)
——THE NONSINGULAR CASE

Abstract The characterizations of nonsingular Hankel and Bezout matrices are demonstrated precisely, and the inversion problem of the two classes of matrices is discussed.

Keywords Bezout matrix; Hankel matrix; compatible polynomial; Vandermonde matrix; inversion of Bezout matrix

Linear Algebra and Its Applications
1993, 186:37－44

通过 Vandermonde 矩阵的 Bezout 矩阵表示①

Bezoutian Representation via Vandermonde Matrices

Abstract The result on Bezoutian representation via Vandermonde matrices is provided with a proof which is based on polynomial models.

§ 1. Introduction

Given a pair $f(z)$, $g(z)$ of polynomials of degree at most n, the Bezoutian $\boldsymbol{B}=\boldsymbol{B}(f, g)=(b_{ij})$ of f and g is a symmetric matrix of order n defined through the generating polynomial

$$\frac{f(z)g(w)-f(w)g(z)}{z-w}=\sum_{i,j=0}^{n-1} b_{ij} z^i w^j.$$

The Barnett factorization formula is fundamental for the study of Bezoutians:

$$\boldsymbol{B}(f,g)=\boldsymbol{S}(f)g(\boldsymbol{C}_f), \tag{1}$$

Where

$$f(z)=\sum_{j=0}^{n} f_j z^j \quad (f_n \neq 0), \quad g(z)=\sum_{j=0}^{n} g_j z^j, \tag{2}$$

and

① Received 22 May 1991; final manuscript accepted 18 June 1992.
本文与杨正宏合作.

$$\boldsymbol{S}(f)=\begin{bmatrix} f_1 & f_2 & \cdot & \cdot & \cdot & \cdot & f_n \\ f_2 & f_3 & \cdot & \cdot & \cdot & f_n & \\ \cdot & & & & \cdot & & \\ \cdot & & & \cdot & & & \\ \cdot & & \cdot & & & & \\ \cdot & \cdot & & & & 0 & \\ f_n & & & & & & \end{bmatrix} \quad \text{and}$$

$$\boldsymbol{C}_f=\begin{bmatrix} 0 & 1 & & & & \\ & & \cdot & & & \\ & \cdot & & \cdot & & \\ & & \cdot & & \cdot & \\ & & & \cdot & & 1 \\ & & & & \cdot & \\ \dfrac{-f_0}{f_n} & \dfrac{-f_1}{f_n} & \cdot & \cdot & \cdot & \dfrac{-f_{n-1}}{f_n} \end{bmatrix} \tag{3}$$

stand for the symmetrizer and the companion matrix of f, respectively (see [8]).

The next result on diagonalization of the Bezoutian by congruence via a Vandermonde matrix is well known (Lander [9], also Datta [1] and Furhmann and Datta [5]), and is usually shown for the case that $f(z)$ has simple roots.

Theorem 1 *Let $f(z)$ and $g(z)$ be of the form (2) with $f_n=1$. If f has simple roots $t_1, t_2, \cdots, t_n$, then the Bezoutian $B(f, g)$ satisfies*

$$\boldsymbol{V}^{\mathrm{T}}\boldsymbol{B}(f, g)\boldsymbol{V}=\mathrm{diag}[f'(t_i)g(t_i)]_{i=1}^{n},$$

Where $\boldsymbol{V}$ is the Vandermonde matrix $\boldsymbol{V}(t_1, t_2, \cdots, t_n)$.

The Bezoutian can be brought into block diagonal form also in the general case (see [4]). Recently Sansigre and Alvarez [11] obtained a generalization of Theorem 1 as follows.

Theorem 2 *Let $f(z)$ be of the form (2) and*

$$f(z)=\prod_{i=1}^{s}(z-t_i)^{n_i}, \quad t_i \text{ distinct}, \quad n_i \geqslant 1, \tag{4}$$

and let $g(z)$ be a polynomial of degree at most n. Then the Bezoutian

$\boldsymbol{B}(f, g)$ *satisfies*

$$\mathbf{V}_f^{\mathrm{T}}\boldsymbol{B}(f, g)\mathbf{V}_f=\operatorname{diag}[\boldsymbol{R}_k\hat{f}_k(\boldsymbol{J}_{n_k}(t_k))g(\boldsymbol{J}_{n_k}(t_k))]_{k=1}^{s}, \tag{5}$$

where $\mathbf{V}_f$ *is the generalized Vandermonde matrix associated with* f, $\hat{f}_k(z)=f(z)/(z-t_k)^{n_k}$, $\boldsymbol{R}_k$ *the flip matrix of order* n_k, *and* $\boldsymbol{J}_m(t)$ *the Jordan block (with* t *along the diagonal) of order* m.

Here by the generalized Vandermonde matrix associated with f (with ordered roots) we mean the matrix of order n

$$\mathbf{V}_f=(\boldsymbol{S}_{n_1}(t_1), \boldsymbol{S}_{n_2}(t_2), \cdots, \boldsymbol{S}_{n_s}(t_s)), \tag{6}$$

where

$$\boldsymbol{S}_{n_i}(t_i)=\left[\binom{k}{j}t_i^{k-j}\right]_{k, j=0}^{n-1, n_i-1}, \quad i=1, 2, \cdots, s.$$

It is well known that $\mathbf{V}_f$ is invertible [see also (14) below], and that

$$\boldsymbol{C}_f\mathbf{V}_f=\mathbf{V}_f\boldsymbol{J}_f, \tag{7}$$

where

$$\boldsymbol{J}_f=\operatorname{diag}[\boldsymbol{J}_{n_k}(t_k)]_{k=1}^{s}. \tag{8}$$

In the present note a proof to Theorem 2 is given, which is based on polynomial models due to Fuhrmann [6]. This result then is applied to obtain an analogous representation for Hankel matrices.

§ 2. Proof of Theorem 2

In order to describe polynomial models we take up some notation. A rational function $u\in\mathbf{C}(z)$ can be decomposed as $u=v+w$, such that $v\in\mathbf{C}[z]$ (the ring of polynomials over $\mathbf{C}$) and $w=0$ or $w=b/c$, b, $c\in\mathbf{C}[z]$, $\deg b<\deg c$. We define the projections π_+ and π_- by

$$\pi_+u=v \quad \text{and} \quad \pi_-u=w.$$

For a given polynomial $p(z)=\sum_{j=0}^{n}p_jz^j \quad (p_n=1)$, put

$$\pi_ph=p\pi_-(p^{-1}h), \quad h\in\mathbf{C}[z].$$

The range of π_p, $\boldsymbol{X}_p=\operatorname{Im}\pi_p$ is just $\boldsymbol{P}_{n-1}$, the space of all complex polynomials of degree not exceeding $n-1$. Denote by $\boldsymbol{S}_p$ the map on $\boldsymbol{X}_p$ given by

$$S_p f=\pi_p(zf) \quad \forall f\in X_p,$$

and by $B_{st}=\{1, z, \cdots, z^{n-1}\}$ and $B_{co}=\{e_1, e_2, \cdots, e_n\}$ the standard basis and the control basis for X_p, respectively, where $e_i(z)=p_i+p_{i+1}z+\cdots+z^{n-i}$. It has been shown [7] that

$$[S_p]_{st}^{st}=C_p^{T}, \tag{9}$$

$$[I_p]_{co}^{st}=S(p), \tag{10}$$

where I stands for the identity operator on X_p, and $[I]_{co}^{st}$ the representation of I with respect to the ordered pair of bases B_{co} and B_{st}; similarly for the others. Also, B_{st} and B_{co} are dual to each other under the pairing $\langle , \rangle$ defined by

$$\langle u, w\rangle=[p^{-1}u, w], \quad u, w\in \mathbf{C}(z),$$

whereas

$$[u, w]=\sum_{j=-\infty}^{\infty} u_{-j-1}w_j$$

whenever $u(z)=\sum_{j=-\infty}^{n_u} u_j z^j$ and $w(z)=\sum_{j=-\infty}^{n_w} w_j z^j$. Thus by duality

$$[S_p]_{co}^{co}=C_p. \tag{11}$$

It follows from (1)(10) and (11) that

$$B(p, q)=[I]_{co}^{st}[q(S_p)]_{co}^{co}=[q(S_p)]_{co}^{st}. \tag{12}$$

Proof of Theorem 2 Let $G_{kj}\in P_{n-1}$, $k=1, 2, \cdots, s$, $j=0, 1, \cdots, n_k-1$, be such that

$$G_{kj}^{(r)}(t_i)=j!\delta_{ki}\delta_{rj}, \quad i=1, 2, \cdots, s, \quad r=0, 1,\cdots, n_i-1.$$

Then $B_{in}=\{G_{10},\cdots,G_{1,n_1-1},\cdots,G_{s0},\cdots,G_{s,n_s-1}\}$ forms a basis for P_{n-1} by the theory of Hermite interpolation. For the polynomial $f(z)$ in (4), we have

$$f(z)=\sum_{k=1}^{s}\sum_{j=0}^{n_k-1} f_{kj}G_{kj}(z),$$

where $f_{kj}=f^{(j)}(t_k)/j!\ \forall k, j$. A simple calculation leads to

$$[I]_{st}^{in}=V_f^{T}. \tag{13}$$

Now put

$$F_{kj}(z)=\frac{f(z)}{(z-t_k)^{j+1}}, \quad k=1,2,\cdots,s, \quad j=0,1,\cdots,n_k-1.$$

Then $\boldsymbol{B}_{\mathrm{sp}}=\{F_{10},\cdots,F_{1,n_1-1},\cdots,F_{s0},\cdots,F_{s,n_s-1}\}$ is also a basis for $\boldsymbol{P}_{n-1}$, called the spectral basis. Compute the indefinite metric $\langle G_{ki}, F_{qj}\rangle\ \forall i, j, k, q$:

$$\langle G_{ki}, F_{qj}\rangle=[G_{ki}, (z-t_q)^{-j-1}]=\frac{G_{ki}^{(j)}(t_q)}{j!}=\delta_{kq}\delta_{ji}.$$

This means that $\boldsymbol{B}_{\mathrm{sp}}$ is the dual basis of $\boldsymbol{B}_{\mathrm{in}}$, and by (13) and the duality of $\boldsymbol{B}_{\mathrm{st}}$ and $\boldsymbol{B}_{\mathrm{co}}$ we derive

$$[\boldsymbol{I}]_{\mathrm{sp}}^{\mathrm{co}}=\boldsymbol{V}_f. \tag{14}$$

The trivial operator identity $\boldsymbol{S}_f\boldsymbol{I}=\boldsymbol{I}\boldsymbol{S}_f$ implies

$$[\boldsymbol{S}_f]_{\mathrm{co}}^{\mathrm{co}}[\boldsymbol{I}]_{\mathrm{sp}}^{\mathrm{co}}=[\boldsymbol{I}]_{\mathrm{sp}}^{\mathrm{co}}[\boldsymbol{S}_f]_{\mathrm{sp}}^{\mathrm{sp}}.$$

It follows from (11)(14) and (7) that

$$[\boldsymbol{S}_f]_{\mathrm{sp}}^{\mathrm{sp}}=\mathrm{diag}[\boldsymbol{J}_{n_k}(t_k)]_{k=1}^{s},$$

and by duality

$$[\boldsymbol{S}_f]_{\mathrm{in}}^{\mathrm{in}}=\mathrm{diag}[\boldsymbol{J}_{n_k}(t_k)^{\mathrm{T}}]_{k=1}^{s}. \tag{15}$$

Let $\boldsymbol{A}^{[h,k]}=(a_{ij}^{[h,k]})$, $h, k=1, 2, \cdots, s$, be the $n_h\times n_k$ matrix such that

$$[F_{k0},\cdots,F_{k,n_k-1}]=[G_{10},\cdots,G_{s,n_s-1}][\boldsymbol{A}^{[h,k]}]_{h=1}^{s}.$$

Then

$$\begin{aligned}F_{kq}^{(r)}(t_k)\delta_{kv} &= F_{kq}^{(r)}(t_v) = \sum_{j=1}^{s}\sum_{i=0}^{n_j-1}a_{iq}^{[j,k]}G_{ji}^{(r)}(t_v)\\ &=r!a_{rq}^{[v,k]},\quad q=0,1,\cdots,n_k-1,\ r=0,1,\cdots,n_v-1,\end{aligned}$$

whence $a_{rq}^{[k]}\equiv a_{rq}^{[k,k]}=F_{kq}^{(r)}(t_k)/r!\ \forall r, q$, and $\boldsymbol{A}^{[h,k]}=0\quad\forall h\neq k$. But $F_{kq}^{(r)}(t_k)=0$ for $r+q\leqslant n_k-2$, so $a_{rq}^{[k]}=0$ if $r+q\leqslant n_k-2$. For $r+q\geqslant n_k-1$

$$F_{kq}^{(r)}(t_k)=\frac{r!}{(r+q+1-n_k)!}\hat{f}_k(t_k),$$

where $\hat{f}_k(z)=f(z)/(z-t_k)^{n_k}$, so that for $r+q\geqslant n_k-1$

$$a_{rq}^{[k]}=\frac{\hat{f}_k^{(r+q+1-n_k)}(t_k)}{(r+q+1-n_k)!},$$

which implies

$$\boldsymbol{A}_k\equiv\boldsymbol{A}^{[k,k]}=\boldsymbol{R}_k\hat{f}_k(\boldsymbol{J}_{n_k}(t_k)),\quad k=1,2,\cdots,s,$$

where $\boldsymbol{R}_k$ is the flip matrix of order n_k. Thus

$$[\boldsymbol{I}]_{\mathrm{sp}}^{\mathrm{in}}=\mathrm{diag}[\boldsymbol{R}_k\hat{f}_k(\boldsymbol{J}_{n_k}(t_k))]_{k=1}^{s}. \tag{16}$$

From the equality $\boldsymbol{I}g(\boldsymbol{S}_f)\boldsymbol{I}=g(\boldsymbol{S}_f)\boldsymbol{I}$ we get

$$[\boldsymbol{I}]_{st}^{in}[g(\boldsymbol{S}_f)]_{co}^{st}[\boldsymbol{I}]_{sp}^{co}=[g(\boldsymbol{S}_f)]_{in}^{in}[\boldsymbol{I}]_{sp}^{in},$$

and therefore by (13)(12)(14)(15), and (16) the result (5) follows, as $\boldsymbol{R}_k^2=\boldsymbol{I}$ and $\boldsymbol{R}_k\boldsymbol{J}_{n_k}(t_k)\boldsymbol{R}_k=\boldsymbol{J}_{n_k}^{\mathrm{T}}(t_k)$ for each k. ■

Incidentally the representation (14) of $\boldsymbol{V}_f$ is an alternative proof of the nonsingularity of the generalized Vandermonde matrix.

As an immediate corollary we obtain the reduction of Hankel matrices with the generalized Vandermonde matrix.

Corollary *Let $f(z)$, $g(z)$ and $\hat{f}_k(z)$ be as in Theorem 2. Then the Hankel matrix $\boldsymbol{H}=\boldsymbol{H}(g/f)$ satisfies*

$$\boldsymbol{V}_f^{-1}\boldsymbol{H}(\boldsymbol{V}_f^{-1})^{\mathrm{T}}=\mathrm{diag}[g(\boldsymbol{J}_{n_k}(t_k))\hat{f}_k(\boldsymbol{J}_{n_k}(t_k))^{-1}\boldsymbol{R}_k]_{k=1}^{s}. \qquad (17)$$

In the case that $f(z)$ has simple roots $t_1, t_2, \cdots, t_n$,

$$\boldsymbol{V}_f^{-1}\boldsymbol{H}(\boldsymbol{V}_f^{-1})^{\mathrm{T}}=\mathrm{diag}\left[\frac{g(t_k)}{f'(t_k)}\right]_{k=1}^{n}.$$

Proof From $\boldsymbol{H}(g/f)=g(\boldsymbol{C}_f)\boldsymbol{S}(f)^{-1}$ it is obvious. ■

We conclude this note by remarking that the reductions of the Bezoutian and Hankel matrices can be applied to obtain characterizations of these matrices compatible with a polynomial (e. g. [4][2][10][3] and [12]).

Acknowledgments

The authors wish to thank the referee for helpful comments and suggestions.

References

[1] B N. Datta. An elementary proof of the stability criterion of Lienard and Chipart. Linear Algebra Appl. 1978, 22:89—96.

[2] M Fiedler. Polynomials and Hankel matrices. Linear Algebra Appl., 1985, 66:235—248.

[3] M Fiedler. Hankel and Loewner matrices. Linear Algebra Appl., 1984, 58: 75—95.

[4] M Fiedler and V Pták. Bezoutian and intertwining matrices. Linear Algebra Appl., 1987, 86:43—52.

[5] P A Fuhrmann and B N Datta. On Bezoutians, Van der Monde matrices, and

the Lienard-Chipart stability criterion. Linear Algebra Appl., 1989, 120: 23—37.

[6] P A Fuhrmann. Algebraic system theory: An analyst's point of view. J. Franklin Inst. 1976, 301:521—540.

[7] U Helmke and P A Fuhrmann. Bezoutians. Linear Algebra Appl., 1989, 122:1 039—1 097.

[8] P Lancaster and M. Tismenetsky. The Theory of Matrices with Applications. 2nd ed., Academic, New York, 1985.

[9] F I Lander. The Bezoutian and the inversion of Hankel and Toeplitz matrices. Mat. Issled, 1974, 9:69—87.

[10] V Pták, Lyapunov, Bezout, and Hankel. Linear Algebra Appl., 1984, 58: 363—390.

[11] G Sansigre and M Alvarez. On Bezoutian reduction with the Vandermonde matrix. Linear Algebra Appl., 1989, 121:401—408.

[12] S-R Yang, H-P Zhang, and G-N Chen. Bezoutian and Hankel matrices. J. Beijing Normal Univ., 1991, 27(Sup. 2): 20—26.

Linear Algebra and Its Applications，
1994,203－204:265－300

关于 Loewner 矩阵的更多结果①

More on Loewner Matrices

Abstract The study of Loewner matrices is continued in the direction close to the recent work by Fiedler, Pták, and Vavřín. All results are stated for the general case of multiple interpolation nodes, i. e. , for generalized Loewner matrices. Both the "Loewner-Bézoutian" connection of Antoulas and Anderson and the "Loewner-Hankel" connection of Vavřín the general case, which are re-proved in a simple way, provide a good opportunity for finding various properties of Loewner matrices, some of them being generalizations of results published by other authors for simple nodes only. There are also results which are new especially the Barnett-like formulas and intertwining relation characteristic for Loewner matrices, as well as the "Loewner matrix-Hankel vector" connection. In order to handle Loewner matrices further on, the authors introduce and deal with a class of called R-matrices. It turns out that some known results concerning Bezoutian, Hankel matrices, and their products can in a natural way be lifted to the level of R-matrices, and the products of Loewner and R-matrices.

§ 1. Introduction and Preliminaries

Following Fiedler and Pták, [10], and Vavřín[17], in the present

① Received 30 November 1993; final manuscript accepted 2 October 1993. The authors were partially supported by Science Foundation of State Education Committee of China. 本文与张惠品合作. Submitted by A. C. Antoulas.

paper we intend to continue the study of Loewner matrices for the general case of multiple Interpolation nodes, i. e. , for generalized Loewner matrices, which includes the simple nodes as special case.

It is well known that the (generalized) Loewner matrix is a fundamental tool for the study of the general rational interpolation problem (GRIP) in the scalar case (see, e. g. , [1][2][5][17]). The definition of such a matrix is as follows.

Given so-called interpolation vectors (IVs)

$$\boldsymbol{y}=(\underbrace{y_1,\cdots,y_1}_{m_1},\cdots,\underbrace{y_u,\cdots,y_u}_{m_u})^{\mathrm{T}} \textit{ and } \boldsymbol{z}=(\underbrace{z_1,\cdots,z_1}_{n_1},\cdots,\underbrace{z_u,\cdots,z_u}_{n_v})^{\mathrm{T}}, \tag{1.1}$$

$y_1, y_2, \cdots, y_u$ [$z_1, z_2, \cdots, z_v$ resp.] being distinct complex numbers, $|\boldsymbol{y}|=\sum_{i=1}^{u} m_i = m$, the dimension of $\boldsymbol{y}$, $|\boldsymbol{z}|=\sum_{i=1}^{v} n_i = n, m+n=N$, if f and g are two polynomials such that

$$f(y_i)f(z_j)\neq 0,\quad i=1,2,\cdots,u; j=1,2,\cdots,v, \tag{1.2}$$

we define the (generalized) Loewner matrix $\boldsymbol{L}_{y,z}(g/f)$ generated by g/f (and associated with the IVs $\boldsymbol{y}$ and $\boldsymbol{z}$) by the formula

$$\boldsymbol{L}_{y,z}\left(\frac{g}{f}\right)=(\boldsymbol{L}_{i,j})_{i,j=1}^{u,v}, \tag{1.3}$$

in which $\boldsymbol{L}_{i,j}=(\boldsymbol{l}_{i,j}^{k,l})_{k=1,l=0}^{m_i-1,n_j-1}\in M_{m_i,n_j}(\mathbf{C})$ with the elements determined by

$$\boldsymbol{l}_{i,j}^{k,l}=\begin{cases}\dfrac{1}{k!}\dfrac{1}{l!}\dfrac{\partial^{k+l}}{\partial\lambda^k\partial\mu^l}\left[\dfrac{\dfrac{g(\lambda)}{f(\lambda)}-\dfrac{g(\mu)}{f(\mu)}}{\lambda-\mu}\right]_{\lambda=y_i,u=z_j}, & \text{if } y_i\neq z_j,\\[2ex] \dfrac{1}{(k+l+1)!}\dfrac{\partial^{k+l+1}}{\partial\lambda^{k+l+1}}\left[\dfrac{g(\lambda)}{f(\lambda)}\right]_{\lambda=y_i=z_j}, & \text{if } y_i=z_j.\end{cases} \tag{1.4}$$

If f is a given polynomial such that (1.2) holds, we denote by $\mathcal{L}_{y,z}(f)$ the class of all $m\times n$ matrices of the form $\boldsymbol{L}_{y,z}(g/f)$ for polynomials g. Moreover, denote by $\mathcal{L}_{m,n}(\boldsymbol{y},\boldsymbol{z})$ the set of all Loewner matrices from $\mathcal{L}_{y,z}(f)$ for polynomials f satisfying (1.2). In particular, if $m=n=u=v$ and all $y_1, y_2, \cdots, y_u, z_1, z_2, \cdots, z_v$ are mutually distinct, we call $\boldsymbol{y}$ and $\boldsymbol{z}$ in (1.1) a pair of the simplest IVs, and the matrix $\boldsymbol{L}_{y,z}(g/f)$ in (1.3) the classical Loewner matrix generated by g/f (and associated with IVs $\boldsymbol{y}$ and $\boldsymbol{z}$)(see [12]) .

By a GRIP we shall understand the following array of date in $\mathbf{C}^2$:

$$\{(Y_i, C_{ik})\} \mid i=1,2,\cdots,\theta; k=0,1,\cdots,v_i-1\} \tag{1.5}$$

consisting of $N=v_1+v_2+\cdots+v_\theta$ pairs, where $Y_i \neq Y_j$ *for* $i \neq j$, v_i is the multiplicity of the interpolation point Y_i. A rational function $\varphi(\lambda)$ is a solution to GRIP (1. 5) if and only if

$$\frac{1}{k!}\varphi^{(k)}(Y_i)=C_{ik}, i=1,2,\cdots,\theta; k=0,1,\cdots,v_i-1. \tag{1.6}$$

As is well known, there exists a unique polynomial $\omega(\lambda) \in \mathbf{C}_{N-1}(\lambda)$ called the Hermite interpolation polynomial of GRIP(1. 5), which is a solution to GRIP(1. 5). Denote by

$$\boldsymbol{Q}=\{\underbrace{Y_1,\cdots,Y_1}_{v_1},\cdots,\underbrace{Y_\theta,\cdots,Y_\theta}_{v_\theta}\} \tag{1.7}$$

the collection of interpolation nodes. We partition $\boldsymbol{Q}$ into a pair of IVs $\boldsymbol{y}$ and $\boldsymbol{z}$ given as in(1. 1) such that each component of $\boldsymbol{y}$ and $\boldsymbol{z}$ is some Y_i and conversely, and the sum of lengths of sets $\{k \mid y_k=Y_i\}$ and $\{k \mid z_k=Y_i\}$ is equal to $v_i, i=1,2,\cdots,\theta$.

Given a fixed pair of IVs $\boldsymbol{y}$ and $\boldsymbol{z}$ as in(1. 1), which is a partition of $\boldsymbol{Q}$ in(1. 7), the GRIP(1. 5) determines uniquely a (generalized) Loewner matrix $\boldsymbol{L}=\boldsymbol{L}_{y,z}(\omega)$, where $\omega(\lambda)$ stands for the Hermite interpolation polynomial of GRIP(1. 5) in $\mathbf{C}_{N-1}(\lambda)$. It is obvious that $\boldsymbol{L}_{y,z}(\omega)=\boldsymbol{L}_{y,z}(\varphi)$ whenever $\varphi(\lambda)$ is a solution to GRIP(1. 5). Conversely, it is known[4] that each Loewner matrix $\boldsymbol{L}$ in $\mathcal{L}_{m,n}(\boldsymbol{y},\boldsymbol{z})$ generated, say, by g/f specifies uniquely a GRIP of the form(1. 5)(up to a constant summand d for C_{i0}):

$$\{(Y_i, C_{i0}+d, C_{i1}, \cdots, C_{i,v_i-1}) \mid i=1,2,\cdots,\theta\}, \tag{1.8}$$

where

$$C_{ik}=\frac{1}{k!}\left[\frac{g(\lambda)}{f(\lambda)}\right]^{(k)}_{\lambda=Y_i},$$

such that the Loewner matrix $\boldsymbol{L}$ is determined by the array of data(1. 8) and the IVs $\boldsymbol{y}$ and $\boldsymbol{z}$. Thus, in this ease, if $\boldsymbol{L}=\boldsymbol{L}_{y,z}(\omega)=\boldsymbol{L}_{y,z}(\omega_0)$, where both of ω and ω_0 are in $\mathbf{C}_{m+n-1}(\lambda)$, then $\omega(\lambda)-\omega_0(\lambda)=\text{const.}$

The theory of Loewner matrices, especially various relations of such matrices to the other matrices such as Bezoutian and Hankel matrices, as well as mentioning connections with other fields such as rational interpolation, Padè approximation, and inversion of finite matrices have been considered in numerous earlier papers. With restriction to simple nodes this theory was studied by Belevitch[2], Fiedler[6][7], Fiedler and Pták [10], Vavřín[15][16] and others. There are also results investigated in the case of multiple nodes, especially the "(generalized) Loewner-Bezoutian" connection, due to Antoulas and Anderson[1] (and Fiedler [7] for simple nodes only, see also Vavřín[15], and Fiedler and Pták[10]), as well as the "(generalized) Loewner-Hankel" connection, due to Vavřín [17] (and Belevitch[2] for simple nodes only, see also Fiedler[7], and Fiedler and Pták[10]).

In this paper we will show that both of the "Loewner-Bezoutian" and "Loewner-Hankel" connections abovementioned provide a good opportunity for finding various properties of Loewner matrices, some of them being generalizations of previous results by other authors for simple nodes only. However, there are also results which are new and valuable, especially the Barnett-like formulas (Corollary 2.4), the (generalized) intertwining characterization for Loewner matrices in terms of the notion of the left and right companion matrices of a polynomial (Corollary 2.5), and the "Loewner matrix-Hankel vector" connection (Theorem 3.7). Note that the last connection is a crucial observation, and will lead to a new approach for solving the general interpolation problem in the scalar case (see [4][18] for more information). Of importance are the methods of proof suggested in this paper, since many of the proofs, based on the "Loewner-Bezoutian" connection (Theorem 2.2) and on Theorem 2.1, are relatively simple and elegant, using much more matrix multlplication than the difficult elementwise calculation. In order to handle Loewner matrices further on, we introduce and deal with a class of so-called ***R***-

matrices, each of which can be written as the form $\boldsymbol{R}=\boldsymbol{BHA}^{T}$, where $\boldsymbol{H}$ stands for a Hankel matrix, and $\boldsymbol{A}$ and $\boldsymbol{B}$ are invertible matrices depending upon a pair of interpolation vectors and a polynomial compatible with $\boldsymbol{H}$. The concept of the $\boldsymbol{R}$-matrix seems to have aesthetic significance since the formulas on $\boldsymbol{R}$-matrices(see. e. g. , Theorems 4. 1, 4. 3, 5. 1 and 5. 3) get a form similar to the formulas concerning Bezoutian and Hankel matrices, as well as their products. Thus, roughly speaking, the advantage of introducing the $\boldsymbol{R}$-matrix class is that many of known results on Bezoutian and Hankel matrices, as well as their products, could be in a natural way lifted to the level of $\boldsymbol{R}$-matrices and the products of Loewner and $\boldsymbol{R}$-matrices, respectively. As an application, an extension of the result on inverses of Loewner matrices is deduced (Corollary 5. 2).

The paper is organized as follows. In Section 2 we set forth with a full proof of the "Loewner-Bezoutian" connection. This key connection is then used to obtain in a natural way some properties of Loewner matrices, especially the Barnett-like factorization formulas and the intertwining characterization for Loewner matrices. Section 3 is devoted to the study of the mutual relations between Loewner and Hankel matrices as well as further connection to the rational interpolation. The "Loewner-Hankel" connection of Vavřín is re-proved in a simple way. Some deeper results on this aspect are also considered, all of them being of special interest in the study of rational interpolation. In Section 4 we introduce and deal with the $\boldsymbol{R}$-matrix class. A Barnett-like formula and intertwining relation for $\boldsymbol{R}$-matrices(Theorems 4. 1 and 4. 3) are investigated. We show also an interesting version of the well known Lander's theorem, that is, a nonsingular matrix is Loewner if and only if its inverse is an $\boldsymbol{R}$-matrix. Finally, in Section 5 a qualitative result and its quantitative version on products of Loewner and $\boldsymbol{R}$-matrices are formulated. As applications, we obtain a generalization of the result due to Vavřín[15], and Fiedler and pták[10] on inverses of Loewner matrices, as well as an extension of a result of Vavřín[16] on products of so-called $\mathcal{L}d$ matrices.

For our discussion we shall need some notions and assertions. Let m and n be fixed positive integers throughout. Denote by $M_{m,n}(\mathbf{C})$ [$M_n(\mathbf{C})$ resp.]the set of all $m\times n$[$n\times n$ resp.]complex matrices, by $\mathbf{C}_n^0(\lambda)$[$\mathbf{C}_n(\lambda)$ resp.]the class of all complex polynomials of degree n[at most n resp.]. It is convenient to adopt the conventions: the degree of zero polynomial equals -1 and $\gcd(f,g)$ is monic where f and g are polynomials.

Given $f,g\in\mathbf{C}_n(\lambda)$, the Bezout matrix (or Bezoutian) of f and g, $\boldsymbol{B}_n(f,g)=(b_{i,j})_{i,j=0}^{n-1}\in M_n(\mathbf{C})$ is a symmetric matrix for which

$$\frac{f(\lambda)g(\mu)-f(\mu)g(\lambda)}{\lambda-\mu}=\sum_{i,j=0}^{n-1}b_{i,j}\lambda^i\mu^j.$$

In this case, the Bezoutian $\boldsymbol{B}_n(f,g)$ is said to be compatible with f and g, written as $\boldsymbol{B}_n(f,g)\in\mathcal{B}_n(f)$ or $\boldsymbol{B}_n(f,g)\in\mathcal{B}_n(g)$.

Lemma Ⅰ[13] *Let* $\boldsymbol{B}\in M_n(\mathbf{C})$ *and* f $\in\mathbf{C}_n^0(\lambda)$, *and let* $\boldsymbol{C}_f\in M_n(\mathbf{C})$ *be the (first) companion matrix of* f. *Then the following are equivalent*:

(*a*)$\boldsymbol{B}=\boldsymbol{B}_n(f,g)$ *for a suitable* $g\in\mathbf{C}_n(\lambda)$.

(*b*)$\boldsymbol{B}=\boldsymbol{B}_n(f,1)g(\boldsymbol{C}_f)=g(\boldsymbol{C}_f^{\mathrm{T}})\boldsymbol{B}_n(f,1)$ *for a suitable* $g\in\boldsymbol{C}_n(\lambda)$. (*Barnett's factorization formula.*)

(*c*)$\boldsymbol{B}\boldsymbol{C}_f=\boldsymbol{C}_f^{\mathrm{T}}\boldsymbol{B}$.

As is well known, $\boldsymbol{B}_n(f,g)$ is nonsingular if and only if $\gcd(f,g)=1$ and $\max\{\deg f,\deg g\}=n$.

$\boldsymbol{H}\in M_{m,n}(\mathbf{C})$ is a Hankel matrix if $\boldsymbol{H}$ is of the form $\boldsymbol{H}=(h_{i+j})_{i,j=0}^{m-1,n-1}$. Denote by $\mathcal{H}_{m,n}$ the set of all $m\times n$ Hankel matrices. In the case of $m=n$, the Hankel matrix $\boldsymbol{H}=(h_{i+j})_{i,j=0}^{n-1}$ of order n is lower triangular if $h_0=h_1=\cdots=h_{n-2}=0$; $\boldsymbol{H}$ is degenerate if it is a singular and lower triangular Hankel matrix. By definition (see[7]), a Hankel matrix $\boldsymbol{H}=(h_{i+j})_{i,j=0}^{m-1,n-1}\in M_{m,n}(\mathbf{C})$ is compatible with a nonzero polynomial $f(\lambda)=\sum_{j=0}^{n}f_j\lambda^j$ in $\mathbf{C}_n(\lambda)$ if $\hat{\boldsymbol{H}}[f]_n=\boldsymbol{0}$, where $\hat{\boldsymbol{H}}=(h_{i+j})_{i,j=0}^{m-2,n}$ and $[f]_n=(f_0,f_1,\cdots,f_n)^{\mathrm{T}}\in\mathbf{C}^{n+1}$. Denote by $\mathcal{H}_{m,n}(f)$ the set of all $m\times n$ Hankel matrices compatible with $f\in\boldsymbol{C}_n(\lambda)$, for short, write $\mathcal{H}_n(f)\equiv\mathcal{H}_{m,n}(f)$. In the sequel, given a rational function h/f, where h and f are polynomials with $\deg h\leqslant\deg f$, we denote by $\boldsymbol{H}=\boldsymbol{H}_{m,n}(h/f)\in M_{m,n}(\mathbf{C})$ the Hankel matrix generated by the Markow parameters of h/f. In this case, h/f is said to be a generating for $\boldsymbol{H}$ (see [11] for detail).

Lemma Ⅱ [13] *Let $\boldsymbol{H}\in M_n(\mathbf{C})$ and $f\in\mathbf{C}_n^0(\lambda)$, and let $\boldsymbol{C}_f$ be as in Lemma I. Then the following are equivalent:*

(a) $\boldsymbol{H}=\boldsymbol{H}_n(h/f)$ *for a suitable* $h\in\mathbf{C}_{n-1}(\lambda)$.

(b) $\boldsymbol{H}=h(\boldsymbol{C}_f)\boldsymbol{B}_n^{-1}(f,1)=\boldsymbol{B}_n^{-1}(f,1)h(\boldsymbol{C}_f^{\mathrm{T}})$ *for a suitable* $h\in\mathbf{C}_{n-1}(\lambda)$.

(c) $\boldsymbol{C}_f\boldsymbol{H}=\boldsymbol{H}\boldsymbol{C}_f^{\mathrm{T}}$.

Lemma Ⅲ [3] *Let $0\neq f\in\mathbf{C}_n(\lambda)$ and $\boldsymbol{H}\in\mathcal{H}_n(f)$. Then there exist $h\in\mathbf{C}_{n-1}(\lambda)$ with* $\deg h<\deg f$ *and a lower triangular Hankel matrix $\boldsymbol{H}_0$ of order n, which is particularly degenerate if* $\deg f>0$, *such that both of h and $\boldsymbol{H}_0$ are dependent only upon f and $\boldsymbol{H}$, and $\boldsymbol{H}$ can be uniquely written as a quasidirect sum of the form*

$$\boldsymbol{H}=\boldsymbol{H}_n(h/f)+\boldsymbol{H}_0,$$

that is, $\boldsymbol{H}=\boldsymbol{H}_n(h/f)+\boldsymbol{H}_0$ *with* $\operatorname{rank}\boldsymbol{H}=\operatorname{rank}\boldsymbol{H}_n(h/f)+\operatorname{rank}\boldsymbol{H}_0$.

Lemma Ⅳ [9][10][3] *Let k be a fixed positive integer and $0\neq f\in\mathbf{C}_n(\lambda)$. We have*

(a) If $\boldsymbol{A}_{2i-1}\in\mathcal{B}_n(f)$ [$\mathcal{H}_n(f)$ *resp.*], $i=1,2,\cdots,k+1$, *and* $\boldsymbol{A}_{2i}\in\mathcal{H}_n(f)$ [$\mathcal{B}_n(f)$ *resp.*], $i=1,2,\cdots,k$, *then* $\boldsymbol{A}_1\boldsymbol{A}_2\cdots\boldsymbol{A}_{2k+1}\in\mathcal{B}_n(f)$ [$\mathcal{H}_n(f)$ *resp.*].

(b) *Quantitatively, if* $g_1,\cdots,g_{2k+1}\in\boldsymbol{C}_n(\lambda)$, $\gcd(f,g_{2i})=1$ *and* $\max(\deg f,\deg g_{2i})=n$, $i=1,2,\cdots,k$, *then there exists* $g\in\boldsymbol{C}_n(\lambda)$ *such that*

$$\boldsymbol{B}_n(f,g_1)\boldsymbol{B}_n^{-1}(f,g_2)\boldsymbol{B}_n(f,g_3)\boldsymbol{B}_n^{-1}(f,g_4)\cdots\boldsymbol{B}_n(f,g_{2k+1})=\boldsymbol{B}_n(f,g).$$

Moreover, $g\in\boldsymbol{C}_n(\lambda)$ *satisfies this property if and only if*

$$g_1g_3\cdots g_{2k+1}-g_2g_4\cdots g_{2k}g=fp$$

for a suitable $p\in\boldsymbol{C}_{kn}(\lambda)$; *in this case, all* $g\in\boldsymbol{C}_n(\lambda)$ *are of the form* $g_0(\lambda)+cf(\lambda)$, $c=\text{const}$.

(c) *More generally, if* $g_1,g_2,\cdots,g_k\in\boldsymbol{C}_n(\lambda)$ *and if* $\boldsymbol{H}_i\in\mathcal{H}_n(f)$, $i=1,2,\cdots,k-1$, *has a unique quasidirect decomposition of the form as in Lemma Ⅲ*

$$\boldsymbol{H}_i=\boldsymbol{H}_n(h_i/f)+\boldsymbol{H}_{oi},\ i=1,2,\cdots,k-1,$$

then there exists $g\in\mathbf{C}_n(\lambda)$ *such that*

$$\boldsymbol{B}_n(f,g_1)\boldsymbol{H}_1\boldsymbol{B}_n(f,g_2)\boldsymbol{H}_2\cdots\boldsymbol{H}_{k-1}\boldsymbol{B}_n(f,g_k)=\boldsymbol{B}_n(f,g).$$

Moreover, if $g\in\mathbf{C}_n(\lambda)$ *satisfies this property, then*

$$g=f\omega+h_1\cdots h_{k-1}g_1\cdots g_k$$

for a suitable polynomial ω *of degree* $(k-1)(2n-1)$ *at most ; in this case, all* $g\in\mathbf{C}_n(\lambda)$ *are of the form* $g_0(\lambda)+cf(\lambda)$, $c=const.$

§ 2. Generalized Loewner Matrices and Bezoutians

Throughout the rest of the paper, we use the following notation for polynomials and matrices associated with IVs $\boldsymbol{y}$ and $\boldsymbol{z}$ in(1.1):

$$a(\lambda)=\prod_{i=1}^{u}(\lambda-y_i)^{m_i},a_{ik}(\lambda)=\frac{a(\lambda)}{(\lambda-y_i)^{k+1}},$$
$$i=1,2,\cdots,u;k=0,1,\cdots,m_i-1,$$
$$b(\lambda)=\prod_{j=1}^{v}(\lambda-z_j)^{n_j},b_{jl}(\lambda)=\frac{b(\lambda)}{(\lambda-z_j)^{l+1}},\tag{2.1a}$$
$$j=1,2,\cdots,v;l=0,1,\cdots,n_i-1,$$
$$A(\lambda)=a(\lambda)b(\lambda),$$

$$\boldsymbol{J}(y_i)=\begin{pmatrix} y_i & 1 & & & \\ & y_i & 1 & & \\ & & \ddots & \ddots & \\ & & & \ddots & 1 \\ & & & & y_i\end{pmatrix}_{m_i\times m_i},\quad \boldsymbol{J}(z_j)=\begin{pmatrix} z_j & 1 & & & \\ & z_j & 1 & & \\ & & \ddots & \ddots & \\ & & & \ddots & 1 \\ & & & & z_i\end{pmatrix}_{n_j\times n_j}.\tag{2.1b}$$

$$\boldsymbol{W}_i^{[r]}(y_i)=\left(\frac{a_{ik}^{(\mu)}(0)}{\mu!}\right)_{k=0,\mu=0}^{m_i-1,r-1},\boldsymbol{W}_j^{[r]}(z_j)=\left(\frac{b_{jl}^{(v)}(0)}{v!}\right)_{l=0,v=0}^{n_j-1,r-1},\tag{2.1c}$$

$$\boldsymbol{W}^{[r]}(\boldsymbol{y})=\begin{pmatrix}\boldsymbol{W}_1^{[r]}(y_1)\\ \vdots \\ \boldsymbol{W}_u^{[r]}(y_u)\end{pmatrix}\in M_{mr}(\mathbf{C}),\ \boldsymbol{W}^{[r]}(\boldsymbol{z})=\begin{pmatrix}\boldsymbol{W}_1^{[r]}(z_1)\\ \vdots \\ \boldsymbol{W}_v^{[r]}(z_v)\end{pmatrix}\in M_{nr}(\mathbf{C}),\tag{2.1d}$$

Denote by $\mathbf{V}^{[r]}(\boldsymbol{y})$ and $\mathbf{V}^{[r]}(\boldsymbol{z})$ the $r\times m$ and $r\times n$ generalized Vandermonde matrices corresponding to a pair of IVs $\boldsymbol{y}$ *and* $\boldsymbol{z}$ *in* (1.1):

$$\begin{aligned}\mathbf{V}^{[r]}(\boldsymbol{y})&=[\boldsymbol{P}_{rm_1}(y_1)\cdots\boldsymbol{P}_{rm_u}(y_u)]\in M_{rm}(\mathbf{C}),\\ \mathbf{V}^{[r]}(\boldsymbol{z})&=[\boldsymbol{P}_{rn_1}(z_1)\cdots\boldsymbol{P}_{rn_v}(z_v)]\in M_{rn}(\mathbf{C}),\end{aligned}\tag{2.2}$$

where $\boldsymbol{P}_{k,l}(\lambda)$ stands for the $k\times l$ matrix defined by

$$(\boldsymbol{P}_{k,l}(\lambda))_{i,j}=\binom{i}{j}\lambda^{i-j},i=0,1,\cdots,k-1;j=0,1,\cdots,l-1.$$

When $r=m$ [$r=n$ resp.] we shall drop the superscript r in $V^{[r]}(\boldsymbol{y})$, $\boldsymbol{W}^{[r]}(\boldsymbol{y})$, $\boldsymbol{W}_i^{[r]}(y_i)$, [$\boldsymbol{V}^{[r]}(\boldsymbol{z})$, $\boldsymbol{W}^{[r]}(\boldsymbol{z})$, $\boldsymbol{W}_j^{[r]}(z_j)$, resp.] to simplify the notation.

We need first a simple result, which is an extension of Theorem 3 of Fiedler[6].

Theorem 2.1 *Let* $\boldsymbol{y}$ *and* $\boldsymbol{z}$ *be a pair of IVs as in* (1.1), *and* $\boldsymbol{P}_k$ *of order* k *the flip matrix* $(\delta_{i,k-j-1})$, $i,j=0,1,\cdots,k-1$. *Let* $\boldsymbol{J}(y_i)$ [$\boldsymbol{J}(z_j)$ *resp.*] *be the Jordan block of order* m_i [n_j *resp.*] *with* y_i [z_j *resp.*] *along the diagonal and* 1 *along the superdiagonal. Then*

$$\boldsymbol{W}^{[r]}(\boldsymbol{y})\boldsymbol{V}^{[r]}(\boldsymbol{y})=\boldsymbol{W}(\boldsymbol{y})\boldsymbol{V}(\boldsymbol{y})=\mathrm{diag}[\boldsymbol{P}_{m_i}a_{i,m_i-1}(\boldsymbol{J}(y_i))]_{i=1}^{u},\ if\ r\geqslant m, \tag{2.3}$$

$$\boldsymbol{W}^{[r]}(\boldsymbol{z})\boldsymbol{V}^{[r]}(\boldsymbol{z})=\boldsymbol{W}(\boldsymbol{z})\boldsymbol{V}(\boldsymbol{z})=\mathrm{diag}[\boldsymbol{P}_{n_j}b_{j,n_j-1}(\boldsymbol{J}(z_j))]_{j=1}^{v},\ if\ r\geqslant n. \tag{2.4}$$

Consequently, $\boldsymbol{V}(\boldsymbol{y})$, $\boldsymbol{V}(\boldsymbol{z})$, $\boldsymbol{W}(\boldsymbol{y})$ *and* $\boldsymbol{W}(\boldsymbol{z})$ *are all invertible.*

Proof Observe that $\boldsymbol{P}_{r,k}(\lambda)$ in (2.2) can be rewritten as

$$\boldsymbol{P}_{r,k}(\lambda)\begin{bmatrix}\boldsymbol{P}_{r,1}(\lambda) & \dfrac{\boldsymbol{P}_{r1}^{(1)}(\lambda)}{1!} & \cdots & \dfrac{\boldsymbol{P}_{r,1}^{(k-1)}(\lambda)}{(k-1)!}\end{bmatrix} \tag{2.5}$$

(the index (q) denotes the qth derivative). In the case of $r\geqslant m$, we have

$$\begin{aligned}\boldsymbol{W}^{[r]}(\boldsymbol{y})\boldsymbol{P}_{r,1}(\lambda)&=\boldsymbol{W}(\boldsymbol{y})\boldsymbol{P}_{m,1}(\lambda)\\&=(a_{10}(\lambda)\cdots a_{1,m_1-1}(\lambda)\cdots a_{u0}(\lambda)\cdots a_{u,m_u-1}(\lambda))^{\mathrm{T}},\end{aligned} \tag{2.6}$$

so that $\boldsymbol{W}^{[r]}(\boldsymbol{y})\boldsymbol{V}^{[r]}(\boldsymbol{y})=\boldsymbol{W}(\boldsymbol{y})\boldsymbol{V}(\boldsymbol{y})$ if $r\geqslant m$. On the other hand, it follows by (2.5) that

$$\boldsymbol{W}_i(y_i)\boldsymbol{P}_{m,m_j}(y_j)=\begin{bmatrix}a_{i0}(y_j) & \dfrac{a_{i0}^{(1)}(y_j)}{1!} & \cdots & \dfrac{a_{i0}^{m_j-1}(y_j)}{(m_j-1)!}\\ a_{i1}(y_j) & \dfrac{a_{i1}^{(1)}(y_j)}{1!} & \cdots & \dfrac{a_{i1}^{m_j-1}(y_j)}{(m_j-1)!}\\ \vdots & \vdots & \vdots & \vdots\\ a_{i,m_i-1}(y_j) & \dfrac{a_{i,m_i-1}^{(1)}(y_i)}{1!} & \cdots & \dfrac{a_{i,m_i-1}^{m_j-1}(y_j)}{(m_j-1)!}\end{bmatrix} \tag{2.7}$$

Thus ,the matrix on the right hand side of (2. 7) is a zero matrix whenever $i\neq j$, and $\boldsymbol{W}(\boldsymbol{y})\boldsymbol{V}(\boldsymbol{y})=\mathrm{diag}[\boldsymbol{W}_i(y_i)\boldsymbol{P}_{m,m_i}(y_i)]_{i=1}^u$. To complete the proof of (2. 3), it suffices to verify

$$\boldsymbol{W}_i(y_i)\boldsymbol{P}_{m,m_i}(y_i)=\boldsymbol{P}_{m_i}a_{i,m_i-1}(\boldsymbol{J}(y_i)),i=1,2,\cdots,u. \qquad (2.8)$$

Suppose now that k,l,p and q are all nonnegative integers such that $k+l=p+q\leqslant 2(m_i-1)$. If $k>p$, then

$$a_{i,p}(\lambda)=a_{i,k}(\lambda)(\lambda-y_i)^{k-p},$$

$$a_{ip}^{(q)}(\lambda)=\sum_{\mu=0}^{q}a_{ik}^{(\mu)}(\lambda)\begin{pmatrix}q\\ \mu\end{pmatrix}[(\lambda-y_i)^{k-p}]^{(q-\mu)}$$

(the index (q) denotes the qth derivative), whence

$$\frac{a_{i,p}^{(q)}(y_i)}{q!}=\frac{a_{i,k}^{(l)}(y_i)}{l!}.$$

This combined with the definition of $a_{i,k}(\lambda)$ in (2. 1) implies

$$a_{i,k}(y_i)=\frac{a_{i,k-1}^{(1)}(y_i)}{1!}=\cdots=\frac{a_{i0}^{(k)}(y_i)}{k!}=0,\quad k=0,1,\cdots,m_i-2,$$

$$a_{i,m_i-1}(y_I)\neq 0,\frac{a_{i,m_i-1}^{(l)}(y_i)}{l!}=\cdots=\frac{a_{i,l^i}^{(m_i-1)}(y_i)}{(m_i-1)!},\quad l=0,1,\cdots,m_i-1.$$

Thus ,(2. 8)follows by (2. 7) and the matrix defined by (2. 8) is a nonsingular Hankel matrix of order m_i for $i=1,2,\cdots,u$. In a similar manner, we can show (2. 4). ■

The next theorem is fundamental for the study of Loewner matrices without any restriction on the size and IVs, which can be considered as an extension of Proposition 2. 1 in [10], and was sketchily treated in the proof of Lemma 2. 5 of [1] in a somewhat different way.

Theorem 2. 2 *Let* **y** *and* **z** *be a pair of IVs as in* (1. 1) *and* f, $g\in\mathbf{C}_r(\lambda)$ *such that* gcd $(f,ab)=1$, *where* a *and* b *are as in* (2. 1) *and* r *is a fixed natural number. Then*

$$\begin{aligned}&\mathrm{diag}^{\mathrm{T}}[f(\boldsymbol{J}(y_i))]_{i=1}^u\boldsymbol{L}_{y,z}\left(\frac{g}{f}\right)\mathrm{diag}[f(\boldsymbol{J}(z_j))]_{j=1}^v\\ &=\boldsymbol{V}^{[r]}(\boldsymbol{y})^{\mathrm{T}}\boldsymbol{B}_r(g,f)\boldsymbol{V}^{[r]}(\boldsymbol{z}).\end{aligned} \qquad (2.9)$$

Proof Assume that the Loewner matrix $\boldsymbol{L}_{y,z}(g/f)\in\mathcal{L}_{y,z}(f)$ is of the form (1. 3)(1. 4). We shall verify that

$$f(\boldsymbol{J}(y_i))^{\mathrm{T}}\boldsymbol{L}_{ij}f(\boldsymbol{J}(z_j))=\boldsymbol{P}_{rm_i}^{\mathrm{T}}(y_i)\boldsymbol{B}_r(g,f)\boldsymbol{P}_{rn_j}(z_j) \tag{2.10}$$

holds for an arbitrary pair of indices $(i,j), i=1,2,\cdots,u, j=1,2,\cdots,v$. Denote by w_{pq} the (p,q)-element of the matrix given on the right hand side of (2.10), $p=0,1,\cdots,m_i-1$; $q=0,1,\cdots,n_j-1$, and put

$$\boldsymbol{D}(\lambda,\mu)=\frac{\frac{g(\lambda)}{f(\lambda)}-\frac{g(\mu)}{f(\mu)}}{\lambda-\mu},\quad Q(\lambda)=(1\ \lambda\ \cdots\ \ \lambda^{r-1})^{\mathrm{T}}.$$

We have by (2.5) and (1.4) that

$$\begin{aligned}
w_{pq} &= (\boldsymbol{P}_{rm_i}^{\mathrm{T}}(y_i))_p\boldsymbol{B}_r(g,f)(\boldsymbol{P}_{rn_j}(z_j))_q = \left[\frac{Q^{(p)}(y_i)}{p!}\right]^{\mathrm{T}}\boldsymbol{B}_r(g,f)\left[\frac{Q^{(q)}(z_j)}{q!}\right]\\
&= \frac{1}{p!q!}\frac{\partial^{p+q}}{\partial\lambda^p\partial\mu^q}[f(\lambda)D(\lambda,u)f(\mu)]\Big|_{\substack{\lambda=y_i\\ \mu=z_j}}\\
&= \frac{1}{p!q!}\sum_{\alpha=0}^{p}\sum_{\beta=0}^{q}\left[f^{(p-\alpha)}(\lambda)\frac{\partial^{\alpha+\beta}D(\lambda,\mu)}{\partial\lambda^\alpha\partial\mu^\beta}f^{(q-\beta)}(\mu)\binom{p}{\alpha}\binom{q}{\beta}\right]\Big|_{\substack{\lambda=y_i\\ \mu=z_j}}\\
&= \sum_{\alpha=0}^{p}\sum_{\beta=0}^{q}\frac{f^{(p-\alpha)}(y_i)}{(p-\alpha)!}\left[\frac{1}{\alpha!\beta!}\frac{\partial^{\alpha+\beta}D(\lambda,\mu)}{\partial\lambda^\alpha\partial\mu^\beta}\right]\Big|_{\substack{\lambda=y_i\\ \mu=z_j}}\frac{f^{(q-\beta)}(z_j)}{(q-\beta)!}\\
&= [f(\boldsymbol{J}(y_i))^{\mathrm{T}}\boldsymbol{L}_{ij}f(\boldsymbol{J}(z_j))]_{p,q},\ \forall\ p,q.
\end{aligned}$$

The proof is complete. ■

In particular, when $r\leqslant\min\{m,n\}$, it follows from the nonsingularity of $\mathbf{V}(\boldsymbol{y})$ and $\mathbf{V}(\boldsymbol{z})$ that rank $\mathbf{V}^{[r]}(\boldsymbol{y})=$ rank $\mathbf{V}^{[r]}(\boldsymbol{z})=r$. Thus as an immediate corollary of Theorem 2.2, we obtain the following results of importance in view of the fact that both of $f(\boldsymbol{J}(y_i))$ and $f(\boldsymbol{J}(z_j))$ in (2.9) are nonsingular whenever gcd $(f,ab)=1$. Here the last part of (*a*) can be considered as an extension of Corollary 2.2 of [10] and (*b*) is stated in [1] as Main Lemma 2.5 and Corollary 2.6.

Recall that if $\boldsymbol{L}_{y,z}(g/f)$ is as in (1.3)(1.4), then we understand by a Loewner submatrix of $\boldsymbol{L}_{y,z}(g/f)$ any submatrix of the form $(\overline{\boldsymbol{L}}_{i,j})$,

$$\overline{\boldsymbol{L}}_{i,j}=(\boldsymbol{l}_{i,j}^{k,l})_{k,l=0}^{\alpha_i-1,\beta_j-1},$$

where $\boldsymbol{l}_{i,j}^{k,l}$ is defined via (1.4), $0\leqslant\alpha_i\leqslant m_i, 0\leqslant\beta_j\leqslant n_j$ (see [1] and [17]).

Corollary 2.3 *Let* $\boldsymbol{y},\boldsymbol{z}$ *be as in Theorem* 2.2. *Let* $f,g\in\mathbf{C}_r(\lambda)$ *such that* $\max\{\deg f,\deg g\}=r$ *and* $\gcd(f,ab)=1$.

(a) *If* $\max\{\deg f,\deg g\}=r\leqslant\min\{m,n\}$, *then we have that*

$$\operatorname{rank} \boldsymbol{L}_{y,z}\left(\frac{g}{f}\right)=\operatorname{rank} \boldsymbol{B}_r(f,g)\leqslant r,$$

and that rank $\boldsymbol{L}_{y,z}(g/f)=r$ *if and only if* gcd $(f,g)=1$. *In particular, if* $r=m=n$, *then* $\boldsymbol{L}_{y,z}(g/f)\in M_n(\mathbf{C})$ *is nonsingular if and only if* gcd$(f,g)=1$ *and* max $\{\deg, f \deg g\}=n$, *i. e.*, f *and* g *are completely relative prime.*

(b) *If* $f,g\in \mathbf{C}_r(\lambda)$ *are such that* gcd $(f,g)=1$ *and* $r=\deg g/f=\max\{\deg f, \deg g\}$, *the McMillan degree of* g/f, *and* $\min\{m,n\}\geqslant r$, *then* rank $\boldsymbol{L}_{y,z}(g/f)=r$, *and all* $r\times r$ *Loewner submatrices of* $\boldsymbol{L}_{y,z}(g/f)$ *are invertible.*

Proof (a): An immediate consequence of Theorem 2. 2 combined with the nonsingularity criterion on Bezoutians. (b): Follows from (a) and (2. 9). ■

Corollary 2. 4 *Let* $\boldsymbol{y},\boldsymbol{z}$ *be as in Theorem* 2. 2, *and let* $f\in \mathbf{C}_r^0(\lambda)$ *be such that* gcd $(f,ab)=1$, *where* a, b *are defined in* (2. 1). *Denote by* e *the linear polynomial* $e(\lambda)=\lambda$. *If* $g\in\mathbf{C}_r(\lambda)$, *then*

$$\boldsymbol{L}_{y,z}\left(\frac{g}{f}\right)=\boldsymbol{L}_{y,z}\left(\frac{1}{f}\right)g(\boldsymbol{C}_z(f)),\quad if\ r=n, \tag{2. 11}$$

$$\boldsymbol{L}_{y,z}\left(\frac{g}{f}\right)=g(\boldsymbol{C}_y^{\mathrm{T}}(f))\boldsymbol{L}_{y,z}\left(\frac{1}{f}\right),\quad if\ r=m, \tag{2. 12}$$

$$\begin{aligned}\boldsymbol{L}_{y,z}\left(\frac{g}{f}\right)&=\boldsymbol{L}_{y,z}\left(\frac{1}{f}\right)g\left(\boldsymbol{L}_{y,z}^{-1}\left(\frac{1}{f}\right)\boldsymbol{L}_{y,z}\left(\frac{e}{f}\right)\right)\\&=g\left(\boldsymbol{L}_{y,z}\left(\frac{e}{f}\right)\boldsymbol{L}_{y,z}^{-1}\left(\frac{1}{f}\right)\right)\boldsymbol{L}_{y,z}\left(\frac{1}{f}\right),\quad if\ r=m=n,\end{aligned} \tag{2. 13}$$

where

$$\boldsymbol{C}_y(f)=\boldsymbol{P}^{-1}\boldsymbol{C}_f\boldsymbol{P},\quad \boldsymbol{C}_z(f)=\boldsymbol{Q}^{-1}\boldsymbol{C}_f\boldsymbol{Q}, \tag{2. 14}$$

$$\boldsymbol{P}^{[r]}=\boldsymbol{V}^{[r]}(\boldsymbol{y})\operatorname{diag}^{-1}[f(\boldsymbol{J}(y_i))]_{i=1}^{u}\in M_{rm}(\mathbf{C}),\quad \boldsymbol{P}=\boldsymbol{P}^{[m]}, \tag{2. 15a}$$

$$\boldsymbol{Q}^{[r]}=\boldsymbol{V}^{[r]}(\boldsymbol{z})\operatorname{diag}^{-1}[f(\boldsymbol{J}(z_j))]_{j=1}^{v}\in M_{rn}(\mathbf{C}),\quad \boldsymbol{Q}=\boldsymbol{Q}^{[n]}. \tag{2. 15b}$$

Proof By (2. 9) and (2. 15), we have

$$\boldsymbol{L}_{y,z}\left(\frac{g}{f}\right)=(\boldsymbol{P}^{[r]})^{\mathrm{T}}\boldsymbol{B}_r(g,f)\boldsymbol{Q}^{[r]}. \tag{2. 16}$$

If $r=n$, then $\boldsymbol{Q}$ in (2. 15) is invertible by Lemma 2. 1, so that it follows from the Barnett formula for $\boldsymbol{B}_r(g,f)$ given in Lemma I with the fact

(2.16) that

$$\boldsymbol{L}_{y,z}\left(\frac{g}{f}\right)=(\boldsymbol{P}^{[r]})^{\mathrm{T}}\boldsymbol{B}_r(1,f)g(\boldsymbol{C}_f)\boldsymbol{Q}$$

$$=(\boldsymbol{P}^{[r]})^{\mathrm{T}}\boldsymbol{B}_r(1,f)\boldsymbol{Q}(\boldsymbol{Q}^{-1}g(\boldsymbol{C}_f)\boldsymbol{Q})=\boldsymbol{L}_{y,z}\left(\frac{1}{f}\right)g(\boldsymbol{C}_z(f)).$$

In a similar manner ,we can show that (2.12) holds if $r=m$, In the case $r=m=n$, since $\boldsymbol{L}_{y,z}(1/f)$ is invertible, we have from (2.11) and (2.12) with $g=e$ that

$$\boldsymbol{C}_y^{\mathrm{T}}(f)=\boldsymbol{L}_{y,z}\left(\frac{e}{f}\right)\boldsymbol{L}_{y,z}^{-1}\left(\frac{1}{f}\right)\ \textit{and}\ \boldsymbol{C}_z(f)=\boldsymbol{L}_{y,z}^{-1}\left(\frac{1}{f}\right)\boldsymbol{L}_{y,z}\left(\frac{e}{f}\right), \tag{2.17}$$

and (2.13) follows. ■

We remark that the results (2.11)～(2.13) can be considered as the Barnett-like formulas for Loewner matrix $\boldsymbol{L}_{y,z}(g/f)$. Here it is natural to call $\boldsymbol{C}_y(f)$ and $\boldsymbol{C}_z(f)$ the left and right companion matrices if $r=m$ and if $r=n$ respectively, which, by(2.14), are similar to the companion $\boldsymbol{C}_f$ of the polynomial f, if $r=m$ and if $r=n$, respectively. Fiedler and Pták[10, Theorem 3.7] also proved the first formula of (2.13) in the case of the classical Loewner matrices.

From the intertwining relation for Bezoutians given in (c) of Lemma I, we can characterize Loewner matrices by means of intertwining relation whenever they are square. Here we shall use the notion of the left and right companion matrices instead of the one of the companion of a polynomial.

Corollary 2.5 *Let* $\mathbf{y}$, $\mathbf{z}$ *be as in* (1.1) *with* $m=n$ *and* $f\in\mathbf{C}_n^0(\lambda)$ *such that* $\gcd(f,ab)=1$. *Then for any* $\mathbf{L}\in M_n(\mathbf{C})$ *the following are equivalent*:

(a) *There exists* $g\in\mathbf{C}_n(\lambda)$ *such that* $\mathbf{L}=\mathbf{L}_{y,z}(g/f)$.

(b) $\boldsymbol{L}\boldsymbol{C}_z(f)=\boldsymbol{C}_y^{\mathrm{T}}(f)L$.

Proof Since both of $\boldsymbol{P}$ and $\boldsymbol{Q}$ in (2.15) are invertible ,the theorem follows from (2.16) and Lemma I. ■

Corollary 2.6 *Let* $\mathbf{y}$, $\mathbf{z}$ *be as in Corollary* 2.5. *Let* $\mathbf{L}\in M_n(\mathbf{C})$ *and* f, $g\in\mathbf{C}_n^0(\lambda)$ *be such that* $\gcd(f,g)=1$ *and* $\gcd(f,ab)=1$. *Then the* *following are equivalent*:

(a) *There exists a constant* $\rho\in\mathbf{C}$ *such that* $\mathbf{L}=\rho\,\mathbf{L}_{y,z}(g/f)$.

(b) $\boldsymbol{L}\boldsymbol{C}_z(f)=\boldsymbol{C}_y^{\mathrm{T}}(f)\boldsymbol{L}$, $\boldsymbol{L}\boldsymbol{C}_z(g)=\boldsymbol{C}_y^{\mathrm{T}}(g)\boldsymbol{L}$, *where* $\mathbf{C}_z(g)$ *and* $\mathbf{c}_y(g)$ *are defined via* (2.14) *but* $\boldsymbol{P},\boldsymbol{Q}$ *therein correspond to the polynomial* f.

Proof Follows from(2.14)and Theorem 4 of Fiedler[6] combined with a stronger result of Lander (see Ptak[13,Proposition 5.6]). ■

§3. (Generalized) Loewner Matrices and Hankel Matrices

This section is devoted to the study of the mutual relations between Loewner and Hankel matrices as well as further connections to GRIP. We shall use the notation of (2.1) and (2.2).

Let us start with an assertion which establishes a relation between Loewner matrices (not necessarily square) and the square Hankel matrices.

Theorem 3.1 *Let* $\mathbf{y},\mathbf{z}$ *be as in* (1.1), *and* $f\in\mathbf{C}_r(\lambda)$ *such that* $(f,ab)=1$. *If* $r\leqslant\min\{m,n\}$ *and* $\mathbf{L}=\mathbf{L}_{y,z}(g/f)$ *for some* $g\in\mathbf{C}_r(\lambda)$, *then there exists* $\boldsymbol{H}\in\mathcal{H}_r(f)$ *such that*

$$\boldsymbol{L}=\boldsymbol{W}^{[r]}(\boldsymbol{y})\boldsymbol{H}\boldsymbol{W}^{[r]}(\boldsymbol{z})^{\mathrm{T}},\tag{3.1}$$

where $\mathrm{W}^{[r]}(\boldsymbol{y})$ *and* $\mathrm{W}^{[r]}(\boldsymbol{z})$ *are as in* (2.1). *In particular, if, further,* $r=m=n$, *then*

$$\boldsymbol{L}=\boldsymbol{W}(\boldsymbol{y})\boldsymbol{H}\boldsymbol{W}(\boldsymbol{z})^{\mathrm{T}}\tag{3.2}$$

Proof By(2.16) and Theorem 2.1, we have

$$\boldsymbol{L}=\boldsymbol{W}^{[r]}(\boldsymbol{y})\hat{\boldsymbol{P}}\boldsymbol{B}_r(g,f)\hat{\boldsymbol{Q}}\boldsymbol{W}^{[r]}(\boldsymbol{z})^{\mathrm{T}},$$

where

$$\hat{\boldsymbol{P}}=\boldsymbol{V}^{(r)}(\boldsymbol{y})\mathrm{diag}^{-1}[\boldsymbol{P}_{m_i}a_{i,m_i-1}(\boldsymbol{J}(y_i))]_{i=1}^{u}(\boldsymbol{P}^{[r]})^{\mathrm{T}}\in M_r(\mathbf{C}),$$

$$\hat{\boldsymbol{Q}}=\boldsymbol{Q}^{[r]}\mathrm{diag}^{-\mathrm{T}}[\boldsymbol{P}_{n_j}b_{j,n_j-1}(\boldsymbol{J}(z_j))]_{j=1}^{v}\boldsymbol{V}^{[r]}(\boldsymbol{z})^{\mathrm{T}}\in M_r(\mathbf{C}).$$

It suffices to verify that $\hat{\boldsymbol{P}}\boldsymbol{B}_r(g,f)\hat{\boldsymbol{Q}}$ belongs to $\mathcal{H}_r(f)$. This will follow from(a) of Lemma Ⅳ if we can show that both $\hat{\boldsymbol{P}}$ and $\hat{\boldsymbol{Q}}$ are in $\mathcal{H}_r(f)$. By(2.15), we may rewrite $\hat{\boldsymbol{P}}$ *in the form*

$$\begin{aligned}\hat{\boldsymbol{P}}&=\boldsymbol{V}^{[r]}(\boldsymbol{y})\mathrm{diag}^{-1}[\boldsymbol{P}_{m_i}a_{i,m_i-1}(\boldsymbol{J}(y_i))]_{i=1}^{u}\mathrm{diag}^{-\mathrm{T}}[f(\boldsymbol{J}(y_i))]_{i=1}^{u}\boldsymbol{V}^{[r]}(\boldsymbol{y})^{\mathrm{T}}\\&=\boldsymbol{V}^{[r]}(\boldsymbol{y})\mathrm{diag}^{-1}[f(\boldsymbol{J}(y_i))^{\mathrm{T}}\boldsymbol{P}_{m_i}a_{i,m_i-1}(\boldsymbol{J}(y_i))]_{i=1}^{u}\boldsymbol{V}^{[r]}(\boldsymbol{y})^{\mathrm{T}}\end{aligned}$$

$$=\boldsymbol{V}^{[r]}(\boldsymbol{y})\mathrm{diag}^{-1}[\boldsymbol{P}_{m_i}a_{i,m_i-1}(\boldsymbol{J}(y_i))f(\boldsymbol{J}(y_i))]_{i=1}^{u}\boldsymbol{V}^{[r]}(\boldsymbol{y})^{\mathrm{T}}.$$

Applying a result on Bezoution representation via the generalized Vandermonde matrix (see [9],[14]), which states that

$$\boldsymbol{V}(\boldsymbol{y})^{\mathrm{T}}\boldsymbol{B}_m(a,f)\boldsymbol{V}(\boldsymbol{y})=\mathrm{diag}[\boldsymbol{P}_{m_i}a_{i,m_i-1}(\boldsymbol{J}(y_i))f(\boldsymbol{J}(y_i))]_{i=1}^{u},\tag{3.3}$$

we obtain

$$\boldsymbol{V}(\boldsymbol{y})\mathrm{diag}^{-1}[\boldsymbol{P}_{m_i}a_{i,m_i-1}(\boldsymbol{J}(y_i))f(\boldsymbol{J}(y_i))]_{i=1}^{u}\boldsymbol{V}(\boldsymbol{y})^{\mathrm{T}}=\boldsymbol{B}_n^{-1}(a,f)\in\mathcal{H}_m(f),$$

so that $\hat{\boldsymbol{P}}\in\mathcal{H}_r(f)$, since $\hat{\boldsymbol{P}}$ is the rth leading submatrix of the Hankel matrix $\boldsymbol{B}_m^{-1}(a,f)$ and $f\in\mathbf{C}_r(\lambda)$. Similarly, we can show that $\hat{\boldsymbol{Q}}\in\mathcal{H}_r(f)$. ■

Corollary 3.2 (*Compare with Theorem* 2.4 *of* [10]) *Let* $\boldsymbol{y},\boldsymbol{z}$ *be a pair of IVs as in Corollary* 2.5, *and* $f\in\mathbf{C}_n(\lambda)$ *such that* $\gcd(f,ab)=1$. *If* $\boldsymbol{L}\in M_n(\mathbf{C})$, *then the following are equivalent*:

(a) $\boldsymbol{L}=\boldsymbol{L}_{y,z}(g/f)$ *for some polynomial* $g\in\mathbf{C}_n(\lambda)$.

(b) *There exists* $\boldsymbol{H}\in\mathcal{H}_n(f)$ *such that*

$$\boldsymbol{L}=\boldsymbol{W}(\boldsymbol{y})\boldsymbol{H}\boldsymbol{W}(\boldsymbol{z})^{\mathrm{T}}.$$

Proof (a)⇒(b): See Theorem 3.1.

(b)⇒(a): Assume that $\boldsymbol{L}=\boldsymbol{W}(\boldsymbol{y})\boldsymbol{H}\boldsymbol{W}(\boldsymbol{z})^{\mathrm{T}}$, $\boldsymbol{H}\in\mathcal{H}_n(f)$. By (2.15) we have

$$\boldsymbol{L}=\boldsymbol{P}^{\mathrm{T}}\boldsymbol{P}^{-\mathrm{T}}\boldsymbol{W}(\boldsymbol{y})\boldsymbol{H}\boldsymbol{W}(\boldsymbol{z})^{\mathrm{T}}\boldsymbol{Q}^{-1}\boldsymbol{Q}.$$

Thus, it suffices to verify $\boldsymbol{P}^{-\mathrm{T}}\boldsymbol{W}(\boldsymbol{y})\boldsymbol{H}\boldsymbol{W}(\boldsymbol{z})^{\mathrm{T}}\boldsymbol{Q}^{-1}=B_n(g,f)$ for some $g\in\mathbf{C}_n(\lambda)$ in terms of Theorem 2.2. This will follow from (a) of Lemma Ⅳ if we show that both $\boldsymbol{P}^{-1}\boldsymbol{W}(\boldsymbol{y})$ and $\boldsymbol{W}(\boldsymbol{z})^{\mathrm{T}}\boldsymbol{Q}^{-1}$ belong to $\mathcal{B}_n(f)$. To prove that, using Theorem 2.1 together with (2.15) and (3.3) we obtain

$$\begin{aligned}\boldsymbol{P}^{-\mathrm{T}}\boldsymbol{W}(\boldsymbol{y})&=\boldsymbol{V}(\boldsymbol{y})^{-\mathrm{T}}\mathrm{diag}^{\mathrm{T}}[f(\boldsymbol{J}(y_i))]_{i=1}^{u}\mathrm{diag}[\boldsymbol{P}_{m_i}a_{i,m_i-1}(\boldsymbol{J}(y_i))]_{i=1}^{u}\boldsymbol{V}(\boldsymbol{y})^{-1}\\&=\boldsymbol{V}(\boldsymbol{y})^{-\mathrm{T}}\mathrm{diag}[\boldsymbol{P}_{m_i}a_{i,m_i-1}(\boldsymbol{J}(y_i))f(\boldsymbol{J}(y_i))]_{i=1}^{u}\boldsymbol{V}(\boldsymbol{y})^{-1}\\&=\boldsymbol{B}_n(a,f)\in\mathcal{B}_n(f).\end{aligned}\tag{3.4a}$$

In a simillar manner, we can show that

$$\boldsymbol{W}(\boldsymbol{z})^{\mathrm{T}}\boldsymbol{Q}^{-1}=\boldsymbol{B}_n(b,f)\in\mathcal{B}_n(f).\tag{3.4b}$$ ■

In the general case without the restriction on the size of Loewner and Hankel matrices, the following strengthening of Corollary 3.2 is due to Vavřin[17]. We present a new proof by a computational method.

Theorem 3.3 [17] *Let* $\boldsymbol{y},\boldsymbol{z}$ *be as in* (1.2), *then the mapping de-*

fined by

$$\boldsymbol{L}=\boldsymbol{W}(\boldsymbol{y})\boldsymbol{H}\boldsymbol{W}(\boldsymbol{z})^{\mathrm{T}} \tag{3.5}$$

is an isomorphism of the classes $\mathcal{H}_{m,n}$ *and* $\mathcal{L}_{m,n}(\boldsymbol{y},\boldsymbol{z})$.

Before proving Theorem 3.3 it will be convenient to have the following Lemma.

Lemma 3,4 *Let* m,n,k,l *be positive integers, and* p,q *nonnegative integers such that* $l+q\leqslant n$ *and* $k+p\leqslant m$. *Let* $\boldsymbol{S}_{km}^{l}(t)$ *be the* $k\times m$ *Toeplitz matrix defined by* $(q_{j-i}(t))_{i,j=0}^{k-1,m-1}$, *where* $q_k(t)=\begin{pmatrix}l\\k\end{pmatrix}t^{l-k}$ *for* $t\in$ **C**. *Then*

$$\boldsymbol{S}=\boldsymbol{S}_{km}^{p}(t_1)\boldsymbol{H}\boldsymbol{S}_{ln}^{q}(t_2)^{\mathrm{T}},\ \forall t_2,t_2\in\mathbf{C}$$

is a $k\times l$ *Hankel matrix whenever* $\boldsymbol{H}=(h_{i+j})_{i,j=0}^{m-1,n-1}$ *is a Hankel matrix. Moreover, the strictly proper rational part of* $\dfrac{b(\lambda)(\lambda+t_1)^p(\lambda+t_2)^q}{a(\lambda)}$ *is a generating for* $\boldsymbol{S}$, *provided* $b(\lambda)/a(\lambda)$ *is a generating for* $\boldsymbol{H}$.

Proof A simple calculation leads to

$$\begin{aligned}(\boldsymbol{S}_{km}^{p}(t)\boldsymbol{H})_{i,j} &= \left(\underbrace{0\cdots0}_{i}\ t^{p}\begin{pmatrix}p\\1\end{pmatrix}t^{p-1}\cdots1\quad\underbrace{0\cdots0}_{m-p-i-1}\right)(h_j\ h_{j+1}\cdots h_{j+m-1})^{\mathrm{T}}\\ &=\sum_{\alpha=0}^{p}\begin{pmatrix}p\\\alpha\end{pmatrix}t^{p-\alpha}h_{i+j+\alpha},\end{aligned}$$

so that $\boldsymbol{S}_{km}^{p}(t)\boldsymbol{H}$ is a $k\times n$ Hankel for any $t\in\mathbf{C}$. Let $\boldsymbol{H}=\boldsymbol{H}_{mn}(b/a)$, and

$$\frac{(\lambda+t)^p b(\lambda)}{a(\lambda)}=\alpha(\lambda)+\frac{\beta(\lambda)}{a(\lambda)} \tag{3.6}$$

where $\alpha(\lambda)$ is a polynomial and $\beta(\lambda)/a(\lambda)$ a strictly proper rational function. If (in some neighbourhood of infinity)

$$\frac{\beta(\lambda)}{a(\lambda)}=\frac{s_0}{\lambda}+\frac{s_1}{\lambda^2}+\cdots+\frac{s_u}{\lambda^{u+1}}+\cdots,$$

comparing the coefficients of $1/\lambda^{u+1}$ on the both sides of (3.6), we obtain

$$s_u=\sum_{\alpha=0}^{p}\begin{pmatrix}p\\\alpha\end{pmatrix}t^{p-\alpha}h_{u+\alpha}.$$

Thus, $s_{i+j}=(\boldsymbol{S}_{km}^{p}(t)\boldsymbol{H})_{i,j}$, and $\beta(\lambda)/a(\lambda)$ generates the Hankel matrix

$S_{km}^{p}(t)H$. Write now

$$S=(S_{ln}^{q}(t_2)(S_{km}^{p}(t_1)H)^{T})^{T},$$

and thus the desired conclusion follows directly by using the fact just proved twice. ■

The Proof of Theorem 3. 3

Suppose that $L\in\mathcal{L}_{m,n}(y,z)$. Then, by definition, there exists a rational function g/f with gcd $(f,ab)=1$ and $l=\max\{\deg f,\deg g\}>\max\{m,n\}$ such that $L=L_{y,z}(g/f)$. Let $y_0,z_0\in\mathbf{C}$ be new interpolation points with multiplicities $m_0=l-m$ and $n_0=l-n$ respectively such that

$$y_i\neq y_0,z_j\neq z_0,f(y_0)f(z_0)\neq 0,\quad i=1,2,\cdots,u,j=1,2,\cdots,v.$$

Put now

$$\bar{y}=(\underbrace{y_0,\cdots,y_0}_{m_0},\cdots,\underbrace{y_u,\cdots,y_u}_{m_u})^{T}\ \textit{and}\ \bar{z}=(\underbrace{z_0,\cdots,z_0}_{n_0},\cdots,\underbrace{z_v,\cdots,z_v}_{n_v})^{T}.$$

Similar to (2. 1), we can define the matrices $W(\bar{y})$ and $W(\bar{z})$ in $M_l(\mathbf{C})$ associated with IVs $\bar{y}$ and $\bar{z}$. It is readily checked that

$$(O_{m\times m_0}\quad I_m)W(\bar{y})=W(y)S_{ml}^{m_0}(-y_0),$$

$$(O_{n\times n_0}\quad I_n)W(\bar{z})=W(z)S_{nl}^{n_0}(-z_0),\tag{3. 7}$$

where $S_{ml}^{m_0}(-y_0)$ and $S_{nl}^{n_0}(-z_0)$ are Toeplitz matrices defined in Lemma 3. 4. Also, we have by Theorem 3. 1 that

$$L_{\bar{y},\bar{z}}\left(\frac{g}{f}\right)=W(\bar{y})\bar{H}W(\bar{z})^{T}$$

for a suitable Hankel matrix $\bar{H}\in M_l(\mathbf{C})$ compatible with $f\in\mathbf{C}_l(\lambda)$. Thus, from the definition of Loewner matrices, (3. 7) and Lemma 3. 4, it follows that

$$L=(O_{m\times m_0}\quad I_m)L_{\bar{y},\bar{z}}\left(\frac{g}{f}\right)(O_{n\times n_0}\quad I_n)^{T}$$

$$=W(y)(S_{ml}^{m_0}(-y_0)\bar{H}S_{nl}^{n_0}(-z_0)^{T})W(z)^{T}=W(y)HW(z)^{T}$$

for some $m\times n$ Hankel matrix H, and therefore the mapping defined by (3. 5) maps $\mathcal{L}_{m,n}(y,z)$ into $\mathcal{H}_{m,n}$. Since the matrices $W(y)$ and $W(z)$ are nonsingular and the dimensions of both spaces $\mathcal{L}_{m,n}(y,z)$ and $\mathcal{H}_{m,n}$ are $m+n-1$, the mapping defined by (3. 5) is an isomorphism. ■

Theorem3. 5 *Let* y,z *be a pair of IVs as in Corollary* 2. 5. *Let*

$\boldsymbol{H}\in\mathcal{H}_{m,n}$ and $\boldsymbol{L}=\boldsymbol{L}_{y,z}(\omega)\in\mathcal{L}_{n,n}(\boldsymbol{y},\boldsymbol{z})$ be such that (3.5) holds, where $\omega(\lambda)\in\mathbf{C}_{2n-1}(\lambda)$. If $\boldsymbol{H}=\boldsymbol{H}_n(h/f)$, $\deg h<\deg f\leqslant n$, then there exists a unique $g\in\mathbf{C}_n(\lambda)$, $\deg g\leqslant\deg f$, such that

$$g+hab=f\omega. \tag{3.8}$$

In particular, if $\omega(\lambda)\in\mathbf{C}_{2n-1}(\lambda)$ is the (unique) Hermite interpolation polynomial of GRIP (1.5) with $N=2n$, then the pair of polynomials $[g,f]$ is a generalized solution to GRIP (1.5) in the sense that

$$[g(\lambda)-f(\lambda)\omega(\lambda)]_{\lambda=Y_i}^{(k)}=0,\quad k=0,1,\cdots,v_i-1;\ i=1,2,\cdots,\theta.$$

Proof Assume first that $\gcd(f,ab)=1$. By (c) of Lemma Ⅳ, there exist $g_0\in\mathbf{C}_n(\lambda)$ and $\omega_0\in\mathbf{C}_{2n-1}(\lambda)$, $\deg g_0\leqslant\deg f\leqslant n$, such that

$$\boldsymbol{B}_n(a,f)\boldsymbol{H}\,\boldsymbol{B}_n(b,f)=\boldsymbol{B}_n(g_0,f),$$

$$g_0+hab=f\omega_0.$$

Thus, in view of (3.4)

$$\boldsymbol{L}=\boldsymbol{L}_{y,z}(\omega)=\boldsymbol{W}(\boldsymbol{y})\boldsymbol{H}\boldsymbol{W}(\boldsymbol{z})^{\mathrm{T}}=\boldsymbol{P}^{\mathrm{T}}\boldsymbol{B}_n(a,f)\boldsymbol{H}_n\left(\frac{h}{f}\right)\boldsymbol{B}_n(b,f)\boldsymbol{Q}$$

$$=\boldsymbol{P}^{\mathrm{T}}\boldsymbol{B}_n(g_0,f)\boldsymbol{Q}=\boldsymbol{L}_{y,z}\left(\frac{g_0}{f}\right)=\boldsymbol{L}_{y,z}(\omega_0),$$

whence $\omega(\lambda)=\omega_0(\lambda)+c$ for some $c\in\mathbf{C}$. Now put $g=g_o+cf$. Then g satisfies $\deg g\leqslant\deg f$ and (3.8). The last assertion follows from (3.8).

To prove the theorem in the case when $\gcd(f,ab)\neq1$, we change y_i and z_j for $y_i(\varepsilon)$ and $z_j(\varepsilon)$ such that all $y_i(\varepsilon)$'s and all $z_j(\varepsilon)$'s are also distinct, respectively for $\varepsilon\in(0,\delta)$, where δ is a suitable positive number, and

$$f(y_i(\varepsilon))f(z_j(\varepsilon))\neq0,\ \lim_{\varepsilon\to0}y_i(\varepsilon)=y_i,\quad \lim_{\varepsilon\to0}z_j(\varepsilon)=z_j,$$

$$i=1,2,\cdots,u;\ j=1,2,\cdots,v.$$

Consequently, the desired conclusion follows from the result just proved by using a limit process. ■

Our next objective is to establish a generalization of Theorem 3.5 to Loewner matrices with an arbitrary size, which was proved in [17, Lemma 4.4] for the minimal generalized solution to GRIP (1.5) only under the assumption that Leowner matrices do not have full rank.

Theorem 3.6 *Let $\boldsymbol{y},\boldsymbol{z}$ be a pair of* IV*s (not necessarily $|\boldsymbol{y}|=|\boldsymbol{z}|$)*

as in (1.1). *Let* $\mathbf{L}=\mathbf{L}_{y,z}(\omega)$ *and* $\mathbf{H}\in\mathcal{H}_{m,n}$ *be such that* (3.5) *holds, where* $\omega\in\mathbf{C}_{N-1}(\lambda)$. *If* $\mathbf{H}=\mathbf{H}_{mn}(h/f)$, $\deg h<\deg f$, *then there exists uniquely a polynomial* g, $\deg g\leqslant\deg f$, *such that*

$$g+hab=f\omega. \tag{3.9}$$

In particular, if further, $\omega(\lambda)\in\mathbf{C}_{N-1}(\lambda)$ *is the (unique) Hermite interpolation polynomial of GRIP* (1.5), *then the pair of polynomials* $[g,f]$ *is a generalized solution to GRIP* (1.5).

Proof Similar to the proof of Theorem 3.5, it suffices to show the theorem under the assumption that $\gcd(f,ab)=1$. Since f and h are not limited to be coprime, we can assume that $l=\deg f>\max\{m,n\}$. Let y_0, z_0, $\bar{\mathbf{y}}$, and $\bar{\mathbf{z}}$ be as in the proof of Theorem 3.3. By the theory of the Hermite interpolation, there exists a polynomial $\alpha(\lambda)$ of degree $\leqslant m_0+n_0-1$ such that

$$f(\lambda)\alpha(\lambda)+h(\lambda)=\bar{h}(\lambda)(\lambda-y_0)^{m_0}(\lambda-z_0)^{n_0} \tag{3.10}$$

holds for a suitable polynomial $\bar{h}$ with $\deg\bar{h}<\deg f$. Set

$$\bar{a}(\lambda)=(\lambda-y_0)^{m_0}a(\lambda) \text{ and } \bar{b}(\lambda)=(\lambda-z_0)^{n_0}b(\lambda).$$

Then, there exists uniquely a pair of polynomials $\bar{g}(\lambda)$ and $\bar{\omega}(\lambda)$ such that

$$\bar{g}+\bar{h}\bar{a}\bar{b}=f\bar{\omega}, \tag{3.11}$$

$\deg\bar{g}<\deg f$ and $\deg\bar{\omega}\leqslant 2l-1$, and therefore, by (3.10)

$$\bar{g}+hab=f(\bar{\omega}-\alpha ab). \tag{3.12}$$

It follows from (3.11) and Theorem 3.1 that

$$\mathbf{L}_{\bar{y},\bar{z}}(\bar{\omega})=\mathbf{L}_{\bar{y},\bar{z}}\left(\frac{\bar{g}}{f}\right)=\mathbf{W}(\bar{\mathbf{y}})\bar{\boldsymbol{H}}\mathbf{W}(\bar{\mathbf{z}})^{\mathrm{T}},$$

where $\bar{\boldsymbol{H}}$ of order l is a Hankel matrix compatible with $f\in\mathbf{C}_l^0(\lambda)$. Applying a well known result on Hankel matrices (see e.g. [13]), we have that

$$\bar{\boldsymbol{H}}=\boldsymbol{H}_l\left(\frac{h_0}{f}\right)$$

for a suitable polynomial h_0, $\deg h_0<\deg f$, so that it is seen easily by Theorem 3.4 that

$$\bar{g}+h_0\bar{a}\bar{b}=f\bar{\omega}.$$

This with (3.11) implies that $\bar{h}(\lambda)=h_0(\lambda)$. Thus, if follows from Lemma 3.4, (3.10), Theorem 3.5 and the proof of Theorem 3.3 that

$$\boldsymbol{L}_{y,z}(\bar{\omega}-\alpha ab)=\boldsymbol{L}_{y,z}(\bar{\omega})=\boldsymbol{W}(\boldsymbol{y})\left(\boldsymbol{S}_{ml}^{m_0}(-y_0)\boldsymbol{H}_l\left(\frac{\bar{h}}{f}\right)\boldsymbol{S}_{nl}^{n_0}(-z_0)^{\mathrm{T}}\right)\boldsymbol{W}(\boldsymbol{z})^{\mathrm{T}}$$

$$=\boldsymbol{W}(\boldsymbol{y})\boldsymbol{H}_{mn}\left(\frac{h}{f}\right)\boldsymbol{W}(\boldsymbol{z})^{\mathrm{T}}=\boldsymbol{L}_{y,z}(\omega).$$

As deg $(\bar{\omega}-\alpha ab)\leqslant N-1$ by (3.12), this implies that there exists a constant $c\in\mathbf{C}$ such that $\bar{\omega}-\alpha ab=\omega-c$. Now put $g(\lambda)=\bar{g}(\lambda)+cf(\lambda)$. Then, (3.9) follows from (3.12), and the last assertion follows in turn from (3.9). ■

We remark that more can be said about the relations between Loewner matrices (or GRIP) and Hankel matrices. For example, we can present the following useful theorems without proofs (see [4], [18]).

Theorem 3.7 *Given GRIP (1.5), there exists a unique vector* $\boldsymbol{h}=(h_0,\cdots,h_{N-2})^{\mathrm{T}}\in\mathbf{C}^{N-1}$ *such that for any* IVs $\boldsymbol{y}$ and $\boldsymbol{z}$ of the collection $\boldsymbol{Q}$ defined by (1.7),

$$\boldsymbol{L}_{y,z}(\omega)=\boldsymbol{W}(\boldsymbol{y})\boldsymbol{H}_{m,n}[\boldsymbol{h}]\boldsymbol{W}(\boldsymbol{z})^{\mathrm{T}}, \tag{3.13}$$

where ω *stands for the Hermite interpolation polynomial in* $\mathbf{C}_{N-1}(\lambda)$ *of GRIP(1.5),* $m=|\boldsymbol{y}|, n=|\boldsymbol{z}|, m+n=N$ *and* $\boldsymbol{H}_{m,n}[\boldsymbol{h}]=(h_{i+j})_{i,j=0}^{m-1,n-1}$ *is a Hankel in* $\mathcal{H}_{m,n}$.

Theorem 3.8 *Let* $\boldsymbol{h}=(h_0,\cdots,h_{N-2})^{\mathrm{T}}\in\mathbf{C}^{N-1}$ *be the vector defined uniquely by GRIP(1.5) as in Theorem 3.7,* $\omega(\lambda)$ *the Hermite interpolation polynomial in* $\mathbf{C}_{N-1}(\lambda)$ *of GRIP(1.5). If* g, f, h *are polynomials such that* deg $h<$deg $f<N$ *and*

$$g(\lambda)=f(\lambda)\omega(\lambda)-h(\lambda)a(\lambda)b(\lambda) \tag{3.14}$$

then

(a) *The pair of polynomials* $[g,f]$ *is a generalized solution to GRIP(1.5).*

(b) deg $g\leqslant$deg f *if and only if* h/f *generates the Hankel matrix* $\boldsymbol{H}_{m,n}[\boldsymbol{h}]=(h_{i+j})_{i,j=0}^{m-1,n-1}$ *for any* m, n *such that* $m+n=N$.

Corollary 3.9 *Let* $\boldsymbol{h}=(h_0,\cdots,h_{N-2})^{\mathrm{T}}\in\mathbf{C}^{N-1}$ *and* $\omega(\lambda)$ *be as in Theorem 3.8. If* $\boldsymbol{y}$ *and* $\boldsymbol{z}$ *are a pair of* IVs *of* $\boldsymbol{Q}$ *in* (1.7), *and* r *is the integer part of* $N/2$, *then*

$$\operatorname{rank} L_{y,z}(\omega)=\min\{|y|,|z|,n_1\} \tag{3.15}$$

where $n_1=\operatorname{rank}\,(h_{i+j})_{i,j=0}^{r-1,N-r-1}$.

This corollary is in fact an immediate consequence of Theorem 3.7 with a known result on the rank of Hankel matrices in [11]. Moreover, this corollary includes Corollary 2.24 of [1] as special case.

The last two Theorems lead to a new approach for solving GRIP in the scalar case, which allows the problem of parametzization of all solutions to GRIP(1.5) in an explicit from to be essentially reduced to that of finding all generatings of the so-called Hankel vector $\boldsymbol{h}$ specified uniquely by GRIP(1.5) as in Theorem 3.7. For detailed discussions of many of these developments we refer to [4][18].

§ 4. Loewner and R-Matrices

In Corollary 2.5, given $f\in \mathbf{C}_n^0(\lambda)$, we have characterized the class of Loewner matrices associated with a pair of IVs $\boldsymbol{y}$ and $\boldsymbol{z}$ in the case when $|\boldsymbol{y}|=|\boldsymbol{z}|=n$ by intertwining relation

$$\boldsymbol{L}\boldsymbol{C}_z(f)=\boldsymbol{C}_y^{\mathrm{T}}(f)\boldsymbol{L}, \tag{4.1}$$

where $\boldsymbol{C}_y(f)$ and $\boldsymbol{C}_z(f)$ stand for the left and right companion matrices of f, respectively. Comparing to intertwining relations given in (c) of Lemma I we see that the roles of $\boldsymbol{C}_f$ and the Bezoutian $\boldsymbol{B}$ are played by $\boldsymbol{C}_z(f)$, $\boldsymbol{C}_y(f)$ and Loewner matrix $\boldsymbol{L}$. Similarly, in parallel with (c) of Lemma II, it is natural to introduce a class of matrices, such that $\boldsymbol{R}$ is in this class if and only if the intertwining relation

$$\boldsymbol{R}\boldsymbol{C}_y^{\mathrm{T}}(f)=\boldsymbol{C}_z(f)\boldsymbol{R} \tag{4.2}$$

holds. In the section, it is out purpose to give a complete description of such a class of matrices in the case when $f\in \mathbf{C}_n(\lambda)$ as well as close connections to Bezoutians, Hankel and Loewner matrices.

As a first step towards this aim we shall establish its characterization of intertwining relation under the assumption that $f\in \mathbf{C}_n^0(\lambda)$.

Theorem 4.1 *Let* $\boldsymbol{y}, \boldsymbol{z}$ *be as in Corollary* 2.5, *and* $f\in \mathbf{C}_n^0(\lambda)$ *such that* $\gcd(f,ab)=1$. *Denote by* $\boldsymbol{P}$ *and* $\boldsymbol{Q}$ *the matrices in* (2.15) *with* $r=$

$m=n$. If $\mathbf{R}\in M_n(\mathbf{C})$ then the following are equivalent:

(a) *There exists $h\in \mathbf{C}_{n-1}(\lambda)$ such that $\mathbf{R}=\boldsymbol{Q}^{-1}\boldsymbol{H}_n(h/f)\mathbf{P}^{-\mathrm{T}}$.*

(b) $\boldsymbol{R}\boldsymbol{C}_y^{\mathrm{T}}(f)=\boldsymbol{C}_z(f)\boldsymbol{R}$.

Proof The equivalence of (a) and (b) follows from Lemma Ⅱ and (2.16). ■

Without the restriction on the size of matrices, we define the $\boldsymbol{R}$-matrices as follows.

Definition 4.2 Let $\mathbf{y},\mathbf{z}$ be a fixed pair of IVs as in (1.2) with $|\mathbf{y}|=m$ and $|\mathbf{z}|=n$, and $0\neq f\in \mathbf{C}_m(\lambda)$. A matrix $\boldsymbol{R}\in M_{nm}(\mathbf{C})$ is said to be an $\boldsymbol{R}$-matrix compatible with f (and associated with IVs $\mathbf{z}$ and $\mathbf{y}$), if it is of the form

$$\boldsymbol{R}=\boldsymbol{B}\boldsymbol{H}\boldsymbol{A}^{\mathrm{T}} \tag{4.3}$$

where $\boldsymbol{H}$ stands for an n×m Hankel matrix compatible with f, and

$$\begin{aligned}\boldsymbol{A}&=\operatorname{diag}\left[f(\boldsymbol{J}(y_i))\right]_{i=1}^{u}\mathbf{V}(\mathbf{y})^{-1}\in M_m(\mathbf{C}),\\ \boldsymbol{B}&=\operatorname{diag}\left[\mathrm{f}(\boldsymbol{J}(z_j))\right]_{j=1}^{v}\mathbf{V}(\mathbf{z})^{-1}\in M_m(\mathbf{C}),\end{aligned} \tag{4.4}$$

Denote by $\mathcal{R}_{z,y}(f)$ the class of the $\boldsymbol{R}$-matrices compatible with $f\in\mathbf{C}_m(\lambda)$, If, further, in (4.3) $\boldsymbol{H}=\boldsymbol{H}_{nm}(h/f)$, where $\deg h\leqslant \deg f\leqslant m$, the proper rational function h/f will be called a generating for the $\boldsymbol{R}$-matrix $\boldsymbol{R}=\boldsymbol{B}\boldsymbol{H}\boldsymbol{A}^{\mathrm{T}}$, written as $\boldsymbol{R}=\boldsymbol{R}_{z,y}(h/f)$.

Remark from Theorem 4.1 that each of matrices $\boldsymbol{R}$ satisfying the property (4.2) belongs to $\boldsymbol{R}_{z,y}(f)$ and is of the form $\boldsymbol{R}=\boldsymbol{B}\boldsymbol{H}_n(h/f)\mathbf{A}^{\mathrm{T}}$ for some $h\in\mathbf{C}_{n-1}(\lambda)$ whenever $f\in\mathbf{C}_n^0(\lambda)$ and $|\mathbf{z}|=|\mathbf{y}|=n$.

The part (a) of next theorem contains an interesting Barnett-like factorization formula for the $\boldsymbol{R}$-matrix of the form $\boldsymbol{R}_{z,y}(h/f)$ with $\deg h\leqslant\deg f=n$ under the assumption that $|\mathbf{z}|=|y|=n$. Also, the part (c) can be considered as a version of the well known Lander's theorem.

Theorem 4.3 *Let $\mathbf{y},\mathbf{z}$ be as in Corollary 2.5, a and b as in (2.1). Let $f,h\in\mathbf{C}_n(\lambda)$ be such that $\deg h\leqslant\deg f\leqslant n$ and $\gcd(f,ab)=1$.*

(a) *If $\deg f=n$, then*

$$\boldsymbol{R}_{z,y}\left(\frac{1}{f}\right)=-\boldsymbol{L}_{y,z}^{-1}\left(\frac{1}{f}\right), \tag{4.5}$$

$$\boldsymbol{R}_{z,y}\left(\frac{h}{f}\right)=h(\boldsymbol{C}_z(f))\boldsymbol{R}_{z,y}\left(\frac{1}{f}\right)=\boldsymbol{R}_{z,y}\left(\frac{1}{f}\right)h(\boldsymbol{C}_y^{\mathrm{T}}(f)). \tag{4.6}$$

(b) *If* $\gcd(f,h)=1$, *then* $\operatorname{rank}\boldsymbol{R}_{z,y}(h/f)=\deg f$. *In particular*, $\boldsymbol{R}_{z,y}(h/f)$ *is nonsingular if and only if* $\gcd(f,h)=1$ *and* $f\in\mathbf{C}_n^0(\lambda)$.

(c) *The inverse of a nonsingular* **R**-*matrix is a Loewner matrix and conversely. More precisely, if* $f\in\mathbf{C}_n^0(\lambda)$, *then a nonsingular* **R**-*matrix* **R** *belongs to* $\mathcal{R}_{z,y}(f)$ *if and only if* $\boldsymbol{R}^{-1}=\boldsymbol{L}_{y,z}(g/f)$ *for some* $g\in\boldsymbol{C}_n(\lambda)$.

Proof (a): Observe from (2.15) and (4.4) that $\boldsymbol{B}=\boldsymbol{Q}^{-1}$ and $\boldsymbol{A}=\boldsymbol{P}^{-1}$ if $r=m=n$. Thus the part (a) follows from Lemma Ⅱ by using a similar argument to that given in the proof of Corollary 2.4.

(b): An immediate consequence of the nonsingularity criterion for Hankel matrices of the form $\boldsymbol{H}=\boldsymbol{H}_n(h/f)$. Finally, (c) follows from Theorem 4.1 and Corollary 2.5. ■

Theorem 4.4 *Let* **y**, **z** *be as in Corollary* 2.4. *Let* $0\neq f\in\mathbf{C}_n(\lambda)$ *and* $\boldsymbol{R}=\boldsymbol{BHA}^{\mathrm{T}}\in\mathcal{R}_{z,y}(f)$, *where the Hankel matrix* $\boldsymbol{H}\in\mathcal{H}_n(f)$ *has uniquely a quasidirect composition of the form* $\mathbf{H}=\mathbf{H}_n(h/f)+\mathbf{H}_0$ *in the sense of Lemma* Ⅲ. *Then*

(a) *There exists* $g\in\mathbf{C}_n(\lambda)$ *such that*

$$\boldsymbol{W}(\boldsymbol{z})^{\mathrm{T}}\boldsymbol{R}\boldsymbol{W}(\boldsymbol{y})=\boldsymbol{B}_n(g,f). \tag{4.7}$$

Moreover, if $g\in\mathbf{C}_n(\lambda)$ *satisfies the property* (4.7), *then*

$$g=f\omega+hab \tag{4.8}$$

for a suitable $\omega\in\mathbf{C}_{2n-1}(\lambda)$.

(b) *In the case when* $\gcd(f,ab)=1$, *there exists* $g\in\mathbf{C}_n(\lambda)$ *such that*

$$\boldsymbol{R}=\boldsymbol{D}(\boldsymbol{z})\boldsymbol{L}_{z,y}\left(\frac{g}{f}\right)\boldsymbol{D}(\boldsymbol{y}), \tag{4.9}$$

where

$$\begin{aligned}\boldsymbol{D}(\boldsymbol{y})&=\operatorname{diag}[f(\boldsymbol{J}(y_i))a_{i,m_i-1}^{-1}(\boldsymbol{J}(y_i))\boldsymbol{P}_{m_i}]_{i=1}^{u}\\ \boldsymbol{D}(\boldsymbol{z})&=\operatorname{diag}[f(\boldsymbol{J}(z_j))b_{j,n_j-1}^{-1}(\boldsymbol{J}(z_j))\boldsymbol{P}_{n_j}]_{j=1}^{v}\end{aligned} \tag{4.10}$$

in which each diagonal block of $\boldsymbol{D}(\boldsymbol{y})$ [$\boldsymbol{D}(\boldsymbol{z})$ *resp.*] *is a nonsingular and upper triangular Hankel matrix of order* $m_i, i=1,2,\cdots,u$ [$n_j, j=1,2,\cdots,v$, *resp.*].

Proof (a): It follows from Lemma 2.1 and the formula (3.3) that

$$\begin{aligned}\boldsymbol{V}(\boldsymbol{y})^{\mathrm{T}}\boldsymbol{W}(\boldsymbol{y})^{\mathrm{T}}\boldsymbol{A}\boldsymbol{V}(\boldsymbol{y})&=[\boldsymbol{W}(\boldsymbol{y})\boldsymbol{V}(\boldsymbol{y})]^{\mathrm{T}}\boldsymbol{A}\boldsymbol{V}(\boldsymbol{y})\\&=\mathrm{diag}[\boldsymbol{P}_{m_i}f(\boldsymbol{J}(y_i))a_{i,m_i-1}(\boldsymbol{J}(y_i))]_{i=1}^{u}\\&=\boldsymbol{V}(\boldsymbol{y})^{\mathrm{T}}\boldsymbol{B}_n(a,f)\boldsymbol{V}(\boldsymbol{y}),\end{aligned}$$

so that $\boldsymbol{B}_n(a,f)=\boldsymbol{W}(\boldsymbol{y})^{\mathrm{T}}\boldsymbol{A}$. Similarly, we can show that $\boldsymbol{B}_n(b,f)=\boldsymbol{W}(\boldsymbol{z})^{\mathrm{T}}\boldsymbol{B}$. Thus, by (c) *of* Lemma Ⅳ, we have in turn that

$$\begin{aligned}\boldsymbol{W}(\boldsymbol{z})^{\mathrm{T}}\boldsymbol{R}\boldsymbol{W}(\boldsymbol{y})&=\boldsymbol{W}(\boldsymbol{z})^{\mathrm{T}}\boldsymbol{B}\boldsymbol{H}\boldsymbol{A}^{\mathrm{T}}\boldsymbol{W}(\boldsymbol{y})\\&=\boldsymbol{B}_n(b,f)\boldsymbol{H}\boldsymbol{B}_n(a,f)=\boldsymbol{B}_n(g,f)\end{aligned}$$

for some $g\in\mathbf{C}_n(\lambda)$. Again by (c) of Lemma Ⅳ, we have (4.8) holds for some $\omega\in\mathbf{C}_{2n-1}(\lambda)$ if $g\in\mathbf{C}_n(\lambda)$ satisfies the property (4.7).

(b): If, further, $\gcd(f,ab)=1$, then, by Lemma 2.1,

$$\boldsymbol{D}(\boldsymbol{y})^{\mathrm{T}}=\boldsymbol{D}(\boldsymbol{y})=\boldsymbol{P}^{-1}\boldsymbol{W}(\boldsymbol{y})^{-1},\quad \boldsymbol{D}(\boldsymbol{z})^{\mathrm{T}}=\boldsymbol{D}(\boldsymbol{z})=\boldsymbol{Q}^{-1}\boldsymbol{W}(\boldsymbol{z})^{-1},$$

so that, in view of (3.7) and (2.16),

$$\boldsymbol{R}=\boldsymbol{W}(\boldsymbol{z})^{-\mathrm{T}}\boldsymbol{Q}^{-\mathrm{T}}[\boldsymbol{Q}^{\mathrm{T}}\boldsymbol{B}_n(g,f)\boldsymbol{P}]\boldsymbol{P}^{-1}\boldsymbol{W}(\boldsymbol{y})^{-1}=\boldsymbol{D}(\boldsymbol{z})\boldsymbol{L}_{y,z}^{\mathrm{T}}\left(\frac{g}{f}\right)\boldsymbol{D}(\boldsymbol{y}),$$

as deserved. ■

The next result gives a complete description of the class $\mathcal{R}_{z,y}(f)$ in the case where $f\in\mathbf{C}_n(\lambda)$ and $|\boldsymbol{y}|=|\boldsymbol{z}|=n$, which is a simple consequence of the proof of Theorem 4.4.

Corollary 4.5 *Let* $\boldsymbol{y},\boldsymbol{z}$ *be as in Corollary* 2.5 *and* $f\in\mathbf{C}_n(\lambda)$ *such that* $\gcd(f,ab)=1$. *If* $\boldsymbol{R}\in M_n(\mathbf{C})$ *then the following are equivalent*:

(a) $\boldsymbol{R}\in\mathcal{R}_{z,y}(f)$.

(b) *There exists* $g\in\mathbf{C}_n(\lambda)$ *such that* (4.7) *holds.*

(c) *There exists* $g\in\mathbf{C}_n(\lambda)$ *such that* (4.9) *holds where* $\boldsymbol{D}(\boldsymbol{y})$ *and* $\boldsymbol{D}(\boldsymbol{z})$ *stand for the matrices given in* (4.10).

Corollary 4.6 *Let* $\boldsymbol{y},\boldsymbol{z}$ *be as in Corollary* 2.5, *and let* $f\in\mathbf{C}_n(\lambda)$ *be such that* $\gcd(f,ab)=1$. *If* $\mathbf{L}\in M_n(\mathbf{C})$ *then the following are equivalent*:

(a) *There exists* $g\in\mathbf{C}_n(\lambda)$ *such that* $\mathbf{L}=\mathbf{L}_{y,z}(g/f)$.

(b) *There exists* $\boldsymbol{R}\in\mathcal{R}_{y,z}(f)$ *such that*

$$\boldsymbol{L}=\boldsymbol{D}(\boldsymbol{y})^{-1}\boldsymbol{R}\boldsymbol{D}(\boldsymbol{z})^{-1},\tag{4.11}$$

where $\boldsymbol{D}(\boldsymbol{y})$ *and* $\boldsymbol{D}(\boldsymbol{z})$ *are defined by* (4.10).

Proof By Corollary 3.2 and (4.10), the equivalence of (a) and (b) follows from the following fact:

$$\begin{aligned} \boldsymbol{L} &= \boldsymbol{W}(\boldsymbol{y})\boldsymbol{H}\boldsymbol{W}(\boldsymbol{z})^{\mathrm{T}} \\ &= \boldsymbol{D}(\boldsymbol{y})^{-1}\operatorname{diag}[f(\boldsymbol{J}(y_i))]_{i=1}^{u}\boldsymbol{V}(\boldsymbol{y})^{-1}\boldsymbol{H}\boldsymbol{V}(\boldsymbol{z})^{-\mathrm{T}}\operatorname{diag}[f(\boldsymbol{J}(z_j))]_{j=1}^{v}\boldsymbol{D}(\boldsymbol{z})^{-1} \\ &= \boldsymbol{D}(\boldsymbol{y})^{-1}[\boldsymbol{A}\boldsymbol{H}\boldsymbol{B}^{\mathrm{T}}]\boldsymbol{D}(\boldsymbol{z})^{-1}. \end{aligned}$$ ■

§ 5. Products of Loewner and *R*-Matrices

This section is devoted to an analysis of products of Loewner and **R**-matrices, in terms of the result established in Lemma Ⅳ combined with Theorem 2.2. Our main interest is to describe some special cases of products just mentioned. In the following discussion we shall need more than two IVs, written as $\boldsymbol{y}_i, \boldsymbol{z}_i, i, j=0,1,\cdots$ instead of $\boldsymbol{y}, \boldsymbol{z}$ in (1.1). As in Section 1 given a pair of IVs $\boldsymbol{y}_i$ and $\boldsymbol{z}_j$ with $|\boldsymbol{y}_i|=|\boldsymbol{z}_j|=n$ and $f\in \mathbf{C}_n(\lambda)$, denote by $\mathcal{L}_{y_i,z_j}(f)$ the class of all matrices of the form $\boldsymbol{L}_{y_i,z_j}(g/f)$ for polynomials $g\in \mathbf{C}_n(\lambda)$ (rather than polynomials $g(\lambda)$ of any degree), and by $a_i(\lambda)$ and $b_j(\lambda)$ the $a(\lambda)$ and $b(\lambda)$ in (2.1) with $\boldsymbol{y}$ and $\boldsymbol{z}$ *replaced by* $\boldsymbol{y}_i$ and $\boldsymbol{z}_j$, respectively, similarly, for $\boldsymbol{P}(\boldsymbol{y}_i), \boldsymbol{Q}(\boldsymbol{z}_j)$ in (2.15), $\boldsymbol{A}_i, \boldsymbol{B}_j$ in (4.4), etc.

Theorem 5.1 (*Compare with Theorem* 3.6 *in* [10]) *Let k be a fixed positive integer. Let* $\boldsymbol{y}_i, \boldsymbol{z}_j$, *be* IVs *such that* $|\boldsymbol{y}_i|=|\boldsymbol{z}_j|=n, i, j=1,2,\cdots,k+1$, *which have consecutive indices therein, and* $f\in \mathbf{C}_n(\lambda)$ *such that* $\gcd(f, a_i b_j)=1, \forall i, j$.

(a) *If* $\boldsymbol{E}_{2i-1}\in \mathcal{L}_{y_i,z_i}(f)[\mathcal{R}_{z_i,y_i}(f)$ *resp.*$], i=1,2,\cdots,k+1, \boldsymbol{E}_{2i}\in \mathcal{R}_{z_i,y_{i+1}}(f)[\mathcal{L}_{y_{i+1},z_i}(f)$ *resp.*$], i=1,2,\cdots,k$, *then*

$$\boldsymbol{E}_1\boldsymbol{E}_2\cdots\boldsymbol{E}_{2k+1}\in \mathcal{L}_{y_i,z_{k+1}}(f)\ [\mathcal{R}_{z_1,y_{k+1}}(f)\ \textit{resp.}]. \tag{5.1}$$

(b) *If* $g_1, g_2, \cdots, g_k, h_1, h_2, \cdots, h_{k-1}\in \mathbf{C}_n(\lambda)$ *such that* $\gcd(f, h_i)=1$ *and* $\max\{\deg f, \deg h_i\}=n, i=1,2,\cdots,k-1$, *then*

$$\boldsymbol{L}_{y_1,z_1}\left(\frac{g_1}{f}\right)\boldsymbol{L}_{y_2,z_1}^{-1}\left(\frac{h_1}{f}\right)\boldsymbol{L}_{y_2,z_2}\left(\frac{g_2}{f}\right)\boldsymbol{L}_{y_3,z_2}^{-1}\left(\frac{h_2}{f}\right)\cdots\boldsymbol{L}_{y_k,z_k}\left(\frac{g_k}{f}\right)=\boldsymbol{L}_{y_1,z_k}\left(\frac{g}{f}\right) \tag{5.2}$$

holds for some $g \in \mathbf{C}_n(\lambda)$. *Moreover*, $g \in \mathbf{C}_n(\lambda)$ *satisfies the property* (5.2) *if and only if*

$$g_1 g_2 \cdots g_k - h_1 h_2 \cdots h_{k-1} g = fp \tag{5.3}$$

for a suitable $p \in \mathbf{C}_{(k-1)n}(\lambda)$. *If* $g = g_0$ *satisfies the property* (5.3), *then the set of all polynomials in* $\mathbf{C}_n(\lambda)$ *satisfying* (5.3) *consists of all polynomials of the form* $g_0 + \beta f$ *where* β *is a constant*.

Proof (a): We prove first that (5.1) holds for $k=1$. Let $\boldsymbol{E}_1 = \boldsymbol{L}_{y_1, z_1}(g_1/f)$ and $\boldsymbol{E}_3 = \boldsymbol{L}_{y_2, z_2}(g_2/f)$ for some $g_1, g_2 \in \mathbf{C}_n(\lambda)$, and $\boldsymbol{R} = \boldsymbol{B}_1 \boldsymbol{H} \boldsymbol{A}_2^{\mathrm{T}} \in \mathcal{R}_{z_1, y_2}(f)$ for some $\boldsymbol{H} \in \mathcal{H}_n(f)$. Then, by (2.15)(2.16)(4.4) and (a) of Lemma Ⅳ, we have that there exists $g \in \mathbf{C}_n(\lambda)$ such that

$$\begin{aligned}\boldsymbol{E}_1\boldsymbol{E}_2\boldsymbol{E}_3 &= \boldsymbol{P}^{\mathrm{T}}(\boldsymbol{y}_1)\boldsymbol{B}_n(g_1, f)\boldsymbol{H}\boldsymbol{B}_n(g_2, f)\boldsymbol{Q}(\boldsymbol{z}_2)\\ &= \boldsymbol{P}^{\mathrm{T}}(\boldsymbol{y}_1)\boldsymbol{B}_n(g, f)\,\boldsymbol{Q}(\boldsymbol{z}_2) = \boldsymbol{L}_{y_1, z_2}\left(\frac{g}{f}\right).\end{aligned}$$

In a similar manner, we can show that if $\boldsymbol{E}_1 \in \mathcal{R}_{z_1, y_1}(f)$, $\boldsymbol{E}_3 \in \mathcal{R}_{z_2, y_2}(f)$ and $\boldsymbol{E}_2 \in \mathcal{L}_{y_2, z_1}(f)$, then $\boldsymbol{E}_1\boldsymbol{E}_2\boldsymbol{E}_3 \in \mathcal{R}_{z_1, y_2}(f)$. (5.1) follows from the previous results by induction form k.

(b): Let us verify the results (5.2) and (5.3) for $k=2$. Owing to (b) of Lemma Ⅳ and (2.16), we have that

$$\begin{aligned}\boldsymbol{L}_{y_1, z_1}\left(\frac{g_1}{f}\right)\boldsymbol{L}_{y_2, z_1}^{-1}\left(\frac{h_1}{f}\right)\boldsymbol{L}_{y_2, z_2}\left(\frac{g_2}{f}\right) &= \boldsymbol{P}^{\mathrm{T}}(\boldsymbol{y}_1)\boldsymbol{B}_n(g_1, f)\boldsymbol{B}_n^{-1}(h_1, f)\boldsymbol{B}_n(g_2, f)\boldsymbol{Q}(\boldsymbol{z}_2)\\ &= \boldsymbol{P}^{\mathrm{T}}(\boldsymbol{y}_1)\boldsymbol{B}_n(g, f)\,\boldsymbol{Q}(\boldsymbol{z}_2) = \boldsymbol{L}_{y_1, z_2}\left(\frac{g}{f}\right)\end{aligned}$$

for some $g \in \mathbf{C}_n(\lambda)$, and $g \in \mathbf{C}_n(\lambda)$ satisfies this property if and only if

$$g_1 g_2 - h_1 g = fp$$

for a suitable $p \in \mathbf{C}_n(\lambda)$. The conclusion in the general case follows by induction. The rest is obvious. ■

The following corollary of Theorem 5.1 was proved first by Z. Vavřín[15] under the assumption that $f \in \mathbf{C}_n^0(\lambda)$ and $\boldsymbol{y}, \boldsymbol{z}$ are a pair of the simples IVs, and then by M. Fiedler and V. Pták[10] under the assumption $f \in \mathbf{C}_n(\lambda)$ and $\boldsymbol{y}, \boldsymbol{z}$ are a pair of the simplest IVs only.

Corollary 5.2 *Let* $\boldsymbol{y}, \boldsymbol{z}$ *be as in Corollary* 2.5, *and let* $f, g, h \in \mathbf{C}_n(\lambda)$ *such that* $\gcd(f, ab) = 1$ *and* $gh - ab = fp$ *for some* $p \in \mathbf{C}_n(\lambda)$. *Then*

$$\boldsymbol{L}_{y,z}^{-1}\left(\frac{g}{f}\right)=\boldsymbol{D}(\boldsymbol{z})\boldsymbol{L}_{z,y}\left(\frac{h}{f}\right)\boldsymbol{D}(\boldsymbol{y}), \tag{5.4}$$

where $\boldsymbol{D}(\boldsymbol{y})$ *and* $\boldsymbol{D}(\boldsymbol{z})$ *are as in* (4.10).

Proof By (b) of Theorem 5.1 with $k=2$, $\boldsymbol{y}_1=\boldsymbol{z}_2=\boldsymbol{y}$ and $\boldsymbol{y}_2=\boldsymbol{z}_1=\boldsymbol{z}$, we have

$$\boldsymbol{L}_{y,z}\left(\frac{g}{f}\right)\boldsymbol{L}_{z,z}^{-1}\left(\frac{b}{f}\right)\boldsymbol{L}_{z,y}\left(\frac{h}{f}\right)=\boldsymbol{L}_{y,y}\left(\frac{a}{f}\right), \tag{5.5}$$

whence both $\boldsymbol{L}_{y,z}(g/f)$ and $\boldsymbol{L}_{z,y}(h/f)$ are nonsingular under the condition that $\gcd(f,ab)=1$. Thus, (5.4) will follow from (5.5) if we can show that

$$\boldsymbol{L}_{y,y}^{-1}\left(\frac{a}{f}\right)=\boldsymbol{D}(\boldsymbol{y}),\quad \boldsymbol{L}_{z,z}^{-1}\left(\frac{b}{f}\right)=\boldsymbol{D}(\boldsymbol{z}). \tag{5.6}$$

Now by (2.15)(4.10) and (3.3),

$$\begin{aligned}\boldsymbol{B}_n^{-1}(a,f)&=\boldsymbol{V}(\boldsymbol{y})\operatorname{diag}[f^{-1}(\boldsymbol{J}(y_i))a_{i,m_i-1}^{-1}(\boldsymbol{J}(y_i))\boldsymbol{P}_{m_i}]_{i=1}^{u}\boldsymbol{V}(\boldsymbol{y})^{\mathrm{T}}\\&=\boldsymbol{P}\boldsymbol{D}(\boldsymbol{y})\boldsymbol{P}^{\mathrm{T}}.\end{aligned}$$

whence, by (2.16)

$$\boldsymbol{L}_{y,y}^{-1}\left(\frac{a}{f}\right)=\boldsymbol{P}^{-1}\boldsymbol{B}_n(a,f)^{-1}\boldsymbol{P}^{-\mathrm{T}}=\boldsymbol{D}(\boldsymbol{y}).$$

In a similar manner, we can show that $\boldsymbol{L}_{z,z}^{-1}(b/f)=\boldsymbol{D}(\boldsymbol{z})$. ■

Theorem 5.3 *Let* $\boldsymbol{y}_i$, $\boldsymbol{z}_i$, *and* f *be as in Theorem* 5.1, $i=1,2,\cdots,k$, *and* $g_1,g_2,\cdots,g_k\in\mathbf{C}_n(\lambda)$. *Let* $\boldsymbol{R}_i=\boldsymbol{B}_i\boldsymbol{H}_i\boldsymbol{A}_{i+1}^{\mathrm{T}}\in M_n(\mathbf{C})$ *be an* $\boldsymbol{R}$*-matrix compatible with* f (*and associated with a pair of* IV*s* $\boldsymbol{z}_i$, $\boldsymbol{y}_{i+1}$), $i=1,2,\cdots,k-1$, *where* $\boldsymbol{H}_i\in\mathcal{H}_n(f)$. *Then there exists* $g\in\mathbf{C}_n(\lambda)$ *such that*

$$\boldsymbol{L}_{y_1,z_1}\left(\frac{g_1}{f}\right)\boldsymbol{R}_1\boldsymbol{L}_{y_2,z_2}\left(\frac{g_2}{f}\right)\boldsymbol{R}_2\cdots\boldsymbol{R}_{k-1}\boldsymbol{L}_{y_k,z_k}\left(\frac{g_k}{f}\right)=\boldsymbol{L}_{y_1,z_k}\left(\frac{g}{f}\right). \tag{5.7}$$

If, further, $\boldsymbol{H}_i$ *has a unique quasidirect decomposition of the form* $\boldsymbol{H}_i=\boldsymbol{H}_n(h_i/f)+\boldsymbol{H}_{0i}$ *in the sense of Lemma* Ⅲ, *then the fact that* $g\in\mathbf{C}_n(\lambda)$ *satisfies the property* (5.7) *implies*

$$g=f\omega+(-1)^{k+1}h_1h_2\cdots h_{k-1}g_1g_2\cdots g_k \tag{5.8}$$

for a suitable polynomial $\omega(\lambda)$ *of degree* $(k-1)(2n-1)$ *at most.*

Proof Observe that each Loewner matrix of the form $\boldsymbol{L}_{y_i,z_i}(g_i/f)$ can be rewritten in a similar form to (2.16), $i=1,2,\cdots,k$, and thus the conclusion follows from Definition 4.2 and (c) of Lemma Ⅳ at once. ■

Corollary 5.4 *Under the same assumption as in Theorem* 5.3, *if*, *further*, $\boldsymbol{R}_0=\boldsymbol{B}_0\boldsymbol{H}_0\boldsymbol{A}_1^{\mathrm{T}}$ *and* $\boldsymbol{R}_k=\boldsymbol{B}_k\boldsymbol{H}_k\boldsymbol{A}_{k+1}^{\mathrm{T}}$ *are* $\boldsymbol{R}$-*matrices compatible with* f *and associated with the* IVs $\boldsymbol{z}_0, \boldsymbol{y}_1$ *and* $\boldsymbol{z}_k, \boldsymbol{y}_{k+1}$ *respectively*, *where* $\boldsymbol{H}_0$, $\boldsymbol{H}_k\in\mathcal{H}_n(f)$ *and* $|\boldsymbol{z}_0|=|\boldsymbol{y}_1|=|\boldsymbol{z}_k|=|\boldsymbol{y}_{k+1}|=n$, *then*

(a)

$$\boldsymbol{R}\equiv\boldsymbol{R}_0\boldsymbol{L}_{y_1,z_1}\left(\frac{g_1}{f}\right)\boldsymbol{R}_1\boldsymbol{L}_{y_2,z_2}\left(\frac{g_2}{f}\right)\boldsymbol{R}_2\cdots\boldsymbol{L}_{y_k,z_k}\left(\frac{g_k}{f}\right)\boldsymbol{R}_k\in\mathcal{R}_{z_0,y_{k+1}}(f). \tag{5.9}$$

Moreover,

$$\boldsymbol{M}_R\equiv\boldsymbol{R}_0\boldsymbol{L}_{y_1,z_1}\left(\frac{g_1}{f}\right)\boldsymbol{R}_1\boldsymbol{L}_{y_2,z_2}\left(\frac{g_2}{f}\right)\boldsymbol{R}_2\cdots\boldsymbol{L}_{y_k,z_k}\left(\frac{g_k}{f}\right)=\boldsymbol{L}_{z_0,z_0}^{-1}\left(\frac{b_0}{f}\right)\boldsymbol{L}_{z_0,z_k}\left(\frac{g}{f}\right) \tag{5.10}$$

for a suitable $g\in\mathbf{C}_n(\lambda)$, *and*

$$\begin{aligned}\boldsymbol{M}_L^{\mathrm{T}}&\equiv\boldsymbol{L}_{y_1,z_1}\left(\frac{g_1}{f}\right)\boldsymbol{R}_1\boldsymbol{L}_{y_2,z_2}\left(\frac{g_2}{f}\right)\boldsymbol{R}_2\cdots\boldsymbol{L}_{y_k,z_k}\left(\frac{g_k}{f}\right)\boldsymbol{R}_k\\&=\boldsymbol{L}_{y_1,y_{k+1}}\left(\frac{\widetilde{g}}{f}\right)\boldsymbol{L}_{y_{k+1},y_{k+1}}^{-1}\left(\frac{a_{k+1}}{f}\right)\end{aligned} \tag{5.11}$$

for a suitable $\widetilde{g}\in\mathbf{C}_n(\lambda)$.

(b) *If*, *further*, $\boldsymbol{y}_1=\boldsymbol{y}_{k+1}=\boldsymbol{y}$, $\boldsymbol{z}_0=\boldsymbol{z}_k=\boldsymbol{z}$, $f\in\mathbf{C}_n^0(\lambda)$, *and* $\boldsymbol{R}_0=\boldsymbol{R}_k$, *then*, *for any* $\boldsymbol{L}\in\mathcal{L}_{y,z}(f)$, $\boldsymbol{R}\in\mathcal{R}_{z,y}(f)$

$$\boldsymbol{L}\boldsymbol{M}_R=\boldsymbol{M}_L^{\mathrm{T}}\boldsymbol{L},\ \boldsymbol{R}\boldsymbol{M}_L^{\mathrm{T}}=\boldsymbol{M}_R\boldsymbol{R}.$$

Moreover, $\boldsymbol{M}_R$ *and* $\boldsymbol{M}_L^{\mathrm{T}}$ *can be written as polynomials of matrices* $\mathbf{C}_y(f)$ *and* $\mathbf{C}_z^{\mathrm{T}}(f)$ *respectively*.

Proof (a): In view of (b) of Theorem 4.4, there exist $g_0, g_{k+1}\in\mathbf{C}_n(\lambda)$ such that

$$\boldsymbol{R}_0=\boldsymbol{B}_0\boldsymbol{H}_0\boldsymbol{A}_1^{\mathrm{T}}=\boldsymbol{D}(\boldsymbol{z}_0)\boldsymbol{L}_{z_0,y_1}\left(\frac{g_0}{f}\right)\boldsymbol{D}(\boldsymbol{y}_1)$$

$$\boldsymbol{R}_k=\boldsymbol{B}_k\boldsymbol{H}_k\boldsymbol{A}_{k+1}^{\mathrm{T}}=\boldsymbol{D}(\boldsymbol{z}_k)\boldsymbol{L}_{z_k,y_{k+1}}\left(\frac{g_{k+1}}{f}\right)\boldsymbol{D}(\boldsymbol{y}_{k+1}).$$

By (5.7) and Corollary 4.5, we get

$$\begin{aligned}\boldsymbol{R}&=\boldsymbol{D}(\boldsymbol{z}_0)\boldsymbol{L}_{z_0,y_1}\left(\frac{g_0}{f}\right)\boldsymbol{D}(\boldsymbol{y}_1)\boldsymbol{L}_{y_1,z_k}\left(\frac{g}{f}\right)\boldsymbol{D}(\boldsymbol{z}_k)\boldsymbol{L}_{z_k,y_{k+1}}\left(\frac{g_{k+1}}{f}\right)\boldsymbol{D}(\boldsymbol{y}_{k+1})\\&=\boldsymbol{D}(\boldsymbol{z}_0)\boldsymbol{L}_{z_0,y_1}\left(\frac{g_0}{f}\right)\widetilde{\boldsymbol{R}}\boldsymbol{L}_{z_k,y_{k+1}}\left(\frac{g_{k+1}}{f}\right)\boldsymbol{D}(\boldsymbol{y}_{k+1})\end{aligned}$$

$$=\boldsymbol{D}(\boldsymbol{z}_0)\boldsymbol{L}_{z_0,y_{k+1}}\left(\frac{\widetilde{g}}{f}\right)\boldsymbol{D}(\boldsymbol{y}_{k+1})\in\mathcal{R}_{z_0,y_{k+1}}(f),$$

where $g,\widetilde{g}$ are suitable polynomils in $\mathbf{C}_n(\lambda)$ and

$$\widetilde{\boldsymbol{R}}=\boldsymbol{D}(\boldsymbol{y}_1)\,\boldsymbol{L}_{y_1,z_k}\left(\frac{g}{f}\right)\boldsymbol{D}(\boldsymbol{z}_k)\in\mathcal{R}_{y_1,z_k}(f).$$

By Theorem 5.3, (5.6)(5.2), there exist, $g,\hat{g}\in\mathbf{C}_n(\lambda)$ such that for some $p\in\mathbf{C}_n(\lambda)$

$$g_0\hat{g}-a_1g=fp \tag{5.12}$$

and

$$\boldsymbol{M}_R=\boldsymbol{D}(\boldsymbol{z}_0)\boldsymbol{L}_{z_0,y_1}\left(\frac{g_0}{f}\right)\boldsymbol{D}(\boldsymbol{y}_1)\boldsymbol{L}_{y_1,z_k}\left(\frac{\hat{g}}{f}\right)$$
$$=\boldsymbol{L}_{z_0,z_0}^{-1}\left(\frac{b_0}{f}\right)\boldsymbol{L}_{z_0,y_1}\left(\frac{g_0}{f}\right)\boldsymbol{L}_{y_1,y_1}^{-1}\left(\frac{a_1}{f}\right)\boldsymbol{L}_{y_1,z_k}\left(\frac{\hat{g}}{f}\right)$$
$$=\boldsymbol{L}_{z_0,z_0}^{-1}\left(\frac{b_0}{f}\right)\boldsymbol{L}_{z_0,z_k}\left(\frac{g}{f}\right),$$

so that (5.10) follows. In a similar manner we can show (5.11) where

$$g_{k+1}\hat{g}-b_k\widetilde{g}=f\widetilde{p} \tag{5.13}$$

for a suitable $\widetilde{p}\in\mathbf{C}_n(\lambda)$.

(b): Under the assumption of (b), we have by (5.10)(2.11) that

$$\boldsymbol{M}_R=\boldsymbol{L}_{z,z}^{-1}\left(\frac{b}{f}\right)\boldsymbol{L}_{z,z}\left(\frac{g}{f}\right)=\boldsymbol{b}^{-1}(\boldsymbol{C}_z(f))g(\boldsymbol{C}_z(f)),$$

and, similarly,

$$\boldsymbol{M}_L^{\mathrm{T}}=\widetilde{g}(\boldsymbol{C}_y^{\mathrm{T}}(f))\boldsymbol{a}^{-1}(\boldsymbol{C}_y^{\mathrm{T}}(f)),$$

where $g,\widetilde{g}\in\mathbf{C}_n(\lambda)$ satisfy

$$g_0\hat{g}-ag=f\,p \text{ and } g_0\hat{g}=b\,\widetilde{g}=f\,\widetilde{p},$$

for some suitable polynomials $p,\widetilde{p}\in\mathbf{C}_n(\lambda)$ by (5.12) and (5.13), since we can assume that $g_0=g_{k+1}$ in the case of $\boldsymbol{R}_0=\boldsymbol{R}_{k+1}$. Thus, $\boldsymbol{M}_R$ and $\boldsymbol{M}_L^{\mathrm{T}}$ can be considered as polynomials of the matrices $\boldsymbol{C}_z(f)$ and $\boldsymbol{C}_y^{\mathrm{T}}(f)$ respectively by the theory of functions on matrices. Also

$$[b^{-1}g-\widetilde{g}a^{-1}](\boldsymbol{C}_z(f))=[a^{-1}b^{-1}(\widetilde{p}-p)](\boldsymbol{C}_z(f))f(\boldsymbol{C}_z(f))=\boldsymbol{O}, \tag{5.14}$$

since $f(\boldsymbol{C}_z(f))=\boldsymbol{Q}^{-1}f(\boldsymbol{C}_f)\boldsymbol{Q}=\boldsymbol{O}$ by (2.14) and the Cayley-Hamilton theorem. Thus, by Corollary 2.5 and (5.14),

$$\boldsymbol{L}\boldsymbol{M}_R=\boldsymbol{L}b^{-1}(\boldsymbol{C}_z(f))=\boldsymbol{L}\tilde{g}\,a^{-1}(\boldsymbol{C}_z(f))=\tilde{g}\,a^{-1}(\boldsymbol{C}_y^{\mathrm{T}}(f))\boldsymbol{L}=\boldsymbol{M}_L^{\mathrm{T}}\boldsymbol{L},$$

for all Loewner matrices $\boldsymbol{L}\in\mathcal{L}_{y,z}(f)$, each of which has the form $\boldsymbol{L}=\boldsymbol{L}_{y,z}(g/f)$ for some $g\in\mathbf{C}_n(\lambda)$ in the case of $f\in\mathbf{C}_n^0(\lambda)$. In a similar manner, we can show

$$\boldsymbol{R}\boldsymbol{M}_L^{\mathrm{T}}=\boldsymbol{M}_R\boldsymbol{R},\quad \forall\,\boldsymbol{R}\in\mathcal{R}_{z,y}(f). \qquad\blacksquare$$

To conclude the paper we remark that in the case of the simplest IVs Z. Vavřín[16] introduced the notion of $\mathcal{L}d$ matrices, each of which is defined as an appropriate diagonal transformation from a classical Loewner matrix, and has proved that products of such $\mathcal{L}d$ matrices can be written as a single $\mathcal{L}d$ matrix. Indeed, each $\mathcal{L}d$ matrix being of the form $\boldsymbol{L}\boldsymbol{D}$ where $\boldsymbol{L}\in\mathcal{L}_{n,n}(\boldsymbol{y},\boldsymbol{z})$ and $\boldsymbol{D}$ is the diagonal matrix $\boldsymbol{D}(\boldsymbol{y})$ defined by (4.10) with all $m_i=1$, we see by (4.9) that the product of two $\mathcal{L}d$ matrices can be expressed as one product of an $\boldsymbol{R}$-matrix and a Loewner matrix. Thus, Vavřín's result on products of $\mathcal{L}d$ matrices follows from (5.10) immediately, and we obtain a generalization to the notion of $\mathcal{L}d$ matrices in the more general case as by-product.

Acknowledgments

The authors acknowledge with thanks the referee for helpful comments and suggestions.

References

[1] A C Antoulas and B D O Anderson. On the scalar rational interpolation problem. IMA J. Math. Control Inform., 1986, 3: 61—88.

[2] V Belevitch. Interpolation Matrices. Philips Res. Rep., 1970, 25: 337—369.

[3] G-N Chen and H-P Zhang. Note on products of Bezoutians and Hankel matrices. Linear Algebra Appl., 1995, 225: 23—35.

[4] G-N Chen, B Zhao and H-P Zhang. The general rational interpolation problems in the scalar case and its Hankel vector. Linear Algebra Appl., 1996, 244: 165—201.

[5] W F Donoghue. Monotone Matrix Functions and Analytic Continuation. Springer-Verlag, Berlin, 1974.

[6] M Fiedler. Hankel and Loewner matrices. Linear ALgebra Appl., 1984, 58: 75—95.

[7] M. Fiedler. Polynomials and Hankel matrices. Linear Algebra Appl. , 1985, 66:235—248.

[8]M Fiedler and V Pták. Intertwining and testing matrices corresponding to a polynomial, Linear Algebra Appl. ,1987,86:53—74.

[9] M Fiedler and V Pták. Bezoutians and Intertwining matrices. Linear Algebra Appl. ,1987,86:43—51.

[10] M Fiedler and V Pták. Loewner and Bézout matrices. Linear Algebra Appl. , 1988,101:187—220.

[11] G Heinig and U Jungnickel. Hankel matrices generated by the Markov parameters, Hankel matrix extension, partial realization and Padè-approximation. Operator Theory: Advances and Applications, Vol. 19, 1986, Birkhäuser Verlag Basel, 231—253.

[12] C Löwner. Über monotone Matrixfunktionen. Math. Z. ,1934,38:177—216.

[13] V Pták. . Lyapunov, Bézout and Hankel. Linear Algebra Appl. , 1984, 58: 363—390.

[14] G Sansigre and M Alvarez. On Bezoutian reduction with Vandermonde matrix . Linear Algebra Appl. ,1989,121:401—408.

[15] Z Vavřín. Inverses of Loewner matrices. Linear Algebra Appl. , 1984, 63: 227—236.

[16] Z Vavřín. Multiplication of diagonal transforms of Loewner matrices. Linear Algebra Appl. ,1988,99:1—40.

[17] Z Vavřín. A unified approach to Loewner and Hankel matrices. Linear Algebra Appl. ,1991:143:171—222.

[18] G. N. Chen, H-P Zhang. Complete solution to the general rational interpolation problem in scalar case. J. Beijing Normal University (Natural Science) 1993,29(4):465—466.

Linear Algebra and Its Applications,
1995,225:23－35

Bezout 与 Hankel 矩阵乘积注记[①]

Note on Products of Bezoutians and Hankel Matrices

Abstract A quantitative theorem due to Fiedler and Pták, on products of Bezoutians and their inverses, is extended to a more general case, using as the main tool an alternative to Fiedler's assertion on the quasi-direct decomposition of Hankel matrices.

§ 1. Introduction and Preliminaries

This paper works out quantitative connections concerning products of Bezoutians and Hankel matrices, a topic which was briefly described by Fiedler and Pták[5,6], but more from a qualitative point of view. The development of the result mentioned was motivated by the desire to give simpler proofs in the theory of Loewner matrices and further by the rational interpolation problems in the scalar case, including multiple nodes.

Various interesting relations of this sort established in [5] and [6] are partially represented as follows. Here, $\mathbf{C}_n(\lambda)$ denotes the class of polynomials of degree n at most; given $f\in\mathbf{C}_n(\lambda)$, by $\mathcal{B}_n(f)[\mathcal{H}_n(f)]$ we denote the set of Bezoutians $\boldsymbol{B}_n(f,g)$ of order n for polynomials $g\in\mathbf{C}_n(\lambda)$ [the set of Hankel matrices of order n compatible with f](see the latter part of this section of the precise definition).

---

① Received 25 February 1993; final manuscript accepted 17 October 1993. The authors were partially supported by the Science Foundation of State Education Committee of China. 本文与张惠品合作, Submitted by Hans Schneider.

Lemma Ⅰ *Let* $0\neq f\in \mathbf{C}_n(\lambda)$, *and let k be a fixed positive integer.*

(a) *If* $\mathbf{A}_{2i-1}\in \mathcal{B}_n(f)$ [$\mathcal{H}_n(f)$ *respectively*], $i=1,2,\cdots,k+1$, *and* $\mathbf{A}_{2i}\in \mathcal{H}_n(f)[\mathcal{B}_n(f)]$, $i=1,2,\cdots,k$, *then*

$$\mathbf{A}_1\mathbf{A}_2\cdots\mathbf{A}_{2k+1}\in \mathcal{B}_n(f)[\mathcal{H}_n(f)].$$

(b) *In particular, if* $g_1, g_2, \cdots, g_{2k+1}\in \mathbf{C}_n(\lambda)$, $\gcd(f, g_{2i})=1$ *and* $\max\{\deg f, \deg g_{2i}\}=n, i=1,2,\cdots,k$, *then there exists* $g\in \mathbf{C}_n(\lambda)$ *such that*

$$\boldsymbol{B}_n(f,g_1)\boldsymbol{B}_n^{-1}(f,g_2)\boldsymbol{B}_n(f,g_3)\boldsymbol{B}_n^{-1}(f,g_4)\cdots\boldsymbol{B}_n^{-1}(f,g_{2k})\boldsymbol{B}_n(f,g_{2k+1})=\boldsymbol{B}_n(f,g). \tag{1.1}$$

Moreover, $g\in \mathbf{C}_n(\lambda)$ *satisfies the property* (1.1) *if and only if*

$$g_1g_3\cdots g_{2k+1}-g_2g_4\cdots g_{2k}g=fp \tag{1.2}$$

for a suitable polynomial $p\in \mathbf{C}_{kn}(\lambda)$.

If $g=g_0\in \mathbf{C}_n(\lambda)$ *is a polynomial having the property* (1.2), *then the set of all polynomials in* $\mathbf{C}_n(\lambda)$ *having the property* (1.2) *consists of all polynomials of the form* $g_0+\beta f, \beta=const.$

In the present note we pursue the investigation of quantitative results concerning the interlacing products of Bezoutians and their inverses of the type given in (b) of Lemma Ⅰ. Moreover, we focus on the more general case of products of such matrices where the inverses of Bezoutians are replaced by Hankel matrices compatible with a polynomial $f\in \mathbf{C}_n(\lambda)$. Some of our quantitative results, e. g., the polynomial congruence equations (2.18) and (2.23) below, are especially important in the case when the compatible polynomial f has degree less than n. In this case, an alternative of Fiedler's assertion [4, Theorem 2.5] on the quasidirect decomposition of Hankel matrices is used as a tool. It turns out that each $\boldsymbol{H}\in \mathcal{H}_n(f)$ (not necessarily singular) can be uniquely written as a quasidirect sum of the form

$$\boldsymbol{H}=\boldsymbol{H}_n\left(\frac{h}{f}\right)+\boldsymbol{H}_0, \tag{1.3}$$

where $h\in C_{n-1}(\lambda)$ with $\deg h\leqslant \deg f$, and $\boldsymbol{H}_0$ of order n is a lower triangular Hankel matrix.

By means of the decomposition (1.3) on Hankel matrices, we establish the properties of products of Bezoutians in $\mathcal{B}_n(f)$ and Hankel matrices in $\mathcal{H}_n(f)$, including mutual relations between the compatible polynomial of the products, which are Bezoutians, and the generating functions, of Hankel matrices therein (see Theorems 4 and 6 below). For applications of these results we refer to [2, Theorems 3.5, 5.3].

In the case deg $f=n$ the result on products of matrices mentioned is not very surprising, since it is an almost immediate consequence of the well-known Barnett formulas (the calculation of the products of Bezoutians and Hankel matrices can be reduced to multiplication of polynomials in the companion matrix $\boldsymbol{C}_f$).

The following notation and preliminaries will be used in the sequel. Let n be a fixed natural number throughout. Denote by $M_n(\mathbf{C})$ the set of complex matrices of order n, by $\mathbf{C}_n^0(\lambda)$ the subclass of $\mathbf{C}_n(\lambda)$ consisting of all polynomials of degree n. We shall adopt the following conventions: the degree of the zero polynomial equals $-\infty$, so that deg $0<$ deg f whenever f is a nonzero polynomial ; gcd(f,g) is always monic where f,g are polynomials.

By a Hankel matrix we shall mean a square matrix of order n which has the form $\boldsymbol{H}=(h_{i+j}), i,j=0,1,\cdots,n-1$. Given two polynomials f and g with deg $g\leqslant$deg f, if $g(\lambda)/f(\lambda)$ has the expansion at infinity

$$\frac{g(\lambda)}{f(\lambda)}=h_{-1}+\frac{h_0}{\lambda}+\frac{h_1}{\lambda^2}+\cdots, \tag{1.4}$$

then $\boldsymbol{H}=(h_{i+j}), i,j=0,1,\cdots,n-1$, is called the Hankel matrix generated by the rational function $g(\lambda)/f(\lambda)$, written as $\boldsymbol{H}=\boldsymbol{H}_n(g/f)$. A Hankel matrix $\boldsymbol{H}=(h_{i+j}), i,j=0,1,\cdots,n-1$, is lower triangular if $h_0=h_1=\cdots=h_{n-2}=0$; a singular and lower triangular Hankel matrix is said to be degenerate. By definition, a Hankel matrix $\boldsymbol{H}=(h_{i+j}), i,j=0,1,\cdots,n-1$, is compatible with a nonzero polynomial $f(\lambda)=f_0+f_1\lambda+\cdots+f_n\lambda^n$ if $\hat{H}[f]_n=\mathbf{0}$, in which $[f]_n=(f_0,f_1,\cdots,f_n)^{\mathrm{T}}\in\mathbf{C}^{n+1}$ and $\hat{\boldsymbol{H}}=(h_{i+j}), i=0,1,\cdots,n-2, j=0,1,\cdots,n$; in this case, write $\boldsymbol{H}\in\mathcal{H}_n(f)$. Given a pair of polynomials g and f in $\mathbf{C}_n(\lambda)$, the Bezoutian $\boldsymbol{B}_n(f,g)=$

$(b_{ij}), i, j=0,1,\cdots,n-1$, of f and g is a symmetric matrix of order n defined by

$$\frac{f(\lambda)g(\mu)-f(\mu)g(\lambda)}{\lambda-\mu}=\sum_{i,j=0}^{n-1}b_{i,j}\lambda^i\mu^j.$$

In this case, the Bezoutian $\boldsymbol{B}_n(f,g)$ is said to be compatible with f and g, written as $\boldsymbol{B}_n(f,g)\in\mathcal{B}_n(f)$ or $\boldsymbol{B}_n(f,g)\in\mathcal{B}_n(g)$.

Let

$$f(\lambda)=\sum_{j=0}^{n}f_j\lambda^j(f_n\neq 0),\quad g(\lambda)=\sum_{j=0}^{n}g_j\lambda^j. \tag{1.5}$$

As is well known, the following factorization formula due to S. Barnett is fundamental for the study of the Bezoutians:

$$\boldsymbol{B}_n(f,g)=\boldsymbol{B}_n(f,1)g(C_f)=g(\boldsymbol{C}_f^{\mathrm{T}})\boldsymbol{B}_n(f,1) \tag{1.6}$$

where $\boldsymbol{C}_f$ of order n stands for the (first) companion matrix of f. Thus, the Bezoutian $\boldsymbol{B}_n(f,g)$ is nonsingular if and only if $\gcd(f,g)=1$ and $\max\{\deg f, \deg g\}=n$. Similarly, we have

$$\boldsymbol{H}_n\left(\frac{h}{f}\right)=h(\boldsymbol{C}_f)\boldsymbol{B}_n^{-1}(f,1)=\boldsymbol{B}_n^{-1}(f,1)h(\boldsymbol{C}_f^{\mathrm{T}}), \tag{1.7}$$

where $f\in\mathbf{C}_n^0(\lambda)$ and $h\in\mathbf{C}_n(\lambda)$, and therefore $\boldsymbol{H}_n(h/f)$ is nonsingular if and only if $\gcd(f,h)=1$, and $\deg f=n$. It follows from (1.6)(1.7) or from Theorem 0.1 of Heinig and Jungnickel [7] that the inverse of $\boldsymbol{H}_n(h/f)$ is given by

$$\mathbf{H}_n^{-1}\left(\frac{h}{f}\right)=\boldsymbol{B}_n(f,g) \tag{1.8}$$

where $g(\lambda)\in\mathbf{C}_n(\lambda)$ is the solution of the congruence

$$h(\lambda)g(\lambda)=1 \bmod f(\lambda). \tag{1.9}$$

§ 2. Main Results

First of all, we need to establish an alternative to Fiedler's assertion on the quasidirect decomposition of Hankel matrices. In contrast to Fiedler's in [4, Theorem 2.5], our decomposition is stated for an *a priori* chosen polynomial f and is always unique whenever the Hankel matrix $\boldsymbol{H}$ (not necessarily singular) is in $\mathcal{H}_n(f)$ for some $0\neq f\in\mathbf{C}_n(\lambda)$. In the rest of the note we shall frequently quote this result rather than Fiedler's.

Theorem 1 *Let* $0\neq f\in\mathbf{C}_n(\lambda)$ *and* $\boldsymbol{H}\in\mathcal{H}_n(f)$. *Then there exist* $h\in\mathbf{C}_{n-1}(\lambda)$ *with* $\deg h<\deg f$ *and a lower triangular Hankel matrix* $\boldsymbol{H}_0$ *of order* n, *which is degenerate if* $\deg f>0$, *where* h *and* $\boldsymbol{H}_0$ *are determined uniquely by* f *and* $\boldsymbol{H}$, *such that* $\boldsymbol{H}$ *can be uniquely written as a quasidirect sum of the form*

$$\boldsymbol{H}=\boldsymbol{H}_n\left(\frac{h}{f}\right)+\boldsymbol{H}_0, \tag{2.1}$$

that is, $\boldsymbol{H}$ *is the sum of both matrices* $\boldsymbol{H}_n(h/f)$ *and* $\boldsymbol{H}_0$ *with* $\operatorname{rank}\boldsymbol{H}=\operatorname{rank}\boldsymbol{H}_n(h/f)+\operatorname{rank}\boldsymbol{H}_0$. (*We shall call* $\boldsymbol{H}_n(h/f)$ *and* $\boldsymbol{H}_0$ *the proper and the lower triangular part of* $\boldsymbol{H}$, *respectively.*)

Proof Observe that if $\boldsymbol{H}\in\mathcal{H}_n(f)$ is a singular Hankel matrix, then the theorem follows from Theorem 2.5 of Fiedler [4] together with Theorems 4 and 1 of Chen and Zhang [1]. Assume now that $\boldsymbol{H}$ is nonsingular. We have then by Theorem 3 of Fiedler [3] that $\boldsymbol{H}\in\mathcal{H}_n(g)$ *for some* $g\in\mathbf{C}_n^0(\lambda)$ *such that* $\gcd(f,g)=1$ *and* $\boldsymbol{H}^{-1}=\boldsymbol{B}_n(f,g)$. Thus, there exists a pair of p,h of polynomials in $\mathbf{C}_n(\lambda)$ with $\deg h<\deg f$ satisfying $gh-fp=1$, whence, by (1.8),

$$\boldsymbol{H}=\boldsymbol{B}_n^{-1}(f,g)=\boldsymbol{H}_n\left(\frac{p}{g}\right)=\boldsymbol{H}_n\left(\frac{h}{f}\right)-\boldsymbol{H}_n\left(\frac{1}{fg}\right).$$

Put now $\boldsymbol{H}_0=-\boldsymbol{H}_n(1/(fg))$. Clearly, $\boldsymbol{H}_0$ is a lower triangular Hankel matrix in $\mathcal{H}_n(f)$ with $\operatorname{rank}\boldsymbol{H}_0=n-\deg f$, so that $\boldsymbol{H}=\boldsymbol{H}_n(h/f)+\boldsymbol{H}_0$ is a quasidirect sum, since $\gcd(f,h)=1$ and $\operatorname{rank}\boldsymbol{H}_n(h/f)=\deg f$, and $\boldsymbol{H}_0$ is degenerate if $\deg f>0$. In the case of $\deg f=0$, we have $\boldsymbol{H}_n(h/f)=\boldsymbol{O}$, $\boldsymbol{H}=\boldsymbol{O}+\boldsymbol{H}_0$, and the theorem follows obviously. In the case of $\deg f>0$, if $\boldsymbol{H}$ can be rewritten as a quasidirect sum of the form

$$\boldsymbol{H}=\boldsymbol{H}_n\left(\frac{h'}{f}\right)+\boldsymbol{H}'_0$$

where $h'\in\mathbf{C}_{n-1}(\lambda)$ such that $\deg h'<\deg f$ and $\boldsymbol{H}'_0$ is a degenerate Hankel matrix, then it is obvious that the Hankel matrix defined by

$$\boldsymbol{H}_n\left(\frac{h-h'}{f}\right)=\boldsymbol{H}'_0-\boldsymbol{H}_0$$

is degenerate as well. Since $f \in \mathbf{C}_n(\lambda)$, this implies that $h = h'$ and in turn $\boldsymbol{H}_0 = \boldsymbol{H}'_0$. Thus, the uniqueness of the decomposition (2.1) does hold. ■

Note that if $\boldsymbol{H}$ of order n is a singular Hankel matrix [and therefore $\boldsymbol{H} \in \mathcal{H}_n(f)$ for a suitable $0 \neq f \in \mathbf{C}_n(\lambda)$], and $\boldsymbol{H} \in \mathcal{H}_n(f) \cap \mathcal{H}_n(\hat{f})$, where both f and $\hat{f}$ are nonzero polynomials in $\mathbf{C}_n(\lambda)$, then the decompositions in the sense of Theorem 1

$$\boldsymbol{H} = \boldsymbol{H}_n\left(\frac{h}{f}\right) + \boldsymbol{H}_0,$$

$$\boldsymbol{H} = \boldsymbol{H}_n\left(\frac{\hat{h}}{\hat{f}}\right) + \widetilde{\boldsymbol{H}}_0$$

will ensure, by Theorem 2.5 of Fiedler [4], that $\boldsymbol{H}_n(h/f) = \boldsymbol{H}_n(\hat{h}/\hat{f})$, the proper part of $\boldsymbol{H}$, and $\boldsymbol{H}_0 = \widetilde{\boldsymbol{H}}_0$, the degenerate part of $\boldsymbol{H}$. Thus, in the case when $\boldsymbol{H}$ is singular Hankel, Theorem 1 is in fact independent of the choice of the polynomial f therein, and is reduced to Theorem 2.5 of Fiedler [4].

Corollary 1 *Let $0 \neq f \in \mathbf{C}_n(\lambda)$. Then $\boldsymbol{H} \in \mathcal{H}_n(f)$ if and only if there exist $h \in \mathbf{C}_{n-1}(\lambda)$ with $\deg h < \deg f$ and a lower triangular Hankel matrix $\boldsymbol{H}_0 \in \mathcal{H}_n(f)$ such that $\boldsymbol{H}$ can be written as a quasidirect sum of the form* (2.1).

Proof Observe by Theorem 3 of [8] that in the case of $f \in \mathbf{C}_n^0(\lambda)$, $\boldsymbol{H} \in \mathcal{H}_n(f)$ if and only if $\boldsymbol{H} = \boldsymbol{H}_n(h/f)$ for a suitable $h \in \mathbf{C}_{n-1}(\lambda)$, *and the conclusion follows with* $\boldsymbol{H}_0 = \boldsymbol{O}$. In the case of $\deg f < n$, the "only if" part follows from Theorem 1, and the "if" part is an immediate consequence of Lemma 2 of [1], since in this case $\boldsymbol{H}_n(h/f)$ is a singular Hankel matrix compatible with f. ■

Theorem 2 *Let $0 \neq f, a, b \in \mathbf{C}_n(\lambda)$, and let $\boldsymbol{H} = (h_{i+j})_{i,j=0}^{n-1} \in \mathcal{H}_n(f)$ be degenerate*, rank $\boldsymbol{H} = k$. *Then*

$$\boldsymbol{B}_n(f, a)\boldsymbol{H}\boldsymbol{B}_n(f, b) = \boldsymbol{B}_n(f, fp) \tag{2.2}$$

holds for a suitable polynomial $p \in \mathbf{C}_k(\lambda)$ depending only upon a, b, and $\boldsymbol{H}$. Moreover, all $p \in \mathbf{C}_k(\lambda)$ satisfying the property (2.2) *are of the form $p_0 + c$, $c =$ const, where $p_0 \in \mathbf{C}_k(\lambda)$ satisfies* (2.2). *If, further* ,

$\boldsymbol{H}\neq\boldsymbol{O}$ *and* $p\in\mathbf{C}_k(\lambda)$ *has the property* (2.2), *then* $\deg p=k$ *if and only if* $a,b\in\mathbf{C}_n^0(\lambda)$.

Proof The theorem is trivial when $\boldsymbol{H}=\boldsymbol{O}$. It remains to prove it in the case of $\boldsymbol{H}\neq\boldsymbol{O}$. A simple calculation leads to

$$\boldsymbol{B}_n(a,1)\boldsymbol{H}\boldsymbol{B}_n(b,1)=\boldsymbol{B}_n(1,p), \tag{2.3}$$

where $p(\lambda)=\sum_{j=0}^{k}p_j\lambda^j$ is a suitable polynomial in $\mathbf{C}_k(\lambda)$ having

$$p_k=-a_nh_{2n-k-1}b_n, \tag{2.4}$$

in which a_n and b_n are the coefficients at λ^n in $a(\lambda)$ and $b(\lambda)$, respectively. Thus, $\deg p=\operatorname{rank}\boldsymbol{H}=k$ if and only if $a_nb_n\neq 0$, that is, $a,b\in\mathbf{C}_n^0(\lambda)$.

Since $\boldsymbol{H}\in\mathcal{H}_n(f)$ is degenerate with $\operatorname{rank}\boldsymbol{H}=k$, we have by Chen and Zhang [1, Theorem 2] that $\deg(fp)\leqslant\deg f+k\leqslant n$. Now we first prove (2.2) under the constraint that $a,b\in\mathbf{C}_n^0(\lambda)$ having n simple roots each:

$$a(\lambda)=a_n\prod_{i=1}^{n}(\lambda-y_i),\quad b(\lambda)=b_n\prod_{i=1}^{n}(\lambda-z_i).$$

Denote by $\boldsymbol{V}(a)=(y_{j+1}^i)_{i,j=0}^{n-1}$ and $\boldsymbol{V}(b)=(z_{j+1}^i)_{i,j=0}^{n-1}$ the Vandermonde matrices associated with $a(\lambda)$ and $b(\lambda)$, respectively. By straightforward algebraic manipulations, it is easily seen that

$$\operatorname{diag}[f(y_i)]_{i=1}^n\boldsymbol{V}(a)^{\mathrm{T}}\boldsymbol{B}_n(1,p)\boldsymbol{V}(b)\operatorname{diag}[f(z_i)]_{i=1}^n=\boldsymbol{V}(a)^{\mathrm{T}}\boldsymbol{B}_n(f,fp)\boldsymbol{V}(b).$$

This, combined with (1.6)(2.3) and the well-known result that

$$\boldsymbol{V}(a)^{-1}\boldsymbol{C}_a\boldsymbol{V}(a)=\operatorname{diag}[y_1,y_2,\cdots,y_n],$$

implies that

$$\begin{aligned}
&\boldsymbol{V}(a)^{\mathrm{T}}\boldsymbol{B}_n(f,a)\boldsymbol{H}\boldsymbol{B}_n(f,b)\boldsymbol{V}(b)\\
&\quad=\boldsymbol{V}(a)^{\mathrm{T}}f(\boldsymbol{C}_a^{\mathrm{T}})[\boldsymbol{B}_n(a,1)\boldsymbol{H}\boldsymbol{B}_n(b,1)]f(\boldsymbol{C}_b)\boldsymbol{V}(b)\\
&\quad=f(\boldsymbol{V}(a)^{\mathrm{T}}\boldsymbol{C}_a^{\mathrm{T}}\boldsymbol{V}(a)^{-\mathrm{T}})[\boldsymbol{V}(a)^{\mathrm{T}}\boldsymbol{B}_n(1,p)\boldsymbol{V}(b)]f(\boldsymbol{V}(b)^{-1}\boldsymbol{C}_b\boldsymbol{V}(b))\\
&\quad=\operatorname{diag}[f(y_i)]_{i=1}^n[\boldsymbol{V}(a)^{\mathrm{T}}\boldsymbol{B}_n(1,p)\boldsymbol{V}(b)]\operatorname{diag}[f(z_j)]_{j=1}^n\\
&\quad=\boldsymbol{V}(a)^{\mathrm{T}}\boldsymbol{B}_n(f,fp)\boldsymbol{V}(b).
\end{aligned}$$

Since both of $\boldsymbol{V}(a)$ and $\boldsymbol{V}(b)$ are nonsingular, (2.2) holds for some polynomial $p\in\mathbf{C}_k(\lambda)$. Now we show (2.2) in the general case when a and b are in $\mathbf{C}_n(\lambda)$. As is well known, any polynomial in $\mathbf{C}_n(\lambda)$ can be written as a sum of a pair of polynomials in $\mathbf{C}_n^0(\lambda)$, each of which can be in turn written as a sum of a pair of polynomials in $\mathbf{C}_n^0(\lambda)$ having n simple roots each. Thus, (2.2) holds for any $a, b \in \mathbf{C}_n(\lambda)$ in terms of the result just proved. The rest is obvious, since the polynomials $p \in \mathbf{C}_k(\lambda)$ having the property (2.2) are independent of the choice of $f \in \mathbf{C}_n(\lambda)$. ■

Theorem 3 *Let* $0 \neq f, h, a, b$ *be all in* $\mathbf{C}_n(\lambda)$, $\deg h < \deg f$. *Then there exists* $g \in \mathbf{C}_n(\lambda)$ *such that*

$$\boldsymbol{B}_n(f,a)\boldsymbol{H}_n\left(\frac{h}{f}\right)\boldsymbol{B}_n(f,b)=\boldsymbol{B}_n(f,g). \tag{2.5}$$

Moreover, $g \in \mathbf{C}_n(\lambda)$ *satisfies the property* (2.5) *if and only if*

$$\deg g \leqslant \deg f \tag{2.6a}$$

and

$$g = hab + f\omega \tag{2.6b}$$

for a suitable polynomial ω *of degree* $2n-1$ *at most. The set of all polynomials satisfying* (2.6) *consists of all polynomials of the form* $g_0 + \beta f$, *where* β *is a constant.*

Proof If $0 \neq f = \text{const}$, then the theorem follows immediately, since in this case $g \in \mathbf{C}_n(\lambda)$ satisfies (2.5) or (2.6) if and only if $g = \text{const}$. Assume now that $\deg f > 0$. We shall distinguish two cases according to the degree of $\gcd(f, h)$. Suppose first that $\gcd(f, h) = 1$. Then there exist polynomials $p \in \mathbf{C}_n^0(\lambda)$ and $q \in \mathbf{C}_{n-1}(\lambda)$ such that

$$hp - fq = 1, \tag{2.7}$$

so that by (1.8),

$$\boldsymbol{H}_n\left(\frac{h}{f}\right)=\boldsymbol{H}_n\left(\frac{q}{p}\right)+\boldsymbol{H}_n\left(\frac{1}{fp}\right)=\boldsymbol{B}_n^{-1}(f,p)+\boldsymbol{H}_n\left(\frac{1}{fp}\right), \tag{2.8}$$

in which $\boldsymbol{H}_n(1/(fp))$ belongs to $\mathcal{H}_n(f)$ and is degenerate, since $\deg(fp) > n$ and $n - \deg f = \operatorname{rank} \boldsymbol{H}_n(1/(fp))$. It follows from Theorem 2, Lemma Ⅰ, and (2.7) that

$$\boldsymbol{B}_n(f,a)\boldsymbol{H}_n\left(\frac{1}{fp}\right)\boldsymbol{B}_n(f,b)=\boldsymbol{B}_n(f,f\tilde{p}) \tag{2.9}$$

for a suitable polynomial $\tilde{p}$ of degree $n - deg\ f$ at most, and

$$\boldsymbol{B}_n(f,a)\boldsymbol{B}_n(f,p)^{-1}\boldsymbol{B}_n(f,b)=\boldsymbol{B}_n(f,\tilde{g}) \tag{2.10}$$

for a suitable $\widetilde{g}\in \mathbf{C}_n(\lambda)$ satisfying

$$ab-p\widetilde{g}=f\alpha \tag{2.11}$$

where $\alpha(\lambda)$ is a polynomial in $\mathbf{C}_n(\lambda)$. Put now

$$g_0(\lambda)=\widetilde{g}(\lambda)+f(\lambda)\widetilde{p}(\lambda). \tag{2.12}$$

Then $g_0(\lambda)\in \mathbf{C}_n(\lambda)$ satisfies (2.5) with $g:=g_0$ by (2.9) (2.10) and (2.8). If $g\in \mathbf{C}_n(\lambda)$ is such that

$$\boldsymbol{B}_n(f,a)\boldsymbol{H}_n\left(\frac{h}{f}\right)\boldsymbol{B}_n(f,b)=\boldsymbol{B}_n(f,g), \tag{2.13}$$

then $\boldsymbol{B}_n(f,g_0)=\boldsymbol{B}_n(f,g)$, g_0 being as in (2.12), so that $g=g_0+\beta f$, where β= const. By (2.11) (2.12) and (2.7), we have

$$g(\lambda)=\widetilde{g}(\lambda)+f(\lambda)[\widetilde{p}(\lambda)+\beta]=h(\lambda)a(\lambda)b(\lambda)+f(\lambda)\omega(\lambda), \tag{2.14}$$

in which $\omega\in \mathbf{C}_{2n-1}(\lambda)$ is defined by

$$\omega(\lambda)=\widetilde{p}(\lambda)-\widetilde{q}(\lambda)\widetilde{g}(\lambda)-h(\lambda)\alpha(\lambda)+\beta, \tag{2.15}$$

and (2.6b) follows. As for (2.6a), if gcd $(f,ab)=1$, then, from (2.14) and the condition that $\gcd(f,h)=1$, we have that $\gcd(f,g)=1$, and therefore, by (2.13), deg f = rank $\boldsymbol{H}_n(h/f)\geqslant$ rank $\boldsymbol{B}_n(f,g)$ = max {deg f, deg g}. Thus, (2.6a) follows under the assumption that $\gcd(f,ab)=1$.

In the case when $\gcd(f,ab)\neq 1$, we have that there exist $a_1, a_2, b_1, b_2\in \mathbf{C}_n(\lambda)$ such that $a=a_1+a_2$, $b=b_2+b_2$, and gcd $(f,a_ib_j)=1$ for $i,j=1,2$. By the result just proved, there exist $g_{11}, g_{12}, g_{21}, g_{22}\in \mathbf{C}_n(\lambda)$ such that

$$\boldsymbol{B}_n(f,a_i)\boldsymbol{H}_n\left(\frac{h}{f}\right)\boldsymbol{B}_n(f,b_j)=\boldsymbol{B}_n(f,g_{ij}),\quad i,,j=1,2,$$

and

$$\deg g_{ij}\leqslant \deg f,\quad i,j=1,2,$$

so that $\deg(g_{11}+g_{12}+g_{21}+g_{22})\leqslant \deg f$ and

$$\boldsymbol{B}_n(f,a)\boldsymbol{H}_n\left(\frac{h}{f}\right)\boldsymbol{B}_n(f,b)=\boldsymbol{B}_n(f,g_{11}+g_{12}+g_{21}+g_{22}).$$

This together with (2.13) obtains that $g=g_{11}+g_{12}+g_{21}+g_{22}+f\gamma$ for some $\gamma\in \mathbf{C}$, whence $\deg g\leqslant \deg f$, i. e., (2.6a) follows again when $\gcd(f,ab)\neq 1$.

Conversely, if (2.6) is fulfilled for some g, that is, $\deg g\leqslant \deg f$

and $g = hab + f\omega$ where $\omega \in \mathbf{C}_{2n-1}(\lambda)$, then $g - g_0 = f(\omega - \omega_0)$, where g_0 is as in (2.12), and $\omega_0(\lambda) \in \mathbf{C}_{2n-1}(\lambda)$ is such that $g_0 + hab = f\omega_0$, so that $\omega - \omega_0$ is same constant δ, and $g = g_0 + f\delta$. Thus, (2.5) follows.

Assume now that $\gcd(f, h) \neq 1$. Then there exist $h_1, h_2 \in \mathbf{C}_n(\lambda)$ such that $h = h_1 + h_2$, $\deg h_i < \deg f$ and $\gcd(f, h_i) = 1$ for $i = 1, 2$. According to what we already proved, we can conclude that the theorem is also true if $\gcd(f, h) \neq 1$. ■

As an application of Corollary 1 and the proof of Theorem 3 we have

Corollary 2 *Let* $\boldsymbol{H} \in M_n(\mathbf{C})$ *and* $0 \neq f \in \mathbf{C}_n(\lambda)$. *Then* $\boldsymbol{H} \in \mathcal{H}_n(f)$ *if and only if there exist two polynomials* $g_1, g_2 \in \mathbf{C}_n^0(\lambda)$ *with* $\gcd(f, g_i) = 1$, *for* $i = 1, 2$, *and a lower triangular Hankel matrix* $\boldsymbol{H}_0 \in \mathcal{H}_n(f)$ *such that*

$$\boldsymbol{H} = \boldsymbol{B}_n^{-1}(f, g_1) + \boldsymbol{B}_n^{-1}(f, g_2) + \boldsymbol{H}_0. \tag{2.16}$$

Proof The "if" part is obvious. By Corollary 1, if $\boldsymbol{H} \in \mathcal{H}_n(f)$ then $\boldsymbol{H} = \boldsymbol{H}_n(h/f) + \boldsymbol{H}_0$ for some $h \in \mathbf{C}_n(\lambda)$ with $\deg h < \deg f$ and some lower Hankel matrix $\boldsymbol{H}_0 \in \mathcal{H}_n(f)$. Thus, if $\gcd(f, h) = 1$, then $\boldsymbol{H}_n(h/f) = \boldsymbol{B}_n^{-1}(f, 2p) + \boldsymbol{B}_n^{-1}(f, 2p) + \boldsymbol{H}'_0$ for some $p \in \mathbf{C}_n^0(\lambda)$ and some lower Hankel matrix $\boldsymbol{H}'_0 \in \mathcal{H}_n(f)$ by (2.8), and (2.16) follows with $g_1 := g_2 := 2p$, $\boldsymbol{H}_0 := \boldsymbol{H}_0 + \boldsymbol{H}'_0$. When $\gcd(f, h) \neq 1$ we can assume that $\deg f > 0$, since in case $0 \neq f = \text{const}$, (2.16) holds obviously with $g_1(\lambda) = \lambda^n$ and $g_2(\lambda) = -\lambda^n$. Then there exist $h_1, h_2 \in \mathbf{C}_n(\lambda)$ such that $h = h_1 + h_2$, $\deg h_i < \deg f$, and $\gcd(f, h_i) = 1$ for $i = 1, 2$, so that $\boldsymbol{H}_n(h_i/f) = \boldsymbol{B}_n^{-1}(f, p_i) + \boldsymbol{H}_{0i}$ for $i = 1, 2$, where p_i is some polynomial in $\mathbf{C}_n^0(\lambda)$ and $\boldsymbol{H}_{i0}$ *is some lower Hankel matrix in* $\mathcal{H}_n(f)$ by (2.8) again. Thus, (2.16) follows with $g_1 := p_1$, $g_2 := p_2$, and $\boldsymbol{H}_0 := \boldsymbol{H}_0 + \boldsymbol{H}_{01} + \boldsymbol{H}_{02}$. ■

Theorems 2 and 3 yield a useful result for Hankel matrices in $\mathcal{H}_n(f)$ when combined with the quasidirect decomposition given in Theorem 1.

Theorem 4 *Let* $0 \neq f, a, b \in \mathbf{C}_n(\lambda)$, *and* $\boldsymbol{H} \in \mathcal{H}_n(f)$. *if* $\boldsymbol{H}$ *has the unique quasidirect decomposition of the form* $\boldsymbol{H} = \boldsymbol{H}_n(h/f) + \boldsymbol{H}_0$ *as in Theorem* 1, *then*:

(a) *There exists* $g \in \mathbf{C}_n(\lambda)$ *such that*

$$\boldsymbol{B}_n(f,a)\boldsymbol{H}\boldsymbol{B}_n(f,b)=\boldsymbol{B}_n(f,g). \tag{2.17}$$

(b) *If* $g \in \mathbf{C}_n(\lambda)$ *satisfies the property* (2.17), *then*

$$g=hab+f\omega \tag{2.18}$$

for a suitable polynomial $\omega \in \mathbf{C}_{2n-1}(\lambda)$, *and* $\deg g-\deg f \leqslant \operatorname{rank} \boldsymbol{H}_0$. *In particular*, $\deg g-\deg f=\operatorname{rank} \boldsymbol{H}_0>0$ *if* $\boldsymbol{H}_0 \neq \boldsymbol{O}$ *and* $a, b \in \mathbf{C}_n^0(\lambda)$; $\max\{\deg f, \deg g\}=\operatorname{rank} \boldsymbol{H}$ *if* $a, b \in \mathbf{C}_n^0(\lambda)$ *and* $\gcd(f,h)=1$.

Proof (a): If $f(\lambda)$ is a nonzero constant, we have $h=0$, $\boldsymbol{H}=\boldsymbol{H}_0$, and thus the theorem follows by (2.3) and Theorems 2 and 3. Suppose now that $\deg f>0$. It follows from Theorem 1 and $\boldsymbol{H}_0$ is degenerate. By Theorems 2 and 3, there exist $p_0, g_0 \in \mathbf{C}_n(\lambda)$ with $\deg p_0 \leqslant \operatorname{rank} \boldsymbol{H}_0$ and $\deg g_0 \leqslant \deg f$ such that $g_0=hab+f\omega_0$ for a suitable $\omega_0 \in \mathbf{C}_{2n-1}(\lambda)$,

$$\boldsymbol{B}_n(f,a)\boldsymbol{H}_n\left(\frac{h}{f}\right)\boldsymbol{B}_n(f,b)=\boldsymbol{B}_n(f,g_0),$$

and

$$\boldsymbol{B}_n(f,a)\boldsymbol{H}_0\boldsymbol{B}_n(f,b)=\boldsymbol{B}_n(f,fp_0).$$

Put now $g=g_0+fp_0$. Then $g \in \mathbf{C}_n(\lambda)$, and (2.17) follows from the last two relations.

(b): If $\hat{g} \in \mathbf{C}_n(\lambda)$ satisfies (2.17) with $g:=\hat{g}$, then, since $\boldsymbol{B}_n(f,g)=\boldsymbol{B}_n(f,\hat{g})$ and $g=g_0+fp_0$, there exists a constant β such that $\hat{g}=g+\beta f=g_0+fp_0+\beta f$, and therefore, by Theorem 3, $\hat{g}=hab+f\omega_0+fp_0+\beta f=hab+f\tilde{\omega}$, where ω_0 is a suitable polynomial in $\mathbf{C}_{2n-1}(\lambda)$ and $\tilde{\omega}(\lambda)=\omega_0(\lambda)+p_0(\lambda)+\beta \in \mathbf{C}_{2n-1}(\lambda)$. *Thus*, (2.18) *follows with* $g:=\hat{g}$, *and* $\omega:=\tilde{\omega}$. Also, the fact that $\hat{g}=g_0+f(p_0+\beta)$, $\beta=$const, together with $\deg g_o \leqslant \deg f$ and $\deg p_0 \leqslant \operatorname{rank} H_0$, ensures that

$$\begin{aligned}\deg \hat{g}=\deg[g_0+f(p_0+\beta)] &\leqslant \deg f+\max\{\deg(p_0+\beta), 0\} \\ &\leqslant \deg f+\operatorname{rank} \boldsymbol{H}_0,\end{aligned}$$

so that $\deg \hat{g}-\deg f \leqslant \operatorname{rank} \boldsymbol{H}_0$. In particular, if $\boldsymbol{H}_0 \neq \boldsymbol{O}$ and $a, b \in \mathbf{C}_n^0(\lambda)$, then $\deg p_0=\operatorname{rank} \boldsymbol{H}_0>0$ in view of Theorem 2, and thus

$$\deg \hat{g}-\deg f=\deg p_0=\operatorname{rank} \boldsymbol{H}_0.$$

In the case when $\gcd(f,h)=1$ and $a, b \in \mathbf{C}_n^0(\lambda)$, we have that $\deg \hat{g}(\lambda) \leqslant$

$\deg f = \operatorname{rank} \boldsymbol{H}$ if $\boldsymbol{H}_0 = \boldsymbol{O}$, and $\operatorname{rank} \boldsymbol{H} = \deg f + \operatorname{rank} \boldsymbol{H}_0 = \deg \hat{g}$ if $\boldsymbol{H}_0 \neq \boldsymbol{O}$. Thus, $\max\{\deg \hat{g}, \deg f\} = \operatorname{rank} \boldsymbol{H}$. ■

Theorem 5 *Let* $0 = f, g, a, b \in \mathbf{C}_n(\lambda)$ *and let* $l = \deg g - \deg f$. *If both of* $\boldsymbol{B}_n(f,a)$ *and* $\boldsymbol{B}_n(f,b)$ *are nonsingular, then*

(a) *The matrix* $\boldsymbol{H}$ *defined by*

$$\boldsymbol{H} = \boldsymbol{B}_n^{-1}(f,a)\boldsymbol{B}_n(f,g)\boldsymbol{B}_n^{-1}(f,b) \tag{2.19}$$

is in $\mathcal{H}_n(f)$.

(b) *There exists uniquely* $h \in \mathbf{C}_{n-1}(\lambda)$ *with* $\deg h < \deg f$ *such that* $\boldsymbol{H}$ *given in* (2.19) *has uniquely the quasidirect decomposition of the form*

$$\boldsymbol{H} = \boldsymbol{H}_n\left(\frac{h}{f}\right) + \boldsymbol{H}_0, \tag{2.20}$$

where $\boldsymbol{H}_0$ *is the lower triangular part of* $\boldsymbol{H}$ *with* $\operatorname{rank} \boldsymbol{H}_0 = \max\{l, 0\}$.

Moreover, if $h \in \mathbf{C}_n(\lambda)$ *is the unique polynomial in* (2.20), *then*

$$g = hab + f\omega \tag{2.21}$$

for a suitable $\omega \in \mathbf{C}_{2n-1}(\lambda)$.

Proof Observe by Theorem 4 of Fiedler [3] that both of $\boldsymbol{B}_n^{-1}(f,a)$ and $\boldsymbol{B}_n^{-1}(f,b)$ are Hankel matrices in $\mathcal{H}_n(f)$. Thus, (a) follows from (a) of Lemma Ⅰ. As for (b), by Theorem 1 it suffices to verity (2.21) and the fact that $\operatorname{rank} \boldsymbol{H}_0 = \max\{l, 0\}$. We have from (2.19) that

$$\boldsymbol{B}_n(f,a)\boldsymbol{H}\boldsymbol{B}_n(f,b) = \boldsymbol{B}_n(f,g),$$

and (2.21) follows from (2.20) and Theorem 4. If $\boldsymbol{H}_0 = \boldsymbol{O}$, then $l = \deg g - \deg f \leqslant 0$ by Theorem 4, whence $\operatorname{rank} \boldsymbol{H}_0 = \max\{l, 0\}$. Conversely, if $l \leqslant 0$, then $\boldsymbol{H}_0 = \boldsymbol{O}$ when $\deg f = n$ in view of Theorem 3 of [8] and Theorem 1 and when $\deg f < n$ in view of Theorem 4. Finally, if $l > 0$ (and therefore $\deg f < n$), we have that $\boldsymbol{H}_0 \neq \boldsymbol{O}$ and $a, b \in \boldsymbol{C}_n^0(\lambda)$ under the assumption that both $\boldsymbol{B}_n(f,a)$ and $\boldsymbol{B}_n(f,b)$ are invertible, whence $\operatorname{rank} \boldsymbol{H}_0 = l$ by Theorem 4 once more. ■

As immediate corollaries, Theorems 4 and 5 have the following generalized results.

Theorem 6 *Let* $m \geqslant 2$ *be a fixed integer, and* $0 \neq f, g_1, g_2, \cdots, g_m \in \mathbf{C}_n(\lambda)$. *Let* $\boldsymbol{H}_i \in \mathcal{H}_n(f)$ *have a unique quasidirect decomposition of the* *form* $\mathbf{H}_i=\mathbf{H}_n(h_i/f)+\mathbf{H}_{0i}$, *where* $h_i\in\mathbf{C}_{n-1}(\lambda)$ *with* $\deg h_i<\deg f$ *and* $\mathbf{H}_{0i}$ *is the lower triangular part of* $\mathbf{H}_i$, $i=1,2,\cdots,m-1$. *Then there exists* $g\in\mathbf{C}_n(\lambda)$ *such that*

$$\mathbf{B}_n(f,g_1)\mathbf{H}_1\mathbf{B}_n(f,g_2)\mathbf{H}_2\cdots\mathbf{H}_{m-1}\mathbf{B}_n(f,g_m)=\mathbf{B}_n(f,g). \tag{2.22}$$

Moreover, if $g\in\mathbf{C}_n(\lambda)$ *satisfies the property* (2.22), *then*

$$g=h_1h_2\cdots h_{m-1}g_1g_2\cdots g_m+f\omega \tag{2.23}$$

for a suitable polynomial ω *of degree* $(m-1)(2n-1)$ *at most; in this case, all* $g\in\mathbf{C}_n(\lambda)$ *satisfying the property* (2.22) *are of the form* $g_0(\lambda)+cf(\lambda)$, $c=\text{const}$.

Proof A simple induction argument based on Theorem 4. ■

Theorem 7 *Let k be a fixed positive integer. Let* $0\neq f,g_1,g_2,\cdots,g_{2k+1}\in\mathbf{C}_n(\lambda)$ *be such that* $\max\{\deg f,\deg g_{2i-1}\}=n$ *and* $\gcd(f,g_{2i-1})=1$, $i=1,2,\cdots,k+1$. *Then*:

(a) *The products*

$$\mathbf{B}_n^{-1}(f,g_1)\mathbf{B}_n(f,g_2)\mathbf{B}_n^{-1}(f,g_3)\mathbf{B}_n(f,g_4)\cdots\mathbf{B}_n(f,g_{2k})\mathbf{B}_n^{-1}(f,g_{2k+1}) \tag{2.24}$$

can be written as a single Hankel matrix, say $\mathbf{H}$, *in* $\mathcal{H}_n(f)$.

(b) *There exist* $h\in\mathbf{C}_n(\lambda)$ *with* $\deg h<\deg f$ *and a lower triangular Hankel matrix* $\mathbf{H}_0$ *depending only upon* f *and* $\mathbf{H}$ *defined by* (2.24), *such that* $\mathbf{H}$ *has a unique quasidirect decomposition of the form*

$$\mathbf{H}=\mathbf{H}_n\left(\frac{h}{f}\right)+\mathbf{H}_0, \tag{2.25}$$

in which $\mathbf{H}_0$ *is the lower triangular part of* $\mathbf{H}$ *with* $\operatorname{rank}\mathbf{H}_0=\max\{\deg g_0-\deg f,0\}$. *Moreover, the polynomial h in* (2.25) *satisfies*

$$g_2g_4\cdots g_{2k}=hg_1g_3\cdots g_{2k+1}+f\omega$$

for a suitable polynomial ω *of degree* $(k+1)n-1$ *at most.*

Proof (a) follows easily from Lemma Ⅰ, since $\mathbf{B}_n^{-1}(f,g_{2i-1})\in\mathcal{H}_n(f)$, $i=1,2,\cdots,k+1$. The result (b) follows from Theorem 5 by induction with respect to k. ■

Acknowledgments

The authors thank the referee for helpful comments and suggestions.

References

[1]G -N Chen and H-P Zhang. Singular Hankel matrices(in chinese). J. Beijing Normal Univ. (Natur. Sci),1992,28(3):292—294.

[2] G -N Chen and H-P Zhang, More on Loewner matrices, Linear Algebra Appl. 1994,203—204:265—300.

[3] M Fiedler. Hankel and Loewner matrices. Linear Algebra Appl. 1984,58:75—95.

[4] M Fiedler. Quasidirect decomposition of Hankel and Toeplitz matrices. Linear Algebra Appl. 1984,61:155—174.

[5] M Fiedler and V Pták. Intertwining and testing matrices corresponding to a polynomial. Linear Algebra Appl. 1987,86:53—74.

[6] M Fiedler and V Pták. Loewner and Bézout matrices. Linear Algebra Appl., 1988,101:187—220.

[7] G Heinig and U Jungnickel. Hankel matrices generated by the Markov parameters of rational functions. Linear Algebra Appl., 1986,76:121—135.

[8] S-R Yang, H-P Zhang, and G-N Chen. Bezoutians and Hankel matrices (in Chinese). J. Beijing Normal Univ. (Natur. Sci.),1991,27(Sup. 2):20—26.

Linear Algebra and Its Applications,
1996,244:165－201

数值情形的一般有理插值问题及其 Hankel 向量[①]

The General Rational Interpolation Problem in the Scalar Case and Its Hankel Vector

Abstract　This paper deals with the general rational interpolation problem (GRIP) in the scalar case. In the recent work of Antoulas, Ball, Kang, and Willems the general solution to the GRIP has been derived in the framework of linear fractions using the so-called generating matrix as the main tool. Within this framework, the contribution here consists in a new approach of computing a particular 2×2 polynomial generating matrix, based on the deep connection between Loewner and Hankel matrices. It turns out that given a fixed GRIP, we can readily obtain a unique vector, called the Hankel vector, such that both of the allowable McMillan degrees of interpolants to this GRIP and the corresponding polynomial generating matrix are completely determined by this Hankel vector combined with its characteristic degrees and characteristic polynomials.

① Support was provided by the Science Foundation of the State Education Committee of China. 本文与赵斌和张惠品合作. Submitted by Hans Schneider

§ 1. Introduction

The present paper is concerned with the general unconstrained rational interpolation problem in the scalar case and its connections with special matrices, which is one of the popular topics connected with the mathematical theory of linear systems. Accordingly, all vectors, matrices, polynomials, etc. are assumed to be complex throughout.

By a general rational interpolation problem (GRIP) we shall understand the following array of data:

$$Y_i, C_{ik}, \quad k = 0,1,\cdots,\tau_i - 1, i = 1,2,\cdots,\theta, \tag{1.1}$$

consisting of $N=\tau_1+\tau_2+\cdots+\tau_\theta$ pairs of points, where $Y_i \neq Y_j$ for $i \neq j$ and τ_i is the multiplicity of the interpolation point Y_i. For simplicity in the sequel, both the Y_i's and C_{ik}'s are assumed to be finite. In the case when all $\tau_i=1$ in (1.1), we have the so-called simplest rational interpolation problem. A scalar function $\phi(\lambda)$ (with no poles in $Y_1, Y_2\cdots, Y_\theta$) is said to be a solution or an interpolant to the GRIP(1.1) if

$$\frac{1}{k!}\varphi^{(k)}(Y_i) = C_{ik}, \quad k = 0,1,\cdots,\tau_i - 1, i = 1,2,\cdots,\theta \tag{1.2}$$

where the index (k) stands for the kth derivative with respect to λ.

Since the number of conditions (1.2) is finite, it makes sense to find among all rational functions the solutions (rational interpolants) to the GRIP (1.1). As is well known, there exists a unique polynomial $\omega(\lambda)$ of degree at most $N-1$, referred to as the Hermite (interpolation) polynomial of the GRIP(1.1), such that $\omega(\lambda)$ is a solution to (1.1). Given a rational function

$$\varphi(\lambda) = \frac{r(\lambda)}{f(\lambda)}, \quad r, f \text{ coprime}, \tag{1.3}$$

the complexity of $\varphi(\lambda)$ is defined to be its McMillan degree:

$$\deg \varphi(\lambda) = \max\{\deg r(\lambda), \deg f(\lambda)\}.$$

The basic object of the rational interpolation is to look for all corresponding rational interpolants to the GRIP(1.1) with an allowable McMillan degree and to investigate the algebraic structure of the problem of parametrizing such interpolants. Here by the allowable McMillan degrees we shall mean those positive integers l for which there exist rational interpolants $\varphi(\lambda)$ with deg $\varphi(\lambda)=l$.

The rational interpolation problem has a long history in mathematics. Its origin can be traced back to the work of Cauchy and Jacobi in the first half of the last century. Recently the GRIP has been solved in the generalization of the so-called state space framework by Antoulas and Anderson[2,3,8], using the (generalized) Loewner matrix as the main too1. A polynomial approach to rational interpolation was put forward in Antoulas and Willems [6] and Antoulas, Ball, Kang, and Willems [5]. For a recent, general account of rational interpolation, see Ball, Gohberg, and Rodman [9,10].

The main tool in the polynomial approach is the so-called generating matrix, which in the scalar case is a 2×2 nonsingular polynomial matrix $\boldsymbol{\Theta}$ computed from the data (1.1), and is characterized in [5,6]. It turns out there that once the generator $\boldsymbol{\Theta}$ is obtained, the parametric expression of all interpolants $r(\lambda)/f(\lambda)$ of a given McMillan degree can be described in terms of linear fractions

$$\begin{bmatrix} r \\ f \end{bmatrix} = \boldsymbol{\Theta} \begin{bmatrix} u \\ v \end{bmatrix},$$

where u, v is a pair of parameter polynomials. It is important to realize that the generator $\boldsymbol{\Theta}$ need not be polynomial, but is in general rational. Using this freedom in the choice of $\boldsymbol{\Theta}$, one can perform rational interpolation with additional constraints: stability (poles in given regions of the complex plane), norm-boundedness, etc. For detailed discussions of many of these developments and their applications, we refer to [1,4,14,19,17,7] and references therein.

In this paper, a solution to the GRIP in the scalar case is presented in the framework of linear fractions. Since the general solution in this framework has been worked out in [5,6], we focus on a novel approach of computing a particular polynomial generating matrix from the data

(1.1). The basis for the present paper is given by the deep connection between Loewner and Hankel matrices as well as the mentioning connection with the GRIP. These enable us to exploit the results known for Hankel matrices. The main tool adopted in our analysis is the notion of what we shall call the Hankel vector of a GRIP, which will lead to a fairly simple and more concrete formula for computing the generating matrix using only standard linear algebra manipulations on finite matrices. More precisely, it is shown, surprisingly, that with each fixed GRIP we may associate a unique vector, the Hankel vector of the GRIP, obtained from a solution to a specific system of linear equations, such that both of the allowable McMillan degrees of interpolants to this GRIP and the corresponding generating matrix are completely determined by that vector combined with its characteristic degrees and a pair of characteristic polynomials. A quantitative connection between the Hankel vector and the Hermite interpolation polynomial of a GRIP is established as well. Thus, roughly speaking, the main idea of the paper is to reduce the GRIP to what amounts to the partial realization problem of linear system theory. This latter problem is a special case of the former in that one can view it as a high order interpolation at a single point (infinity). The solution of the partial realization problem is identical with the set of rational generators of the Hankel vector, which has been formulated by Heinig and Jungnickel [15] in detail. This idea, in particular, would doubtless be very advantageous numerically. Also, such a treatment yields some new results, and simpler proofs and a better understanding of several known properties of the GRIP and of Loewner matrices, as by-products.

Our paper is laid out as follows. Section 2 is of preliminary character; we collect some definitions and the necessary background for the generalized Loewner matrices and for rational generators of Hankel matrices. In Section 3 we discuss in detail the existence and computational aspects of the Hankel vector of a GRIP. The main result of this paper consists of finding a particular polynomial generating matrix and an explicit parametrization of all rational interpolants to a GRIP in terms of the Hankel vector. A complete analysis of these results forms the subject matter of Section 4, and simplified proofs for some previous known results on GRIP and Loewner matrices are derived as by-products. The final section, Section 5, contains the procedure to follow when solving a GRIP, and three numerical examples.

§ 2. The Generalized Loewner Matrices and Rational Generators of Hankel Matrices

To set up the framework, we first need some machinery of the generalized Loewner matrices and rational generators of Hankel matrices (see, e.g., [2,20] and [15,16], respectively).

Let

$$\boldsymbol{y}=(\underbrace{y_1,\cdots,y_1}_{m_1},\cdots,\underbrace{y_u,\cdots,y_u}_{m_u})^{\mathrm{T}} \text{ and } \boldsymbol{z}=(\underbrace{z_1,\cdots,z_1}_{n_1},\cdots,\underbrace{z_v,\cdots,z_v}_{n_v})^{\mathrm{T}} \tag{2.1}$$

be a pair of so-called interpolation vectors (IVs), where $y_1, y_2, \cdots, y_u$ (and $z_1, z_2, \cdots, z_v$) are distinct; $|\boldsymbol{y}|=\sum_{i=1}^{u} m_i = m$, the dimension of $\boldsymbol{y}$; and $|\boldsymbol{z}|=\sum_{i=1}^{v} n_i = n$, $m+n=N$. Given polynomials f and g such that

$$f(y_i)f(z_j)\neq 0,\quad i=1,2,\cdots,u,\quad j=1,2,\cdots,v, \tag{2.2}$$

we define the (generalized) Loewner matrices $\boldsymbol{L}=\boldsymbol{L}_{y,z}(g/f)$ generated by g/f (and associated with the IVs $\boldsymbol{y}$ and $\boldsymbol{z}$) via the formula

$$\boldsymbol{L}_{y,z}\left(\frac{g}{f}\right)=(\boldsymbol{L}_{ij})_{i,j=1}^{u,v}, \tag{2.3a}$$

whose (i,j)th block $\boldsymbol{L}_{ij}=(l_{ij}^{kl})_{k,l=0}^{m_i-1,n_j-1}\in \mathbf{C}^{m_i\times m_j}$ with entries l_{ij}^{kl} determined by the divided difference for multiple points with respect to the rational function g/f:

$$l_{ij}^{kl}=[\underbrace{y_i,\cdots,y_i}_{k+1},\underbrace{z_j,\cdots,z_j}_{l+1}]_{g/f}=\frac{1}{k!\ l!}\frac{\partial^{k+l}}{\partial\lambda^k\partial\mu^l}\left[\frac{\dfrac{g(\lambda)}{f(\lambda)}-\dfrac{g(\mu)}{f(\mu)}}{\lambda-\mu}\right]_{\lambda=y_i,\mu=z_j}; \tag{2.3b}$$

if $y_i = z_j$, the value on the right hand side of (2.3b) must be replaced with

$$\frac{1}{(k+l+1)!}\frac{\partial^{k+l+1}}{\partial \lambda^{k+l+1}}\left[\frac{g(\lambda)}{f(\lambda)}\right]_{\lambda=y_i=z_j}.$$

Denote by $\mathscr{L}_{m,n}(\boldsymbol{y},\boldsymbol{z})$ the class of such Loewner matrices associated with a fixed pair of IVs $\boldsymbol{y}$ and $\boldsymbol{z}$ as in (2.1) in advance.

In particular, if $y_1, y_2, \cdots, y_u, z_1, z_2, \cdots, z_v$ are all distinct and $m=u$, $n=v$, we call $\boldsymbol{y}$ and $\boldsymbol{z}$ in (2.1) a pair of the simplest IVs, and the matrix $\boldsymbol{L}_{y,z}(g/f)$ in (2.3) a classical Loewner matrix generated by g/f (and associated with the IVs $\boldsymbol{y}$ and $\boldsymbol{z}$).

Notice that the (generalized) Loewner matrix $\boldsymbol{L}_{y,z}(g/f)$ in (2.3) is built only on the N values and derivatives of the rational function g/f:

$$\frac{1}{k!}\left(\frac{g(\lambda)}{f(\lambda)}\right)^{(k)}_{\lambda=y_i}, i=1,2,\cdots,u, k=0,1,\cdots,m_i-1, \tag{2.4a}$$

$$\frac{1}{k!}\left(\frac{g(\lambda)}{f(\lambda)}\right)^{(k)}_{\lambda=z_j}, j=1,2,\cdots,v, k=0,1,\cdots,n_j-1, \tag{2.4b}$$

In the sequel when we write $\boldsymbol{L}\in\mathscr{L}_{m,n}(\boldsymbol{y},\boldsymbol{z})$ we shall mean that either that $\boldsymbol{L}=\boldsymbol{L}_{y,z}(g/f)$ for a suitable g/f or that $\boldsymbol{L}$ is built on the finite values as C_{ik} in (1.1), which can be considered as the values and derivatives of some rational function.

Let now $\boldsymbol{Q}$ denote the collection of Y_i's in (1.1):

$$\boldsymbol{Q}=\{\underbrace{Y_1,\cdots,Y_1}_{\tau_1},\cdots,\underbrace{Y_\theta,\cdots,Y_\theta}_{\tau_\theta}\} \text{ (with length } N\text{)}, \tag{2.5}$$

where Y_i is listed τ_i times. We assign to each $\boldsymbol{Q}$ as in (2.5) a polynomial $A(\lambda)$ of degree N:

$$A(\lambda)=\prod_{i=1}^{\theta}(\lambda-Y_i)^{\tau_i}. \tag{2.6}$$

Let us partition $\boldsymbol{Q}$ arbitrarily into a pair of IVs $\boldsymbol{y}$ and $\boldsymbol{z}$, called a pair of IVs of $\boldsymbol{Q}$, such that in case of multiple points equal Y_i's have consecutive indices in both $\boldsymbol{y}$ and $\boldsymbol{z}$, and the sum of lengths of sets $\{k: y_k=Y_i\}$ and $\{k: z_k=Y_i\}$ is equal to $\tau_i, i=1,2,\cdots,\theta$. Given a fixed pair of IVs $\boldsymbol{y}$ and $\boldsymbol{z}$ as in (2.1) of Q in (2.5), the GPIP(1.1) determines uniquely a (generalized) Loewner matrix $\boldsymbol{L}=\boldsymbol{L}_{y,z}(\omega)\in\mathscr{L}_{m,n}(\boldsymbol{y},\boldsymbol{z})$, where $\omega(\lambda)$, of degree $N-1$ at most, is the Hermite interpolation polynomial of (1.1). It is obvious that $\boldsymbol{L}_{y,z}(\omega)=\boldsymbol{L}_{y,z}(\phi)$ whenever $\phi(\lambda)$ is a solution to (1.1). Thus, we have the first part of the following lemma.

Lemma 2.1 *Let* $\mathbf{y}$ *and* $\mathbf{z}$ *be a fixed pair of IVs of Q in* (2.5) *with* $|\mathbf{y}|=m$ *and* $|\mathbf{z}|=n$. *To each fixed GRIP with Q as the collection of interpolation points, we associate a unique* $m\times n$ *Loewner matrix* $\mathbf{L}\in\mathscr{L}_{m,n}(\mathbf{y},\mathbf{z})$ *such that* $\mathbf{L}=\mathbf{L}_{y,z}(\phi)$ *is independent of the solution* ϕ *chosen for this GRIP. Conversely, each Loewner matrix* $\mathbf{L}$ *in* $\mathscr{L}_{m,n}(\mathbf{y},\mathbf{z})$ *determines uniquely a GRIP of the form* (1.1) *up to a constant summand d for* C_{i0}:

$$Y_i, C_{i0}+d, C_{i1}, \cdots, C_{i,r_i-1}, \quad i=1,2,\cdots,\theta, \tag{2.7}$$

such that the Loewner matrix $\mathbf{L}$ *is to be built on the values of the date* (2.7) *and associated with the IVs* $\mathbf{y}$ *and* $\mathbf{z}$. *Consequently, if* $\mathbf{L}=\mathbf{L}_{y,z}(\omega)=\mathbf{L}_{y,z}(\omega_0)$, *where both of* $\omega(\lambda)$ *and* $\omega_0(\lambda)$ *are polynomials of degree* $|\mathbf{y}|+|\mathbf{z}|-1$ *at most, then* $\omega(\lambda)-\omega_0(\lambda)=$const.

The proof of the second part of Lemma 2.1 involves straightforward but rather tedious manipulations and will be omitted.

For additional information about Loewner matrices we refer to [13] and [12].

As mentioned earlier, we also need to recall some results concerning an explicit parameterization of the rational generators for a Hankel matrix, due to Heinig and Jungnickel[15](see also [16]). By an $m\times n$ Hankel matrix $\boldsymbol{H}$, we shall mean a matrix of the form $\boldsymbol{H}=(h_{i+j})_{i,j=0}^{m-1,n-1}$. Denote by $\mathscr{H}_{m,n}$ the class of all $m\times n$ Hankel matrices. Let $g(\lambda)$ and $f(\lambda)$ be a pair of polynomials such that $\deg g<\deg f$. If the strictly proper rational function $\phi(\lambda)=g(\lambda)/f(\lambda)$ has the Laurent expansion at infinity

$$\phi(\lambda)=\frac{g(\lambda)}{f(\lambda)}=\sum_{j=1}^{\infty}h_{j-1}\lambda^{-j}, \tag{2.8}$$

then the $k\times l$ Hankel matrix generated by g/f is, by definition, the matrix $\boldsymbol{H}_{kl}(g/f)=(h_{i+j})_{i,j=0}^{k-1,l-1}$, where both k and l are arbitrary positive integers. Suppose now we are given a fixed positive integer N and $\boldsymbol{b}=(h_0,h_1,\cdots,h_{N-2})^{\mathrm{T}}\in\mathbf{C}^{N-1}$. We introduce the $k\times l$ Hankel matrices, $k+l=N$,

$$\boldsymbol{H}_{kl}[\boldsymbol{b}] = (h_{i+j})_{i,j=0}^{k-1,l-1}. \qquad (2.9)$$

If for any k and l such that $k+l=N$ we have $\boldsymbol{H}_{kl}[\boldsymbol{b}]=\boldsymbol{H}_{kl}(g/f)$, $\deg g<\deg f$, then we call the strictly proper rational function g/f a rational generator to $\boldsymbol{b}\in\mathbf{C}^{N-1}$ or to the Hankel matrix $\boldsymbol{H}_{kl}[\boldsymbol{b}]$ in (2.9). Clearly, this is equivalent to

$$\boldsymbol{b}=\boldsymbol{H}_{N-1,1}\left(\frac{g}{f}\right),$$

or, what is the same,

$$\frac{g(\lambda)}{f(\lambda)} = \sum_{j=1}^{N-1} h_{j-1}\lambda^{-j} + o(\lambda^{-N+1}). \qquad (2.10)$$

Now we review the notion of the characteristic degrees and characteristic polynomials of a vector. Let $\mathbf{0}\neq\boldsymbol{b}=(h_0,h_1,\cdots,h_{N-2})^{\mathrm{T}}\in\mathbf{C}^{N-1}$ and $m=[N/2]$, the integer part of $N/2$. Then the characteristic degrees $n_1=n_1(\boldsymbol{b})$ and $n_2=n_2(\boldsymbol{b})$ of $\boldsymbol{b}\in\mathbf{C}^{N-1}$ are defined by

$$n_1=\begin{cases}\operatorname{rank} \boldsymbol{H}_{mm}[\boldsymbol{b}], & \text{if } N \text{ is even}\\ \operatorname{rank} \boldsymbol{H}_{m,m+1}[\boldsymbol{b}], & \text{if } N \text{ is odd}\end{cases} \quad \text{and} \quad n_2=N-n_1. \qquad (2.11)$$

In general, it is known [16] that

$$n_1\leqslant n_2 \text{ and } \operatorname{rank} \boldsymbol{H}_{kl}[\boldsymbol{b}]=\min\{k,l,n_1\}, k+l=N. \qquad (2.12)$$

Denote by $\mathbf{C}_n(\lambda)$ the class of polynomials of degree at most n, and by $\mathbf{C}_n^0(\lambda)$ the subclass consisting of all monic polynomials of degree n. Put

$$\mathscr{A}_l=(1,\lambda,\cdots,\lambda^{l-1})\operatorname{Ker} \boldsymbol{H}_{kl}[\boldsymbol{b}], \quad k+l=N, \qquad (2.13)$$

a subspace of $\mathbf{C}_{l-1}(\lambda)$. It follows from (2.12) that $n_1=n_2=N/2$ iff N is even and the Hankel matrix $\boldsymbol{H}_{n_1,n_2}[\boldsymbol{b}]$ is regular. In this case, $\dim \mathscr{A}_{n_i+1}=n_1+1-(n_1-1)=2$ by (2.13), so that there exists a pair of polynomials $p(\lambda)$ and $q(z)$ in $\mathbf{C}_{n_1}(\lambda)$ forming a basis of $\mathscr{A}_{n_1+1}$, which will be called a pair of characteristic polynomials of the vector $\mathbf{b}\in\mathbf{C}^{N-1}$. In the case $n_1<n_2$, we have that $\dim \mathscr{A}_{n_1+1}=1$ by (2.12)(2.13) again, so that there exists $p(\lambda)\in\mathbf{C}_{n_1}(\lambda)$ forming a basis of $\mathscr{A}_{n_1+1}$. Similarly, if $1<n_1<n_2$, we have that $\dim \mathscr{A}_{n_2+1}=n_2-n_1+2$ and $p(\lambda),\lambda p(\lambda),\cdots,\lambda^{n_2-n_1}p(\lambda)\in\mathscr{A}_{n_2+1}$, so that there exists $q(\lambda)\in\mathbf{C}_{n_2}(\lambda)$ such that $p(\lambda),\lambda p(\lambda),\cdots,\lambda^{n_2-n_1}p(\lambda)$, $q(\lambda)$ form a basis of $\mathscr{A}_{n_2+1}$. In case $1=n_1<n_2$, choose an arbitrary $q(\lambda)\in$

$\mathbf{C}_{n_2}(\lambda)$ such that $p(\lambda), \lambda p(\lambda), \cdots, \lambda^{n_2-1} p(\lambda), q(\lambda)$ are linearly independent. Such $(p(\lambda), q(\lambda))$ will also be called a pair of characteristic polynomials of $\mathbf{b} \in \mathbf{C}^{N-1}$ in the case $n_1 < n_2$. For our purpose it is convenient to choose pairs of so-called canonical characteristic polynomials of $\mathbf{b} \in \mathbf{C}^{N-1}$ such that either $p(\lambda) \in \mathbf{C}^0_{n_1}(\lambda)$ and $\deg q(\lambda) < n_2$ or $q(\lambda) \in \mathbf{C}^0_{n_2}(\lambda)$ and $\deg p(\lambda) < n_1$. Such canonical pairs of characteristic polynomials do always exist.

Given now $\mathbf{0} \neq \boldsymbol{b} = (h_0, h_1, \cdots, h_{N-2})^{\mathrm{T}} \in \mathbf{C}^{N-1}$ with characteristic degrees n_1 and n_2 and a canonical pair of characteristic polynomials $(p(\lambda), q(\lambda))$:

$$p(\lambda) = \sum_{j=0}^{n_1} p_j \lambda^j \text{ and } q(\lambda) = \sum_{j=0}^{n_2} q_j \lambda^j,$$

we denote by $\boldsymbol{S}(p)$ and $\boldsymbol{S}(q)$ the symmetrizers of $p(\lambda)$ and $q(\lambda)$, respectively:

$$\boldsymbol{S}(p) = \begin{bmatrix} p_1 & p_2 \cdots p_{n_1} \\ p_2 & \\ \vdots & \iddots \\ p_{n_1} & \end{bmatrix} \in \mathbf{C}^{n_1 \times n_1}, \quad \boldsymbol{S}(q) = \begin{bmatrix} q_1 & q_2 \cdots q_{n_2} \\ q_2 & \\ \vdots & \iddots \\ q_{n_1} & \end{bmatrix} \in \mathbf{C}^{n_2 \times n_2},$$

Define in turn polynomials $\gamma(\lambda) \in \mathbf{C}_{n_1-1}(\lambda)$ and $\delta(\lambda) \in \mathbf{C}_{n_2-1}(\lambda)$ via

$$\begin{aligned} \gamma(\lambda) &= (1, \lambda, \cdots, \lambda^{n_1-1}) \boldsymbol{S}(p) (h_0, \cdots, h_{n_1-1})^{\mathrm{T}}, \\ \delta(\lambda) &= (1, \lambda, \cdots, \lambda^{n_2-1}) \boldsymbol{S}(q) (h_0, \cdots, h_{n_2-1})^{\mathrm{T}}. \end{aligned} \tag{2.14}$$

It is known [15] that the constant $\sigma = q(0)\gamma(0) - p(0)\delta(0)$ does not vanish, and

$$q(\lambda)\gamma(\lambda) - p(\lambda)\delta(\lambda) = \sigma, \tag{2.15}$$

so that the pairs of polynomials $(p(\lambda), q(\lambda))$, $(q(\lambda), \delta(\lambda))$, and $(p(\lambda), \gamma(\lambda))$ are all coprime, respectively.

The next result is concerned with a complete parametrization of rational generators to a vector $\boldsymbol{b} \in \mathbf{C}^{N-1}$.

Theorem 2.2 [15] *Let $\mathbf{0} \neq \boldsymbol{b} \in \mathbf{C}^{N-1}$, n_1, n_2, $(p(\lambda), q(\lambda))$, and $(\gamma(\lambda), \delta(\lambda))$ be as before, and let $\Delta(\mathbf{b}) = n_1 - \deg p(\lambda)$. Then the complete parametrization of the class $\mathscr{F}_n$ of rational generators to $\boldsymbol{b} \in \mathbf{C}^{N-1}$ with every McMillan degree $n < N$ can be described in the following way:*

(a) *If* $n_1 = n_2$, *then*

$$\mathcal{F}_n = \begin{cases} \varnothing, & if \quad n < n_1, \\ \left\{ \dfrac{u(\lambda)\gamma(\lambda)+v(\lambda)\delta(\lambda)}{u(\lambda)p(\lambda)+v(\lambda)q(\lambda)} : \gcd(u,v)=1, \right. \\ \left. \quad u \in \mathbf{C}^0_{n-n_1}(\lambda), v \in \mathbf{C}_{n-n_2}(\lambda) \right\}, & if \quad n \geqslant n_1. \end{cases}$$

(b) *If* $n_1 < n_2$ *and* $\Delta(\boldsymbol{b}) = 0$, *then*

$$\mathcal{F}_n = \begin{cases} \varnothing, & if \quad n < n_2 \text{ and } n \neq n_1, \\ \left\{ \dfrac{\gamma(\lambda)}{p(\lambda)} \right\}, & if\ n = n_1, \\ \left\{ \dfrac{u(\lambda)\gamma(\lambda)+v(\lambda)\delta(\lambda)}{u(\lambda)p(\lambda)+v(\lambda)q(\lambda)} : \gcd(u,v)=1, \right. \\ \left. \quad u \in \mathbf{C}^0_{n-n_1}(\lambda), v \in \mathbf{C}_{n-n_2}(\lambda) \right\}, & if \quad n \geqslant n_2. \end{cases}$$

(c) *If* $n_1 < n_2$ *and* $\Delta(\boldsymbol{b}) > 0$, *then*

$$\mathcal{F}_n = \begin{cases} \varnothing, & if \quad n < n_2, \\ \left\{ \dfrac{u(\lambda)\gamma(\lambda)+v(\lambda)\delta(\lambda)}{u(\lambda)p(\lambda)+v(\lambda)q(\lambda)} : \gcd(u,v)=1, \right. \\ \left. \quad u \in \mathbf{C}_{n-n_1}(\lambda), v \in \mathbf{C}^0_{n-n_2}(\lambda) \right\}, & if \quad n \geqslant n_2. \end{cases}$$

We remark that in case $\boldsymbol{b} = \mathbf{0} \in \mathbf{C}^{N-1}$ the set of its rational generators is obviously identical with

$$\{0\} \cup \left\{ \frac{h}{f} : \deg f - \deg h \geqslant N \right\}, \tag{2.16}$$

which says that Theorem 2. 2 is also true for the case $\boldsymbol{b} = \mathbf{0} \in \mathbf{C}^{N-1}$ if we take $n_1 = 0, n_2 = N, p(\lambda) = \delta(\lambda) = 1$, and $q(\lambda) = \gamma(\lambda) = 0$ therein. This clearly displays the fact that the corresponding set of allowable McMillan degrees of rational generators consists of $0, N, N+1, \cdots$.

§ 3. The Hankel Vector of a GRIP

The main work of the section lies in showing the existence and discussing in detail the computational aspects of the Hankel vector of a GRIP. The basic results here are stated and proved in Theorems 3. 3 and 3. 6.

We use the following notation throughout this section for polynomials and matrices associated with a pair of IVs $\boldsymbol{y}$ and $\boldsymbol{z}$ defined via (2. 1):

$$a(\lambda)=\prod_{i=1}^{u}(\lambda-y_1)^{m_i}, \quad a_{ik}(\lambda)=\frac{a(\lambda)}{(\lambda-y_i)^{k+1}},$$

$$i=1,2,\cdots,u, \quad k=0,1,\cdots,m_i-1, \tag{3.1a}$$

$$b(\lambda)=\prod_{j=1}^{v}(\lambda-z_j)^{n_j}, \quad b_{jl}(\lambda)=\frac{b(\lambda)}{(\lambda-z_j)^{l+1}},$$

$$j=1,2,\cdots,v, \quad l=0,1,\cdots,n_j-1, \tag{3.1b}$$

$$A(\lambda)=a(\lambda)b(\lambda), \tag{3.2}$$

$$\boldsymbol{W}_y=\begin{bmatrix}\boldsymbol{W}_{y_1}\\ \boldsymbol{W}_{y_2}\\ \vdots\\ \boldsymbol{W}_{y_u}\end{bmatrix}\in\mathbf{C}^{m\times m}, \quad \boldsymbol{W}_z=\begin{bmatrix}\boldsymbol{W}_{z_1}\\ \boldsymbol{W}_{z_2}\\ \vdots\\ \boldsymbol{W}_{z_v}\end{bmatrix}\in\mathbf{C}^{n\times n}, \tag{3.3a}$$

$$\boldsymbol{W}_{y_i}=\left(\frac{a_{ik}^{(j)}(0)}{j!}\right)_{k=0,j=0}^{m_i-1,m-1}, \quad \boldsymbol{W}_{z_j}=\left(\frac{b_{jl}^{(k)}(0)}{k!}\right)_{l=0,k=0}^{n_j-1,n-1}. \tag{3.3b}$$

It will be convenient to have the following lemmas before proving Theorem 3.3.

Lemma 3.1[12] *Let* $\mathbf{y}$ *and* $\mathbf{z}$ *be a pair of IVs as in* (2.1), *and* $\boldsymbol{Q}_k$ *of order* k *be the flip matrix* $(\delta_{i,k-j-1})$, $i,j=0,1,\cdots,k-1$. *Let* $\boldsymbol{J}(y_i)$ $[\boldsymbol{J}(z_j)]$ *be the Jordan block of order* $m_i[n_j]$ *with* $y_i[z_j]$ *along the diagonal and* 1 *along the superdiagonal. Then*

$$\boldsymbol{W}_y\boldsymbol{V}_y=\operatorname{diag}[\boldsymbol{Q}_{m_i}a_{i,m_i-1}(\boldsymbol{J}(y_i))]_{i=1}^{u}, \tag{3.4a}$$

$$W_z\boldsymbol{V}_z=\operatorname{diag}[\boldsymbol{Q}_{n_j}b_{j,n_j-1}(\boldsymbol{J}(z_j))]_{j=1}^{v}, \tag{3.4b}$$

where $\boldsymbol{V}_y$ *and* $\boldsymbol{V}_z$ *stand for the generalized Vandermonde matrices corresponding to* $\mathbf{y}$ *and* $\mathbf{z}$, *respectively*:

$$\begin{aligned}\boldsymbol{V}_y&=[\boldsymbol{P}_{mm_1}(y_1)\cdots\boldsymbol{P}_{mm_u}(y_u)]\in\mathbf{C}^{m\times m},\\ \boldsymbol{V}_z&=[\boldsymbol{P}_{nn_1}(z_1)\cdots\boldsymbol{P}_{nn_v}(z_v)]\in\mathbf{C}^{n\times n},\end{aligned} \tag{3.5}$$

whereas

$$(\boldsymbol{P}_{k,l}(\lambda))_{i,j}=\binom{i}{j}\lambda^{i-j}, i=0,1,\cdots,k-1, \quad j=0,1,\cdots,l-1.$$

Lemma 3.2[20] *Let* $\mathbf{y}$ *and* $\mathbf{z}$ *be a pair of IVs as in* (2.1), $|\mathbf{y}|=m$, $|\mathbf{z}|=n$, *and let* $\boldsymbol{W}_y$ *and* $\boldsymbol{W}_z$ *be as in* (3.3). *Then the mapping* $\mathscr{F}$ *defined by*

$$\mathscr{F}(\boldsymbol{H})=\boldsymbol{W}_y\boldsymbol{H}\boldsymbol{W}_z^{\mathrm{T}}, \quad \forall\,\boldsymbol{H}\in\mathscr{H}_{m,n} \tag{3.6}$$

is an isomorphism of the classes $\mathscr{H}_{m,n}$ *and* $\mathscr{L}_{m,n}(\mathbf{y},\mathbf{z})$.

For the proof of Lemma 3.2 we could simply refer to [20] (for a different proof of this result, see [12, Theorem 3.3]), but the idea of the proof, which carries on the one of [13] or [20], will be very useful in the proofs of Theorems 3.3 and 3.6 below. Given a rational function $Q(\lambda)$ having the expansion at infinity

$$Q(\lambda) = Q_d\lambda^d + Q_{d-1}\lambda^{d-1} + \cdots + Q_0 + \sum_{k=1}^{\infty} Q_{-k}\lambda^{-k}, d \leqslant m+n-2,$$

and the fixed parameters $h_0, h_1, \cdots, h_{m+n-2}$, we define a functional via

$$[Q(\lambda)]\downarrow = \sum_{k=0}^{d} Q_k h_k. \tag{3.7}$$

Note that for an arbitrary but fixed mumber h_{m+n-1}, the functional in (3.7) can be extended to $Q(\lambda)$ with $d \leqslant m+n-1$ in its expansion at infinity. Now let $\boldsymbol{H} = (h_{i+j})_{i,j=0}^{m-1,n-1} \in \mathscr{H}_{m,n}$ have a rational generator $H(\lambda)/F(\lambda)$ of McMillan degree less than N (by Theorem 2.2) such that $N(\lambda)/F(\lambda) = \sum_{j=0}^{\infty} h_j\lambda^{-j-1}$ (expansion at infinity) for some numbers $h_{m+n-1}, h_{m+n}, \cdots$. Then there exists a unique pair of polynomials $W(\lambda)$ and $R(\lambda)$ such that $H(\lambda)A(\lambda) = F(\lambda)W(\lambda) - R(\lambda)$ and $\deg R(\lambda) < \deg F(\lambda)$, where $A(\lambda)$ is defined via (3.2). It is shown in [20] that with the given parameters $h_0, \cdots, h_{m+n-2}, h_{m+n-1}$, the value $[A(\lambda)/(\lambda - Y)]\downarrow$ makes sense and is defined as a function of the variable Y, and that $W(Y) = [A(\lambda)/(\lambda - Y)]\downarrow$ as well, which is a rational generator (polynomial) for the Loewner matrix $\boldsymbol{L} = \boldsymbol{W}_y \boldsymbol{H} \boldsymbol{W}_z^{\mathrm{T}}$, that is, $\boldsymbol{L} = \boldsymbol{L}_{y,z}(W)$.

Taking into account the remark on Lemma 3.2 above, we have the following significant result.

Theorem 3.3 *Let the GRIP (1.1) be given, and $\omega(\lambda) \in \mathbf{C}_{N-1}(\lambda)$ be its Hermite polynomial. Then there exists a unique vector $\boldsymbol{b} = (h_0, h_1, \cdots, h_{N-2})^{\mathrm{T}} \in \mathbf{C}^{N-1}$ such that*

$$\boldsymbol{L}_{y,z}(\omega) = \boldsymbol{W}_y \boldsymbol{H}_{|y|,|z|}[\boldsymbol{b}]\boldsymbol{W}_z^{\mathrm{T}} \tag{3.8}$$

whenever $\mathbf{y}, \mathbf{z}$ are an arbitrary pair of IVs *of Q in (2.5) with dimensions $|\mathbf{y}|$ and $|\mathbf{z}|$, respectively.*

Proof Let $\mathbf{y}'$ and $\mathbf{z}'$ be a fixed pair of IVs of Q in (2.5), $|\mathbf{y}'| = m'$,

$|z'|=n', m'+n'=N$. Write $\boldsymbol{L}'=\boldsymbol{L}_{y',z'}(\omega)$. In view of Lemma 3.2, there exists a unique $m'\times n'$ Hankel matrix $\boldsymbol{H}=(h_{i+j})_{i,j=0}^{m'-1,n'-1}$ such that

$$\boldsymbol{L}'=\boldsymbol{W}_{y'}\boldsymbol{H}\boldsymbol{W}_{z'}^{\mathrm{T}}.$$

Let us denote by $\boldsymbol{b}$ the vector $(h_0, h_1, \cdots, h_{N-2})^{\mathrm{T}}\in \mathbf{C}^{N-1}$ corresponding to $\boldsymbol{H}$. We shall prove that with this chosen $\boldsymbol{b}\in \mathbf{C}^{N-1}$, (3.8) is valid for an arbitrary pair of IVs $\boldsymbol{y}$ and $\boldsymbol{z}$ of Q in (2.5). Indeed, it follows from Theorem 2.2 that there exists a rational generator $H(\lambda)/F(\lambda)$ to $\boldsymbol{b}\in \mathbf{C}^{N-1}$, $\deg H(\lambda)<\deg F(\lambda)<N$, such that the expansion of $H(\lambda)/F(\lambda)$ at infinity is of the form

$$\frac{H(\lambda)}{F(\lambda)}=\sum_{j=0}^{\infty}h_j\lambda^{-j-1}$$

for some suitable numbers $h_{N-1}, h_N, \cdots$. Now let $W(\lambda), R(\lambda)$ be the unique pair of polynomials such that $H(\lambda)A(\lambda)=F(\lambda)W(\lambda)-R(\lambda)$ and $\deg R(\lambda)<\deg F(\lambda)$. By the exposition following Lemma 3.2, with the fixed parameters $h_0, h_1, \cdots, h_{N-2}, h_{N-1}$, the value $[A(\lambda)/(\lambda-Y)]\downarrow$ makes sense as a function of the variable Y, and is identical with the polynomial $W(Y)$, which is a generating polynomial in $\mathbf{C}_{N-1}(\lambda)$ to the Loewner matrix $\boldsymbol{L}'=\boldsymbol{L}_{y',z'}(\omega)$, i.e., $\boldsymbol{L}'=\boldsymbol{L}_{y',z'}(\omega)=\boldsymbol{L}_{y',z'}(W)$. Thus by the last part of Lemma 2.1, there exists a constant d such that $\omega(\lambda)=W(\lambda)+d$. Consider now an arbitrary pair of IVs $\boldsymbol{y}$ and $\boldsymbol{z}$ of Q in (2.5), determined by, say, (2.1) and the Loewner matrix $\boldsymbol{L}=\boldsymbol{L}_{y,z}(\omega(\lambda))=\boldsymbol{L}_{y,z}(W(\lambda)+d)$. Denote by $l_{ik,jl}$ the (k,l)th entry in the (i,j)th block of $\boldsymbol{L}$. A similar manipulation to that used in the proof of [20, Lemma 3.2] leads to

$$\begin{aligned}l_{ik,jl}&=[\underbrace{y_i,\cdots,y_i}_{k+1},\underbrace{z_j,\cdots,z_j}_{l+1}]_{W+d}=[\underbrace{y_i,\cdots,y_i}_{k+1},\underbrace{z_j,\cdots,z_j}_{l+1}]_W\\&=\frac{1}{k!\ l!}\frac{\partial^{k+l}}{\partial\lambda^k\partial\mu^l}\left[\frac{W(\lambda)-W(\mu)}{\lambda-\mu}\right]_{\lambda=y_i,\mu=z_j}\\&=\left[\frac{a(x)}{(x-\lambda)^{k+1}}\frac{b(x)}{(x-\mu)^{l+1}}\right]\downarrow_{\lambda=y_i,\mu=z_j}=[a_{ik}(x)b_{jl}(x)]\downarrow\\&=(\boldsymbol{W}_y\boldsymbol{H}_{|y|,|z|}[\boldsymbol{b}]\boldsymbol{W}_z^{\mathrm{T}})_{ik,jl},\end{aligned}$$

where $a(\lambda), a_{ik}(\lambda)$ and $b(\lambda), b_{jl}(\lambda)$ are defined via (3.1a) and (3.1b), respectively. Hence (3.8) is valid for an arbitrary pair of IVs $\boldsymbol{y}$ and $\boldsymbol{z}$ of

$\boldsymbol{Q}$ in (2. 5). It remains to show the uniqueness of the vector of $\boldsymbol{b}\in \mathbf{C}^{N-1}$. But that is an immediate consequence of Lemma 3. 2 and (3. 8). ■

The above observation is a crucial one with respect to the GRIPs to be studied. In the sequel, instead of working with the Loewner matrices determined by the GRIP(1. 1), we shall work directly with the unique vector $\boldsymbol{b}$ given in Theorem 3. 3. With the help of Theorem 3. 3, it is natural to introduce the following concept:

Definition 3. 4 Given a GRIP(1. 1), the unique vector $\mathbf{b}\in \mathbf{C}^{N-1}$ constructed in Theorem 3. 3 such that(3. 8)holds for an arbitrary pair of IVs $\boldsymbol{y}$ and $\boldsymbol{z}$ of $\boldsymbol{Q}$ in(2. 5) is called the Hankel vector of(1. 1). Accordingly, the characteristic degrees and characteristic polynomials of this Hanker vector can be referred to as those of the GRIP(1. 1).

As an application of Theorem 3. 3, we have readily the following corollary with a simpler proof, whose part (b) has been proved in [2, Corollary 2. 24] based on the main Lemma 2. 5 therein.

Corollary 3. 5 *Given a GRIP* (1. 1), *let* $\omega(\lambda)$ *be its Hermite polynomial and let* n_1, n_2 *be the characteristic degrees of its Hankel vector* $\boldsymbol{b}$. *If* $\mathbf{y}$ *and* $\mathbf{z}$ *are a pair of IVs of* $\boldsymbol{Q}$ *in* (2. 5), *then*

$$\operatorname{rank} \boldsymbol{L}_{y,z}(\omega)=\min\{|y|,|z|,n_1\}. \tag{3. 9}$$

In particular:

(a) *For arbitrary but fixed natural numbers* m *and* n *such that* $m+n=N$,

$$\operatorname{rank} \boldsymbol{L}_{y,z}(\omega)=\min\{m,n,n_1\}=\text{const}$$

provided $\mathbf{y}$ *and* $\mathbf{z}$ *are a pair of IVs of* $\boldsymbol{Q}$ *in* (2. 5) *such that* $|\mathbf{y}|=m$ *and* $|\mathbf{z}|=n$. *Accordingly, if* $m=[N/2]$, *then all* $m\times m$ (*if* N *is even*) *or all* $m\times(m+1)$ (*if* N *is odd*) *Loewner matrices built on the data* (1. 1) *have the same rank* n_1.

(b) *For a pair of IVs* $\mathbf{y}$ *and* $\mathbf{z}$ *of* $\boldsymbol{Q}$ *in* (2. 5), *if* $|\mathbf{y}|, |\mathbf{z}|\geqslant n_1$,

$$\operatorname{rank} \boldsymbol{L}_{y,z}(\omega)=n_1.$$

Proof The main result (3. 9) follows from Theorem 3. 3 and (2. 12), both of $\boldsymbol{W}_y$ and $\boldsymbol{W}_z$ in (3. 3) being invertible. The remainder is obvious by this main result. ■

For practical applications one would like to be able to compute the Hankel vector $\boldsymbol{b}\in\mathbf{C}^{N-1}$ of the GRIP(1. 1) directly from the original interpolation data. The next theorem deals with this computation, and (3. 10) below can be considered as an extreme case of the matrix equation (3. 8).

Theorem 3. 6 *Let $\boldsymbol{b}=(h_0,h_1,\cdots,h_{N-2})^{\mathrm{T}}\in\mathbf{C}^{N-1}$ be the Hankel vector of the GRIP (1. 1). Then there exists a unique complex number d such that*

$$(C_{10},\cdots,C_{1,r_1-1},\cdots,C_{\theta 0},\cdots,C_{\theta,r_\theta-1})^{\mathrm{T}}=\boldsymbol{W}_Q(h_0,h_1,\cdots,h_{N-2},d)^{\mathrm{T}} \tag{3.10}$$

where $\boldsymbol{W}_Q\in\mathbf{C}^{N\times N}$ is invertible and is defined by

$$\boldsymbol{W}_Q=\begin{bmatrix}\boldsymbol{W}_{Q_1}\\ \boldsymbol{W}_{Q_2}\\ \vdots\\ \boldsymbol{W}_{Q_\theta}\end{bmatrix},\quad \boldsymbol{W}_{Q_i}=\left(\frac{A_{ik}^{(j)}(0)}{j!}\right)_{k=0,j=0}^{r_i-1,N-1},\quad A_{ik}(\lambda)=\frac{A(\lambda)}{(\lambda-Y_i)^{k+1}}. \tag{3.11}$$

Proof In view of the proof of Theorem 3. 3, for a given sequence of numbers of $h_0,h_1,\cdots,h_{N-2},h_{N-1}$ (where h_{N-1} is a fixed complex number), there exists a constant d' such that $f(Y)=[A(\lambda)/(\lambda-Y)]\downarrow+d'$ is a polynomial interpolant to the GRIP (1. 1). Thus, for $i=1,2,\cdots,\theta$, we have by (3. 7) (3. 11) and (1. 2) that

$$\begin{aligned}C_{i0}&=\left[\frac{A(\lambda)}{\lambda-Y_i}\right]\downarrow+d'=[A_{i0}(\lambda)]\downarrow+d'\\&=\left[\lambda^{N-1}+\frac{A_{i0}^{(N-2)}(0)}{(N-2)!}\lambda^{N-2}+\cdots+A_{i0}(0)\right]\downarrow+d'\\&=A_{i0}(0)h_0+\frac{A_{i0}^{(1)}(0)}{1!}h_1+\cdots+\frac{A_{i0}^{(N-2)}(0)}{(N-2)!}h_{N-2}+h_{N-1}+d',\\C_{ik}&=\frac{1}{k!}f^{(k)}(Y)\Big|_{Y=Y_i}=\left[\frac{A(\lambda)}{(\lambda-Y_i)^{k+1}}\right]\downarrow=[A_{ik}(\lambda)]\downarrow\\&=A_{ik}(0)h_0+\frac{A_{ik}^{(1)}(0)}{1!}h!+\cdots+\frac{A_{ik}^{(N-k-1)}(0)}{(N-k-1)!}h_{N-k-1},\end{aligned}$$

$$k=1,2,\cdots,\tau_i-1.$$

Put now $d=h_{N-1}+d'$. Then it is obvious that all the equations above are equivalent to the system (3.10). Finally, the uniqueness of d is an immediate consequence of the nonsingularity of the $\boldsymbol{W}_Q$, which is derived from Lemma 3.1.

We remark that Fiedler [13] has also considered the equation (3.10) in the case of the simplest rational interpolation problem only.

Note that the matrix $\boldsymbol{W}_Q \in \mathbf{C}^{N\times N}$ given in (3.10)(3.11) depends only on $\boldsymbol{Q}$ in (2.5), and not on the values of the C_{ik}'s in (1.1).

Combining Theorem 3.6 with Lemma 3.1, we obtain readily a closed form explicit representation formula for the Hankel vector in terms of the original interpolation data.

Corollary 3.7 *Given the GRIP* (1.1), *let* $\boldsymbol{b}=(h_0, h_1, \cdots, h_{N-2})^{\mathrm{T}} \in \mathbf{C}^{N-1}$ *be the Hankel vector. With the notation in Theorem* 3.6, *we have*

$$\boldsymbol{W}_Q=\operatorname{diag}[\boldsymbol{Q}_{\tau_i} A_{i,r_{i-1}}(\boldsymbol{J}_{\tau_i}(Y_i))]_{i=1}^{\theta} \boldsymbol{V}_Q^{-1}, \tag{3.12}$$

$$\boldsymbol{W}_Q^{-1}=\boldsymbol{V}_Q \operatorname{diag}\Big[\prod_{j\neq i} \boldsymbol{J}_{\tau_i}(Y_i-Y_j)^{-\tau_j} \boldsymbol{Q}_{\tau_i}\Big]_{i=1}^{\theta}, \tag{3.13}$$

where $\boldsymbol{Q}_{\tau_i}$ *is the flip matrix of order* τ_i, $\boldsymbol{J}_{\tau_i}(Y_i)$ *the Jordan block of order* τ_i, *and* $\mathbf{V}_Q$ *the generalized Vandermonde matrix associated with* $A(\lambda)$:

$$\mathbf{V}_Q=[\boldsymbol{P}_{N,\tau_1}(Y_1) \cdots \boldsymbol{P}_{N,\tau_\theta}(Y_\theta)] \in \mathbf{C}^{N\times N},$$

$$(\boldsymbol{P}_{N,l}(t))_{ij}=\binom{i}{j} t^{i-j}, i=0,1,\cdots,N-1, \quad j=0,1,\cdots,l-1.$$

Moreover,

$$\begin{bmatrix} h_0 \\ h_1 \\ \vdots \\ h_{N-2} \end{bmatrix}=\sum_{k=1}^{\theta} \boldsymbol{P}_{N-1,\tau_k}(Y_k) \prod_{j\neq k} \boldsymbol{J}_{\tau_k}(Y_k-Y_j)^{-\tau_j} \begin{bmatrix} C_{k,\tau_k-1} \\ \vdots \\ C_{k0} \end{bmatrix}. \tag{3.14}$$

In particular, if all $\tau_i=1$ (*and thus* $\theta=N, Y_1, Y_2, \cdots, Y_N$ *distinct and* $\boldsymbol{Q}=\{Y_1, Y_2, \cdots, Y_N\}$), *then* (3.14) *can be reduced to*

$$h_l=\sum_{k=1}^{N} Y_k^l \prod_{j\neq k} (Y_k-Y_j)^{-1} C_{k0}, \quad 0 \leqslant l \leqslant N-2. \tag{3.15}$$

We remark that the inverse of the matrix $\boldsymbol{J}_{\tau_k}(Y_k-Y_j)(j\neq k)$ occurring in (3.14) can be expressed in terms of the explicit formula

$$J_{\tau_k}(Y_k-Y_j)^{-1}=\begin{bmatrix} \alpha_{kj}^{-1} & -\alpha_{kj}^{-2} & \cdots & (-1)^{\tau_k-1}\alpha_{kj}^{-\tau_k} \\ & \alpha_{kj}^{-1} & \ddots & \vdots \\ & & \ddots & -\alpha_{kj}^{-2} \\ & & & \alpha_{kj}^{-1} \end{bmatrix} \tag{3.16}$$

where $\alpha_{kj}=Y_k-Y_j$, $k\neq j$. Furthermore, in case of necessity, the matrix $\prod_{j\neq k}J_{\tau_k}(Y_k-Y_j)^{-\tau_j}$ $(j\neq k)$ in (3.14) can also be formulated by a well-known result given in function theory of matrices that

$$\prod_{j\neq k}J_{\tau_k}(Y_k-Y_j)^{-\tau_j}=A_{k,\tau_k-1}^{-1}(J_{\tau_k}(Y_k))$$

$$=\begin{bmatrix} f(Y_k) & f^{(1)}(Y_k) & \cdots & \frac{1}{(\tau_k-1)!}f^{(\tau_k-1)}(Y_k) \\ & \ddots & \ddots & \vdots \\ & & \ddots & f^{(1)}(Y_k) \\ & & & f(Y_k) \end{bmatrix} \tag{3.17}$$

where $f(\lambda)$ stands for the complex rational function $1/A_{k,\tau_k-1}(\lambda)$.

The above considerations show that the problem of determining the Hankel vector of a GRIP can be readily achieved with the aid of the explicit formula (3.14). Moreover, the description of W_Q^{-1} in (3.13) together with (3.14) can be used, on one hand, to derive a numerical algorithm to find the Hankel vector of a GRIP, and, on the other hand, to develop a fast recursive algorithm to update the Hankel vector in an explicit manner when some C_{ik} are changed or some new data (Y_i, C_{ik}), $i\geqslant\theta+1$, $k=0,1,\cdots,\tau_i-1$, are provided, without having to start from scratch. Detailed discussions of these developments will be treated elsewhere.

An example will now illustrate the recursiveness aspects of determining the Hankel vector.

Example 3.8 Let $Y_1=0, Y_2=1, Y_3=2$; $C_{10}=0, C_{20}=0, C_{21}=\frac{1}{2}$, $C_{22}=0, C_{23}=3, C_{30}=1$. Then, after a simple computation, the formula (3.14) with $N=6$, $\tau_1=\tau_3=1$, $\tau_2=4$, and $Q=\{0,1,1,1,1,2\}$ produces the Hankel vector

$$\begin{aligned}\boldsymbol{b}=&P_{51}(Y_1)(Y_1-Y_2)^{-4}(Y_1-Y_3)^{-1}C_{10}+\\&P_{54}(Y_2)\boldsymbol{J}_4(Y_2-Y_1)^{-1}\boldsymbol{J}_4(Y_2-Y_3)^{-1}(C_{23},C_{22},C_{21},C_{20})^{\mathrm{T}}+\\&P_{51}(Y_3)(Y_3-Y_1)^{-1}(Y_3-Y_2)^{-4}C_{30}\end{aligned}$$

$$=\begin{bmatrix}1&&&\\1&1&&\\1&2&1&\\1&3&3&1\\1&4&6&4\end{bmatrix}\begin{bmatrix}-1&0&-1&0\\&-1&0&-1\\&&-1&0\\&&&-1\end{bmatrix}\begin{bmatrix}3\\0\\\frac{1}{2}\\0\end{bmatrix}+\frac{1}{2}\begin{bmatrix}1\\2\\4\\8\\16\end{bmatrix}$$

$$=\left(-3,-\frac{5}{2},-2,-1,\frac{3}{2}\right)^{\mathrm{T}}.$$

If now we choose the same GRIP but $C_{23}=\frac{1}{2}$, the corresponding Hankel vector $\bar{\boldsymbol{b}}\in\mathbf{C}^6$ can be achieved with the aid of the original Hankel vector $\boldsymbol{b}$ without having to start from scratch. Indeed, by (3.14),

$$\bar{\boldsymbol{b}}-\boldsymbol{b}=P_{54}(Y_2)\boldsymbol{J}_4(Y_2-Y_1)^{-1}\boldsymbol{J}_4(Y_2-Y_3)^{-1}\left(\frac{1}{2},-3,0,0,0\right)^{\mathrm{T}}.$$

Thus,

$$\begin{aligned}\bar{\boldsymbol{b}}&=\left(-3,\frac{5}{2},-2,-1,\frac{3}{2}\right)^{\mathrm{T}}-\frac{5}{2}(-1,-1,-1,-1,-1)^{\mathrm{T}}\\&=\left(-\frac{1}{2},0,\frac{1}{2},\frac{3}{2},4\right)^{\mathrm{T}}.\end{aligned}$$

In addition, if new data, say, $Y_4=3$, $C_{40}=1$, are provided for the original GRIP, then (3.14) with

$$N=7,\tau_1=\tau_3=\tau_4=1,\tau_2=4,\text{and }\boldsymbol{Q}=\{0,1,1,1,1,2,3\}$$

yields the corresponding Hankel vector:

$$\begin{aligned}\boldsymbol{b}'=&P_{64}(Y_2)\boldsymbol{J}_4(Y_2-Y_1)^{-1}\boldsymbol{J}_4(Y_2-Y_3)^{-1}\boldsymbol{J}_4(Y_2-Y_4)^{-1}(C_{23},C_{22},C_{21},C_{20})^{\mathrm{T}}+\\&P_{61}(Y_3)(Y_3-Y_1)^{-1}(Y_3-Y_2)^{-4}(Y_3-Y_4)^{-1}C_{30}+\\&P_{61}(Y_4)(Y_4-Y_1)^{-1}(Y_4-Y_2)^{-4}(Y_4-Y_3)^{-1}C_{40}\end{aligned}$$

$$=\begin{bmatrix}P_{54}(Y_2)\boldsymbol{J}_4(Y_2-Y_1)^{-1}\boldsymbol{J}_4(Y_2-Y_3)^{-1}\\(15\ 10\ 10)\boldsymbol{J}_4(Y_2-Y_1)^{-1}\boldsymbol{J}_4(Y_2-Y_3)^{-1}\end{bmatrix}\boldsymbol{J}_4(Y_2-Y_4)^{-1}\begin{bmatrix}C_{23}\\C_{22}\\C_{21}\\C_{20}\end{bmatrix}+$$

$$\begin{bmatrix}P_{51}(Y_3)(Y_3-Y_1)^{-1}(Y_3-Y_2)^{-4}\\32(Y_3-Y_1)^{-1}(Y_3-Y_2)^{-4}\end{bmatrix}(Y_3-Y_4)^{-1}C_{30}+$$

$$P_{61}(Y_4)(Y_4-Y_1)^{-1}(Y_4-Y_2)^{-4}(Y_4-Y_3)^{-1}C_{40}$$

$$=\begin{bmatrix}-1&0&-1&0\\-1&-1&-1&-1\\-1&-2&-2&-2\\-1&-3&-4&-4\\-1&-4&-7&-8\\-1&-5&-11&-15\end{bmatrix}\begin{bmatrix}-\frac{1}{2}&-\frac{1}{4}&-\frac{1}{8}&-\frac{1}{16}\\&\ddots&\ddots&\vdots\\&&\ddots&-\frac{1}{4}\\&&&-\frac{1}{2}\end{bmatrix}\begin{bmatrix}3\\0\\\frac{1}{2}\\0\end{bmatrix}-$$

$$\frac{1}{2}\begin{bmatrix}1\\2\\4\\8\\16\\32\end{bmatrix}+\frac{1}{3}\left(\frac{1}{2}\right)^4\begin{bmatrix}1\\3\\9\\27\\81\\243\end{bmatrix}=\left(\frac{4}{3},1,\frac{1}{2},-\frac{1}{2},-\frac{5}{2},-6\right)^{\mathrm{T}}.$$

We note that a formula for the Hankel vector will appear in Corollary 4. 2 later on, which is related to the Hermite interpolation polynomial of a GRIP.

§ 4. The Parametrization of all Rational Interpolants to a GRIP: Main Results

In this section our goal is to construct a particular 2×2 polynomial generating matrix and further all corresponding rational interpolants to a GRIP with a given allowable McMillan degree, using the Hankel vector described in the previous section as the main tool. Going from the Hankel vector to GRIP setting has a very big advantage. This is that once the vector is obtained, the corresponding generating matrix will follow using only simple linear algebra manipulation on finite matrices related to the Hankel vector.

The first step toward this goal is to investigate the close relation between the proper rational interpolants to the GRIP(1. 1) with degree at most $N-1$ and the rational generators to the Hankel vector $\boldsymbol{b}\in\mathbf{C}^{N-1}$

of this GRIP, which can be considered as an improvement of Lemma 4. 4 in [20]. Here a rational function $r(\lambda)/f(\lambda)$ is said to be proper if $\deg r(\lambda)\leqslant \deg f(\lambda)$.

Theorem 4. 1 *Let* $\boldsymbol{b}=(h_0, h_1, \cdots, h_{N-2})\in \mathbf{C}^{N-1}$ *and* $\omega(\lambda)$ *be the Hankel vector and the Hermite polynomial of the GRIP* (1. 1), *respectively, and* $A(\lambda)$ *be as in* (2. 6). *Then there exists a one to one correspondence between the class of proper rational interpolants to the GRIP* (1. 1) *with degree at most* $N-1$, $\mathcal{R}=\{r/f: r/f$ *is an interpolant to the GRIP* (1. 1), $\deg r\leqslant \deg f<N, \gcd(f,r)=1\}$, *and the subclass of the rational generators to* $\boldsymbol{b}\in \mathbf{C}^{N-1}$ *with degree at most* $N-1$, $\mathcal{H}=\{h/f: h/f$ *is a rational generator to* $\boldsymbol{b}$, $\deg h<\deg f<N, \gcd(f,h)=\gcd(f,A)=1\}$; *this means that given* $h/f\in\mathcal{H}$, *there exists a unique polynomial* $r(\lambda)$ *fulfilling the equation*

$$h(\lambda)A(\lambda)=f(\lambda)\omega(\lambda)-r(\lambda) \qquad (4.1)$$

such that $r/f\in\mathcal{R}$, *and conversely*.

Proof Suppose first $h(\lambda)/f(\lambda)\in\mathcal{H}$. Then, by definition, $\gcd(h,f)=\gcd(f,A)=1$ and $\deg h<\deg f<N$, so that there exists a unique pair of polynomials $W(\lambda)$ and $r_1(\lambda)$ such that

$$h(\lambda)A(\lambda)=W(\lambda)f(\lambda)-r_1(\lambda),\quad \deg r_1(\lambda)<\deg f(\lambda). \qquad (4.2)$$

These imply that $\gcd(f,r_1)=1$ and

$$\frac{h(\lambda)}{f(\lambda)}A(\lambda)=W(\lambda)-\frac{r_1(\lambda)}{f(\lambda)},$$

and therefore, in view of $\gcd(f,A)=1$ and $A^{(k)}(Y_i)=0\ \forall\, i,k$,

$$W^{(k)}(\lambda)|_{\lambda=Y_i}=\left(\frac{r_1(\lambda)}{f(\lambda)}\right)^{(k)}_{\lambda=Y_i}\quad \forall\, i,k. \qquad (4.3)$$

According to the proof of Theorem 3. 3, we have that $W(\lambda)+d$ is a polynomial interpolant to the GRIP(1. 1) for a suitable constant d, where the degree of $W(\lambda)$ is at most $N-1$ by (4. 2). Due to the uniqueness of the Hermite polynomial in $\mathbf{C}_{N-1}(\lambda)$, we derive that $W(\lambda)+d=\omega(\lambda)$, and thus $r_1(\lambda)/f(\lambda)+d$ is a rational interpolant to the GRIP (1. 1) as well by (4. 3). Putting now $r(\lambda)=r_1(\lambda)+df(\lambda)$, we obtain that $r(\lambda)/f(\lambda)\in\mathcal{R}$, since $\deg r(\lambda)\leqslant\deg f(\lambda)$ and $\gcd(f,r)=1$ by (4. 2). Now (4. 1) follows immediately from (4.2) with $r_1(\lambda)$ and $W(\lambda)$ therein replaced by $r(\lambda)-df(\lambda)$ and $\omega(\lambda)+d$, respectively.

Conversely, assume that $r(\lambda)/f(\lambda)\in\mathcal{R}$. Then $\gcd(r,f)=\gcd(f,A)=1$ and $\deg r\leqslant\deg f<N$. Let $h_0(\lambda)/f_0(\lambda)$ be a fixed member of $\mathcal{H}$, noting that $\mathcal{H}$ is nonempty by Theorem 2.2. According to what we have already proved in the first part, we have that there exist $\eta\in\mathbf{C}$ and a polynomial $r_0(\lambda)$, $\deg r_0<\deg f$, such that

$$\frac{h_0(\lambda)}{f_0(\lambda)}A(\lambda)=\omega(\lambda)-\frac{r_0(\lambda)+\eta f_0(\lambda)}{f_0(\lambda)}. \tag{4.4}$$

On the other hand, the fact that $r(\lambda)/f(\lambda)\in\mathcal{R}$ implies that

$$[\omega(\lambda)f(\lambda)-r(\lambda)]^{(k)}_{\lambda=Y_i}=\sum_{j=0}^{k}\binom{k}{j}f^{(j)}(\lambda)\left[\omega(\lambda)-\frac{r(\lambda)}{f(\lambda)}\right]^{(k-j)}\Bigg|_{\lambda=Y_i}=0$$

for $i=1,2,\cdots,\theta$, $k=0,1,\cdots,\tau_i-1$, and therefore (4.1) is valid for a suitable but unique polynomial $h(\lambda)$ with $\deg h<\deg f$. It follows then from (4.4) and (4.1) that

$$\left[\frac{h_0(\lambda)}{f_0(\lambda)}-\frac{h(\lambda)}{f(\lambda)}\right]A(\lambda)=\frac{r(\lambda)}{f(\lambda)}-\frac{r_0(\lambda)+\eta f_0(\lambda)}{f_0(\lambda)}.$$

Thus we have that either $h_0(\lambda)/f_0(\lambda)=h(\lambda)/f(\lambda)\in\mathcal{H}$ or

$$\begin{aligned}&N+\deg[h_0(\lambda)f(\lambda)-h(\lambda)f_0(\lambda)]\\&=\deg[r(\lambda)f_0(\lambda)-f(\lambda)r_0(\lambda)-\eta f(\lambda)f_0(\lambda)]\leqslant\deg[f(\lambda)f_0(\lambda)],\end{aligned}$$

which is equivalent to

$$\deg[f(\lambda)f_0(\lambda)]-\deg[h_0(\lambda)f(\lambda)-h(\lambda)f_0(\lambda)]\geqslant N.$$

This last inequality ensures that the Laurent expansions of $h_0(\lambda)/f_0(\lambda)$ and of $h(\lambda)/f(\lambda)$ at infinity have the same coefficients in λ^{-k} for $k=1,2,\cdots,N-1$, so that $h(\lambda)/f(\lambda)\in\mathcal{H}$ as well, because $\gcd(h,f)=\mathrm{gc}(f,A)=1$ and $\deg h<\deg f$. ■

Corollary 4.2 *Let* $\boldsymbol{b}=(h_0,h_1,\cdots,h_{N-2})^{\mathrm{T}}\in\mathbf{C}^{N-1}$, $\omega(\lambda)=\sum_{j=0}^{N-1}w_j\lambda^j$, *and* $A(\lambda)=\sum_{j=0}^{N}A_j\lambda^j\,(A_N=1)$ *be as in Theorem* 4.1. *Then*

$$\boldsymbol{b}=\boldsymbol{H}_{N-1,1}\left(\frac{\omega(\lambda)}{A(\lambda)}\right), \tag{4.5a}$$

and

$$\begin{aligned} w_{N-1} &= A_N h_0, \\ w_{N-2} &= A_{N-1} h_0 + A_N h_1, \\ &\vdots \\ w_1 &= A_2 h_0 + A_3 h_1 + \cdots + A_N h_{N-2}, \\ w_0 &= C_{10} - w_1 Y_1 - \cdots - w_{N-1} Y_1^{N-1}. \end{aligned} \tag{4.5b}$$

Proof (4.5a) follows from (4.1) immediately, and (4.5b) follows from(4.5a) combined with the fact that $\omega(Y_1)=C_{10}$. ■

Corollary 4.2 declares in fact that if $\omega(\lambda)$ is known, then we can obtain the Hankel vector $\boldsymbol{b}$ by means of (4.5a). Conversely, $\omega(\lambda)$ can be computed via (4.5b) from the Hankel vector $\boldsymbol{b}$ combined with one of the conditions (1.2) as well.

Example 4.3 Given a GRIP as in the first part of Example 3.8, find the Hermite polynomial $\omega(\lambda)$. Since $\boldsymbol{b}=\left(-3,-\frac{5}{2},-2,-1,\frac{3}{2}\right)^{\mathrm{T}}\in\mathbf{C}^5$ and $A(\lambda)=\lambda(\lambda-1)^4(\lambda-2)$, we have by (4.5b)

$$\omega(\lambda)=-\frac{1}{2}\lambda(6\lambda^4-31\lambda^3+58\lambda^2-48\lambda+15).$$

Making use of Theorem 4.1 combined with Theorem 2.2. we will obtain directly a complete parametrization of all proper rational interpolants to the GRIP (1.1) with every McMillan degree less than N in terms of the characteristic degrees n_1, n_2 and characteristic polynomials $p(\lambda), q(\lambda)$ of the Hankel vector $\boldsymbol{b}\in\mathbf{C}^{N-1}$. Note that the coprimeness constraint on $r(\lambda)$ and $f(\lambda)$ in the following theorem has been expressed equivalently as a coprimeness constraint on the parameter polynomials $u(\lambda)$ and $v(\lambda)$ together with the condition that $\gcd(f,A)=1$.

Theorem 4.4 *Let $\mathbf{0}\neq\boldsymbol{b}=(h_0,h_1,\cdots,h_{N-2})\in\mathbf{C}^{N-1}$ and $\omega(\lambda)$ be the Hankel vector and the Hermite polynomial of the GRIP (1.1), respectively. Let (n_1,n_2) and $(p(\lambda),q(\lambda))$ be the characteristic degrees and a canonical pair of characteristic polynomials of $\boldsymbol{b}\in\mathbf{C}^{N-1}$. Let $\mathcal{R}$ be as in Theorem 4.1. Further, define the polynomials $\alpha(\lambda)$ and $\beta(\lambda)$ via*

$$\alpha(\lambda)=p(\lambda)\omega(\lambda)-\gamma(\lambda)A(\lambda),\quad \beta(\lambda)=q(\lambda)\omega(\lambda)-\delta(\lambda)A(\lambda), \tag{4.6}$$

where $\gamma(\lambda)$ and $\delta(\lambda)$ are as in (2.14), and $A(\lambda)$ as in (2.6). Then the *subclass* $\mathcal{G}_n$ *of* $\mathcal{R}$ *consisting of proper rational interpolants to the GRIP* (1.1) *with McMillan degree* n *no greater than* $N-1$ *can be described in the following way*:

(a) *If* $n_1=n_2$ (*and therefore* $\deg p(\lambda)=n_1>\deg q(\lambda)$), *then*

$$\mathcal{G}_n=\begin{cases}\varnothing, & n<n_1,\\ \left\{\dfrac{u(\lambda)\alpha(\lambda)+v(\lambda)\beta(\lambda)}{u(\lambda)p(\lambda)+v(\lambda)q(\lambda)}: u(\lambda)\in \mathbf{C}^0_{n-n_1}(\lambda), v(\lambda)\in\mathbf{C}_{n-n_2}(\lambda),\right. & \\ \qquad \left.\gcd(u,v)=\gcd(up+vq,A)=1\right\}, & n\geqslant n_2.\end{cases}$$

(b) *If* $n_1<n_2$ *and* $\deg p(\lambda)=n_1$ (*and therefore* $\deg q(\lambda)<n_2$), *then*

$$\mathcal{G}_n=\begin{cases}\varnothing, & n<n_2,\quad n\neq n_1,\\ \left\{\dfrac{\alpha(\lambda)}{p(\lambda)}:\gcd(p,A)=1\right\}, & n=n_1,\\ \left\{\dfrac{u(\lambda)\alpha(\lambda)+v(\lambda)\beta(\lambda)}{u(\lambda)p(\lambda)+v(\lambda)q(\lambda)}: u(\lambda)\in \mathbf{C}^0_{n-n_1}(\lambda), v(\lambda)\in\mathbf{C}_{n-n_2}(\lambda),\right. & \\ \qquad \left.\gcd(u,v)=\gcd(up+vq,A)=1\right\}, & n\geqslant n_2.\end{cases}$$

(c) *If* $n_1<n_2$ *and* $\deg p(\lambda)<n_1$ (*and therefore* $\deg q(\lambda)=n_2$), *then*

$$\mathcal{G}_n=\begin{cases}\varnothing, & n_1<n_2,\\ \left\{\dfrac{u(\lambda)\alpha(\lambda)+v(\lambda)\beta(\lambda)}{u(\lambda)p(\lambda)+v(\lambda)q(\lambda)}: u(\lambda)\in \mathbf{C}_{n-n_1}(\lambda), v(\lambda)\in\mathbf{C}^0_{n-n_2}(\lambda),\right. & \\ \qquad \left.\gcd(u,v)=\gcd(up+vq,A)=1\right\}, & n\geqslant n_2.\end{cases}$$

We point out that the restriction on the McMillan degree n to be no greater than $N-1$ in Theorem 4.4 is not essential for the theorem to be valid. This will become apparent later on.

As an immediate consequence of Theorem 4.4, we have the following

Corollary 4.5 *Under the assumption of Theorem* 4.4, *we have*

$$n_1=\max\{\deg\alpha(\lambda),\deg p(\lambda)\}\leqslant\max\{\deg\beta(\lambda),\deg q(\lambda)\}=n_2. \tag{4.7}$$

Proof Observe by (2.15) and (4.6) that

$$\det\begin{bmatrix}\alpha(\lambda) & \beta(\lambda)\\ p(\lambda) & q(\lambda)\end{bmatrix}=\alpha(\lambda)q(\lambda)-\beta(\lambda)p(\lambda)=-\sigma A(\lambda), \tag{4.8}$$

where $\sigma\neq0$ is as in (2.15). We prove first (4.7) in case (a) of Theorem

4.4 [i.e., when $n_1=n_2$ and therefore $\deg p(\lambda)=n_1>\deg q(\lambda)$]. To do this, it suffices to show that $\deg \alpha(\lambda)\leqslant n_1$ by (4.8). If $\deg \alpha(\lambda)>n_1$, according to (a) of Theorem 4.4 with $n=n_1$, then $\mathscr{G}_{n_1}$ would be the set of the form

$$\left\{\frac{\alpha(\lambda)+\gamma\beta(\lambda)}{p(\lambda)+\gamma q(\lambda)}: 0\neq\gamma\in\mathbf{C}, \gcd(p+\gamma q, A)=1\right\}\subset\mathscr{R},$$

where the class $\mathscr{R}$ is defined as in Theorem 4.1, and the numerator $\alpha(\lambda)+\gamma\beta(\lambda)$ therein would be of degree m, $m\geqslant\deg \alpha(\lambda)$, for a suitable $0\neq\gamma\in\mathbf{C}$ such that $\gcd(p(\lambda)+\gamma q(\lambda), A(\lambda))=1$, whence $\deg[\alpha(\lambda)+\gamma\beta(\lambda)]>\deg[p(\lambda)+\gamma q(\lambda)]=n_1$, a contradiction with the fact that $\mathscr{G}_{n_1}\subset\mathscr{R}$. Thus, $\max\{\deg \alpha(\lambda), \deg p(\lambda)\}=n_1$. This together with (4.8) ensures in turn that $\deg \beta(\lambda)=n_2$ and $\max\{\deg \beta(\lambda), \deg q(\lambda)\}=n_2$ in the case $n_1=n_2$. In a similar manner, we can verify that (4.7) is valid as well in cases (b) and (c) of Theorem 4.4. ■

In the case of all $\tau_i=1$ in (1.1), the polynomial matrix of order 2

$$\begin{bmatrix}\alpha(\lambda) & \beta(\lambda)\\ p(\lambda) & q(\lambda)\end{bmatrix} \tag{4.9}$$

constructed in Equation (4.8) is identical with the matrix $\boldsymbol{\Theta}(\lambda)$ of the so-called column reduced form in [6,5] after a possible normalization by a nonzero constant, and therefore the characteristic degrees n_1, n_2 are, of course, the same as the so-called reachability indices given in [5].

In the second part of this section we will expand on the approach to the GRIP based on the Hankel vector and show how the result given in Theorem 4.4 can be extended to the general case.

Motivated by Theorem 4.4, it is natural to ask if the 2×2 polynomial matrix in (4.9) is a generating matrix of the GRIP (1.1), or equivalently, if every rational interpolant to (1.1) can be expressed in terms of the linear fractional representation formula

$$\phi(\lambda)=\frac{u(\lambda)\alpha(\lambda)+v(\lambda)\beta(\lambda)}{u(\lambda)p(\lambda)+v(\lambda)q(\lambda)}, \quad \gcd(up+vq, A)=1, \tag{4.10}$$

where $(p(\lambda), q(\lambda))$ and $(\alpha(\lambda), \beta(\lambda))$ are as in Theorem 4.4, and $u(\lambda)$ and $v(\lambda)$ are two parameter polynomials, and vice versa. In the next theorem we show that the answer to this question is yes .

Theorem 4.6 *Under the assumption of Theorem* 4.4, *the rational function* $\phi(\lambda)=r(\lambda)/f(\lambda)$ *with* $\gcd(f,r)=1$ *is an interpolant to the GRIP* (1.1) *if , and only if , there exist coprime polynomials* $u(\lambda)$, $v(\lambda)$ *such that* (4.10) *holds.*

Proof Suppose first that $\phi(\lambda)=r(\lambda)/f(\lambda)$ is determined by (4.10), where $\gcd(u,v)=1$, Then we have by (4.6) that

$$\omega(\lambda)-\phi(\lambda)=\frac{u(\lambda)\gamma(\lambda)+v(\lambda)\delta(\lambda)}{u(\lambda)p(\lambda)+v(\lambda)q(\lambda)}A(\lambda),\quad \gcd(up+vq,A)=1,$$

so that $\phi(\lambda)$ satisfies the interpolation conditions (1.2), that is, $\phi(\lambda)$ is a rational interpolant to the GRIP (1.1). In addition, it follows from (4.8) that

$$q(\lambda)[u(\lambda)\alpha(\lambda)+v(\lambda)\beta(\lambda)]-\beta(\lambda)[u(\lambda)p(\lambda)+v(\lambda)q(\lambda)]=-\sigma u(\lambda)A(\lambda),$$
$$-p(\lambda)[u(\lambda)\alpha(\lambda)+v(\lambda)\beta(\lambda)]+\alpha(\lambda)[u(\lambda)p(\lambda)+v(\lambda)q(\lambda)]=-\sigma v(\lambda)A(\lambda).$$

These last relations ensure in turn that $r(\lambda)=u(\lambda)\alpha(\lambda)+v(\lambda)\beta(\lambda)$ and $f(\lambda)=u(\lambda)p(\lambda)+v(\lambda)q(\lambda)$ are coprime under the coprimeness constraints that $\gcd(u,v)=1$ and $\gcd(up+vq,A)=1$.

Conversely, let $\phi(\lambda)=r(\lambda)/f(\lambda)$ with $\gcd(f,r)=1$ be an interpolant to the GRIP (1.1). It is easy to see that (4.10) holds for a suitable pair of polynomials $u(\lambda)$ and $v(\lambda)$ if and only if there exist two polynomials $u(\lambda)$ and $v(\lambda)$ such that

$$\begin{bmatrix}\alpha(\lambda) & \beta(\lambda)\\ p(\lambda) & q(\lambda)\end{bmatrix}\begin{bmatrix}u(\lambda)\\ v(\lambda)\end{bmatrix}=\begin{bmatrix}r(\lambda)\\ f(\lambda)\end{bmatrix}. \tag{4.11}$$

To verify that this condition is valid , using (4.8), where $0\neq\sigma\in\mathbf{C}$, we obtain from (4.11) that

$$\begin{bmatrix}u(\lambda)\\ v(\lambda)\end{bmatrix}=-\frac{1}{\sigma A(\lambda)}\begin{bmatrix}q(\lambda) & -\beta(\lambda)\\ -p(\lambda) & \alpha(\lambda)\end{bmatrix}\begin{bmatrix}r(\lambda)\\ f(\lambda)\end{bmatrix}. \tag{4.12}$$

$\phi(\lambda)=r(\lambda)/f(\lambda)$ with $\gcd(f,r)=1$ being an interpolant to the GRIP (1.1), there exists a unique polynomial $h(\lambda)$ (not necessarily with $\deg h<\deg f$) such that $f(\lambda)\omega(\lambda)-r(\lambda)=h(\lambda)A(\lambda)$. Thus, by (4.6),

$$\begin{aligned}q(\lambda)r(\lambda)-\beta(\lambda)f(\lambda)&=q(\lambda)r(\lambda)-[q(\lambda)\omega(\lambda)-\delta(\lambda)A(\lambda)]f(\lambda)\\&=[f(\lambda)\delta(\lambda)-q(\lambda)h(\lambda)]A(\lambda),\end{aligned}$$

so that $u(\lambda)=-\sigma^{-1}[f(\lambda)\delta(\lambda)-q(\lambda)h(\lambda)]$ is a polynomial in terms of (4.12). In a similar manner, we can show that $v(\lambda)$ determined by (4.12) is also a polynomial. Furthermore, the coprimeness of $f(\lambda)$ and $r(\lambda)$ together with (4.11) implies the coprimeness of $u(\lambda)$ and $v(\lambda)$, as desired. ■

Theorem 4.6 and Corollary 4.5 now assert that the 2×2 nonsingular polynomial matrix of the form (4.9) is a generating matrix of column reduced form (this means no zeros at infinity). It follows in turn that on leaving out the restriction on the McMillan degree $n<N$ for rational interpolants, Theorem 4.4 will still be valid. According to the so-called predictable-degree property of column reduced polynomial matrices (see e.g. [18, Theorem 6.3~13]) and the Main Theorem of [6] combined with Theorems 4.4 and 4.6, we have the following main result immediately.

Theorem 4.7 *Under the assumption of Theorem* 4.4, *we have*:

(a) $\gcd(\alpha(\lambda),p(\lambda))=1$ *if and only if* $\gcd(p,A)=1$; *in this case*, $\alpha(\lambda)/p(\lambda)$ *of (McMillan) degree* n_1 *is an interpolant to the GRIP* (1.1).

(b) *If* $n_1<n_2$ *and* $\gcd(p,A)=1$, *then* $\alpha(\lambda)/p(\lambda)$ *is the only interpolant of least degree* n_1, *and the allowable McMillan degrees consist of* $n_1, n_2, n_2+1, n_2+2, \cdots$.

(c) *If* $n_1=n_2$ *or* $n_1<n_2$ *but* $p(\lambda)$ *and* $A(\lambda)$ *are not coprime*, *then there exists a family of interpolants of least degree* n_2, *and the allowable McMillan degrees consist of* $n_2, n_2+1, n_2+2, \cdots$.

Combining Theorem 4.7 with Theorems 4.4 and 4.6, we are now ready to write down all rational interpolants to the GRIP (1.1) using the explicit linear fractional parameterization. It is convenient to adopt the convention that $\mathbf{C}_{-1}(\lambda)=\mathbf{C}_0(\lambda)$ in the subsequent formulas.

Theorem 4.8 *Under the assumption of Theorem* 4.4 *on the GRIP* (1.1), *denote further by* $\mathscr{G}_n^*$ [$\mathscr{G}'_n$] *the set of all* [*all nonproper*] *rational interpolants to* (1.1) *with every allowable McMillan degree n. Then*

$$\mathscr{G}_n^*=\mathscr{G}_n\cup\mathscr{G}'_n,$$

where $\mathscr{G}_n$ *is as in Theorem* 4.4 *but without the restraint that the McMillan degree n therein must be less than N. Moreover*, $\mathscr{G}'_n$ *can be parametrized in terms of the linear fractional representation formula as follows.*

(a) *If* $n_1=n_2$ *(and therefore* $\deg p(\lambda)=n_1>\deg q(\lambda)$*), then*

$$\mathscr{G}'_n=\begin{cases}\varnothing, & n<n_1,\\ \left\{\dfrac{u(\lambda)\alpha(\lambda)+v(\lambda)\beta(\lambda)}{u(\lambda)p(\lambda)+v(\lambda)q(\lambda)}: u(\lambda)\in\mathbf{C}_{n-n_1-1}(\lambda), v(\lambda)\in\mathbf{C}^0_{n-n_2}(\lambda),\right.\\ \left.\quad \gcd(u,v)=\gcd(up+vq,A)=1\right\}, & n\geq n_2.\end{cases}$$

(b) *If* $n_1<n_2$ *and* $\deg p(\lambda)=n_1$ *(and therefore* $\deg q(\lambda)<n_2$*), then*

$$\mathscr{G}'_n=\begin{cases}\varnothing, & n<n_2,\\ \left\{\dfrac{u(\lambda)\alpha(\lambda)+v(\lambda)\beta(\lambda)}{u(\lambda)p(\lambda)+v(\lambda)q(\lambda)}: u(\lambda)\in\mathbf{C}_{n-n_1-1}(\lambda), v(\lambda)\in\mathbf{C}^0_{n-n_2}(\lambda).\right.\\ \left.\quad \gcd(u,v)=\gcd(up+vq,A)=1\right\}, & n\geq n_2.\end{cases}$$

(c) *If* $n_1<n_2$*, and* $\deg p(\lambda)<n_1$ *(and therefore* $\deg q(\lambda)=n_2$*), then*

$$\mathscr{G}'_n=\begin{cases}\varnothing, & n<n_2, n\neq n_1,\\ \left\{\dfrac{\alpha(\lambda)}{p(\lambda)}: gcd(p,A)=1\right\}, & n=n_1,\\ \left\{\dfrac{u(\lambda)\alpha(\lambda)+v(\lambda)\beta(\lambda)}{u(\lambda)p(\lambda)+v(\lambda)q(\lambda)}: u(\lambda)\in\mathbf{C}^0_{n-n_1}(\lambda), v(\lambda)\in\mathbf{C}_{n-n_2-1}(\lambda),\right.\\ \left.\quad \gcd(u,v)=\gcd(up+vq,A)=1\right\}, & n\geq n_2.\end{cases}$$

An immediate corollary of the previous theorem is

Corollary 4.9 *Each rational interpolant to the GRIP* (1.1) *with degree n greater than or equal to* n_2 *can be parametrized in terms of a linear fractional function with at most* $2n-N+1$ *parameters.*

Recall [20] that a pair of polynomials $[r(\lambda), f(\lambda)]$ is said to be a generalized solution to the GRIP (1.1) if

$$r(\lambda)\equiv f(\lambda)\omega(\lambda) \bmod A(\lambda),$$

where $\omega(\lambda)$ and $A(\lambda)$ are as before. Also, define the degree of such a solution $[r,f]$ via $\deg[r,f]=\max\{\deg r, \deg f\}$.

Corollary 4.10 *Under the assumption of Theorem* 4.4*, the set of generalized solutions* $[r(\lambda), f(\lambda)]$ *to the GPIP* (1.1) *with every allowable degree n has the form*

$$\{[u(\lambda)\alpha(\lambda)+v(\lambda)\beta(\lambda), u(\lambda)p(\lambda)+v(\lambda)q(\lambda)]:$$
$$u(\lambda), v(\lambda) \text{ arbitrary such that } \deg[u\alpha+v\beta, up+vq]=n\},$$

whereas the allowable degrees of generalized solutions to (1.1) *consist of* $n_1, n_1+1, n_1+2, \cdots$.

In particular, if $n_1 < n_2$, *the set of generalized solutions to* (1.1) *with every degree* n, $n_1 \leqslant n < n_2$, *is nonempty and identical with*

$$\{[u(\lambda)\alpha(\lambda), u(\lambda)p(\lambda)]: u \neq 0 \text{ of degree } n-n_1\}.$$

Proof Follows from Theorem 4.8 and similar arguments to those in the proofs of Theorems 4.6 and 4.7. ■

We point out that in case $\boldsymbol{b}=\boldsymbol{0}\in \mathbf{C}^{N-1}$, it follows from Lemma 2.1 and Theorem 3.3 that a rational function $g(\lambda)$ with no poles in $Y_1, Y_2, \cdots, Y_N$ is an interpolant to the GRIP (1.1) if and only if it is of the form

$$g(\lambda)=h(\lambda)A(\lambda)+C_{10}$$

where C_{10} is as in (1.1), and $h(\lambda)$ is a rational function with no poles in $Y_1, Y_2, \cdots, Y_N$. Therefore Theorem 4.8 is also true for the case of $\boldsymbol{b}=\boldsymbol{0}$ if we take $n_1=0, n_2=N, p(\lambda)=\delta(\lambda)=1, q(\lambda)=\gamma(\lambda)=0$ [and therefore, by (4.6), $\alpha(\lambda)=\omega(\lambda)=C_{10}$ and $\beta(\lambda)=-A(\lambda)$] therein.

We remark that in [2, Theorem 2.26], Antoulas and Anderson have given a parametrization of all rational interpolants to the GRIP (1.1) of every allowable McMillan degree n by using the Loewner matrix approach, but these n's must be less than N. In [6, Main Theorem], Antoulas and Willems deal with the simplest rational interpolation problem, and provide a parametrization of all interpolants without the restriction on the McMillan degrees by using an elegant method based on the reachability of an appropriately defined pair of matrices (for the matrix version see [5, 9, 10]). In [19, Theorem 6.11] Van Barel and Bultheel have developed an algorithm solving the GRIP to write down all interpolants or generalized solutions using a parametrization, but only in an implicit, recursive way. However, for some applications one may want a closed form, explicit formula in terms of the original interpolation data.

Since our analysis here is independent to a great extent of previously found results on Loewner matrices and on GRIPs, it is sometimes possible to achieve new results and simplified proofs for some of the well-known results mentioned on Loewner matrices and on GRIPs as by-products. In addition to Corollaries 3.5 and 4.10, we now give another basic example on the imaginary line, which was established in Belevitch [11] for the simplest IVs, and in Antoulas and Anderson [2] as the main Lemma 2.5 with a different proof.

Recall that if the (generalized) Loewner matrix $\boldsymbol{L}$ is of the form (2.3), then any submatrix of the form

$$\bar{\boldsymbol{L}}=(\bar{\boldsymbol{L}}_{ij}),\quad \bar{\boldsymbol{L}}_{ij}=(l_{ij}^{kl})_{k,l=0}^{\rho_i-1,\sigma_j-1}\ (0\leqslant\rho_i\leqslant m_i,0\leqslant\sigma_j\leqslant n_j), \tag{4.13}$$

is said to be a (generalized) Loewner submatrix. It is clear that if $\boldsymbol{L}$ associates with a GRIP then $\bar{\boldsymbol{L}}$ in (4.13) associates accordingly with a sub-GRIP of this GRIP in terms of Lemma 2.1.

Corollary 4.11 *Let $\phi(\lambda)$ be a rational interpolant to the GRIP* (1.1), *and let $\mathbf{L}=\mathbf{L}_{y,z}(\omega)$, where $\omega(\lambda)$ stands for the Hermite interpolation polynomial of* (1.1), *and $\mathbf{y},\mathbf{z}$ are a pair of* IVs *of $\boldsymbol{Q}$ in* (2.5), *$|\mathbf{y}|=m$ and $|\mathbf{z}|=n$. Then $\mathbf{L}$ has the property*

$$\operatorname{rank}\mathbf{L}=\deg\phi(\lambda)\quad if\ m,n\geqslant\deg\phi(\lambda). \tag{4.14}$$

Moreover, $\mathbf{L}$ is proper, i.e., all $n_1\times n_1$ (generalized) Loewner submatrices of $\mathbf{L}$ are nonsingular.

Proof Observe by Theorem 4.7 that either $\deg\phi(\lambda)=n_1$ or $\deg\phi(\lambda)\geqslant n_2$. In case $\deg\phi(\lambda)=n_1$, the condition that $m,n\geqslant\deg\phi(\lambda)=n_1$ $(m+n=N)$ implies that rank $\boldsymbol{L}=\deg\phi(\lambda)=n_1$ in view of (b) of Corollary 3.5. In case $\deg\phi(\lambda)\geqslant n_2$, the condition that $m,n\geqslant\deg\phi(\lambda)$ $(m+n=N)$ ensures that

$$n_2\leqslant\deg\phi(\lambda)\leqslant n\leqslant N-\deg\phi(\lambda)\leqslant N-n_2=n_1,$$

so that, since $n_1\leqslant n_2$, we have $\deg\phi(\lambda)=n=n_1=n_2$. Thus, the desired result (4.14) follows from (b) of Corollary 3.5 again. Finally, if $\bar{\boldsymbol{L}}$ is an $n_1\times n_1$ (generalized) Loewner submatrix of $\boldsymbol{L}$, then, using (4.14) with $\boldsymbol{L},m,n$ replaced by $\bar{\boldsymbol{L}},n_1,n_1$, respectively, we have

$$\operatorname{rank}\bar{\boldsymbol{L}}=\deg\phi(\lambda)=n_1.$$

■

§ 5. Computational Procedure and Examples

We will first summarize the nonrecursive procedure to follow, according to the results covered in previous sections, when solving completely the GRIP (1. 1). Our main tool here is the Hankel vector rather than the Loewner matrix or other structured matrices in interpolation problems. The steps are as follows.

Step 1 Compute $A(\lambda)$ by (2. 6).

Step 2 Determine the Hankel vector $\boldsymbol{b}=(h_0, h_1, \cdots, h_{N-2})^{\mathrm{T}} \in \mathbf{C}^{N-1}$ of the GRIP (1. 1) from the unique solution to Equation (3. 10) or directly from the formula (3. 14). In case all $\tau_i=1$, the formula (3. 15) can apply.

Step 3 Determine the characteristic degrees n_1 and n_2 of $\boldsymbol{b}\in\mathbf{C}^{N-1}$ by (2. 11).

Step 4 Construct the characteristic polynomials $p(\lambda)$ and $q(\lambda)$ of $\boldsymbol{b}\in\mathbf{C}^{N-1}$ using the procedure described in the second part of Section 2, distinguishing two cases according to $n_1=n_2$ or $n_1<n_2$ — or rather, if $n_1=n_2$, choose $p(\lambda)$ and $q(\lambda)$ as a basis of $\mathscr{A}_{n_1+1}$, and if $n_1<n_2$, choose $p(\lambda)$ as a basis of $\mathscr{A}_{n_1+1}$ and then $q(\lambda)$ such that $p(\lambda), \lambda p(\lambda), \cdots, \lambda^{n_2-n_1} p(\lambda), q(\lambda)$ form a basis of $\mathscr{A}_{n_2+1}$ [in case $n_1=1$, choose $q(\lambda)\in\mathbf{C}_{n_2}(\lambda)$ arbitrarily such that $p(\lambda)$, $\lambda p(\lambda), \cdots, \lambda^{n_2-1} p(\lambda), q(\lambda)$, are linearly independent]. Bring $(p(\lambda), q(\lambda))$ to a canonical pair of characteristic polynomials of $\boldsymbol{b}\in\mathbf{C}^{N-1}$ such that either $p(\lambda)\in\mathbf{C}_{n_1}^0(\lambda)$ and $\deg q(\lambda)<n_2$ or $\deg p(\lambda)<n_1$ and $q(\lambda)\in\mathbf{C}_{n_2}^0(\lambda)$.

Step 5 Compute the Hermite interpolation polynomial $\omega(\lambda)$ by (4. 5b). Define $\alpha(\lambda)$ and $\beta(\lambda)$ via (2. 14) and (4. 6).

Step 6 Find the allowable McMillan degrees of interpolants by Theorem 4. 7.

Step 7 Construct all corresponding rational interpolants to the GRIP (1. 1) for every given allowable McMillan degree in terms of Theorem 4. 8.

Notice that if $\omega(\lambda)$ is known (it is in the case when a simplest rational interpolation problem is considered), then in step 2, the vector $\boldsymbol{b}$ will be determined by (4.5a).

The procedure above will now be illustrated in terms of numerical examples.

Example 5.1[6]　Let $Y_1=0, Y_2=1, Y_3=2, Y_4=\frac{1}{2}$; $C_{10}=0, C_{20}=3, C_{30}=4, C_{40}=2$. This is a simplest rational interpolation problem by definition. The corresponding polynomials $A(\lambda)$ and $\omega(\lambda)$ are

$$A(\lambda)=\lambda(\lambda-1)(\lambda-2)\left(\lambda-\frac{1}{2}\right),\quad \omega(\lambda)=\frac{1}{3}\lambda(2\lambda^2-9\lambda+16).$$

The Hankel vector $\boldsymbol{b}\in\mathbf{C}^3$ is

$$\boldsymbol{b}=\left(\frac{2}{3},-\frac{2}{3},\frac{2}{3}\right)^{\mathrm{T}}$$

determined by (4.5a), and by (2.11)

$$n_1=\operatorname{rank}\begin{bmatrix}\frac{2}{3} & -\frac{2}{3}\\ -\frac{2}{3} & \frac{2}{3}\end{bmatrix}=1,\quad n_2=4-1=3.$$

Solving

$$\begin{bmatrix}\frac{2}{3} & -\frac{2}{3}\\ -\frac{2}{3} & \frac{2}{3}\end{bmatrix}\begin{bmatrix}p_0\\ p_1\end{bmatrix}=\mathbf{0},$$

we have $p(\lambda)=\lambda+1$. Since $n_1=1$, we can choose $q(\lambda)=-2\lambda+1$ such that $p(\lambda), \lambda p(\lambda), \lambda^2 p(\lambda), q(\lambda)$ are linearly independent, and thus $p(\lambda)$ and $q(\lambda)$ form a canonical pair of characteristic polynomials for $\boldsymbol{b}\in\mathbf{C}^3$.

Define $\alpha(\lambda)$ and $\beta(\lambda)$ via (2.14) and (4.6):

$$\alpha(\lambda)=6\lambda,\quad \beta(\lambda)=\lambda(2\lambda-1)(\lambda-4).$$

In view of Theorem 4.7, since $p(Y_i)\neq 0, i=1,2,3,4$, there exist a unique minimal interpolant of degree 1, namely

$$\phi^{\min}(\lambda)=\frac{6\lambda}{\lambda+1},\quad \deg\phi^{\min}(\lambda)=n_1=1,$$

and interpolants with degree at least $n_2=3$. For an example, the family of interpolants with degree 4 can be parametrized in terms of the linear fractional representation formula

$$\phi(\lambda)=\frac{(u_3\lambda^3+u_2\lambda^2+u_1\lambda+u_0)6\lambda+(v_1\lambda+v_0)\lambda(2\lambda-1)(\lambda-4)}{(u_3\lambda^3+u_2\lambda^2+u_1\lambda+u_0)(\lambda+1)+(v_1\lambda+v_0)(-2\lambda+1)},$$

where firstly u,v must be coprime, which is equivalent to

$$u_3v_0^3-u_2v_0^2v_1+u_1v_0v_1^2-u_0v_1^3\neq 0$$

if $v_1\neq 0$, or $u\in \mathbf{C}_3^0(\lambda)$ arbitrary if $v_1=0(v_0\neq 0)$. The constraint that

$$\gcd(up+vq,A)=1$$

must also be satisfied:

$$u_0+v_0\neq 0,\quad 2u_3+2u_2+2u_1+2u_0\neq v_1+v_0,$$
$$8u_3+4u_2+2u_1+u_0\neq 2v_1+v_0,\quad u_3+2u_2+4u_1+8u_0\neq 0.$$

Moreover, $\phi(\lambda)$ is proper if $u\in \mathbf{C}_3^0(\lambda)$ and $v\in \mathbf{C}_1(\lambda)$; it is nonproper if $u\in \mathbf{C}_2(\lambda)$ and $v\in \mathbf{C}_1^0(\lambda)$.

Example 5.2 Let $Y_1=0, Y_2=1, Y_3=2; C_{10}=0, C_{20}=0, C_{21}=\frac{1}{2}$, $C_{22}=0, C_{23}=3, C_{30}=1$. Then

$$A(\lambda)=\lambda(\lambda-1)^4(\lambda-2).$$

The Hankel vector $\boldsymbol{b}\in \mathbf{C}^5$ and the Hermite polynomial $\omega(\lambda)$ are

$$\boldsymbol{b}=\left(-3,-\frac{5}{2},-2,-1,\frac{3}{2}\right)^{\mathrm{T}} \text{ and}$$

$$\omega(\lambda)=-\frac{1}{2}\lambda(6\lambda^4-31\lambda^3+58\lambda^2-48\lambda+15),$$

which were formulated in Examples 3.8 and 4.3, and

$$n_1=\operatorname{rank}\boldsymbol{H}_{33}[\boldsymbol{b}]=\operatorname{rank}\begin{bmatrix}-3 & -\frac{5}{2} & -2\\ -\frac{5}{2} & -2 & -1\\ -2 & -1 & \frac{3}{2}\end{bmatrix}=3,\quad n_2=6-3=3.$$

Solving

$$\boldsymbol{H}_{24}[\boldsymbol{b}]\boldsymbol{x}=\begin{bmatrix}-3 & -\frac{5}{2} & -2 & -1\\ -\frac{5}{2} & -2 & -1 & \frac{3}{2}\end{bmatrix}[x_1\quad x_2\quad x_3\quad x_4]^{\mathrm{T}}=\mathbf{0},$$

we can obtain a canonical pair of characteristic polynomials $p(\lambda), q(\lambda)$:

$$p(\lambda)=(\lambda-1)(\lambda^2-3\lambda+1),\quad q(\lambda)=\lambda^2-8\lambda+6.$$

In view of Theorem 4.7, there exist interpolants with every degree at least 3. Define $\alpha(\lambda)$ and $\beta(\lambda)$ via (2.14) and (4.6):

$$\alpha(\lambda)=-\frac{1}{2}\lambda(\lambda-1)^2,\quad \beta(\lambda)=-\frac{1}{2}\lambda(\lambda-1)(5\lambda-4).$$

Now consider only the minimal rational interpolants of degree $n=3$. Such family can be parametrized in terms of the linear fractional representation formula

$$\phi(\lambda)=-\frac{1}{2}\lambda(\lambda-1)\frac{u_0(\lambda-1)+(5\lambda+4)}{u_0(\lambda-1)(\lambda^2-3\lambda+1)+(\lambda^2-8\lambda+6)},$$

$$u_0=\text{const},$$

where the constraint that

$$\gcd(up+vq,A)=1$$

must be satisfied, which is equivalent to $u_0\neq\pm 6$.

Example 5.3 Let $Y_1=0, Y_2=1, Y_3=2; C_{10}=0, C_{20}=0, C_{21}=\frac{1}{2}$, $C_{22}=0, C_{23}=\frac{1}{2}, C_{30}=1$. (The data are the same as those in Example 5.2 but with $C_{23}=3$ replaced by $C_{23}=\frac{1}{2}$.) Then $A(\lambda)=\lambda(\lambda-1)^4(\lambda-2)$.

The Hankel vector $\boldsymbol{b}\in\mathbf{C}^5$ is $\boldsymbol{b}=\left(-\frac{1}{2},0,\frac{1}{2},\frac{3}{2},4\right)^{\mathrm{T}}$,

which was found in Example 3.8 as the vector $\bar{\boldsymbol{b}}$, and

$$n_1=\operatorname{rank}\boldsymbol{H}_{33}[\boldsymbol{b}]=\operatorname{rank}\begin{bmatrix}-\frac{1}{2} & 0 & \frac{1}{2}\\ 0 & \frac{1}{2} & \frac{3}{2}\\ \frac{1}{2} & \frac{3}{2} & 4\end{bmatrix}=2,\quad n_2=6-2=4.$$

Solving $\boldsymbol{H}_{33}[\boldsymbol{b}]\boldsymbol{x}=\boldsymbol{0}$ and $\boldsymbol{H}_{15}[\boldsymbol{b}]\boldsymbol{x}=\boldsymbol{0}$, we obtain

$$p(\lambda)=\lambda^2-3\lambda+1,\quad q(\lambda)=\lambda,$$

forming a canonical pair of characteristic polynomials for $\boldsymbol{b}\in\mathbf{C}^5$.

Compute $\omega(\lambda)$, $\alpha(\lambda)$, and $\beta(\lambda)$ via (4.5b)(2.14) and (4.6):

$$\omega(\lambda)=\frac{1}{2}\lambda(-\lambda^4+6\lambda^3-13\lambda^2+13\lambda-5),$$

$$\alpha(\lambda)=-\frac{1}{2}\lambda(\lambda-1),\quad \beta(\lambda)=\frac{1}{2}\lambda(\lambda-1)(\lambda^2-2\lambda+2).$$

In view of Theorem 4.7, since $\gcd(p,A)=1$ and $\deg p(\lambda)=n_1=2$, there exists a unique minimal interpolant

$$\phi^{\min}(\lambda)=-\frac{1}{2}\frac{\lambda(\lambda-1)}{\lambda^2-3\lambda+1},\quad \deg\phi^{\min}(\lambda)=n_1=2,$$

and interpolants with every degree at least $n_2=4$. Consider first the family of interpolants with degree $n=4$. Such family can be parametrized in terms of the linear fractional representation formula

$$\phi(\lambda)=-\frac{1}{2}\lambda(\lambda-1)\frac{(u_2\lambda^2+u_1\lambda+u_0)-v_0(\lambda^2-2\lambda+2)}{(u_2\lambda^2+u_1\lambda+u_0)(\lambda^2-3\lambda+1)+v_0\lambda},$$

$$v_0=\text{const},$$

where $u(\lambda)$, v_0 must be coprime, i.e., $v_0\neq 0$, and the constraint that

$$\gcd(up+vq,A)=1$$

must also be satisfied: $u_0\neq 0$, $u_2+u_1+u_0\neq v_0$, $4u_2+2u_1+u_0\neq 2v_0$. The family of the interpolants with every degree $n\geqslant 5$ can be parametrized analogously.

Acknowledgments

The authors acknowledge with many thanks the helpful comments and suggestions of the referees.

References

[1]A C Antoulas. Rational interpolation and the Euclidean algorithm. Linear Algebra Appl., 1988, 108: 157—171.

[2]A C Antoulas and B D O Anderson. On the scalar rational interpolation problem. IMA J. Math. Control Inform., 1986, 3: 61—88.

[3]A C Antoulas and B D O Anderson. On the problem of stable rational interpolation. Linear Algebra Appl. 1989, 122—124: 301—329.

[4]A C Antoulas and B D O Anderson. State-space and polynomial approaches to rational interpolation. In Progress in Systems and Control Theory: Realization and Modelling in System Theory (M A Kaashoek, J H van Schuppen, and A C M Ran, Eds.), Vol. I, Birkhäuser, Boston, 1990: 73—82.

[5]A C Antoulas, J A Ball, J Kang, and J C Willems. On the solution of the minimal rational interpolation problem. Linear Algebra Appl., 1990, 137—138: 511—573.

[6]A C Antoulas and J C Willems. Minimal rational interpolation and Prony's method. in Analysis and Optimization of Systems (A Bensoussan and J L Lions, Eds.), Lecture Notes in Control and Inform. Sci. 144, Springer-Verlag, 1990:297—306.

[7]A C Antoulas and J C Willems. A behavioral approach to linear exact modeling. IEEE Trans. Automat. Control, 1993, 38(12):1 776—1 802.

[8]B D O Anderson and A C Antoulas. Rational interpolation and state-variable realization. Linear Algebra Appl., 1990, 137—138:479—510.

[9]J A Ball, I Gohberg, and L Rodman. Two-sided Lagrange-Sylvester interpolation problem for rational matrix functions. in Proc. Symp. Pure Math. 51, Part I(W B Arveson and R G Douglas, Eds.), Amer. Math. Soc., 1990:17—83.

[10]J A Ball, I Gohberg, and L Rodman. Interpolation of Rational Matrix Functions. Birkhäuser, 1990.

[11]V Belevitch. Interpolation matrices. Philips Res. Rep., 1970, 25:337—369.

[12] G -N Chen and H-P Zhang. More on Loewner matrices. Linear Algebra Appl., 1994, 203—204:265—300.

[13]M Fideler. Hankel and Löwner matrices. Linear Algebra Appl., 1984, 58:75—95.

[14]M H Gutknecht, The rational interpolation problem revisited. Rocky Mountain J. Math., 1990, 21(1):263—280.

[15] G Heinig and U Jungnickel. Hankel matrices generated by Markov parameters, Hankel matrix extension, partial realization and Padé-approximation in Oper. Theory Adv. Appl. 19, Birkhäuser, Basel, 1986:231—253.

[16]G Heinig and K Rost. Algebraic Methods for Toeplitz-like Matrices and Operators. Oper. Theory Adv. Appl. 13, Birkhäuser, Basel, 1984.

[17] B Hoehn. Algorithmen zur rationalen Interpolation und Loewner-Matrizen. Thesis, Univ. Leipzig, 1991.

[18]T Kailath. Linear Systems. Prentice-Hall, 1980.

[19]M Van Barel and A Bultheel. A new formal approach to the rational interpolation problem. Numer. Math., 1992, 62:87—122.

[20]Z Vavřin. A unified approach to Loewner and Hankel matrices. Linear Algebra Appl., 1991, 143:171—222.

Linear Algebra and its Applications,
1999,286:223－259

广义块 Loewner 矩阵的求逆，最小部分实现与有理矩阵插值问题①

Inversion of a Generalized Block Loewner Matrix, the Minimal Partial Realization, and Matrix Rational Interpolation Problem

Abstract Inversion theorems for generalized block Loewner matrices with non-square blocks are presented. Connections are made to the general (i. e. Hermite) matrix rational interpolation, theory of partial realization and its extensions, theoretical and computational aspects of many other special types of matrices such as block Hankel, block Toeplitz, Vandermonde, block Bezout, etc.

Keywords Generalized block Loewner matrix; Matrix rational interpolation; Block Hankel vector; Minimal partial realization; Sylvester matrix equation; Fundamental matrix equations

§ 1. Introduction

It is well known that the scalar, block, and generalized block

① Received 14 March 1998; accepted 1 August 1998. Project supported by NSFC and Doctoral Programme Foundation of Institution of Higher Education of China. 本文与胡永建合作. Submitted by R. A. Brualdi.

Loewner matrices play a significant role in some specific interpolation problems, such as the scalar and matrix unconstrained rational interpolation problems (see[36, 9, 15～17, 1, 3, 4, 39, 12, 6, 7]), the minimal state-variable realization from interpolation data (see[5, 7, 6]), and the Nevanlinna-Pick interpolation in the class of Nevanlinna matrix functions [36, 17, 13, 14, 28, 33, 10]. Starting from such matrices, we can study the connections to Hankel, Toeplitz, Bezout matrices, and their extensions of block matrix form [18, 39, 15, 4].

A few inversion results are given for Loewner matrices [40, 37, 15], and for block Loewner matrices [12].

Motivated by a preceding work of Van Barel and Vavřín [12], in the present paper we investigate the inversion problem for generalized block Loewner matrices with non-square blocks and its wide connections with the matrix rational interpolation with multiple nodes, theory of partial realization and its extensions, theoretical and computational aspects of many other special types of matrices, such as block Hankel, block Toeplitz, block Bezout, Vandermonde, etc. in a comparatively understandable way. In this way, the suggested work helps to generalize old and to establish new connections and to build a deeper basis of the theory of all mutually connected topics. Besides that, the used methods may yield as an inspiration in further development in this field.

To a given general matrix rational interpolation problem (GRMIP) with multiple nodes, there corresponds a class of the so-called generalized block Loewner matrices (not necessarily square, but with the same sum of numbers of the rows and columns), denoted by $\mathscr{L}$ (see (2.3a) and (2.3b) for definition). For a matrix $\boldsymbol{L}\in\mathscr{L}$, we have a Sylvester matrix equation of the form

$$\boldsymbol{A}_{\xi}\boldsymbol{L}-\boldsymbol{L}\boldsymbol{A}_{\pi}=\boldsymbol{B}_{-}\boldsymbol{C}_{-}-\boldsymbol{B}_{+}\boldsymbol{C}_{+}$$

which encodes all the information of the given interpolation problem (see Lemma 2.1), and further four linear matrix equations with the same $\boldsymbol{L}\in\mathscr{L}$ as their coefficient matrices: $\boldsymbol{P}\boldsymbol{L}=\boldsymbol{C}_{-}$, $\boldsymbol{U}\boldsymbol{L}=\boldsymbol{C}_{+}$; $\boldsymbol{L}\widetilde{\boldsymbol{P}}=\boldsymbol{B}_{+}$ and

$L\widetilde{U}=B_{-}$. These tools are particularly relevant to the study of the inversion problem for a square matrix $L\in\mathscr{L}$. On the one hand, the solvability of the half of four matrix equations mentioned just amounts to a criterion of non-singularity for $L\in\mathscr{L}$, and on the other hand, in terms of the solutions of all or the half of these four matrix equations, we can get three formulas for the inverse L^{-1}, respectively (see Theorems 2.2 and 2.4), the last of which is a generalization of the work of [12].

Based on the fact that the class of block Hankel matrices can be considered as a special type of the generalized block Loewner matrices, as applications of aforesaid inversion theorems for $L\in\mathscr{L}$, we derive, in Section 3, the corresponding inversion theorems for block Hankel matrices (Theorems 3.1 and 3.2) in the Loewner matrix framework. Conversely, thanks to a deep connection between the generalized (block) Loewner and (block) Hankel matrices (Lemma 4.1), we may well reduce the inversion problem for the former to what amounts to the same problem but for the latter by using the so-called block Hankel vector for a GRMIP as the main tool. This enables us to exploit the related results known for (block) Hankel and even for (block) Toeplitz matrices, on which a considerable literature exists (see e.g., [24,23,21] for the scalar matrix case, and [35,11,26,34,20,25] and references therein for the block matrix case), for the (block) Loewner matrix case. Such a result appears in Theorem 4.3. A new and simpler derivation is given as a by-product for a known result which says that the inverse of a block Hankel matrix with non-square blocks is identical with a generalized Bezoutian of a quadruple of certain polynomial matrices (see [35] for the square block case, and [26] for the non-square block case). In the block Hankel matrix framework, we have also a Sylvester matrix equation (see Eq. (4.7)) and four matrix equations (see Eqs. (4.8)~(4.11)) with the same block Hankel matrix as coefficient matrices, which are equivalent to and simpler than those for block Loewner case. They provide the key for inversing the block Hankel matrices, and further the generalized block Loewner matrices.

Finally, in Section 5, we consider the connection between the inversion problem for a matrix $\boldsymbol{L}\in\mathscr{L}$ and the solution of the minimal partial realization, and the solution of the GRMIP. Here the matrix equations $\boldsymbol{PL}=\boldsymbol{C}_-$, $\boldsymbol{UL}=\boldsymbol{C}_+$; $\boldsymbol{L}\widetilde{\boldsymbol{P}}=\boldsymbol{B}_+$, $\boldsymbol{L}\widetilde{\boldsymbol{U}}=\boldsymbol{B}_-$, and the analogues for the block Hankel case play also a key part. It turns out that, under the condition that a matrix $\boldsymbol{L}\in\mathscr{L}$ is invertible, there exists an explicit one-to-one correspondence between the set of the minimal, proper rational matrix interpolants to the GRMIP and a certain subset of the minimal partial realizations for the block Hankel vector of the GRMIP (see Theorem 5.4). This clarifies that not only the GRMIP provides a generalization of the partial realization problem, but also it can be treated in the (block) Hankel matrix setting, or equivalently, in the realization problem framework.

§ 2. Two inversion theorems for generalized block Loewner matrices with nonsquare blocks

In this section we present two inversion results for the generalized block Loewner matrices.

§ 2.1 *The class $\mathscr{L}$ of generalized block Loewner matrices associated with a given GRMIP*

We first recall the definition of a generalized block Loewner matrix (see [17, 39, 4, 15] for the scalar case, the extension to the block matrix case is immediate). Given a file of data

$$\{(x_i, \boldsymbol{w}_{ik}): i=1,2,\cdots,\rho, \quad k=0,1,\cdots,\tau_i-1\} \tag{2.1}$$

consisting of $N=\tau_i+\tau_2+\cdots+\tau_\rho(\geqslant 2)$ pairs of points, where the $x_i\in\mathbf{C}$ are different elements with respective multiplicities $\tau_i(\geqslant 1)$, and $\boldsymbol{w}_{ik}\in\mathbf{C}^{p\times q}$. Partition the collection of nodes $x_1, x_2, \cdots, x_\rho$, each x_i taking τ_i times, into two sequences

$$\mathscr{Y}=\{\underbrace{y_1,\cdots,y_1}_{m_1\text{ times}}\cdots,\underbrace{y_u,\cdots,y_u}_{m_u\text{ times}}\},$$

$$\mathscr{Z}=\{\underbrace{z_1,\cdots,z_1}_{n_1\text{ times}}\cdots,\underbrace{z_v,\cdots,z_v}_{n_v\text{ times}}\},\tag{2.2}$$

such that $y_1,y_2,\cdots,y_u[z_1,z_2,\cdots,z_v$, resp.]are distinct with multiplicities $m_1,m_2,\cdots,m_u[n_1,n_2,\cdots,n_v]$, $\sum_{i=1}^{u}m_i=m\ [\sum_{j=1}^{v}n_j=n]$, $m+n=N$, and in case $y_i=z_j$, then $m_i+n_j=\tau_k$ if $y_i(=z_j)$ is equal to some x_k. We introduce the corresponding generalized block Loewner matrix as the $mp\times nq$ matrix

$$\boldsymbol{L}=\boldsymbol{L}_{\mathscr{Y}\mathscr{Z}}=[\boldsymbol{L}_{ij}]_{i,j=1}^{u,v},\tag{2.3a}$$

where $\boldsymbol{L}_{ij}=[\boldsymbol{l}_{ij}^{kl}]_{k=0,l=0}^{m_i-1,n_j-1}\in\mathbf{C}^{m_ip\times n_jq}$ with entries $\boldsymbol{l}_{ij}^{kl}\in\mathbf{C}^{p\times q}$ determined via

$$\boldsymbol{l}_{ij}^{kl}=\begin{cases}\dfrac{1}{k!\ l!}\dfrac{\partial^{k+l}}{\partial\lambda^k\partial\mu^l}\left[\dfrac{\Omega(\lambda)-\Omega(\mu)}{\lambda-\mu}\right]\Bigg|_{\lambda=y_i,\mu=z_j}, & \text{if } y_i\neq z_j,\\ \dfrac{1}{(k+l+1)!}\dfrac{\partial^{k+l+1}}{\partial\lambda^{k+l+1}}\Omega(\lambda)\Bigg|_{\lambda=y_i}, & \text{if } y_i=z_j.\end{cases}\tag{2.3b}$$

Here (and in the sequel) by $\Omega(\lambda)\in\mathbf{C}^{p\times q}[\lambda]$ we denote the unique Hermitian(interpolation) polynomial matrix, which is of formal degree at most $N-1$, of the so-called GRMIP. The last problem reads as follows: Given the data(2.1), find all rational matrix interpolants $F(\lambda)$ subject to

$$\frac{1}{k!}\frac{\mathrm{d}^k}{\mathrm{d}\lambda^k}F(\lambda)\Bigg|_{\lambda=x_i}=\boldsymbol{w}_{ik},i=1,2,\cdots,\rho,\quad k=0,1,\cdots,\tau_i-1.\tag{2.4}$$

In terms of $\mathscr{Y}$ and $\mathscr{Z}$ in Eq. (2.2), we set

$$\frac{1}{k!}\Omega^{(k)}(y_i)=\boldsymbol{C}_{ik},i=1,2,\cdots,u,k=0,1,\cdots,m_i-1,\tag{2.4a}$$

$$\frac{1}{l!}\Omega^{(l)}(z_j)=\boldsymbol{D}_{jl},j=1,2,\cdots,v,l=0,1,\cdots,n_j-1,\tag{2.4b}$$

where $\boldsymbol{C}_{ik}[\boldsymbol{D}_{jl}$, resp.] is equal to some $\boldsymbol{w}_{pk}[\boldsymbol{w}_{pl}]$ if $y_i[z_j]$ coincides with some x_p. In the case of all $\tau_i=1$, the generalized block Loewner matrix of Eqs. (2.3a) and (2.3b) is reduced to a certain block Loewner matrix, which is of the form

$$\boldsymbol{L}_{\mathscr{Y}\mathscr{Z}}=\left[\frac{\boldsymbol{C}_{i0}-\boldsymbol{D}_{j0}}{y_i-z_j}\right]_{i=1,j=1}^{m,n}\tag{2.5}$$

with $\boldsymbol{C}_{i0}, \boldsymbol{D}_{j0} \in \mathbf{C}^{p\times q}$ and $y_i \neq z_j$, $\forall i, j$.

In general, we call the matrix $\boldsymbol{L}_{\mathscr{Y}\mathscr{Z}}$ of Eqs. (2.3a) and (2.3b) the generalized block Loewner matrix associated with the GRMIP (2.4) and the sequences $\mathscr{Y}$ and $\mathscr{Z}$, and denote by $\mathscr{L}$ the class of all the generalized block Loewner matrices associated with the GRMIP (2.4), which are not necessarily square, but with the same sum of numbers of the rows and columns, i. e., $mp+nq$.

In the present paper we deal primarily with the square generalized block Loewner matrices in the class $\mathscr{L}$, which means that $mp=nq$.

Denote by $\boldsymbol{J}_{m_i}(y_i)$ the upper Jordan cell of order m_i with y_i along the main diagonal, and by $\boldsymbol{I}_p$ the identity matrix of order p, by row $(\boldsymbol{B}_1, \boldsymbol{B}_2, \cdots, \boldsymbol{B}_s)$ or row $[\boldsymbol{B}_i]_{i=1}^s$ the block row matrix with blocks $\boldsymbol{B}_1$, $\boldsymbol{B}_2, \cdots, \boldsymbol{B}_s$, and by col $(\boldsymbol{C}_1, \boldsymbol{C}_2, \cdots, \boldsymbol{C}_s)$ or col$[\boldsymbol{C}_j]_{j=1}^s$ the block column matrix with blocks $\boldsymbol{C}_1, \boldsymbol{C}_2, \cdots, \boldsymbol{C}_s$.

A fundamental role is played in our analysis by the following Lemma.

Lemma 2.1 *Let $\mathscr{Y}$ and $\mathscr{Z}$ be the sequences of interpolation nodes given in Eq.* (2.2), *and let $\mathbf{L}\in\mathscr{L}$ be of the form* (2.3a) *and* (2.3b) (*not necessarily square*). *Then*

$$(\mathrm{diag}[\boldsymbol{J}_{m_i}(y_i)]_{i=1}^u \otimes \boldsymbol{I}_p)^{\mathrm{T}}\boldsymbol{L}-\boldsymbol{L}(\mathrm{diag}[J_{n_j}(z_j)]_{j=1}^v \otimes \boldsymbol{I}_q)$$
$$=\mathrm{col}[\mathrm{col}(\boldsymbol{C}_{i0}, \cdots, \boldsymbol{C}_{i,m_i-1})]_{i=1}^u \cdot \mathrm{row}[\mathrm{row}(\underbrace{\boldsymbol{I}_q, \boldsymbol{0}, \cdots, \boldsymbol{0}}_{n_j \text{ times}})]_{j=1}^v -$$
$$\mathrm{col}[\mathrm{col}(\underbrace{\boldsymbol{I}_p, \boldsymbol{0}, \cdots, \boldsymbol{0}}_{m_i \text{ times}})]_{i=1}^u \cdot \mathrm{row}[\mathrm{row}(\boldsymbol{D}_{j0}, \cdots, \boldsymbol{D}_{j,n_j-1})]_{j=1}^v, \tag{2.6}$$

where $\boldsymbol{C}_{ik}$ and $\boldsymbol{D}_{jl}$ as in Eqs. (2.4a) *and* (2.4b) *and the notation $\boldsymbol{A}\otimes\boldsymbol{B}$ indicates the tensor product of two matrices $\boldsymbol{A}$ and $\boldsymbol{B}$.*

We note that the Sylvester matrix Eq. (2.6) can be found in the literature (see, e. g., [10], but the idea has an older origin and can be traced back to the work in [27,31,32]) as a special case of the more general result. Here, however, we give a direct verification for completeness.

Proof of Lemma 2.1 Let $\Omega(\lambda)$ be the Hermitian (interpolation) polynomital matrix of the GRMIP (2.4). From the identity

$$\lambda \frac{\Omega(\lambda)-\Omega(\mu)}{\lambda-\mu}-\frac{\Omega(\lambda)-\Omega(\mu)}{\lambda-\mu}\mu=\Omega(\lambda)-\Omega(\mu)$$

we deduce by computing the partial derivatives $\frac{1}{k!\ l!}\frac{\partial^{k+1}}{\partial\lambda^k\partial\mu^l}$ of both sides

$$\frac{1}{k!l!}\left\{\sum_{\alpha=0}^{k}\frac{k!}{(k-\alpha)!\alpha!}\frac{\mathrm{d}^{k-\alpha}\lambda}{\mathrm{d}\lambda^{k-\alpha}}\cdot\frac{\partial^{\alpha+l}}{\partial\lambda^{\alpha}\partial\mu^{l}}\left[\frac{\Omega(\lambda)-\Omega(\mu)}{\lambda-\mu}\right]-\right.$$

$$\left.\sum_{\beta=0}^{l}\frac{l!}{(l-\beta)!\beta!}\frac{\partial^{k+l-\beta}}{\partial\lambda^{k}\partial\mu^{l-\beta}}\left[\frac{\Omega(\lambda)-\Omega(\mu)}{\lambda-\mu}\right]\cdot\frac{\mathrm{d}^{\beta}\mu}{\mathrm{d}\mu^{\beta}}\right\}$$

$$=\sum_{\alpha=0}^{k}\frac{1}{(k-\alpha)!}\frac{\mathrm{d}^{k-\alpha}\lambda}{\mathrm{d}\lambda^{\lambda-\alpha}}\cdot\frac{1}{\alpha!l!}\frac{\partial^{\alpha+l}}{\partial\lambda^{\alpha}\partial\mu^{l}}\left[\frac{\Omega(\lambda)-\Omega(\mu)}{\lambda-\mu}\right]-$$

$$\sum_{\beta=0}^{l}\frac{1}{k!(l-\beta)!}\frac{\partial^{k+l-\beta}}{\partial\lambda^{k}\partial\mu^{l-\beta}}\left[\frac{\Omega(\lambda)-\Omega(\mu)}{\lambda-\mu}\right]\cdot\frac{1}{\beta!}\frac{\mathrm{d}^{\beta}\mu}{\mathrm{d}\mu^{\beta}}$$

$$=\frac{1}{k!}\Omega^{(k)}(\lambda)\delta_{0l}-\frac{1}{l!}\Omega^{(l)}(\mu)\delta_{0k},k=0,1,\cdots,m_i-1,l=0,1,\cdots,n_j-1$$

in which the symbol δ_{ij} is the Kronecker δ. By choosing $\lambda=y_i$ and $\mu=z_j$ in the last equations (in the case of $y_i=z_j$ we take the limit as $z_j\to y_i$) we get from Eqs. (2. 4a) and (2. 4b) that

$$(\boldsymbol{J}_{m_i}(y_i)^{\mathrm{T}}\otimes\boldsymbol{I}_p)\boldsymbol{L}_{ij}-\boldsymbol{L}_{ij}(\boldsymbol{J}_{n_j}(z_j)\otimes\boldsymbol{I}_q)$$

$$=\mathrm{col}(\boldsymbol{C}_{i0},\cdots,\boldsymbol{C}_{i,m_i-1})\mathrm{row}(\underbrace{\boldsymbol{I}_q,\boldsymbol{0},\cdots,\boldsymbol{0}}_{n_j\text{ times}})-$$

$$\mathrm{col}(\underbrace{\boldsymbol{I}_p,\boldsymbol{0},\cdots,\boldsymbol{0}}_{m_i\text{ times}})\mathrm{row}(\boldsymbol{D}_{j0},\cdots,\boldsymbol{D}_{j,n_j-1})$$

for $1\leqslant i\leqslant u,1\leqslant j\leqslant v$. These are obviously equivalent to Eq. (2. 6). ■

As a matter of convenience we use the following notation throughout this paper

$$\begin{aligned}
\boldsymbol{A}_{\xi}&=\mathrm{diag}[\boldsymbol{J}_{m_i}(y_i)^{\mathrm{T}}]_{i=1}^{u}\otimes\boldsymbol{I}_p\in\mathbf{C}^{mp\times mp},\\
\boldsymbol{A}_{\pi}&=\mathrm{diag}[\boldsymbol{J}_{n_j}(z_j)]_{j=1}^{v}\otimes\boldsymbol{I}_q\in\mathbf{C}^{nq\times nq},\\
\boldsymbol{B}_{-}&=\mathrm{col}[\mathrm{col}(\boldsymbol{C}_{i0},\cdots,\boldsymbol{C}_{i,m_i-1})]_{i=1}^{u}\in\mathbf{C}^{mp\times q},\\
\boldsymbol{B}_{+}&=\mathrm{col}[\mathrm{col}(\underbrace{\boldsymbol{I}_p,\boldsymbol{0},\cdots,\boldsymbol{0}}_{m_i\text{ times}})]_{i=1}^{u}\in\mathbf{C}^{mp\times p},\qquad(2.7)\\
\boldsymbol{C}_{-}&=\mathrm{row}[\mathrm{row}(\underbrace{\boldsymbol{I}_p,\boldsymbol{0},\cdots,\boldsymbol{0}}_{n_j\text{ times}})]_{j=1}^{v}\in\mathbf{C}^{q\times nq},\\
\boldsymbol{C}_{+}&=\mathrm{row}[\mathrm{row}(\boldsymbol{D}_{j0},\cdots,\boldsymbol{D}_{j,n_j-1})]_{j=1}^{v}\in\mathbf{C}^{p\times nq}.
\end{aligned}$$

Then, by Eq. (2.6), $A_\xi, A_\pi, L, B_+, B_-, C_+, C_-$ are connected by the matrix identity

$$A_\xi L - L A_\pi = B_- C_- - B_+ C_+. \tag{2.6'}$$

We thus say that $L \in \mathscr{L}$ has displacement rank two with respect to the displacement defined by $A_\xi L - L A_\pi$ in a sense first defined in [30,31].

Further on, we consider the following four linear matrix equations for $L \in \mathscr{L}$ (not necessarily square):

$$PL = C_-, \tag{2.8}$$

$$UL = C_+, \tag{2.9}$$

$$L\widetilde{P} = B_+, \tag{2.10}$$

$$L\widetilde{U} = B_-, \tag{2.11}$$

where $P \in \mathbf{C}^{q\times mp}$, $U \in \mathbf{C}^{p\times mp}$, $\widetilde{P} \in \mathbf{C}^{nq\times p}$ and $\widetilde{U} \in \mathbf{C}^{nq\times q}$ are unknown, and of the block form

$$P = \text{row}[\text{row}(P_{i0}, P_{i1}, \cdots, P_{i,m_i-1})]_{i=1}^{u}, \ \forall P_{ik} \in \mathbf{C}^{q\times p},$$

$$U = \text{row}[\text{row}(U_{i0}, U_{i1}, \cdots, U_{i,m_i-1})]_{i=1}^{u}, \ \forall U_{ik} \in \mathbf{C}^{p\times p},$$

$$\widetilde{P} = \text{col}[\text{col}(\widetilde{P}_{j0}, \widetilde{P}_{j1}, \cdots, \widetilde{P}_{j,n_j-1})]_{j=1}^{v}, \ \forall \widetilde{P}_{jl} \in \mathbf{C}^{q\times p},$$

$$\widetilde{U} = \text{col}[\text{col}(\widetilde{U}_{j0}, \widetilde{U}_{j1}, \cdots, \widetilde{U}_{j,n_j-1})]_{j=1}^{v}, \ \forall \widetilde{U}_{jl} \in \mathbf{C}^{q\times q}.$$

In the case of simple nodes, i. e., of all $\tau_i = 1$, L is reduced to a block Loewner matrix of the form (2.5), and Eqs. (2.8)~(2.11) become

$$\text{row}(P_{10}, P_{20}, \cdots, P_{m0})L = \text{row}(I_q, I_q, \cdots, I_q), \tag{2.12}$$

$$\text{row}(U_{10}, U_{20}, \cdots, U_{m0})L = \text{row}(D_{10}, D_{20}, \cdots, D_{n0}), \tag{2.13}$$

$$L\ \text{col}(\widetilde{P}_{10}, \widetilde{P}_{20}, \cdots, \widetilde{P}_{n0}) = \text{col}(I_p, I_p, \cdots, I_p), \tag{2.14}$$

$$L\ \text{col}(\widetilde{U}_{10}, \widetilde{U}_{20}, \cdots, \widetilde{U}_{n0}) = \text{col}(C_{10}, C_{20}, \cdots, C_{m0}). \tag{2.15}$$

§ 2.2 *The first inversion theorem for a square* $L \in \mathscr{L}$

Now we proceed to prove an inversion theorem for a square $L \in \mathscr{L}$, on the basis of the use of the strategy given in [25].

Theorem 2.2 *Let* $\mathbf{L} \in \mathscr{L}$ *be of the form* (2.3a) *and* (2.3b) *with*

$$mp = nq.$$

(a) *If there exist* $\widetilde{\mathbf{P}}$ *and* $\widetilde{\mathbf{U}}$ *solving Eqs.* (2.10) *and* (2.11), *respectively, then and only then* $\mathbf{L}$ *is invertible. In this case, the inverse* $\mathbf{L}^{-1}$ *has the form*

$$L^{-1}=\mathrm{row}[\mathcal{U}^{i}\widetilde{P}]_{i=0}^{m-1}(V_{\mathcal{Y}}^{T}\otimes I_{p})^{-1}, \tag{2.16}$$

where

$$\mathcal{U}=A_{\pi}+\widetilde{U}C_{-}-\widetilde{P}C_{+} \tag{2.17}$$

and $V_{\mathcal{Y}}$ *stands for the generalized Vandermonde matrix associated with the sequence* $\mathcal{Y}$ *in* Eq. (2.2), *and defined via*

$$V_{\mathcal{Y}}=\mathrm{row}[P_{m,m_i}(y_i)]_{i=1}^{u}\in \mathbf{C}^{m\times m}, \tag{2.18}$$

whereas

$$P_{kl}(\lambda)_{ij}=\frac{i!}{j!\ (i-j)!}\lambda^{i-j} \tag{2.19}$$

for $0\leqslant i\leqslant k-1, 0\leqslant j\leqslant l-1$.

(b) *If there exist* P, U *solving Eqs.* (2.8) *and* (2.9), *respectively, then and only then* L *is invertible. In this case, the inverse* L^{-1} *has the form*

$$L^{-1}=(V_{\mathcal{Z}}\otimes I_{q})^{-1}\mathrm{col}[P\mathcal{B}^{j}]_{j=0}^{n-1}, \tag{2.20}$$

where

$$\mathcal{B}=A_{\xi}-B_{-}P+B_{+}U \tag{2.21}$$

and $V_{\mathcal{Z}}$ *stands for the generalized Vandermonde matrix associated with the sequence* $\mathcal{Z}$ *in* Eq. (2.2):

$$V_{\mathcal{Z}}=\mathrm{row}[P_{nn_j}(z_j)]_{j=1}^{v}, \tag{2.22}$$

whereas the $P_{nn_j}(\lambda)$ *are defined as in Eq.* (2.19).

Proof To prove (a), we assume that $\widetilde{P}$ and $\widetilde{U}$ solve Eqs. (2.10) and (2.11), respectively. It follows from Eq. (2.6) that

$$A_{\xi}L-LA_{\pi}=L\widetilde{U}C_{-}-L\widetilde{P}C_{+}$$

so that

$$A_{\xi}L=L\mathcal{U},$$

where $\mathcal{U}$ is defined by Eq. (2.17). A simple calculation proves that row $[A_{\xi}^{i}B_{+}]_{i=0}^{m-1}=V_{\mathcal{Y}}^{T}\otimes I_{p}$, where $V_{\mathcal{Y}}$ is as in Eq. (2.18). Further on, from $A_{\xi}L=L\mathcal{U}$, we have naturally that $A_{\xi}^{i}L=L\mathcal{U}^{i}$ for any non-negative integer i, and therefore

$$V_{\mathcal{Y}}^{T}\otimes I_{p}=\mathrm{row}[A_{\xi}^{i}L\widetilde{P}]_{i=0}^{m-1}=\mathrm{row}[L\mathcal{U}^{i}\widetilde{P}]_{i=0}^{m-1}=L\ \mathrm{row}[\mathcal{U}^{i}\widetilde{P}]_{i=0}^{m-1}.$$

Since $\boldsymbol{V}_{\mathscr{Y}}$ is invertible, this implies the invertiblity of the matrix $\boldsymbol{L}$ and Eq. (2.16). The rest is plain.

The proof of (b) is similar to that just given, except that we have to apply $\mathrm{col}[\boldsymbol{C}_-\boldsymbol{A}_\pi^j]_{j=0}^{n-1}=\boldsymbol{V}_{\mathscr{Z}}\otimes\boldsymbol{I}_q$ instead of using now $[\boldsymbol{A}_\xi^i\boldsymbol{B}_+]_{i=0}^{m-1}=\boldsymbol{V}_{\mathscr{Y}}^{\mathrm{T}}\otimes\boldsymbol{I}_p$. ■

It is to be noted that a method of finding the inverse of a generalized Vandermonde matrix $\boldsymbol{V}_{\mathscr{Y}}$ and $\boldsymbol{V}_{\mathscr{Z}}$ may be employed in the inversion of $\boldsymbol{L}$ of Eqs. (2.16) and (2.20). (See also [22] for another method in the simple node case.) To do this, we need the following notation for polynomials and matrices associated with the sequences $\mathscr{Y}$ and $\mathscr{Z}$ in Eq. (2.2):

$$a(\lambda)=\prod_{i=1}^{u}(\lambda-y_i)^{m_i}=\sum_{i=0}^{m-1}a_i\lambda^i+\lambda^m, b(\lambda)=\prod_{j=1}^{v}(\lambda-z_j)^{n_j}=\sum_{j=0}^{n-1}b_j\lambda^j+\lambda^n,$$

$$a_{ik}(\lambda)=\frac{a(\lambda)}{(\lambda-y_i)^{k+1}}, i=1,2,\cdots,u, k=0,1,\cdots,m_i-1,$$

$$b_{jl}(\lambda)=\frac{b(\lambda)}{(\lambda-z_j)^{l+1}}, j=1,2,\cdots,v, l=0,1,\cdots,n_j-1, \tag{2.23}$$

$$\boldsymbol{W}_{y_i}=\left(\frac{1}{\mu!}a_{ik}^{(\mu)}(0)\right)_{k=0,\mu=0}^{m_i-1,m-1}\in\mathbf{C}^{m_i\times m}, \boldsymbol{W}_{z_j}=\left(\frac{1}{\mu!}b_{jl}^{(\mu)}(0)\right)_{l=0,\mu=0}^{n_j-1,n-1}\in\mathbf{C}^{n_j\times n},$$

$$\boldsymbol{W}_{\mathscr{Y}}=\mathrm{col}[\boldsymbol{W}_{y_i}]_{i=1}^{u}\in\mathbf{C}^{m\times m}, \boldsymbol{W}_{\mathscr{Z}}=\mathrm{col}[\boldsymbol{W}_{z_j}]_{j=1}^{v}\in\mathbf{C}^{n\times n},$$

where the index (μ) stands for the μth derivate with respect to λ at $\lambda=0$.

It is known (see e. g., [16]) that

$$\boldsymbol{W}_{\mathscr{Y}}\boldsymbol{V}_{\mathscr{Y}}=\mathrm{diag}[\boldsymbol{Q}_{m_i}a_{i,m_i-1}(\boldsymbol{J}_{m_i}(y_i))]_{i=1}^{u},$$

$$\boldsymbol{W}_{\mathscr{Z}}\boldsymbol{V}_{\mathscr{Z}}=\mathrm{diag}[\boldsymbol{Q}_{n_j}b_{j,n_j-1}(\boldsymbol{J}_{n_j}(z_j))]_{j=1}^{v}, \tag{2.24}$$

where $\boldsymbol{Q}_k$ is the flip matrix of order k: $(\delta_{i,k-j-1})_{i,j=0}^{k-1}$, and $\boldsymbol{V}_{\mathscr{Y}}, \boldsymbol{V}_{\mathscr{Z}}$ are as in Eqs. (2.18) and (2.22) respectively. Thus we have

$$\boldsymbol{V}_{\mathscr{Y}}^{-1}=\mathrm{diag}[a_{i,m_i-1}^{-1}(\boldsymbol{J}_{m_i}(y_i))\boldsymbol{Q}_{m_i}]_{i=1}^{u}\boldsymbol{W}_{\mathscr{Y}},$$

$$\boldsymbol{V}_{\mathscr{Z}}^{-1}=\mathrm{diag}[b_{j,n_j-1}^{-1}(\boldsymbol{J}_{n_j}(z_j))\boldsymbol{Q}_{n_j}]_{j=1}^{v}\boldsymbol{W}_{\mathscr{Z}}, \tag{2.25}$$

whereas

$$a_{i,m_i-1}^{-1}(\boldsymbol{J}_{m_i}(y_i))=\prod_{k\neq i}(\boldsymbol{J}_{m_i}(y_i)-y_k\boldsymbol{I}_{m_i})^{-m_k},$$

$$b_{j,n_j-1}^{-1}(\boldsymbol{J}_{n_j}(z_j))=\prod_{l\neq j}(\boldsymbol{J}_{n_j}(z_j)-z_l\boldsymbol{I}_{n_j})^{-n_l}. \tag{2.26}$$

This process carried out for $\boldsymbol{V}_{\mathscr{Y}}^{-1}$ or $\boldsymbol{V}_{\mathscr{Z}}^{-1}$ yields a somewhat more precise result than Eq. (2.16) or Eq. (2.20).

Corollary 2.3 *Let* $\mathbf{L}\in\mathscr{L}$ *be of the form* (2.3a) *and* (2.3b) *with*

$$mp=nq.$$

(a) *If* $\widetilde{\mathbf{P}}, \widetilde{\mathbf{U}}$ *are the solutions to Eps.* (2.10) *and* (2.11), *respectively, then* $\mathbf{L}^{-1}$ *exists and* $\mathbf{L}^{-1}$ *is of the form*

$$\mathbf{L}^{-1}=\mathrm{row}[(\boldsymbol{A}_\pi+\widetilde{\boldsymbol{U}}\boldsymbol{C}_- - \widetilde{\boldsymbol{P}}\boldsymbol{C}_+)^i\,\widetilde{\boldsymbol{P}}]_{i=0}^{m-1}\times$$
$$\{(\boldsymbol{W}_{\mathscr{Y}}^{\mathrm{T}}\mathrm{diag}[\boldsymbol{Q}_{m_i}a_{i,m_i-1}^{-1}(\boldsymbol{J}_{m_i}^{\mathrm{T}}(y_i))]_{i=1}^{u})\otimes\boldsymbol{I}_p\}. \tag{2.16}$$

(b) *If* $\mathbf{P}, \mathbf{U}$ *are the solutions to Eqs.* (2.8) *and* (2.9), *respectively, then* $\mathbf{L}^{-1}$ *exists and* $\mathbf{L}^{-1}$ *is of the form*

$$\mathbf{L}^{-1}=\{(\mathrm{diag}[b_{j,n_j-1}^{-1}(\boldsymbol{J}_{n_j}(z_j))\boldsymbol{Q}_{n_j}]_{j=1}^{v}\boldsymbol{W}_{\mathscr{Z}})\otimes\boldsymbol{I}_q\}\,\mathrm{col}[\boldsymbol{P}(\boldsymbol{A}_\xi-\boldsymbol{B}_-\boldsymbol{P}+\boldsymbol{B}_+\boldsymbol{U})^j]_{j=0}^{n-1}. \tag{2.20^1}$$

§ 2.3 *The second inversion theorem for a square* $\mathbf{L}\in\mathscr{L}$

Still another inversion result is expressed as follows, which can be considered as the second inversion theorem of this paper, and as a direct extension of an inversion theorem of [12]. We will modify the scheme given in([12], Theorem 2.2) in a recursive fashion by using the original data and solutions of all the Eqs. (2.8)~(2.11).

Theorem 2.4 *Let* $\mathbf{L}\in\mathscr{L}$ *be square. If* $\widetilde{\mathbf{P}}, \widetilde{\mathbf{U}}$ *solve Eqs.* (2.10) *and* (2.11)(*similarly*, $\mathbf{P}, \mathbf{U}$ *solve Eqs.* (2.8) *and* (2.9)) *respectively, then and only then* $\mathbf{L}^{-1}$ *exists. In this case, setting*

$$\boldsymbol{\Gamma}_{ji}^{lk}=\widetilde{\boldsymbol{U}}_{jl}\boldsymbol{P}_{lk}-\widetilde{\boldsymbol{P}}_{jl}\boldsymbol{U}_{ik},\ \forall\, i,j,k,l \tag{2.27}$$

$\mathbf{L}^{-1}$ *can be written in the block form*

$$\boldsymbol{L}^{-1}=(\widetilde{\boldsymbol{L}}_{ji})_{j,i=1}^{v,u}, \tag{2.28}$$

where $\widetilde{\boldsymbol{L}}_{ji}=(\widetilde{\boldsymbol{L}}_{ji}^{lk})_{l,k=0}^{n_j-1,m_i-1}\in\mathbf{C}^{n_jq\times m_ip}$ *with entries* $\widetilde{\boldsymbol{L}}_{ji}^{lk}\in\mathbf{C}^{q\times p}$ *determined by*

$$\widetilde{\boldsymbol{L}}_{ji}^{lk}=\sum_{\alpha=0}^{n_j-1-l}\sum_{\beta=0}^{m_i-1-k}\boldsymbol{\Gamma}_{ji}^{l+\alpha,k+\beta}(-1)^\beta\binom{\alpha+\beta}{\beta}(y_i-z_j)^{-\beta-\alpha-1},\ if\ y_i\neq z_j \tag{2.29}$$

and

$$\widetilde{\boldsymbol{L}}_{ji}^{lk}=-\sum_{\beta=k}^{m_i-1}\boldsymbol{\Gamma}_{ji}^{l+k-1-\beta,\beta},\ if\ y_i=z_j\ and\ l+k>m_i-1,$$

$$\widetilde{\boldsymbol{L}}_{ji}^{lk}=\sum_{\alpha=l}^{n_j-1}\boldsymbol{\Gamma}_{ji}^{\alpha,l+k-1-\alpha},\ if\ y_i=z_j\ and\ l+k>n_i-1,$$

$$\widetilde{\boldsymbol{L}}_{ji}^{lk} = -\sum_{\beta=k}^{l+k-1}\Gamma_{ji}^{l+k-1-\beta,\beta} + \widetilde{\boldsymbol{L}}_{ji}^{0,l+k}, \text{if } y_i = z_j \text{ and } l+k \leqslant m_i - 1,$$

$$\widetilde{\boldsymbol{L}}_{ji}^{lk} = \sum_{\alpha=l}^{l+k-1}\Gamma_{ji}^{\alpha,l+k-1-\alpha} + \widetilde{\boldsymbol{L}}_{ji}^{l+k,0}, \text{if } y_i = z_j \text{ and } l+k \leqslant n_j - 1, \tag{2.30}$$

in which

$$\widetilde{\boldsymbol{L}}_{ji}^{0k} = \boldsymbol{P}_{ik} - \sum_{\alpha\neq j}\widetilde{\boldsymbol{L}}_{\alpha i}^{0k}, \quad \widetilde{\boldsymbol{L}}_{ji}^{l0} = \widetilde{\boldsymbol{P}}_{jl} - \sum_{\beta\neq i}\widetilde{\boldsymbol{L}}_{j\beta}^{l0}, \tag{2.31}$$

whereas $\widetilde{\boldsymbol{L}}_{\alpha i}^{0k}(\alpha\neq j)$ *and* $\widetilde{\boldsymbol{L}}_{j\beta}^{l0}(\beta\neq i)$ *are determined by Eq.* (2.29) *in advance.*

Proof The first part of the theorem has been proved in Theorem 2.2. Only the second part remains to be proved. Assume now that $\boldsymbol{L}^{-1}$ exists and is of the block form (2.28), and that, $\boldsymbol{P}, \boldsymbol{U}, \widetilde{\boldsymbol{P}}, \widetilde{\boldsymbol{U}}$ are solution of Eqs. (2.8)~(2.11), respectively. Multiplying Eq. (2.6′) on the left and the right by $\boldsymbol{L}^{-1}$, and taking into consideration Eqs. (2.8)~(2.11), we recover

$$\boldsymbol{L}^{-1}\boldsymbol{A}_{\xi} - \boldsymbol{A}_{\pi}\boldsymbol{L}^{-1} = \widetilde{\boldsymbol{U}}\boldsymbol{P} - \widetilde{\boldsymbol{P}}\boldsymbol{U}. \tag{2.32}$$

Comparing the $(j,i)(n_j q\times m_i p)$-blocks of both sides of Eq. (2.32) we obtain

$$\begin{aligned}&\widetilde{\boldsymbol{L}}_{ji}(\boldsymbol{J}_{m_i}^{\mathrm{T}}(y_i)\otimes\boldsymbol{I}_p) - (\boldsymbol{J}_{n_j}(z_j)\otimes\boldsymbol{I}_q)\widetilde{\boldsymbol{L}}_{ji}\\ &\quad = \mathrm{col}[\boldsymbol{U}_{jl}]_{l=0}^{n_j-1}\mathrm{row}[\boldsymbol{P}_{ik}]_{k=0}^{m_i-1} - \mathrm{col}[\widetilde{\boldsymbol{P}}_{jl}]_{l=0}^{n_j-1}\mathrm{row}[\boldsymbol{U}_{ik}]_{k=0}^{m_i-1}\end{aligned} \tag{2.33}$$

for $1\leqslant j\leqslant v$ and $1\leqslant i\leqslant u$. The last equation is clearly equivalent to

$$(y_i - z_j)\widetilde{\boldsymbol{L}}_{ji}^{lk} + (\widetilde{\boldsymbol{L}}_{ji}^{l,k+1} - \widetilde{\boldsymbol{L}}_{ji}^{l+1,k}) = \boldsymbol{\Gamma}_{ji}^{lk}, \quad 0\leqslant k\leqslant m_i - 1, 0\leqslant l\leqslant n_j - 1, \tag{2.34}$$

where we set $\widetilde{\boldsymbol{L}}_{jl}^{n_j,k} = \boldsymbol{O}$ and $\widetilde{\boldsymbol{L}}_{jl}^{l,m_i} = \boldsymbol{O}$.

For a fixed pair of (j,i), if $y_i\neq z_j$, the matrix Eq. (2.33) has clearly a unique solution $\widetilde{\boldsymbol{L}}_{ji}$, and the validity of Eq. (2.29) can be verified directly as follows. From Eq. (2.34) we deduce

$$\widetilde{\boldsymbol{L}}_{ji}^{lk} = \frac{\boldsymbol{\Gamma}_{ji}^{lk}}{y_i - z_j} - \frac{\widetilde{\boldsymbol{L}}_{ji}^{l,k+1} - \widetilde{\boldsymbol{L}}_{ji}^{l+1,k}}{y_i - z_j}, 0\leqslant k\leqslant m_i - 1, 0\leqslant l\leqslant n_j - 1. \tag{2.35}$$

This gives a recurrence relation for $\widetilde{\boldsymbol{L}}_{ji}^{lk}$ with the initializations $\widetilde{\boldsymbol{L}}_{jl}^{n_j,k} = \boldsymbol{O}$

and $\widetilde{\boldsymbol{L}}_{ji}^{l,m_i}=\boldsymbol{O}$. It is not difficult to check that each entry $\widetilde{\boldsymbol{L}}_{ji}^{lk}$ of $\widetilde{\boldsymbol{L}}_{ji}$ is uniquely determined by the $\boldsymbol{\Gamma}_{ji}^{\alpha,\beta}$ for $0\leqslant\alpha\leqslant n_j-1$ and $0\leqslant\beta\leqslant m_i-1$ via Eq. (2.29). To do this, let $\widetilde{\boldsymbol{L}}_{ji}^{lk}$ be as in Eq. (2.29) for the moment. We then have

$$\widetilde{\boldsymbol{L}}_{ji}^{lk}=(y_i-z_j)^{-1}\boldsymbol{\Gamma}_{ji}^{lk}+\sum_{\alpha=1}^{n_j-1-l}\sum_{\beta=0}^{m_i-1-k}\boldsymbol{\Gamma}_{ji}^{l+\alpha,k+\beta}(-1)^{\beta}\binom{\alpha+\beta}{\beta}(y_i-z_j)^{-\beta-\alpha-1}+$$
$$\sum_{\beta=1}^{m_i-1-k}\boldsymbol{\Gamma}_{ji}^{l,k+\beta}(-1)^{\beta}(y_i-z_j)^{-\beta-1}.$$

Thus the relation (2.35) amounts to

$$\sum_{\alpha=1}^{n_j-1-l}\sum_{\beta=0}^{m_i-1-k}\boldsymbol{\Gamma}_{ji}^{l+\alpha,k+\beta}(-1)^{\beta}\binom{\alpha+\beta}{\beta}(y_i-z_j)^{-\beta-\alpha}+$$
$$\sum_{\beta=1}^{m_i-1-k}\boldsymbol{\Gamma}_{ji}^{l,k+\beta}(-1)^{\beta}(y_i-z_j)^{-\beta}=-\widetilde{\boldsymbol{L}}_{ji}^{l,k+1}+\widetilde{\boldsymbol{L}}_{ji}^{l+1,k}. \tag{2.36}$$

By Eq. (2.29) again, we may write $\widetilde{\boldsymbol{L}}_{ji}^{l,k+1}$ and $\widetilde{\boldsymbol{L}}_{ji}^{l+1,k}$ as

$$\widetilde{\boldsymbol{L}}_{ji}^{l,k+1}=\sum_{\alpha=0}^{n_j-1-l}\sum_{\beta=1}^{m_i-1-k}\boldsymbol{\Gamma}_{ji}^{l+\alpha,k+\beta}(-1)^{\beta-1}\binom{\alpha+\beta-1}{\beta-1}(y_i-z_j)^{-\beta-\alpha},$$

$$\widetilde{\boldsymbol{L}}_{ji}^{l+1,k}=\sum_{\alpha=1}^{n_j-1-l}\sum_{\beta=0}^{m_i-1-k}\boldsymbol{\Gamma}_{ji}^{l+\alpha,k+\beta}(-1)^{\beta}\binom{\alpha+\beta-1}{\beta}(y_i-z_j)^{-\beta-\alpha}$$

then plugging into Eq. (2.36) gives

$$\sum_{\alpha=1}^{n_j-1-l}\sum_{\beta=1}^{m_i-1-k}\boldsymbol{\Gamma}_{ji}^{l+\alpha,k+\beta}\left\{(-1)^{\beta}\binom{\alpha+\beta}{\beta}+(-1)^{\beta-1}\binom{\alpha+\beta-1}{\beta-1}-\right.$$
$$\left.(-1)^{\beta}\binom{\alpha+\beta-1}{\beta}\right\}(y_i-z_j)^{-\beta-\alpha}+$$
$$\sum_{\alpha=1}^{n_j-1-l}\boldsymbol{\Gamma}_{ji}^{l+\alpha,k}\{(-1)^{0}-(-1)^{0}\}(y_i-z_j)^{-\alpha}+$$
$$\sum_{\beta=1}^{m_i-1-k}\boldsymbol{\Gamma}_{ji}^{l,k+\beta}\{(-1)^{\beta}+(-1)^{\beta-1}\}(y_i-z_j)^{-\beta}=\boldsymbol{O},$$

where we use the identity

$$\binom{\alpha+\beta}{\beta}-\binom{\alpha+\beta-1}{\beta}=\binom{\alpha+\beta-1}{\beta-1}.$$

Further on, in the case of $y_i=z_j$, we then get from Eq. (2.34)

$$\widetilde{\boldsymbol{L}}_{ji}^{l,k+1}-\widetilde{\boldsymbol{L}}_{ji}^{l+1,k}=\boldsymbol{\Gamma}_{ji}^{lk}, \tag{2.37}$$

for $0 \leqslant k \leqslant m_i - 1, 0 \leqslant l \leqslant n_j - 1$. It follows from Eqs. (2.8) and (2.10) that

$$\sum_{\beta=1}^{u} \widetilde{L}_{j\beta}^{l0} = \widetilde{P}_{jl}, \quad \sum_{\alpha=1}^{v} \widetilde{L}_{\alpha j}^{0k} = P_{ik}, 0 \leqslant l \leqslant n_j - 1, \quad 0 \leqslant k \leqslant m_i - 1.$$

Since $y_i = z_j$, we have that $y_\beta \neq z_j$ for all $\beta \neq i$, and $y_i \neq z_\alpha$ for all $\alpha \neq j$, and therefore all the $\widetilde{L}_{j\beta}^{l0} (\beta \neq i)$ and $\widetilde{L}_{\alpha i}^{0k} (\alpha \neq j)$ can be uniquely determined by the formula (2.29) in advance. Thus, both $\widetilde{L}_{ji}^{l0}$ and $\widetilde{L}_{ji}^{0k}$ are well defined via

$$\widetilde{L}_{ji}^{l0} = \widetilde{P}_{jl} - \sum_{\beta \neq i} \widetilde{L}_{j\beta}^{l0}, \quad \widetilde{L}_{ji}^{0k} = P_{ik} - \sum_{\alpha \neq j} \widetilde{L}_{\alpha i}^{0k} \tag{2.38}$$

for $0 \leqslant l \leqslant n_j - 1$ and $0 \leqslant k \leqslant m_i - 1$. Then the validity of Eq. (2.30) for general l, k can be verified by an inductive argument via Eq. (2.37) together with the boundary conditions (2.38). Thus we derive Eq. (2.30) directly, where $\widetilde{L}_{ji}^{0,l+k}$ and $\widetilde{L}_{ji}^{l+k,0}$ are defined via Eq. (2.38) in advance. This completes the proof. ■

It is to be noted that if $L \in \mathscr{L}$ is invertible and if $y_i = z_j$, the proof of Theorem 2.4 says that the formula (2.30) determines completely the block entry $\widetilde{L}_{ji}$ of L^{-1}, even if the matrix Eq. (2.33) has infinitely many solutions.

In particular, if all $\tau_i = 1$ (and therefore all $m_i = n_j = 1$ and $y_i \neq z_j$, $\forall i, j$) we have, by Theorem 2.4,

Corollary 2.5[12] *Let L be a block Loewner matrix of the form* (2.5) *with $mp = nq$ and let $P, U, \widetilde{P}, \widetilde{U}$ be solutions of Eqs.* (2.12) ~ (2.15), *respectively. Then L^{-1} exists and is of the form*

$$L^{-1} = \left[\frac{\widetilde{P}_{j0} U_{i0} - \widetilde{U}_{j0} P_{i0}}{z_j - y_i} \right]_{j,i=1}^{n,m}. \tag{2.39}$$

§ 3. Two inversion theorems for block Hankel matrices with non-square blocks

Let us consider a special case of generalized block Loewner matrices that can be partitioned into blocks in such a way that each matrix H has Hankel structure, i. e., the matrix H has the form

$$\boldsymbol{H}=(\boldsymbol{H}_{i+j})_{i,j=0}^{m-1,n-1}\in\mathbf{C}^{mp\times nq} \tag{3.1}$$

with entries $\boldsymbol{H}_{i+j}\in\mathbf{C}^{p\times q}$. Such a matrix will be called a block Hankel matrix (generally speaking, with non-square blocks). The aim of this section is to specify Theorems 2.2 and 2.4 to the block Hankel case.

§3.1 *The block Hankel matrices in the block Loewner matrix framework*

To begin with, let us formulate the following matrix rational interpolation problem. Given $\boldsymbol{H}$ as in Eq. (3.1), find a rational $p\times q$ matrix function $G(\lambda)$ subject to

$$G(0)=\boldsymbol{R},\quad \frac{1}{k!}G^{(k)}(0)=\boldsymbol{H}_{m+n-1-k},\quad k=1,2,\cdots,m+n-1, \tag{3.2}$$

where $\boldsymbol{R}$ is an arbitrary $p\times q$ matrix. By choosing $u=v=1, m_1=m, n_1=n$, and the sequences

$$\mathscr{Y}=\{\underbrace{0,\cdots,0}_{m\text{ times}}\},\quad \mathscr{Z}=\{\underbrace{0,\cdots,0}_{n\text{ times}}\},$$

and by using the formulas (2.3a) and (2.3b) for $y_1=z_1=0$, we obtain a corresponding generalized block Loewner matrix

$$\boldsymbol{L}_H=\boldsymbol{L}_{\mathscr{Y}\mathscr{Z}}=(\boldsymbol{H}_{m+n-2-i-j})_{i,j=0}^{m-1,n-1}\in\mathbf{C}^{mp\times nq} \tag{3.3}$$

which is exactly a block Hankel matrix. Hence, by Lemma 2.1, the following relation is valid for $\boldsymbol{L}_H$

$$\begin{aligned}&(\boldsymbol{J}_m^{\mathrm{T}}(0)\otimes\boldsymbol{I}_p)\boldsymbol{L}_H-\boldsymbol{L}_H(\boldsymbol{J}_n(0)\otimes\boldsymbol{I}_q)\\&\quad=\mathrm{col}(\boldsymbol{R},\boldsymbol{H}_{m+n-2},\cdots,\boldsymbol{H}_n)\mathrm{row}(\underbrace{\boldsymbol{I}_q,\boldsymbol{0},\cdots,\boldsymbol{0}}_{n\text{ times}})-\\&\quad\mathrm{col}(\underbrace{\boldsymbol{I}_p,\boldsymbol{0},\cdots,\boldsymbol{0}}_{m\text{ times}})\mathrm{row}(\boldsymbol{R},\boldsymbol{H}_{m+n-2},\cdots,\boldsymbol{H}_m).\end{aligned} \tag{3.4}$$

It is clear that

$$\boldsymbol{L}_H=(\boldsymbol{Q}_m\otimes\boldsymbol{I}_p)\boldsymbol{H}(\boldsymbol{Q}_n\otimes\boldsymbol{I}_q),\quad \boldsymbol{J}_m(0)=\boldsymbol{Q}_m\boldsymbol{J}_m^{\mathrm{T}}(0)\boldsymbol{Q}_m, \tag{3.5}$$

where $\boldsymbol{Q}_m$ is the flip matrix of order m. Therefore, if $\boldsymbol{H}$ is invertible, then and only then $\boldsymbol{L}_H$ is invertible. Multiplying Eq. (3.4) on the left by $\boldsymbol{Q}_m\otimes\boldsymbol{I}_p$ and on the right by $\boldsymbol{Q}_n\otimes\boldsymbol{I}_q$, we recover a certain Sylvester equation for the block Hankel matrix $\boldsymbol{H}$

$$(\boldsymbol{J}_m(0)\otimes\boldsymbol{I}_p)\boldsymbol{H}-\boldsymbol{H}(\boldsymbol{J}_n^{\mathrm{T}}(0)\otimes\boldsymbol{I}_q)=\boldsymbol{B}_-^H\boldsymbol{C}_-^H-\boldsymbol{B}_+^H\boldsymbol{C}_+^H, \tag{3.6}$$

where

$$\boldsymbol{B}_{-}^{H}=\operatorname{col}(\boldsymbol{H}_{n},\cdots,\boldsymbol{H}_{m+n-2},\boldsymbol{R})\in\mathbf{C}^{mp\times q},\boldsymbol{C}_{-}^{H}=\operatorname{row}(\underbrace{\mathbf{0},\cdots,\mathbf{0}}_{n\text{ times}},\boldsymbol{I}_{q})\in\mathbf{C}^{q\times nq},$$

$$\boldsymbol{B}_{+}^{H}=\operatorname{col}(\underbrace{\mathbf{0},\cdots,\mathbf{0}}_{m\text{ times}},\boldsymbol{I}_{p})\in\mathbf{C}^{mp\times p},\boldsymbol{C}_{+}^{H}=\operatorname{row}(\boldsymbol{H}_{m},\cdots,\boldsymbol{H}_{m+n-2},\boldsymbol{R})\in\mathbf{C}^{p\times nq}.\tag{3.7}$$

This shows that $\boldsymbol{H}$ has displacement rank two with respect to the structure defined by $(\boldsymbol{J}_m(0)\otimes\boldsymbol{I}_p)\boldsymbol{H}-\boldsymbol{H}(\boldsymbol{J}_n^{\mathrm{T}}(0)\otimes\boldsymbol{I}_q)$. Further on, the corresponding Eqs. (2.8)～(2.11) have the forms

$$\operatorname{row}(\boldsymbol{P}_0,\boldsymbol{P}_1,\cdots,\boldsymbol{P}_{m-1})\boldsymbol{H}=\boldsymbol{C}_{-}^{H},\tag{3.8}$$

$$\operatorname{row}(\boldsymbol{U}_0,\boldsymbol{U}_1,\cdots,\boldsymbol{U}_{m-1})\boldsymbol{H}=\boldsymbol{C}_{+}^{H},\tag{3.9}$$

$$\boldsymbol{H}\operatorname{col}(\widetilde{\boldsymbol{P}}_0,\widetilde{\boldsymbol{P}}_1,\cdots,\widetilde{\boldsymbol{P}}_{n-1})=\boldsymbol{B}_{+}^{H},\tag{3.10}$$

$$\boldsymbol{H}\operatorname{col}(\widetilde{\boldsymbol{U}}_0,\widetilde{\boldsymbol{U}}_1,\cdots,\widetilde{\boldsymbol{U}}_{n-1})=\boldsymbol{B}_{-}^{H},\tag{3.11}$$

where $\boldsymbol{P}_i\in\mathbf{C}^{q\times p},\boldsymbol{U}_i\in\mathbf{C}^{p\times p},i=1,2,\cdots,m-1$ and $\widetilde{\boldsymbol{P}}_j\in\mathbf{C}^{q\times p},\widetilde{\boldsymbol{U}}_j\in\mathbf{C}^{q\times q},j=1,2,\cdots,n-1$.

§ 3.2 *Inversion theorems for block Hankel matrices*

The next result is a special case of Theorem 2.2, in which the inverse matrix $\boldsymbol{H}^{-1}$ is constructed via the solutions of Eqs. (3.10) and (3.11) (or similarly, of Eqs. (3.8) and (3.9)). Compare Section 3 of [25].

Theorem 3.1 *Let* $\boldsymbol{H}$ *be a block Hankel matrix of the form* (3.1) *with* $mp=nq$, *and let* $\boldsymbol{Q}_m$ *be the flip matrix of order* m.

(a) *If there exist* $\widetilde{\boldsymbol{P}}_H$ *and* $\widetilde{\boldsymbol{U}}_H$ *solving Eqs.* (3.10) *and* (3.11), *respectively, then and only then* $\boldsymbol{H}$ *is invertible. In this case*, $\boldsymbol{H}^{-1}$ *is of the form*

$$\boldsymbol{H}^{-1}=\operatorname{row}[(\boldsymbol{J}_n^{\mathrm{T}}(0)\otimes\boldsymbol{I}_q+\widetilde{\boldsymbol{U}}_H\boldsymbol{C}_{-}^{H}-\widetilde{\boldsymbol{P}}_H\boldsymbol{C}_{+}^{H})^i\widetilde{\boldsymbol{P}}_H]_{i=0}^{m-1}(\boldsymbol{Q}_m\otimes\boldsymbol{I}_p).\tag{3.12}$$

(b) *If there exist* $\boldsymbol{P}_H$ *and* $\boldsymbol{U}_H$ *solving Eqs.* (3.8) *and* (3.9), *respectively, then and only then* $\boldsymbol{H}$ *is invertible. In this case*, H^{-1} *is of the form*

$$\boldsymbol{H}^{-1}=(\boldsymbol{Q}_n\otimes\boldsymbol{I}_q)\operatorname{col}[\boldsymbol{P}_H(\boldsymbol{J}_m(0)\otimes\boldsymbol{I}_p-\boldsymbol{B}_{-}^{H}\boldsymbol{P}_H+\boldsymbol{B}_{+}^{H}\boldsymbol{U}_H)^j]_{j=0}^{n-1}.\tag{3.13}$$

Proof Assume that Eqs. (3.10) and (3.11) have the solutions $\widetilde{\boldsymbol{P}}_H$ and $\widetilde{\boldsymbol{U}}_H$, respectively. From Eqs. (3.6)(3.10)(3.11) we deduce

$$(J_m(0)\otimes I_p)H=H(J_n^{\mathrm{T}}(0)\otimes I_q+\widetilde{U}_H C_-^H-\widetilde{P}_H C_+^H).$$

For $0\leqslant i\leqslant m-1$, we get

$$H(J_n^{\mathrm{T}}(0)\otimes I_q+\widetilde{U}_H C_-^H-\widetilde{P}_H C_+^H)^i\widetilde{P}_H=(J_m(0)\otimes I_p)^i H\,\widetilde{P}_H$$
$$=(J_m(0)\otimes I_p)^i B_+^H=\mathrm{col}(\underbrace{0,\cdots,0}_{m-i\text{ times}},I_p,0,\cdots,0)$$

and therefore

$$Q_m\otimes I_p=H\ \mathrm{row}[(J_n^{\mathrm{T}}(0)\otimes I_q+\widetilde{U}_H C_-^H-\widetilde{P}_H C_+^H)^i\widetilde{P}_H]_{i=0}^{m-1},$$

so that H^{-1} exists and is of the form (3.12).

Similarly, we can prove (b). ■

We may similarly obtain the specific of Theorem 2.4 to the block Hankel case (see [38], p. 58, for the scalar case: $p=q=1$).

Theorem 3.2 *Let H be as in Eq. (3.1), $mp=nq$. If there exist $P_H=\mathrm{row}[P_i]_{i=0}^{m-1}$, $U_H=\mathrm{row}[U_i]_{i=0}^{m-1}$ [$\widetilde{P}_H=\mathrm{col}[\widetilde{P}_j]_{j=0}^{n-1}$ and $\widetilde{U}_H=\mathrm{col}[\widetilde{U}_j]_{j=0}^{n-1}$, resp.] solving Eqs. (3.8) ~ (3.11), respectively, then and only then H is invertible. In this case, H^{-1} is of the form*

$$H^{-1}=(\widetilde{H}_{ji})_{j,i=0}^{n-1,m-1},\tag{3.14}$$

where the $\widetilde{H}_{ji}\in\mathbf{C}^{q\times p}$ are defined via

$$\widetilde{H}_{ji}=\sum_{\beta=i}^{m-2}\Gamma_{i+j-\beta,\beta+1}+\widetilde{P}_{i+j+1-m},\quad \text{if } i+j\geqslant m-1(i\neq m-1),$$
$$\widetilde{H}_{ji}=-\sum_{\alpha=j}^{n-2}\Gamma_{\alpha+1,i+j-\alpha}+\widetilde{P}_{i+j+1-n},\quad \text{if } i+j\geqslant n-1(j\neq n-1),$$
$$\widetilde{H}_{ji}=\sum_{\beta=i}^{i+j}\Gamma_{i+j-\beta,\beta+1},\quad \text{if } i+j\leqslant m-1,\widetilde{H}_{n-1,i}=P_i,$$
$$\widetilde{H}_{ji}=-\sum_{\alpha=j}^{i+j}\Gamma_{\alpha+1,i+j-\alpha},\quad \text{if } i+j\leqslant n-1,\widetilde{H}_{j,m-1}=\widetilde{P}_j,\tag{3.15}$$

in which

$$\Gamma_{ji}=\widetilde{U}_jP_i-\widetilde{P}_jU_i,\quad j=0,1,\cdots,n-1,i=0,1\cdots,m-1.$$

Proof If H is invertible, from Eq. (3.6) we deduce

$$H^{-1}(J_m(0)\otimes I_p)-(J_m^{\mathrm{T}}(0)\otimes I_p)H^{-1}=\widetilde{U}_HP_H-\widetilde{P}_HU_H.$$

By comparing the (j,i)-block entries of both sides we have

$$\widetilde{H}_{j,i-1}-\widetilde{H}_{j-1,i}=\widetilde{U}_jP_i-\widetilde{P}_jU_i,\tag{3.16}$$

for $0 \leqslant i \leqslant m-1$ and $0 \leqslant j \leqslant n-1$, where we set $\widetilde{\boldsymbol{H}}_{j,-1}=\boldsymbol{O}$ and $\widetilde{\boldsymbol{H}}_{-1,i}=\boldsymbol{O}$. On the other hand, by Eqs. (3.8) and (3.10), $\widetilde{\boldsymbol{H}}_{n-1,i}=\boldsymbol{P}_i$ for $0 \leqslant i \leqslant m-1$, and $\widetilde{\boldsymbol{H}}_{j,m-1}=\widetilde{\boldsymbol{P}}_j$ for $0 \leqslant j \leqslant n-1$. Thus Eq. (3.15) follows from Eq. (3.16) together with these boundary conditions. ■

We emphasize that the result of Theorem 3.2 consists indeed of nothing more than the representation for a certain generalized Bezoutian. We shall prove this in Remark (c) following Theorem 4.3 (see Eqs. (4.16)(4.28a) and (4.28b) below).

Also, it is to be noted that if $\boldsymbol{H}$ is a block Hankel matrix of the form (3.1) then $(\boldsymbol{Q}_m \otimes \boldsymbol{I}_p)\boldsymbol{H}$ and $\boldsymbol{H}(\boldsymbol{Q}_n \otimes \boldsymbol{I}_q)$ are both the block Toeplitz matrices, so that the inversion problem for block Toeplitz matrices can be reduced what amounts to the same problem but for block Hankel matrices.

§4. The connection between inversions for generalized block Loewner matrices and for block Hankel matrices

The question arises as to whether inversions for $\boldsymbol{L} \in \mathscr{L}$ and for block Hankel matrices can be treated in a unifying framework. In this section, we shall prove that this is indeed the case by establishing a very rigid relationship between these two inversion problems in terms of the so-called block Hankel vector of a GRMIP. It turns out that the former problem can be reduced to what amounts to the study of the latter with little additional effort. This key observation thereby allows us to exploit all the known and more-studied results on inversion of block Hankel (or block Toeplitz) matrices for the block Loewner case.

Here we will use the notation given in Sections 2 and 3, especially, those given in Eq. (2.23) for polynomials and matrices associated with the sequences $\mathscr{Y}$ and $\mathscr{X}$ in Eq. (2.2).

§4.1 *The block Hankel vector for a GRMIP*

Let $\Omega(\lambda) \in \mathbf{C}^{p \times q}[\lambda]$ be the Hermitian (interpolation) polynomial matrix of the GRMIP (2.4), and let

$$A(\lambda) = a(\lambda)b(\lambda) = \prod_{i=1}^{\rho}(\lambda - x_i)^{\tau_i} = \prod_{i=1}^{u}(\lambda - y_i)^{m_i}\prod_{j=1}^{v}(\lambda - z_j)^{n_j}, \tag{4.1}$$

where $\tau_1 + \tau_2 \cdots + \tau_\rho = N = m+n$. Consider the first $N-1$ Markov parameters $\boldsymbol{H}_0, \boldsymbol{H}_1, \cdots, \boldsymbol{H}_{N-2}$, of size $p \times q$, of the strictly proper rational matrix function $\Omega(\lambda)/A(\lambda)$, that is, for sufficiently large values of $|\lambda|$,

$$\frac{\Omega(\lambda)}{A(\lambda)} = \sum_{j=0}^{\infty}\boldsymbol{H}_j\lambda^{-j-1}. \tag{4.2}$$

As in the scalar case [16], we call $\boldsymbol{b} = (\boldsymbol{H}_0, \boldsymbol{H}_1, \cdots, \boldsymbol{H}_{N-2})$ the *block Hankel vector* of the GRMIP (2.4).

The following result is useful for inverting generalized block Loewner matrices $\boldsymbol{L} \in \mathscr{L}$. Here we should mention that though we had introduced the concept of the (block) Hankel vector independently and the decomposition Eq. (4.3) (when $p=q=1$) in [16], it was Fiedler who first made a detailed study of the Loewner-Hankel matrix connection in [18] for scalar, simple-node case (see also [39] for the scalar, multiple-node case).

Lemma 4.1 [16] *Let the GRMIP (2.4) be given, and let* $\boldsymbol{b} = (\boldsymbol{H}_0, \boldsymbol{H}_1, \cdots, \boldsymbol{H}_{N-2})$ *be its block Hankel vector. If* $\mathscr{Y}$ *and* $\mathscr{X}$ *are arbitrarily given as in Eq. (2.2), then the corresponding* $\boldsymbol{L}_{\mathscr{Y}\mathscr{X}} \in \mathscr{L}$ *(not necessarily square), defined via Eqs. (2.3a) and (2.3b), admits the decomposition*

$$\boldsymbol{L}_{\mathscr{Y}\mathscr{X}} = (\boldsymbol{W}_{\mathscr{Y}} \otimes \boldsymbol{I}_p)\boldsymbol{H}_{mp,nq}[\boldsymbol{b}](\boldsymbol{W}_{\mathscr{X}}^{\mathrm{T}} \otimes \boldsymbol{I}_q), \tag{4.3}$$

in which $\boldsymbol{W}_{\mathscr{Y}}, \boldsymbol{W}_{\mathscr{X}}$ *are non-singular and defined as in Eq. (2.23), and* $\boldsymbol{H}_{mp,nq}[\boldsymbol{b}]$ *is a block Hankel matrix of the form*

$$\boldsymbol{H}_{mp,nq}[\boldsymbol{b}] = (\boldsymbol{H}_{i+j})_{i,j=0}^{m-1,n-1} \in \mathbf{C}^{mp \times nq}.$$

Moreover, an extreme case of Eq. (4.3) holds true

$$\mathrm{col}[\mathrm{col}(\boldsymbol{w}_{i0}, \cdots, \boldsymbol{w}_{i,\tau_i-1})]_{i=1}^{\rho} = (\boldsymbol{W}_A \otimes \boldsymbol{I}_p)\mathrm{col}(\boldsymbol{H}_0, \boldsymbol{H}_1, \cdots, \boldsymbol{H}_{m+n-2}, \boldsymbol{H}_{m+n-1}), \tag{4.4}$$

where $\boldsymbol{W}_A \in \mathbf{C}^{N \times N}$ *is non-singular and defined as* $\boldsymbol{W}_{\mathscr{Y}}$ *in Eq. (2.23) but with* $A(\lambda)$ *in place of* $a(\lambda)$ *therein, and* $\boldsymbol{H}_{m+n-1}$ *is the* $(m+n)$*th Markov parameter of* $\Omega(\lambda)/A(\lambda)$ *in Eq. (4.2).*

Remarks (a) We may prove[16] that for the GRMIP(2.4), the different pairs of the sequences $\mathscr{Y}$ and $\mathscr{X}$ as in Eq. (2.2) correspond to the different $\boldsymbol{W}_{\mathscr{Y}}$, $\boldsymbol{W}_{\mathscr{X}}$ and $\boldsymbol{L}_{\mathscr{Y}\mathscr{X}}$; however, they always give rise to one and the same block vector $\boldsymbol{b}$ such that Eq. (4.3) holds. It was the block Hankel vector $\boldsymbol{b}=(\boldsymbol{H}_0, \boldsymbol{H}_1, \cdots, \boldsymbol{H}_{N-2})$ that plays this key role. Thus, in the case of $mp=nq$, all the matrices $\boldsymbol{L}\in\mathscr{L}$ are invertible if, and only if, one of them is invertible, or equivalently, $\boldsymbol{H}_{mp,nq}[\boldsymbol{b}]$ is invertible.

Conversely, to a block vector $\boldsymbol{b}=(\boldsymbol{H}_0, \boldsymbol{H}_1, \cdots, \boldsymbol{H}_{m+n-2})$ with each $\boldsymbol{H}_k\in\mathbf{C}^{p\times q}$, and to a pair of sequences $\mathscr{Y}$ and $\mathscr{X}$ defined as in Eq. (2.2), there corresponds uniquely a GRMIP of the form (2.4) (up to a constant matrix summand for $\boldsymbol{w}_{i0}$) such that $\boldsymbol{b}$ is its block Hankel vector [16].

(b) The result of Eq. (4.4) says that $\mathrm{col}(\boldsymbol{H}_0, \boldsymbol{H}_1, \cdots, \boldsymbol{H}_{m+n-2}, \boldsymbol{H}_{m+n-1})$, consisting of the first $m+n$ Markov parameters of $\Omega(\lambda)/A(\lambda)$, solves uniquely the matrix equation

$$(\boldsymbol{W}_A\otimes\boldsymbol{I}_p)\boldsymbol{X}=\mathrm{col}[\mathrm{col}(\boldsymbol{w}_{i0}, \cdots, \boldsymbol{w}_{i,\tau_i-1})]_{i=1}^{\rho}.$$

Similarly, we have

$$\begin{aligned}&(\boldsymbol{W}_{\mathscr{Y}}\otimes\boldsymbol{I}_p)\{\boldsymbol{H}_{mp,nq}[\boldsymbol{b}]\mathrm{col}(b_0\boldsymbol{I}_q, \cdots, b_{n-1}\boldsymbol{I}_q)+\\&\qquad\mathrm{col}(\boldsymbol{H}_n, \cdots, \boldsymbol{H}_{m+n-2}, \boldsymbol{H}_{m+n-1})\}=\boldsymbol{B}_-,\\&(\boldsymbol{W}_{\mathscr{X}}\otimes\boldsymbol{I}_q)\{\mathrm{row}(a_0\boldsymbol{I}_p, \cdots, a_{m-1}\boldsymbol{I}_p)\boldsymbol{H}_{mp,nq}[\boldsymbol{b}]+\\&\qquad\mathrm{row}(\boldsymbol{H}_m, \cdots, \boldsymbol{H}_{m+n-2}, \boldsymbol{H}_{m+n-1})\}=\boldsymbol{C}_+.\end{aligned}\tag{4.5}$$

Indeed, the first amounts to

$$\begin{aligned}\frac{\Omega_a(\lambda)}{a(\lambda)}=&\frac{1}{\lambda}(\boldsymbol{H}_0b_0+\cdots+\boldsymbol{H}_{n-1}b_{n-1}+\boldsymbol{H}_n)+\\&\frac{1}{\lambda^2}(\boldsymbol{H}_1b_0+\cdots+\boldsymbol{H}_nb_{n-1}+\boldsymbol{H}_{n+1})+\cdots+\\&\frac{1}{\lambda^m}(\boldsymbol{H}_{m-1}b_0+\cdots+\boldsymbol{H}_{m+n-2}b_{n-1}+\boldsymbol{H}_{m+n-1})+\cdots,\end{aligned}$$

where $\Omega_a(\lambda)\in\mathbf{C}^{p\times q}[\lambda]$ is the unique Hermitian interpolation polynomial matrix, subject to the condition (2.4a), of degree $\leqslant m-1$. But, the last expression follows readily from

$$\Omega_a(\lambda)-\Omega(\lambda)=a(\lambda)h(\lambda),$$

where $h(\lambda)\in\mathbf{C}^{p\times q}[\lambda]$ is a suitable polynomial matrix of degree at most

$n-1$. Similarly, we can prove another result of Eq. (4.5) by considering the unique Hermitian interpolation polynomial matrix, $\Omega_b(\lambda)$, subject to Eq. (2.4b), of degree $\leqslant n-1$.

§ 4.2 *The fundamental matrix equations associated with a block Hankel matrix*

Further on, under the transformation (4.3), Eqs. (2.8) ~ (2.11) with the same $\boldsymbol{L}$ as coefficient matrices will be reduced to what amounts to the corresponding matrix equations with the same $\boldsymbol{H}=\boldsymbol{H}_{mp,nq}[\boldsymbol{b}]$ as coefficient matrices. To prove this, we need the following lemma.

Lemma 4.2 *Let $a(\lambda)$ and $b(\lambda)$ be as in Eq. (2.23), and let $\boldsymbol{C}_a$ and $\boldsymbol{C}_b$ be the (first) companion matrices of $a(\lambda)$ and $b(\lambda)$, respectively. Then*

$$\boldsymbol{W}_{\mathscr{Y}}\boldsymbol{C}_a=\operatorname{diag}[\boldsymbol{J}_{m_i}^{\mathrm{T}}(y_i)]_{i=1}^{u}\boldsymbol{W}_{\mathscr{Y}},\quad \boldsymbol{W}_{\mathscr{Z}}\boldsymbol{C}_b=\operatorname{diag}[\boldsymbol{J}_{n_j}^{\mathrm{T}}(z_j)]_{j=1}^{v}\boldsymbol{W}_{\mathscr{Z}},$$

Proof It is well known that

$$\boldsymbol{V}_{\mathscr{Y}}\operatorname{diag}[\boldsymbol{J}_{m_i}(y_i)]_{i=1}^{u}=\boldsymbol{C}_a\boldsymbol{V}_{\mathscr{Y}}$$

so that, by Eq. (2.24),

$$\begin{aligned}\boldsymbol{W}_{\mathscr{Y}}\boldsymbol{C}_a&=\boldsymbol{W}_{\mathscr{Y}}\boldsymbol{V}_{\mathscr{Y}}\operatorname{diag}[\boldsymbol{J}_{m_i}(y_i)]_{i=1}^{u}\boldsymbol{V}_{\mathscr{Y}}^{-1}\\&=\operatorname{diag}[\boldsymbol{Q}_{m_i}a_{i,m_i-1}(\boldsymbol{J}_{m_i}(y_i))\boldsymbol{J}_{m_i}(y_i)]_{i=1}^{u}\boldsymbol{V}_{\mathscr{Y}}^{-1}\\&=\operatorname{diag}[\boldsymbol{J}_{m_i}^{\mathrm{T}}(y_i)\boldsymbol{Q}_{m_i}a_{i,m_i-1}(\boldsymbol{J}_{m_i}(y_i))]_{i=1}^{u}\boldsymbol{V}_{\mathscr{Y}}^{-1}\\&=\operatorname{diag}[\boldsymbol{J}_{m_i}^{\mathrm{T}}(y_i)]_{i=1}^{u}\boldsymbol{W}_{\mathscr{Y}}.\end{aligned}$$

We can similarly obtain the second result. ■

Due to Lemmas 4.1 and 4.2 the relation (2.6′) amounts to

$$\begin{aligned}&(\boldsymbol{C}_a\otimes\boldsymbol{I}_p)\boldsymbol{H}_{mp,nq}[\boldsymbol{b}]-\boldsymbol{H}_{mp,nq}[\boldsymbol{b}](\boldsymbol{C}_b^{\mathrm{T}}\otimes\boldsymbol{I}_q)\\&=(\boldsymbol{W}_{\mathscr{Y}}\otimes\boldsymbol{I}_p)^{-1}(\boldsymbol{B}_-\boldsymbol{C}_--\boldsymbol{B}_+\boldsymbol{C}_+)(\boldsymbol{W}_{\mathscr{Z}}^{\mathrm{T}}\otimes\boldsymbol{I}_q)^{-1}.\end{aligned}\tag{4.6}$$

It is easy to verify by Remark (b) following Lemma 4.1 that

$$(\boldsymbol{W}_{\mathscr{Y}}\otimes\boldsymbol{I}_p)^{-1}\boldsymbol{B}_+=\boldsymbol{B}_+^H\quad\text{and}\quad\boldsymbol{C}_-(\boldsymbol{W}_{\mathscr{Z}}^{\mathrm{T}}\otimes\boldsymbol{I}_q)^{-1}=\boldsymbol{C}_-^H,$$

where $\boldsymbol{B}_+^H$, $\boldsymbol{C}_-^H$ are as in Eq. (3.7). Further on, we have by Eq. (4.5) that

$$(\boldsymbol{W}_{\mathscr{Y}}\otimes\boldsymbol{I}_p)^{-1}\boldsymbol{B}_-=\boldsymbol{H}_{mp,nq}[\boldsymbol{b}]\operatorname{col}(b_0\boldsymbol{I}_q,b_1\boldsymbol{I}_q,\cdots,b_{n-1}\boldsymbol{I}_q)+\boldsymbol{B}_-^H,$$

$$\boldsymbol{C}_+(\boldsymbol{W}_{\mathscr{Z}}^{\mathrm{T}}\otimes\boldsymbol{I}_q)^{-1}=\operatorname{row}(a_0\boldsymbol{I}_p,a_1\boldsymbol{I}_p,\cdots,a_{m-1}\boldsymbol{I}_p)\boldsymbol{H}_{mp,nq}[\boldsymbol{b}]+\boldsymbol{C}_+^H,$$

where $\boldsymbol{B}_-^H$, $\boldsymbol{C}_+^H$ are as in Eq. (3.7) but with $\boldsymbol{H}=\boldsymbol{H}_{mp,nq}[\boldsymbol{b}]$ and $\boldsymbol{R}=$

$\boldsymbol{H}_{m+n-1}$, the $(m+n)$th Markov parameter of $\Omega(\lambda)/A(\lambda)$ (see Eq. (4.2)). Thus, Eq. (4.6) can be rewritten as

$$(\boldsymbol{J}_m(0)\otimes\boldsymbol{I}_p)\boldsymbol{H}_{mp,nq}[\boldsymbol{b}]-\boldsymbol{H}_{mp,nq}[\boldsymbol{b}](\boldsymbol{J}_n^{\mathrm{T}}(0)\otimes\boldsymbol{I}_q)=\boldsymbol{B}_-^H\boldsymbol{C}_-^H-\boldsymbol{B}_+^H\boldsymbol{C}_+^H. \tag{4.7}$$

Similarly, Eqs. (2.8) ~ (2.11) have the becoming form

$$\boldsymbol{P}_H\boldsymbol{H}_{mp,nq}[\boldsymbol{b}]=\boldsymbol{C}_-^H, \tag{4.8}$$

$$\boldsymbol{U}_H\boldsymbol{H}_{mp,nq}[\boldsymbol{b}]=\boldsymbol{C}_+^H, \tag{4.9}$$

$$\boldsymbol{H}_{mp,nq}[\boldsymbol{b}]\widetilde{\boldsymbol{P}}_H=\boldsymbol{B}_+^H, \tag{4.10}$$

$$\boldsymbol{H}_{mp,nq}[\boldsymbol{b}]\widetilde{\boldsymbol{U}}_H=\boldsymbol{B}_-^H, \tag{4.11}$$

where $\boldsymbol{C}_+^H$ and $\boldsymbol{B}_-^H$ are as in Eq. (4.7) with $\boldsymbol{R}=\boldsymbol{H}_{m+n-1}$ therein, and

$$\begin{aligned}
&\boldsymbol{P}_H=\boldsymbol{P}(\boldsymbol{W}_{\mathscr{Y}}\otimes\boldsymbol{I}_p)=\mathrm{row}(\boldsymbol{P}_0,\boldsymbol{P}_1,\cdots,\boldsymbol{P}_{m-1}),\\
&\boldsymbol{U}_H=\boldsymbol{U}(\boldsymbol{W}_{\mathscr{Y}}\otimes\boldsymbol{I}_p)-\mathrm{row}[a_i\boldsymbol{I}_p]_{i=0}^{m-1}=\mathrm{row}(\boldsymbol{U}_0,\boldsymbol{U}_1,\cdots,\boldsymbol{U}_{m-1}),\\
&\widetilde{\boldsymbol{P}}_H=(\boldsymbol{W}_{\mathscr{Z}}^{\mathrm{T}}\otimes\boldsymbol{I}_q)\widetilde{\boldsymbol{P}}=\mathrm{col}(\widetilde{\boldsymbol{P}}_0,\widetilde{\boldsymbol{P}}_1,\cdots,\widetilde{\boldsymbol{P}}_{n-1}),\\
&\widetilde{\boldsymbol{U}}_H=(\boldsymbol{W}_{\mathscr{Z}}^{\mathrm{T}}\otimes\boldsymbol{I}_q)\widetilde{\boldsymbol{U}}-\mathrm{col}[b_j\boldsymbol{I}_q]_{j=0.}^{n-1}=\mathrm{col}(\widetilde{\boldsymbol{U}}_0,\widetilde{\boldsymbol{U}}_1,\cdots,\widetilde{\boldsymbol{U}}_{n-1}).
\end{aligned} \tag{4.12}$$

These mean that if we put

$$\begin{aligned}
&\widetilde{\boldsymbol{P}}_H(\lambda)=\mathrm{row}(\boldsymbol{I}_q,\lambda\boldsymbol{I}_q,\cdots,\lambda^{n-1}\boldsymbol{I}_q)\widetilde{\boldsymbol{P}}_H\in\mathbf{C}^{q\times p}[\lambda],\\
&\widetilde{\boldsymbol{U}}_H(\lambda)=-\lambda^n\boldsymbol{I}_q+\mathrm{row}(\boldsymbol{I}_q,\lambda\boldsymbol{I}_q,\cdots,\lambda^{n-1}\boldsymbol{I}_q)\widetilde{\boldsymbol{U}}_H\in\mathbf{C}^{q\times q}[\lambda],\\
&\boldsymbol{P}_H(\lambda)=\boldsymbol{P}_H\mathrm{col}(\boldsymbol{I}_p,\lambda\boldsymbol{I}_p,\cdots,\lambda^{m-1}\boldsymbol{I}_p)\in\mathbf{C}^{q\times p}[\lambda],\\
&\boldsymbol{U}_H(\lambda)=-\lambda^m\boldsymbol{I}_p+\boldsymbol{U}_H\mathrm{col}(\boldsymbol{I}_p,\lambda\boldsymbol{I}_p,\cdots,\lambda^{m-1}\boldsymbol{I}_p)\in\mathbf{C}^{p\times p}[\lambda]
\end{aligned} \tag{4.13}$$

where $\boldsymbol{P}_H,\boldsymbol{U}_H,\widetilde{\boldsymbol{P}}_H,\widetilde{\boldsymbol{U}}_H$ are the solutions to Eps. (4.8) ~ (4.11), respectively, then

$$\begin{aligned}
\widetilde{\boldsymbol{P}}_H(\lambda)&=\sum_{j=1}^{v}\sum_{l=0}^{n_j-1}\widetilde{\boldsymbol{P}}_{jl}b_{jl}(\lambda),\\
\widetilde{\boldsymbol{U}}_H(\lambda)&=-b(\lambda)\boldsymbol{I}_q+\sum_{j=1}^{v}\sum_{l=0}^{n_j-1}\widetilde{\boldsymbol{U}}_{jl}b_{jl}(\lambda),\\
\boldsymbol{P}_H(\lambda)&=\sum_{i=1}^{u}\sum_{k=0}^{m_i-1}\boldsymbol{P}_{ik}a_{ik}(\lambda),\\
\boldsymbol{U}_H(\lambda)&=-a(\lambda)\boldsymbol{I}_p+\sum_{i=1}^{u}\sum_{k=0}^{m_i-1}\boldsymbol{U}_{ik}a_{ik}(\lambda),
\end{aligned} \tag{4.14}$$

where $\widetilde{\boldsymbol{P}}_{jl}, \widetilde{\boldsymbol{U}}_{jl}, \boldsymbol{P}_{ik}, \boldsymbol{U}_{ik}$ are determined by the solutions to Eqs. (2.8) ~ (2.11), and $a_{ik}(\lambda), b_{jl}(\lambda)$ are as in Eq. (2.23).

Theoretically, the fundamental matrix Eqs. (4.8) ~ (4.11) associated with $\boldsymbol{H}_{mp,nq}[\boldsymbol{b}]$ are simpler and easier to comprehend than Eps. (2.8) ~ (2.11). For instance, in the case of $m=n$ and $p=q=1$, we have that the four Eqs. (4.8) ~ (4.11) amount to two Eqs. (4.8) and (4.9) only (but Eqs. (2.8) ~ (2.11) do not always amount to Eqs. (2.8) and (2.9)). Computationally, many of the advantageous properties of different methods of solving Hankel and Toeplitz systems carry over without modification to Eqs. (4.8) ~ (4.11).

§ 4.3 *The third inversion theorem for* $\mathbf{L} \in \mathscr{L}$

In summary, we have the third inversion theorem for generalized block Loewner matrices in this paper as follows.

Theorem 4.3 *Let the GRMIP (2.4) be given, and let* $\boldsymbol{b}=(\boldsymbol{H}_0, \boldsymbol{H}_1, \cdots, \boldsymbol{H}_{m+n-2})$ *be its block Hankel vector. Let* $\mathbf{L}=\mathbf{L}_{\mathscr{Y}\mathscr{Z}} \in \mathscr{L}$ *be of the form (2.3) with* $mp=nq$*. Then the following are equivalent:*

(a) $\boldsymbol{L}_{\mathscr{Y}\mathscr{Z}}$ *is invertible.*

(b) $\boldsymbol{H}=(\boldsymbol{H}_{i+j})_{i,j=0}^{m-1,n-1}$ *is invertible.*

(c) *Eqs.* (4.8) *and* (4.9) *are solvable.*

(d) *Eqs.* (4.10) *and* (4.11) *are solvable.*

In this case, $\boldsymbol{L}_{\mathscr{Y}\mathscr{Z}}^{-1}$ *has the form*

$$\boldsymbol{L}_{\mathscr{Y}\mathscr{Z}}^{-1}=(\mathbf{W}_{\mathscr{Z}}^{\mathrm{T}} \otimes \boldsymbol{I}_q)^{-1} \boldsymbol{H}^{-1} (\mathbf{W}_{\mathscr{Y}} \otimes \boldsymbol{I}_p)^{-1}, \tag{4.15}$$

where $\mathbf{W}_{\mathscr{Y}}, \mathbf{W}_{\mathscr{Z}}$ *are as in Eq.* (2.23). *Moreover,* $\boldsymbol{H}^{-1}$ *is identical with the generalized Bezoutian associated with the quadruple* $\{\widetilde{\boldsymbol{P}}_H(\lambda), \widetilde{\boldsymbol{U}}_H(\lambda); \boldsymbol{U}_H(\lambda), \boldsymbol{P}_H(\lambda)\}$

$$\boldsymbol{H}^{-1}=\mathbb{B}(\widetilde{\boldsymbol{P}}_H(\lambda), \widetilde{\boldsymbol{U}}_H(\lambda); \boldsymbol{U}_H(\lambda), \boldsymbol{P}_H(\lambda)), \tag{4.16}$$

where $\widetilde{\boldsymbol{P}}_H(\lambda), \widetilde{\boldsymbol{U}}_H(\lambda), \boldsymbol{P}_H(\lambda), \boldsymbol{U}_H(\lambda)$ *are as in Eq.* (4.13).

To prove Theorem 4.3, it is sufficient to justfy Eq. (4.16). It was first recognized in [35] that the inverse of an invertible block Hankel matrix with square blocks can be represented as the generalized Bezoutian of quadruple of certain matrix polynomials (see also [19] for another proof). This result for non-square block case can be found in [26]. In the following proof, we shall present a simpler derivation of Eq. (4.16) by considering it as an immediate consequence of Lemma 4.1 and Corollary 2.5.

Proof of Theorem 4.3 Let $\boldsymbol{b}=(\boldsymbol{H}_0,\boldsymbol{H}_1,\cdots,\boldsymbol{H}_{m+n-2})$ be the block Hankel vector of the GRMIP (2.4). By choosing $\mathscr{Y}_0$ and $\mathscr{Z}_0$ as in Eq. (2.2) but with $y_i\neq z_j$ for all $i\neq j$, and $u=m, v=n, m_1=\cdots=m_u=n_1=\cdots=n_v=1$

$$\mathscr{Y}_0=\{y_1,y_2,\cdots,y_m\},\quad \mathscr{Z}_0=\{z_1,z_2,\cdots,z_n\} \tag{4.17}$$

as was mentioned in the Remark (a) following Lemma 4.1, we obtain a unique GRMIP of the form (up to a constant matrix summand $\boldsymbol{E}$ for all $\boldsymbol{C}_{i0}$ and $\boldsymbol{D}_{j0}$)

$$F(y_i)=\boldsymbol{C}_{i0}, i=1,2,\cdots,m,\quad F(z_j)=\boldsymbol{D}_{j0}, j=1,2,\cdots,n, \tag{4.18}$$

for which $\boldsymbol{b}=(\boldsymbol{H}_0,\boldsymbol{H}_1,\cdots,\boldsymbol{H}_{m+n-2})$ is the block Hankel vector.

Then, by Lemma 4.1, we have

$$\boldsymbol{L}_{\mathscr{Y}_0\mathscr{Z}_0}=\left[\frac{\boldsymbol{C}_{i0}-\boldsymbol{D}_{j0}}{y_i-z_j}\right]_{i,j=1}^{m,n}=(\boldsymbol{W}_{\mathscr{Y}_0}\otimes\boldsymbol{I}_p)(\boldsymbol{H}_{i+j})_{i,j=0}^{m-1,n-1}(\boldsymbol{W}_{\mathscr{Z}_0}^{\mathrm{T}}\otimes\boldsymbol{I}_q). \tag{4.19}$$

If, in addition, $mp=nq$, and if $\boldsymbol{H}=(\boldsymbol{H}_{i+j})_{i,j=0}^{m-1,n-1}$ is invertible (and so is $\boldsymbol{L}_{\mathscr{Y}_0\mathscr{Z}_0}$), then, by Corollary 2.5,

$$\boldsymbol{H}^{-1}=(\boldsymbol{W}_{\mathscr{Z}_0}^{\mathrm{T}}\otimes\boldsymbol{I}_q)\left[\frac{\widetilde{\boldsymbol{P}}_{j0}\boldsymbol{U}_{i0}-\widetilde{\boldsymbol{U}}_{j0}\boldsymbol{P}_{i0}}{z_j-y_i}\right]_{j,i=1}^{n,m}(\boldsymbol{W}_{\mathscr{Y}_0}\otimes\boldsymbol{I}_p), \tag{4.20}$$

where $\mathrm{row}(\boldsymbol{P}_{10},\boldsymbol{P}_{20},\cdots,\boldsymbol{P}_{m0})$, $\mathrm{row}(\boldsymbol{U}_{10},\boldsymbol{U}_{20},\cdots,\boldsymbol{U}_{m0})$, $\mathrm{col}(\widetilde{\boldsymbol{P}}_{10},\widetilde{\boldsymbol{P}}_{20},\cdots,\widetilde{\boldsymbol{P}}_{n0})$ and $\mathrm{col}(\widetilde{\boldsymbol{U}}_{10},\widetilde{\boldsymbol{U}}_{20},\cdots,\widetilde{\boldsymbol{U}}_{n0})$ are the solutions of Eqs. (2.12)~(2.15) with $\boldsymbol{C}_{i0}$, $\boldsymbol{D}_{j0}$ determined by Eq. (4.18). Let

$$\hat{a}(\lambda)=\prod_{i=1}^{m}(\lambda-y_i), a_i(\lambda)=\hat{a}(\lambda)/(\lambda-y_i), i=1,2,\cdots,m,$$

$$\hat{b}(\lambda)=\prod_{j=1}^{v}(\lambda-z_j), b_j(\lambda)=\hat{b}(\lambda)/(\lambda-z_j), j=1,2,\cdots,n, \tag{4.21}$$

and

$$\widetilde{U}(\lambda) = -\hat{b}(\lambda)I_q + \sum_{j=1}^{n}\widetilde{U}_j b_j(\lambda),\quad P(\lambda) = \sum_{j=1}^{n}P_i a_i(\lambda),$$
$$\widetilde{P}(\lambda) = \sum_{j=1}^{n}\widetilde{P}_j b_j(\lambda),\quad U(\lambda) = -\hat{a}(\lambda)I_p + \sum_{i=1}^{m}U_i a_i(\lambda). \tag{4.22}$$

Then Eq. (4.20) amounts to

$$H^{-1} = (W_{\mathscr{Z}_0}^{\mathrm{T}} \otimes I_q)\operatorname{diag}\left[\frac{1}{b_j(z_j)}I_q\right]_{j=1}^{n} \times$$
$$\left[\frac{\widetilde{P}_{j0}b_j(z_j)U_{i0}a_i(y_i) - \widetilde{U}_{j0}b_j(z_j)P_{i0}a_i(y_i)}{z_j - y_i}\right]_{j,i=1}^{n,m} \times$$
$$\operatorname{diag}\left[\frac{1}{a_i(y_i)}I_p\right]_{i=1}^{m}(W_{\mathscr{Y}_0} \otimes I_p)$$
$$= (V_{\mathscr{Z}_0}^{\mathrm{T}} \otimes I_q)^{-1}\left[\frac{\widetilde{P}(z_j)U(y_i) - \widetilde{U}(z_j)P(y_i)}{z_j - y_i}\right]_{j,i=1}^{n,m}(V_{\mathscr{Y}_0} \otimes I_p)^{-1}.$$

But we have ([12], Theorem 4.1) that $\widetilde{P}(\lambda)U(\lambda) = \widetilde{U}(\lambda)P(\lambda)$ (which can be obtained from Eqs. (4.24) and (5.7) as well), so that, by definition [2] of a generalized Bezoutian of the quadruple $\{\widetilde{P}(\lambda), \widetilde{U}(\lambda); U(\lambda), P(\lambda)\}$,

$$\left[\frac{\widetilde{P}(z_j)U(y_i) - \widetilde{U}(z_j)P(y_i)}{z_j - y_i}\right]_{j,i=1}^{n,m}$$
$$= (V_{\mathscr{Z}_0}^{\mathrm{T}} \otimes I_q)\mathbb{B}(\widetilde{P}(\lambda), \widetilde{U}(\lambda); U(\lambda), P(\lambda))(V_{\mathscr{Y}_0} \otimes I_p).$$

Thus we have shown that

$$H^{-1} = \mathbb{B}(\widetilde{P}(\lambda), \widetilde{U}(\lambda); U(\lambda), P(\lambda)). \tag{4.23}$$

On the other hand, Eqs. (4.12)～(4.14) with $u=m$, $v=n$, all $m_i = n_j = 1$ imply

$$\widetilde{P}(\lambda) = \widetilde{P}_H(\lambda), \widetilde{U}(\lambda) = \widetilde{U}_H(\lambda), P(\lambda) = P_H(\lambda), U(\lambda) = U_H(\lambda), \tag{4.24}$$

where $\widetilde{P}_H(\lambda)$, $\widetilde{U}_H(\lambda)$, $P_H(\lambda)$, $U_H(\lambda)$ are as in Eq. (4.13). Thus, Eq. (4.16) follows from Eqs. (4.23) and (4.24). ■

Remarks (a) The choice of $R = H_{m+n-1}$ in C_+^H and B_-^H of Eq. (3.7) is not significant for computing H^{-1}. Indeed, if, for instance, we take $R = O$ in place of H_{m+n-1} and denote the corresponding solutions of Eqs. (4.9) and (4.11) by U_H^0 and $\widetilde{U}_H^0$, respectively, it is obvious by the construction of Eqs. (4.8)～(4.11) that

$$\boldsymbol{U}_H=\boldsymbol{U}_{\boldsymbol{H}}^0+\boldsymbol{H}_{m+n-1}\boldsymbol{P}_H \quad \text{and} \quad \widetilde{\boldsymbol{U}}_H=\boldsymbol{U}_H^0+\widetilde{\boldsymbol{P}}_H\boldsymbol{H}_{m+n-1}$$

or equivalently,

$$\begin{aligned}\boldsymbol{U}_H(\lambda)&=\boldsymbol{U}_{\boldsymbol{H}}^0(\lambda)+\boldsymbol{H}_{m+n-1}\boldsymbol{P}_H(\lambda),\\ \widetilde{\boldsymbol{U}}_H(\lambda)&=\widetilde{\boldsymbol{U}}_H^0(\lambda)+\widetilde{\boldsymbol{P}}_{\boldsymbol{H}}(\lambda)\boldsymbol{H}_{m+n-1}.\end{aligned} \tag{4.25}$$

Thus, by Eq. (4.16) and by Remark 2.3 of [2], we have also

$$\begin{aligned}\boldsymbol{H}^{-1}&=\mathbb{B}(\widetilde{\boldsymbol{P}}_H(\lambda),\widetilde{\boldsymbol{U}}_H^0(\lambda)+\widetilde{\boldsymbol{P}}_{\boldsymbol{H}}(\lambda)\boldsymbol{H}_{m+n-1};\boldsymbol{U}_{\boldsymbol{H}}^0(\lambda)+\boldsymbol{H}_{m+n-1}\boldsymbol{P}_H(\lambda),\boldsymbol{P}_H(\lambda))\\ &=\mathbb{B}(\widetilde{\boldsymbol{P}}_H(\lambda),\widetilde{\boldsymbol{U}}_H^0(\lambda);\boldsymbol{U}_H^0(\lambda)\boldsymbol{P}_H(\lambda)).\end{aligned} \tag{4.26}$$

This means that Eq. (4.16) is independent of the choice of $\boldsymbol{R}\in\mathbf{C}^{p\times q}$, more precisely, Eq. (4.16) holds for each quadruple $\{\widetilde{\boldsymbol{P}}_H(\lambda),\widetilde{\boldsymbol{U}}_H(\lambda);\boldsymbol{U}_H(\lambda),\boldsymbol{P}_H(\lambda)\}$ of polynomial matrices as long as $\boldsymbol{P}_H,\boldsymbol{U}_H,\widetilde{\boldsymbol{P}}_H,\widetilde{\boldsymbol{U}}_H$ solve Eqs. (4.8)～(4.11) but with an arbitrary $\boldsymbol{R}\in\mathbf{C}^{p\times q}$ in place of $\boldsymbol{H}_{m+n-1}$ appearing in $\boldsymbol{C}_+^H$ and $\boldsymbol{B}_-^H$ of Eq. (3.7). Compare Section 4 of [19] for the square block case: $p=q$.

(b) Since $\boldsymbol{H}=\boldsymbol{H}_{mp,nq}[\boldsymbol{b}]$ is independent of the choice of sequences $\mathscr{Y}$, $\mathscr{Z}$ in Eq. (2.2), the formula (4.15) gives indeed the inverses of all invertible $\boldsymbol{L}\in\mathscr{L}$ in which $\boldsymbol{H}^{-1}$ can be viewed as an invariant, a striking and useful fact. Further on, by Eq. (2.24), $\boldsymbol{W}_{\mathscr{Y}}^{-1}$ and $\boldsymbol{W}_{\mathscr{Z}}^{-1}$ can be expressed as

$$\begin{aligned}\boldsymbol{W}_{\mathscr{Y}}^{-1}&=\boldsymbol{V}_{\mathscr{Y}}\mathrm{diag}[a_{i,m_i-1}^{-1}(\boldsymbol{J}_{m_i}(y_i))\boldsymbol{Q}_{m_i}]_{i=1}^{u},\\ \boldsymbol{W}_{\mathscr{Z}}^{-1}&=\boldsymbol{V}_{\mathscr{Z}}\mathrm{diag}[b_{j,n_j-1}^{-1}(\boldsymbol{J}_{n_j}(z_j))\boldsymbol{Q}_{n_j}]_{j=1}^{v},\end{aligned} \tag{4.27}$$

where $\boldsymbol{V}_{\mathscr{Y}},\boldsymbol{V}_{\mathscr{Z}}$ are as in Eqs. (2.18) and (2.22), respectively.

(c) From Eq. (3.15) we deduce that

$$\boldsymbol{H}^{-1}=\begin{bmatrix}\widetilde{\boldsymbol{U}}_0 & \boldsymbol{O} & \cdots & \boldsymbol{O}\\ \widetilde{\boldsymbol{U}}_1 & \widetilde{\boldsymbol{U}}_0 & \ddots & \vdots\\ \vdots & \ddots & \ddots & \boldsymbol{O}\\ \widetilde{\boldsymbol{U}}_{n-1} & \cdots & \widetilde{\boldsymbol{U}}_1 & \widetilde{\boldsymbol{U}}_0\end{bmatrix}\begin{bmatrix}\boldsymbol{P}_1 & \cdots & \boldsymbol{P}_{m-1} & \boldsymbol{O}\\ \vdots & \cdot^{\cdot^{\cdot}} & \cdot^{\cdot^{\cdot}} & \vdots\\ \boldsymbol{P}_{n-1} & \boldsymbol{O} & \cdots & \boldsymbol{O}\\ \boldsymbol{O} & \boldsymbol{O} & \cdots & \boldsymbol{O}\end{bmatrix}-$$

$$\begin{bmatrix} \widetilde{P}_0 & O & \cdots & O \\ \widetilde{P}_1 & \widetilde{P}_0 & \cdots & O \\ \vdots & \ddots & \ddots & \vdots \\ \widetilde{P}_{n-1} & \widetilde{P}_{n-2} & \cdots & \widetilde{P}_{n-m} \end{bmatrix} \begin{bmatrix} U_1 & \cdots & U_{m-1} & -I_p \\ \vdots & \iddots & \iddots & O \\ U_{m-1} & -I_p & \iddots & \vdots \\ -I_p & O & \cdots & O \end{bmatrix}, \text{if } n \geqslant m, \tag{4.28a}$$

$$H^{-1} = \begin{bmatrix} \widetilde{U}_0 & O & \cdots & O \\ \widetilde{U}_1 & \widetilde{U}_0 & \ddots & \vdots \\ \vdots & \ddots & \ddots & O \\ \widetilde{U}_{n-1} & \cdots & \widetilde{U}_1 & \widetilde{U}_0 \end{bmatrix} \begin{bmatrix} P_1 & \cdots & \cdots & P_{m-1} & O & \cdots & O \\ \vdots & & \iddots & \iddots & \vdots & \iddots & \vdots \\ P_{n-1} & \cdots & P_{m-1} & O & O & \cdots & O \\ O & \cdots & O & \cdots & O & \cdots & O \end{bmatrix} -$$

$$\begin{bmatrix} \widetilde{P}_0 & O & \cdots & O & \cdots & O \\ \widetilde{P}_1 & \widetilde{P}_0 & \ddots & O & \cdots & O \\ \vdots & & & \vdots & & \vdots \\ \widetilde{P}_{n-1} & \cdots & \widetilde{P}_0 & O & \cdots & O \end{bmatrix} \begin{bmatrix} U_1 & \cdots & U_{m-1} & -I_p \\ \vdots & \iddots & \iddots & O \\ U_{m-1} & -I_p & \iddots & \vdots \\ -I_p & O & \cdots & O \end{bmatrix}, \text{if } n < m, \tag{4.28b}$$

which consists of nothing more than an expression for the generalized Bezoution $\mathbb{B}$ with non-square blocks in Eq. (4.16).

(d) When we deal with the matrix Nevanlinna-Pick interpolation in the upper half plane, the Hermitian positive-definite generalized block Loewner matrices are encountered. The approach we described so far can be used smoothly to this case. In particular, we then have that $p=q$ and

$$\mathscr{X} = \overline{\mathscr{Y}} = \{\underbrace{\overline{y}_1, \cdots, \overline{y}_1}_{m_1 \text{ times}}, \cdots, \underbrace{\overline{y}_u, \cdots, \overline{y}_u}_{m_u \text{ times}}\}$$

in Eq. (2.2) (i. e., $m=n=N/2$) and thus Eq. (4.15) is of the form

$$L_{\mathscr{Y}\overline{\mathscr{Y}}}^{-1} = (W_{\mathscr{Y}}^* \otimes I_p)^{-1} H_{mp,mp}^{-1}[b](W_{\mathscr{Y}} \otimes I_p)^{-1},$$

where $H_{mp,mp}[b]$ is a Hermitian block Hankel matrix: $H_i = H_i^* \ \forall i$, and its inverse of Eq. (4.16) can be rewritten as

$$H_{mp,mp}^{-1}[b] = \mathbb{B}(P_H^*(\lambda), U_H^*(\lambda); U_H(\lambda), P_H(\lambda)).$$

Then there will be a computational reduction considerably.

§ 5. The connections with minimal partial realization and matrix rational interpolation

In this section we proceed to give a parametrization of the set of all the minimal, proper rational interpolants to the GRMIP (2.4) under the assumption that a matrix $L \in \mathscr{L}$ is invertible. Here (and in the sequel) a matrix whose elements are all rational functions is called a rational matrix function. A rational matrix function is said proper [strictly proper, resp.] if each element is proper [strictly proper]. A rational interpolant to the GRMIP (2.4) is minimal if it is of least McMillan degree. By definition, the McMillan degree of a proper rational matrix function $\psi(\lambda)$, denoted by $\delta(\psi)$, is defined as the dimension of the state space of a minimal realization of $\psi(\lambda)$ (see, e.g., [29] for further details).

§ 5.1 *The minimal partial realizations of the block Hankel vector b*

Let now the GRMIP (2.4) be given, and let $\boldsymbol{b} = (\boldsymbol{H}_0, \boldsymbol{H}_1, \cdots, \boldsymbol{H}_{m+n-2})$ be its block Hankel vector with each $\boldsymbol{H}_j \in \mathbf{C}^{p\times q}$. Suppose that $L_{\mathscr{Y}\mathscr{X}} \in \mathscr{L}$ of the form (2.3) is invertible, so that, by Lemma 4.1, then and only then the block Hankel matrix

$$\boldsymbol{H} \equiv \boldsymbol{H}_{mp,nq}[\boldsymbol{b}] = (\boldsymbol{H}_{i+j})_{i,j=0}^{m-1,n-1} \in \mathbf{C}^{mp\times nq} \tag{5.1}$$

is also invertible. In order to obtain the set of aforementioned interpolants to GRMIP (2.4), we first treat the minimal partial realization problem of the block Hankel vector $\boldsymbol{b}$, that is, to reconstruct all the strictly proper rational matrix functions, of least McMillan degree, which have $\boldsymbol{H}_0, \boldsymbol{H}_1, \cdots, \boldsymbol{H}_{m+n-2}$ as the first $m+n-1$ Markov parameters. In the case when $\boldsymbol{H}$ given in Eq. (5.1) is invertible, this least McMillan degree is exactly $mp(=nq)$. Let $\widetilde{\boldsymbol{P}}_H(\lambda)$, $\widetilde{\boldsymbol{U}}_H(\lambda)$, $\boldsymbol{P}_H(\lambda)$ and $\boldsymbol{U}_H(\lambda)$ be as in Eq. (4.13), and be rewritten in the form (by Eq. (4.12))

$$\widetilde{\boldsymbol{P}}_H(\lambda) = \sum_{j=0}^{n-1} \widetilde{\boldsymbol{P}}_j \lambda^j, \quad \widetilde{\boldsymbol{P}}_j \in \mathbf{C}^{q\times p},$$

$$\widetilde{\boldsymbol{U}}_H(\lambda) = -\lambda^n \boldsymbol{I}_q + \sum_{j=0}^{n-1} \widetilde{\boldsymbol{U}}_j \lambda^j, \quad \widetilde{\boldsymbol{U}}_j \in \mathbf{C}^{q\times q},$$

$$P_H(\lambda) = \sum_{i=0}^{m-1} P_i\lambda^i, P_i \in \mathbf{C}^{q\times p},$$

$$U_H(\lambda) = -\lambda^m I_p + \sum_{i=0}^{m-1} U_i\lambda^j, U_i \in \mathbf{C}^{p\times p}. \quad (5.2)$$

It is easy to verify that there exist polynomial matrices $\widetilde{M}_H(\lambda)$, $\widetilde{N}_H(\lambda)$, $M_H(\lambda)$ and $N_H(\lambda)$ such that

$$U_H^{-1}(\lambda)N_H(\lambda) = \widetilde{N}_H(\lambda)\widetilde{U}_H^{-1}(\lambda) = \frac{H_0}{\lambda} + \cdots + \frac{H_{m+n-2}}{\lambda^{m+n-1}} + \frac{H_{m+n-1}}{\lambda^{m+n}} + \cdots,$$

$$M_H(\lambda) = P_H(\lambda)\left(\frac{H_0}{\lambda} + \cdots + \frac{H_{m+n-2}}{\lambda^{m+n-1}} + \frac{H_{m+n-1}}{\lambda^{m+n}} + \cdots\right),$$

$$\widetilde{M}_H(\lambda) = \left(\frac{H_0}{\lambda} + \cdots + \frac{H_{m+n-2}}{\lambda^{m+n-1}} + \frac{H_{m+n-1}}{\lambda^{m+n}} + \cdots\right)\widetilde{P}_H(\lambda).$$

Moreover, these polynomial matrices are defined via

$$\widetilde{M}_H(\lambda) = [H_0, \cdots, H_{n-2}] \begin{bmatrix} \widetilde{P}_1 & \cdots & \widetilde{P}_{n-1} \\ \vdots & \iddots & \\ \widetilde{P}_{n-1} & & O \end{bmatrix} \mathrm{col}(I_p, \lambda I_p, \cdots, \lambda^{n-2} I_p),$$

$$\widetilde{N}_H(\lambda) = [H_0, \cdots, H_{n-1}] \begin{bmatrix} \widetilde{U}_1 & \cdots & \widetilde{U}_{n-1} & -I_q \\ \vdots & \iddots & \iddots & \\ \widetilde{U}_{n-1} & \iddots & & \\ I_q & & & O \end{bmatrix} \mathrm{col}(I_q, \lambda I_q, \cdots, \lambda^{n-1} I_q),$$

$$M_H(\lambda) = \mathrm{row}(I_q, \lambda I_q, \cdots, \lambda^{m-2} I_q) \begin{bmatrix} P_1 & \cdots & P_{m-1} \\ \vdots & \iddots & \\ P_{m-1} & & O \end{bmatrix} \begin{bmatrix} H_0 \\ \vdots \\ H_{m-2} \end{bmatrix},$$

$$N_H(\lambda) = \mathrm{row}(I_p, \lambda I_p, \cdots, \lambda^{m-1} I_p) \begin{bmatrix} U_1 & \cdots & U_{m-1} & -I_p \\ \vdots & \iddots & \iddots & \\ U_{m-1} & \iddots & & \\ -I_p & & & O \end{bmatrix} \begin{bmatrix} H_0 \\ \vdots \\ H_{m-1} \end{bmatrix}.$$

(5.3)

In particular, from these we deduce

$$\begin{aligned}&\boldsymbol{N}_H(\lambda)\widetilde{\boldsymbol{U}}_H(\lambda)=\boldsymbol{U}_H(\lambda)\widetilde{\boldsymbol{N}}_H(\lambda),\\&\boldsymbol{P}_H(\lambda)\widetilde{\boldsymbol{M}}_H(\lambda)=\boldsymbol{M}_H(\lambda)\widetilde{\boldsymbol{P}}_H(\lambda),\\&-\boldsymbol{N}_H(\lambda)\widetilde{\boldsymbol{P}}_H(\lambda)+\boldsymbol{U}_H(\lambda)\widetilde{\boldsymbol{M}}_H(\lambda)=\boldsymbol{I}_p,\\&-\boldsymbol{P}_H(\lambda)\widetilde{\boldsymbol{N}}_H(\lambda)+\boldsymbol{M}_H(\lambda)\widetilde{\boldsymbol{U}}_H(\lambda)=\boldsymbol{I}_q\end{aligned}\tag{5.4}$$

or equivalently,

$$\begin{bmatrix}\boldsymbol{U}_H(\lambda) & -\boldsymbol{N}_H(\lambda)\\ \boldsymbol{P}_H(\lambda) & -\boldsymbol{M}_H(\lambda)\end{bmatrix}\begin{bmatrix}\widetilde{\boldsymbol{M}}_H(\lambda) & -\widetilde{\boldsymbol{N}}_H(\lambda)\\ \widetilde{\boldsymbol{P}}_H(\lambda) & -\widetilde{\boldsymbol{U}}_H(\lambda)\end{bmatrix}=\begin{bmatrix}\boldsymbol{I}_p & \boldsymbol{O}\\ \boldsymbol{O} & \boldsymbol{I}_q\end{bmatrix}.\tag{5.5}$$

Since clearly Eq. (5.5) implies

$$\begin{bmatrix}\widetilde{\boldsymbol{M}}_H(\lambda) & -\widetilde{\boldsymbol{N}}_H(\lambda)\\ \widetilde{\boldsymbol{P}}_H(\lambda) & -\widetilde{\boldsymbol{U}}_H(\lambda)\end{bmatrix}\begin{bmatrix}\boldsymbol{U}_H(\lambda) & -\boldsymbol{N}_H(\lambda)\\ \boldsymbol{P}_H(\lambda) & -\boldsymbol{M}_H(\lambda)\end{bmatrix}=\begin{bmatrix}\boldsymbol{I}_p & \boldsymbol{O}\\ \boldsymbol{O} & \boldsymbol{I}_q\end{bmatrix},\tag{5.6}$$

we have

$$\begin{aligned}&\widetilde{\boldsymbol{N}}_H(\lambda)\boldsymbol{M}_H(\lambda)=\widetilde{\boldsymbol{M}}_H(\lambda)\boldsymbol{N}_H(\lambda),\\&\widetilde{\boldsymbol{U}}_H(\lambda)\boldsymbol{P}_H(\lambda)=\widetilde{\boldsymbol{P}}_H(\lambda)\boldsymbol{U}_H(\lambda),\\&-\widetilde{\boldsymbol{N}}_H(\lambda)\boldsymbol{P}_H(\lambda)+\widetilde{\boldsymbol{M}}_H(\lambda)\boldsymbol{U}_H(\lambda)=\boldsymbol{I}_p,\\&\widetilde{\boldsymbol{U}}_H(\lambda)\boldsymbol{M}_H(\lambda)-\widetilde{\boldsymbol{P}}_H(\lambda)\boldsymbol{N}_H(\lambda)=\boldsymbol{I}_q.\end{aligned}\tag{5.7}$$

Further on, it follows from Eq. (5.4) that $\boldsymbol{U}_H(\lambda)$ and $\boldsymbol{N}_H(\lambda)$ is left coprime, and $\widetilde{\boldsymbol{N}}_H(\lambda)$ and $\widetilde{\boldsymbol{U}}_H(\lambda)$ is right coprime. Since $\delta(\boldsymbol{U}_H^{-1}\boldsymbol{N}_H)=\delta(\widetilde{\boldsymbol{N}}_H\widetilde{\boldsymbol{U}}_H^{-1})=mp=nq$, $\boldsymbol{U}_H^{-1}(\lambda)\boldsymbol{N}_H(\lambda)$ and $\widetilde{\boldsymbol{N}}_H(\lambda)\widetilde{\boldsymbol{U}}_H^{-1}(\lambda)$ are respectively the left coprime and the right coprime polynomial matrix fractions of some minimal partial realization of $\boldsymbol{b}=(\boldsymbol{H}_0,\boldsymbol{H}_1,\cdots,\boldsymbol{H}_{m+n-2})$.

Notice [8] that given $\widetilde{\boldsymbol{P}}_H(\lambda)$, $\widetilde{\boldsymbol{U}}_H(\lambda)$, $\boldsymbol{P}_H(\lambda)$ and $\boldsymbol{U}_H(\lambda)$ as in Eq. (5.2), which are such that both $\widetilde{\boldsymbol{U}}_H^{-1}\widetilde{\boldsymbol{P}}_H$ and $\boldsymbol{P}_H\boldsymbol{U}_H^{-1}$ are strictly proper rational, then $\widetilde{\boldsymbol{M}}_H(\lambda)$, $\widetilde{\boldsymbol{N}}_H(\lambda)$, $\boldsymbol{M}_H(\lambda)$, $\boldsymbol{N}_H(\lambda)$ defined by Eq. (5.3) form the only polynomial matrix sequence such that Eq. (5.4) or Eq. (5.5) holds true.

By summarizing the proceeding analysis and by applying Theorem 3.5 of [8], we can state the following result (a precursor thereof for the square block case: $p=q$, can be found in ([19], Theorem 4.1)).

Theorem 5.1 *Let* $\boldsymbol{b}=(\boldsymbol{H}_0,\boldsymbol{H}_1,\cdots,\boldsymbol{H}_{m+n-2})$ *be the block Hankel vector of the GRMIP (2.4) such that* $\boldsymbol{H}$ *of Eq. (5.1) is invertible. Let* $\boldsymbol{P}_H(\lambda),\boldsymbol{U}_H(\lambda),\widetilde{\boldsymbol{P}}_H(\lambda)$ *and* $\widetilde{\boldsymbol{U}}_H(\lambda)$ *be given by Eq. (5.2), and the quantities* $\boldsymbol{M}_H(\lambda),\boldsymbol{N}_H(\lambda),\widetilde{\boldsymbol{M}}_H(\lambda)$*, and* $\widetilde{\boldsymbol{N}}_H(\lambda)$ *be given by Eq. (5.3). Then the collection of all the minimal partial realizations of* $\boldsymbol{b}$ *admits a parametrized description by means of a linear fractional transformation*

$$\begin{aligned}\boldsymbol{G}_X(\lambda)&=(\widetilde{\boldsymbol{N}}_H(\lambda)+\widetilde{\boldsymbol{M}}_H(\lambda)\boldsymbol{X})(\widetilde{\boldsymbol{U}}_H(\lambda)+\widetilde{\boldsymbol{P}}_H(\lambda)\boldsymbol{X})^{-1}\\&=(\boldsymbol{U}_H(\lambda)+\boldsymbol{X}\boldsymbol{P}_H(\lambda))^{-1}(\boldsymbol{N}_H(\lambda)+\boldsymbol{X}\boldsymbol{M}_H(\lambda)),\end{aligned}\tag{5.8}$$

where $\boldsymbol{X}$*, as the free parameter, runs over all the* $p\times q$ *constant matrices. Furthermore,* $\delta(\boldsymbol{G}_X)=mp=nq$ *for each* $\boldsymbol{X}\in\mathbf{C}^{p\times q}$.

Remarks (a) If desired, we can take $\boldsymbol{U}_H^0(\lambda),\widetilde{\boldsymbol{U}}_H^0(\lambda)$, defined in the Remark (a) following Theorem 4.3, in place of $\boldsymbol{U}_H(\lambda),\widetilde{\boldsymbol{U}}_H(\lambda)$ in Theorem 5.1.

(b) By Eq. (5.4) we see readily that $\widetilde{\boldsymbol{N}}_H(\lambda)+\widetilde{\boldsymbol{M}}_H(\lambda)\boldsymbol{X}$ and $\widetilde{\boldsymbol{U}}_H(\lambda)+\widetilde{\boldsymbol{P}}_H(\lambda)\boldsymbol{X}$ are right coprime for any $\boldsymbol{X}\in\mathbf{C}^{p\times q}$, while $\boldsymbol{N}_H(\lambda)+\boldsymbol{X}\boldsymbol{M}_H(\lambda)$ and $\boldsymbol{U}_H(\lambda)+\boldsymbol{X}\boldsymbol{P}_H(\lambda)$ are left coprime for any $\boldsymbol{X}\in\mathbf{C}^{p\times q}$.

(c) Since both $\boldsymbol{G}_X(\lambda)$ and $\boldsymbol{G}_0(\lambda)(=\widetilde{\boldsymbol{N}}_H(\lambda)\widetilde{\boldsymbol{U}}_H^{-1}(\lambda)=\boldsymbol{U}_H^{-1}(\lambda)\boldsymbol{N}_H(\lambda))$ with $\boldsymbol{X}=\boldsymbol{O}$ are the minimal partial realizations of $\mathbf{b}$, we have

$$\boldsymbol{G}_X(\lambda)-\boldsymbol{G}_0(\lambda)=(\boldsymbol{U}_H(\lambda)+\boldsymbol{X}\boldsymbol{P}_H(\lambda))^{-1}\boldsymbol{X}\widetilde{\boldsymbol{U}}_H^{-1}(\lambda)=\frac{\boldsymbol{X}}{\lambda^{m+n}}+\cdots.\tag{5.9}$$

Indeed, by Eqs. (5.8) and (5.4),

$$\begin{aligned}\boldsymbol{G}_X(\lambda)-\boldsymbol{G}_0(\lambda)&=(\boldsymbol{U}_H(\lambda)+\boldsymbol{X}\boldsymbol{P}_H(\lambda))^{-1}(\boldsymbol{N}_H(\lambda)+\boldsymbol{X}\boldsymbol{M}_H(\lambda))-\widetilde{\boldsymbol{N}}_H(\lambda)\widetilde{\boldsymbol{U}}_H^{-1}(\lambda)\\&=(\boldsymbol{U}_H(\lambda)+\boldsymbol{X}\boldsymbol{P}_H(\lambda))^{-1}[(\boldsymbol{N}_H(\lambda)+\boldsymbol{X}\boldsymbol{M}_H(\lambda))\widetilde{\boldsymbol{U}}_H(\lambda)-\\&\quad(\boldsymbol{U}_H(\lambda)+\boldsymbol{X}\boldsymbol{P}_H(\lambda))\widetilde{\boldsymbol{N}}_H(\lambda)]\widetilde{\boldsymbol{U}}_H^{-1}(\lambda)\\&=(\boldsymbol{U}_H(\lambda)+\boldsymbol{X}\boldsymbol{P}_H(\lambda))^{-1}[(\boldsymbol{N}_H(\lambda)\widetilde{\boldsymbol{U}}_H(\lambda)-\boldsymbol{U}_H(\lambda)\widetilde{\boldsymbol{N}}_H(\lambda))+\\&\quad\boldsymbol{X}(\boldsymbol{M}_H(\lambda)\widetilde{\boldsymbol{U}}_H(\lambda)-\boldsymbol{P}_H(\lambda)\widetilde{\boldsymbol{N}}_H(\lambda))]\widetilde{\boldsymbol{U}}_H^{-1}(\lambda)\\&=(\boldsymbol{U}_H(\lambda)+\boldsymbol{X}\boldsymbol{P}_H(\lambda))^{-1}\boldsymbol{X}\widetilde{\boldsymbol{U}}_H^{-1}(\lambda)\end{aligned}$$

so that Eq. (5.9) follows from the fact that $-(\boldsymbol{U}_H(\lambda)+\boldsymbol{X}\boldsymbol{P}_H(\lambda))$ and $-\widetilde{\boldsymbol{U}}_H(\lambda)$ are monic, of degrees m and n, respectively, for any $\boldsymbol{X}\in\mathbf{C}^{p\times q}$.

(d) Theorem 5.1 shows that a minimal partial realization of $\boldsymbol{b}$, with the matrix $\boldsymbol{H}$ of Eq. (5.1) non-singular, is completely determined by that $\boldsymbol{X}\in\mathbf{C}^{p\times q}$ given in Eq. (5.8), and

$$\boldsymbol{G}_X(\lambda)=\frac{\boldsymbol{H}_0}{\lambda}+\frac{\boldsymbol{H}_1}{\lambda^2}+\cdots+\frac{\boldsymbol{H}_{m+n-2}}{\lambda^{m+n-1}}+\frac{\boldsymbol{H}_{m+n-1}+\boldsymbol{X}}{\lambda^{m+n}}+\cdots.$$

Actually, there exists a one-to-one correspondence between all minimal continuations $(\boldsymbol{H}_0,\boldsymbol{H}_1,\cdots,\boldsymbol{H}_{m+n-2},\boldsymbol{R},\cdots)$ of $\boldsymbol{b}$, that is, all the sequences of the Markov parameters of all the minimal partial realizations of $\boldsymbol{b}$, and $\boldsymbol{X}\in\mathbf{C}^{p\times q}$ via Eq. (5.8). Each $\boldsymbol{X}\in\mathbf{C}^{p\times q}$ determines only one minimal continuation $(\boldsymbol{H}_0,\boldsymbol{H}_1,\cdots,\boldsymbol{H}_{m+n-2},\boldsymbol{H}_{m+n-1}+\boldsymbol{X},\cdots)$ of $\boldsymbol{b}$, and conversely, given any minimal continuation of $\boldsymbol{b}$, there exists some $\boldsymbol{X}\in\mathbf{C}^{p\times q}$ such that $\boldsymbol{G}_X(\lambda)$, defined by Eq. (5.8), is a minimal partial realization of the infinite sequence $(\boldsymbol{H}_0,\boldsymbol{H}_1,\cdots,\boldsymbol{H}_{m+n-2},\boldsymbol{H}_{m+n-1}+\boldsymbol{X},\cdots)$. For further details, see ([8], Sections Ⅲ, Ⅴ).

§ 5.2 *The minimal, proper rational matrix interpolants to the GRMIP* (2.4)

The next step is to parametrize the set of proper rational matrix interpolants, of least McMillan degree, to the GRMIP(2.4) under the assumption that a matrix $\boldsymbol{L}_{\mathscr{Y}\mathscr{X}}\in\mathscr{L}$ with $mp=nq$ is invertible. To this end, we shall show that this parametrization problem can be reduced to a special case, that is, the solution of the minimal partial realization problem of the block Hankel vector $\boldsymbol{b}=(\boldsymbol{H}_0,\boldsymbol{H}_1,\cdots,\boldsymbol{H}_{m+n-2})$ mentioned just. (In the scalar case, this attempt has been made in ([16], Theorem 4.1).) Here, the polynomial matrices $\boldsymbol{P}_H(\lambda)$, $\boldsymbol{U}_H(\lambda)$, $\widetilde{\boldsymbol{P}}_H(\lambda)$, $\widetilde{\boldsymbol{U}}_H(\lambda)$ given as in Eq. (4.13) or Eq. (5.12) play again a key role.

Let $\boldsymbol{\Omega}(\lambda)$ and $A(\lambda)$ be as before. We say that a pair of polynomial matrices $[\widetilde{\boldsymbol{Q}}(\lambda),\widetilde{\boldsymbol{F}}(\lambda)]_r$ $[[\boldsymbol{Q}(\lambda),\boldsymbol{F}(\lambda)]_l$, resp.$]$ with $\widetilde{\boldsymbol{Q}}(\lambda)\in\mathbf{C}^{p\times q}[\lambda]$ and $\widetilde{\boldsymbol{F}}(\lambda)\in\mathbf{C}^{q\times q}[\lambda]$ [with $\boldsymbol{Q}(\lambda)\in\mathbf{C}^{p\times q}[\lambda]$ and $\boldsymbol{F}(\lambda)\in\mathbf{C}^{p\times p}[\lambda]$], is a right [left] generalized solution to the GRMIP (2.4), if

$$\frac{\mathrm{d}^k}{\mathrm{d}\lambda^k}(\widetilde{\boldsymbol{Q}}(\lambda)-\boldsymbol{\Omega}(\lambda)\widetilde{\boldsymbol{F}}(\lambda))\bigg|_{\lambda=x_i}=\boldsymbol{O},$$

$$i=1,2,\cdots,\rho,\quad k=0,1,\cdots,\tau_i-1.$$

$$\left[\frac{\mathrm{d}^k}{\mathrm{d}\lambda^k}(\boldsymbol{Q}(\lambda)-\boldsymbol{F}(\lambda)\boldsymbol{\Omega}(\lambda))\bigg|_{\lambda=x_i}=\boldsymbol{O},\right. \tag{5.10}$$

$$\left. i=1,2,\cdots,\rho,\quad k=0,1,\cdots,\tau_i-1,\text{resp.}\right]$$

We say that the interpolation node x_i is attainable for $[\widetilde{\boldsymbol{Q}}(\lambda),\widetilde{\boldsymbol{F}}(\lambda)]_r$ (or $[\boldsymbol{Q}(\lambda),\boldsymbol{F}(\lambda)]_l$) if $\det\widetilde{\boldsymbol{F}}(x_i)\neq\boldsymbol{O}$ (or $\det\boldsymbol{F}(x_i)\neq\boldsymbol{O}$). If, further, $\widetilde{\boldsymbol{F}}(\lambda)[\boldsymbol{F}(\lambda)$, resp.] is invertible at all $\lambda=x_i$, then and only then $\widetilde{\boldsymbol{Q}}(\lambda)\widetilde{\boldsymbol{F}}^{-1}(\lambda)[\boldsymbol{F}^{-1}(\lambda)\boldsymbol{Q}(\lambda)]$ is a right [left] rational matrix interpolant to the GPMIP (2.4).

Theorem 5.2 *Let $\boldsymbol{b}=(\boldsymbol{H}_0,\boldsymbol{H}_1,\cdots,\boldsymbol{H}_{m+n-2})$ be the block Hankel vector of the GRMIP (2.4). Let $\widetilde{\boldsymbol{P}}_H(\lambda),\widetilde{\boldsymbol{U}}_H(\lambda),\boldsymbol{P}_H(\lambda),\boldsymbol{U}_H(\lambda),\widetilde{\boldsymbol{M}}_H(\lambda)$, $\widetilde{\boldsymbol{N}}_H(\lambda),\boldsymbol{M}_H(\lambda),\boldsymbol{N}_H(\lambda)$ be as before. Define*

$$\widetilde{\boldsymbol{Q}}_X(\lambda)=\boldsymbol{\Omega}(\lambda)[\widetilde{\boldsymbol{U}}_H(\lambda)+\widetilde{\boldsymbol{P}}_H(\lambda)\boldsymbol{X}]-A(\lambda)[\widetilde{\boldsymbol{N}}_H(\lambda)+\widetilde{\boldsymbol{M}}_H(\lambda)\boldsymbol{X}], \tag{5.11}$$

$$\boldsymbol{Q}_X(\lambda)=[\boldsymbol{U}_H(\lambda)+\boldsymbol{X}\boldsymbol{P}_H(\lambda)]\boldsymbol{\Omega}(\lambda)-[\boldsymbol{N}_H(\lambda)+\boldsymbol{X}\boldsymbol{M}_H(\lambda)]A(\lambda), \tag{5.12}$$

where $\boldsymbol{X}\in\mathbf{C}^{p\times q}$ is a constant matrix. Then

(a) *Each column [row, resp.] degree of $\widetilde{\boldsymbol{Q}}_X(\lambda)$ [$\boldsymbol{Q}_X(\lambda)$] is less than or equal to the corresponding column [row] degree of*

$$\widetilde{\boldsymbol{U}}_H(\lambda)+\widetilde{\boldsymbol{P}}_H(\lambda)\boldsymbol{X}\ [\boldsymbol{U}_H(\lambda)+\boldsymbol{X}\boldsymbol{P}_H(\lambda)].$$

(b) $[\widetilde{\boldsymbol{Q}}_X(\lambda),\widetilde{\boldsymbol{U}}_H(\lambda)+\widetilde{\boldsymbol{P}}_H(\lambda)\boldsymbol{X}]_r,[[\boldsymbol{Q}_X(\lambda),\widetilde{\boldsymbol{U}}_H(\lambda)+\boldsymbol{X}\boldsymbol{P}_H(\lambda)]_l$, *resp.*] *is a right [left] generalized solution to the GRMIP (2.4).*

(c) *If $\widetilde{\boldsymbol{U}}_H(\lambda)+\widetilde{\boldsymbol{P}}_H(\lambda)\boldsymbol{X}$ and $A(\lambda)\boldsymbol{I}_q[\boldsymbol{U}_H(\lambda)+\boldsymbol{X}\boldsymbol{P}_H(\lambda)$ and $A(\lambda)\boldsymbol{I}_p$, resp.] are right [left] coprime for some $\boldsymbol{X}\in\mathbf{C}^{p\times q}$, then and only then $\boldsymbol{Q}_X(\lambda)(\boldsymbol{U}_H(\lambda)+\widetilde{\boldsymbol{P}}_H(\lambda)\boldsymbol{X})^{-1}[(\boldsymbol{U}_H(\lambda)+\boldsymbol{X}\boldsymbol{P}_H(\lambda))^{-1}\boldsymbol{Q}_X(\lambda)$, resp.] for the same $\boldsymbol{X}\in\mathbf{C}^{p\times q}$ is a right [left] proper rational matrix solution, of the McMillan degree $mp(=nq)$, to the GRMIP (2.4).*

Proof (a) Let

$$(\widetilde{\boldsymbol{N}}_H(\lambda)+\widetilde{\boldsymbol{M}}_H(\lambda)\boldsymbol{X})(\widetilde{\boldsymbol{U}}_H(\lambda)+\widetilde{\boldsymbol{P}}_H(\lambda)\boldsymbol{X})^{-1}$$

be of the matrix form $[h_{ij}(\lambda)/f_{ij}(\lambda)]_{i,j=1}^{p,q}$, where both $h_{ij}(\lambda)$ and $f_{ij}(\lambda)$ are coprime scalar polynomials with $\deg h_{ij}(\lambda)<\deg f_{ij}(\lambda)$ for each i,j, and let, similarly,

$$\widetilde{\boldsymbol{N}}_H(\lambda)+\widetilde{\boldsymbol{M}}_H(\lambda)\boldsymbol{X}=[H_{kl}(\lambda)]_{k,l=1}^{p,q},$$

$$\widetilde{\boldsymbol{U}}_H(\lambda)+\widetilde{\boldsymbol{P}}_H(\lambda)\boldsymbol{X}=[F_{kl}(\lambda)]_{k,l=1}^{q},\ \boldsymbol{\Omega}(\lambda)=[\omega_{ij}(\lambda)]_{i,j=1}^{p,q}.$$

Then we can express $\boldsymbol{H}_{mp,nq}[\boldsymbol{b}]$, generated by $(\widetilde{\boldsymbol{N}}_H(\lambda)+\widetilde{\boldsymbol{M}}_H(\lambda))(\widetilde{\boldsymbol{U}}_h(\lambda)+\widetilde{\boldsymbol{P}}_H(\lambda)\boldsymbol{X})^{-1}$, as $\sum_{i=1}^{p}\sum_{j=1}^{q}\boldsymbol{H}_{m,n}\left(\frac{h_{ij}(\lambda)}{f_{ij}(\lambda)}\right)\otimes\boldsymbol{E}_{ij}$, in which $\boldsymbol{E}_{ij}$ is the $p\times q$ matrix with one as the (i,j) element and zero, elsewhere. From Lemma 4.1 we deduce clearly that

$$\begin{aligned}\boldsymbol{L}_{\mathscr{Y}\mathscr{X}}(\boldsymbol{\Omega}(\lambda)) &= \sum_{i=1}^{p}\sum_{j=1}^{q}\boldsymbol{L}_{\mathscr{Y}\mathscr{X}}(\omega_{ij}(\lambda))\otimes\boldsymbol{E}_{ij}\\ &= (\boldsymbol{W}_{\mathscr{Y}}\otimes\boldsymbol{I}_p)\boldsymbol{H}_{mp,nq}[\boldsymbol{b}](\boldsymbol{W}_{\mathscr{X}}^{\mathrm{T}}\otimes\boldsymbol{I}_q)\\ &= \sum_{i=1}^{p}\sum_{j=1}^{q}\left[\boldsymbol{W}_{\mathscr{Y}}\boldsymbol{H}_{mn}\left(\frac{h_{ij}(\lambda)}{f_{ij}(\lambda)}\right)\boldsymbol{W}_{\mathscr{X}}^{\mathrm{T}}\right]\otimes\boldsymbol{E}_{ij},\end{aligned}$$

so that

$$\boldsymbol{L}_{\mathscr{Y}\mathscr{X}}(\omega_{ij}(\lambda))=\boldsymbol{W}_{\mathscr{Y}}\boldsymbol{H}_{mn}\left(\frac{h_{ij}(\lambda)}{f_{ij}(\lambda)}\right)\boldsymbol{W}_{\mathscr{X}}^{\mathrm{T}},\ \forall\, i,j.$$

Set now

$$g_{ij}(\lambda)\equiv\omega_{ij}(\lambda)f_{ij}(\lambda)-A(\lambda)h_{ij}(\lambda),\ i=1,2,\cdots,p,\ j=1,2,\cdots,q,$$

$$\hat{\boldsymbol{G}}(\lambda)=[\hat{G}_{ij}(\lambda)]_{i,j=1}^{p,q}\equiv\left[\frac{g_{ij}(\lambda)}{f_{ij}(\lambda)}\right]_{i,j=1}^{p,q}(\widetilde{\boldsymbol{U}}_H(\lambda)+\widetilde{\boldsymbol{P}}_H(\lambda)\boldsymbol{X}).$$

Then

$$\begin{aligned}\hat{\boldsymbol{G}}_{ij}(\lambda) &= \sum_{s=1}^{q}\left[\frac{g_{is}(\lambda)}{f_{is}(\lambda)}F_{sj}(\lambda)\right]\\ &= \sum_{s=1}^{q}[\omega_{is}(\lambda)F_{sj}(\lambda)]-A(\lambda)\sum_{s=1}^{q}\left[\frac{h_{is}(\lambda)}{f_{is}(\lambda)}F_{sj}(\lambda)\right]\\ &= \sum_{s=1}^{q}[\omega_{is}(\lambda)F_{sj}(\lambda)]-A(\lambda)H_{ij}(\lambda),\\ & i=1,2,\cdots,p,\quad j=1,2,\cdots,q,\end{aligned}$$

which amount to, by Eq. (5.11),

$$\hat{\boldsymbol{G}}(\lambda)=\boldsymbol{\Omega}(\lambda)(\widetilde{\boldsymbol{U}}_H(\lambda)+\widetilde{\boldsymbol{P}}_H(\lambda)\boldsymbol{X})-A(\lambda)(\widetilde{\boldsymbol{N}}_H(\lambda)+\widetilde{\boldsymbol{M}}_H(\lambda)\boldsymbol{X})=\widetilde{\boldsymbol{Q}}_X(\lambda).$$

Thus, each column degree of $\widetilde{\boldsymbol{Q}}_X(\lambda)$ is less than or equal to the corresponding column degree of $\widetilde{\boldsymbol{U}}_H(\lambda)+\widetilde{\boldsymbol{P}}_H(\lambda)\boldsymbol{X}$, since each $g_{ij}(\lambda)/f_{ij}(\lambda)$ is proper rational (see Theorem 4.1 of [16]). The proof for another half of (a) is carried out in much the same way, but with $\hat{\boldsymbol{G}}(\lambda)$ replaced by

$$\boldsymbol{G}(\lambda)=(\boldsymbol{U}_H(\lambda)+\boldsymbol{X}\boldsymbol{P}_H(\lambda))\left[\frac{g_{ij}(\lambda)}{f_{ij}(\lambda)}\right]_{i,j=1}^{p,q}.$$

(b) This follows from Eqs. (5.11) and (5.12) directly.

(c) Let $\widetilde{\boldsymbol{U}}_H(\lambda)+\widetilde{\boldsymbol{P}}_H(\lambda)\boldsymbol{X}$ and $A(\lambda)\boldsymbol{I}_q$ be right coprime for some $\boldsymbol{X}\in\mathbf{C}^{p\times q}$. Then there exist $\boldsymbol{Z}(\lambda),\boldsymbol{Y}(\lambda)\in\mathbf{C}^{q\times q}[\lambda]$ such that

$$\boldsymbol{Z}(\lambda)A(\lambda)+\boldsymbol{Y}(\lambda)(\widetilde{\boldsymbol{U}}_H(\lambda)+\widetilde{\boldsymbol{P}}_H(\lambda)\boldsymbol{X})=\boldsymbol{I}_q,$$

so that $\widetilde{\boldsymbol{U}}_H(\lambda)+\widetilde{\boldsymbol{P}}_H(\lambda)\boldsymbol{X}$ is invertible at each $\lambda=x_i, i=1,2,\cdots,\rho$. Conversely, if $\widetilde{\boldsymbol{U}}_H(\lambda)+\widetilde{\boldsymbol{P}}_H(\lambda)\boldsymbol{X}$ and $A(\lambda)\boldsymbol{I}_q$ are not right coprime for some $\boldsymbol{X}\in\mathbf{C}^{p\times q}$, then there exists at least one x_i such that $(\lambda-x_i)$ is a divisor of each element of the polynomial matrix $\widetilde{\boldsymbol{U}}_H(\lambda)+\widetilde{\boldsymbol{P}}_H(\lambda)\boldsymbol{X}$, so that $\widetilde{\boldsymbol{U}}_H(\lambda)+\widetilde{\boldsymbol{P}}_H(\lambda)\boldsymbol{X}$ is not invertible for $\lambda=x_i$. Thus, from (a), (b) together with the fact $\widetilde{\boldsymbol{U}}_H(\lambda)+\widetilde{\boldsymbol{P}}_H(\lambda)\boldsymbol{X}$ is column reduced, we deduce that $\widetilde{\boldsymbol{Q}}_X(\lambda)\cdot(\widetilde{\boldsymbol{U}}_H(\lambda)+\widetilde{\boldsymbol{P}}_H(\lambda)\boldsymbol{X})^{-1}$ is a right, proper rational matrix interpolant to the GRMIP (2.4) for the same $\boldsymbol{X}\in\mathbf{C}^{p\times q}$. In order to show that $\widetilde{\boldsymbol{Q}}_X(\lambda)\cdot(\widetilde{\boldsymbol{U}}_H(\lambda)+\widetilde{\boldsymbol{P}}_H(\lambda)\boldsymbol{X})^{-1}$ is of the McMillan degree $mp(=nq)$, it suffices to verify that $\widetilde{\boldsymbol{Q}}_X(\lambda)$ and $\widetilde{\boldsymbol{U}}_H(\lambda)+\widetilde{\boldsymbol{P}}_H(\lambda)\boldsymbol{X}$ are right coprime for the same $\boldsymbol{X}\in\mathbf{C}^{p\times q}$, i.e., all the greatest common right divisors of them are unimodular. Suppose now that $\boldsymbol{R}(\lambda)\in\mathbf{C}^{q\times q}[\lambda]$ is such a divisor, so that there exist two suitable right coprime polynomial matrices $\boldsymbol{F}_1(\lambda)\in\mathbf{C}^{p\times q}[\lambda]$ and $\boldsymbol{G}_1(\lambda)\in\mathbf{C}^{q\times q}[\lambda]$ such that

$$\widetilde{\boldsymbol{Q}}_X(\lambda)=\boldsymbol{F}_1(\lambda)\boldsymbol{R}(\lambda),\quad \widetilde{\boldsymbol{U}}_H(\lambda)+\widetilde{\boldsymbol{P}}_H(\lambda)X=\boldsymbol{G}_1(\lambda)\boldsymbol{R}(\lambda).$$

Then $\boldsymbol{R}(\lambda)$ is clearly invertible at all $\lambda=x_i$, and, by Eq. (5.11)

$$A(\lambda)(\widetilde{\boldsymbol{N}}_H(\lambda)+\widetilde{\boldsymbol{M}}_H(\lambda)\boldsymbol{X})=(\boldsymbol{\Omega}(\lambda)\boldsymbol{G}_1(\lambda)-\boldsymbol{F}_1(\lambda))\boldsymbol{R}(\lambda)$$

so that $\widetilde{\boldsymbol{N}}_H(\lambda)+\widetilde{\boldsymbol{M}}_H(\lambda)\boldsymbol{X}=\boldsymbol{H}_1(\lambda)\boldsymbol{R}(\lambda)$ in which

$$\boldsymbol{H}_1(\lambda)=(\widetilde{\boldsymbol{N}}_H(\lambda)+\widetilde{\boldsymbol{M}}_H(\lambda)\boldsymbol{X})\boldsymbol{R}^{-1}(\lambda)=(\boldsymbol{\Omega}(\lambda)\boldsymbol{G}_1(\lambda)-\boldsymbol{F}_1(\lambda))/A(\lambda)$$

is a $p\times q$ polynomial matrix. Hence, $\boldsymbol{R}(\lambda)$ is in turn a (greatest) common right divisor of $\widetilde{\boldsymbol{U}}_H(\lambda)+\widetilde{\boldsymbol{P}}_H(\lambda)\boldsymbol{X}$ and $\widetilde{\boldsymbol{N}}_H(\lambda)+\widetilde{\boldsymbol{M}}_H(\lambda)\boldsymbol{X}$, and therefore it is unimodular by the irreducibility of $(\widetilde{\boldsymbol{U}}_H(\lambda)+\widetilde{\boldsymbol{M}}_H(\lambda)\boldsymbol{X})(\widetilde{\boldsymbol{U}}_H(\lambda)+\widetilde{\boldsymbol{P}}_H(\lambda)\boldsymbol{X})^{-1}$ (see Remark (b) after Theorem 5. 1).

Similarly, we can prove another assertion of (c). ■

Theorem 5. 2 is naturally supplemented with the converse assertion.

Theorem 5. 3 *Let $\boldsymbol{b}$, $\boldsymbol{\Omega}(\lambda)$, $A(\lambda)$, $\widetilde{\boldsymbol{U}}_H(\lambda)$, $\widetilde{\boldsymbol{P}}_H(\lambda)$, $\boldsymbol{U}_H(\lambda)$, $\boldsymbol{P}_H(\lambda)$, $\widetilde{\boldsymbol{N}}_H(\lambda)$, $\widetilde{\boldsymbol{M}}_H(\lambda)$, $\boldsymbol{N}_H(\lambda)$, $\boldsymbol{M}_H(\lambda)$ be as before. Let $\widetilde{\boldsymbol{U}}_H(\lambda)+\widetilde{\boldsymbol{P}}_H(\lambda)\boldsymbol{X}\in\mathbf{C}^{q\times q}[\lambda]$ and $\widetilde{\boldsymbol{Q}}_X(\lambda)\in\mathbf{C}^{p\times q}[\lambda]$ $[\boldsymbol{U}_H(\lambda)+\boldsymbol{X}\boldsymbol{P}_H(\lambda)\in\mathbf{C}^{p\times p}[\lambda]$ and $\boldsymbol{Q}_X(\lambda)\in\mathbf{C}^{p\times q}[\lambda]$, resp.] be right [left] coprime for some $\boldsymbol{X}\in\mathbf{C}^{p\times q}$ such that $\widetilde{\boldsymbol{Q}}_X(\lambda)(\widetilde{\boldsymbol{U}}_H(\lambda)+\widetilde{\boldsymbol{P}}_H(\lambda)X)^{-1}$ $[(\boldsymbol{U}_H(\lambda)+\boldsymbol{X}\boldsymbol{P}_H(\lambda))^{-1}\boldsymbol{Q}_X(\lambda)$, resp.] is a proper, right [left] rational matrix interpolant to the GRMIP (2. 4). Then there must exist a unique polynomial matrix $\widetilde{\boldsymbol{H}}_X(\lambda)\in\mathbf{C}^{p\times q}[\lambda]$ $[\boldsymbol{H}_X(\lambda)\in\mathbf{C}^{p\times q}[\lambda]]$ such that for the same $\boldsymbol{X}\in\mathbf{C}^{p\times q}$*

$$\begin{aligned}&A(\lambda)\widetilde{\boldsymbol{H}}_X(\lambda)=\boldsymbol{\Omega}(\lambda)(\widetilde{\boldsymbol{U}}_H(\lambda)+\widetilde{\boldsymbol{P}}_H(\lambda)\boldsymbol{X})-\widetilde{\boldsymbol{Q}}_X(\lambda),\\&[A(\lambda)\boldsymbol{H}_X(\lambda)=(\boldsymbol{U}_H(\lambda)+\boldsymbol{X}\boldsymbol{P}_H(\lambda))\boldsymbol{\Omega}(\lambda)-\boldsymbol{Q}_X(\lambda),resp.]\end{aligned}\tag{5.13}$$

and $\widetilde{\boldsymbol{H}}_X(\lambda)(\widetilde{\boldsymbol{U}}_H(\lambda)+\widetilde{\boldsymbol{P}}_H(\lambda)\boldsymbol{X})^{-1}$ $[(\boldsymbol{U}_H(\lambda)+\boldsymbol{X}\boldsymbol{P}_H(\lambda))^{-1}\boldsymbol{H}_X(\lambda)$, resp.] is a minimal partial realization of the block Hankel vector $\boldsymbol{b}$. Moreover, $\widetilde{\boldsymbol{H}}_X(\lambda)=\widetilde{\boldsymbol{N}}_H(\lambda)+\widetilde{\boldsymbol{M}}_H(\lambda)\boldsymbol{X}$ $[\boldsymbol{H}_X(\lambda)=\boldsymbol{N}_H(\lambda)+\boldsymbol{X}\boldsymbol{M}_H(\lambda)]$ for the same $\boldsymbol{X}\in\mathbf{C}^{p\times q}$.

Proof By hypothesis, there exists a unique $\widetilde{\boldsymbol{H}}_X(\lambda)\in\mathbf{C}^{p\times q}[\lambda]$ such that Eq. (5. 13) holds for $\boldsymbol{X}\in\mathbf{C}^{p\times q}$ given. Let us rewrite $\widetilde{\boldsymbol{Q}}_X(\lambda)(\widetilde{\boldsymbol{U}}_H(\lambda)+\widetilde{\boldsymbol{P}}_H(\lambda)\boldsymbol{X})^{-1}$ in the matrix form

$$\widetilde{\boldsymbol{Q}}_X(\lambda)(\widetilde{\boldsymbol{U}}_H(\lambda)+\widetilde{\boldsymbol{P}}_H(\lambda)\boldsymbol{X})^{-1}=\left[\frac{g_{ij}(\lambda)}{f_{ij}(\lambda)}\right]_{i,j=1}^{p,q},$$

where $g_{ij}(\lambda)$ and $f_{ij}(\lambda)$ are coprime scalar polynomials with $\deg g_{ij}(\lambda)\leqslant\deg f_{ij}(\lambda)$ for all i,j. Then it is easy to see by Eq. (5. 13) that $\widetilde{\boldsymbol{H}}(\lambda)\cdot(\widetilde{\boldsymbol{U}}_H(\lambda)+\widetilde{\boldsymbol{P}}_H(\lambda)\boldsymbol{X})^{-1}$ has the matrix form $\left[\frac{h_{ij}(\lambda)}{f_{ij}(\lambda)}\right]_{i,j=1}^{p,q}$ for certain scalar polynomials $h_{ij}(\lambda)$ such that

$$\frac{g_{ij}(\lambda)}{f_{ij}(\lambda)}+A(\lambda)\frac{h_{ij}(\lambda)}{f_{ij}(\lambda)}=\omega_{ij}(\lambda),\quad i=1,2,\cdots,p,\quad j=1,2,\cdots,q,$$

and therefore

$$g_{ij}(\lambda)+A(\lambda)h_{ij}(\lambda)=\omega_{ij}(\lambda)f_{ij}(\lambda),\forall i,j.$$

It follows from the known result for the scalar case [16,15] that $h_{ij}(\lambda)$ and $f_{ij}(\lambda)$ are coprime for all i,j with $\deg h_{ij}(\lambda)<\deg f_{ij}(\lambda),\forall i,j$, and that

$$\begin{aligned}\boldsymbol{L}_{\mathscr{Y}\mathscr{X}}(\boldsymbol{\Omega}(\lambda)) &= \sum_{i=1}^{p}\sum_{j=1}^{q}[\boldsymbol{W}_{\mathscr{Y}}\boldsymbol{H}_{m,n}\left(\frac{h_{ij}(\lambda)}{f_{ij}(\lambda)}\right)\boldsymbol{W}_{\mathscr{X}}^{\mathrm{T}}]\otimes\boldsymbol{E}_{ij}\\ &= [\boldsymbol{W}_{\mathscr{Y}}\otimes\boldsymbol{I}_p][\sum_{i=1}^{p}\sum_{j=1}^{q}\boldsymbol{H}_{m,n}\left(\frac{h_{ij}(\lambda)}{f_{ij}(\lambda)}\right)\otimes\boldsymbol{E}_{ij}][\boldsymbol{W}_{\mathscr{X}}^{\mathrm{T}}\otimes\boldsymbol{I}_q]\\ &= [\boldsymbol{W}_{\mathscr{Y}}\otimes\boldsymbol{I}_p]\boldsymbol{H}_{mp,nq}(\widetilde{\boldsymbol{H}}_X(\lambda)(\widetilde{\boldsymbol{U}}_H(\lambda)+\widetilde{\boldsymbol{P}}_H(\lambda)\boldsymbol{X})^{-1})[\boldsymbol{W}_{\mathscr{X}}^{\mathrm{T}}\otimes\boldsymbol{I}_q].\end{aligned}$$

Thus, by Lemma 4.1,

$$\boldsymbol{H}_{mp,nq}[\boldsymbol{b}]=\boldsymbol{H}_{mp,nq}(\widetilde{H}_X(\lambda)(\widetilde{\boldsymbol{U}}(\lambda)+\widetilde{\boldsymbol{P}}(\lambda)\boldsymbol{X})^{-1})$$

which means that the strictly proper rational matrix function $\widetilde{\boldsymbol{H}}_X(\lambda)(\widetilde{\boldsymbol{U}}_H(\lambda)+\widetilde{\boldsymbol{P}}_H(\lambda)\boldsymbol{X})^{-1}$ has $\boldsymbol{H}_0,\boldsymbol{H}_1,\cdots,\boldsymbol{H}_{m+n-2}$ as its first $m+n-1$ Markov parameters.

It remains to verify the right-coprimeness of $\widetilde{\boldsymbol{H}}_X(\lambda)$ and $\widetilde{\boldsymbol{U}}_H(\lambda)+\widetilde{\boldsymbol{P}}_H(\lambda)\boldsymbol{X}$. But this is carried out in much the same way as the proof of the last part of (c) of Theorem 5.2. Thus, since $\widetilde{\boldsymbol{U}}_H(\lambda)+\widetilde{\boldsymbol{P}}_H(\lambda)\boldsymbol{X}$ is clearly column-reduced, we have[28] that $\delta(\widetilde{\boldsymbol{H}}_X(\lambda)(\widetilde{\boldsymbol{U}}_H(\lambda)+\widetilde{\boldsymbol{P}}_H(\lambda)\boldsymbol{X})^{-1})=\delta(\widetilde{\boldsymbol{U}}_H(\lambda)+\widetilde{\boldsymbol{P}}_H(\lambda)\boldsymbol{X})=mp$. Applying Theorem 5.1 we derive that $\widetilde{\boldsymbol{H}}_X(\lambda)=\widetilde{\boldsymbol{N}}_H(\lambda)+\widetilde{\boldsymbol{M}}_H(\lambda)\boldsymbol{X}$ for the same $\boldsymbol{X}$. Similarly, we can prove the other part. ■

Due to Theorems 5.2 and 5.3 we obtain the fundamental result of this section.

Theorem 5.4 *Let the GRMIP* (2.4) *be given, and let* $\boldsymbol{b}=(\boldsymbol{H}_0,\boldsymbol{H}_1,\cdots,\boldsymbol{H}_{m+n-2})$ *be its block Hankel vector. Let* $\boldsymbol{P}_H(\lambda),\boldsymbol{U}_H(\lambda),\widetilde{\boldsymbol{P}}_H(\lambda),\widetilde{\boldsymbol{U}}_H(\lambda),\boldsymbol{N}_H(\lambda),\boldsymbol{M}_H(\lambda),\widetilde{\boldsymbol{N}}_H(\lambda),\widetilde{\boldsymbol{M}}_H(\lambda)$ *be as before. Define* $\widetilde{\boldsymbol{Q}}_X(\lambda)$ *and* $\boldsymbol{Q}_X(\lambda)$ *for* $\boldsymbol{X}\in\mathbf{C}^{p\times q}$ *as in Eqs.* (5.11) *and* (5.12), *respectively. Then the collection of all minimal, proper and right* [*left, resp.*] *rational matrix interpolants to the*

GRMIP (2.4) *admits a parametrized description by means of a linear fractional transformation*

$$\begin{aligned}\boldsymbol{F}_X(\lambda)&=\widetilde{\boldsymbol{Q}}_X(\lambda)(\widetilde{\boldsymbol{U}}_H(\lambda)+\widetilde{\boldsymbol{P}}_H(\lambda)\boldsymbol{X})^{-1}\\&=\boldsymbol{\Omega}(\lambda)-A(\lambda)(\widetilde{\boldsymbol{N}}(\lambda)+\widetilde{\boldsymbol{M}}_H(\lambda)\boldsymbol{X})(\widetilde{\boldsymbol{U}}_H(\lambda)+\widetilde{\boldsymbol{P}}_H(\lambda)\boldsymbol{X})^{-1},\end{aligned}\tag{5.14}$$

$$\begin{aligned}[\boldsymbol{F}_X(\lambda)&=(\boldsymbol{U}_H(\lambda)+\boldsymbol{X}\boldsymbol{P}_H(\lambda))^{-1}\boldsymbol{Q}_X(\lambda)\\&=\boldsymbol{\Omega}(\lambda)-A(\lambda)(\boldsymbol{U}_H(\lambda)+\boldsymbol{X}\boldsymbol{P}_H(\lambda))^{-1}(\boldsymbol{N}_H(\lambda)+\boldsymbol{X}\boldsymbol{M}_H(\lambda)),]\end{aligned}$$

where the free parameter $\boldsymbol{X}$ *runs over all the* $p\times q$ *constant matrices such that*

$$\widetilde{\boldsymbol{U}}_H(\lambda)+\widetilde{\boldsymbol{P}}_H(\lambda)\boldsymbol{X}[\boldsymbol{U}_H(\lambda)+\boldsymbol{X}\boldsymbol{P}_H(\lambda),resp.]\tag{5.15}$$

is invertible at each node x_i. *Moreover*, $\delta(\boldsymbol{F}_X(\lambda))=mp=nq$ *for all* $\boldsymbol{X}\in\mathbf{C}^{p\times q}$ *subject to Eq.* (5.15).

§ 5.3 Remarks

(a) Theorem 5.4 establishes the connection between the minimal partial realizations of the block Hankel vector $\boldsymbol{b}$ of the GRMIP (2.4) and the minimal, proper rational matrix interpolants to the GRMIP (2.4). Actually, there exists a one-to-one correspondence between $\boldsymbol{G}_X(\lambda)$ in Eq. (5.8) with those $\boldsymbol{X}\in\mathbf{C}^{p\times q}$ for which Eq. (5.15) holds, and the minimal, proper rational matrix interpolants $\boldsymbol{F}_X(\lambda)$ in Eq. (5.14). Thus, the problem of finding the latter can be reduced to what amounts to the more-studied problem of finding the former.

(b) From Eqs. (5.4)～(5.7), (5.11) and (5.12) we deduce that the coefficient matrices, $\widetilde{\boldsymbol{T}}(\lambda)$, $\boldsymbol{T}(\lambda)\in\mathbf{C}^{(p+q)\times(p+q)}[\lambda]$, of Eq. (5.14) given by

$$\widetilde{\boldsymbol{T}}(\lambda)=\begin{bmatrix}\boldsymbol{\Omega}(\lambda)\widetilde{\boldsymbol{P}}_H(\lambda)-A(\lambda)\widetilde{\boldsymbol{M}}_H(\lambda) & -\boldsymbol{\Omega}(\lambda)\widetilde{\boldsymbol{U}}_H(\lambda)+A(\lambda)\widetilde{\boldsymbol{N}}_H(\lambda)\\ \boldsymbol{P}_H(\lambda) & -\widetilde{\boldsymbol{U}}_H(\lambda)\end{bmatrix},$$

$$\boldsymbol{T}(\lambda)=\begin{bmatrix}\boldsymbol{U}_H(\lambda) & -\boldsymbol{U}_H(\lambda)\boldsymbol{\Omega}(\lambda)+A(\lambda)\boldsymbol{N}_H(\lambda)\\ \boldsymbol{P}_H(\lambda) & -\boldsymbol{P}_H(\lambda)\boldsymbol{\Omega}(\lambda)+A(\lambda)\boldsymbol{M}_H(\lambda)\end{bmatrix}$$

satisfy

$$\widetilde{\boldsymbol{T}}(\lambda)\boldsymbol{T}(\lambda)=\boldsymbol{T}(\lambda)\widetilde{\boldsymbol{T}}(\lambda)=-A(\lambda)\boldsymbol{I}_{p+q},$$

$$\det\widetilde{\boldsymbol{T}}(\lambda)=(-A(\lambda))^p,\quad \det\boldsymbol{T}(\lambda)=(-A(\lambda))^q.$$

Compare Theorem 4.1 of [12] for the case of simple interpolation nodes. Let now

$$\widetilde{\boldsymbol{K}}(\lambda)=\sum_{j=1}^{v}\sum_{l=0}^{n_j-1}\sum_{k=0}^{l}\boldsymbol{D}_{jk}\widetilde{\boldsymbol{U}}_{jk}b_{jl}(\lambda)\in\mathbf{C}^{p\times q}[\lambda],$$

$$\widetilde{\boldsymbol{V}}(\lambda)=b(\lambda)\boldsymbol{I}_p+\sum_{j=1}^{v}\sum_{l=0}^{n_j-1}\boldsymbol{D}_{jl}\widetilde{\boldsymbol{P}}_{jl}b_{jl}(\lambda)\in\mathbf{C}^{p\times p}[\lambda],$$

$$\boldsymbol{K}(\lambda)=\sum_{i=1}^{u}\sum_{k=0}^{m_i-1}\sum_{l=0}^{k}\boldsymbol{U}_{il}\boldsymbol{C}_{il}a_{ik}(\lambda)\in\mathbf{C}^{p\times q}[\lambda],$$

$$\boldsymbol{V}(\lambda)=a(\lambda)\boldsymbol{I}_q+\sum_{i=1}^{u}\sum_{k=0}^{m_i}\boldsymbol{P}_{ik}\boldsymbol{C}_{ik}a_{ik}(\lambda)\in\mathbf{C}^{q\times q}[\lambda],$$

where $a_{ik}(\lambda), b_{jl}(\lambda)$ are as in Eq. (2.23), $\boldsymbol{C}_{ik}, \boldsymbol{D}_{jl}\in\mathbf{C}^{p\times q}$ are as in Eqs. (2.4a) and (2.4b), and $\boldsymbol{P}_{ik}, \boldsymbol{U}_{ik}, \widetilde{\boldsymbol{P}}_{jl}, \widetilde{\boldsymbol{U}}_{jl}$ are defined via Eqs. (2.8) ~ (2.11). It can be readily verified that

$$[\widetilde{\boldsymbol{K}}(\lambda), \widetilde{\boldsymbol{U}}_H(\lambda)]_r \text{ and } [\boldsymbol{K}(\lambda), \boldsymbol{U}_H(\lambda)]_l$$

are respectively the unique right and the unique left generalized solution to the GRMIP (2.4) such that $\deg\widetilde{\boldsymbol{K}}(\lambda)<\deg\widetilde{\boldsymbol{U}}_H(\lambda)=n$ with $-\widetilde{\boldsymbol{U}}_H(\lambda)$ monic, and $\deg\boldsymbol{K}(\lambda)\leqslant\deg\boldsymbol{U}_H(\lambda)=m$ with $-\boldsymbol{U}_H(\lambda)$ monic, respectively. Similarly, $[\widetilde{\boldsymbol{V}}(\lambda), \widetilde{\boldsymbol{P}}_H(\lambda)]$ and $[\boldsymbol{V}(\lambda), \boldsymbol{P}_H(\lambda)]$ are respectively the unique polynomial matrix pair with $\deg\widetilde{\boldsymbol{P}}(\lambda)<\deg\widetilde{\boldsymbol{V}}(\lambda)=n$ ($\widetilde{\boldsymbol{V}}(\lambda)$ is monic), and $\deg\boldsymbol{P}_H(\lambda)\leqslant\deg\boldsymbol{V}(\lambda)=m$ ($\boldsymbol{V}(\lambda)$ is monic) such that

$$\frac{\mathrm{d}^k}{\mathrm{d}\lambda^k}[\widetilde{\boldsymbol{V}}(\lambda)-\boldsymbol{\Omega}(\lambda)\widetilde{\boldsymbol{P}}_H(\lambda)]\bigg|_{\lambda=x_i}=\boldsymbol{O}, i=1,2,\cdots,\rho, k=0,1,2,\cdots,\tau_i-1,$$

$$\frac{\mathrm{d}^k}{\mathrm{d}\lambda^k}[\boldsymbol{V}(\lambda)-\boldsymbol{P}_H(\lambda)\boldsymbol{\Omega}(\lambda)]\bigg|_{\lambda=x_i}=\boldsymbol{O}, i=1,2,\cdots,\rho, k=0,1,2,\cdots,\tau_i-1,$$

(Compare Theorems 3.1 and 3.2 of [12] for the case of simple interpolation nodes.) Thus, we obtain from Theorem 5.4 that

$$\widetilde{\boldsymbol{K}}(\lambda)=\boldsymbol{\Omega}(\lambda)\widetilde{\boldsymbol{U}}_H(\lambda)-A(\lambda)\widetilde{\boldsymbol{N}}_H(\lambda)\equiv\widetilde{\boldsymbol{Q}}_0(\lambda),$$

$$\widetilde{\boldsymbol{V}}(\lambda)=\boldsymbol{\Omega}(\lambda)\widetilde{\boldsymbol{P}}_H(\lambda)-A(\lambda)\widetilde{\boldsymbol{M}}_H(\lambda)\equiv\widetilde{\boldsymbol{Q}}_\infty(\lambda),$$

$$\boldsymbol{K}(\lambda)=\boldsymbol{U}_H(\lambda)\boldsymbol{\Omega}(\lambda)-A(\lambda)N_H(\lambda)\equiv\boldsymbol{Q}_0(\lambda),$$

$$\boldsymbol{V}(\lambda)=\boldsymbol{P}_H(\lambda)\boldsymbol{\Omega}(\lambda)-A(\lambda)\boldsymbol{M}_H(\lambda)\equiv\boldsymbol{Q}_\infty(\lambda).$$

(c) As a consequence of Eqs. (5.11) and (5.12) and Theorem 5.1 we have that for any $\boldsymbol{X}\in\mathbf{C}^{p\times q}$ a common left divisor of $\boldsymbol{Q}_X(\lambda)$ and $\widetilde{\boldsymbol{U}}_H(\lambda)+\widetilde{\boldsymbol{P}}_H(\lambda)\boldsymbol{X}$ divides $A(\lambda)I_q$; similarly for a common right divisor of $\boldsymbol{Q}_X(\lambda)$ and $\boldsymbol{U}_H(\lambda)+\boldsymbol{X}\boldsymbol{P}_H(\lambda)$, but with $A(\lambda)I_p$ in place of $A(\lambda)I_q$. Thus, we may generalize the results of [12] on the attainability to the multiple node case as well.

(d) A much more general parametrization of rational matrix interpolants to the GRMIP (2.4), which does not the very restrictive assumption that a generalized block Loewner matrix in Eqs. (2.3a) and (2.3b) is non-singular, and related problems will be dealt with in separate investigations by adopting the so-called block Hankel vector approach to the GRMIP (2.4).

Acknowledgements

The authors acknowledge with many thanks helpful comments and suggestions of the referee.

References

[1]B D O Anderson, A C Antoulas. Rational interpolation and state-variable realization. Linear Algebra Appl., 1990, 137—138: 479—510.

[2]B D O Anderson, E I Jury. Generalized Bezoutian and Sylvester matrices in multivarible linear control. IEEE Trans. Automat. Control, 1976, AC—21: 551—556.

[3]A C Antoulas, B D O Anderson. On the problem of state rational interpolation. Linear Algebra Appl., 1989, 122—124: 301—329.

[4]A C Antoulas, B D O Anderson. On the scalar rational interpolation problem. IMA J. Math. Control Inform., 1986, 3: 61—88.

[5]A C Antoulas, B D O Anderson. State-space and polynomial approaches to rational interpolation. in: M A Kaashoek, J H van Schuppen, A C M Ran (Eds.), Progress in Systems and Control Theory: Relaization and Modelling in System Theory, vol. I. Birkhäuser, Boston, 1990: 73—82.

[6]A C Antoulas, J A Ball, J Kang, J C Willems. On the solution of the minimal rational interpolation problem. Linear Algebra Appl., 1990, 137—138: 511—573.

[7]A C Antoulas, J C Willems. Minimal rational interpolation and Prony's method. in: A Bensoussan, J L Lions (Eds.), Analysis and Optimization of systems. Lecture Notes in Control and Information Science 144, Springer, Berlin,

1990:297—306.
[8]A C Antoulas. On recursiveness and related topics in Linear systems. IEEE Trans. Automat. Control,1986,AC—31:1 121—1 135.
[9]V Belevitch. Interpolation matrices. Philips Res. Rep. ,1970,25:337—369.
[10]J A Ball,I Gohberg,L Rodman. Interpolation of Rational Matrix Functions. Birkhäuser,Basel,1990.
[11]A Ben-Artzi,T Shalom. On inversion of block Toeplitz matrices. Integral Equations Operator Theory,1985,8:751—779.
[12]M Van Barel,Z Vavřín. Inverse of a block Löwner matrix J. Comput. Appl. Math. 1996,69:285—301.
[13]G -N Chen. The general rational interpolation and its connection with the Nevanlinna Pick interpolation and power moment problem. Linear Algebra Appl. ,1998,273:83—117.
[14]G -N Chen,Y-J Hu. The truncated Hamburger moment problem in both the nondegenerate and degenerate cases,and matrix continued fractions. Linear Algebra Appl. ,1998,277:199—236.
[15]G -N Chen,H-P Zhang. More on Loewner matrices. Linear Algebra Appl. , 1994,203—204:265—300.
[16]G -N Chen,B Zhao,H-P Zhang. The general rational interpolation problem in the scalar case and its Hankel vector. Linear Algebra Appl. ,1996,244:165—201.
[17]W F Donoghue Jr. Monotone Matrix Functions. Springer,New York,1974.
[18]M Fiedler. Hankel and Loewner matrices. Linear Algebra Appl. ,1984,58:75—95.
[19]P A Fuhrmann. Block Hankel matrix inversion—the polynomial approach. Oper. Theory Adv. Appl. vol. 19,Birkhäuser,Basel,1986.
[20]L Gemignani. Schur complements of Bezoutians and the inversion of block Hankel and block Toeplitz matrices. Linear Algebra. ,Appl. ,1997,253:39—59.
[21]I Gohberg,G Heining. Inversion of finite-section Toeplitz matrices consisting of elements of noncommutative algebra (in Russian). Rev. Roumaine Math. Pures Appl. ,1974,19(5):623—663.
[22]I Gohberg,I Koltracht. Triangular factors of Cauchy and Vandermonde matrices. Integral Equations Operator Theory,1996,26:46—59.
[23]I Gohberg,N Y Krupnik. A formula for inverting finite Toeplitz matrices(in Russian). Mat. Issled. ,1972,7(2):272—283.
[24]I Gohberg,T Shalom. On Bezoutian of nonsquare matrix polynomials and inversion of matrices with nonsquare blocks. Linear Algebra Appl. ,1990,137—138: 249—323.
[25]I Gohberg,T Shalom. On inversion of square matrices partitioned into nonsquare blocks. Integral Equations Operator Theory,1989,12:539—566.
[26] M I Gekhtman, M Shmoish. On invertibility of nonsquare generalized

Bezoutians. Linerar Algebra Appl. ,1995,223－224:205－241.

[27]G Heinig,K Rost. Algebraic Methods for Toeplitz-like Matrices and Operators. Birkhäuser,Basel,Boston,Stuttgart,1984.

[28]T S Ivanchenko, L A Sakhnovich. An operator approach to the Potapov scheme for the solution of interpolation problems. in:I Gohberg,L A Sakhnovich (Eds.),Matrix and Operator Valued Functions. Operator Theory:Advances and Applications. Vol. 72,Birkhäuser,London,1994:48－86.

[29]T Kailath. Linear Systems. Prentice-Hall,Englewood Cliffs,NJ,1980.

[30]T Kailath,S Kung,M Morf. Displacement ranks of matrices and linear equations. J. Math. Anal. Appl. ,1979,68:395－407.

[31] T Kailath, S Kung, M Morf. Displacement ranks of a matrix. Bull. Amer. Soc. ,1979,1:769－773.

[32] T Kailath, A H Sayed. Displacement structure: theory and applications. SIAM Review,1995,37:297－386.

[33]I V Kovalishina, V P Potapov. An indefinite metric in the Nevanlinna-Pick problem. Amer. Math. Soc. Transl. ,1988,138(2):15－19.

[34]G Labahn,T Shalom. Inversion of Toeplitz matrices with only two standard equations. Linear Algebra Appl. ,1992,175:143－158.

[35]L Lerer, M Tismenetsky. Generalized Bezoutian and the inversion problem for block matrices, I. General Scheme. Integral Equations Operator Theory, 1986,9:790－819.

[36]K Löwner. Über monotone Matrixfunktionen. Math. Z. ,1934,38:177－216.

[37]V Pták, Z Vavřín. Bezout, Hankel, and Loewner matrices. Linear Algebra Appl. ,1993,184:13－36.

[38]W F Trench. A note on a Toeplitz inversion formula. Linear Algebra Appl. , 1990,129:55－61.

[39]Z Vavřín. A unified approach to Loewner and Hankel matrices. Linear Algebra Appl. ,1991,143:171－222.

[40]Z Vavřín. Inverses of Löwner matrices. Linear Algebra Appl. ,1984,63:227－236.

陈公宁文集

四、解析函数插值和矩量问题以及有关分析问题

Ⅳ. Interpolations for Analytic Functions and Moment Problems, and Some Related Questions in Analysis

数学物理学报
2010,30A(3):577－583

完全不确定 Hamburger 矩阵矩量问题的有限阶解①

Solutions of Finite Order to a Completely Indeterminate Hamburger Matrix Moment Problem

摘要 该文描述带有矩量序列$\{\boldsymbol{v}_m\}_0^\infty \subset \mathbf{C}^{q\times q}$的完全不确定 Hamburger 矩阵矩量问题：$\boldsymbol{v}_m = \int_{-\infty}^{\infty} x^m \mathrm{d}\rho(x), m = 0,1,\cdots$ 的有限阶解，即该问题的那些解 ρ，使得 $\mathbf{C}^{q\times q}$-值多项式的线性空间 P 在对应的空间 $L^2\left(\mathbf{R}, \frac{\mathrm{d}\rho}{E(x)}\right)$ 内稠密，这里 $E(x)$ 为在实轴 $\mathbf{R}$ 上取正值的某个数值多项式. 作为预备知识，作者考虑所谓广义 Akhiezer 插值的矩阵情形变种与它的相关矩阵矩量问题之间的一种关系.

关键词 矩阵测度；Hamburger 矩阵矩量问题；广义 Akhiezer 矩阵插值；N-极端矩阵测度；Riesz 定理；有限阶解

§1. 引言

设 $\boldsymbol{v} := \{\boldsymbol{v}_m\}_0^\infty$ 是给定的 Hamburger $q\times q$ 埃尔米特矩阵矩量序列，亦即：至少存在一个实轴 $\mathbf{R}$ 上的正 $q\times q$ 矩阵测度 ρ，求解如下的 Hamburger 矩阵矩量问题(HMMP)

$$\boldsymbol{v}_m = \int_{-\infty}^{\infty} x^m \mathrm{d}\rho(x), \quad m = 0,1,\cdots \tag{1}$$

用 $\mu^H(\boldsymbol{v})$ 记问题(1)的解 ρ 的集合. Riesz 定理的一种矩阵变种断言，对于$\rho\in$

① 收稿日期：2007-12-11；修订日期：2009-07-06. 基金项目：国家自然科学基金(10271018)资助. 本文与秦建国合作.

$\mu^H(\boldsymbol{v})$,$\mathbf{C}^{q\times q}$-值多项式的线性空间 P 在 $L^2(\mathbf{R},\mathrm{d}\rho)$ 内稠密,当且仅当对某个(因而对任意)$z\in\mathbf{C}\backslash\mathbf{R}$,$\int_{-\infty}^{\infty}\frac{\mathrm{d}\rho(x)}{x-z}$ 是矩阵凸集 $\left\{\int_{-\infty}^{\infty}\frac{\mathrm{d}\sigma(x)}{x-z}:\sigma\in\mu^H(\boldsymbol{v})\right\}$ 的一个极端点(按凸性意义)(见 Duran 和 Lopez-Rodriguez[4]). 这样的 $\rho\in\mu^H(\boldsymbol{v})$ 称为 N-极端矩阵测度. 问题(1)称为确定的,假如 $\mu^H(\boldsymbol{v})$ 中仅含有一个元素,否则该问题是不确定的. 当问题(1)确定时,$\mu^H(\boldsymbol{v})$ 中的唯一元素 ρ(对于它,P 自动地在 $L^2(\mathbf{R},\mathrm{d}\rho)$ 内稠密)是一个 N-极端矩阵测度. 本文后面总假定问题(1)是完全不确定的,换言之,它有最大可能的不确定度,或等价地,定义域为 P 的右移位算子 A_0 在 Hilbert 空间 $\overline{P}$(P 的闭包)有亏指数(q,q). 因此,块 Hankel 矩阵$(\boldsymbol{v}_{i+j})_{i,j=0}^{l}$ 对 $l=0,1,\cdots$ 均为埃尔米特正定的(见 Krein[8,第10节]).

如果 $P(z)$ 为 $\mathbf{C}^{q\times q}$-值整函数,那么令 $P^*(\bar{z}):=(P(z))^*$,$\forall z\in\mathbf{C}$. 与一个完全不确定 HMMP(1)相对应的有 4 个 $\mathbf{C}^{q\times q}$-值整函数 $A(z)$,$B(z)$,$C(z)$ 与 $D(z)$. 如同数值情形,它们使得解集 $\mu^H(\boldsymbol{v})$ 的 Nevanlinna 参数化表示由下式以及 Stieltjes 反演公式给出

$$G_\rho(z)=\int_{-\infty}^{\infty}\frac{\mathrm{d}\rho(x)}{x-z}=-\{C^*(z)[I_q+V_\rho(z)]-\mathrm{i}A^*(z)[I_q-V_\rho(z)]\}\times$$
$$\{D^*(z)[I_q+V_\rho(z)]-\mathrm{i}B^*(z)[I_q-V_\rho(z)]\}^{-1},\quad \forall z\in\mathbf{C}^+,\qquad(2)$$

其中,$\mathbf{C}^+:=\{z\in\mathbf{C}:\mathrm{Im}\,z>0\}$,而 $V_\rho(z)$ 遍历 $\mathbf{C}^+$ 内 $\mathbf{C}^{q\times q}$-值解析压缩函数集合 $\mathbf{S}_q$:$(V_\rho(z))^*V_\rho(z)\leqslant I_q$,$\forall z\in\mathbf{C}^+$. 公式(2)提供集合 $\mathbf{S}_q$ 与集合 $\mu^H(\boldsymbol{v})$ 之间一种同胚映射. 而且,N-极端矩阵测度恰对应到 $V_\rho(z)\equiv\boldsymbol{U}$ 的情形,这里 $\boldsymbol{U}$ 为常数 $q\times q$ 酉矩阵. 它们组成 $\mu^H(\boldsymbol{v})$ 内一个特异类,这是因为它们对应于前面提及的 $\overline{A}_0$ 的自伴扩张,或者对应于有关 $\boldsymbol{v}$ 的无限阶块 Jacobi 矩阵确定的算子,其谱包含在某个极小类型整函数的零点集合内(见 Krein[8,第10节]).

$\rho\in\mu^H(\boldsymbol{v})$ 称为有限阶的,譬如至多为 l 阶的,假如空间 P 在对应的 $L^2\left(\mathbf{R},\frac{\mathrm{d}\rho}{E(x)}\right)$ 内稠密,这里 $E(z)$ 为某个在实轴 $\mathbf{R}$ 上取正值的次数为 $2l$ 的数值多项式. 于是,$\mu^H(\boldsymbol{v})$ 内 N-极端矩阵测度恰为 0 阶的. 本文旨在描述这些有限阶 $\rho\in\mu^H(\boldsymbol{v})$,应用的主要工具是广义 Akhiezer 插值的一种矩阵变种与它的相关矩阵矩量问题之间的深刻的关系. 我们的结果是:$\rho\in\mu^H(\boldsymbol{v})$ 具有有限阶,当且仅当对某个在 $\mathbf{R}$ 上取正值的数值多项式 $E(z)$,

$\frac{\rho}{E(z)}\in\mu^H(\boldsymbol{\eta})$是 N-极端矩阵测度,其中,$\boldsymbol{\eta}$ 是某个广义 Akhiezer 插值相关矩量问题的 Hamburger 矩阵矩量序列(见第 2 节). 本文的主要结果展现在定理 3.1 中,它可视为 Akhiezer[1],胡永建和陈公宁[6],Simon[9,第6节]相关结论的对矩阵情形的深刻的推广,同时,本文的证明也大为简化了 Akhiezer[1],Simon[9]的有关推证.

§2. 广义 Akhiezer 矩阵插值与相关的矩阵矩量问题

给定 $\rho\in\mu^H(\boldsymbol{\nu})$,令 $G_\rho:\mathbf{C}\backslash\mathbf{R}\to\mathbf{C}^{q\times q}$ 为由下式确定的 ρ 的 Stieltjes 变换

$$G_\rho(z):=\int_{-\infty}^{\infty}\frac{\mathrm{d}\rho(x)}{x-z},\quad z\in\mathbf{C}\backslash\mathbf{R},\tag{3}$$

它属于 $\mathbf{N}_q$ 类,即在 $\mathbf{C}^+$ 内解析并有非负埃尔米特虚部的 $\mathbf{C}^{q\times q}$-值函数的集合. 首先,我们需要如下的 Hamburger-Nevanlinna 引理的一种矩阵推广.

引理 2.1(Akhiezer[1],Kovalishina[7]) 设 $\tilde{\boldsymbol{\nu}}=\{\tilde{\boldsymbol{\nu}}_m\}_0^\infty$ 为 Hamburger $q\times q$ 埃尔米特矩阵矩量序列. 如果 $\rho\in\mu^H(\tilde{\boldsymbol{\nu}})$,那么 $G_\rho(z)=\int_{-\infty}^{\infty}\frac{\mathrm{d}\rho(x)}{x-z}\in\mathbf{N}_q$,且可渐近表示为在每个扇形 $\pi_\varepsilon:=\left\{z\in\mathbf{C}^+:\varepsilon\leqslant\arg z\leqslant\frac{\pi}{2}-\varepsilon\right\}$,$\forall\varepsilon\in\left(0,\frac{\pi}{2}\right)$内一致收敛的级数:$G_\rho(z)=-\sum_{j=0}^{\infty}\frac{\tilde{\nu}_j}{z^{j+1}}(z\to\infty)$. 反之,如果 $G(z)\in\mathbf{N}_q$ 在 $z=\mathrm{i}y(y>0)$有展开式:$G(\mathrm{i}y)=-\sum_{j=0}^{\infty}\frac{\tilde{\nu}_j}{(\mathrm{i}y)^{j+1}}(y\to\infty)$,那么存在 $\mu^H(\tilde{\nu})$内唯一的 ρ,使得 $G(z)=G_\rho(z)=\int_{-\infty}^{\infty}\frac{\mathrm{d}\rho(x)}{x-z}$.

给定 $\mathbf{C}^+$ 内长度为 n 的数据$\triangle:=\{(z_j,\tau_j)_{j=1}^s\}$(此意味着,诸 z_j 在 $\mathbf{C}^+$ 内互异且有重数 τ_j,$j=1,2,\cdots,s$,而 $n=\tau_1+\tau_2+\cdots+\tau_s$),与 q 阶矩阵序列$\{\boldsymbol{W}_{jk}\}$,$j=1,2,\cdots,s$;$k=0,1,\cdots,\tau_j-1$. 这里的广义 Akhiezer 矩阵插值(GAMI)需要寻求对应于 $\rho\in\mu^H(\boldsymbol{\nu})$的形如(3)式的所有 G_ρ,使得

$$\frac{1}{k!}G_\rho^{(k)}(z_j)=\int_{-\infty}^{\infty}\frac{\mathrm{d}\rho(x)}{(x-z_j)^{k+1}}=\boldsymbol{W}_{jk},$$
$$j=1,2,\cdots,s,\ k=0,1,\cdots,\tau_j-1.\tag{4}$$

设 $\Omega(z)$是满足条件$\frac{1}{k!}\Omega^{(k)}(z_j)=\boldsymbol{W}_{jk}$与$\frac{1}{k!}\Omega^{(k)}(\bar{z}_j)=\boldsymbol{W}_{jk}^*$,$j=1,2,\cdots,s$,$k=0,1,\cdots,\tau_j-1$ 的至多 $2n-1$ 次的(带有埃尔米特矩阵系数的)$\mathbf{C}^{q\times q}$-值

Sylvester-Lagrange 多项式，又设 $E(z)$ 是 $2n$ 次实系数的数值多项式

$$E(z)=\prod_{j=1}^{s}(z-z_j)^{\tau_j}(z-\bar{z}_j)^{\tau_j}=:\sum_{j=0}^{2n}a_jz^j\quad(a_{2n}=1),\tag{5}$$

它有性质：$E(x)>0,\forall x\in R$. 这时对充分大的 $|z|$，我们有

$$\frac{\Omega(z)}{E(z)}:=\sum_{j=0}^{\infty}\frac{\boldsymbol{S}_j}{z^{j+1}},\quad \boldsymbol{S}_j=\boldsymbol{S}_j^*\in\mathbf{C}^{q\times q},\forall j.\tag{6}$$

因此

$$\Omega(z)=\sum_{k=0}^{2n-1}\sum_{j=0}^{2n-k-1}a_{j+k+1}\boldsymbol{S}_jz^k.\tag{7}$$

现设 $\{\boldsymbol{H}_m\}_0^\infty$ 为满足由下式给出的 $2n+1$ 项递推关系的埃尔米特矩阵序列

$$\boldsymbol{H}_{2n+m}:=\boldsymbol{v}_m-\sum_{k=O}^{2n-1}a_k\boldsymbol{H}_{m+k},\quad m\geqslant 0,\tag{8a}$$

原始条件为

$$\boldsymbol{H}_j:=\boldsymbol{S}_j,\ j=0,1,\cdots,2n-1.\tag{8b}$$

此矩阵序列 $\{\boldsymbol{H}_m\}_0^\infty$ 对应着一个相关的 Hamburger 矩阵矩量问题：寻求 $\mathbf{R}$ 上的埃尔米特半正定 $q\times q$ 矩阵测度 σ，使得

$$\boldsymbol{H}_m=\int_{-\infty}^{\infty}x^m\,\mathrm{d}\sigma(x),\ m=0,1,\cdots.\tag{9}$$

但要注意，问题(9)不必是完全不确定的，即使它是可解的.

如下精细的定理是胡永建和陈公宁[6]对应结论的一种推广.

定理 2.1　设 $\Omega(z)$ 与 $E(z)$ 如前，GAMI 问题(4)可解，当且仅当矩阵矩量问题(9)可解. 此时，公式

$$G_\rho(z)=\int_{-\infty}^{\infty}\frac{1}{x-z}\mathrm{d}\rho(x)=\Omega(z)+E(z)\int_{-\infty}^{\infty}\frac{1}{x-z}\mathrm{d}\sigma(x)\tag{10}$$

实现 GAMI 问题(4)的解 $G_\rho(\rho\in\mu^H(\nu))$ 与问题(9)的解 σ 之间的一一对应，其中，ρ 与 σ 二者由下式关系连接

$$\mathrm{d}\sigma(x)=\frac{\mathrm{d}\rho(x)}{E(x)},\ x\in\mathbf{R}.$$

证　设对 $\rho\in\mu^H(\nu)$，$G_\rho(z)=\int_{-\infty}^{\infty}\frac{1}{x-z}\mathrm{d}\rho(x)\in\mathbf{N}_q$ 是 GAMI 问题(4)的一个解. $G_{\rho E}(z)$ 按下式确定

$$G_\rho(z)=\Omega(z)+E(z)G_{\rho E}(z),\ z\in\mathbf{C}\backslash\mathbf{R}.\tag{11}$$

则 $G_{\rho E}(z)$ 有明确的定义，且在 $\mathbf{C}\backslash\mathbf{R}$ 内解析. 从 Chen 和 Li[3] 我们推出，$G_{\rho E}(z)$ 属于 $\mathbf{N}_q$ 类，且有表示

$$G_{\rho E}(z)=\int_{-\infty}^{\infty}\frac{1}{x-z}\frac{\mathrm{d}\rho(x)}{E(x)},\ z\in\mathbf{C}\backslash\mathbf{R}.$$

另一方面，从引理 2.1，(6)与(11)两式，对充分大的$|z|$，我们有

$$G_{\rho E}(z)=\int_{-\infty}^{\infty}\frac{1}{x-z}\frac{\mathrm{d}\rho(x)}{E(x)}=-\frac{1}{E(z)}\sum_{j=0}^{\infty}\frac{\boldsymbol{v}_j}{z^{j+1}}-\sum_{j=0}^{\infty}\frac{\boldsymbol{S}_j}{z^{j+1}},$$

于是，在每个扇形 π_ε 内，$G_{\rho E}(z)=-\sum\limits_{j=0}^{\infty}\frac{\widetilde{\boldsymbol{H}}_j}{z^{j+1}}(z\to\infty)$ 一致地成立，这里诸 $\widetilde{\boldsymbol{H}}_j$ 由下面确定

$$\widetilde{\boldsymbol{H}}_j=\boldsymbol{S}_j,\ j=0,1,\cdots,2n-1,$$

$$\widetilde{\boldsymbol{H}}_{2n+m}=\boldsymbol{v}_m-\sum_{k=0}^{2n-1}a_k\widetilde{\boldsymbol{H}}_{m+k},m=0,1,\cdots,$$

进而按(8)式，$\widetilde{\boldsymbol{H}}_j=\boldsymbol{H}_j$，$j=0,1,\cdots$. 对 $G_{\rho E}(z)$，应用引理 2.1，我们看出，$\frac{\mathrm{d}\rho(x)}{E(x)}$求解矩阵矩量问题(9).

反之，假定 σ 求解问题(9)，亦即：$\sigma\in\mu^H(\{\boldsymbol{H}_m\}_0^\infty)$. 再按 Chen 和 Li[3]，由下式确定的 $\mathbf{C}^{q\times q}$-值函数 G

$$G(z):=\Omega(z)+E(z)\int_{-\infty}^{\infty}\frac{1}{x-z}\mathrm{d}\sigma(x),\quad z\in\mathbf{C}\backslash\mathbf{R}$$

属于 $\mathbf{N}_q$ 类，其积分表示中对应的矩阵测度是 $E(x)\mathrm{d}\sigma(x)$，$x\in\mathbf{R}$. 于是，对充分大的$|z|$，从(7)(8)两式与引理 2.1 推出

$$G(z)=\sum_{k=0}^{2n-1}\sum_{j=0}^{2n-k-1}a_{j+k+1}\boldsymbol{S}_jz^k-\sum_{j=0}^{2n}a_jz^j\left(\sum_{j=0}^{\infty}\frac{\boldsymbol{H}_j}{z^{j+1}}\right)=-\sum_{j=0}^{\infty}\frac{\boldsymbol{v}_j}{z^{j+1}}.$$

因此，再次应用引理 2.1，$G(z)$应重合于

$$G_\rho(z)=\int_{-\infty}^{\infty}\frac{1}{x-z}\mathrm{d}\rho(x),$$

式中 $\mathrm{d}\rho(x)=E(x)\mathrm{d}\sigma(x)$，$x\in\mathbf{R}$. 显然，$G(z)$满足条件(4). ■

推论 2.1 带有 $\rho\in\mu^H(\boldsymbol{v})$的 GAMI 问题(4)有唯一解 $G_\rho(z)\in\mathbf{N}_q$，当且仅当相关的 Hamburger 矩阵矩量问题(9)是确定的.

推论 2.2 $\rho\in\mu^H(\boldsymbol{v})$使得 $G_\rho(z)$满足条件(4)，当且仅为$\frac{\rho}{E(x)}\in\mu^H(\{\boldsymbol{H}_m\}_0^\infty)$，亦即

$$\boldsymbol{H}_m=\int_{-\infty}^{\infty}x^m\frac{\mathrm{d}\rho(x)}{E(x)},\ m=0,1,\cdots.$$

§3. $\mu^H(\boldsymbol{\nu})$内有限阶解

现在我们转到本文的主题:描述问题(1)的有限阶解.如果 $\rho\in\mu^H(\boldsymbol{\nu})$ 使得空间 $\overline{P}$(P 在 $L^2(\mathbf{R},\mathrm{d}\rho)$内的闭包)重合于对应的 $L^2\left(\mathbf{R},\dfrac{\mathrm{d}\rho}{\widetilde{E}(x)}\right)$,这里 $\widetilde{E}(x)$是某个次数为 $2l$,且在 $\mathbf{R}$ 上取正值的多项式,那么我们将称 ρ 至多为 l 阶的.在后面定理 3.1 中,我们要证明:$\rho\in\mu^H(\boldsymbol{\nu})$的阶与这样的 $\widetilde{E}(z)$ 的选取无关.这个描述的关键处在于我们必须将对应的空间$L^2(\mathbf{R},\mathrm{d}\rho)$与 $L^2\left(\mathbf{R},\dfrac{\mathrm{d}\rho}{\widetilde{E}(x)}\right)$(其中 $\rho\in\mu^H(\boldsymbol{\nu})$)想象为左的或右的 $\mathbf{C}^{q\times q}$-模.对 $m=1,2,\cdots$, 所有的形如$\dfrac{\boldsymbol{I}_q}{(x-z)^m}$(这里 $z\in\mathbf{C}\backslash\mathbf{R}$)的有理 $\mathbf{C}^{q\times q}$-值函数均属于 $L^2(\mathbf{R},\mathrm{d}\rho)$. 设 $E(x)$ 如(5)式所示,而 U 为下式确定的从 $L^2(\mathbf{R},\mathrm{d}\rho)$ 到 $L^2\left(\mathbf{R},\dfrac{\mathrm{d}\rho}{E(x)}\right)$的酉变换

$$(Uf)(x)=\prod_{j=1}^{s}(x-z_j)^{\tau_j}f(x).$$

则 U 将 $P\cup\left\{\dfrac{\boldsymbol{I}_q}{(x-z_j)^{k+1}}\right\}_{j=1,2,\cdots,s,\ k=0,1,\cdots,\tau_j-1}$ 的生成映射到 P 上.因此,其稠密性的结论是等价的,换言之,空间 P 在 $L^2\left(\mathbf{R},\dfrac{\mathrm{d}\rho}{E(x)}\right)$内稠密,即 $\overline{P}=L^2\left(\mathbf{R},\dfrac{\mathrm{d}\rho}{E(x)}\right)$,这里 $E(x)$ 如(5)式所示,当且仅当 $\overline{P}\cup\left\{\dfrac{\boldsymbol{I}_q}{(x-z_j)^{k+1}}\right\}_{j=1,2,\cdots,s;\ k=0,1,\cdots,\tau_j-1}$ 的生成重合于 $L^2(\mathbf{R},\mathrm{d}\rho)$.因此我们有

引理 3.1　设 $\rho\in\mu^H(\boldsymbol{\nu})$的阶至多等于 $l(\leqslant n)$.则此时且仅在此时,对于 $\mathbf{C}^+$ 内长度为 n 的某数据$\{(z_j,\tau_j)_{j=1}^s\}$,有

$$\overline{P}\cup\left\{\frac{\boldsymbol{I}_q}{(x-z_j)^{k+1}}\right\}_{j=1,2,\cdots,s;\ k=0,1,\cdots,\tau_j-1}$$

生成 $L^2(\mathbf{R},\mathrm{d}\rho)$,或等价地,$P$ 在 $L^2\left(\mathbf{R},\dfrac{\mathrm{d}\rho}{E(x)}\right)$内稠密,这里 $E(x)$如(5)式.

设 $\rho\in\mu^H(\boldsymbol{\nu})$由(2)式给定,其中 $\mathbf{V}_\rho\in\mathbf{S}_q$.对 $\mathbf{C}^+$ 内长度为 $m:=n+1$ 的某数据$\triangle=\{(z_j,\tau_j)_{j=1}^{s+1}\}$,其中 $z_{s+1}\neq z_j(1\leqslant j\leqslant s)$,且 $\tau_{s+1}=1$,我们定义所谓的块 Pick 矩阵如下

$$\boldsymbol{P}_m^{(V_\rho)} := (\boldsymbol{P}_{ij})_{i,j=1}^{s+1} \in \mathbf{C}^{mq\times mq}, \tag{12}$$

式中 $\boldsymbol{P}_{ij} := (\boldsymbol{P}_{ij}^{kl})_{k=0,l=0}^{\tau_i-1,\tau_j-1}$，其元素 $\boldsymbol{P}_{ij}^{kl} \in \mathbf{C}^{q\times q}$ 由下式确定

$$\boldsymbol{P}_{ij}^{kl} = \frac{1}{k!\,l!}\frac{\partial^{k+l}}{\partial\lambda^k\partial\mu^l}\left[\frac{I_q - V_\rho(\lambda)V_\rho(\bar{\mu})^*}{\lambda-\mu}\right]_{\lambda=z_i,\ \mu=\bar{z}_j}. \tag{13}$$

引理 3.2 设 $\rho\in\mu^H(\boldsymbol{\nu})$ 由(2)式给定，其中 $V_\rho\in\mathbf{S}_q$. 则对 $\mathbf{C}^+$ 内长度为 $m=n+1$ 的某数据 $\triangle=\{(z_j,\tau_j)_{j=1}^{s+1}\}$，不存在其他的 $\xi\in\mu^H(\boldsymbol{\nu})$ 使得 $G_\xi^{(k)}(z_j)=G_\rho^{(k)}(z_j)$，$j=1,2,\cdots,s+1$；$k=0,1,\cdots,\tau_j-1$，当且仅当 $V_\rho(z)\in\mathbf{S}_q$ 满足

$$\mathrm{rank}\ \boldsymbol{P}_m^{(V_\rho)} = \mathrm{rank}\ \boldsymbol{P}_n^{(V_\rho)}, \tag{14}$$

这里，$\boldsymbol{P}_n^{(V_\rho)}$ 为 $\boldsymbol{P}_m^{(V_\rho)}$ 的左上角 nq 阶子阵. 此时，(2)式中的 $V_\rho(z)\in\mathbf{S}_q$ 是形式次数至多为 n 的有理 $\mathbf{C}^{q\times q}$-值函数.

证 按前面的分析，带有 $\mathbf{C}^+$ 中长度为 m 的数据 $\triangle=\{(z_j,\tau_j)_{j=0}^{s+1}\}$ 与 $\boldsymbol{W}_{jk} := \frac{1}{k!}G_\rho^{(k)}(z_j)$，$\forall\, j,k$ 的 GAMI 问题(4)有唯一解 $G_\rho(z)$ ($\rho\in\mu^H(\boldsymbol{\nu})$)，当且仅当不存在其他的 $V\in\mathbf{S}_q$ 满足条件 $V^{(k)}(z_j)=V_\rho^{(k)}(z_j)$，$j=1,2,\cdots,s+1$；$k=0,1,\cdots,\tau_j-1$. 然而后者成立当且仅当由(12)(13)式定义的对应块 Pick 矩阵 $\boldsymbol{P}_m^{(V_\rho)}$ 是埃尔米特半正定的，且拥有性质(14). 这个结果是基于对 $\mathbf{S}_q$ 类内给定函数与 $\mathbf{C}^+$ 内给定长度为 m 的数据△的经典 Schur-Nevanlinna 算法的矩阵变种(见文献[5，推论 5.6]对于 $p\times q$ 矩阵值 Schur 函数的一个类似结论). 在这种情形下，按类似的理由，对应的 $V_\rho(z)\in\mathbf{S}_q$ 必须是形式次数不超过 n 的有理 $\mathbf{C}^{q\times q}$-值函数. ■

注 3.1 从类似于文献[5，推论 5.6]的推证，我们得出，$\boldsymbol{P}_m^{(V_\rho)}$ 的秩不依赖于 $\mathbf{C}^+$ 内长度为 m 的数据△的选取，于是，当数据 $\triangle=\{(z_j,\tau_j)_{j=1}^{s+1}\}$ 换为 $\mathbf{C}^+$ 内长度为 m 的任意数据△′时，引理 3.2 仍然成立. 此结论将用于后面分析.

推论 3.1 设 $\rho\in\mu^H(\boldsymbol{\nu})$. 如果 ρ 是一个 N-极端矩阵测度，那么此时且仅在此时，对每个 $z_0\in\mathbf{C}^+$，不存在 $\mu^H(\boldsymbol{\nu})$ 内其他的 ξ，使得

$$G_\xi(z_0)=G_\rho(z_0).$$

证 如果 ρ 是一个 N-极端矩阵测度，那么存在一个常数 $q\times q$ 酉矩阵 $\boldsymbol{U}$ 使得 $G_\rho(z)$ 采纳表示(2)，但其中 $V_\rho(z) := \boldsymbol{U}$，因而所要的唯一性结果成立. 这是因为对每个 $z_0\in\mathbf{C}^+$，$G_\xi(z_0)=G_\rho(z_0)$，当且仅当对每个 $z_0\in$

$\mathbf{C}^+$，$V_\rho(z_0)=V_\xi(z_0)$，因而 $V_\rho(z_0)$ 此时等于 $\boldsymbol{U}$. 反之，假定对每个 $z_0\in\mathbf{C}^+$，不存在其他的 $\xi\in\mu^H(\boldsymbol{\nu})$ 满足 $G_\xi(z_0)=G_\rho(z_0)$. 这意味着带有单个单重节点 $z_1:=z_0$ 与 $\boldsymbol{W}_{10}:=G_\rho(z_0)$ 的 GAMI 问题(4)只有唯一解，这恰是 $G_\rho(z)$，或等价地，不存在 $\mathbf{S}_q$ 内其他的 $\widetilde{V}$ 使得 $\widetilde{V}(z_0)=V_\rho(z_0)$. 此时且仅在此时，引理 3.2 连同注 3.1 断定，$V_\rho(z)V_\rho(z)^*=I_q$，$\forall z\in\mathbf{C}^+$，或等价地，$\operatorname{Im}G_\rho(z)=O$，$\forall z\in\mathbf{C}^+$，于是，$G_\rho(z)$ 重合于某个常数 $q\times q$ 埃尔米特矩阵 $\boldsymbol{G}$，因而 $V(z)$ 重合于某个常数 $q\times q$ 酉矩阵. 因此，按定义，$\rho\in\mu^H(\boldsymbol{\nu})$ 是一个 N-极端矩阵测度. ■

综上所述，我们可以讨论本文的主要结论.

定理 3.1　设 $\rho_0\in\mu^H(\boldsymbol{\nu})$ 且 $G_{\rho_0}(z)$ 为它的 Stieltjes 变换并有形如(2)式的表示($\rho:=\rho_0$). 固定 $n\in\{0,1,\cdots\}$. 则下列诸命题彼此等价

(a) ρ_0 的阶 $\leqslant n$.

(b) 对 $\mathbf{C}^+$ 内长度为 n 的某数据 $\{(z_j,\tau_j)_{j=1}^s\}$，$\overline{P}\cup\left\{\dfrac{\boldsymbol{I}_q}{(x-z_j)^{k+1}}\right\}_{j=1,k=0}^{s,\tau_j-1}$ 生成 $L^2(\mathbf{R},\mathrm{d}\rho_0)$.

(c) 对 $\mathbf{C}^+$ 内长度为 n 的任意数据 $\{(z_j,\tau_j)_{j=1}^s\}$，$\overline{P}\cup\left\{\dfrac{\boldsymbol{I}_q}{(x-z_j)^{k+1}}\right\}_{j=1,k=0}^{s,\tau_j-1}$ 生成 $L^2(\mathbf{R},\mathrm{d}\rho_0)$.

(d) 对 $\mathbf{C}^+$ 内长度为 $m=n+1$ 的某数据 $\{(z_j,\tau_j)_{j=1}^{s+1}\}$(其中 z_{s+1} 单重)，不存在其他的 $\rho\in\mu^H(\boldsymbol{\nu})$ 使得 $G_\rho^{(k)}(z_j)=G_{\rho_0}^{(k)}(z_j)$，$j=1,2,\cdots,s+1$，$k=0,1,\cdots,\tau_j-1$.

(e) 对 $\mathbf{C}^+$ 内长度为 $m=n+1$ 的某数据 $\{(z_j,\tau_j)_{j=1}^{s+1}\}$(其中 z_{s+1} 单重)，$V_{\rho_0}\in\mathbf{S}_q$ 满足(14)式，其中 ρ 被 ρ_0 代替.

(f) 对 $\mathbf{C}^+$ 内长度为 $m=n+1$ 的任意数据 $\{(z_j,\tau_j)_{j=1}^{s+1}\}$(其中 z_{s+1} 单重)，不存在其他的 $\rho\in\mu^H(\boldsymbol{\nu})$ 使得 $G_\rho^{(k)}(z_j)=G_{\rho_0}^{(k)}(z_j)$，$j=1,2,\cdots,s+1$，$k=0,1,\cdots,\tau_j-1$.

(g) 对 $\mathbf{C}^+$ 内长度为 $m=n+1$ 的任意数据 $\{(z_j,\tau_j)_{j=1}^{s+1}\}$(其中 z_{s+1} 单重)，$V_{\rho_0}\in\mathbf{S}_q$ 满足(14)式，其中 $\rho:=\rho_0$.

(h) 对 $\mathbf{C}^+$ 内长度为 $m=n+1$ 的某数据 $\{(z_j,\tau_j)_{j=1}^{s+1}\}$(其中 z_{s+1} 单重)与 $\boldsymbol{W}_{jk}:=\dfrac{1}{k!}G_{\rho_0}^{(k)}(z_j)$，$j=1,2,\cdots,s+1$，$k=0,1,\cdots,\tau_j-1$，相关的矩阵矩

量问题(8)(9)为确定的,其唯一解为 $d\sigma(x)=\dfrac{d\rho_0(x)}{\prod_{j=1}^{s+1}|x-z_j|^{2\tau_j}}$.

(i) 对 $\mathbf{C}^+$ 内长度为 $m=n+1$ 的任意数据 $\{(z_j,\tau_j)_{j=1}^{s+1}\}$(其中 z_{s+1} 单重)与 $\boldsymbol{W}_{jk}:=\dfrac{1}{k!}G_{\rho_0}^{(k)}(z_j),j=1,2,\cdots,s+1,k=0,1,\cdots,\tau_j-1$,相关的矩阵矩量问题(8)(9)为确定的,其唯一解为 $d\sigma(x)=\dfrac{d\rho_0(x)}{\prod_{j=1}^{s+1}|x-z_j|^{2\tau_j}}$.

(j)对 $\mathbf{C}^+$ 内长度为 n 的某数据 $\{(z_j,\tau_j)_{j=1}^{s}\}$,$\dfrac{\rho_0}{E(x)}$是 $\mu^H(\{\boldsymbol{H}_m\}_{m=0}^{\infty})$内一个 N-极端矩阵测度,其中 $E(x)$与$\{\boldsymbol{H}_m\}_{m=0}^{\infty}$分别如(5)式与(8)式.

(k)对 $\mathbf{C}^+$ 内长度为 n 的任意数据 $\{(z_j,\tau_j)_{j=1}^{s}\}$,$\dfrac{\rho_0}{E(x)}$是 $\mu^H(\{\boldsymbol{H}_m\}_{m=0}^{\infty})$内一个 N-极端矩阵测度,其中 $E(x)$与$\{\boldsymbol{H}_m\}_{m=0}^{\infty}$同(j).

在这种情形下我们有

(l) (2)式中的 $V_{\rho_0}(z)\in\mathbf{S}_q$ 是形式次数至多为 n 的有理 $\mathbf{C}^{q\times q}$-值函数.

证 (a)⇔(b)⇔(j):根据引理 3.1.(d)⇔(h)⇔(e);(f)⇔(i)⇔(g):根据定理 2.1 与引理 3.2.(e)⇔(g):见引理 3.2 与注 3.1.(j)⇔(h);(k)⇔(i):按推论 3.1.(b)⇔(c):因为(b)⇔(d)⇔(f)⇔(c).(e)⇒(l):见引理 3.2 最后. ■

注 3.2 在 $q=1$ 的数值情形,公式(2)被下式替代

$$G_\rho(z)=\int_{-\infty}^{\infty}\frac{d\rho(x)}{x-z}=-\{C^*(z)\Phi_\rho(z)+A^*(z)\}\times\{D^*(z)\Phi_\rho(z)+B^*(z)\}^{-1},\forall z\in\mathbf{C}^+,\tag{2'}$$

其中,代替对 $q>1$ 的 $V_\rho(z)$,$\Phi_\rho(z)$遍历 $\mathbf{N}_1\cup\{\infty\}$.$N$-极端测度 $\rho\in\mu^H(\boldsymbol{v})$ 确切地对应到 $\Phi_\rho(z)\equiv t\in\mathbf{R}\cup\{\infty\}$ 的情形.另外,公式(14)必须更换为 rank $\boldsymbol{P}_n^{(v)}=$ rank $\boldsymbol{P}_{n+1}^{(v)}$,这里,$\boldsymbol{P}_n^{(v)}$ 和 $\boldsymbol{P}_{n+1}^{(v)}$ 分别是关于 $\mathbf{N}_1$ 类的 $\Phi_{\rho_0}(z)$,带有 $\mathbf{C}^+$ 内长度为 n 的数据 $\{(z_j,\tau_j)_1^s\}$ 与长度为 $n+1$ 的数据 $\{(z_j,\tau_j)_1^{s+1}\}$ 的 Pick 矩阵(见 Chen 和 Li[3] 的有关定义).这时,(a)~(l)(但要用 $\Phi_{\rho_0}(z)$ 代替(e),(g)与(l)中的 $V_{\rho_0}(z)$)是全部彼此等价的,并且特别地,ρ_0 的阶$=n$⇔$\Phi_{\rho_0}(z)$是一个 McMillan 次数为 n 的有理函数⇔rank $\boldsymbol{P}_n^{(v)}=$ rank $\boldsymbol{P}_{n+1}^{(v)}=n$.此外,假如这些命题中一个因而全部成立,则 ρ_0 是 $\mu^H(\boldsymbol{v})$内极端点,且亦是一个纯点测度.

注 3.3 在确定情形,对 Hamburger 矩阵矩量问题的仔细的研究成果,我们推荐读者参见 Kovalishina[7],Chen 和 Hu[2] 以及其中的参考文献.

参考文献

[1] Akhiezer N I. The Classical Moment Problem and Some Related Questions in the Analysis. London: Oliver and Boyd, 1965.

[2] Chen G N, Hu Y J. The Nevanlinna-Pick interpolation problems and power moment problems for matrix valued functions Ⅲ: the infinitely many data case. Linear Alg Appl, 2000, 306:59—87.

[3] Chen G N, Li X Q. The Nevanlinna-Pick interpolation problems and power moment problems for matrix valued functions. Linear Alg Appl, 1999, 288: 123—148.

[4] Duran A J, Lopez-Rodriguez P. *N*-extremal matrices for an indeterminate matrix moment problem. J Funct Anal, 2000, 174: 301—321.

[5] Fritzsche B, Kirstein B, et al. On rank invariance of Schwarz-Pick-Potapov block matrices of matricial Schur functions. Integr Equ Oper Theory, 2004, 48: 305—330.

[6] 胡永建,陈公宁. Akhiezer 型插值问题的矩阵变型. 北京师范大学学报(自然科学版), 2000, 36: 722—726.

[7] Kovalishina I V. Analytic theory of a class of interpolation problems. Math USSR Izv, 1984, 22(3): 419—463.

[8] Krein M G. Fundamental aspects of the representation theory of Hermitian operators with deficiency index (m, m). Ukrain Mat Z, 1949, 1: 3—66; Amer Math Soc Transl, 1970, 97: 75—143.

[9] Simom B. The classical moment problem as a self-adjoint finite difference operator. Adv in Math, 1998, 137: 82—203.

Solutions of Finite Order to a Completely Indeterminate Hamburger Matrix Moment Problem

Abstract In this paper we describe solutions ρ of finite order to a completely indeterminate Hamburger matrix moment problem with moments $\{\boldsymbol{v}_m\}_0^\infty \subset \mathbf{C}^{q\times q}$: $\boldsymbol{v}_m = \int_{-\infty}^{\infty} x^m \mathrm{d}\rho(x)$, $m=0,1,\cdots$, i. e., those solutions ρ for which the linear space P of $\mathbf{C}^{q\times q}$-valued polynomials is dense in the corresponding $L^2\left(\mathbf{R}, \frac{d\rho}{E(x)}\right)$ for some scalar polynomial $E(x)$ positive on the real line **R**. As preliminaries we consider a certain connection of a matrix version of the so-called generalized Akhiezer interpolation with its related matrix moment problem.

Keywords Matrix measure; Hamburger matrix moment problem; Generalized Akhiezer matrix interpolation; N-extremal matrix measure; Riesz's theorem; Solution of finite order

北京师范大学学报(自然科学版),
2002,36(5):588－596

对矩阵值函数的 Nevanlinna-Pick 插值问题与幂矩量问题(Ⅱ)①

The Nevanlinna-Pick Interpolation Problems and Power Moment Problems for Matrix-valued Functions(Ⅱ)

摘要 继续作者最近得到的 NP 问题(在 $\mathbf{N}_p$ 类或 $\mathbf{S}_p$ 类)与幂矩量问题的内在对应的研究成果,主要考虑这 2 个问题的解的具体描述.

关键词 NP 插值;幂矩量问题;Nevanlinna 类;Stieltjes 类;块 Hankel 向量;基本矩阵不等式

§1. 问题的提出

作者在文献[1]中建立了 NP 问题(在 $\mathbf{N}_p$ 类或 $\mathbf{S}_p$ 类)与幂矩量问题的一种崭新的对应关系,应用它从此 2 个问题之一的可解性可推断另一个的可解性,并且对它们的解的描述也有类似的情形.本文主要考虑幂矩量问题与 Stieltjes 矩量问题的可解性与解的描述.作为推论得到 NP 问题的相关结果.

有关的记号与定义见文献[1].本文涉及的内容可见文献[2～5].

NP 问题(在 $\mathbf{N}_p$ 类或 $\mathbf{S}_p$ 类) 给定 $\mathbf{C}^+$ 内 θ 个不同点 $\lambda_1,\lambda_2,\cdots,\lambda_\theta$,其重数分别为 $\tau_1,\tau_2,\cdots,\tau_\theta$, $N=\sum\limits_{k=1}^{\theta}\tau_k\geqslant 2$;给定 $p\times p$ 矩阵 $\boldsymbol{c}_{ik}$, $i=1,2,\cdots,\theta$, $k=0,1,\cdots,\tau_i-1$.求解所有的 $F\in\mathbf{N}_p$(或 $\mathbf{S}_p$)使其满足插值条件:

① 收稿日期:2000-01-10.
国家自然科学基金资助项目(19971009),本文与李小青合作.

$$(1/k!)F^{(k)}(\lambda_i)=\boldsymbol{c}_{ik},i=1,2,\cdots,\theta,k=0,1,\cdots,\tau_i-1. \tag{1}$$

设 $\omega(\lambda)$ 为满足(1) 与

$$(1/k!)F^{(k)}(\bar{\lambda}_i)=\boldsymbol{c}_{ik}^{*},i=1,2,\cdots,\theta,k=0,1,\cdots,\tau_i-1 \tag{2}$$

的次数不超过 $2N-1$ 的唯一的矩阵多项式,并且,

$$A(\lambda)=\prod_{k=1}^{\theta}(\lambda-\lambda_k)^{\tau_k}(\lambda-\bar{\lambda}_k)^{\tau_k}. \tag{3}$$

如果 $\omega(\lambda)/A(\lambda)$ 在 ∞ 处的罗朗展开为

$$\frac{\omega(\lambda)}{A(\lambda)}=\frac{\boldsymbol{h}_0}{\lambda}+\frac{\boldsymbol{h}_1}{\lambda^2}+\cdots+\frac{\boldsymbol{h}_{2N-2}}{\lambda^{2N-1}}+\frac{\boldsymbol{h}_{2N-1}}{\lambda^{2N}}+o(\lambda^{-2N}), \tag{4}$$

其中各 $\boldsymbol{h}_i$ 显然是 p 阶 Hermite 矩阵,那么称

$$\boldsymbol{b}=(\boldsymbol{h}_0,\boldsymbol{h}_1,\cdots,\boldsymbol{h}_{2N-2})\text{ 与 }\tilde{\boldsymbol{b}}=(\boldsymbol{h}_0,\boldsymbol{h}_1,\cdots,\boldsymbol{h}_{2N-1}) \tag{5}$$

分别为 $\mathrm{NP}(\mathbf{N}_p)$ 与 $\mathrm{NP}(\mathbf{S}_p)$ 问题的块 Hankel 向量.

与式(5) 中向量相对应的有如下 2 个矩量问题:

PM 问题　求在实轴上满足 $\tau(-\infty)=\boldsymbol{O}$ 的所有不减的右连续 $\mathbf{C}^{p\times p}$-值函数 $\tau(u)$ 使得

$$\boldsymbol{h}_j=\int_{-\infty}^{+\infty}u^j\mathrm{d}\tau(u),j=0,1,\cdots,2N-3;\boldsymbol{h}_{2N-2}\geqslant\int_{-\infty}^{+\infty}u^{2N-2}\mathrm{d}\tau(u). \tag{6}$$

SM 问题　求在正半实轴上的所有不减的 $\mathbf{C}^{p\times p}$- 值函数 $\tau(u)(0<u<+\infty)$ 使得

$$\boldsymbol{h}_j=\int_{0}^{+\infty}u^j\mathrm{d}\tau(u),j=0,1,\cdots,2N-2;\boldsymbol{h}_{2N-1}\geqslant\int_{-\infty}^{+\infty}u^{2N-1}\mathrm{d}\tau(u). \tag{7}$$

应用 Hamburger-Nevanlinna(H-N) 定理(见文献[3]) 可断定上述 PM 问题的解 $\tau(u)$ 与下面 H_N 问题的解 $\Phi(\lambda)\in\mathbf{N}_p$ 存在着 1-1 对应关系:

$$\Phi(\lambda)=\int_{-\infty}^{+\infty}(u-\lambda)^{-1}\mathrm{d}\tau(u), \tag{8}$$

这里,H_N 问题指的是求所有 $\Phi(\lambda)\in\mathbf{N}_p$ 使得对某个 $\boldsymbol{\beta}\geqslant\boldsymbol{O}$,渐近表示式

$$\lim_{\lambda\to\infty}\lambda^{2N-1}(\Phi(\lambda)+\boldsymbol{h}_0/\lambda+\boldsymbol{h}_1/\lambda^2+\cdots+\boldsymbol{h}_{2N-2}/\lambda^{2N-3})=-\boldsymbol{h}_{2N-2}+\boldsymbol{\beta} \tag{9}$$

在每个扇形区域 $s_\varepsilon:\varepsilon\leqslant\arg\lambda\leqslant\pi-\varepsilon,\quad 0<\varepsilon<\pi/2$　(10)

内一致成立.类似地,SM 问题解 $\tau(u)$ 与寻求所有 $\Phi(\lambda)\in\mathbf{S}_p$ 使得对某个 $\boldsymbol{\gamma}\geqslant\boldsymbol{O}$,渐近表示式

$$\lim_{\lambda\to\infty}\lambda^{2N}(\Phi(\lambda)+\boldsymbol{h}_0/\lambda+\boldsymbol{h}_1/\lambda^2+\cdots+\boldsymbol{h}_{2N-2}/\lambda^{2N-1})=-\boldsymbol{h}_{2N-1}+\boldsymbol{\gamma} \tag{11}$$

在每个扇形(10) 内一致成立的问题的解 $\Phi(\lambda)$ 也存在着 1-1 对应关系:

$$\Phi(\lambda)=\int_0^{+\infty}(u-\lambda)^{-1}\mathrm{d}\tau(u),\tag{12}$$

这里，$\boldsymbol{\gamma}=\boldsymbol{h}_{2N-1}-\int_0^{+\infty}u^{2N-1}\mathrm{d}\tau(u)$.

定理 Ⅰ[1] NP($\mathbf{N}_p$)问题的解 $F(\lambda)$ 与 PM 问题(6)的解 $\tau(u)$ 之间有下列 1-1 对应关系：

$$F(\lambda)=\omega(\lambda)+A(\lambda)\int_{-\infty}^{+\infty}(u-\lambda)^{-1}\mathrm{d}\tau(u).\tag{13}$$

定理 Ⅱ[1] NP($\mathbf{S}_p$)问题的解 $F(\lambda)$ 与 SM 问题(7)的解 $\tau(u)$ 之间有下列 1-1 对应关系：

$$F(\lambda)=\omega(\lambda)+A(\lambda)\int_0^{+\infty}(u-\lambda)^{-1}\mathrm{d}\tau(u).\tag{14}$$

定理 Ⅲ[1] 下列 4 个命题彼此等价：

1)NP($\mathbf{N}_p$)问题是可解的；

2)PM 问题(6)是可解的；

3)由 NP($\mathbf{N}_p$)问题的块 Hankel 向量 $\boldsymbol{b}$ 生成的块 Hankel 矩阵 $\boldsymbol{H}_{N-1}=(\boldsymbol{h}_{i+j})_{i,j=0}^{N-1}\geqslant\boldsymbol{O}$;

4)由 NP($\mathbf{N}_p$)问题的数据生成的块 Loewner 矩阵 $\boldsymbol{L}(\omega(\lambda))\geqslant\boldsymbol{O}$.

定理 Ⅳ[1] 下列 4 个命题彼此等价：

1)NP($\mathbf{S}_p$)问题是可解的；

2)SM 问题(7)是可解的；

3)$\boldsymbol{H}_{N-1}=(\boldsymbol{h}_{i+j})_{i,j=0}^{N-1}\geqslant\boldsymbol{O}$ 与 $\widetilde{\boldsymbol{H}}_{N-1}=(\boldsymbol{h}_{i+j+1})_{i,j=0}^{N-1}\geqslant\boldsymbol{O}$;

4)$\boldsymbol{L}(\omega(\lambda))\geqslant\boldsymbol{O}$ 与 $\boldsymbol{L}(\lambda\omega(\lambda))\geqslant\boldsymbol{O}$.

§2. PM 问题与 NP($\mathbf{N}_p$)问题解的参数化描述

设 PM 问题(6)可解，且总假定 $\boldsymbol{H}_{N-1}=(\boldsymbol{h}_{i+j})_{i,j=0}^{N-1}>\boldsymbol{O}$. 这时存在 2 个 p 阶 Hermite 矩阵 $\boldsymbol{h}_{2N-1}$ 与 $\boldsymbol{h}_{2N}$ 使得 $\boldsymbol{H}_N=(\boldsymbol{h}_{i+j})_{i,j=0}^{N}>\boldsymbol{O}$. 于是，按定理 Ⅲ 存在某个 $\tau(u)(-\infty<u<+\infty)$ 使得

$$\boldsymbol{h}_j=\int_{-\infty}^{+\infty}u^j\mathrm{d}\tau(u),j=0,1,\cdots,2N-1;\boldsymbol{h}_{2N}\geqslant\int_{-\infty}^{+\infty}u^{2N}\mathrm{d}\tau(u).\tag{15}$$

现令 $\boldsymbol{H}_k=(\boldsymbol{h}_{i+j})_{i,j=0}^{k}$，$k=0,1,\cdots,N$，它们是 Hermite 正定的：$\boldsymbol{H}_k>\boldsymbol{O}$，$k=0,1,\cdots,N$. 又令 $\boldsymbol{\Gamma}_k=\boldsymbol{H}_k^{-1}=(\boldsymbol{v}_{ij}^k)_{i,j=0}^{k}$，$k=0,1,\cdots,N$. 接着定义矩阵多项式 $P_k(\lambda)$ 与 $P_k^0(\lambda)$ 如下[6]：

$$P_k(\lambda)=\Big(\sum_{j=0}^{k}\lambda^j \boldsymbol{v}_{jk}^k\Big)(\boldsymbol{v}_{kk}^k)^{-1/2},\quad k=0,1,\cdots,N,$$

$$P_k^0(\lambda)=\int_{-\infty}^{+\infty}\mathrm{d}\tau(u)\,\frac{P_k(u)-P_k(\lambda)}{u-\lambda},\quad k=0,1,\cdots,N,\tag{16}$$

其中 $\tau(u)$ 为问题(15) 的任意解(注意,上述定义中 $P_k^0(\lambda)$ 不依赖于这样 $\tau(u)$ 的选取).容易验明以下的直交关系成立:

$$\int_{-\infty}^{+\infty}P_j(u)^*\,\mathrm{d}\tau(u)P_k(u)=\begin{cases}\boldsymbol{O}, & k\neq j\leqslant N,\\ \boldsymbol{I}_p, & k=j<N,\\ \boldsymbol{I}_p-(\boldsymbol{v}_{NN}^N)^{1/2}\boldsymbol{\beta}_\tau(\boldsymbol{v}_{NN}^N)^{1/2}, & k=j=N,\end{cases}\tag{17}$$

式中,$\boldsymbol{\beta}_\tau=\boldsymbol{h}_{2N}-\int_{-\infty}^{+\infty}u^{2N}\mathrm{d}\tau(u)$ 与 $\tau(u)$ 的选取有关.特别有,如果 $\tau(u)$ 满足式(15),且其中最后不等式变为等式,那么 $\boldsymbol{\beta}_\tau=\boldsymbol{O}$,且对 $k,j=0,1,\cdots,N$,

$$\int_{-\infty}^{+\infty}P_j(u)^*\,\mathrm{d}\tau(u)P_k(u)=\begin{cases}\boldsymbol{O}, & k\neq j,\\ \boldsymbol{I}_p, & k=j.\end{cases}\tag{17'}$$

对于这 2 类矩阵多项式 $P_k(\lambda)$ 与 $P_k^0(\lambda)$,我们有

$$\lambda P_k(\lambda)=P_{k-1}(\lambda)\boldsymbol{b}_{k-1}^*+P_k(\lambda)\boldsymbol{a}_k+P_{k+1}(\lambda)\boldsymbol{b}_k,\quad k=0,1,\cdots,N-1,$$

$$\lambda P_k^0(\lambda)=P_{k-1}^0(\lambda)\boldsymbol{b}_{k-1}^*+P_k^0(\lambda)\boldsymbol{a}_k+P_{k+1}^0(\lambda)\boldsymbol{b}_k,\quad k=1,2,\cdots,N-1,$$

这里,$\boldsymbol{b}_{-1}=\boldsymbol{O}$,且对 $k=0,1,\cdots,N-1$ 与式(17′) 中的 $\tau(u)$,

$$\boldsymbol{a}_k=\int_{-\infty}^{+\infty}P_k(u)^*\,\mathrm{d}\tau(u)uP_k(u)=\boldsymbol{a}_k^*,\ \boldsymbol{b}_k=(\boldsymbol{v}_{k+1,k+1}^{k+1})^{-1/2}(\boldsymbol{v}_{k,k}^k)^{1/2}.$$

这表明 $P_k(\lambda)$ 与 $P_k^0(\lambda)$ 满足差分方程(带有某些关于 $\boldsymbol{y}_0$ 与 $\boldsymbol{y}_1$ 的初始条件):

$$\boldsymbol{y}_{k-1}\boldsymbol{b}_{k-1}^*+\boldsymbol{y}_k\boldsymbol{a}_k+\boldsymbol{y}_{k+1}\boldsymbol{b}_k=\lambda\boldsymbol{y}_k,\quad k=1,2,\cdots.\tag{18}$$

令 $\boldsymbol{y}_k$ 是式(18) 的一个解,而 $\boldsymbol{z}_k^*$ 是同一方程(但用 μ 代替 λ) 的一个解,则对 $l=1,2,\cdots,N-1$,

$$(\lambda-\bar{\mu})\sum_{k=l}^{N-1}\boldsymbol{y}_k\boldsymbol{z}_k=(\boldsymbol{y}_N\boldsymbol{b}_{N-1}\boldsymbol{z}_{N-1}-\boldsymbol{y}_{N-1}\boldsymbol{b}_{N-1}^*\boldsymbol{z}_N)-(\boldsymbol{y}_l\boldsymbol{b}_{l-1}\boldsymbol{z}_{l-1}-\boldsymbol{y}_{l-1}\boldsymbol{b}_{l-1}^*\boldsymbol{z}_l).$$

特殊地取 $\boldsymbol{y}_k=P_k(\lambda),\boldsymbol{z}_k=P_k(\lambda)^*,l=0$ 与 $\mu=\lambda$,我们有

$$(\lambda-\bar{\lambda})\sum_{k=0}^{N-1}P_k(\lambda)P_k(\lambda)^*=P_N(\lambda)\boldsymbol{b}_{N-1}P_{N-1}(\lambda)^*-P_{N-1}(\lambda)\boldsymbol{b}_{N-1}^*P_N(\lambda)^*.\tag{19}$$

同理可推出

$$(\lambda-\bar{\lambda})\sum_{k=0}^{N-1}P_k^0(\lambda)P_k^0(\lambda)^* = P_N^0(\lambda)\boldsymbol{b}_{N-1}P_{N-1}^0(\lambda)^* - P_{N-1}^0(\lambda)\boldsymbol{b}_{N-1}^*P_N^0(\lambda)^*, \tag{20}$$

$$(\lambda-\bar{\lambda})\sum_{k=0}^{N-1}P_k^0(\lambda)P_k(\lambda)^* = P_N^0(\lambda)\boldsymbol{b}_{N-1}P_{N-1}(\lambda)^* - P_{N-1}^0(\lambda)\boldsymbol{b}_{N-1}^*P_N(\lambda)^* - \boldsymbol{I}_p. \tag{21}$$

定理 1 如果 $\boldsymbol{H}_{N-1}>\boldsymbol{O}$,$\tau(u)$ 为 PM 问题的解,且 $\Phi(\lambda)=\int_{-\infty}^{+\infty}(u-\lambda)^{-1}\mathrm{d}\tau(u)(\mathrm{Im}\,\lambda\neq 0)$,那么 $\Phi(\lambda)$ 满足

$$\frac{\Phi(\lambda)-\Phi(\lambda)^*}{\lambda-\bar{\lambda}}-\sum_{k=0}^{N-1}\psi_k(\lambda)\psi_k(\lambda)^*\geqslant\boldsymbol{O},\quad \mathrm{Im}\,\lambda\neq 0, \tag{22}$$

其中 $\psi_k(\lambda)=\Phi(\lambda)^*P_k(\bar{\lambda})+P_k^0(\bar{\lambda}),k=0,1,\cdots,N-1.$

证 设 $\tau(u)$ 为 PM 问题的解,则有唯一的 $\boldsymbol{\beta}\geqslant\boldsymbol{O}$ 使得式(6) 成立,且 $\boldsymbol{h}_{2N-2}-\boldsymbol{\beta}=\int_{-\infty}^{+\infty}u^{2N-2}\mathrm{d}\tau(u)$. 这时,

$$\int_{-\infty}^{+\infty}|u-\bar{\lambda}|^{-2}\mathrm{d}\tau(u)=\sum_{k=0}^{N-1}\psi_k(\lambda)\psi_k(\lambda)^*+E(\bar{\lambda}), \tag{23}$$

式中 $E(\bar{\lambda})=\int_{-\infty}^{+\infty}[(u-\bar{\lambda})^{-1}\boldsymbol{I}_p-\sum_{k=0}^{N-1}\psi_k(\lambda)P_k(u)^*]\mathrm{d}\tau(u)[(u-\lambda)^{-1}\boldsymbol{I}_p-\sum_{k=0}^{N-1}P_k(u)\psi_k(\lambda)^*]+\psi_{N-1}(\lambda)(\boldsymbol{v}_{N-1,N-1}^{N-1})^{1/2}\boldsymbol{\beta}(\boldsymbol{v}_{N-1,N-1}^{N-1})^{1/2}\psi_{N-1}(\lambda)^*$.

事实上,式(23) 为式(17)(但用 $N-1$ 代替其中的 N) 与 $\int_{-\infty}^{+\infty}(u-\bar{\lambda})^{-1}\mathrm{d}\tau(u)P_k(u)=P_k^0(\bar{\lambda})+\Phi(\lambda)^*P_k(\bar{\lambda})=\psi_k(\lambda),\forall k$ 的直接推论. 因为 $E(\bar{\lambda})\geqslant\boldsymbol{O}$ $(\mathrm{Im}\,\lambda\neq 0)$ 与 $\int_{-\infty}^{+\infty}|u-\bar{\lambda}|^{-2}\mathrm{d}\tau(u)=\int_{-\infty}^{+\infty}(\lambda-\bar{\lambda})^{-1}[(u-\lambda)^{-1}-(u-\bar{\lambda})^{-1}]\mathrm{d}\tau(u)=\dfrac{\Phi(\lambda)-\Phi(\lambda)^*}{\lambda-\bar{\lambda}}$,式(22) 由式(23) 得出. 定理证毕. ■

令

$$\boldsymbol{J}=\begin{pmatrix}\boldsymbol{O} & \mathrm{i}\boldsymbol{I}_p\\ -\mathrm{i}\boldsymbol{I}_p & \boldsymbol{O}\end{pmatrix},\quad \Omega(\bar{\lambda})=\begin{pmatrix}\sum_{k=0}^{N-1}P_k(\bar{\lambda})P_k(\bar{\lambda})^* & \sum_{k=0}^{N-1}P_k(\bar{\lambda})P_k^0(\bar{\lambda})^*\\ \sum_{k=0}^{N-1}P_k^0(\bar{\lambda})P_k(\bar{\lambda})^* & \sum_{k=0}^{N-1}P_k^0(\bar{\lambda})P_k^0(\bar{\lambda})^*\end{pmatrix},$$

则 $\Omega(\lambda)\geqslant\boldsymbol{O},\forall\lambda\in\mathbf{C}$,而不等式(22) 可改写为

$$(\Phi(\lambda)^* \quad I_p)\frac{J-\mathrm{i}(\bar{\lambda}-\lambda)\Omega(\bar{\lambda})}{\mathrm{i}(\bar{\lambda}-\lambda)}\begin{pmatrix}\Phi(\lambda)\\ I_p\end{pmatrix}\geqslant O,\quad \operatorname{Im}\lambda\neq 0. \tag{22'}$$

定义 $\mathbf{C}^{2p\times 2p}$- 值函数 $\boldsymbol{\Theta}(\lambda)$ 如下：

$$\boldsymbol{\Theta}(\lambda)=\begin{pmatrix}C^0(\lambda) & D^0(\lambda)\\ C(\lambda) & D(\lambda)\end{pmatrix}$$

$$=\begin{pmatrix}P_{N-1}^0(\lambda)(\boldsymbol{v}_{N-1,N-1}^{N-1})^{-1/2} & P_N^0(\lambda)\boldsymbol{b}_{N-1}(\boldsymbol{v}_{N-1,N-1}^{N-1})^{-1/2}\\ -P_{N-1}(\lambda)(\boldsymbol{v}_{N-1,N-1}^{N-1})^{-1/2} & -P_N(\lambda)\boldsymbol{b}_{N-1}(\boldsymbol{v}_{N-1,N-1}^{N-1})^{-1/2}\end{pmatrix}. \tag{24}$$

根据式(19) ～ (21)，

$$\boldsymbol{\Theta}(\bar{\lambda})\boldsymbol{J}\boldsymbol{\Theta}(\bar{\lambda})^*=\boldsymbol{J}-\mathrm{i}(\bar{\lambda}-\lambda)\boldsymbol{J}\Omega(\bar{\lambda})\boldsymbol{J}, \tag{25}$$

因而 $\boldsymbol{\Theta}(\lambda)$ 在 $\mathbf{C}^+$ 内是 $\boldsymbol{J}$- 扩张的，而在实轴上是 $\boldsymbol{J}$- 酉的，亦即

$$\boldsymbol{\Theta}(\lambda)\boldsymbol{J}\boldsymbol{\Theta}(\lambda)^*\geqslant \boldsymbol{J},\forall\lambda\in\mathbf{C}^+ \text{ 与 } \boldsymbol{\Theta}(\lambda)\boldsymbol{J}\boldsymbol{\Theta}(\lambda)^*=\boldsymbol{J},\forall\lambda\in\mathbf{R}. \tag{26}$$

从 $\boldsymbol{J}$- 理论可得到对称关系：$\boldsymbol{\Theta}^{-1}(\lambda)=\boldsymbol{J}\boldsymbol{\Theta}(\bar{\lambda})^*\boldsymbol{J},\forall\lambda\in\mathbf{C}$，这与式(25) 推得

$$\boldsymbol{J}-\boldsymbol{\Theta}^{-1}(\lambda)^*\boldsymbol{J}\boldsymbol{\Theta}^{-1}(\lambda)=\boldsymbol{J}[\boldsymbol{J}-\boldsymbol{\Theta}(\bar{\lambda})\boldsymbol{J}\boldsymbol{\Theta}(\bar{\lambda})^*]\boldsymbol{J}=\mathrm{i}(\bar{\lambda}-\lambda)\Omega(\bar{\lambda}),$$

因而 $\boldsymbol{\Theta}^{-1}(\bar{\lambda})$ 是一个内函数，也就是说，

$$\boldsymbol{\Theta}^{-1}(\lambda)^*\boldsymbol{J}\boldsymbol{\Theta}^{-1}(\lambda)\leqslant \boldsymbol{J},\forall\lambda\in\mathbf{C}^+ \text{ 与 } \boldsymbol{\Theta}^{-1}(\lambda)^*\boldsymbol{J}\boldsymbol{\Theta}^{-1}(\lambda)=\boldsymbol{J},\forall\lambda\in\mathbf{R}. \tag{27}$$

因此，不等式(22) 或(22′) 可表示为

$$(\Phi(\lambda)^* \quad I_p)\frac{\boldsymbol{\Theta}^{-1}(\lambda)^*\boldsymbol{J}\boldsymbol{\Theta}^{-1}(\lambda)}{\mathrm{i}(\bar{\lambda}-\lambda)}\begin{pmatrix}\Phi(\lambda)\\ I_p\end{pmatrix}\geqslant O,\quad \operatorname{Im}\lambda\neq 0. \tag{22''}$$

定理 1 有下列的逆命题.

定理 2 如果 $\mathbf{C}^{p\times p}$- 值函数 $\Phi(\lambda)$ 在 $\mathbf{C}^+$ 内解析且满足不等式(22)($\lambda\in\mathbf{C}^+$)，那么 $\Phi(\lambda)$ 为 H_N 问题(9) 的一个解.

证 首先假定 $\Phi(\lambda)$ 满足不等式(22)($\lambda\in\mathbf{C}^+$) 且不等式中"$\geqslant$"换为"$>$"，或等价地，严格不等式(22″) 对 $\lambda\in\mathbf{C}^+$ 成立. 这时，显然有 $\Phi(\lambda)\in\mathbf{N}_p$. 令

$$\begin{pmatrix}p(\lambda)\\ q(\lambda)\end{pmatrix}=\boldsymbol{\Theta}^{-1}(\lambda)\begin{pmatrix}\Phi(\lambda)\\ I_p\end{pmatrix},\ \forall\lambda\in\mathbf{C}^+. \tag{28}$$

容易看出，它是满足 $\det q(\lambda)\neq 0,\forall\lambda\in\mathbf{C}^+$ 的解析的 Nevanlinna 对，于是对 $\lambda\in\mathbf{C}^+$，由式(28) 推出 $\begin{pmatrix}\Phi(\lambda)q^{-1}(\lambda)\\ q^{-1}(\lambda)\end{pmatrix}=\boldsymbol{\Theta}(\lambda)\begin{pmatrix}G(\lambda)\\ I_p\end{pmatrix}$，其中 $G(\lambda)=$

$p(\lambda)q^{-1}(\lambda)$,因而

$$\Phi(\lambda)=[C^0(\lambda)G(\lambda)+D^0(\lambda)][C(\lambda)G(\lambda)+D(\lambda)]^{-1}, \qquad (29)$$

式中,$G(\lambda)\in \mathbf{N}_p$.现假定 $G(\lambda)$ 有积分表示:

$$G(\lambda)=\boldsymbol{\alpha}+\boldsymbol{\beta}\lambda+\int_{-\infty}^{+\infty}[(u-\lambda)^{-1}-u(1+u^2)^{-1}]\mathrm{d}\sigma(u),\mathrm{Im}\,\lambda\neq 0,$$

其中,$\boldsymbol{\alpha}=\boldsymbol{\alpha}^*$,$\boldsymbol{\beta}\geqslant \boldsymbol{O}$,$\mathrm{d}\sigma(u)\geqslant \boldsymbol{O}$,且$\int_{-\infty}^{+\infty}(1+u^2)^{-1}\mathrm{tr}\,\mathrm{d}\,\sigma(u)<\infty$,则经过计算便知在每个扇形区域 s_ε 内一致地有

$$\lim_{\lambda\to\infty}\lambda^{2N-1}[\Phi(\lambda)+\boldsymbol{h}_0/\lambda+\boldsymbol{h}_1/\lambda^2+\cdots+\boldsymbol{h}_{2N-3}/\lambda^{2N-2}]$$
$$=-\boldsymbol{h}_{2N-2}-\boldsymbol{\beta}_0\geqslant-\boldsymbol{h}_{2N-2},$$

其中,$\boldsymbol{\beta}_0=-(\boldsymbol{v}_{N-1,N-1}^{N-1})^{-1}\boldsymbol{\beta}(\boldsymbol{\beta}+\boldsymbol{v}_{N-1,N-1}^{N-1})^{-1}\leqslant \boldsymbol{O}$,这表示 $\Phi(\lambda)$ 为 H_N 问题(9)的解.

其次,假定 $\Phi(\lambda)$ 满足一般不等式(22)($\lambda\in\mathbf{C}^+$).则 $\Phi(\lambda)\in\mathbf{N}_p$.将 $W(\lambda)=\boldsymbol{\Theta}^{-1}(\lambda)^*\boldsymbol{J}\boldsymbol{\Theta}^{-1}(\lambda)$(Weyl 矩阵)写成分块形式:

$$W(\lambda)=\begin{pmatrix}-R(\lambda) & S(\lambda)\\ S^*(\lambda) & -T(\lambda)\end{pmatrix},$$

式中 $R(\lambda)=\mathrm{i}(\bar{\lambda}-\lambda)\sum_{k=0}^{N-1}P_k(\bar{\lambda})P_k(\bar{\lambda})^*>\boldsymbol{O},\forall\lambda\in\mathbf{C}^+$,等.通过变换:

$$u(\lambda)=[R(\lambda)^{1/2}\Phi(\lambda)-R(\lambda)^{-1/2}S(\lambda)][S(\lambda)^*R(\lambda)^{-1}S(\lambda)-T(\lambda)]^{-1/2},$$

因而

$$\Phi(\lambda)=R(\lambda)^{-1}S(\lambda)+R(\lambda)^{-1/2}u(\lambda)[S(\lambda)^*R(\lambda)^{-1}S(\lambda)-T(\lambda)]^{1/2},$$

不等式(22)变为等价形式:

$$u(\lambda)^*u(\lambda)\leqslant \boldsymbol{I}_p,\ \forall\lambda\in\mathbf{C}^+.$$

现定义与 $\Phi(\lambda)$ 有关的函数列 $\Phi_k(\lambda)$ 如下:

$$\Phi_k(\lambda)=R(\lambda)^{-1}S(\lambda)+R(\lambda)^{-1/2}(1-k^{-1})u(\lambda)\cdot$$
$$[S(\lambda)^*R(\lambda)^{-1}S(\lambda)-T(\lambda)]^{-1/2},k=1,2,\cdots,$$

则 $\lim_{k\to\infty}\Phi_k(\lambda)=\Phi(\lambda),\forall\lambda\in\mathbf{C}^+$.因为$[(1-k^{-1})u(\lambda)^*][(1-k^{-1})u(\lambda)]<\boldsymbol{I}_p,\forall\lambda\in\mathbf{C}^+,k=1,2,\cdots$,所以 $\Phi_k(\lambda)$ 满足严格的不等式(22),因此,由前一部分的推证可得每个 $\Phi_k(\lambda)$ 为 H_N 问题的解.这时

$$\begin{pmatrix}p_k(\lambda)\\ q_k(\lambda)\end{pmatrix}=\boldsymbol{\Theta}^{-1}(\lambda)\begin{pmatrix}\Phi_k(\lambda)\\ I_p\end{pmatrix},\quad k=1,2,\cdots$$

是 Nevanlinna 对,$\det q_k(\lambda)\neq 0,\forall\lambda\in\mathbf{C}^+$.

假定某个 NP 问题(在 $\mathbf{N}_p$ 类)有 N 个单重的插值点,且以 $\boldsymbol{b}=(\boldsymbol{h}_0,\boldsymbol{h}_1,\cdots,\boldsymbol{h}_{2N-2})$ 为其块 Hankel 向量. 用 $\omega^0(\lambda)$ 与 $A^0(\lambda)$ 分别记它的对应插值多项式 $\omega(\lambda)$ 与 $A(\lambda)$,则从定理 Ⅰ 推出 $F_k(\lambda)=\omega^0(\lambda)+\Phi_k(\lambda)A^0(\lambda),k=1,2,\cdots$,是刚才引入的新 NP 问题的解,因而

$$F(\lambda)=\omega^0(\lambda)+\Phi(\lambda)A^0(\lambda)=\lim_{k\to\infty}F_k(\lambda)$$

是 H_N 问题的一个解(这里用到 $\mathbf{N}_p$ 类的紧性). ■

定理 3 设 $\boldsymbol{H}_{N-1}>\boldsymbol{O}$,则 PM 问题(6)的一般解有形式(8),其中

$$\Phi(\lambda)=[C^0(\lambda)p(\lambda)+D^0(\lambda)q(\lambda)][C(\lambda)p(\lambda)+D(\lambda)q(\lambda)]^{-1},\forall\lambda\in\mathbf{C}^+,\tag{30}$$

式中 $\begin{pmatrix}p(\lambda)\\q(\lambda)\end{pmatrix}$ 为任意的 Nevanlinna 对.

定理 4 设 $\boldsymbol{H}_{N-1}>\boldsymbol{O}$,则 NP($\mathbf{N}_p$) 问题的一般解可以表示为关于任意 Nevanlinna 对 $\begin{pmatrix}p(\lambda)\\q(\lambda)\end{pmatrix}$ 的线性分式变换:

$$F(\lambda)=\omega(\lambda)+A(\lambda)[C^0(\lambda)p(\lambda)+D^0(\lambda)q(\lambda)][C(\lambda)p(\lambda)+D(\lambda)q(\lambda)]^{-1}.\tag{31}$$

下面定理告诉我们,不等式(22) 或(22′)(在 $\boldsymbol{H}_{N-1}>\boldsymbol{O}$ 前提下) 重合于文献[3] 中基本矩阵不等式 FMI(**H**).

定理 5 设 $\boldsymbol{H}_{N-1}>\boldsymbol{O}$,这里 $\boldsymbol{h}_j=\boldsymbol{h}_j^*\in\mathbf{C}^{p\times p}$,则对于 $\operatorname{Im}\lambda>0$,不等式(22) 或(22′) 可以改写成 FMI(**H**) 形式:

$$\begin{bmatrix}\boldsymbol{h}_0 & \cdots & \boldsymbol{h}_{N-1} & \Phi(\lambda)\\ \boldsymbol{h}_1 & \cdots & \boldsymbol{h}_N & \lambda(\Phi(\lambda)+\boldsymbol{h}_0/\lambda)\\ \vdots & & \vdots & \vdots\\ \boldsymbol{h}_{N-1} & \cdots & \boldsymbol{h}_{2N-2} & \lambda^{N-1}(\Phi(\lambda)+\boldsymbol{h}_0/\lambda+\cdots+\boldsymbol{h}_{N-2}/\lambda^{N-1})\\ & * & & (\Phi(\lambda)-\Phi(\lambda)^*)/(\lambda-\bar{\lambda})\end{bmatrix}\geqslant\boldsymbol{O},\operatorname{Im}\lambda>0.\tag{32}$$

证 令 $b(\lambda)=\operatorname{col}[\boldsymbol{I}_p,\lambda\boldsymbol{I}_p,\cdots,\lambda^{N-1}\boldsymbol{I}_p]$ 与 $-c(\lambda)=\operatorname{col}[\boldsymbol{O},\boldsymbol{h}_0,\boldsymbol{h}_1+\lambda\boldsymbol{h}_0,\cdots,\boldsymbol{h}_{N-2}+\lambda\boldsymbol{h}_{N-3}+\cdots+\lambda^{N-2}\boldsymbol{h}_0]$. 为证明本定理,只需验证对 $\lambda\in\mathbf{C}^+$ 均有

$$\Omega(\bar{\lambda})=\begin{bmatrix}b(\lambda)^*\\-c(\lambda)^*\end{bmatrix}\boldsymbol{H}_{N-1}^{-1}(b(\lambda)\quad-c(\lambda)),$$

(这里 $\Omega(\bar{\lambda})$ 同于式(22′)中的,)或者

$$\sum_{k=0}^{N-1} P_k(\bar{\lambda})P_k(\bar{\lambda})^* = b(\lambda)^* \boldsymbol{H}_{N-1}^{-1} b(\lambda),$$
$$\sum_{k=0}^{N-1} P_k(\bar{\lambda})P_k^0(\bar{\lambda})^* = -b(\lambda)^* \boldsymbol{H}_{N-1}^{-1} c(\lambda), \tag{33}$$
$$\sum_{k=0}^{N-1} P_k^0(\bar{\lambda})P_k^0(\bar{\lambda})^* = c(\lambda)^* \boldsymbol{H}_{N-1}^{-1} c(\lambda).$$

为此,令 $\boldsymbol{Y}_k = [\boldsymbol{I}_p, \bar{\lambda}\boldsymbol{I}_p, \cdots, \bar{\lambda}^k \boldsymbol{I}_p], k = 0,1,\cdots,N-1, \boldsymbol{B}_k = [\boldsymbol{h}_k, \boldsymbol{h}_{k+1}, \cdots, \boldsymbol{h}_{2k-1}], k = 0,1,\cdots,N-1$. 应用 $\boldsymbol{H}_k$ 的分块形式:

$$\boldsymbol{H}_k = (\boldsymbol{h}_{i+j})_{i,j=0}^k = \begin{pmatrix} \boldsymbol{H}_{k-1} & \boldsymbol{B}_k^* \\ \boldsymbol{B}_k & \boldsymbol{h}_{2k} \end{pmatrix}, k = 1,2,\cdots,N-1,$$

可得递归公式

$$\begin{aligned}\boldsymbol{Y}_k \boldsymbol{H}_k^{-1} \boldsymbol{Y}_k^* &= \boldsymbol{Y}_{k-1} \boldsymbol{H}_{k-1}^{-1} \boldsymbol{Y}_{k-1}^* + (\boldsymbol{Y}_{k-1} \quad \bar{\lambda}_k \boldsymbol{I}_p) \begin{pmatrix} -\boldsymbol{H}_{k-1}^{-1} \boldsymbol{B}_k^* \\ \boldsymbol{I}_p \end{pmatrix} \boldsymbol{v}_{kk}^k (-\boldsymbol{B}_k \boldsymbol{H}_{k-1}^{-1} \quad \boldsymbol{I}_p) \begin{pmatrix} \boldsymbol{Y}_{k-1}^* \\ \lambda^k \boldsymbol{I}_p \end{pmatrix} \\ &= \boldsymbol{Y}_{k-1} \boldsymbol{H}_{k-1}^{-1} \boldsymbol{Y}_{k-1}^* + P_k(\bar{\lambda}) P_k(\bar{\lambda})^*, k = 1,2,\cdots,N-1,\end{aligned}$$

因此,

$$b(\lambda)^* \boldsymbol{H}_{N-1}^{-1} b(\lambda) = \boldsymbol{Y}_{N-1} \boldsymbol{H}_{N-1}^{-1} \boldsymbol{Y}_{N-1} = \sum_{k=0}^{N-1} P_k(\bar{\lambda}) P_k(\bar{\lambda})^*.$$

类似地可以验明式(33)中其他对应的块分量也彼此相等. 定理证毕. ■

推论 1 设 $\boldsymbol{H}_{N-1} > \boldsymbol{O}$,则 $\Phi(\lambda) \in \mathbf{N}_p$ 求解 H_N 问题(9)当且仅当它满足不等式 FMI(**H**),即不等式(32).

推论 2 设 $F(\lambda)$ 在 $\mathbf{C}^+$ 内解析,则 $F(\lambda)$ 为 NP($\mathbf{N}_p$)问题的解当且仅当它满足不等式:

$$(F(\lambda)^* \quad \boldsymbol{I}_p) \frac{\hat{\boldsymbol{\Theta}}^{-1}(\lambda)^* \boldsymbol{J} \hat{\boldsymbol{\Theta}}^{-1}(\lambda)}{\mathrm{i}(\bar{\lambda} - \lambda)} \begin{pmatrix} F(\lambda) \\ \boldsymbol{I}_p \end{pmatrix} \geqslant \boldsymbol{O}, \quad \operatorname{Im} \lambda > 0, \tag{34}$$

式中 $\hat{\boldsymbol{\Theta}}(\lambda) = \begin{pmatrix} A(\lambda)\boldsymbol{I}_p & \omega(\lambda) \\ \boldsymbol{O} & \boldsymbol{I}_p \end{pmatrix} \boldsymbol{\Theta}(\lambda)$.

§3. SM 问题与 NP($\mathbf{S}_p$)问题解的参数化描述

只考虑完全非退化的情形,即有

$$\boldsymbol{H}_{N-1} > \boldsymbol{O} \text{ 与 } \widetilde{\boldsymbol{H}}_{N-1} > \boldsymbol{O}. \tag{35}$$

按定理Ⅱ我们主要讨论 SM 问题的解集,它显然可视为分别带有矩量 $\boldsymbol{h}_0$,

$\boldsymbol{h}_1,\cdots,\boldsymbol{h}_{2N-2}$ 与 $\boldsymbol{h}_1,\boldsymbol{h}_2,\cdots,\boldsymbol{h}_{2N-1}$ 的 SM 问题解集之交集. 在条件(35) 下,存在 p 阶矩阵 $\boldsymbol{h}_{2N}$ 与 $\boldsymbol{h}_{2N+1}$ 使得

$$\boldsymbol{H}_N=(\boldsymbol{h}_{i+j})_{i,j=0}^{N}>\boldsymbol{O} \text{ 与 } \widetilde{\boldsymbol{H}}_N=(\boldsymbol{h}_{i+j+1})_{i,j=0}^{N}>\boldsymbol{O}. \tag{36}$$

设 $P_k(\lambda)$ 与 $P_k^0(\lambda)$ 如前,而 $\widetilde{P}_k(\lambda)$ 与 $\widetilde{P}_k^0(\lambda)$ 为对应于 $\widetilde{\boldsymbol{H}}_N$ 的矩阵多项式. 同样地,可以引用记号 $\widetilde{\boldsymbol{H}}_k,\widetilde{\boldsymbol{\Gamma}}_k,\tilde{\boldsymbol{v}}_{ij}^k$,等. 照此规定,$\widetilde{\mathrm{H}}_N$ 问题指的是寻求所有 $\Phi(\lambda)\in\mathbf{N}_p$ 满足:对某个 $\boldsymbol{v}\geqslant\boldsymbol{O}$,渐近式

$$\lim_{\lambda\to\infty}\lambda^{2N-1}\left(\Phi(\lambda)+\frac{\boldsymbol{h}_1}{\lambda}+\cdots+\frac{\boldsymbol{h}_{2N-2}}{\lambda^{2N-2}}\right)=-\boldsymbol{h}_{2N-1}+\boldsymbol{v} \tag{37}$$

在每个扇形(10) 内一致成立. 这时,它的一般解 $\Phi(\lambda)$ 有参数化表示

$$\Phi(\lambda)=[\widetilde{C}^0(\lambda)p(\lambda)+\widetilde{D}^0(\lambda)q(\lambda)][\widetilde{C}(\lambda)p(\lambda)+\widetilde{D}(\lambda)q(\lambda)]^{-1}, \tag{38}$$

其中,$\begin{pmatrix}p(\lambda)\\q(\lambda)\end{pmatrix}$ 是任意的 Nevanlinna 对,而其系数矩阵 $\widetilde{\boldsymbol{\Theta}}(\lambda)=\begin{pmatrix}\widetilde{C}^0(\lambda)&\widetilde{D}^0(\lambda)\\\widetilde{C}(\lambda)&\widetilde{D}(\lambda)\end{pmatrix}$ 中,$\widetilde{C}^0(\lambda)=P_{N-1}^0(\lambda)(\tilde{\boldsymbol{v}}_{N-1,N-1}^{N-1})^{-1/2}$ 等. 但要注意,表达式(30) 与(38) 中的系数矩阵 $\boldsymbol{\Theta}(\lambda)$ 与 $\widetilde{\boldsymbol{\Theta}}(\lambda)$ 都可以分别地右乘上任意的 $\boldsymbol{J}$-酉常数矩阵而不影响解的表示. 现令

$$P(\lambda)=\begin{pmatrix}\lambda\boldsymbol{I}_p&\boldsymbol{h}_0\\\boldsymbol{O}&\boldsymbol{I}_p\end{pmatrix},$$

$$\mathbf{V}=\begin{pmatrix}\boldsymbol{v}_{N-1,N-1}^{N-1}(\tilde{\boldsymbol{v}}_{N-1,N-1}^{N-1})^{-1}\boldsymbol{v}_{N-1,N-1}^{N-1}&\boldsymbol{O}\\\boldsymbol{I}_p&(\boldsymbol{v}_{N-1,N-1}^{N-1})^{-1}\tilde{\boldsymbol{v}}_{N-1,N-1}^{N-1}(\boldsymbol{v}_{N-1,N-1}^{N-1})^{-1}\end{pmatrix},$$

$$\widetilde{\mathbf{V}}=\begin{pmatrix}\boldsymbol{v}_{N-1,N-1}^{N-1}&\tilde{\boldsymbol{v}}_{N-1,N-1}^{N-1}(\boldsymbol{v}_{N,N}^{N})^{-1}\tilde{\boldsymbol{v}}_{N-1,N-1}^{N-1}(\boldsymbol{v}_{N-1,N-1}^{N-1})^{-1}-\boldsymbol{v}_{N-1,N-1}^{N-1}h_0\\\boldsymbol{O}&(\boldsymbol{v}_{N-1,N-1}^{N-1})^{-1}\end{pmatrix}.$$

容易验证,$\mathbf{V}$ 与 $\widetilde{\mathbf{V}}$ 均为 $2p$ 阶 $\boldsymbol{J}$-酉矩阵,并且,

$$\widetilde{\boldsymbol{\Theta}}(\lambda)\widetilde{\mathbf{V}}=P(\lambda)\boldsymbol{\Theta}(\lambda)\mathbf{V}P^{-1}(\lambda),\ \forall\lambda\neq 0. \tag{39}$$

(式(39) 的证明要涉及 $P_N(\lambda),P_N^0(\lambda),\widetilde{P}_N(\lambda),\widetilde{P}_N^0(\lambda),P_{N-1}(\lambda),P_{N-1}^0(\lambda),\widetilde{P}_{N-1}(\lambda),\widetilde{P}_{N-1}^0(\lambda)$ 之间的关系. 这里从略.)

下面用 $\boldsymbol{\Theta}(\lambda)\mathbf{V}$ 与 $\widetilde{\boldsymbol{\Theta}}(\lambda)\widetilde{\mathbf{V}}$ 分别取代式(30) 与(38) 中的系数矩阵 $\boldsymbol{\Theta}(\lambda)$ 与 $\widetilde{\boldsymbol{\Theta}}(\lambda)$,并写成分块形式:

$$\boldsymbol{\Theta}(\lambda)\mathbf{V}=\begin{pmatrix}K_{11}(\lambda)&K_{12}(\lambda)\\K_{21}(\lambda)&K_{22}(\lambda)\end{pmatrix},\ \widetilde{\boldsymbol{\Theta}}(\lambda)\widetilde{\mathbf{V}}=\begin{pmatrix}\overline{K}_{11}(\lambda)&\overline{K}_{12}(\lambda)\\\overline{K}_{21}(\lambda)&\overline{K}_{22}(\lambda)\end{pmatrix}. \tag{40}$$

定理 6 在条件(35) 下,SM 问题(7) 的一般解有形式(12),其中

$\Phi(\lambda)$ 有表达式：

$$\Phi(\lambda)=[K_{11}(\lambda)p(\lambda)+K_{12}(\lambda)q(\lambda)][K_{21}(\lambda)p(\lambda)+K_{22}(\lambda)q(\lambda)]^{-1}, \tag{41}$$

这里 $\begin{pmatrix}p(\lambda)\\q(\lambda)\end{pmatrix}$ 为任意的 Stieltjes 对.

证 只要验证表达式(41) 为问题(11) 的一般解. 首先设 $\Phi(\lambda)\in\mathbf{S}_p$ 为问题(11) 的解(对应于 $\boldsymbol{v}\geqslant\boldsymbol{O}$). 这时，$\Phi(\lambda)$，$\lambda\Phi(\lambda)$ 均属于 $\mathbf{N}_p$. 因为 $\Phi(\lambda)$ 为 H_N 问题的解，所以按定理 3(用 $\boldsymbol{\Theta}(\lambda)\mathbf{V}$ 取代其中的 $\boldsymbol{\Theta}(\lambda)$)，

$$\begin{pmatrix}\Phi(\lambda)\\I_p\end{pmatrix}=\boldsymbol{\Theta}(\lambda)\mathbf{V}\begin{pmatrix}p(\lambda)\\q(\lambda)\end{pmatrix}[K_{21}(\lambda)p(\lambda)+K_{22}(\lambda)q(\lambda)]^{-1}, \tag{42}$$

这里 $\begin{pmatrix}p(\lambda)\\q(\lambda)\end{pmatrix}$ 为某个 Nevanlinna 对. 因为 $\lambda\Phi(\lambda)+\boldsymbol{h}_0\in\mathbf{N}_p$，且对某个 $\boldsymbol{\beta}\geqslant\boldsymbol{O}$，下式在每个扇形(10) 内一致成立：

$$\lim_{\lambda\to\infty}\lambda^{2N-1}(\lambda\Phi(\lambda)+\boldsymbol{h}_0+\boldsymbol{h}_1/\lambda+\cdots+\boldsymbol{h}_{2N-2}/\lambda^{2N-2})=-\boldsymbol{h}_{2N-1}+\boldsymbol{\beta},$$

$\lambda\Phi(\lambda)+\boldsymbol{h}_0$ 为 $\overline{\mathrm{H}}_N$ 问题的解，于是，对某个 Nevanlinna 对 $\begin{pmatrix}\tilde{p}_0(\lambda)\\\tilde{q}_0(\lambda)\end{pmatrix}$ 有

$$\lambda\Phi(\lambda)+\boldsymbol{h}_0=[\overline{K}_{11}(\lambda)\tilde{p}_0(\lambda)+\overline{K}_{12}(\lambda)\tilde{q}_0(\lambda)][\overline{K}_{21}(\lambda)\tilde{p}_0(\lambda)+\overline{K}_{22}(\lambda)\tilde{q}_0(\lambda)]^{-1}$$

或者

$$P(\lambda)\begin{pmatrix}\Phi(\lambda)\\I_p\end{pmatrix}=\widetilde{\boldsymbol{\Theta}}(\lambda)\widetilde{\mathbf{V}}\begin{pmatrix}\tilde{p}_0(\lambda)\\\tilde{q}_0(\lambda)\end{pmatrix}[\overline{K}_{21}(\lambda)\tilde{p}_0(\lambda)+\overline{K}_{22}(\lambda)\tilde{q}_0(\lambda)]^{-1}. \tag{43}$$

用 $P(\lambda)$ 左乘式(42)，且记 $\begin{pmatrix}\tilde{p}(\lambda)\\\tilde{q}(\lambda)\end{pmatrix}=P(\lambda)\begin{pmatrix}p(\lambda)\\q(\lambda)\end{pmatrix}$，按式(39)，

$$\overline{K}_{21}(\lambda)\tilde{p}(\lambda)+\overline{K}_{22}(\lambda)\tilde{q}(\lambda)=K_{21}(\lambda)p(\lambda)+K_{22}(\lambda)q(\lambda),$$

于是，式(43) 可变换为

$$P(\lambda)\begin{pmatrix}\Phi(\lambda)\\I_p\end{pmatrix}=\widetilde{\boldsymbol{\Theta}}(\lambda)\widetilde{\mathbf{V}}\begin{pmatrix}\tilde{p}(\lambda)\\\tilde{q}(\lambda)\end{pmatrix}[\overline{K}_{21}(\lambda)\tilde{p}(\lambda)+\overline{K}_{22}(\lambda)\tilde{q}(\lambda)]^{-1},$$

其中 $\begin{pmatrix}\tilde{p}(\lambda)\\\tilde{q}(\lambda)\end{pmatrix}$ 也是某个 Nevanlinna 对. 为了证明 $\Phi(\lambda)$ 有形式(41)，我们只要验证式(42) 中的 $\begin{pmatrix}p(\lambda)\\q(\lambda)\end{pmatrix}$ 为某个 Stieltjes 对，但这是 $\begin{pmatrix}p(\lambda)\\q(\lambda)\end{pmatrix}$ 与 $\begin{pmatrix}\tilde{p}(\lambda)\\\tilde{q}(\lambda)\end{pmatrix}$ 均

为 Nevanlinna 对的直接推论.

反之,设 $\Phi(\lambda)$ 有形式(41),其中 $\begin{pmatrix} p(\lambda) \\ q(\lambda) \end{pmatrix}$ 为某个 Stieltjes 对,这时 $\begin{pmatrix} p(\lambda) \\ q(\lambda) \end{pmatrix}$ 显然为 Nevanlinna 对,于是 $\Phi(\lambda)$ 为 H_N 问题的一个解,因而 $\Phi(\lambda) \in \mathbf{N}_p$. 易证 $\begin{pmatrix} \tilde{p}(\lambda) \\ \tilde{q}(\lambda) \end{pmatrix} = P(\lambda) \begin{pmatrix} p(\lambda) \\ q(\lambda) \end{pmatrix}$ 也是 Nevanlinna 对. 这时,应用公式(39),我们有

$$P(\lambda) \begin{pmatrix} \Phi(\lambda) \\ I_p \end{pmatrix} = \widetilde{\boldsymbol{\Theta}}(\lambda) \widetilde{\mathbf{V}} \begin{pmatrix} \tilde{p}(\lambda) \\ \tilde{q}(\lambda) \end{pmatrix} [\overline{K}_{21}(\lambda) \tilde{p}(\lambda) + \overline{K}_{22}(\lambda) \tilde{q}(\lambda)]^{-1}.$$

这表明 $\lambda\Phi(\lambda) + \boldsymbol{h}_0$ 为 $\overline{\mathrm{H}}_N$ 问题的解. 因此,$\lambda\Phi(\lambda) \in \mathbf{N}_p$,且对某个 $\boldsymbol{\beta} \geqslant \boldsymbol{O}$ 渐近式

$$\lim_{\lambda \to \infty} \lambda^{2N-1} (\lambda\Phi(\lambda) + \boldsymbol{h}_0 + \boldsymbol{h}_1/\lambda + \cdots + \boldsymbol{h}_{2N-2}/\lambda^{2N-2}) = -\boldsymbol{h}_{2N-1} + \boldsymbol{\beta}$$

在每一个扇形(10)内一致成立. 由此得出 $\Phi(\lambda) \in \mathbf{S}_p$,因而 $\Phi(\lambda)$ 为问题(11)的一个解,证毕. ■

定理 7 $NP(\mathbf{S}_p)$ 问题的一般解 $F(\lambda)$ 可以表示为任意 Stieltjes 对 $\begin{pmatrix} p(\lambda) \\ q(\lambda) \end{pmatrix}$ 的线性分式变换形式:

$$F(\lambda) = [S_{11}(\lambda) p(\lambda) + S_{12}(\lambda) q(\lambda)][S_{21}(\lambda) p(\lambda) + S_{22}(\lambda) q(\lambda)]^{-1},$$

其中系数矩阵有分块表示:

$$\begin{pmatrix} S_{11}(\lambda) & S_{12}(\lambda) \\ S_{21}(\lambda) & S_{22}(\lambda) \end{pmatrix} = \begin{pmatrix} A(\lambda)\boldsymbol{I}_p & \omega(\lambda) \\ \boldsymbol{O} & \boldsymbol{I}_p \end{pmatrix} \boldsymbol{\Theta}(\lambda) \mathbf{V}.$$

§4. 参考文献

[1] Chen G N, Li X Q. The Nevanlinna-Pick interpolation problems and power moment problems for matrix-valued functions. Linear Algebra Appl, 1999,288:123.

[2] Chen G N. The general rational interpolation problem and its connection with the Nevanlinna-Pick interpolation and power moment problem. Linear Algebra Appl, 1998,273:83.

[3] Kovalishina I. Analytic theory of a class of interpolation problems. Math USSR Izv, 1984,22:419.

[4]Ball J A, Cohberg I, Rodman L. Interpolation of Rational Matrix Functions. Basel: Birkhäuser-Verlag, 1990.

[5]Foias C, Frazo A E. The Commutant Lifting Approach to the Interpolation Problems. Basel: Birkhäuser-Verlag, 1990.

[6]Dym H. On hermitian block Hankel matrices, matrix polynomials, the Hamburger moment problem, interpolation and maximum entropy. Integr Equ Oper Theory, 1989, 12:757.

THE NEVANLINNA-PICK INTERPOLATION PROBLEMS AND POWER MOMENT PROBLEMS FOR MATRIX-VALUED FUNCTIONS (Ⅱ)

Abstract The authors pursue previous investigations of establishing an intrinsic correspondence between a Nevanlinna-Pick interpolation in the class $\mathbf{N}_p$ or $\mathbf{S}_p$ and a certain power moment problem. The attention is mainly turned to descriptions of solutions to these two problems.

Keywords NP interpolation; power moment problem; Nevanlinna class; Stieltjes class; block Hankel vector; fundamental matrix inequality

Linear Algebra and Its Applications
1998, 273: 83－117

一般有理插值问题及其与 Nevanlinna-Pick 插值和幂矩量问题的联系①

The General Rational Interpolation Problem and Its Connection with the Nevanlinna-Pick Interpolation and Power Moment Problem

Abstract The Hankel vector approach in a recent work of the author with Zhao and Zhang on the general rational interpolation problem (GRIP) allows the problem to be reduced to what amounts to a limiting case, i. e. a high order rational interpolation at only a single node (infinity). In particular, it gives rise to a new method to find the coefficient matrix of the linear fractional parametrization of interpolants to the GRIP. This approach is extended in the present paper to the Nevanlinna-Pick (NP) interpolation problem with multiple nodes in the class of Nevanlinna functions, which can be in essence considered as a GRIP with certain symmetries. The Hankel vector, suitably adapted to the present situation, is in fact a finite nonnegative sequence relative to an axis if the problem is solvable—in particular, a finite positive sequence in the indeterminate case. This development leads to an intrinsic connec-

① Received 4 February 1997; final manuscript accepted 13 July 1997.

Partially supported by the National Natural Science Foundation of China. Submitted by Hans Schneider. A part of the work on this paper was done during the author's stay at the Department of Mathematics, the University of Michigan.

tion between the NP problem and problems of the truncated power moment and reconstructing generators of the Hankel vector in question, and therefore opens up certain ways for the discussion and solution of the NP problem and its relatives on the basis of the theories of the rational interpolation and algebraic moment problems.

§ 1. Introduction

In the present paper, we continue our investigations [7] of the general rational interpolation problem (GRIP) and its applications. The main concern is several components in classic analysis—the scalar Nevanlinna-Pick (NP) interpolation problem with multiple nodes in the class $\mathcal{N}$ of Nevanlinna functions, the GRIP of certain symmetries, the truncated power moment problems, and the problem of reconstructing functions with a given asymptotic expansion. We shall establish a very rigid relationship between them in terms of the so-called Hankel vector of a GRIP. The NP problem is in essence thought of as a special GRIP of certain symmetries, and has actually good GRIP analogs, provided it is solvable. On the other hand, the power moment problem is equivalent to the problem of finding functions which belong to $\mathcal{N}$ with a given asymptotic expansion. The latter could be viewed as part of the study of the problem of finding rational generators to a given real vector if we kept all functions rational. This leads to a method of solving the NP problem and some related problems. In the indeterminate case, the solutions to each of these problems considered are represented as a linear fractional transformation with a polynomial coefficient matrix built from the interpolation data via the Hankel vector approach.

In the analogy to [7], we shall show in the present paper that to the statement of a NP problem of order N there corresponds a unique real vector $\boldsymbol{b}=(h_0, h_1, \cdots, h_{2N-2})$ of dimension $2N-1$, referred to as the Hankel vector of that NP problem, which is a finite nonnegative sequence relative to an axis if the problem is solvable—in particular, a finite positive sequence in the indeterminate case. The key observation thereby allows the NP problem to be reduced to the study of the power moment problem and/or the problem of reconstructing functions in $\mathcal{N}$ having a given asymptotic expansion associated with that Hankel vector. Furthermore, it gives rise, with little additional effort, to an extension of the basic theorem of the NP problem theory and its more constructive proof, and to a new structure of the solutions, in a simpler way based heavily on the use of the theory of GRIPs combined with some well-known results of the power moment problems.

The mathematical apparatus of the Hankel vector used to solve the NP problem here is not greatly complicated in transition from the simple node to the multiple node case. It seems also highly appropriate for the NP problem in the case of matrix-valued functions (for details see [8]). Furthermore, the method presented here admits a precise algorithmization, which may be useful in numerical solutions of a class of interpolation problems.

In spite of many different approaches to and viewpoints on the NP problem (see, e. e. g, [1,3~5,9,10,12,13,16~26,28] and references therein), we have not found an approach such as the one stated here. It seems that the connection between the GRIP or NP problem and power moment problems described in this paper offers some new insights and makes the results relative to the NP problem and the power moment problem more transparent.

Explicit formulations of the above-mentioned NP problem and some related problems in the class $\mathcal{N}$ of Nevanlinna functions are as follows. Here the class $\mathcal{N}$ consists of all functions $w(\lambda)$ holomorphic in the open upper half plane $\mathbf{C}^+ = \{\lambda \in \mathbf{C}: \operatorname{Im} \lambda > 0\}$ that have the property that $\operatorname{Im} w(\lambda) \geqslant 0$ for any $\lambda \in \mathbf{C}^+$. Each $w(\lambda) \in \mathcal{N}$ admits a continuation into the lower half plane by reflection:

$$w(\lambda) = \overline{w(\bar{\lambda})}, \quad \operatorname{Im} \lambda < 0, \tag{1.1}$$

in which the bar denotes complex conjugation.

Ⅰ. General NP Problem in $\mathcal{N}$. Suppose that we are given θ distinct interpolation nodes in $\mathbf{C}^+$,

$$Y_1, Y_2, \cdots, Y_\theta, \tag{1.2}$$

with given multiplicities $\tau_1, \tau_2, \cdots, \tau_\theta, \tau_i \geqslant 1$, where $\sum_{i=1}^{\theta} \tau_i = N > 1$ is the number of interpolation nodes counting with multiplicities, and are given N complex numbers C_{ik} (with Im $C_{i0} \geqslant 0$) for $k=0,1,\cdots,\tau_i-1, i=1,2,\cdots,\theta$. It is required to find necessary and sufficient conditions for a function $w(\lambda) \in \mathcal{N}$ to exist which satisfies the equations

$$\frac{1}{k!}\frac{\mathrm{d}^k}{\mathrm{d}\lambda^k} w(\lambda)\Big|_{\lambda=Y_i} = C_{ik}, \quad k=0,1,\cdots,\tau_i-1, i=1,2,\cdots,\theta, \tag{1.3}$$

and to describe all the solutions to the NP problem (1.3) of order N if these conditions are met.

As is well known, the investigation is based on the positivity of the so-called (generalized) Pick matrix defined by

$$\boldsymbol{\Lambda} = [\boldsymbol{\Lambda}_{ij}]_{i,j=1}^{\theta} \in \mathbf{C}^{N \times N}, \tag{1.4a}$$

where $\boldsymbol{\Lambda}_{ij} \in \mathbf{C}^{\tau_i \times \tau_j}$ with entries $K_{kl}(Y_i, \overline{Y}_j)$ determined by the divided differences of a suitable function $w(\lambda)$ [see (2.4b) below]:

$$K_{kl}(Y_i, \overline{Y}_j) = [Y_i, \cdots, Y_i, \overline{Y}_j, \cdots, \overline{Y}_j]_{w(\lambda)}. \tag{1.4b}$$

To the NP problem (1.3) there is related an interpolation problem of order $2N$, which is considered as a certain GRIP later on unless otherwise mentioned.

Ⅱ. GRIP: of Certain Symmetries. Given the interpolation data as in (1.3), find all rational functions $\varphi(\lambda)$ satisfying the equations

$$\left.\begin{aligned} &\frac{1}{k!}\frac{d^k}{d\lambda^k}\varphi(\lambda)\Big|_{\lambda=Y_i} = C_{ik}, \\ &\frac{1}{k!}\frac{d^k}{d\lambda^k}\varphi(\lambda)\Big|_{\lambda=\overline{Y}_i} = \overline{C}_{ik}, \end{aligned}\right\} \quad k=0,1,\cdots,\tau_i-1,\ i=1,2,\cdots,\theta. \tag{1.5}$$

We should emphasize that if $w(\lambda)$ (not necessarily rational) is in $\mathcal{N}$, then it is a solution to the problem (1.3) if, and only if, it satisfies the equations (1.5). Thus if we keep all functions rational (this constraint is not essential in most of the results cited here), the theory of the NP interpolation could be viewed as part of the study of the GRIP. In this case, the Pick matrix $\boldsymbol{\Lambda}$ of order N defined by (1.4) is always Hermitian and coincides with the (generalized) Loewner matrix $\boldsymbol{L}_{y\bar{y}}(w)$ associated with the GRIP (1.5) and with a special pair of interpolation vectors $\boldsymbol{y}$ and $\bar{\boldsymbol{y}}$:

$$\boldsymbol{y}=(\underbrace{Y_1,\cdots,Y_1}_{\tau_1},\cdots,\underbrace{Y_\theta,\cdots,Y_\theta}_{\tau_\theta})^{\mathrm{T}}\in\mathbf{C}^N,$$
$$\bar{\boldsymbol{y}}=(\underbrace{\overline{Y}_1,\cdots,\overline{Y}_1}_{\tau_1},\cdots,\underbrace{\overline{Y}_\theta,\cdots,\overline{Y}_\theta}_{\tau_\theta})^{\mathrm{T}}\in\mathbf{C}^N \tag{1.6}$$

(see Section 2 below for details).

Note that the positivity of a Hermitian matrix $\boldsymbol{\Lambda}$ is equivalent to the positivity of $\boldsymbol{M\Lambda M}^*$ for any invertible $\boldsymbol{M}$, so there exist many matrices one might use in the test for determining the solvability of the problem (1.3). In particular, it follows from Lemma 3.1 below that the positivity of the Pick matrix (1.4) is equivalent to the positivity of the real Hankel vector $\boldsymbol{b}=(h_0,h_1,\cdots,h_{2N-2})$ of the GRIP (1.5), that is, $\{h_i\}_0^{2N-2}$ is in fact a finite nonnegative sequence relative to an axis in the sense that the real Hankel matrix generated by $\boldsymbol{b}$ is Hermitian nonnegative:

$$\boldsymbol{H}_{NN}[\boldsymbol{b}]=(h_{i+j})_{i,j=0}^{N-1}\geqslant\boldsymbol{O}. \tag{1.7}$$

Further, $\boldsymbol{\Lambda}>\boldsymbol{O}$ (Hermitian positive) if and only if $\boldsymbol{H}_{NN}[\boldsymbol{b}]>\boldsymbol{O}$, namely, $\{h_i\}_0^{2N-2}$ is a finite positive sequence relative to an axis.

It has been well recognized that there exists a close connection between the positivity of real Hankel matrix and the algebraic moment problems [1, 18, 27]. A theorem of M. Riesz (see, e.g., [1]), of which we shall give an alternative proof in Section 2, says that the condition (1.7) is necessary and sufficient for solvability of the following moment problem.

Ⅲ. Truncated Power Moment Problem. Given a real vector $\boldsymbol{b}=(h_0,h_1,\cdots,h_{2N-2})$, find all nondecreasing functions $\sigma(u)\,(-\infty<u<\infty)$ such that

$$h_j = \int_{-\infty}^{\infty} u^j \mathrm{d}\sigma(u) \quad (j = 0,1,\cdots,2N-3),$$

$$h_{2N-2} \geqslant \int_{-\infty}^{\infty} u^{2N-2} \mathrm{d}\sigma(u). \tag{1.8}$$

Note that in the case of $\boldsymbol{H}_{NN}[\boldsymbol{b}]>\boldsymbol{O}$ we can choose appropriate real numbers h_{2N-1} and h_{2N} such that $\boldsymbol{H}_{N+1,N+1}[\widetilde{\boldsymbol{b}}]>\boldsymbol{O}$ in which $\widetilde{\boldsymbol{b}}=(h_0,h_1,\cdots,h_{2N-2},h_{2N-1},h_{2N})$ of dimension $2N+1$, and vice versa. Hence, in this case, the solvability of the problem (1.8) is equivalent to that of the following problem, and the set of solutions of the latter clearly forms a subset of the solutions of the former.

Ⅳ. Truncated Hamburger Moment Problem.　Given a real $\boldsymbol{b}=(h_0,h_1,\cdots,h_{2N-2})$, find all nondecreasing functions $\sigma(u)$ $(-\infty<u<\infty)$ such that

$$h_j = \int_{-\infty}^{\infty} u^j \mathrm{d}\sigma(u), \quad j = 0,1,\cdots,2N-2. \tag{1.9}$$

In what follow, we call a real vector $\boldsymbol{b}=(h_0,h_1,\cdots,h_{2N-2})$ a Hamburger moment vector if there exists a $\sigma(u)$ such that (1.9) holds.

Thanks to the Hamburger-Nevanlinna theorem (see, e.g., [1, p. 95]), there exists a one-to-one correspondence between the class of solutions $\sigma(u)$ of the problem (1.9) and the class of functions $F(\lambda)$ in $\mathscr{N}$ satisfying the condition (1.10) in any sector (1.11) given in the problem (1.10) below. As a rule, in that correspondence and in the subsequent discussion all functions $\sigma(u)$ $(-\infty<u<\infty)$ which are substantially equal are not considered as distinct.

Ⅴ. Problem of Reconstructing Functions in $\mathscr{N}$ With a Given Asymptotic Expansion.　Given a real $\boldsymbol{b}=(h_0,h_1,\cdots,h_{2N-2})$, find all functions $F(\lambda)$ in $\mathscr{N}$ such that

$$F(\lambda)=-\frac{h_0}{\lambda}-\frac{h_1}{\lambda^2}-\cdots-\frac{h_{2N-2}}{\lambda^{2N-1}}-R_{2N-1}(\lambda), \tag{1.10}$$

where $R_{2N-1}(\lambda)=o(\lambda^{-2N+1})$ as $\lambda\to\infty$ uniformly in any sector

$$\varepsilon\leqslant\arg\lambda\leqslant\pi-\varepsilon, \quad 0<\varepsilon<\frac{\pi}{2}. \tag{1.11}$$

Thus the solvability of the problems (1.3)(1.8)(1.9), and (1.10) is equivalent if the vector $\boldsymbol{b}$ coincides with the Hankel vector of the GRIP (1.5) and if $\boldsymbol{H}_{NN}[\boldsymbol{b}]>\boldsymbol{O}$. Further, a necessary and sufficient condition for the existence of infinitely many solutions to the problem (1.3) (and therefore to all of problems mentioned above) is just the positiveness of the Hankel matrix $\boldsymbol{H}_{NN}[\boldsymbol{b}]$:

$$\boldsymbol{H}_{NN}[\boldsymbol{b}]>\boldsymbol{O} \tag{1.12}$$

(see Corollary 3.10 below). This observation is by no means trivial.

If (1.12) holds, we have [18, 27] that the generators $F(\lambda)$ in $\mathcal{N}$ of the vector $\boldsymbol{b}$ given in the problem (1.10) are parametrized by a linear fractional transformation of an arbitrary function $\psi(\lambda)$ in $\mathcal{N}$ such that $\lim\limits_{\lambda\to\infty}\psi(\lambda)/\lambda=0$ uniformly in any sector (1.11):

$$-\frac{\gamma(\lambda)+\psi(\lambda)\delta(\lambda)}{p(\lambda)+\psi(\lambda)q(\lambda)},\quad \operatorname{Im}\lambda>0, \tag{1.13}$$

where $p(\lambda)$ and $q(\lambda)$ are respectively the Nth and $(N-1)$th orthogonal polynomials with respect to $\boldsymbol{b}$, considered as a finite positive sequence relative to an axis, while $\gamma(\lambda)$ and $\delta(\lambda)$ are the corresponding polynomials of the second kind. Taking into account the statements of the GRIP (1.5), we can in turn obtain a description of all the solutions $w(\lambda)$ of the NP problem (1.3) in the indeterminate case by means of the formula (1.13) with the exception of a different range of the parameter function $\psi(\lambda)$:

$$w(\lambda)=\omega(\lambda)-\frac{\gamma(\lambda)+\psi(\lambda)\delta(\lambda)}{p(\lambda)+\psi(\lambda)q(\lambda)}A(\lambda),\qquad \psi(\lambda)\in\mathcal{N}\cup\{\infty\}, \tag{1.14}$$

in which $\omega(\lambda)$ stands for the unique Hermite interpolation polynomial, of degree at most $2N-1$, of the GRIP (1.5), and $A(\lambda)$ is the polynomial of degree $2N$ with real coefficients defined by

$$A(\lambda)=\prod_{j=1}^{\theta}(\lambda-Y_j)^{\tau_j}(\lambda-\overline{Y}_j)^{\tau_j} \tag{1.15}$$

(see Theorem 3.11 below for further details). It is obvious that (1.14) can be rewritten as a linear fractional transformation, whose coefficient matrix has the form

$$\begin{bmatrix} A(\lambda) & \omega(\lambda) \\ 0 & 1 \end{bmatrix}\begin{bmatrix} -\gamma(\lambda) & -\delta(\lambda) \\ p(\lambda) & q(\lambda) \end{bmatrix}.$$

Furthermore, it will be proved in Theorem 3.9 below that the formula (1.13) with $\psi(\lambda)\in\mathcal{N}\cup\{\infty\}$ in place of the range of $\psi(\lambda)$ therein can be represented in the form $\int_{-\infty}^{\infty}\mathrm{d}\sigma(u)/(u-\lambda)$, which describes the solutions of the problem (1.8) [to determine the function $\sigma(u)$ from the function $\psi(\lambda)$ one can apply the Stieltjes-Perron inversion formula]. Thus, if $\boldsymbol{H}_{NN}[\boldsymbol{b}]>\boldsymbol{O}$ (1.14) defines in fact a one-to-one correspondence between the class of solutions to the NP problem (1.3) and that to the problem (1.8)

$$w(\lambda)=\omega(\lambda)+\int_{-\infty}^{\infty}\frac{\mathrm{d}\sigma(u)}{u-\lambda}A(\lambda). \tag{1.16a}$$

Suppose now that $\boldsymbol{H}_{NN}[\boldsymbol{b}]\geqslant\boldsymbol{O}$ is singular. We shall show that the two problems (1.3) and (1.8) have only a single solution, respectively, namely, these two problems are determined in this case. Their unique solutions $\omega(\lambda)$ and $\sigma(u)$ $(-\infty<u<\infty)$ satisfy the relation

$$w(\lambda)=\omega(\lambda)+\int_{-\infty}^{\infty}\frac{\mathrm{d}\sigma(u)}{u-\lambda}A(\lambda). \tag{1.16b}$$

where $\omega(\lambda)$ and $A(\lambda)$ are as in (1.14) and (1.15), and $\sigma(u)$ has only a finite number of points of increase (not more than $N-1$). Moreover, with the help of the statement of the quasidirect decomposition of singular Hankel matrices one can write down a more explicit expression of (1.16b), and can amplify these results to some extent.

It is thereby shown that the solution of the NP problem (1.3) is adequate to the study of the power moment problem (1.8) associated with the Hankel vector $\boldsymbol{b}$, and the formula (1.16) defines a one-to-one correspondence between the solutions of these two problems.

This paper is divided into three sections. In Section 2, restricted to the case of the GRIP (1.5), we first recall statements from the theory of GRIPs given mainly in [7, 14], and then handle that problem under the condition that $\boldsymbol{H}_{NN}[\boldsymbol{b}]\geqslant\boldsymbol{O}$ is not degenerate (that is, $\boldsymbol{H}_{NN}[\boldsymbol{b}]\neq\boldsymbol{H}_d$ in (2.20) below). Section 3 is devoted to an instructive alternative proof and extension of criteria for the NP problem (1.3) to be solvable, and to be determinate or indeterminate; to the intrinsic connection between the NP problem and the problems of the power moment and reconstructing functions in $\mathscr{N}$ such that (1.10) holds in which $\boldsymbol{b}$ is the Hankel vector; and also to an effective construction of the solutions to (1.3)(1.8)(1.9) and (1.10) via the Hankel vector approach.

§ 2. The General Rational Interpolation Problem of a Suitable Symmetric Form

The NP problem studied here involves naturally the consideration of certain results of the rational interpolation problem (1.5) and the problem (1.10), in particular, the rational solution(s) of least degree to the problems (1.5) and (1.10). Suppose now that we are given the interpolation data as in (1.5). It is required to determine all complex-valued rational functions $\varphi(\lambda)$ (with no poles at any Y_i or $\overline{Y}_i$), called interpolants to the GRIP (1.5), such that the equations (1.5) hold.

It is well known that there always exists a unique polynomial solution $\omega(\lambda)$ of degree at most $2N-1$ to the GRIP (1.5), referred to as the Hermite (interpolation) polynomial of that problem:

$$\omega(\lambda) = \sum_{i=1}^{\theta} \frac{A(\lambda)}{(\tau_i-1)!}\left\{\left(\frac{\mathrm{d}}{\mathrm{d}\zeta}\right)^{\tau_i-1}\left[\sum_{j=0}^{\tau_i-1}\frac{C_{ij}(\zeta-Y_i)^{\tau_i+j}}{(\lambda-\zeta)A(\zeta)}\right]_{\zeta=Y_i}+\right.$$
$$\left.\left(\frac{\mathrm{d}}{\mathrm{d}\zeta}\right)^{\tau_i-1}\left[\sum_{j=0}^{\tau_i-1}\frac{\overline{C}_{ij}(\zeta-\overline{Y}_i)^{\tau_i+j}}{(\lambda-\zeta)A(\zeta)}\right]_{\zeta=\overline{Y}_i}\right\}. \quad (2.1)$$

where $A(\lambda)$ is as in (1.15).

As usual, we define the complexity of a rational function $g(\lambda)/f(\lambda)$ as its McMillan degree:

$$\deg\frac{g(\lambda)}{f(\lambda)}=\max\{\deg g(\lambda),\ \deg f(\lambda)\},\quad g \text{ and } f \text{ coprime.} \quad (2.2)$$

Put now

$$Q=\{Y_1,\cdots,Y_1,\cdots,Y_\theta,\cdots,Y_\theta,\overline{Y}_1,\cdots,\overline{Y}_1,\cdots,\overline{Y}_\theta,\cdots,\overline{Y}_\theta\}, \quad (2.3)$$

the interpolation node collection of $2N$ numbers, taking each Y_i or $\overline{Y}_i$ τ_i times. Let us partition Q into an arbitrary pair of interpolation vectors (IVs for short) $\hat{\boldsymbol{y}}$ and $\hat{\boldsymbol{z}}$ with dimension $|\hat{\boldsymbol{y}}|$ and $|\hat{\boldsymbol{z}}|$, $|\hat{\boldsymbol{y}}|+|\hat{\boldsymbol{z}}|=2N$. Denote by $\boldsymbol{L}_{\hat{y}\hat{z}}(\omega(\lambda))$ the (generalized) Loewner matrix associated with the GRIP (1.5) and the IVs $\hat{\boldsymbol{y}}$ and $\hat{\boldsymbol{z}}$(see, e.g. [6] for details), In particular, to the pair of IVs $\mathbf{y}$ and $\bar{\mathbf{y}}$ given in (1.6) there is related an $N\times N$ Hermitian Loewner matrix

$$L_{y\bar{y}}(\omega)=[L_{ij}]_{i,j=1}^{\theta} \tag{2.4a}$$

where the (i,j)th block L_{ij} is in $\mathbf{C}^{\tau_i\times\tau_j}$ with entries $K_{kl}(Y_i,\bar{Y}_j)$ determined by the divided difference with respect to the Hermite polynomial $\omega(\lambda)$:

$$\begin{aligned}K_{kl}(Y_i,\bar{Y}_j)&=[Y_i,\cdots,Y_i,\bar{Y}_j,\cdots,\bar{Y}_j]_{\omega(\lambda)}\\&=\frac{1}{k!l!}\frac{\partial^{k+l}}{\partial\zeta^k\partial\eta^l}\left[\frac{\omega(\zeta)-\omega(\eta)}{\zeta-\eta}\right]_{\zeta=Y_i,\eta=\bar{Y}_j}\end{aligned} \tag{2.4b}$$

Theorem 2.1 *Let the GRIP (1.5) be given, and let $\omega(\lambda)$ be its Hermite polynomial. Then there exists a unique real vector $\boldsymbol{b}=(h_0,h_1,\cdots,h_{2N-2})$ such that*

$$L_{\hat{y},\hat{z}}(\omega)=W_{\hat{y}}H_{|\hat{y}|,|\hat{z}|}[\boldsymbol{b}]W_{\hat{z}}^{\mathrm{T}} \tag{2.5}$$

whenever $\hat{\mathbf{y}},\hat{\mathbf{z}}$ are an arbitrary pair of IVs of Q in (2.3) with dimensions $|\hat{\mathbf{y}}|$ and $|\hat{\mathbf{z}}|$, respectively, where $W_{\hat{y}}(W_{\hat{z}})$ is some appropriately defined nonsingular matrix of order $|\hat{\mathbf{y}}|(|\hat{\mathbf{z}}|)$ dependent of the interpolation vector $\hat{\mathbf{y}}(\hat{\mathbf{z}})$ only, and $H_{|\hat{y}|,|\hat{z}|}[\boldsymbol{b}]=(h_{i+j})_{i,j=0}^{|\hat{y}|-1,|\hat{z}|-1}$, $|\hat{\mathbf{y}}|+|\hat{\mathbf{z}}|=2N$. In particular, for the IVs $\mathbf{y}$ and $\bar{\mathbf{y}}$ given in (1.6) we have

$$L_{y,\bar{y}}(\omega)=W_yH_{NN}[\boldsymbol{b}]W_y^*. \tag{2.6}$$

where the asterisk denotes the hermitian transpose of the indicated matrix.

Proof (2.5)(2.6), and the fact that $\boldsymbol{b}$ is real follow from Theorem 3.3 of [7], noting that if $\hat{\mathbf{y}}=\mathbf{y}$ and $\hat{\mathbf{z}}=\bar{\mathbf{y}}$ then $W_{\bar{y}}^{\mathrm{T}}=W_y^*$, and therefore $H_{NN}[\boldsymbol{b}]=H_{NN}[\boldsymbol{b}]^{\mathrm{T}}=H_{NN}[\boldsymbol{b}]^*$. ■

Theorem 2.1 says that given the GRIP (1.5), the corresponding Loewner matrices always give rise to one and the same vector $\boldsymbol{b}$ such that (2.5) holds. We refer to $\boldsymbol{b}$ of dimension $2N-1$ defined uniquely by Theorem 2.1 as the Hankel vector of the GRIP (1.5); it plays a key role in full analysis of the GRIP. Conversely, each real vector $\boldsymbol{b}=(h_0,h_1,\cdots,h_{2N-2})$ of dimension $2N-1$ generates uniquely a certain GRIP of the form (1.5) up to a constant summand d for C_{i0}:

$$Y_i, C_{i0}+d,\ C_{il}, \cdots, C_{i,\tau_i-1},\quad i=1,2,\cdots,\theta, \tag{2.7}$$

such that $\boldsymbol{b}$ is the Hankel vector of the GRIP with the interpolation data (2.7) (see [7, Lemma 2.1]).

The Hankel vector of a GRIP has certain properties of interest. For instance, we have

Theorem 2.2[7] *With the notation as in Theorem 2.1, let* $\boldsymbol{b}=(h_0, h_1, \cdots, h_{2N-2})$ *be the Hankel vector of the GRIP* (1.5). *Then* $\boldsymbol{b}$ *coincides with the first* $2N-1$ *Markov parameters of the strictly proper rational function* $\omega(\lambda)/A(\lambda)$, *that is, for sufficiently large values of* $|\lambda|$

$$\frac{\omega(\lambda)}{A(\lambda)}=\frac{h_0}{\lambda}+\frac{h_1}{\lambda^2}+\cdots+\frac{h_{2N-2}}{\lambda^{2N-1}}+o(\lambda^{-2N+1}). \tag{2.8}$$

By definition, a strictly proper rational function $\varphi(\lambda)$ is called a rational generator to a vector $\boldsymbol{b}=(h_0, h_1, \cdots, h_{2N-2})\neq\boldsymbol{0}$ if it has as its Laurent expansion at infinity the series on the right side of (2.8). So, in particular, $\omega(\lambda)/A(\lambda)$ is a rational generator to $\boldsymbol{b}$. It is obvious that if $\varphi(\lambda)$ is a rational generator to $\boldsymbol{b}\neq\boldsymbol{0}$ such that $-\varphi(\lambda)\in\mathcal{N}$, then $-\varphi(\lambda)$ is actually a solution to the problem (1.10). It is seen from (2.8) and (2.1) that

$$h_l=\sum_{i=1}^{\theta}\frac{2}{(\tau_i-1)!}\left(\frac{\mathrm{d}}{\mathrm{d}\zeta}\right)^{\tau_i-1}\sum_{j=0}^{\tau_i-1}\mathrm{Re}\left.\frac{C_{ij}(\zeta-Y_i)^{\tau_i+j}\zeta^l}{A(\zeta)}\right|_{\zeta=Y_i},$$

$$l=0,1,\cdots. \tag{2.9}$$

In particular, if all $\tau_i=1$ $(\theta=N)$, then the problem (1.5) is reduced to the classical rational interpolation problem, and the formula (2.9) will be replaced by

$$h_l=2\sum_{i=1}^{N}\mathrm{Re}\frac{C_{i0}Y_i^l}{\prod_{j\neq i}(Y_i-Y_j)\prod_{j=1}^{N}(Y_i-\bar{Y}_j)},\quad l=0,1,\cdots. \tag{2.9'}$$

It is obvious that (2.9) and (2.8) imply in turn that $\omega(\lambda)$ has real coefficients.

Let us now make the real Hankel vector $\boldsymbol{b}$ of the GRIP (1.5) our starting point. If $\boldsymbol{b}\neq\boldsymbol{0}$, the characteristic degrees $n_1=n_1(\boldsymbol{b})$ and $n_2=n_2(\boldsymbol{b})$ of $\boldsymbol{b}$ are, by definition,

$$n_1=\mathrm{rank}\ \boldsymbol{H}_{NN}[\boldsymbol{b}],\quad\text{and}\quad n_2=2N-n_1. \tag{2.10}$$

It is known [14] that

$$n_1 \leqslant n_2 \quad \text{and} \quad \operatorname{rank} \boldsymbol{H}_{k,l}[\boldsymbol{b}] = \min\{k, l, n_1\}, \quad k+l=2N.$$

Denote by $\mathbf{C}_n(\lambda)$ the class of polynomials of degree at most n, and by $\mathbf{C}_n^0(\lambda)$ the subclass consisting of all monic polynomials of degree exact n. Put

$$\mathscr{A}_l = (1, \lambda, \cdots, \lambda^{l-1}) \ker \boldsymbol{H}_{kl}[\boldsymbol{b}], \quad k+l=2N, \tag{2.11}$$

a subspace of $\mathbf{C}_{l-1}(\lambda)$. There always exists a canonical pair of the so-called characteristic polynomials $(p(\lambda), q(\lambda))$, with real coefficients, of the vector $\boldsymbol{b}$ such that (1) either $p(\lambda) \in \mathbf{C}_{n_1}^0(\lambda)$ and $q(\lambda) \in \mathbf{C}_{n_2-1}(\lambda)$ or $P(\lambda) \in \mathbf{C}_{n_1-1}(\lambda)$ and $q(\lambda) \in \mathbf{C}_{n_2}^0(\lambda)$, and (2) $p(\lambda), q(\lambda)$ form a basis of $\mathscr{A}_{n_1+1}$ if $n_1 = n_2 = N$ ($\boldsymbol{H}_{NN}[\boldsymbol{b}]$ is then nonsingular), and $p(\lambda)$ (forming a basis of $\mathscr{A}_{n_1+1}$), $\lambda p(\lambda), \cdots, \lambda^{n_2-n_1} p(\lambda)$, $q(\lambda)$ form a basis of $\mathscr{A}_{n_2+1}$ if $n_1 < n_2$. [In case $1 = n_1 < n_2$, choose an arbitrary $q(\lambda) \in \mathbf{C}_{n_2}(\lambda)$ such that $p(\lambda)$, $\lambda p(\lambda), \cdots, \lambda^{n_2-1} p(\lambda)$, $q(\lambda)$ are linearly independent]. Set now

$$p(\lambda) = \sum_{j=0}^{n_1} p_j \lambda^j \quad \text{and} \quad q(\lambda) = \sum_{j=0}^{n_2} q_j \lambda^j, \tag{2.12}$$

and denote by $S(p(\lambda))$ and $S(q(\lambda))$ the symmetrizers of $p(\lambda)$ and $q(\lambda)$, respectively:

$$S(p(\lambda)) = \begin{bmatrix} p_1 & \cdots & p_{n_1-1} & p_{n_1} \\ \vdots & \iddots & \iddots & 0 \\ p_{n_1-1} & \iddots & \iddots & \vdots \\ p_{n_1} & 0 & \cdots & 0 \end{bmatrix} \in \mathbf{R}^{n_1 \times n_1},$$

$$S(q(\lambda)) = \begin{bmatrix} q_1 & \cdots & q_{n_2-1} & q_{n_2} \\ \vdots & \iddots & \iddots & 0 \\ q_{n_2-1} & \iddots & \iddots & \vdots \\ q_{n_2} & 0 & \cdots & 0 \end{bmatrix} \in \mathbf{R}^{n_2 \times n_2}.$$

Define in turn $\gamma(\lambda) \in \mathbf{C}_{n_1-1}(\lambda)$ and $\delta(\lambda) \in \mathbf{C}_{n_2-1}(\lambda)$, both with real coefficients, by

$$\gamma(\lambda) = (1, \lambda, \cdots, \lambda^{n_1-1}) S(p(\lambda)) (h_0, h_1, \cdots, h_{n_1-1})^{\mathrm{T}},$$
$$\delta(\lambda) = (1, \lambda, \cdots, \lambda^{n_2-1}) S(q(\lambda)) (h_0, h_1, \cdots, h_{n_2-1})^{\mathrm{T}}. \tag{2.13}$$

We point out that if we introduce a linear functional in the space of $\mathbf{C}_{2N-1}(\lambda)$ associated with the vector $\boldsymbol{b}$ and a suitable real h_{2N-1},

$$S_\lambda\{R(\lambda)\}=r_0h_0+r_1h_1+\cdots+r_{2N-2}h_{2N-2}+r_{2N-1}h_{2N-1} \quad (2.14)$$

for $R(\lambda)=\sum_{j=0}^{2N-1} r_j\lambda^j\in \mathbf{C}_{2N-1}(\lambda)$, then (2.13) can be represented as

$$\gamma(\lambda)=S_u\left(\frac{p(\lambda)-p(u)}{\lambda-u}\right) \quad \text{and} \quad \delta(\lambda)=S_u\left(\frac{q(\lambda)-q(u)}{\lambda-u}\right).$$

Further, we conclude that the real number $\sigma=q(0)\gamma(0)-p(0)\delta(0)$ does not vanish, and

$$\det\begin{bmatrix}\gamma(\lambda) & \delta(\lambda)\\ p(\lambda) & q(\lambda)\end{bmatrix}=q(\lambda)\gamma(\lambda)-p(\lambda)\delta(\lambda)=\sigma, \quad (2.15)$$

so that the pairs of polynomials $(p(\lambda),q(\lambda))$, $(p(\lambda),\gamma(\lambda))$, and $(q(\lambda),\delta(\lambda))$ are coprime, respectively; namely, $\gcd(p,q)=\gcd(p,\gamma)=\gcd(q,\delta)=1$. In what follows, without loss of generality, we select $\sigma>0$.

Theorem 2.3[14, 7] *With the above notation, the complete parametrization of the class $\mathscr{F}_n$ of rational generators of exact degree n to the Hankel vector $\mathbf{0}\neq \boldsymbol{b}=(h_0,h_1,\cdots,h_{2N-2})$ can be described in the following way:*

(a) *If $n_1=n_2=N$, then*

$$\mathscr{F}_n=\begin{cases}\varnothing, & \text{if } n<n_1,\\ \left\{\dfrac{\gamma(\lambda)u(\lambda)+\delta(\lambda)v(\lambda)}{p(\lambda)u(\lambda)+q(\lambda)v(\lambda)}:\right. \\ \quad \left.\gcd(u,v)=1, u\in \mathbf{C}^0_{n-n_1}(\lambda), v\in \mathbf{C}_{n-n_2}(\lambda)\right\}, & \text{if } n\geqslant n_1.\end{cases}$$

(b) *If $n_1<n_2$ and* deg $p(\lambda)=n_1$, *then*

$$\mathscr{F}_n=\begin{cases}\varnothing, & \text{if } n<n_2,\ n\neq n_1,\\ \{\gamma(\lambda)/p(\lambda)\} & \text{if } n=n_1,\\ \left\{\dfrac{\gamma(\lambda)u(\lambda)+\delta(\lambda)v(\lambda)}{p(\lambda)u(\lambda)+q(\lambda)v(\lambda)}:\right. \\ \quad \left.\gcd(u,v)=1, u\in \mathbf{C}^0_{n-n_1}(\lambda), v\in \mathbf{C}_{n-n_2}(\lambda)\right\}, \\ & \text{if } n\geqslant n_2.\end{cases}$$

(c) *If $n_1<n_2$ and* deg $p(\lambda)<n_1$, *then*

$$\mathscr{F}_n=\begin{cases}\varnothing, & \text{if } n<n_2,\\ \left\{\dfrac{\gamma(\lambda)u(\lambda)+\delta(\lambda)v(\lambda)}{p(\lambda)u(\lambda)+q(\lambda)v(\lambda)}:\right. \\ \quad \left.\gcd(u,v)=1, u\in \mathbf{C}_{n-n_1}(\lambda), v\in \mathbf{C}^0_{n-n_2}(\lambda)\right\}, & \text{if } n\geqslant n_2.\end{cases}$$

We remark that Theorem 2.3 is valid even for $\boldsymbol{b}=\mathbf{0}$, if we take the conventions that

$$n_1=0,\quad n_2=2N,\quad p(\lambda)=\delta(\lambda)=1,\quad q(\lambda)=\gamma(\lambda)=0,\tag{2.16}$$

and $\deg u(\lambda)-\deg v(\lambda)\geqslant 2N$ therein.

Theorem 2.4[7] *Let $\mathbf{0}\neq\boldsymbol{b}$ be the Hankel vector of the GRIP* (1.5), *and let (n_1,n_2), $p(\lambda)$, $q(\lambda)$, $\gamma(\lambda)$, $\delta(\lambda)$, $\omega(\lambda)$, and $A(\lambda)$ be as before. Define $\alpha(\lambda)$ and $\beta(\lambda)$ via*

$$\alpha(\lambda)=p(\lambda)\omega(\lambda)-\gamma(\lambda)A(\lambda),\quad \beta(\lambda)=q(\lambda)\omega(\lambda)-\delta(\lambda)A(\lambda).\tag{2.17}$$

Then

$$\det\begin{bmatrix}\alpha(\lambda) & \beta(\lambda)\\ p(\lambda) & q(\lambda)\end{bmatrix}=q(\lambda)\alpha(\lambda)-p(\lambda)\beta(\lambda)=-\sigma A(\lambda),\tag{2.18}$$

where $\sigma(>0)$ is the same as in (2.15). *Moreover, the complete parametrization of the class $\mathscr{G}_n$ of all the rational interpolants of degree exact n to the GRIP* (1.5) *can be described in the following way:*

(a) *If $n_1=n_2$ (and therefore $\deg p(\lambda)=N>\deg q(\lambda)$), then*

$$\mathscr{G}_n=\begin{cases}\varnothing, & if\ \ n<n_1,\\ \left\{\dfrac{\alpha(\lambda)u(\lambda)+\beta(\lambda)v(\lambda)}{p(\lambda)u(\lambda)+q(\lambda)v(\lambda)}:\right. & \\ \quad \gcd(u,v)=\gcd(pu+qv,A)=1; & \\ \quad u\in\mathbf{C}^0_{n-n_1}(\lambda),v\in\mathbf{C}_{n-n_2}(\lambda), & \\ \quad \left. or\ u\in\mathbf{C}_{n-n_1-1}(\lambda),\ v\in\mathbf{C}^0_{n-n_2}(\lambda)\right\}, & if\ \ n\geqslant n_1.\end{cases}$$

(b) *If $n_1<n_2$ $\deg p(\lambda)=n_1$, then*

$$\mathscr{G}_n=\begin{cases}\varnothing, & if\ \ n<n_2, n\neq n_1,\\ \{\alpha(\lambda)/p(\lambda):\gcd(p,A)=1\}, & if\ \ n=n_1,\\ \left\{\dfrac{\alpha(\lambda)u(\lambda)+\beta(\lambda)v(\lambda)}{p(\lambda)u(\lambda)+q(\lambda)v(\lambda)}:\right. & \\ \quad \gcd(u,v)=\gcd(pu+qv,A)=1; & \\ \quad u\in\mathbf{C}^0_{n-n_1}(\lambda),v\in\mathbf{C}_{n-n_2}(\lambda), & \\ \quad \left. or\ u\in\mathbf{C}_{n-n_1-1}(\lambda),\ v\in\mathbf{C}^0_{n-n_2}(\lambda)\right\}, & if\ \ n\geqslant n_2.\end{cases}$$

(c) *If* $n_1 < n_2$ deg $p(\lambda) < n_1$, *then*

$$\mathcal{G}_n = \begin{cases} \varnothing, & if \ \ n < n_2, n \neq n_1, \\ \{\alpha(\lambda)/p(\lambda) : \gcd(p, A) = 1\}, & if \ \ n = n_1, \\ \Big\{ \dfrac{\alpha(\lambda)u(\lambda) + \beta(\lambda)v(\lambda)}{p(\lambda)u(\lambda) + q(\lambda)v(\lambda)} : & \\ \gcd(u, v) = \gcd(pu + qv, A) = 1; & \\ u \in \mathbf{C}_{n-n_1}(\lambda), v \in \mathbf{C}^0_{n-n_2}(\lambda) & \\ or \ u \in \mathbf{C}^0_{n-n_1}(\lambda),\ v \in \mathbf{C}_{n-n_2-1}(\lambda) \Big\}, & if \ \ n \geqslant n_2. \end{cases}$$

We note that it is trivial to verify that in case $\boldsymbol{b} = \mathbf{0}$, Theorem 2. 4 is also true if we agree to the conventions (2. 16) [and therefore $\alpha(\lambda) = \omega(\lambda) = C_{10}$ and $\beta(\lambda) = -A(\lambda)$ by (2. 17)].

Thus, to the statement of the GRIP (1. 5) of order $2N$ there corresponds a unique real Hankel vector $\boldsymbol{b} = (h_0, h_1, \cdots, h_{2N-2})$ of dimension $2N-1$, whose pair of characteristic polynomials $p(\lambda)$ and $q(\lambda)$ combined with the Hermite polynomial $\omega(\lambda)$ in question determines fully all the solutions of that problem.

Corollary 2. 5[7] *With the notation of Theorem* 2. 4, *we have*

(a) $n_1 = \max\{\deg \alpha(\lambda), \deg p(\lambda)\} \leqslant n_2 = \max\{\deg \beta(\lambda), \deg q(\lambda)\}$.

(b) $\gcd(\alpha(\lambda), p(\lambda)) = 1$ *if*, *and only if*, $\gcd(p(\lambda), A(\lambda)) = 1$; *in this case*, $\alpha(\lambda)/p(\lambda)$ *is of degree* n_1, *is real on the real axis*, *and is an interpolant to the GRIP* (1. 5). *If*, *further*, $n_1 < n_2$, *then* $\alpha(\lambda)/p(\lambda)$ *is the only rational interpolant of degree* $< N$ *to the GRIP* (1. 5).

(c) *if* $n_1 = n_2 = N$, *there exists a family of interpolants of least degree N to the GRIP* (1. 5), *which are real on the real axis and of the form*

$$\frac{\alpha(\lambda) + t\beta(\lambda)}{p(\lambda) + tq(\lambda)}, \tag{2. 19}$$

where $t \in \mathbf{R} \cup \{\infty\}$ *is the free parameter such that*

$$\gcd(p(\lambda) + tq(\lambda), A(\lambda)) = 1.$$

Corollary 2. 6[7] *Let* $\varphi(\lambda)$ *be an interpolant to the GRIP* (1. 5), *and let* $\omega(\lambda)$ *be as before*, *and let* $\mathbf{y}$, $\bar{\mathbf{y}}$ *be IVs of Q as in* (1. 6). *Then*

$$\text{rank } \boldsymbol{L}_{yy}(\omega(\lambda)) = \deg \varphi(\lambda) \quad if\ N \geqslant \deg \varphi(\lambda).$$

In the rest of this section, we continue the study of the GRIP (1.5), but in contrast with the previous part, we do it under the restriction (1.7), where $\boldsymbol{b}=(h_0, h_1, \cdots, h_{2N-2})$ is the (real) Hanekl vector, $\boldsymbol{b} \neq \boldsymbol{0}$. These results will be used in the next section over and again. It is convenient to have the following lemma before proving the main Theorems 2.3′ and 2.4′. It is a simple consequence of the statement of the so-called quasidirect decomposition of singular Hankel matrices [11].

Lemma 2.7 *If $\boldsymbol{H}_{NN}[\boldsymbol{b}] \geqslant \boldsymbol{O}$ is singular, and if its (unique) quasidirect decomposition has the form*

$$\boldsymbol{H}_{NN}[\boldsymbol{b}] = \boldsymbol{H}_p + \boldsymbol{H}_d, \tag{2.20}$$

where $\boldsymbol{H}_p$ and $\boldsymbol{H}_d$ are the proper and degenerate parts of $\boldsymbol{H}_{NN}[\boldsymbol{b}]$, respectively, such that $\text{rank } \boldsymbol{H}_{NN}[\boldsymbol{b}] = \text{rank } \boldsymbol{H}_p + \text{rank } \boldsymbol{H}_d$, *then $\boldsymbol{H}_p \geqslant \boldsymbol{O}$ and $\boldsymbol{H}_d \geqslant \boldsymbol{O}$, and there exists a unique real $\hat{\alpha}$ with $0 \leqslant \hat{\alpha} \leqslant h_{2N-2}$ such that*

$$\boldsymbol{H}_p = \boldsymbol{H}_{NN}[\boldsymbol{b}_p], \quad \boldsymbol{H}_d = \boldsymbol{H}_{NN}[\boldsymbol{b}_d], \tag{2.21}$$

in which $\boldsymbol{b}_p + \boldsymbol{b}_d = \boldsymbol{b}$, $\boldsymbol{b}_p = (h_0, h_1, \cdots, h_{2N-3}, h_{2N-2} - \hat{\alpha})$ and $\boldsymbol{b}_d = (0, \cdots, 0, \hat{\alpha})$.

Proof It follows from [11] that if $\boldsymbol{H}_{NN}[\boldsymbol{b}] \geqslant \boldsymbol{O}$ is written in the block form

$$\boldsymbol{H}_{NN}[\boldsymbol{b}] = \begin{bmatrix} \boldsymbol{A} & \boldsymbol{B} \\ \boldsymbol{B}^* & \boldsymbol{C} \end{bmatrix},$$

where the order of $\boldsymbol{A}$ is equal to the maximum of the first linearly independent columns of $\boldsymbol{H}_{NN}[\boldsymbol{b}]$, then its quasidirect decomposition is of the form

$$\boldsymbol{H}_{NN}[\boldsymbol{b}] = \boldsymbol{H}_p + \boldsymbol{H}_d = \begin{bmatrix} \boldsymbol{A} & \boldsymbol{B} \\ \boldsymbol{B}^* & \boldsymbol{B}^* \boldsymbol{A}^{-1} \boldsymbol{B} \end{bmatrix} + \begin{bmatrix} \boldsymbol{O} & \boldsymbol{O} \\ \boldsymbol{O} & \boldsymbol{C} - \boldsymbol{B}^* \boldsymbol{A}^{-1} \boldsymbol{B} \end{bmatrix}.$$

Since $\boldsymbol{H}_{NN}[\boldsymbol{b}] \geqslant \boldsymbol{O}$ is equivalent to $\boldsymbol{A} > \boldsymbol{O}$ and $\boldsymbol{C} - \boldsymbol{B}^* \boldsymbol{A}^{-1} \boldsymbol{B} \geqslant \boldsymbol{O}$, we have that $\boldsymbol{H}_p \geqslant \boldsymbol{O}$ and $\boldsymbol{H}_d \geqslant \boldsymbol{O}$, so that $\boldsymbol{C} - \boldsymbol{B}^* \boldsymbol{A}^{-1} \boldsymbol{B}$ is a certain diagonal matrix of the form $\text{diag}(0, \cdots, 0, \hat{\alpha})$ where $\hat{\alpha} \geqslant 0$, as desired. ■

Observe that if $\boldsymbol{H}_{NN}[\boldsymbol{b}] \geqslant \boldsymbol{O}$ is proper, that is, either $\boldsymbol{H}_{NN}[\boldsymbol{b}] > \boldsymbol{O}$ or $\boldsymbol{H}_{NN}[\boldsymbol{b}] \geqslant \boldsymbol{O}$ is singular and proper (i. e., $\boldsymbol{H}_{NN}[\boldsymbol{b}] = \boldsymbol{H}_p$ in (2.20)), and $n_1 = \text{rank } \boldsymbol{H}_{NN}[\boldsymbol{b}]$, the aforementioned canonical pair of characteristic *polynomials* $p(\lambda)$ and $q(\lambda)$ of the vector $\boldsymbol{b}$ can be chosen to be the n_1th and (n_1-1)th orthogonal polynomials with respect to the finite positive sequence $\{h_j\}_0^{2n_1-2}$ truncated form the Hankel vector $\boldsymbol{b}$. Indeed, under the hypothesis on $\boldsymbol{H}_{NN}[\boldsymbol{b}]$, the finite sequence $\{h_j\}_0^{2n_1-2}$ for which the Hankel matrix of order n_1 is positive definite $[(h_{i+j})_{i,j=1}^{n_1-1}>\boldsymbol{O}]$ can be always continued in a suitable way to an infinite real sequence $h_0, h_1, \cdots, h_{2n_1-2}, h'_{2n_1-1}, h'_{2n_1}, \cdots$, such that the latter is also positive relative to an axis, where $h'_{2n_1-1}=h_{2n_1-1}$ if $n_1<n_2$ and h'_{2n_1-1} arbitrary, otherwise. Denote by $\hat{p}(\lambda)$ and $\hat{q}(\lambda)$ the n_1th and (n_1-1)th orthogonal polynomials of that infinite positive sequence, or, for short, of the Hankel vector $\boldsymbol{b}$, respectively. Then we have [1] that

$$\hat{p}(\lambda)=\alpha \det\begin{bmatrix} h_0 & h_1 & \cdots & h_{n_1} \\ h_1 & h_2 & \cdots & h_{n_1+1} \\ \vdots & \vdots & & \vdots \\ h_{n_1-1} & h_{n_1} & \cdots & h'_{2n_1-1} \\ 1 & \lambda & \cdots & \lambda^{n_1} \end{bmatrix}, \tag{2.22}$$

in which $\alpha\neq 0$ is a suitable real constant such that $\hat{p}(\lambda)$ is monic, and that $S_\lambda\{\lambda^k\hat{p}(\lambda)\}=0$ for $k=0,1,\cdots,n_1-1$, where $S_\lambda\{\cdot\}$ is defined by (2.14), whence $\hat{p}(\lambda)\in(1,\lambda,\cdots,\lambda^{n_1})\ker(h_{i+j})_{i,j=0}^{n_1,n_1+1}\subset\mathscr{A}_{n_1+1}$, where $\mathscr{A}_{n_1+1}$ is defined by (2.11) with $l=n_1+1$. By analogy, we can prove that $\hat{q}(\lambda)\in\mathscr{A}_{n_2+1}$.

In the case when $\boldsymbol{H}_{NN}[\boldsymbol{b}]\geqslant\boldsymbol{O}$ $(\boldsymbol{b}\neq\boldsymbol{0})$ is singular and of the form (2.20) with $\boldsymbol{H}_p\neq\boldsymbol{O}$ and $\boldsymbol{H}_d\neq\boldsymbol{O}$, we have by Lemma 2.7 that rank $\boldsymbol{H}_p=n_1-1$ and therefore the sequence $\{h_j\}_0^{2n_1-4}$ is in turn positive. Then it is obvious that the canonical pair of characteristic polynomials $p(\lambda)$ and $q(\lambda)$ of the vector $\boldsymbol{b}$ can be chosen as the (n_1-1)th orthogonal polynomial with respect to that positive sequence and an arbitrary $q(\lambda)\in\mathbf{C}_{n_2}^0(\lambda)$ such that $q(\lambda)\in\mathscr{A}_{n_2+1}$, respectively.

In the sequel, if $\boldsymbol{H}_{NN}[\boldsymbol{b}]\geqslant\boldsymbol{O}$ is not degenerate, we call a pair of polynomials $p(\lambda)$ and $q(\lambda)$ so constructed just a special canonical pair of (characteristic) polynomials of $\boldsymbol{b}$.

From the theory of orthogonal polynomials with respect to a positive sequence we have [1] that all the zeros of the orthogonal polynomial $p(\lambda)$ are real and simple, whence $\gcd(p(\lambda), A(\lambda))=1$. The last condition implies in turn, by Corollary 2.5, that $\alpha(\lambda)/p(\lambda)$ is always a solution of degree exact n_1 to the GRIP (1.5), where $\alpha(\lambda)$ is defined by (2.17). Similarly, we can prove that if $\boldsymbol{H}_{NN}[\boldsymbol{b}]>\boldsymbol{O}$, then for any $t\in \mathbf{R}\cup\{\infty\}$, the rational function $[\alpha(\lambda)+t\beta(\lambda)]/[p(\lambda)+tq(\lambda)]$ of exact degree N is also a solution to (1.5), by Corollary 2.5(c) and by the realness of the zeros of $p(\lambda)+tq(\lambda)$.

Also, it is well known [1] that if $\boldsymbol{H}_{NN}[\boldsymbol{b}]\geqslant\boldsymbol{O}$ is proper, then $-\gamma(\lambda)/p(\lambda)$ belongs to $\mathcal{N}$ and is a rational generator of $-\boldsymbol{b}$, and that if, in particular, $\boldsymbol{H}_{NN}[\boldsymbol{b}]>\boldsymbol{O}$, then $-[\gamma(\lambda)+t\beta(\lambda)]/[p(\lambda)+tq(\lambda)]$ belongs to $\mathcal{N}$ for any real finite t, and is a rational generator of $-\boldsymbol{b}$ as well. Thus we have proved

Theorem 2.3′ *Let $\boldsymbol{0}\neq\boldsymbol{b}=(h_0, h_1, \cdots, h_{2N-2})$ be the Hankel vector of the GRIP (1.5) such that $\boldsymbol{H}_{NN}[\boldsymbol{b}]\geqslant\boldsymbol{O}$ is not degenerate. Then Theorem 2.3 is also valid if $p(\lambda)$ and $q(\lambda)$ therein are chosen to be a special canonical pair of polynomials of $\boldsymbol{b}$, where cases (a)(b), and (c) correspond to the cases when $\boldsymbol{H}_{NN}[\boldsymbol{b}]>\boldsymbol{O}$, when $\boldsymbol{H}_{NN}[\boldsymbol{b}]\geqslant\boldsymbol{O}$ is singular and proper and when $\boldsymbol{H}_{NN}[\boldsymbol{b}]\geqslant\boldsymbol{O}$ is singular and not proper, respectively. If, in particular, $\boldsymbol{H}_{NN}[\boldsymbol{b}]>\boldsymbol{O}$ $\{\boldsymbol{H}_{NN}[\boldsymbol{b}]\geqslant\boldsymbol{O}$ is singular and proper, respectively$\}$, then for any real finite t, the rational function $-[\gamma(\lambda)+t\delta(\lambda)]/[p(\lambda)+tq(\lambda)]$ of exact degree N $\{-\gamma(\lambda)/p(\lambda)$ of exact degree $n_1\}$ belongs to $\mathcal{N}$.*

For the GRIP (1.5) under the condition that $\boldsymbol{H}_{NN}[\boldsymbol{b}]\geqslant\boldsymbol{O}$ is not degenerate we have

Theorem 2.4′ *Let $\boldsymbol{0}\neq\boldsymbol{b}=(h_0, h_1, \cdots, h_{2N-2})$ be as in Theorem 2.3′. Then Theorem 2.4 is also true if $p(\lambda)$ and $q(\lambda)$ therein are chosen to be a special canonical pair of polynomials of the Hankel vector $\boldsymbol{b}$ of the GRIP (1.5). Furthermore, if $\boldsymbol{H}_{NN}[\boldsymbol{b}]>\boldsymbol{O}$, the GRIP (1.5) for any $t\in\mathbf{R}\cap\{\infty\}$ has a solution (with degree N) of the form $[\alpha(\lambda)+t\beta(\lambda)]/$*

$[p(\lambda)+tq(\lambda)]$; *if* $\boldsymbol{H}_{NN}[\boldsymbol{b}]\geqslant\boldsymbol{O}$ *is singular, the GRIP* (1.5) *has a solution (with exact degree* n_1*) of the form* $\alpha(\lambda)/p(\lambda)$.

We remark that restricted to the rational solutions to the problem (1.10) [(1.3)], one needs only to the find the solutions given in Theorem 2.3′ [Theorem 2.4′] belonging to the class $\mathcal{N}$.

§ 3. The General NP Interpolation in the Class $\mathcal{N}$

In this section we pull together the results of the previous section to discuss the NP problem (1.3) in the class $\mathcal{N}$, and its connections with the power moment problems (1.8) and (1.9) as well as the problem (1.10). Here our main tool is the real Hankel vector $\boldsymbol{b}=(h_0,h_1,\cdots,h_{2N-2})$ of the corresponding GRIP (1.5). In the sequel, it is convenient to call that vector $\boldsymbol{b}$ the Hankel vector of the NP problem (1.3). It should be emphasized that the aforementioned polynomials $A(\lambda)$, $\omega(\lambda)$, $p(\lambda)$, $q(\lambda)$, $\gamma(\lambda)$, $\delta(\lambda)$ [and therefore $\alpha(\lambda)$ and $\beta(\lambda)$] have all real coefficients.

Let us start with a basic result, which is an immediate consequence of Theorem 2.1.

Lemma 3.1 *With the notation as before, we have that* $\boldsymbol{L}_{y\bar{y}}(\omega)$ *coincides with the Pick matrix* $\boldsymbol{\Lambda}$ *defined via* (1.4) *with* $w(\lambda)$ *replaced by* $w(\lambda)$ *therein, and that* $\boldsymbol{L}_{y\bar{y}}[\omega]\geqslant\boldsymbol{O}(>\boldsymbol{O})$ *if and only if* $\boldsymbol{H}_{NN}[\boldsymbol{b}]\geqslant\boldsymbol{O}$ $(>\boldsymbol{O})$.

In the case when $\boldsymbol{H}_{NN}[\boldsymbol{b}]\geqslant\boldsymbol{O}$ *is degenerate, i.e.,* $\boldsymbol{H}_{NN}[\boldsymbol{b}]=\boldsymbol{H}_d$ in (2.20), in addition to Theorem 2.4′, we have

Lemma 3.2 *Let* $\boldsymbol{H}_{NN}[\boldsymbol{b}]\geqslant\boldsymbol{O}$ *be degenerate,* $\boldsymbol{b}=\boldsymbol{b}_d=(0,\cdots,\hat{\alpha})$. *Then the Hermite polynomial* $\omega(\lambda)$ *of the GRIP* (1.5) *has the form* $\omega(\lambda)=\hat{\alpha}\lambda+\hat{\beta}$ *for some suitable real* $\hat{\beta}$, *which is also a solution to the NP problem* (1.3).

Proof The fact that $\omega(\lambda)=\hat{\alpha}\lambda+\hat{\beta}$ with $\hat{\alpha}\geqslant 0$ follows form Lemma 2.7 and the remark after Theorem 2.1, while $\omega(\lambda)\in\mathcal{N}$ by definition. ■

It follows from Theorem 2.4′ that $\alpha(\lambda)/p(\lambda)$ is a solution to the problem (1.5) if $\boldsymbol{H}_{NN}[\boldsymbol{b}]\geqslant\boldsymbol{O}$ is proper. At the same time, it is also a solution to the NP problem (1.3). To verify this, it is sufficient to prove $\alpha(\lambda)/p(\lambda)\in\mathcal{N}$. Observe first that if $\boldsymbol{H}_{NN}[\boldsymbol{b}]\geqslant\boldsymbol{O}$ is proper, then the rational function $\alpha(\lambda)/p(\lambda)$ is real on the real λ-axis and regular at infinity, and has only simple poles at the (real) zeros of $p(\lambda)$. A simple calculation shows that $\alpha(\lambda)/p(\lambda)$ at each zero z_0 of $p(\lambda)$ has the negative residue

$$\begin{aligned}\frac{\alpha(z_0)}{p'(z_0)}&=\frac{\alpha(z_0)q(z_0)-\beta(z_0)p(z_0)}{p'(z_0)q(z_0)-q'(z_0)p(z_0)}\\&=\frac{-\sigma A(z_0)}{\sigma[q^2(z_0)+\text{some positive real numbers}]}<0,\end{aligned}\tag{3.1}$$

where we use the formula (2.18) and the limiting case of Christoffel-Daboux formula [1, p. 9]. Thus, $\alpha(\lambda)/p(\lambda)$ belongs to $\mathcal{N}$, and therefore is a solution to the NP problem (1.3). Similarly, we can prove that if $\boldsymbol{H}_{NN}[\boldsymbol{b}]>\boldsymbol{O}$, then for any $t\in\mathbf{R}\cup\{\infty\}$, $[\alpha(\lambda)+t\beta(\lambda)]/[p(\lambda)+tq(\lambda)]$ is a solution to the problem (1.3).

All these results can be stated in

Lemma 3.3 *Let $\boldsymbol{0}\neq\boldsymbol{b}=(h_0,h_1,\cdots,h_{2N-2})$ be the Hankel vector of the GRIP* (1.5) *such that $\boldsymbol{H}_{NN}[\boldsymbol{b}]\geqslant\boldsymbol{O}$ is proper, and let $p(\lambda)$, $q(\lambda)$ be its special canonical pair of (characteristic) polynomials. Let $\alpha(\lambda)$ and $\beta(\lambda)$ be as in* (2.18). *Then $\alpha(\lambda)/p(\lambda)$ is a rational solution of exact degree n_1 to the problem* (1.3). *If, in particular, $\boldsymbol{H}_{NN}[\boldsymbol{b}]>\boldsymbol{O}$, then for any $t\in\mathbf{R}\cup\{\infty\}$, $[\alpha(\lambda)+t\beta(\lambda)]/[p(\lambda)+tq(\lambda)]$ is also a rational solution of exact degree N to the problem* (1.3).

Note that we shall prove in Corollary 3.5 that if $\boldsymbol{H}_{NN}[\boldsymbol{b}]\geqslant\boldsymbol{O}$ is not proper, then $\alpha(\lambda)/p(\lambda)$ is also a rational solution of exact degree n_1 to the problem (1.3).

In what follows, we take as conventions that when the Hankel vector $\boldsymbol{b}\neq\boldsymbol{0}$ has the property that $\boldsymbol{H}_{NN}[\boldsymbol{b}]\geqslant\boldsymbol{O}$ is not degenerate, by $p(\lambda)$ and $q(\lambda)$ we denote always a special pair of polynomials of $\boldsymbol{b}$, by $\gamma(\lambda)$ and $\delta(\lambda)$ the corresponding polynomials defined by (2.13), and by $\alpha(\lambda)$ and $\beta(\lambda)$ that defined by (2.18). In case $\boldsymbol{H}_{NN}[\boldsymbol{b}]=\boldsymbol{H}_d$, we take that $p(\lambda)=1$, $\gamma(\lambda)=0$, and $\alpha(\lambda)=\omega(\lambda)$.

The following is an important result which includes a Pick-type theorem for the problem (1.3) and a brand-new and simpler solvability criterion, namely, $\boldsymbol{H}_{NN}[\boldsymbol{b}]\geqslant\boldsymbol{O}$. At the same time, it brings to light an interesting connection between the NP problem and the power moment problem combined with its relatives associated with the Hankel vector in question.

Theorem 3.4 *Let the NP problem* (1.3) *be given, and* $\mathbf{0}\neq\boldsymbol{b}=(h_0, h_1, \cdots, h_{2N-2})$ *be its Hankel vector with characteristic degrees* n_1 *and* n_2. *Then the following are equivalent:*

(a) *There exists a rational solution* $\varphi(\lambda)\in\mathcal{N}$, *real on the real* λ-*axis, to the NP problem* (1.3).

(b) *There exists a solution* $\varphi(\lambda)\in\mathcal{N}$ *to the problem* (1.3).

(c) *The Pick-Loewner matrix* $\boldsymbol{\Lambda}=\boldsymbol{L}_{y\bar{y}}(\omega)$ *defined by* (2.4) *is Hermitian nonnegative:* $\boldsymbol{\Lambda}=\boldsymbol{L}_{y\bar{y}}(\omega)\geqslant\boldsymbol{O}$.

(d) *The Hankel vector* $\boldsymbol{b}$ *forms a finite nonnegative sequence relative to an axis, in the sense that* (1.7) *holds.*

(e) *The truncated power moment problem* (1.8) *has a solution* $\sigma(u)$ $(-\infty<u<+\infty)$.

(f) *There exists a real* α *with* $0\leqslant\alpha\leqslant h_{2N-2}$ *such that the vector* $(h_0, h_1, \cdots, h_{2N-3}, h_{2N-2}-\alpha)$ *of dimension* $2N-1$ *is a Hamburger vector.*

(g) *There exists a real* α *with* $0\leqslant\alpha\leqslant h_{2N-2}$ *and a rational function* $F(\lambda)\in\mathcal{N}$ *such that* $F(\lambda)$ *is a generator of the vector* $(h_0, h_1, \cdots, h_{2N-3}, h_{2N-2}-\alpha)$.

(h) *There exist real* α *with* $0\leqslant\alpha\leqslant h_{2N-2}$ *and a function* $F(\lambda)\in\mathcal{N}$ *such that*

$$F(\lambda)=-\frac{h_0}{\lambda}-\frac{h_1}{\lambda^2}-\cdots-\frac{h_{2N-2}-\alpha}{\lambda^{2N-1}}-R_{2N-2}(\lambda), \tag{3.2}$$

where $R_{2N-1}(\lambda)=o(\lambda^{-2N+1})$ *as* $\lambda\to\infty$ *uniformly in any sector* (1.11).

In this case, the problem (1.3) *has always a rational solution* $\varphi(\lambda)$ *of the form*

$$\varphi(\lambda)=\omega(\lambda)-\frac{\gamma(\lambda)}{p(\lambda)}A(\lambda)=\frac{\alpha(\lambda)}{\gamma(\lambda)}, \tag{3.3}$$

in with $\gamma(\lambda)/p(\lambda)$ *is a rational generator to the vector* $(h_0, h_1, \cdots, h_{2N-3}, h_{2N-2}-\hat{\alpha})$, *where* $\hat{\alpha}=0$ *if* $\boldsymbol{H}_{NN}[\boldsymbol{b}]\geqslant\boldsymbol{O}$ *is proper, and* $\hat{\alpha}>0$ *is as in* (2.21) *otherwise.*

Proof (a)⇒(b), (e)⇔(f), and (g)⇒(h): Obvious. (f)⇔(h) follows from the Hamburger-Nevanlinna theorem mentioned above.

(c)⇔(d) follows from Lemma 3.1. (e)⇒(d): The proof is very simple. For if the equations (1.8) hold, then for any real $\boldsymbol{x}=(x_0, x_1, \cdots, x_{N-1})^{\mathrm{T}}$ we have

$$\boldsymbol{x}^{\mathrm{T}}\boldsymbol{H}_{NN}[\boldsymbol{b}]\boldsymbol{x}\geqslant\int_{-\infty}^{+\infty}\Big(\sum_{k=0}^{N-1}x_k u^k\Big)^2\mathrm{d}\sigma(u)\geqslant 0. \tag{3.4}$$

It remains to prove the implications of (b)⇒(c), (d)⇒(g), and (d)⇒(a).

(b)⇒(c): Assume that $\varphi(\lambda)\in\mathscr{N}$ is a solution to the problem (1.3), so that, by (1.1), it satisfies the equations (1.5). By virtue of the integral representation of the functions in $\mathscr{N}$ (see e. g. [1, p. 92]),

$$\varphi(\lambda)=\mu\lambda+\gamma+\int_{-\infty}^{+\infty}\Big(\frac{1}{u-\lambda}-\frac{u}{1+u^2}\Big)\mathrm{d}\tau(u), \tag{3.5}$$

where $\mu\geqslant 0$, $\operatorname{Im}\gamma=0$, $\tau(u)$ is a positive Borel measure on the real axis such that $\int_{-\infty}^{+\infty}(1+u^2)^{-1}\mathrm{d}\tau(u)$ is finite, and $\varphi(\lambda)/\lambda\to\mu$ as $\lambda\to\infty$ uniformly in any sector (1.11), we conclude that the entry $K_{kl}(Y_i, \overline{Y}_j)$ of $\boldsymbol{L}_{ij}\in\mathbf{C}^{\tau_i\times\tau_j}$ defined by (2.4) with $\varphi(\lambda)$ in place of $\omega(\lambda)$ has the form

$$K_{kl}(Y_i, \overline{Y}_j)=\frac{1}{k!l!}\frac{\partial^{k+l}}{\partial\xi^k\partial\eta^l}\Big[\mu+\int_{-\infty}^{+\infty}\frac{\mathrm{d}\tau(u)}{(u-\xi)(u-\eta)}\Big]_{\xi=Y_i, \eta=\overline{Y}_j},$$

so that the quadratic form

$$\begin{aligned}\sum_i\sum_j\boldsymbol{x}_i^*\boldsymbol{L}_{ij}\boldsymbol{x}_j&=\sum_{i,j}\sum_{k,l}K_{kl}(Y_i, \overline{Y}_j)\overline{x}_i^k x_j^l\\&=\mu\Big|\sum_i x_i^0\Big|^2+\int_{-\infty}^{+\infty}\Big|\sum_{i,k}\frac{1}{k!}\frac{\partial^k}{\partial\xi^k}\Big[\frac{1}{u-\xi}\Big]_{\xi=Y_i}x_i^k\Big|^2\mathrm{d}\tau(u)\geqslant 0\end{aligned} \tag{3.6}$$

for an arbitrary selection of block vectors $\boldsymbol{x}=(\boldsymbol{x}_1^{\mathrm{T}}, \boldsymbol{x}_2^{\mathrm{T}}, \cdots, \boldsymbol{x}_\theta^{\mathrm{T}})^{\mathrm{T}}\in\mathbf{C}^N$, where

$$\boldsymbol{x}_i=(x_i^0, x_i^1, \cdots, x_i^{\tau_i-1})^{\mathrm{T}},\quad i=1,2,\cdots,\theta.$$

Thus, $\boldsymbol{\Lambda}=\boldsymbol{L}_{y\bar{y}}(\omega)=\boldsymbol{L}_{y\bar{y}}(\varphi)\geqslant\boldsymbol{O}$.

(d)⇒(g): Assume (d). If $\boldsymbol{H}_{NN}[\boldsymbol{b}]\geqslant \boldsymbol{O}$ is proper, (g) follows from Theorem 2. 3′ (with $\alpha=0$ if $\boldsymbol{H}_{NN}[\boldsymbol{b}]>\boldsymbol{O}$, with $\alpha=\hat{\alpha}$ as in (2. 21) if $\boldsymbol{H}_{NN}[\boldsymbol{b}]\geqslant \boldsymbol{O}$ is singular). If $\boldsymbol{H}_{NN}[\boldsymbol{b}]\geqslant \boldsymbol{O}$ is degenerate, we take $\alpha=h_{2N-2}$ and $F(\lambda)=0$. Finally, if $\boldsymbol{H}_{NN}[\boldsymbol{b}]\geqslant \boldsymbol{O}$ is singular and of the form (2. 20) with $\boldsymbol{H}_p=\boldsymbol{H}_{NN}[\boldsymbol{b}_p]\neq \boldsymbol{O}$ and $\boldsymbol{H}_d=\boldsymbol{H}_{NN}[\boldsymbol{b}_d]\neq \boldsymbol{O}$, where $\boldsymbol{b}_p$ and $\boldsymbol{b}_d$ are defined as in (2. 21), then there exists a rational generator $\gamma(\lambda)/p(\lambda)$ of $\boldsymbol{b}_p$ such that $-\gamma(\lambda)/p(\lambda)\in \mathcal{N}$, by Theorem 2. 3′ again, so (g) holds with $\alpha=\hat{\alpha}$.

(d)⇒(a): Assume that $\boldsymbol{H}_{NN}[\boldsymbol{b}]\geqslant \boldsymbol{O}$. Then (a) follows from Lemma 3. 3 (Lemma 3. 2) if $\boldsymbol{H}_{NN}[\boldsymbol{b}]\geqslant \boldsymbol{O}$ is proper (is degenerate). It remains to prove the implication of (d)⇒(a) in the case when $\boldsymbol{H}_{NN}[\boldsymbol{b}]\geqslant \boldsymbol{O}$ is singular of the form (2. 20) with $\boldsymbol{H}_p=\boldsymbol{H}_{NN}[\boldsymbol{b}_p]\neq \boldsymbol{O}$ and $\boldsymbol{H}_d=\boldsymbol{H}_{NN}[\boldsymbol{b}_d]\neq \boldsymbol{O}$. Partition the problem (1. 3) appropriately into a pair of NP problems, denoted by $(1.3)_p$ and $(1.3)_d$, with the same interpolation nodes as in (2. 3), such that $\boldsymbol{b}_p=(h_0, h_1, \cdots, h_{2N-3}, h_{2N-2}-\alpha)$ and $\boldsymbol{b}_d=\boldsymbol{b}-\boldsymbol{b}_p=(0, \cdots, 0, \hat{\alpha})$ are the corresponding Hankel vectors, respectively, and such that $\omega_d(\lambda)=\hat{\alpha}\lambda\in \mathcal{N}$ is a solution to the problem $(1.3)_d$. Such a pair of NP problems always exists uniquely. Due to Lemma 3. 3, there exists a rational solution to $(1.3)_p$ which is of the form $\omega_p(\lambda)-[\gamma(\lambda)/p(\lambda)]\cdot A(\lambda)\in \mathcal{N}$, where $\omega_p(\lambda)$ is the Hermite interpolation polynomial of the GRIP $(1.3)_p$. Thus, $\omega_p(\lambda)-[\gamma(\lambda)/p(\lambda)]A(\lambda)+\omega_d(\lambda)=\omega(\lambda)-[\gamma(\lambda)/p(\lambda)]A(\lambda)$ belongs to the class $\mathcal{N}$ and is obviously a rational solution to the GRIP (1. 5) [and therefore to the problem (1. 3)]. This completes the proof of the theorem. ■

We remark that the above proof of (d)⇔(g) provides in essence an alternative proof of a result of M. Riesz (see, e. g. [1, p. 71]) by making use of the Stieltjes-Perron inversion formula.

In the case when $\boldsymbol{\Lambda}\geqslant \boldsymbol{O}$ (or $\boldsymbol{H}_{NN}[\boldsymbol{b}]\geqslant \boldsymbol{O}$) is singular, more can be said about the class of solutions to the NP problem (1. 3), and to the power moment problem (1. 8) and its relatives. We shall prove in Corollary 3. 5 below that each of these problems with the exception of the problem (1. 9) has only a single solution. This means that these problems are determinate in the case when $\boldsymbol{H}_{NN}[\boldsymbol{b}]\geqslant\boldsymbol{O}$ is singular.

Corollary 3.5 *With the notation as before, the following are equivalent:*

(a) *$\boldsymbol{H}_{NN}[\boldsymbol{b}]\geqslant\boldsymbol{O}$ (or $\boldsymbol{\Lambda}\geqslant\boldsymbol{O}$) is singular with rank $n_1(<N)$, and of the form of the quasidirect decomposition defined as in Lemma 2.7.*

(b) *The class of solutions to the problem (1.3) is nonempty and consists of rational functions of degree at most $N-1$.*

(c) *The problem (1.3) has a unique solution which is rational, real on the real λ-axis, and of degree exact n_1.*

(d) *The problem (1.3) has only a single solution.*

(e) *The problem (1.8) has a solution $\sigma(u)$ $(-\infty<u<+\infty)$ which has at most $N-1$ points of increase.*

(f) *The problem (1.8) has a unique solution $\sigma(u)$ $(-\infty<u<+\infty)$ for which the number of points of increase is n_1 if $\boldsymbol{H}_{NN}[\boldsymbol{b}]$ is proper, and n_1-1 otherwise.*

(g) *The problem (1.8) has a unique solution $\sigma(u)$ $(-\infty<u<+\infty)$.*

(h) *There exists a unique real α with $0\leqslant\alpha\leqslant h_{2N-2}$ such that the Hamburger moment problem*

$$h_j=\int_{-\infty}^{+\infty}u^j\,\mathrm{d}\sigma(u)\quad(j=0,1,\cdots,2N-3),$$

$$h_{2N-2}-\alpha=\int_{-\infty}^{+\infty}u^{2N-2}\,\mathrm{d}\sigma(u)\tag{3.7}$$

has a unique solution $\sigma(u)$ $(-\infty<u<+\infty)$.

(i) *There exists a unique real α with $0\leqslant\alpha\leqslant h_{2N-2}$ such that the problem (1.10) with h_{2N-2} replaced by $h_{2N-2}-\alpha$ has only a single solution $F(\lambda)$.*

In this case, the (unique) solution give in (b)(c), or (d) is of the form (3.3), while $-\gamma(\lambda)/p(\lambda)$ therein is the unique solution given in (i) with $\alpha=\hat{\alpha}$, where $\hat{\alpha}$ is defined uniquely by (2.21); the (unique) solution $\sigma(u)$ $(-\infty<u<+\infty)$ to the problem (1.8) given in (e)(f), or (g) must be concentrated on the real zeros $\lambda_1,\lambda_2,\cdots,\lambda_r$ ($r=n_1$ if $\boldsymbol{H}_{NN}[\boldsymbol{b}]$ is proper, and $r=n_1-1$ otherwise) of $p(\lambda)$, at which its discontinui-

ties are

$$\sigma(\lambda_i+0)-\sigma(\lambda_i-0)=\frac{\gamma(\lambda_i)}{p'(\lambda_i)},\quad i=1,2,\cdots,r$$

(*where* $\sigma(u)=0$ $(-\infty<u<+\infty)$ *if* $\boldsymbol{H}_{NN}[\boldsymbol{b}]=\boldsymbol{H}_d$); *finally, the real* α *given in* (h) *is equal to* $\hat{\alpha}$ *given in* (2. 21), *and the Hamburger moment problem* (3. 7) *has a unique solution* $\sigma(u)$ $(-\infty<u<+\infty)$ *which is the same as that of the problem* (1. 8) *just mentioned*.

Proof (a)⇒(b): Assume (a). Then the problem (1. 3) has at least one solution $\varphi(\lambda)$ by Theorem 3. 5. Going to the proof of rationality of the solution $\varphi(\lambda)$, we have further that the quadratic form in (3. 6) vanishes for a nontrivial choice of the $\boldsymbol{x}=(\boldsymbol{x}_1^{\mathrm{T}},\boldsymbol{x}_2^{\mathrm{T}},\cdots,\boldsymbol{x}_\theta^{\mathrm{T}})^{\mathrm{T}}\in\mathbf{C}^N$ where $\boldsymbol{x}_i=(x_i^0,x_i^1,\cdots,x_i^{\tau_i-1})^{\mathrm{T}}$ for $i=1,2,\cdots,\theta$. The rational function of u defined by

$$\sum_{i,k}\frac{1}{k!}\frac{\partial^k}{\partial\xi^k}\left[\frac{1}{u-\xi}\right]_{\xi=Y_i}x_i^k$$

has the form

$$\sum_{i=1}^{\theta}\sum_{k=0}^{\tau_i-1}\frac{(-1)^k x_i^k}{(u-Y_i)^{k+1}}=\frac{R(u)}{\prod_{i=1}^{\theta}(u-Y_i)^{\tau_i}},$$

in which $R(u)$ is some polynomial of u with leading term $\sum_{i=1}^{\theta}x_i^0u^{N-1}$. Since the measure $\tau(u)$ must be concentrated on the zeros of $R(u)$ and the number of these is at most $N-1$ [it is $\leqslant N-2$ if $\mu\neq 0$, since then $\sum_i x_i^0=0$ by (3. 6)]. It follows from the representation (3. 5) that any solution $\varphi(\lambda)\in\mathcal{N}$ to the problem (1. 3) is always rational, of degree at most $N-1$.

(b)⇒(c): The rational solution $\varphi(\lambda)$ of degree at most $N-1$ to the problem (1. 3) is also an interpolant to the GRIP (1. 5), so that it is unique, and of degree exact n_1 by Corollary 2. 5(b), and has the form (3. 3), which is real on the real λ-axis by Theorem 3. 4.

(c)⇒(d): Obvious.

(d)⇒(a): The condition (d) implies that $\boldsymbol{H}_{NN}[\boldsymbol{b}]\geqslant\boldsymbol{O}$ (or $\boldsymbol{\Lambda}\geqslant\boldsymbol{O}$)

by Theorem 3.4 and that $\boldsymbol{H}_{NN}[\boldsymbol{b}]$ (or $\boldsymbol{\Lambda}$) is singular by Lemma 3.3, since $\boldsymbol{H}_{NN}[\boldsymbol{b}]>\boldsymbol{O}$ (or $\boldsymbol{\Lambda}>\boldsymbol{O}$) implies that (1.3) has infinitely many solutions.

(a)⇒(e): Assume (a). Then (1.8) has a solution $\sigma(u)$ by Theorem 3.4, and the equality applies in (3.4) for some $\boldsymbol{x}\neq\boldsymbol{0}$. Thus, the polynomial $\sum_{k=0}^{N-1} x_k u^k$ must be zero at all points of the increase of the function $\sigma(u)$, the number of which is obviously at most $N-1$.

(e)⇒(f), (i): Assume (e). Then there exists a real α with $0\leqslant\alpha\leqslant h_{2N-2}$ such that the problem (3.7) has a solution $\sigma(u)$ which is the same as that given in (e). It follows from the Hamburger-Nevanlinna theorem that the function

$$F(\lambda)=\int_{-\infty}^{+\infty}\frac{\mathrm{d}\sigma(u)}{u-\lambda} \tag{3.8}$$

belongs to $\mathcal{N}$, and generates the vector $-\boldsymbol{b}'=-(h_0,h_1,\cdots,h_{2N-3},h_{2N-2}-\alpha)$ in the sense that (3.2) holds uniformly in any sector (1.11). The function $F(\lambda)$ is, by (e), rational, of degree at most $N-1$, so that it (and therefore the real α) is unique by Theorem 2.3′ with $\boldsymbol{b}'$ in place of $\boldsymbol{b}$ therein. Thus, we have by Theorem 3.4 that $\boldsymbol{b}'=\boldsymbol{b}_p$, i.e., $\alpha=\hat{\alpha}$, whence the function $\sigma(u)$ $(-\infty<u<+\infty)$ is also unique, with n_1 points of increase if $\hat{\alpha}=0$ or with n_1-1 if $\hat{\alpha}>0$.

(f)⇒(g), and (h)⇒(e): Obvious.

(g)⇒(h): The proof is very simple with $\alpha=h_{2N-2}-\int_{-\infty}^{+\infty}u^{2N-2}\mathrm{d}(u)$.

(i)⇒(a) follows from Theorem 3.4, since $\boldsymbol{H}_{NN}[\boldsymbol{b}]>\boldsymbol{O}$ implies that for $\alpha=0$, the vector $-\boldsymbol{b}$ has infinitely many (rational) generators $F(\lambda)$ in $\mathcal{N}$ by Theorem 2.3′. The last part of the corollary is obvious. ■

An interesting consequence of Theorems 3.4 and 2.3′ and of Corollary 3.5 is the following result (see Theorem 2.3 and the formula (2.38) of [27] or [18, p. 183]; for a generalization to the operator case see [2]). Here we state an alternative proof for it.

Theorem 3.6 *In order for the problem* (1.9) *to have a solution*, *it* *is necessary and sufficient that* $\boldsymbol{H}_{NN}[\boldsymbol{b}]\geqslant\boldsymbol{O}$ *be proper. In this case, if* $\boldsymbol{H}_{NN}[\boldsymbol{b}]$ *is singular, then* (1.9) *has only a single solution* $\sigma(u)$ $(-\infty<u<+\infty)$, *which is such that*

$$\int_{-\infty}^{+\infty}\frac{\mathrm{d}\sigma(u)}{u-\lambda}=-\frac{\gamma(\lambda)}{p(\lambda)},\tag{3.9}$$

where $-\gamma(\lambda)/p(\lambda)$ *is the unique solution to the problem* (1.10); *and if* $\boldsymbol{H}_{NN}[\boldsymbol{b}]>\boldsymbol{O}$, *then the formula*

$$\int_{-\infty}^{+\infty}\frac{\mathrm{d}\sigma(u)}{u-\lambda}=-\frac{\gamma(\lambda)+\varphi(\lambda)\delta(\lambda)}{p(\lambda)+\varphi(\lambda)q(\lambda)}\tag{3.10}$$

establishes a one-to-one correspondence between the class of solutions $\sigma(u)$ *to the problem* (1.9) *and the class of solutions to the problem* (1.10) *when* $\varphi(\lambda)$ *runs over the class* $\mathcal{N}$ *such that* $\varphi(\lambda)/\lambda\to 0$ *as* $\lambda\to\infty$.

Proof The sufficiency of the condition of the first statement of this theorem follows from the last result of Theorem 2.3′ and the Hamburger-Nevanlinna theorem. It is very simple to prove that the condition is necessary. For if (1.9) [and therefore, (1.8)] has a solution $\sigma(u)$, then $\boldsymbol{H}_{NN}[\boldsymbol{b}]\geqslant\boldsymbol{O}$ by Theorem 3.4. if $\boldsymbol{H}_{NN}[\boldsymbol{b}]\geqslant\boldsymbol{O}$ were singular and not proper, Corollary 3.5 would imply that the rational function $-F(\lambda)=-\int_{-\infty}^{+\infty}\mathrm{d}\sigma(u)/(u-\lambda)$ of degree at most $N-1$ was a generator of the vector $\boldsymbol{b}$. But according to Theorem 2.3′ this is impossible. Thus one part of the theorem is proved.

If $\boldsymbol{H}_{NN}[\boldsymbol{b}]\geqslant\boldsymbol{O}$ is singular and proper, the result (3.9) follows from Corollary 3.5 at once. It remains to prove the last result of this theorem. Assume that $\boldsymbol{H}_{NN}[\boldsymbol{b}]>\boldsymbol{O}$ and $\sigma(u)$ is a solution to the problem (1.9), and therefore $\int_{-\infty}^{+\infty}\mathrm{d}\,\sigma(u)/(u-\lambda)$ is a solution to the problem (1.10). Then, by Lemma 3.8 below, we have (3.10) for some $\varphi(\lambda)\in\mathcal{N}$ such that $\varphi(\lambda)/\lambda\to 0$ as $\lambda\to\infty$ uniformly in any sector (1.11). Conversely, if $\varphi(\lambda)\in\mathcal{N}$ is such that the condition just mentioned holds, then there exists a sequence of proper rational functions $\varphi_n(\lambda)$ of the class $\mathcal{N}$ converging to $\varphi(\lambda)$ uniformly on each compact set in $\mathbf{C}^+$. Let now

$$F(\lambda)=\frac{\gamma(\lambda)+\varphi(\lambda)\delta(\lambda)}{p(\lambda)+\varphi(\lambda)q(\lambda)} \quad \text{and} \quad F_n(\lambda)=\frac{\gamma(\lambda)+\varphi_n(\lambda)\delta(\lambda)}{p(\lambda)+\varphi_n(\lambda)q(\lambda)},$$

$$n=1,2,\cdots. \tag{3.11}$$

Then $-F_n(\lambda)\in\mathcal{N}$, and $F_n(\lambda)$ is a rational generator to $\boldsymbol{b}$ by Theorem 2.3′ and Lemma 3.7 below, so that $-F_n(\lambda)$ is a solution to the problem (1.10). We have

$$-F_n(\lambda)=\int_{-\infty}^{+\infty}\frac{d\sigma_n(u)}{u-\lambda}$$

for some solution $\sigma_n(u)$ to the problem (1.9), by the Hamburger-Nevanlinna theorem again. Going to the limit and using Helly's theorem, we have that (3.10) holds for a solution $\sigma(u)$ to (1.9). ■

Lemma 3.7[1] *If $\boldsymbol{H}_{NN}[\boldsymbol{b}]>\boldsymbol{O}$, then for a fixed $\lambda\in\mathbf{C}^+$, the set of values assumed at the point λ by the functions*

$$F(\lambda,\tau)=-\frac{\gamma(\lambda)+\tau\delta(\lambda)}{p(\lambda)+\tau q(\lambda)}, \tag{3.12}$$

where τ varies along the whole real axis, coincides with a nondegenerate circular contour $K(\lambda)$ situated in $\mathbf{C}^+$, and the point $F(\lambda,\tau)$ lies within the circle $K(\lambda)$ or on its boundary if and only if $\operatorname{Im}\tau\geqslant 0$.

Lemma 3.8 *If $\boldsymbol{H}_{NN}[\boldsymbol{b}]>\boldsymbol{O}$, then for each function $F(\lambda)$ such that $-F(\lambda)$ is a solution to the problem (1.10), there exists a $\varphi(\lambda)\in\mathcal{N}$ satisfying $\varphi(\lambda)/\lambda\to 0$ as $\lambda\to\infty$ in any sector (1.11) such that*

$$F(\lambda)=\frac{\gamma(\lambda)+\varphi(\lambda)\delta(\lambda)}{p(\lambda)+\varphi(\lambda)q(\lambda)}. \tag{3.13}$$

Proof Assume that $\boldsymbol{H}_{NN}[\boldsymbol{b}]>\boldsymbol{O}$ and that $F(\lambda)$ is a nonrational function such that $-F(\lambda)$ is a solution to (1.10). Then, by the Hamburger-Nevanlinna theorem there is a solution $\sigma(u)$ to the problem (1.9) such that

$$-F(\lambda)=\int_{-\infty}^{+\infty}\frac{d\sigma(u)}{u-\lambda}.$$

We take $\{\sigma_n(u)\}_{n=1}^{\infty}$ as a sequence of bounded nondecreasing functions $(-\infty<u<+\infty)$ each of which has a finite set of points of increase, such that $\sigma_n(u)\to\sigma(u)$ substantially as $n\to\infty$. Set $h_j^{(n)}=\int_{-\infty}^{+\infty}u^j\, d\sigma_n(u)$,

$0 \leqslant j \leqslant 2N-2$, and $\boldsymbol{b}^{(n)}=(h_0^{(n)},\cdots,h_{2N-2}^{(n)})$. We have that $h_j^{(n)} \to h_j$ ($j=0,1,\cdots,2N-2$) as $n \to \infty$, and $-F_n(\lambda)=\int_{-\infty}^{+\infty} \mathrm{d}\sigma_n(u)/(u-\lambda) \to -F(\lambda)$ as $n \to \infty$ by Helly's theorem. Then by Theorem 3.4 and the condition that $\boldsymbol{H}_{NN}[\boldsymbol{b}]>\boldsymbol{O}$, the real Hankel matrices $\boldsymbol{H}_{NN}^{(n)}=(h_{i+j}^{(n)})_{i,j=0}^{N-1}>\boldsymbol{O}$ for sufficiently large values of n. Denote by $p^{(n)}(\lambda), q^{(n)}(\lambda), \gamma^{(n)}(\lambda), \delta^{(n)}(\lambda)$ the polynomials of $\boldsymbol{b}^{(n)}$ comparable to $p(\lambda), q(\lambda), \gamma(\lambda), \delta(\lambda)$ relative to the vector $\boldsymbol{b}$. We have $p^{(n)}(\lambda) \to p(\lambda)$, etc. Since $-F_n(\lambda) \in \mathcal{N}$ is a rational function such that

$$F_n(\lambda)=\frac{h_0^{(n)}}{\lambda}+\frac{h_1^{(n)}}{\lambda}+\cdots+\frac{h_{2N-2}^{(n)}}{\lambda^{2N-1}}+o(\lambda^{-2N+1})$$

as $\lambda \to \infty$ uniformly in any sector (1.11), $F_n(\lambda)$ is a rational generator to $\boldsymbol{b}^{(n)}$ for large values of n. It follows from Theorem 2.3′ and Lemma 3.7 that $F_n(\lambda)$ has the form

$$F_n(\lambda)=\frac{\gamma^{(n)}(\lambda)+\varphi^{(n)}(\lambda)\delta^{(n)}(\lambda)}{p^{(n)}(\lambda)+\varphi^{(n)}(\lambda)q^{(n)}(\lambda)}$$

for large values of n, where $\varphi^{(n)}(\lambda) \in \mathcal{N}$ is proper rational. Going to the limit, and using the compactness property [15, p. 14] of all Nevanlinna functions in $\mathbf{C}^+$, we have a subsequence $\varphi^{(n_j)}$ of $\varphi^{(n)}(\lambda)$ converging to $\varphi(\lambda) \in \mathcal{N}$ such that $\varphi(\lambda)/\lambda \to 0$ as $\lambda \to \infty$, and therefore $\{F_{n_j}(\lambda)\}$ converging to $\widetilde{F}(\lambda)$ uniformly on each compact set in $\mathbf{C}^+$, which is analytic in $\mathbf{C}^+$, and coincides with $F(\lambda)$ by the uniqueness theorem of analytic functions. Thus, (3.13) holds, where $\varphi(\lambda) \in \mathcal{N}$, and $\varphi(\lambda)/\lambda \to 0$ as $\lambda \to \infty$ uniformly in any sector (1.11). ■

For the problem (1.8) we have by a result of M. Riesz (see [1, p. 71]), or by Theorem 3.4, that it has a solution of $\sigma(u)$ ($-\infty<u<+\infty$) if and only if $\boldsymbol{H}_{NN}[\boldsymbol{b}] \geqslant \boldsymbol{O}$, and that if $\boldsymbol{H}_{NN}[\boldsymbol{b}] \geqslant \boldsymbol{O}$ is singular, (1.8) has only a single solution $\sigma(u)$ such that (3.9) holds where $-\gamma(\lambda)/p(\lambda)$ is the unique generator of $\boldsymbol{b}_p$, the proper part of $\boldsymbol{b}$. In the case of $\boldsymbol{H}_{NN}[\boldsymbol{b}]>\boldsymbol{O}$ we have

Theorem 3.9 *Let* $\boldsymbol{b}=(h_0,h_1,\cdots,h_{2N-2})$ *be real. If* $\boldsymbol{H}_{NN}[\boldsymbol{b}]>\boldsymbol{O}$, *then the formula* (3.10) *establishes a one-to-one correspondence between*

the class of solutions $\sigma(u)$ to the problem (1.8) *and the class of functions $\varphi(\lambda)$ of the class $\mathcal{N}$ augmented by the constant ∞.*

Proof We consider the functions

$$F_\sigma(\lambda)=\int_{-\infty}^{+\infty}\frac{\mathrm{d}\sigma(u)}{u-\lambda},\quad \lambda\in\mathbf{C}^+, \tag{3.14}$$

in which $\sigma(u)$ $(-\infty<u<\infty)$ runs over the solutions to (1.8). It can be shown in a similar way to that given in proof of Theorem 2.24 of [1] that the set of values assumed at the point $\lambda\in\mathbf{C}^+$ by all functions $F_\sigma(\lambda)$ coincides with the point set $K(\lambda)$ defined by Lemma 3.7 and its interior. Thus, by Lemma 3.7,

$$F_\sigma(\lambda)=-\frac{\gamma(\lambda)+\tau(\lambda)\delta(\lambda)}{p(\lambda)+\tau(\lambda)q(\lambda)},\quad \lambda\in\mathbf{C}^+, \tag{3.15}$$

where $\tau(\lambda)$ is such that $\operatorname{Im}\tau(\lambda)\geqslant 0$ for $\lambda\in\mathbf{C}^+$, and is defined by the last equation. If $\tau(\lambda)$ is not an infinite constant, it is meromorphic in $\mathbf{C}^+$, and further is regular in $\mathbf{C}^+$, since the possible singularities are removable, so that $\tau(\lambda)\in\mathcal{N}$.

Conversely, suppose that $\varphi(\lambda)\in\mathcal{N}$. Denote by $F(\lambda)$ the corresponding function given in the right side of (3.10). For a fixed $\lambda\in\mathbf{C}^+$, the point $F(\lambda)$ lies in the point set $K(\lambda)$ or its interior, in view of Lemma 3.7. We draw a chord through the point $F(\lambda)$ and find a pair of points w_1 and w_2 on the boundary of $K(\lambda)$. Clearly by Lemma 3.7.

$$w_i=-\frac{\gamma(\lambda)+\tau_i\delta(\lambda)}{p(\lambda)+\tau_i q(\lambda)}\quad \text{for some}\quad \tau_i\in\mathbf{R}\ \ (i=1,2)$$

and the point $F(\lambda)$ may be expressed as $F(\lambda)=\theta w_1+(1-\theta)w_2$, where $0\leqslant\theta\leqslant 1$. Due to Theorem 3.6, we can find solutions $\sigma_1(u)$ and $\sigma_2(u)$ to the problem (1.9) for which

$$\int_{-\infty}^{+\infty}\frac{\mathrm{d}\sigma_1(u)}{u-\lambda}=w_1,\quad \int_{-\infty}^{+\infty}\frac{\mathrm{d}\sigma_2(u)}{u-\lambda}=w_2.$$

We then introduce the function $\sigma(u)=\theta\sigma_1(u)+(1-\theta)\sigma_2(u)$, which evidently is a solution to the problem (1.8). Also, the equality (3.10) will hold.

It remains to discuss the case of $\varphi(\lambda)\equiv\infty$. In this case, it is sufficient to use the result proved already and to go to the limit, by Helly's theorem. ■

In the case when $\boldsymbol{\Lambda}>\boldsymbol{O}$ (or equivalently, $\boldsymbol{H}_{NN}[\boldsymbol{b}]>\boldsymbol{O}$), due to Lemma 3.3, Theorem 3.4, and Corollary 3.5, each of the problems (1.3) (1.8)(1.9), and (1.10) has infinitely many solutions, so it is indeterminate. More precisely, we have

Corollary 3.10 *With the notation as before, the following are equivalent:*

(a) $\boldsymbol{\Lambda}=\boldsymbol{L}_{y\bar{y}}(\omega)>\boldsymbol{O}$.

(b) $\boldsymbol{H}_{NN}[\boldsymbol{b}]>\boldsymbol{O}$.

(c) *There exists at least one nonrational solution to the problem* (1.3).

(d) *There exist infinitely many solutions* $w(\lambda)$ *to the problem* (1.3). *Among them are infinitely many real rational solutions* $\varphi(\lambda)$ *of exact degree* N.

(e) *There exist infinitely many solutions* $\sigma(u)$ $(-\infty<u<+\infty)$ *to the problem* (1.9). *Among them are solutions with an infinite number of points of increase.*

(f) *There exist infinitely many solutions* $\sigma(u)$ $(-\infty<u<+\infty)$ *to the problem* (1.8). *Among them are solutions with an infinite number of points of increase.*

(g) *There exist infinitely many solutions to the problem* (1.10). *Among them are rational generators, of exact degree* N, *to the vector* $\boldsymbol{b}$.

(h) *There exists at least one nonrational solution to the problem* (1.10).

Comparing with the statements of the GRIP (1.5), in the case of $\boldsymbol{\Lambda}>\boldsymbol{O}$ (or $\boldsymbol{H}_{NN}[\boldsymbol{b}]>\boldsymbol{O}$) we have that the solutions to the problem (1.3) can be described by a linear fractional transformation with a 2×2 coefficient matrix of polynomials. This is shown by the following theorem (see, e.g., [16, 5] for other parametrizations of the solutions to (1.3)).

Theorem 3.11 *Let the NP problem* (1.3) *be given, and* $\boldsymbol{b}=(h_0, h_1, \cdots, h_{2N-2})$ *be its Hankel vector. If* $\boldsymbol{H}_{NN}[\boldsymbol{b}]>\boldsymbol{O}$, *the set of all the solutions* $w(\lambda)$ *to* (1.3) *is described by a linear fractional transformation of an arbitrary function* $\varphi(\lambda)\in\mathcal{N}\cup\{\infty\}$:

$$w(\lambda)=\omega(\lambda)-\frac{\gamma(\lambda)+\varphi(\lambda)\delta(\lambda)}{p(\lambda)+\varphi(\lambda)q(\lambda)}A(\lambda)=\frac{\alpha(\lambda)+\varphi(\lambda)\beta(\lambda)}{p(\lambda)+\varphi(\lambda)q(\lambda)}, \tag{3.16}$$

whose coefficient matrix is of form

$$\begin{bmatrix} A(\lambda) & \omega(\lambda) \\ 0 & 1 \end{bmatrix}\begin{bmatrix} -\gamma(\lambda) & -\delta(\lambda) \\ p(\lambda) & q(\lambda) \end{bmatrix}.$$

In order to prove Theorem 3.11 we need two auxiliary statements.

Lemma 3.12 *If* $\boldsymbol{H}_{NN}[\boldsymbol{b}]>\boldsymbol{O}$, *the set of values assumed at the point* λ *such that* $\operatorname{Im}\lambda>0$ *and* $\lambda\neq Y_i$, $i=1,2,\cdots,\theta$, *by the functions*

$$w(\lambda,\tau)=\omega(\lambda)-\frac{\gamma(\lambda)+\tau\delta(\lambda)}{p(\lambda)+\tau q(\lambda)}A(\lambda)=\frac{\alpha(\lambda)+\tau\beta(\lambda)}{p(\lambda)+\tau q(\lambda)}, \tag{3.17}$$

where τ *varies along the whole real axis, coincides with a nondegenerate circular contour* $\hat{K}(\lambda)$ *situated in* $\mathbf{C}^+$, *and the point* $w(\lambda,\tau)$ *lies within the circle* $\hat{K}(\lambda)$ *or on its boundary if and only if* $\operatorname{Im}\tau\geqslant 0$.

Proof The statement follows from Lemmas 3.3 and 3.7. ■

Lemma 3.13 *If* $\boldsymbol{H}_{NN}[\boldsymbol{b}]>\boldsymbol{O}$, *then for each solution* $w(\lambda)$ *to the problem* (1.3) *there exists a* $\varphi(\lambda)\in\mathcal{N}$ *such that* (3.16) *holds.*

Proof Assume that $w(\lambda)$ is a nonrational solution to (1.3), so that it satisfies the equations (1.5). Consider the integral representation given as in (3.5):

$$w(\lambda)=\mu\lambda+\gamma+\int_{-\infty}^{+\infty}\left(\frac{1}{u-\lambda}-\frac{u}{1+u^2}\right)\mathrm{d}\tau(u). \tag{3.18}$$

We take $\{\tau_n(u)\}$ as a sequence of bounded nondecreasing functions ($-\infty<u<+\infty$), each of which has a finite set of points of increase, such that $\tau_n(u)\to\tau(u)$ substantially as $n\to\infty$. The remainder of the proof follows now by a similar argument to that used in the proof of Lemma 3.8, in view of Theorem 2.4′ and Lemma 3.12. ■

Proof of Theorem 3.11 Assume that $\boldsymbol{H}_{NN}[\boldsymbol{b}]>\boldsymbol{O}$ and that $w(\lambda)$ is a solution to the problem (1.3). Then, by Lemma 3.13, $w(\lambda)$ has the form (3.16) for some $\varphi(\lambda)\in\mathcal{N}\cup\{\infty\}$.

Conversely, for any $\varphi(\lambda)\in\mathcal{N}\cup\{\infty\}$, the function $w(\lambda)$ of the form (3.16) makes sense, and obviously satisfies (1.5). In order to complete the proof it is sufficient to verify that the function $w(\lambda)$ belongs to the class $\mathcal{N}$. But this is an immediate consequence of Lemma 3. 12 by a similar argument to that used in the last part of the proof of Theorem 3. 9 [since $w(\lambda)$ is such that $\operatorname{Im} w(\lambda)\geqslant 0$ for all points λ of holomorphy in $\mathbf{C}^+$, and is meromorphic in $\mathbf{C}^+$]. ■

Form the previous theorems we have

Corollary 3. 14 *Let $\xi\in\mathbf{C}^+$ be fixed such that $\xi\neq Y_i$, $i=1,2,\cdots,\theta$. If $\boldsymbol{H}_{NN}[\boldsymbol{b}]\geqslant\boldsymbol{O}$ and if $w_0(\lambda)$ is a solution to the problem* (1. 3) *for which $|w(\xi)|$ is largest when $w(\lambda)$ runs over all solutions to* (1. 3), *then $w_0(\lambda)$ is uniquely determined by its value $w_0(\xi)$, and $w_0(\lambda)$ is real rational, of degree at most N.*

Proof The statement is trivial in the case when $\boldsymbol{H}_{NN}[\boldsymbol{b}]\geqslant\boldsymbol{O}$ is singular, since the problem (1. 3) has only a single solution, which is real rational, of degree at most $N-1$. In the case of $\boldsymbol{H}_{NN}[\boldsymbol{b}]>\boldsymbol{O}$ the result follows from Theorem 3. 11 combined with Lemma 3. 12. ■

Corollary 3. 15 *There exists a one-to-one correspondence between the class of solutions $\sigma(u)$ of the problem* (1. 8) *and the class of solutions $\omega(\lambda)$ of the NP problem* (1. 3) *defined via*

$$w(\lambda)=\omega(\lambda)+\int_{-\infty}^{+\infty}\frac{\mathrm{d}\sigma(u)}{u-\lambda}A(\lambda).$$

Proof It follows from Theorems 3. 11 and 3. 9 if $\boldsymbol{H}_{NN}[\boldsymbol{b}]>\boldsymbol{O}$. In the case when $\boldsymbol{H}_{NN}[\boldsymbol{b}]\geqslant\boldsymbol{O}$ is singular the result is trivial. ■

Acknowledgments

The author wishes to acknowledge a debt of gratitude to the department for the invitation and hospitality. The author acknowledges with many thanks helpful comments and suggestions of the referees.

References

[1] N I Akhiezer. The Classical Moment Problem. Oliver & Boyd, London, 1965.

[2] T Ando. Truncated moment problems for operators. Acta. Sci. Math. (Szeged), 1970, 31: 319—334.

[3] J A Ball. Interpolation problems for Pick-Nevanlinna and Löwner types for

meromorphis matrix functions. Integral Equations Operator Theory, 1983, 6:804—840.

[4] J A Ball. Nevanlinna-Pick interpolaton: Generalizations and applications. in Survey of Some Recent Results in Operator Theory I (J B Conway and B B Morrell, Eds.). Pitman, Boston, 1988:51—94.

[5] J A Ball, I Gohberg, and L Rodman. Interpolation of Rational Matrix Functions. Birkhäuser, Basel, 1990.

[6] G-N Chen and H-P Zhang. More on Loewner matrices. Linear Algebra Appl., 1994,203—204:265—300.

[7] G-N Chen, B Zhao, and H-P Zhang. The general rational interpolation problem in the scalar case and its Hankel vector. Linear Algebra Appl., 1996, 244:165—201.

[8] G-N Chen and X-Q Li. The Nevanlinna-Pick interpolation problems and the power moment problems for matrix-valued functions. Linear Algebra Appl., 1999,288(1—3):123—148.

[9] W F Donoghue. Monotone Matrix Functions and Analytic Continuation. Springer-Verlag, New York, 1974.

[10] H Dym. *J*-contractive matrix functions, reproducing kernel Hilbert spaces and interpolation. in Regional Conf. Ser. in Math. 71, Amer. Math. Soc., Providence, 1989.

[11] M Fiedler. Quasidirect decomposition of Hankel and Toeplitz matrices. Linear Algebra Appl., 1984,61:155—174.

[12] C Foias and A E Frazho. The Commutant Lifting Approach to Interpolation Problems. Birkhäuser,Basel, 1990.

[13] J B Garnett. Bounded Analytic Functions. Academic, 1981.

[14] G Heinig and U Jungnickel. Hankel matrices generated by the Markov parameters, Hankel matrix extension, partial realization and Padé approximation. in Oper. Theory Adv. Appl. 19, Birkhäuser Basel, 1986:231—253.

[15] V E Katsnel'son. Methods of *J*-theory in continuous interpolation problems of analysis I. transl. by T. Ando, Sapporo, Japan, 1985.

[16] I V Kovalishina. Analytic theory of a class of interpolation problems. Math. USSR Izv., 1984, 22:449—463.

[17] I V Kovalishina and V P Potapov. An indefinite metric in the Nevanlinna-Pick problem. Amer. Math. Soc. Transl. Ser. 2, 1988,138:15—19.

[18] M G Krein and A A Nudelman. The Markov Moment Problem and Extremal Problems. Transl. Math. Monogr. 50, Amer. Math. Soc., Providence, 1977.

[19] M G Krein and P G Rekhtman. On the Nevanlinna-Pick problem (in Ukrainian), Trudy Odess. Gos. Univ. (Travaux Univ. Odessa), 1938, 2: 63—68.

[20] B Sz Nagy and C Foias. Harmonic Analysis of Operators on Hilbert Spaces. North-Holland, Amsterdam, 1970.

[21] B Sz Nagy and A Korańyi. Relations d'un problème de Nevanlinna et Pick avec la théorie des opérateurs de l'espace Hilbertien. Acta. Math. Acad. Sci. Hungar., 1956, 7: 295—302.

[22] R Nevanlinna. Asymptotische Entwicklungen beschränkter Funktionen und Stieltjesche Momentenproblem, Ann. Acad. Sci. Fenn. 1922, 18A(5): 1—53.

[23] R Nevanlinna. Über beschränkte analytische Funktionen. Ann. Acad. Sci. Fenn., 1929, 32A: 1—75.

[24] G Pick. Über die beschränkte analytischer Funktionen welche durch vorgegebene Funktionswerte bewirkt werden. Math. Ann., 1916, 77: 7—23.

[25] M Rosenblum and J Rovnyak. Hardy Classes and Operator Theory. Oxford U. P., 1985.

[26] D Sarason. Generalized interpolation in H^∞. Amer. Math. Soc. Transl., 1967, 127(2): 180—203.

[27] J A Shohat and J D Tamarkin. The Problem of Moment. Amer. Math. Soc., Providence, 1943.

[28] J L Walsh. Interpolation and approximation by rational functions in the complex domain. in Amer. Math. Soc. Colloq. Publ., 20, Providence, 1956.

Linear Algebra and its Applications,
1999,288:123－148

对矩阵值函数的 Nevanlinna-Pick 插值问题与幂矩量问题①

The Nevanlinna-Pick Interpolation Problems and Power Moment Problems for Matrix-valued Functions

Abstract

This paper is concerned with the study of several types of matrix interpolation problems, including

- The matrix Nevanlinna-Pick interpolation problem in the Nevanlinna class $\mathcal{N}_p$,
- The matrix Nevanlinna-Pick interpolation problem in the Stieltjes class $\mathcal{S}_p$,
- The truncated matrix power moment problem on the real axis,
- The Stieltjes matrix power moment problem.

We use the so-called block Hankel vector approach for the treatment of these problems and make explicit the correspondence between solutions to the Nevanlinna-Pick interpolation problem and solutions to an associated power moment problem with the block Hankel vector under consideration. This enables one not only to find existence criteria for the

① Received 4 February 1997; accepted 9 September 1998. Submitted by R. A. Brualdi. Dedicated to Professor Ky Fan on the Occasion of his 84th Birthday. Project supported by NSFC and Doctoral Programme Foundation of Institution of Higher Education of China. 本文与李小青合作.

solutions of the Nevanlinna-Pick interpolation problem, but also to describe the solutions based on the results of power moment problem, and vice versa.

Keywords Nevanlinna-Pick interpolation problem; Power moment problem; Nevanlinna class; Stieltjes class; Block Hankel vector; Stieltjes moment problem

§ 0. Introduction

Both the moment problem and Nevanlinna-Pick interpolation (NP) problem have been extensively studied in the literature using a variety of methods (see, e. g. , [1,3 ~ 11,14 ~ 33] and references therein), and interconnections between these two types of interpolation problems are well recognized.

In 1970s, Krein and Nudel'man ([24], Proposition V. 5. 1, [25]) clarified a clever connection between solutions to the classical NP problem in the Stieltjes class with simple nodes situated in $(-\infty, 0]$ and solutions to a certain Stieltjes moment problem. Thanks to this relation, the study of the Stieltjes moment problem enables one not only to find solvability criteria, but also to parametrize the solutions of that NP problem. (See also [26] for the further developments.)

Some of these same ideas appeared in a different form earlier but implicitly in the study of the power moment problem of [1], Ch. 3, Section 4, and somewhat later in the solvability problem for the NP problem in the Caratheodory class of [14], p. 51, and also recently in the solution of rational interpolation of [12] and in the solution of the NP problem in the Nevanlinna or Caratheodory class of [10,11], and in the solution of the boundary NP problem of [32], etc.

In the present paper, we continue our investigations [10] of both the multiple NP problem and the power moment problem on the real axis. Here we generalize the results obtained earlier to the case of matrix-valued functions via the so-called block Hankel vector approach.

The main concern is the precise connection between the NP problem in the Nevanlinna class $\mathcal{N}_p$ or in the Stieltjes class $\mathcal{S}_p$ on the one hand and the matrix power moment (PM) problem on the real axis or on the semiaxis on the other hand. The so-called block Hankel vector, suitably adapted to the present frainework, which is in fact a finite Hermitian matrix sequence, plays a significant role in our analysis. The essence of the block Hankel vector approach is to transfer the NP problem via the block Hankel vector to the PM problem, and vice versa. More precisely, it is shown that to a given NP problem in the class $\mathcal{N}_p$[in the class $\mathcal{S}_p$, resp.], there corresponds a unique block-vector, referred to as *the block Hankel vector* of that NP problem, such that the NP problem can be reduced to what amounts to the study of a certain truncated matrix power moment problem on real axis [on the semiaxis] associated with that block Hankel vector. The latter, in essence, consists of nothing more than a special case of the former in that one can view it as a high order NP problem in $\mathcal{N}_p$ [in $\mathcal{S}_p$, resp.] at a single point (infinity)(see Theorems 1.3 and 2.2 below).

The paper is divided into two sections besides the introduction. In Section 1 we establish the integral and explicit connection between solutions to the NP problem in the Nevanlinna class $\mathcal{N}_p$ and solutions to an associated PM problem on the real axis with the block Hankel vector of the former. Section 2 is devoted to the same investigation as in Section 1 but for both the NP problem in the Stieltjes class $\mathcal{S}_p$ and the PM problem on the semiaxis. Also, the point of view presented here can be used to understand and to generalize considerably an endearing proposition of [24] mentioned above.

§1. Connection between the matrix Nevanlinna-Pick interpolation in the class $\mathcal{N}_p$ and the matrix power moment problem on the real axis

Let $\mathcal{N}_p$ denote the class of all the $\mathbf{C}^{p\times p}$-valued holomorphic functions

$F(\lambda)$ such that $F^*(\bar{\lambda}) = F(\lambda)$ and $(F(\lambda) - F^*(\lambda))/(\lambda - \bar{\lambda}) \geqslant \boldsymbol{O}$ (nonnegative Hermitian) for all $\lambda \in \mathbf{C}\backslash\mathbf{R}$.

In this section, the notion of the so-called block Hankel vector of a matrix Nevanlinna-Pick interpolation problem with multiple nodes in the class $\mathcal{N}_p$ (NP($\mathcal{N}_p$) problem, for short) is introduced. Then we shall establish a very rigid relationship between the NP($\mathcal{N}_p$) problem and a certain truncated matrix power moment problem on the real axis (PM($\mathbf{R}$) problem) associated with the block Hankel vector of the former.

§ 1.1 The NP($\mathcal{N}_p$) problem

We are given a sequence of $p \times p$ matrices $\boldsymbol{c}_{ik}$ $(1 \leqslant i \leqslant \theta, 0 \leqslant k \leqslant \tau_i - 1)$, and distinct nodes $\lambda_i \in \mathbf{C}^+ \equiv \{\lambda \in \mathbf{C} : \text{Im}\,\lambda > 0\}, 1 \leqslant i \leqslant \theta$, and to each λ_i there corresponds some positive integer τ_i. The objective is to find necessary and sufficient conditions on the data to ensure the existence a function $F(\lambda) \in \mathcal{N}_p$ such that

$$\frac{1}{k!}\frac{\mathrm{d}^k}{\mathrm{d}\lambda^k}F(\lambda)\,|_{\lambda=\lambda_i} = \boldsymbol{c}_{ik}, \quad i = 1,2,\cdots,\theta,\ k = 0,1,\cdots,\tau_i - 1, \tag{1.1}$$

and to describe all the existing solutions when these conditions are met.

It is well known (see, e.g. [2]) that a function $F(\lambda) \in \mathcal{N}_p$ if, and only if, it admits an integral representation

$$F(\lambda) = \boldsymbol{\alpha} + \boldsymbol{\beta}\lambda + \int_{-\infty}^{+\infty}\left(\frac{1}{u-\lambda} - \frac{u}{1+u^2}\right)\mathrm{d}\sigma(u), \quad \text{Im}\,\lambda \neq 0, \tag{1.2}$$

where $\boldsymbol{\alpha} = \boldsymbol{\alpha}^*$, $\boldsymbol{\beta} \geqslant \boldsymbol{O}$, $\mathrm{d}\sigma(u) \geqslant \boldsymbol{O}$ (a nondecreasing $\mathbf{C}^{p\times p}$-valued function) and $\int_{-\infty}^{+\infty}(1+u^2)^{-1}\,\mathrm{tr}\,\mathrm{d}\sigma(u) < \infty$. In this case,

$$\lim_{\lambda\to\infty}\frac{F(\lambda)}{\lambda} = \boldsymbol{\beta}$$

unifromly in each sector S_ε defined by

$$\varepsilon \leqslant \arg\lambda \leqslant \pi - \varepsilon \quad \left(0 < \varepsilon < \frac{\pi}{2}\right). \tag{1.3}$$

Observe that if $F(\lambda)$ is a solution to the NP($\mathcal{N}_p$) problem (1.1), then and only then $F(\lambda) \in \mathcal{N}_p$ is also an interpolant of the condition

$$\frac{1}{k!}\frac{d^k}{d\lambda^k}F(\lambda)\mid_{\lambda=\bar{\lambda}_i}=c_{ik}^*,\quad i=1,2,\cdots,\theta,\ k=0,1,\cdots,\tau_i-1. \tag{1.1'}$$

Let $N=\tau_1+\tau_2+\cdots+\tau_\theta\geqslant 2$. Among $\mathbf{C}^{p\times p}[\lambda]$, there exists a unique polynomial matrix, referred to as *the Hermitian interpolation polynomial matrix* of the NP($\mathcal{N}_p$) problem (1.1), which satisfies both Eqs. (1.1) and (1.1′), and is of degree at most $2N-1$. We denote by $\omega(\lambda)$ such a polynomial matrix throughout. It can be formulated by

$$\omega(\lambda)=\sum_{i=1}^{\theta}\frac{A(\lambda)}{(\tau_i-1)!}\left\{\left(\frac{d}{d\xi}\right)^{\tau_i-1}\left[\sum_{k=0}^{\tau_i-1}\frac{c_{ik}(\xi-\lambda_i)^{\tau_i+k}}{(\lambda-\xi)A(\xi)}\right]_{\xi=\lambda_i}+\left(\frac{d}{d\xi}\right)^{\tau_i-1}\left[\sum_{k=0}^{\tau_i-1}\frac{c_{ik}^*(\xi-\bar{\lambda}_i)^{\tau_i+k}}{(\lambda-\xi)A(\xi)}\right]_{\xi=\bar{\lambda}_i}\right\}, \tag{1.4}$$

where

$$A(\lambda)=\prod_{i=1}^{\theta}(\lambda-\lambda_i)^{\tau_i}(\lambda-\bar{\lambda}_i)^{\tau_i}. \tag{1.5}$$

§ 1.2 The block Hankel vector of the NP($\mathcal{N}_p$) problem

Let $\boldsymbol{y}$ and $\bar{\boldsymbol{y}}$ be the sequences

$$\boldsymbol{y}=(\underbrace{\lambda_1,\cdots,\lambda_1}_{\tau_1\text{ times}},\cdots,\underbrace{\lambda_\theta,\cdots,\lambda_\theta}_{\tau_\theta\text{ times}}),$$

$$\bar{\boldsymbol{y}}=(\underbrace{\bar{\lambda}_1,\cdots,\bar{\lambda}_1}_{\tau_1\text{ times}},\cdots,\underbrace{\bar{\lambda}_\theta,\cdots,\bar{\lambda}_\theta}_{\tau_\theta\text{ times}}), \tag{1.6}$$

which are uniquely defined by the collection of interpolation nodes modulo permutation. Set up the (generalized) block Loewner matrix $\boldsymbol{L}_{y\bar{y}}(\omega(\lambda))$ generated by $\omega(\lambda)$ (and associated with the sequences $\boldsymbol{y}$ and $\bar{\boldsymbol{y}}$) via the formula:

$$\boldsymbol{L}_{y\bar{y}}(\omega(\lambda))=(\boldsymbol{L}_{ij})_{i,j=1}^{\theta}\in\mathbf{C}^{Np\times Np}, \tag{1.7a}$$

Where $\boldsymbol{L}_{ij}=(\boldsymbol{L}_{ij}^{kl})_{k=0,l=0}^{\tau_i-1,\tau_j-1}\in\mathbf{C}^{\tau_ip\times\tau_jp}$ with entries $\boldsymbol{L}_{ij}^{kl}\in\mathbf{C}^{p\times p}$ determined by

$$\boldsymbol{L}_{ij}^{kl}=\frac{1}{k!l!}\frac{\partial^{k+l}}{\partial\lambda^k\partial\mu^l}\left[\frac{\omega(\lambda)-\omega(\mu)}{\lambda-\mu}\right]_{\lambda=\lambda_i,\mu=\bar{\lambda}_j} \tag{1.7b}$$

We need the following notation for polynomials and matrices associated with the sequence $\boldsymbol{y}$ in Eq. (1.6):

$$a(\lambda)=\prod_{i=1}^{\theta}(\lambda-\lambda_i)^{\tau_i}=\sum_{j=0}^{N-1}a_j\lambda^j+\lambda^N,$$

$$a_{ik}(\lambda)=a(\lambda)/(\lambda-\lambda_i)^{k+1}, i=1,2,\cdots,\theta,\ k=0,1,\cdots,\tau_i-1, \tag{1.8}$$

$$W_y=\operatorname{col}\left[\left(\frac{a_{ik}^{(j)}(0)}{j!}\right)_{k=0,j=0}^{\tau_i-1,N-1}\right]_{i=1}^{\theta}\in \mathbf{C}^{N\times N},$$

where the index (j) stands for the jth derivate with respect to λ at $\lambda=0$.

Let now C_a denote the (first) companion matrix, of order N, of $a(\lambda)$ in (1,8), and let V_y be the generalized Vandermonde matrix, of order N, associated with the sequence y in Eq. (1.6):

$$V_y=\operatorname{row}\left[\left[\binom{k}{j}\lambda_i^{k-j}\right]_{k,j=0}^{N-1,\tau_i-1}\right]_{i=1}^{\theta}.$$

Set

$$A_\xi=\operatorname{diag}[J_{\tau_i}^{\mathrm{T}}(\lambda_i)]_{i=1}^{\theta}, \tag{1.9}$$

where $J_{\tau_i}(\lambda_i)$ is the super Jordan cell of order τ_i with λ_i along the main diagonal. It is known (see, e. g. , [12]) that

$$C_aV_y=V_yA_\xi^{\mathrm{T}}$$

and

$$W_yV_y=\operatorname{diag}[Q_{\tau_i}a_{i,\tau_i-1}(J_{\tau_i}(\lambda_i))]_{i=1}^{\theta},$$

where Q_{τ_i} is the flip matrix of order τ_i and $a_{i,\tau_i-1}(\lambda)$ is as in Eq. (1.8). Thus, both W_y and V_y are invertible. Further on, it is readily verified that

$$A_\xi W_y=W_yC_a. \tag{1.10}$$

Theorem 1.1 *Let the* $NP(\mathcal{N}_p)$ *problem* (1.1) *be given, and* $\omega(\lambda)$ *be its Hermitian interpolation polynomial matrix. Then there exists a unique Hermitian block vector* $h=(h_0,h_1,\cdots,h_{2N-2}), h_i=h_i^*\in \mathbf{C}^{p\times p}$, $i=0,1,\cdots,2N-2$, *such that*

$$L_{y\bar{y}}(\omega(\lambda))=(W_y\otimes I_p)H_{Np}[h](W_y\otimes I_p)^*, \tag{1.11}$$

where W_y *is as in Eq.* (1.8), *and* $A\otimes B$ *stands for the tensor product of matrices* A *and* B. *and*

$$H_{Np}[h]=(h_{i+j})_{i,j=0}^{N-1}\in \mathbf{C}^{Np\times Np}$$

is the Hermitian block Hankel matrix generated by h. *The elements* h_0, $h_1,\cdots, h_{2N-2}$ of h *coincide with the first* $2N-1$ *Markov parameters of the rational* $p \times p$ *matrix function* $\omega(\lambda)/A(\lambda)$, *that is*, $\omega(\lambda)/A(\lambda)$ *has the Laurent expansion at infinity.*

$$\frac{\omega(\lambda)}{A(\lambda)} = \sum_{j=0}^{+\infty} \frac{\boldsymbol{h}_j}{\lambda^{j+1}}. \tag{1.12}$$

Moreover, *an extreme case of* (1.11) *holds true*:

$$\begin{aligned}\operatorname{col}[\operatorname{col}[(\boldsymbol{c}_{i0}, \cdots, \boldsymbol{c}_{i,\tau_i-1})]_{i=1}^{\theta}, \operatorname{col}[(\boldsymbol{c}_{i0}^*, \cdots, \boldsymbol{c}_{i,\tau_i-1}^*)]_{i=1}^{\theta}] \\ = (\boldsymbol{W}_A \otimes \boldsymbol{I}_p)\operatorname{col}(\boldsymbol{h}_0, \boldsymbol{h}_1, \cdots, \boldsymbol{h}_{2N-2}, \boldsymbol{h}_{2N-1}),\end{aligned} \tag{1.13}$$

Where $W_A \in \mathbf{C}^{2N\times 2N}$ *is defined as* $\boldsymbol{W}_y$ *in Eq.* (1.8) *but with* $A(\lambda)$ *in place of* $a(\lambda)$ *therein.*

In the case of $p = 1$, the statement of Theorem 1.1 is a particular case of the more general result given in [13], Theorems 3.3, 3.6 and Corollary 4.2 (see also [10], Theorems 2.1 and 2.2). The generalization to the case of $p > 1$ is immediate. Here, however, we give an alternative proof of Theorem 1.1, which will rely heavily on certain fundamental matrix equations.

Proof of Theorem 1.1 From the identity

$$\lambda \frac{\omega(\lambda) - \omega(\mu)}{\lambda - \mu} - \mu \frac{\omega(\lambda) - \omega(\mu)}{\lambda - \mu} = \omega(\lambda) - \omega(\mu),$$

we deduce by computing the partial derivatives $(1/(k!l!))(\partial^{k+1}/(\partial\lambda^k \partial\mu^l))$ of both sides at $\lambda = y_i$ and $\mu = \bar{y}_j$ for any i, j, k, l that

$$(\boldsymbol{A}_\xi \otimes \boldsymbol{I}_p)\boldsymbol{L}_{y\bar{y}}(\omega(\lambda)) - \boldsymbol{L}_{y\bar{y}}(\omega(\lambda))(\boldsymbol{A}_\xi \otimes \boldsymbol{I}_p)^* = \boldsymbol{BC}^* - \boldsymbol{CB}^*, \tag{1.14}$$

Where $\boldsymbol{A}_\xi$ is as in Eq. (1.9), and

$$\boldsymbol{C} = \operatorname{col}[\operatorname{col}(\underbrace{\boldsymbol{I}_p, \boldsymbol{0}, \cdots, \boldsymbol{0}}_{\tau_i \text{ times}})]_{i=1}^{\theta} \in \mathbf{C}^{Np\times p},$$

$$\boldsymbol{B} = \operatorname{col}[\operatorname{col}(\boldsymbol{c}_{i0}, \boldsymbol{c}_{i1}, \cdots, \boldsymbol{c}_{i,\tau_i-1})]_{i=1}^{\theta} \in \mathbf{C}^{Np\times p}. \tag{1.15}$$

Since $\lambda_i \neq \bar{\lambda}_j$ for any i, j, the matrix Eq. (1.14) has only one solution, which is clearly $\boldsymbol{L}_{y\bar{y}}(\omega(\lambda))$ itself.

Multiplying Eq. (1.14) on the left by $(\boldsymbol{W}_y \otimes \boldsymbol{I}_p)^{-1}$ and on the right by $(\boldsymbol{W}_y^* \otimes \boldsymbol{I}_p)^{-1}$ and using Eq. (1.10) we get

$$(\boldsymbol{C}_a \otimes \boldsymbol{I}_p)\boldsymbol{H} - \boldsymbol{H}(\boldsymbol{C}_a \otimes \boldsymbol{I}_p)^* = (\boldsymbol{W}_y \otimes \boldsymbol{I}_p)^{-1}(\boldsymbol{BC}^* - \boldsymbol{CB}^*)(\boldsymbol{W}_y^* \otimes \boldsymbol{I}_p)^{-1}, \tag{1.16}$$

Where

$$H=(W_y\otimes I_p)^{-1}L_{y\bar{y}}(\omega(\lambda))(W_y^*\otimes I_p)^{-1}\in \mathbf{C}^{Np\times Np}. \quad (1.17)$$

On the other hand, if $\omega(\lambda)/A(\lambda)$ has h_0, $h_1,\cdots$ as its Markov parameters, then from Eqs. (1.12) and (1.4) we deduce that

$$h_j=2\sum_{i=1}^{\theta}\frac{1}{(\tau_i-1)!}\left(\frac{d}{d\xi}\right)^{\tau_j-1}\left[\sum_{k=0}^{\tau_i-1}\mathrm{Re}\frac{c_{ik}(\xi-\lambda_i)^{\tau_i+k}\xi^j}{A(\xi)}\right]_{\xi=\lambda_i},$$
$$j=0,1,2,\cdots.$$

which are all Hermitian $p\times p$ matrices. A simple calculation shows that the last expression amounts indeed to that $\mathrm{col}(h_0, h_1,\cdots, h_{2N-2}, h_{2N-1})$ solves uniquely Eq. (1.13). Similarly, we can derive

$$(W_y\otimes I_p)^{-1}B=B_H$$
$$\equiv (h_{i+j})_{i,j=0}^{N-1}\mathrm{col}(\bar{a}_0 I_p,\cdots,\bar{a}_{N-1}I_p)+\mathrm{col}(h_N,\cdots,h_{2N-2},h_{2N-1}),$$
$$(W_y\otimes I_p)^{-1}C=C_H\equiv \mathrm{col}(\underbrace{0,\cdots,0}_{N-1\ \text{times}},I_p).$$

Thus, Eq. (1.16) amounts to

$$(C_a\otimes I_p)H-H(C_a\otimes I_p)^*=B_HC_H^*-C_HB_H^* \quad (1.16')$$

which has only one solution, since $\sigma(C_a)\cap\sigma(C_a^*)=\varnothing$ (or equivalently, $\lambda_i\neq\bar{\lambda}_j$ for any i,j). Taking $H=(h_{i+j})_{i,j=0}^{N-1}$ for the moment, we have

$$(J_N(0)\otimes I_p)H-H(J_N(0)\otimes I_p)^{\mathrm{T}}$$
$$=\mathrm{col}(h_N,\cdots,h_{2N-1})C_H^*-C_H\,\mathrm{row}(h_N,\cdots,h_{2N-1}),$$

which implies that $H=(h_{i+j})_{i,j=0}^{N-1}$ is the unique solution to Eq. (1.16′). Thus, Eq. (1.11) follows from Eq. (1.17). The proof is complete. ■

The preceding theorem implies in fact that to each NP($\mathcal{N}_p$) problem of the form (1.1) there corresponds uniquely a Hermitian block vector h defined via Eqs. (1.12) and (1.13) such that Eq. (1.11) holds. In what follows, as in the scalar case [10], we refer to the Hermitian block vector $h=(h_0,h_1,\cdots,h_{2N-2})$ determined uniquely by Theorem 1.1 as *the block Hankel vector* of the NP($\mathcal{N}_p$) problem (1.1). It will play a key role throughout our investigations. In the other direction, if $h=(h_0,h_1,\cdots,h_{2N-2})$ with each $h_j=h_j^*\in\mathbf{C}^{p\times p}$ is given in advance, then there is associated with each sequence $\boldsymbol{y}$ as in Eq. (1.6) a unique NP($\mathcal{N}_p$) problem of the form (1.1), to within addition of a Hermitian constant matrix of order p for $\boldsymbol{c}_{i0}$ in (1.1), such that $\boldsymbol{h}$ is the block Hankel vector of that NP($\mathcal{N}_p$) problem [13].

In view of the fact that the coefficients of $A(\lambda)$ in Eq. (1.5) are all real. Eq. (1.12) implies in turn that the coefficients of $\omega(\lambda)$ are all Hermitian, a useful result later on.

§ 1.3 The PM(R) problem

Given a finite sequence of $p \times p$ Hermitian matrices $\boldsymbol{h}_0, \boldsymbol{h}_1, \cdots, \boldsymbol{h}_{2n}$, find the conditions on the data which will ensure the existence of a nondecreasing $\mathbf{C}^{p\times p}$-valued function $\tau(u)(-\infty < u < +\infty)$ subject to

$$\begin{aligned} \boldsymbol{h}_j &= \int_{-\infty}^{+\infty} u^j \, \mathrm{d}\tau(u), \quad j = 0, 1, \cdots, 2n-1. \\ \boldsymbol{h}_{2n} &\geqslant \int_{-\infty}^{+\infty} u^{2n} \, \mathrm{d}\tau(u), \end{aligned} \tag{1.18}$$

and describe the set of solutions.

Note that the PM(R) problem (1.18) is different from and closely relative to the truncated Hamburger matrix moment (HM) problem of finding a nondecreasing $\mathbf{C}^{p\times p}$ -valued function $\tau(u)(-\infty < u < +\infty)$ subject to

$$\boldsymbol{h}_j = \int_{-\infty}^{+\infty} u^j \, \mathrm{d}\tau(u), \quad j = 0, 1, \cdots, 2n, \tag{1.19}$$

that is, the HM problem (1.19) is the same as the PM($\mathbf{R}$) problem (1.18) but with $\boldsymbol{h}_{2n} = \int_{-\infty}^{+\infty} u^{2n} \mathrm{d}\tau(u)$ in place of the last one in (1.18).

The following result shows the close relation of the HM problem (1.19) to the problem of reconstructing functions $\Phi(\lambda)$ in $\mathcal{N}_p$ with a given asymptotic expansion

$$\Phi(\lambda) = -\frac{\boldsymbol{h}_0}{\lambda} - \frac{\boldsymbol{h}_1}{\lambda^2} - \cdots - \frac{\boldsymbol{h}_{2n}}{\lambda^{2n+1}} - R_{2n+1}(\lambda), \tag{1.20}$$

where $R_{2n+1}(\lambda) = o(\lambda^{-2n-1})$ as $\lambda \to \infty$ uniformly in each sector (1.3).

Lemma 1.2 (The Hamburger-Nevanlinna theorem) *If $\tau(u)$ is a solution to the HM problem* (1.19), *then there exists $\Phi(\lambda) \in \mathcal{N}_p$*

determined by

$$\Phi(\lambda)=\int_{-\infty}^{+\infty}\frac{\mathrm{d}\tau(u)}{u-\lambda},\tag{1.21}$$

for which the asymptotic representation

$$\lim_{\lambda\to\infty}\lambda^{2n+1}\left[\Phi(\lambda)+\frac{\boldsymbol{h}_0}{\lambda}+\frac{\boldsymbol{h}_1}{\lambda^2}+\cdots+\frac{\boldsymbol{h}_{2n-1}}{\lambda^{2n}}\right]=-\boldsymbol{h}_{2n}\tag{1.22}$$

holds uniformly in each sector (1.3).

Conversely, if Eq. (1.22) *holds, at least for* $\lambda=iy$ ($y\to+\infty$), *for a certain* $\Phi(\lambda)\in\mathcal{N}_p$, *then* $\Phi(\lambda)$ *admits the representation* (1.21) *in which* $\tau(u)$ *is a solution to the HM problem* (1.19).

§ 1.4 A one-to-one correspondence between solutions to the NP($\mathcal{N}_p$) problem and solutions to an associated PM(R) problem

If now we let $n=N-1$ in the PM(**R**) problem (1.18), and further let $(\boldsymbol{h}_0,\boldsymbol{h}_1,\cdots,\boldsymbol{h}_{2N-2})$ coincide with the block Hankel vector $\boldsymbol{h}$ of the NP($\mathcal{N}_p$)problem (1.1), then we can derive an elegant correspondence between the solution sets of these two problems as follows.

Theorem 1.3 *The NP*($\mathcal{N}_p$) *problem* (1.1) *is solvable if, and only if, the PM*(**R**)*problem* (1.18) *associated with the Hankel vector* $\boldsymbol{h}=(\boldsymbol{h}_0,\boldsymbol{h}_1,\cdots,\boldsymbol{h}_{2N-2})$ *is solvable. In this case, there exists a one-to-one correspondence between their solutions* $F(\lambda)$ *and* $\tau(u)$ ($-\infty<u<+\infty$) *determined by the formula*:

$$F(\lambda)=\omega(\lambda)+\int_{-\infty}^{+\infty}\frac{\mathrm{d}\tau(u)}{u-\lambda}A(\lambda).\tag{1.23}$$

where $\omega(\lambda)$ *and* $A(\lambda)$ *are as before. More precisely, if* $F(\lambda)$ *is a solution to the problem* (1.1) *and admits an integral representation of the form* (1.2) *for* $\operatorname{Im}\lambda\neq0$, *then there exists* $\tau(u)$ ($-\infty<u<+\infty$) *such that*

$$\Phi(\lambda)\equiv\frac{-\omega(\lambda)+F(\lambda)}{A(\lambda)}=\int_{-\infty}^{+\infty}\frac{\mathrm{d}\tau(u)}{u-\lambda}\tag{1.24}$$

and

$$\boldsymbol{h}_j = \int_{-\infty}^{+\infty} u^j \mathrm{d}\tau(u), \quad j = 0,1,\cdots,2N-3,$$
$$\boldsymbol{h}_{2N-2} = \int_{-\infty}^{+\infty} u^{2N-2} \mathrm{d}\tau(u) + \boldsymbol{\beta}, \tag{1.25}$$

in which $\boldsymbol{\beta} \geqslant \boldsymbol{O}$ *is identical with that appearing in Eq.* (1.2), *and*

$$\mathrm{d}\tau(u) = \frac{\mathrm{d}\sigma(u)}{A(u)}, \quad -\infty < u < +\infty. \tag{1.26}$$

Conversely, each solution $\tau(u)$ *to the PM*($\mathbf{R}$) *problem* (1.18) *satisfying Eq.* (1.25) *with* $\boldsymbol{\beta} = \boldsymbol{h}_{2N-2} - \int_{-\infty}^{+\infty} u^{2N-2} \mathrm{d}\tau(u)$ *therein is associated with the solution* $F(\lambda)$ *to the problem* (1.1) *defined via Eq.* (1.23), *which admits a representation of form Eq.* (1.2), *where*

$$\boldsymbol{\beta} = \boldsymbol{h}_{2N-2} - \int_{-\infty}^{+\infty} u^{2N-2} \mathrm{d}\tau(u) \geqslant \boldsymbol{O}$$

and

$$\mathrm{d}\sigma(u) = A(u)\mathrm{d}\tau(u), \quad -\infty < u < +\infty.$$

Proof Suppose first that $F(\lambda) \in \mathcal{N}_p$ is a solution to the problem (1.1) and admits a representation of form (1.2) for $\operatorname{Im}\lambda \neq 0$:

$$F(\lambda) = \boldsymbol{\alpha} + \boldsymbol{\beta}\lambda + \int_{-\infty}^{+\infty}\left(\frac{1}{u-\lambda} - \frac{u}{1+u^2}\right)\mathrm{d}\tau(u)$$

where $\boldsymbol{\alpha} = \boldsymbol{\alpha}^*$, $\boldsymbol{\beta} \geqslant \boldsymbol{O}$, $\mathrm{d}\sigma(u) \geqslant \boldsymbol{O}$, and $\int_{-\infty}^{+\infty} (1+u^2)^{-1} \operatorname{tr} \mathrm{d}\sigma(u) < \infty$. Observe that the integral

$$\int_{-\infty}^{+\infty} \frac{\mathrm{d}\sigma(u)}{A(u)(\lambda - u)}$$

makes sense, since the poles of $[A(u)(\lambda - u)]^{-1}$, regarding as a function of u for each fixed λ with $\operatorname{Im}\lambda \neq 0$, are not on the real axis, and $\lim\limits_{u\to\infty}(1+u^2)/(A(u)(1-u)) = 0$. Define in turn

$$\hat{\omega}(\lambda) = F(\lambda) + A(\lambda)\int_{-\infty}^{+\infty} \frac{\mathrm{d}\sigma(u)}{A(u)(\lambda - u)},$$

which can be rewritten as

$$\hat{\omega}(\lambda) = \boldsymbol{\alpha} + \boldsymbol{\beta}\lambda + \int_{-\infty}^{+\infty}\left[\frac{A(u) - A(\lambda)}{A(u)(u-\lambda)} - \frac{u}{1+u^2}\right]\mathrm{d}\sigma(u). \tag{1.27}$$

It is evidently seen from Eq. (1.27) that the last integrand therein must be a certain polynomial of λ of degree at most $2N-1$, whose coefficients

$f_i(u)$ of $\lambda^i (i = 0, 1, \cdots, 2N-1)$ are certain functions of u with no real poles, and satisfy that $\lim\limits_{u\to\infty}(1+u^2)f_i(u) < \infty$. By integrating the polynomial in question term by term, we derive that $\hat{\omega}(\lambda)$ given in Eq. (1.27) is indeed a certain polynomial matrix of $\mathbf{C}^{P\times P}[\lambda]$, of degree at most $2N-1$, subject to Eqs. (1.1)and (1.1′), so that $\hat{\omega}(\lambda)$ has to be identical with $\omega(\lambda)$.

Since the fact that $\lim\limits_{u\to\infty}(1+u^2)/A(u) < \infty$ together with the condition that $\int_{-\infty}^{+\infty}(1+u^2)^{-1}\,\mathrm{tr}\,\mathrm{d}\sigma(u) < +\infty$ implies that $\int_{-\infty}^{+\infty}A(u)^{-1}\,\mathrm{tr}\,\mathrm{d}\sigma(u)$ makes sense, we can set

$$\tau(u) = \int_{-\infty}^{u} A(u)^{-1}\mathrm{d}\sigma(u) \quad \text{for } -\infty < u < +\infty. \tag{1.28}$$

Then we have in turn the relation (1.26) with $\mathrm{d}\tau(u) \geqslant \boldsymbol{O}$ and $\int_{-\infty}^{+\infty}\mathrm{tr}\,\mathrm{d}\tau(u) < +\infty$. Let now

$$\Phi(\lambda) \equiv \int_{-\infty}^{+\infty}\frac{\mathrm{d}\sigma(u)}{A(u)(u-\lambda)} = \int_{-\infty}^{+\infty}\frac{\mathrm{d}\tau(u)}{u-\lambda}.$$

Then $\Phi(\lambda)$ belongs to the class $\mathcal{N}_p$, and is such that $F(\lambda) = \omega(\lambda) + \Phi(\lambda)A(\lambda)$. In virtue of Eq. (1.12) and the fact that

$$\lim_{\lambda\to\infty}\left(\frac{F(\lambda)}{A(\lambda)}\lambda^{2N-1}\right) = \lim_{\lambda\to\infty}\left(\frac{F(\lambda)}{\lambda}\frac{\lambda^{2N}}{A(\lambda)}\right) = \lim_{\lambda\to\infty}\frac{F(\lambda)}{\lambda} = \boldsymbol{\beta}$$

uniformly in each sector (1.3), we derive

$$\lim_{\lambda\to\infty}\lambda^{2N-1}\left[\Phi(\lambda) + \frac{\boldsymbol{h}_0}{\lambda} + \frac{\boldsymbol{h}_1}{\lambda^2} + \cdots + \frac{\boldsymbol{h}_{2N-3}}{\lambda^{2N-2}}\right] = -(\boldsymbol{h}_{2N-2} - \boldsymbol{\beta})$$

uniformly in each sector (1.3). Hence, Lemma 1.2 implies that $\tau(u)$ in Eq. (1.28) satisfies Eq. (1.25), as needed.

Next suppose, conversely, that $\tau(u)$ is a solution to the problem (1.18), so that $\boldsymbol{\beta} = \boldsymbol{h}_{2N-2} - \int_{-\infty}^{+\infty}u^{2N-2}\mathrm{d}\tau(u) \geqslant \boldsymbol{O}$ is given. Let

$$\Phi(\lambda) = \int_{-\infty}^{+\infty}\frac{\mathrm{d}\tau(u)}{u-\lambda},\ \mathrm{Im}\,\lambda \neq 0,$$

$$\sigma(u) = \int_0^u A(u)\mathrm{d}\tau(u) \quad \textit{for} \quad -\infty < u < +\infty.$$

Then $\mathrm{d}\tau(u) = A(u)\mathrm{d}\tau(u) \geqslant \boldsymbol{O}\ (-\infty < u < +\infty)$ and $\int_{-\infty}^{+\infty}(1+u^2)^{-1}\,\mathrm{tr}\,\mathrm{d}\sigma(u)$

$= \int_{-\infty}^{+\infty}(1+u^2)^{-1}A(u)\ \mathrm{tr}\ \mathrm{d}\sigma(u) < \infty$, since $\int_{-\infty}^{+\infty} u^k\ \mathrm{tr}\ \mathrm{d}\tau(u) < \infty$ for $k=0,1,\cdots,2N-2$. It is obvious that

$$\lim_{u\to\infty}(1+u^2)\left(\frac{1}{u-\lambda}-\frac{u}{1+u^2}\right)=\lambda \quad \text{for any } \lambda \notin \mathbf{R}$$

so that

$$\int_{-\infty}^{+\infty}\left(\frac{1}{u-\lambda}-\frac{u}{1+u^2}\right)\mathrm{d}\sigma(u)$$

exists for any $\lambda \notin \mathbf{R}$. Define now

$$\hat{F}(\lambda)=\boldsymbol{\beta}\lambda+\int_{-\infty}^{+\infty}\left(\frac{1}{u-\lambda}-\frac{u}{1+u^2}\right)\mathrm{d}\sigma(u),\ \mathrm{Im}\,\lambda \neq 0. \tag{1.29}$$

Then $\hat{F}(\lambda)$ belongs to $\mathcal{N}_p$, and is such that

$$\lim_{\lambda\to\infty}\frac{\hat{F}(\lambda)}{\lambda}=\boldsymbol{\beta} \tag{1.30}$$

uniformly in each sector (1.3). In a similar way to the proof of the first part of the theorem, we can show that

$$\tilde{\omega}(\lambda)\equiv\hat{F}(\lambda)-A(\lambda)\Phi(\lambda)=\hat{F}(\lambda)+A(\lambda)\int_{-\infty}^{+\infty}\frac{\mathrm{d}\sigma(u)}{A(u)(u-\lambda)} \tag{1.31}$$

is a certain polynomial matrix in $\mathbf{C}^{p\times p}[\lambda]$, of degree at most $2N-1$, subject to

$$\left.\frac{\mathrm{d}^k}{\mathrm{d}\lambda^k}\tilde{\omega}(\lambda)\right|_{\lambda=\lambda_i}=\left.\frac{\mathrm{d}^k}{\mathrm{d}\lambda^k}\hat{F}(\lambda)\right|_{\lambda=\lambda_i},\quad \left.\frac{\mathrm{d}^k}{\mathrm{d}\lambda^k}\tilde{\omega}(\lambda)\right|_{\lambda=\bar{\lambda}_i}=\left.\frac{\mathrm{d}^k}{\mathrm{d}\lambda^k}\hat{F}(\lambda)\right|_{\lambda=\bar{\lambda}_i},\quad i=1,2,\cdots,\theta,\ k=0,1,\cdots,\tau_i-1.$$

On the other hand, we have by Lemma 1.2 that

$$\lim_{\lambda\to\infty}\lambda^{2N-1}\left[\Phi(\lambda)+\frac{\boldsymbol{h}_0}{\lambda}+\frac{\boldsymbol{h}_1}{\lambda^2}+\cdots+\frac{\boldsymbol{h}_{2N-3}}{\lambda^{2N-2}}\right]=-(\boldsymbol{h}_{2N-2}-\boldsymbol{\beta})$$

uniformly in each sector (1.3). This together with Eq. (1.30) implies that for the sufficiently large values of $|\lambda|$,

$$\lim_{\lambda\to\infty}\lambda^{2N-1}\left[-\frac{\tilde{\omega}(\lambda)}{A(\lambda)}+\frac{\boldsymbol{h}_0}{\lambda}+\frac{\boldsymbol{h}_1}{\lambda^2}+\cdots+\frac{\boldsymbol{h}_{2N-3}}{\lambda^{2N-2}}\right]=-\boldsymbol{h}_{2N-2}.$$

But the last expression is also valid with $\omega(\lambda)$ in place of $\tilde{\omega}(\lambda)$ therein, so that there exists a $p\times p$ constant matrix $\boldsymbol{\alpha}$ such that

$$\omega(\lambda) - \tilde{\omega}(\lambda) \equiv \boldsymbol{\alpha}, \tag{1.32}$$

which is evidently Hermitian: $\boldsymbol{\alpha} = \boldsymbol{\alpha}^*$, since the coefficients of both $\omega(\lambda)$ and $\tilde{\omega}(\lambda)$ are all Hermitian. Thus, from Eqs. (1.29) (1.31) (1.32) we deduce

$$F(\lambda) \equiv \boldsymbol{\alpha} + \hat{F}(\lambda) = \omega(\lambda) + A(\lambda)\Phi(\lambda) \in \mathscr{N}_p.$$

The proof is complete. ■

We remark that the preceding theorem says in fact among matrix moment problems the PM(**R**) problem (1.18) with $n = N-1$ rather than the HM problem (1.19) with $n = N-1$ is the closest resemblance to the NP($\mathscr{N}_p$) problem. Theorem 1.3 provides also a very important link between the HM problem (1.19) with $n = N-1$ and a version of the NP($\mathscr{N}_p$) problem (1.1) in which the interpolants $F(\lambda)$ are required further to satisfy that $\lim\limits_{\lambda\to\infty} F(\lambda)/\lambda = \boldsymbol{O}$ uniformly in each sector (1.3).

§ 1.5 The solution of the NP($\mathscr{N}_p$) problem

We first need the following lemma. In [3], the sufficiency of the condition (1.33) was proved, while the necessity follows directly from the fact

$$\boldsymbol{H}_n = (\boldsymbol{h}_{i+j})_{i,j=0}^n \geqslant \left(\int_{-\infty}^{+\infty} u^{i+j}\,\mathrm{d}\tau(u)\right)_{i,j=0}^n = \int_{-\infty}^{+\infty} (u^{i+j})_{i,j=0}^n \otimes \mathrm{d}\tau(u) \geqslant \boldsymbol{O},$$

where $\tau(u)(-\infty < u < +\infty)$ is a solution to the PM(**R**) problem (1.18).

Lemma 1.4[3] *The PM(**R**) problem (1.18) is solvable if, and only if, the following block Hankel matrix $\boldsymbol{H}_n$ is nonnegative Hermitian:*

$$\boldsymbol{H}_n = (\boldsymbol{h}_{i+j})_{i,j=0}^n \geqslant \boldsymbol{O}. \tag{1.33}$$

The following result includes indeed a new solvability criterion for the NP($\mathscr{N}_p$) problem (see (b) of Theorem 1.5), and an alternative proof of the well known solvability criterion for the same problem (see (c) of Theorem 1.5).

Theorem 1.5 *Let $\boldsymbol{h} = (\boldsymbol{h}_0, \boldsymbol{h}_1, \cdots, \boldsymbol{h}_{2N-2})$ be the block Hankel vector of the NP ($\mathscr{N}_p$) problem (1.1), and $\omega(\lambda)$ be as before. Then the following are equivalent:*

(a) *The* $NP(\mathcal{N}_p)$ *problem* (1.1) *is solvable.*

(b) $\boldsymbol{H}_{NP}[\boldsymbol{h}] \equiv (\boldsymbol{h}_{i+j})_{i,j=0}^{N-1} \geqslant \boldsymbol{O}$.

(c) *The generalized block Loewner matrix* $\mathbf{L}_{y\bar{y}}(\omega(\lambda)) \geqslant \boldsymbol{O}$.

Proof Follows from Theorems 1.1, 1.3 and Lemma 1.4 immediately. ■

The Hamburger-Nevanlinna theorem lets us restate the PM($\mathbf{R}$) problem (1.18) as follows. Given a Hermitian $p \times p$ matrix sequence $\boldsymbol{h}_0, \boldsymbol{h}_1, \cdots, \boldsymbol{h}_{2n}$, it is required to reconstruct all $\Phi(\lambda) \in \mathcal{N}_p$ with prescribed asymptotic expansion at infinity

$$\Phi(\lambda) = -\frac{\boldsymbol{h}_0}{\lambda} - \frac{\boldsymbol{h}_1}{\lambda^2} - \cdots - \frac{\boldsymbol{h}_{2n} - \boldsymbol{\beta}}{\lambda^{2n+1}} - \cdots \tag{1.34}$$

for some $\boldsymbol{\beta}$ with $\boldsymbol{O} \leqslant \boldsymbol{\beta} \leqslant \boldsymbol{h}_{2n}$, or equivalently, with the asymptotic representation

$$\lim_{\lambda \to \infty} \lambda^{2n+1}\left[\Phi(\lambda) + \frac{\boldsymbol{h}_0}{\lambda} + \frac{\boldsymbol{h}_1}{\lambda^2} + \cdots + \frac{\boldsymbol{h}_{2n-1}}{\lambda^{2n}}\right] = -(\boldsymbol{h}_{2n} - \boldsymbol{\beta}), \boldsymbol{O} \leqslant \boldsymbol{\beta} \leqslant \boldsymbol{h}_{2n}, \tag{1.35}$$

uniformly in each sector (1.3). In case $\boldsymbol{\beta} = \boldsymbol{O}$, then the above mentioned problem amounts to the HM problem (1.19).

From Theorem 1.3 we see that if the set of solutions $\Phi(\lambda)$ to the problem (1.35) with $n = N - 1$ and $(\boldsymbol{h}_0, \boldsymbol{h}_1, \cdots, \boldsymbol{h}_{2N-2})$ being the block Hankel vector of the problem (1.1) admits a parametrized description by means of a linear fractional transformation

$$\Phi(\lambda) = [a(\lambda)P(\lambda) + b(\lambda)Q(\lambda)][c(\lambda)P(\lambda) + d(\lambda)Q(\lambda)]^{-1}, \tag{1.36}$$

where the free parameter

$$\begin{bmatrix} P(\lambda) \\ Q(\lambda) \end{bmatrix}$$

runs over a suitable set; the matrix of the transformation (1.36) has the form

$$\boldsymbol{\Theta}(\lambda) = \begin{bmatrix} a(\lambda) & b(\lambda) \\ c(\lambda) & d(\lambda) \end{bmatrix} \in \mathbf{C}^{2p \times 2p}, \tag{1.37}$$

then the set of solutions $F(\lambda)$ to the NP($\mathcal{N}_p$) problem (1.1) can be parametrized by means of a linear fractional transformation

$$F(\lambda) = \omega(\lambda) + A(\lambda)[a(\lambda)P(\lambda) + b(\lambda)Q(\lambda)][c(\lambda)P(\lambda) + d(\lambda)Q(\lambda)]^{-1}, \tag{1.38}$$

whose matrix has the form

$$\begin{bmatrix} A(\lambda)\boldsymbol{I}_p & \omega(\lambda) \\ \boldsymbol{O} & \boldsymbol{I}_p \end{bmatrix}\begin{bmatrix} a(\lambda) & b(\lambda) \\ c(\lambda) & d(\lambda) \end{bmatrix}. \tag{1.39}$$

In the nondegenerate situation (i. e. , $\boldsymbol{H}_n = (\boldsymbol{h}_{i+j})_{i,j=0}^{n} > \boldsymbol{O}$), a complete description of the set of solutions to the problem (1. 35) has been given in [22],Theorem H. See also [9,11] for the general case (i. e. ,$\boldsymbol{H}_n \geqslant \boldsymbol{O}$). Each of them leads to a description of solutions to the $\mathrm{NP}(\mathcal{N}_p)$ problem (1. 1) in yet another form. See [11], Theorem 5. 2 for one of such results.

§ 1. 6 On the HM problem and its associated $\mathrm{NP}(\mathcal{N}_p)$ problems

We consider the HM problem (1. 19) with $n = N-1$ and with $\boldsymbol{h} = (\boldsymbol{h}_0, \boldsymbol{h}_1, \cdots, \boldsymbol{h}_{2N-2})$ as its Hermitian moment vector. We define the sequence $\boldsymbol{y}$ as in Eq. (1. 6) at own convenience, even, if desired, considering a simple-node case: $\boldsymbol{y} = (y_1, y_2, \cdots, y_N)$. From the remark following Theorem 1. 1 it follows that to such $\boldsymbol{h}$ and $\boldsymbol{y}$ together with a fixed $\boldsymbol{h}_{2N-1} \in \mathbf{C}^{p\times p}$ there corresponds only one $\mathrm{NP}(\mathcal{N}_p)$ problem of the form (1. 1), such that $\boldsymbol{h}$ is the block Hankel vector thereof, called the $\mathrm{NP}(\mathcal{N}_p)$ problem with data $\boldsymbol{h}$, $\boldsymbol{h}_{2N-1}, \boldsymbol{y}$ Then Theorem 1. 3 provides incidentally an important link between the HM problem with $n = N-1$ and with $\boldsymbol{h}_0, \boldsymbol{h}_1, \cdots, \boldsymbol{h}_{2N-2}$, as moments and a version of the $\mathrm{NP}(\mathcal{N}_p)$ problem with data $\mathbf{h}, \boldsymbol{h}_{2N-1}, \boldsymbol{y}$, in which an extra constraint is added, namely, require that the interpolants $F(\lambda)$ be such that $\lim\limits_{\lambda\to\infty} F(\lambda)/\lambda = \boldsymbol{O}$ uniformly in each sector (1. 3). Thus, the solution of that HM problem may become the point of departure for the statement and investigation of some new problems in the version of the $\mathrm{NP}(\mathcal{N}_p)$ problem with data $\boldsymbol{h}$, $\boldsymbol{h}_{2N-1}, \boldsymbol{y}$ mentioned above (the $\mathrm{VNP}(\mathcal{N}_p)$ problem with data $\boldsymbol{h}, \boldsymbol{h}_{2N-1}, \boldsymbol{y}$, for short), for example, to find the solvability criteria and formulas describing the solutions. The last implication clearly is reversible.

The following result is a simple consequence of [11], Theorem 3.1.

Theorem 1.6 *Let* $\boldsymbol{h} = (\boldsymbol{h}_0, \boldsymbol{h}_1, \cdots, \boldsymbol{h}_{2N-2})$ *be a block rector with each* $\boldsymbol{h}_i = \boldsymbol{h}_i^* \in \mathbf{C}^{p\times p}$. *Then the following are equivalent:*

(a) *The HM problem* (1.19) *with* $n = N-1$ *and with* $\boldsymbol{h} = (\boldsymbol{h}_0, \boldsymbol{h}_1, \cdots, \boldsymbol{h}_{2N-2})$ *as the Hermitian moment vector is solvable.*

(b) *The* VNP$(\mathcal{N}_p)$ *problem with data* $\boldsymbol{h}, \boldsymbol{h}_{2N-1}, \boldsymbol{y}$ *mentioned above is solvable.*

(c) *The block Hankel matrix defined by* $(\boldsymbol{h}_{i+j})_{i,j=0}^{N-1}$ *is nonnegatively extendable, that is, there exist hermitian matrices* $\boldsymbol{h}_{2N-1}$, $\boldsymbol{h}_{2N}$ *of order p such that* $(\boldsymbol{h}_{i+j})_{i,j=0}^{N} \geqslant \boldsymbol{O}$.

Further on, digging a little deeper, one can try to gain an insight into the description of solutions to that VNP$(\mathcal{N}_p)$ problem by examining what become of the interpolants for the case when the last inequality in the PM($\mathbf{R}$) problem (1.18) with $n = N-1$ is replaced by the equality. The equality versus inequality issue can be analyzed via a variety of approaches.

It turns out that if the PM($\mathbf{R}$) problem (1.18) with $n = N-1$ has the solutions $\tau(u)$ defined via the form (1.36), in which

$$\Phi(\lambda) = \int_{-\infty}^{+\infty} \mathrm{d}\tau(u)/(u-\lambda), \text{ and } \begin{bmatrix} P(\lambda) \\ Q(\lambda) \end{bmatrix}$$

is an arbitrary Nevanlinna pair, and if the condition (c) of Theorem 1.6 holds true, then the corresponding HM problem has the solutions $\tilde{\tau}(u)$ defined via

$$\int_{-\infty}^{+\infty} \frac{\mathrm{d}\tilde{\tau}(u)}{u-\lambda} = [a(\lambda)G(\lambda) + b(\lambda)][c(\lambda)G(\lambda) + d(\lambda)]^{-1}, \tag{1.40}$$

where $G(\lambda)$ is an arbitrary member in $\mathcal{N}_p$ such that $\lim\limits_{\lambda\to\infty} G(\lambda)/\lambda = \boldsymbol{O}$ uniformly in each sector (1.3). On the basis of the last result and Theorem 1.3, we obtain immediately the description of solutions for the VNP$(\mathcal{N}_p)$ problem with data $\boldsymbol{h}$, $\boldsymbol{h}_{2N-1}, \boldsymbol{y}$ by means of the formula (1.23) with $\tau(u)$ replaced by $\tilde{\tau}(u)$ of Eq. (1.40). See e.g. [11],

Section 3, for further details.

The preceding reasoning clearly is reversible. We begin with the NP($\mathcal{N}_p$) problem and the VNP($\mathcal{N}_p$) problem both with data $\boldsymbol{h}, \boldsymbol{h}_{2N-1}, \boldsymbol{y}$, where the sequence $\boldsymbol{y}$ consists of simple nodes. It turns out that if the solutions $F(\lambda)$ to the former problem are given by a formula:

$$F(\lambda) = [\alpha(\lambda)P(\lambda) + \beta(\lambda)Q(\lambda)][\gamma(\lambda)P(\lambda) + \delta(\lambda)Q(\lambda)]^{-1}, \tag{1.41}$$

where

$$\begin{bmatrix} P(\lambda) \\ Q(\lambda) \end{bmatrix}$$

is an arbitrary Nevanlinna pair, then, by a straightforward verification, the solutions $\widetilde{F}(\lambda)$ to the latter, if there exist, are given by

$$\widetilde{F}(\lambda) = [\alpha(\lambda)G(\lambda) + \beta(\lambda)][\gamma(\lambda)G(\lambda) + \delta(\lambda)]^{-1}, \tag{1.42}$$

in which $G(\lambda)$ is an arbitrary member of the class $\mathcal{N}_p$ such that $\lim\limits_{\lambda \to \infty} G(\lambda)/\lambda = \boldsymbol{O}$ uniformly in each sector (1.3). By Theorem 1.3, the last result enables us to obtain the description of the solutin set for the HM problem (1.19) with $n = N - 1$ and with $\boldsymbol{h}$ as its Hermitian moment vector, which must have the form (1.40). (The same conclusion is also a result of Dym [16] by a different approach only for the nondegenerate case.)

§ 2. Connection between the NP problem in the class $\mathcal{S}_p$ and the Stieltjes matrix power moment problem

This section is devoted to the same investigations as that in Section 1 but for both the NP problem in the class $\mathcal{S}_p$ (the NP($\mathcal{S}_p$) problem, for short) and the matrix power moment problem on the semiaxis $\mathbf{R}^- = (-\infty, 0)$ (the PM($\mathbf{R}^-$) problem), i.e., the Stieltjes matrix power moment problem. Here the class $\mathcal{S}_p$ consists of all the $\mathbf{C}^{p\times p}$-valued functions $F(\lambda)$ holomorphic in the complex plane $\mathbf{C}$ with a cut along the semiaxis $[0,+\infty)$, satisfying that $(F(\lambda)-F^*(\lambda))/(\lambda-\bar{\lambda})\geqslant \boldsymbol{O}$ for $\operatorname{Im}\lambda\neq 0$, and $F(x)\geqslant \boldsymbol{O}$ for real $x<0$. The fundamental result is a one-to-one correspondence between solutions to the NP($\mathscr{S}_p$) problem and solutions to an associated PM($\mathbf{R}^-$) problem (see Theorem 2.2 below). An extension thereof is also given in Theorem 2.5.

§ 2.1 The NP($\mathscr{S}_p$) problem and its block Hankel vector, and the PM($\mathbf{R}^-$) problem

We are given the interpolation data as in the NP($\mathscr{N}_p$) problem (1.1). The objective is to find necessary and sufficient conditions on the data to ensure the existence a function $F(\lambda)\in\mathscr{S}_p$ such that Eq. (1.1) holds, and describe all the existing solutions when these conditions are met. We call this the NP($\mathscr{S}_p$)problem for short.

It is well known [27,24] that $F(\lambda)\in\mathscr{S}_p$ if, and only if, it admits an integral representation

$$F(\lambda)=\boldsymbol{v}+\int_0^{+\infty}\frac{1}{u-\lambda}\mathrm{d}\sigma(u). \tag{2.1}$$

where $\boldsymbol{v}\geqslant\boldsymbol{O}$, $\mathrm{d}\sigma(u)\geqslant\boldsymbol{O}$ for $u\geqslant 0$ and $\mathrm{d}\sigma(u)=\boldsymbol{O}$ for $u<0$, and

$$\int_0^{+\infty}\frac{1}{1+u}\operatorname{tr}\,\mathrm{d}\sigma(u)<+\infty. \tag{2.2}$$

In this case, $\lim\limits_{\lambda\to\infty}F(\lambda)=\boldsymbol{v}$ uniformly in any sector (1.3).

The following characterization of the Stieltjes class can be obtained from [24],Appendix, for the scalar case, the proof in the matrix case is identical with that of the scalar case (see also [27], Theorem 2.2 with $T(z^{-1})$ in place of $T(z)$ therein).

Theorem 2.1 [24] *$F(\lambda)\in\mathscr{S}_p$ if, and only if, $F(\lambda)\in\mathscr{N}_p$ and $\lambda F(\lambda)\in\mathscr{N}_p$*

Let $N=\sum_{i=1}^{\theta}\tau_i$, and, let $\omega(\lambda)\in\mathbf{C}^{p\times p}[\lambda]$ and $A(\lambda)$ be as before. Denote by $\boldsymbol{h}_0,\boldsymbol{h}_1,\cdots$ the Markov parameters of $\omega(\lambda)/A(\lambda)$ (see Eq. (1.12)). We refer to $\tilde{\boldsymbol{h}}=(\boldsymbol{h}_0,\boldsymbol{h}_1,\cdots,\boldsymbol{h}_{2N-2},\boldsymbol{h}_{2N-1})$ (rather than $(\boldsymbol{h}_0,\boldsymbol{h}_1,\cdots,\boldsymbol{h}_{2N-2})$) as *the block Hankel vector* of the NP($\mathscr{S}_p$) problem.

Since

$$\frac{\lambda\omega(\lambda)}{A(\lambda)}=\boldsymbol{h}_0+\frac{\boldsymbol{h}_1}{\lambda}+\frac{\boldsymbol{h}_2}{\lambda^2}+\cdots+\frac{\boldsymbol{h}_{2N-2}}{\lambda^{2N-2}}+\frac{\boldsymbol{h}_{2N-1}}{\lambda^{2N-1}}+\cdots$$

from Theorem 1.1 we deduce that

$$\boldsymbol{L}_{\bar{yy}}(\lambda\omega(\lambda))=\boldsymbol{L}_{\bar{yy}}(\lambda\omega(\lambda)-A(\lambda)\boldsymbol{h}_0)=(\boldsymbol{W}_y\otimes\boldsymbol{I}_p)(\boldsymbol{h}_{i+j+1})_{i,j=0}^{N-1}(\boldsymbol{W}_y\otimes\boldsymbol{I}_p)^*. \tag{2.3}$$

Note that, as was mentioned in Section 1, the block Hankel vector $\tilde{\boldsymbol{h}}$, of the NP($\mathscr{S}_p$) problem is also Hermitian, i.e., $\boldsymbol{h}_j=\boldsymbol{h}_j^*\in\mathbf{C}^{p\times p}$, $j=0,1,\cdots,2N-1$.

The PM($\mathbf{R}^-$) problem can be formulated as follows. Given a finite sequence of $p\times p$ Hermitian matrices $\boldsymbol{h}_0$, $\boldsymbol{h}_1,\cdots,\boldsymbol{h}_{2n+1}$, find the conditions to ensure the existence of a nondecreasing $\mathbf{C}^{p\times p}$-valued function $\tau(u)$ $(0\leqslant u<+\infty)$ such that

$$\boldsymbol{h}_j=\int_0^{+\infty}u^j\,\mathrm{d}\tau(u),\quad j=0,1,\cdots,2n.$$

$$\boldsymbol{h}_{2n+1}\geqslant\int_0^{+\infty}u^{2n+1}\,\mathrm{d}\tau(u), \tag{2.4}$$

and describe the set of solutions.

It is known [17] that a solvability criterion of the PM($\mathbf{R}^-$) problem (2.4) is the nonnegativity of Hermitian block Hankel matrices $\boldsymbol{H}_n$ and $\tilde{\boldsymbol{H}}_n$:

$$\boldsymbol{H}_n=(\boldsymbol{h}_{i,j})_{i,j=0}^n\geqslant\boldsymbol{O},\quad \tilde{\boldsymbol{H}}_n=(\boldsymbol{h}_{i+j+1})_{i,j=0}^n\geqslant\boldsymbol{O} \tag{2.5}$$

(see also [3, Corollary 6] for a proof of sufficiency of condition (2.5)).

§ 2.2 A one-to-one correspondence between solutions to the NP($\mathscr{S}_p$) problem and solutions to an associated PM($\mathbf{R}^-$) problem

If now we take $n=N-1$ and the block Hankel vector $\tilde{\boldsymbol{h}}=(\boldsymbol{h}_0,\boldsymbol{h}_1,\cdots,\boldsymbol{h}_{2N-1})$ of the NP($\mathscr{S}_p$) problem as the moment vector in (2.4), we can derive a correspondence between solutions to the NP($\mathscr{S}_p$) problem and solutions to the PM($\mathbf{R}^-$) problem (2.4) by using Theorem 1.3.

Theorem 2.2 *The NP($\mathscr{S}_p$) problem is solvable if and only if the PM($\mathbf{R}^-$) problem (2.4) associated with the block Hankel vector $\tilde{\boldsymbol{h}}$ is solvable. In this case, there exists a one-to-one correspondence between their solutions $F(\lambda)$ and $\tau(u)$ $(0\leqslant u<+\infty)$ determined by the* *formula*:

$$F(\lambda)=\omega(\lambda)+\int_0^{+\infty}\frac{\mathrm{d}\tau(u)}{u-\lambda}A(\lambda) \tag{2.6}$$

More precisely, if $F(\lambda)$ *is a solution to the* $NP(\mathscr{S}_p)$ *problem and admits an integral representation of the form* (2.1), *then there exists* $\tau(u)$ *such that*

$$\Phi(\lambda)\equiv\frac{-\omega(\lambda)+F(\lambda)}{A(\lambda)}=\int_0^{+\infty}\frac{1}{u-\lambda}\mathrm{d}\tau(u)$$

and

$$\boldsymbol{h}_j=\int_0^{+\infty}u^j\mathrm{d}\tau(u),j=0,1,\cdots,2N-2,$$

$$\boldsymbol{h}_{2N-1}=\int_0^{+\infty}u^{2N-1}\mathrm{d}\tau(u)+\boldsymbol{v}, \tag{2.7}$$

in which $\boldsymbol{v}\geqslant\boldsymbol{O}$ *is identical with that appearing in Eq.* (2.1), *and*

$$\mathrm{d}\tau(u)=\frac{\mathrm{d}\sigma(u)}{A(u)},\quad 0\leqslant u<+\infty.$$

Conversely, each solution $\tau(u)$ $(0\leqslant u<+\infty)$ *to the* $PM(\mathbf{R}^-)$ *problem* (2.4) *satisfying Eq.* (2.7) *with* $\boldsymbol{v}=\boldsymbol{h}_{2N-1}-\int_0^{+\infty}u^{2N}\mathrm{d}\tau(u)\geqslant\boldsymbol{O}$ *therein is associated with the solution* $F(\lambda)$ *to the* $NP(\mathscr{S}_p)$ *problem defined by Eq.* (2.6), *which admits a representation of the form* (2.1) *in which* $\boldsymbol{v}=\boldsymbol{h}_{2N-1}-\int_0^{+\infty}u^{2N-1}\mathrm{d}\tau(u)\geqslant\boldsymbol{O}$ *and*

$$\mathrm{d}\sigma(u)=A(u)\mathrm{d}\tau(u)\quad(0\leqslant u<+\infty). \tag{2.8}$$

In other words, the aforementioned correspondence can be also described briefly by Eq. (2.8).

Proof Suppose first that $F(\lambda)$ is a solution to the $\mathrm{NP}(\mathscr{S}_p)$ problem, and admits the representation (2.1). Then, since $\lim\limits_{u\to\infty}\frac{u(1+u)}{1+u^2}=1$, $\int_0^{+\infty}u(1+u^2)^{-1}\mathrm{d}\sigma(u)$ makes sense, and hence

$$F(\lambda)=\left[\boldsymbol{v}+\int_0^{+\infty}\frac{u}{1+u^2}\mathrm{d}\sigma(u)\right]+\int_{-\infty}^{+\infty}\left(\frac{1}{u-\lambda}-\frac{u}{1+u^2}\right)\mathrm{d}\sigma(u),$$

$$\operatorname{Im}\lambda\neq0,$$

which means that $F(\lambda)$ admits the representation (1. 2) but with $\boldsymbol{\alpha} = \boldsymbol{\nu} + \int_0^{+\infty} u/(1+u^2)\mathrm{d}\sigma(u)$ and $\boldsymbol{\beta} = \boldsymbol{O}$, and therefore $F(\lambda)$ belongs to $\mathcal{N}_p$. From Theorem 1. 3 it follows that there exists a $\tau(u)(-\infty < u < +\infty)$, satisfying $\mathrm{d}\tau(u) = A(\lambda)^{-1}\mathrm{d}\sigma(u) \geqslant \boldsymbol{O}$, and $\mathrm{d}\tau(u) = 0$ for $u < \boldsymbol{O}$, such that Eq. (2. 6) holds. Define in turn

$$\Phi(\lambda) \equiv \frac{-\omega(\lambda) + F(\lambda)}{A(\lambda)} = \int_0^{+\infty} \frac{1}{u-\lambda}\mathrm{d}\tau(u). \tag{2.9}$$

Since

$$\int_0^{+\infty} \frac{1}{1+u}\mathrm{tr}\ \mathrm{d}\tau(u) = \int_0^{+\infty} \frac{1}{(1+u)A(u)}\mathrm{tr}\ \mathrm{d}\sigma(u) < \infty,$$

$\Phi(\lambda) \in \mathcal{S}_p$ by Eqs. (2. 1) and (2. 2) with $\tau(u)$ in place of $\sigma(u)$ therein. Taking into account Eq. (1. 12) and the fact that

$$\lim_{\lambda\to\infty}\left[\frac{F(\lambda)}{A(\lambda)}\lambda^{2N}\right] = \boldsymbol{\nu}$$

uniformly in each sector (1. 3), we derive from Eq. (2. 9) that

$$\lim_{\lambda\to\infty}\lambda^{2N}\left[\Phi(\lambda) + \frac{\boldsymbol{h}_0}{\lambda} + \cdots + \frac{\boldsymbol{h}_{2N-2}}{\lambda^{2N-1}}\right] = -\boldsymbol{h}_{2N-1} + \boldsymbol{\nu} \tag{2.10}$$

uniformly in each sector (1. 3), so that, by Lemma 1. 2, $\tau(u)$ is a solution to the PM($\mathbf{R}^-$) problem (2. 4) with $n = N-1$ such that Eq. (2. 7) holds true.

Conversely, if $\Phi(\lambda) \in \mathcal{S}_p$ is of the form

$$\Phi(\lambda) = \int_0^{+\infty} \frac{1}{u-\lambda}\mathrm{d}\tau(u).$$

where $\tau(u)$ is a solution to the PM($\mathbf{R}^-$) problem with $n = N-1$ such that (2. 7) holds for $\boldsymbol{\nu} = \boldsymbol{h}_{2N-1} - \int_0^{+\infty} u^{2N-1}\mathrm{d}\tau(u) \geqslant \boldsymbol{O}$, then, by using Theorem 1. 3 again, there exists a $\sigma(u)$, satisfying Eq. (2. 8) (and therefore $\mathrm{d}\sigma(u) \geqslant \boldsymbol{O}$ for $u \geqslant \boldsymbol{O}$ and $\mathrm{d}\sigma(u) = \boldsymbol{O}$ for $u < 0$) such that

$$F(\lambda) \equiv \omega(\lambda) + \Phi(\lambda)A(\lambda) = \boldsymbol{\alpha} + \int_0^{+\infty}\left(\frac{1}{u-\lambda} - \frac{u}{1-u^2}\right)\mathrm{d}\sigma(u) \tag{2.11}$$

for a certain Hermitian $\boldsymbol{\alpha}$, where $\sigma(u)$ is such that

$$\int_0^{+\infty} \frac{1}{1+u}\mathrm{tr}\ \mathrm{d}\sigma(u) = \int_0^{+\infty} \frac{A(u)}{1+u}\mathrm{tr}\ \mathrm{d}\tau(u) < \infty.$$

On the other hand, since Eq. (2.10) holds uniformly in each sector (1.3) by Lemma 1.2, it follows from Eqs. (1.12) and (2.11) that

$$\lim_{\lambda\to\infty}\left[\frac{F(\lambda)}{A(\lambda)}\lambda^{2N}\right]=\boldsymbol{v}=\lim_{\lambda\to\infty}F(\lambda)=\boldsymbol{\alpha}-\int_0^{+\infty}\frac{u}{1+u^2}\mathrm{d}\sigma(u)$$

uniformly in each sector (1.3), so that

$$F(\lambda)=\boldsymbol{v}+\int_0^{+\infty}\frac{1}{u-\lambda}\mathrm{d}\sigma(u)\in\mathscr{S}_p.$$

Thus, as $F(\lambda)$ satisfies clearly Eqs. (1.1) and (1.1′) by Eq. (2.11), $F(\lambda)$ is a solution to the NP ($\mathscr{S}_p$) problem, as desired. ■

From Eqs. (2.5) (2.3) and (1.11) we deduce the following theorem.

Theorem 2.3 *Let $\tilde{\boldsymbol{h}}=(\boldsymbol{h}_0,\boldsymbol{h}_1,\cdots,\boldsymbol{h}_{2N-1})$ be the block Hankel vector of the NP($\mathscr{S}_p$) problem. Then the following are equivalent:*

(a) *The NP($\mathscr{S}_p$) problem is solvable.*

(b) $(\boldsymbol{h}_{i+j})_{i,j=0}^{N-1}\geqslant\boldsymbol{O}$ *and* $(\boldsymbol{h}_{i+j+1})_{i,j=0}^{N-1}\geqslant\boldsymbol{O}$.

(c) $\boldsymbol{L}_{\bar{yy}}(\omega(\lambda))\geqslant\boldsymbol{O}$ *and* $\boldsymbol{L}_{\bar{yy}}(\lambda\omega(\lambda))\geqslant\boldsymbol{O}$.

§ 2.3 The Krein-Nudel'man (KN) problem

In the scalar case: $p=1$, it is interesting to recall an endearing proposition, due to Krein and Nudel'man [24, p. 195]. It shows that the KN problem (that is, the classical NP problem in the class $\mathscr{S}_1$, the class $\mathscr{S}_p$ with $p=1$, with simple nodes situated in $(-\infty,0]$) is equivalent to a certain Stieltjes moment problem. We formulate the KN problem in the following manner.

The KN problem: Given real numbers $(0<)\ x_1<x_2<\cdots<x_n$ and $\{c_k\}_{k=1}^n$, it is required to determine a function $F(-\lambda)$ *in* $\mathscr{S}_1$ such that

$$F(-x_k)=c_k,\quad k=1,2,\cdots,n. \tag{2.12}$$

Let now $\hat{\omega}(\lambda)$ be the Hermitian interpolation polynomial, of degree at most $n-1$, of the KN problem subject to Eq. (2.12), and let $\hat{A}(\lambda)$ be the annihilator polynomial defined by

$$\hat{A}(\lambda)=\prod_{k=1}^{n}(\lambda+x_k). \tag{2.13}$$

As before, we refer to $\hat{\boldsymbol{h}}=(h_0,h_1,\cdots,h_{n-1})$ generated by the first n

Markov parameters of $\hat{\omega}(\lambda)/\hat{A}(\lambda)$ as *the Hankel vector* of the KN problem (2.12). More precisely, it follows from [13, p. 180] that

$$h_k = \sum_{j=1}^{n}(-x_k)^k \prod_{i \neq j}(x_i - x_j)^{-1} c_j, \quad k = 0,1,\cdots, \tag{2.14}$$

which are clearly all real.

The Stieltjes moment problem associated with the Hankel vector $\hat{\boldsymbol{h}}$ of the KN problem is required to determine a nondecreasing function $\tau(u)(0 \leqslant u < +\infty)$ such that

$$\begin{aligned} h_i &= \int_0^{+\infty} u^j \mathrm{d}\tau(u), \quad j = 0,1,\cdots,n-2, \\ h_{n-1} &\geqslant \int_0^{+\infty} u^{n-1} \mathrm{d}\tau(u). \end{aligned} \tag{2.15}$$

Theorem 2.4 [24, Proposition 5.1, p. 195] *A sequence* $\{c_k\}_1^n$ *admits a representation*

$$c_k = \nu + \int_0^{+\infty} \frac{\mathrm{d}\sigma(u)}{x_k + u}, \quad k = 1,2,\cdots,n, \tag{2.16}$$

where $\nu \geqslant 0$ *and* $(0 <)\ x_1 < x_2 < \cdots < x_n$, *if, and only if, the sequence*

$$s_k = \sum_{j=1}^{n} a_{jk} c_j, \quad k = 0,1,\cdots,n-1$$

has a solvable Stieltjes moment problem, where numbers a_{jk} *are defined via*

$$\frac{u^k}{\hat{A}(u)} = \sum_{j=1}^{n} \frac{a_{jk}}{x_j + u}, \quad k = 0,1,\cdots,n-1, \quad 0 < x_1 < \cdots < x_n. \tag{2.17}$$

In this case, each representation (2.16) *is associated with a representation*

$$\begin{aligned} s_j &= \int_0^{+\infty} u^j \mathrm{d}\tau(u), \quad j = 0,1,\cdots,n-2, \\ s_{n-1} &= \int_0^{+\infty} u^{n-1} \mathrm{d}\tau(u) + \nu. \end{aligned} \tag{2.18}$$

where

$$\mathrm{d}\tau(u) = \mathrm{d}\sigma(u)/\hat{A}(u). \tag{2.19}$$

Conversely, each representation (2.18) *is associated with a representation* (2.16) *for which*

$$\mathrm{d}\sigma(u) = \hat{A}(u)\mathrm{d}\tau(u).$$

A simple calculation concludes that the moments $\{s_k\}_0^{n-1}$ given in Theorem 2.4 are nothing more than the Hankel vector $\hat{\boldsymbol{h}} = (h_0, h_1, \cdots, h_{n-1})$ defined via Eq. (2.14), since

$$(-x_j)^k = a_{jk} \prod_{i \neq j} (x_i - x_j), \quad \forall k, j.$$

Thus, the point of view presented in this paper can be used to understand and to rewrite down the correspondence given in Theorem 2.4 via the Hankel vector approach.

Theorem 2.4 says indeed that if $F(-\lambda) \in \mathscr{S}_1$ is a solution to the KN problem (2.12), and admits a representation: $F(\lambda) = \nu + \int_0^{+\infty} \frac{\mathrm{d}\sigma(u)}{u+\lambda}$, then and only then $\tau(u)$, defined via Eq. (2.19), satisfies Eq. (2.18), where $(s_0, s_1, \cdots, s_{n-1})$ is identical with the Hankel vector $\hat{\boldsymbol{h}}$ of the KN problem.

Further on, both Theorems 2.4 and 2.2 can be generalized considerably.

§2.4 The NP($\mathscr{S}_p$) problem with both real and non-real interpolation nodes

We are given real numbers $x_1 < x_2 < \cdots < x_\varepsilon (< 0)$ with multiplicities $v_1, v_2, \cdots, v_\varepsilon$, respectively, and $\lambda_1, \lambda_2, \cdots, \lambda_\theta \in \mathbf{C}^+$ with multiplicities $\tau_1, \tau_2, \cdots, \tau_\theta$, respectively, and a set of $p \times p$ matrices $\boldsymbol{D}_{ik} (1 \leqslant i \leqslant \varepsilon, k = 0, 1, \cdots, v_i - 1)$ and $\boldsymbol{C}_{ik} (1 \leqslant i \leqslant \theta, k = 0, 1, \cdots, \tau_i - 1)$, where each $\boldsymbol{D}_{ik}$ is Hermitian. It is required to find conditions to ensure the existence of a function $F(\lambda) \in \mathscr{S}_p$ such that

$$\begin{aligned} &\frac{1}{k!}\frac{\mathrm{d}^k}{\mathrm{d}\lambda^k}F(\lambda)\Big|_{\lambda=x_i} = \boldsymbol{D}_{ik}, \ i = 1, 2, \cdots, \varepsilon, \quad k = 0, 1, \cdots, v_i - 1, \\ &\frac{1}{k!}\frac{\mathrm{d}^k}{\mathrm{d}\lambda^k}F(\lambda)\Big|_{\lambda=\lambda_i} = \boldsymbol{C}_{ik}, \ i = 1, 2, \cdots, \theta, \quad k = 0, 1, \cdots, \tau_i - 1, \end{aligned} \tag{2.20}$$

and to describe the set of solutions when these conditions are met.

The NP ($\mathscr{S}_p$) problem (2.20) contains clearly the NP($\mathscr{S}_p$) problem considered in Section 2.1 and the KN problem (2.12) as two special cases.

Let $M = \sum_{j=1}^{\varepsilon} v_j + 2\sum_{j=1}^{\theta} \tau_j > 1$. We call the unique polynomial matrix $\Omega(\lambda) \in \mathbf{C}^{p\times p}[\lambda]$, of degree at most $M-1$, subject to Eq. (2.20) and

$$\frac{1}{k!}\frac{\mathrm{d}^k}{\mathrm{d}\lambda^k}F(\lambda)\mid_{\lambda=\bar{\lambda}_i} = \boldsymbol{C}_{ik}^*,\ i = 1,2,\cdots,\theta,\quad k = 0,1,\cdots,\tau_i - 1, \tag{2.20'}$$

the Hermitian interpolation polynomial matrix of the NP($\mathscr{S}_p$) problem (2.20). Further on, let $\hat{A}(\lambda)$ be defined as

$$\hat{A}(\lambda) = \prod_{j=1}^{\theta}(\lambda-\lambda_j)^{\tau_j}(\lambda-\bar{\lambda}_j)^{\tau_j}\prod_{j=1}^{\varepsilon}(\lambda-x_j)^{v_j}. \tag{2.21}$$

We refer to the block vector $\check{\boldsymbol{h}} = (\boldsymbol{h}_0, \boldsymbol{h}_1, \cdots, \boldsymbol{h}_{M-1})$ generated by the first M Markov parameters of $\Omega(\lambda)/\check{A}(\lambda)$ as *the block Hankel vector* of the NP ($\mathscr{S}_p$) problem (2.20). The vector $\check{\boldsymbol{h}}$ is clearly Hermitian, i. e., $\boldsymbol{h}_j = \boldsymbol{h}_j^*$, $\forall j$.

The following theorem presents an explicit one-to-one correspondence between solutions to the problem(2.20) and solutions to an associated Stieltjes moment problem with the block Hankel vector $\check{\boldsymbol{h}}$ of the former.

Theorem 2.5 *Let $\check{\boldsymbol{h}} = (\boldsymbol{h}_0, \boldsymbol{h}_1, \cdots, \boldsymbol{h}_{M-1})$ be the block Hankel vector of the NP ($\mathscr{S}_p$) problem (2.20). Then the NP($\mathscr{S}_p$) problem (2.20) is solvable if, and only if, the Stieltjes matrix moment problem associated with $\check{\boldsymbol{h}}$:*

$$\begin{aligned} \boldsymbol{h}_j &= \int_0^{+\infty} u^j \mathrm{d}\tau(u),\quad j = 0,1,\cdots,M-2, \\ \boldsymbol{h}_{M-1} &\geqslant \int_0^{+\infty} u^{M-1}\mathrm{d}\tau(u) \end{aligned} \tag{2.22}$$

is solvable. In this case, there exists a one-to-one correspondence between their solutions $F(\lambda)$ and $\tau(u)$ $(0 \leqslant u < +\infty)$ determined by the formula:

$$F(\lambda) = \Omega(\lambda) + \int_0^{+\infty}\frac{\mathrm{d}\tau(u)}{u-\lambda}\check{A}(\lambda). \tag{2.23}$$

More precisely, if $F(\lambda)\in\mathscr{S}_p$ *is a solution to the problem* (2.20) *and admits a representation*(2.1), *then and only then*

$$\mathrm{d}\tau(u)=\frac{1}{A(u)}\mathrm{d}\sigma(u)\quad(0\leqslant u<+\infty),\tag{2.24}$$

and

$$\boldsymbol{h}_j=\int_0^{+\infty}u^j\,\mathrm{d}\tau(u),\quad j=0,1,\cdots,M-2,\quad \boldsymbol{h}_{M-1}=\int_0^{+\infty}u^{M-1}\mathrm{d}\tau(u)+\boldsymbol{v},$$

where $\boldsymbol{v}\geqslant\boldsymbol{O}$ *is the same as that appearing in Eq.* (2.1).

The proof Theorem 2.5 is only a little different from the one in Theorems 1.3 and 2.2, and therefore will be omitted.

It is known [3] that the Stieltjes matrix moment problem (2.22) is solvable if

$$\boldsymbol{H}_0=(\boldsymbol{h}_{i+j})_{i,j=0}^{[(m-1)/2]}\geqslant\boldsymbol{O},\quad \boldsymbol{H}_1=(\boldsymbol{h}_{i+j+1})_{i,j=0}^{[(m-2)/2]}\geqslant\boldsymbol{O},\tag{2.25}$$

where by $[r]$ denote the integer part of real r. The proof of necessity of the condition (2.25) is very simple by a direct verification. Thus, from Theorem 2.5 we deduce an extension of Theorem 2.3 to a certain extent.

Theorem 2.6 *Let* $\check{\boldsymbol{h}}=(\boldsymbol{h}_0,\boldsymbol{h}_1,\cdots,\boldsymbol{h}_{M-1})$ *be the block Hankel vector of the NP* ($\mathscr{S}_p$) *problem* (2.20). *Then the NP*($\mathscr{S}_p$) *problem* (2.20) *is solvable if, and only if,* (2.25) *holds true.*

Thanks to the correspondence given in Theorem 2.5, the study of the Stieltjes matrix moment problem (2.22) enables one not only to find solvability conditions, but also to get the solutions of the NP($\mathscr{S}_p$) problem (2.20).

The solution of the Stieltjes matrix moment problem (2.22) with M even was first considered by Dyukarev [17] and by Bolotnikov [8], who gave the description of all the solutions for nondegenerate and degenerate cases, respectively. In the case when M is odd, a more detailed discussion is needed and will be dealt in separate investigations.

Acknowledgements

The authors acknowledge with many thanks the helpful comments and suggestions of the referee.

References

[1] N I Akhiezer. The Classical Moment Problem. Oliver and Boyd, London, 1965.

[2] G D Allen, F J Narcowich. R-operators I. Representation theory and applications. Indiana U. Math. J., 1976,25: 945 — 963.

[3] T Ando. Truncated moment problems for operators. Acta Sci. Math. (Szeged), 1970,31:319 — 334.

[4] J A Ball. Nevanlinna-Pick interpolation Generalizations and applications. in: J B Conway, B B Morrell (Eds.), Surveys of Some Recent Results in Operator Theory 1. Pitman, Boston, 1988: 51 — 94.

[5] J A Ball. I Gohberg, L Rodman. Interpolation of Rational Matrix Functions. Vol. 45, Birkhäuser, Basel. OT, 1990.

[6] J A Ball, J W Helton. Interpolation problems of Pick-Nevanlinna and Loewner types for meromorphic matrix functions parametrization of the set of all solutions. Integral Equations and Operator Theory, 1986,9: 155 — 203.

[7] V A Bolotnikov. The two-sided Nevanlinna-Pick problem in the Stieltjes class. in: Oper. Theory Adv. Appl., Vol. 62, Birkäuser, Basel, 1993: 15 — 37.

[8] V A Bolotnikov. Description of the solutions of a degenerate moment problem on the axis and the semiaxis. Teor. Funktsii. Funftsional. Anal. i Prilozhen, 1988,50: 25 — 31.

[9] V A Bolotnikov. On degenerate Hamburger moment problem and extensions of nonnegative Hankel block matrices, Integral Equations and Operator Theory, 1996,25: 253 — 276.

[10] G-N Chen. The general rational interpolation problem and its connection with the Nevanlinna-Pick interpolation and power moment problem. Linear Algebra Appl., 1998, 273: 83 — 117.

[11] G-N Chen, Y-J Hu. The truncated Hamburger moment problem in both the nondegenerate and degenerate cases, and matrix continued fractions. Linear Algebra Appl., 1998,277: 199 — 236.

[12] G-N Chen, H-P Zhang. More on Loewner matrices. Linear Algebra Appl., 1994, 203 — 204; 265 — 300.

[13] G-N Chen, B Zhao, H-P Zhang. The general rational interpolation problem in the scalar case and its Hankel vector. Linear Algebra Appl.,1996. 244: 165 — 201.

[14] Ph Delsarte, Y Genin, Y Kamp. The Nevanlinna-Pick problem for

matrix-valued functions. SIAM J. Applied Math., 1979,36: 47－61.

[15] H Dym. J contractive matrix functions, reproducing Kernel Hilbert spaces and interpolation. CBMS Regional Conference Series in Math., Vol. 71, American Mathematical Society, Providence, RI, 1989.

[16] H Dym. On Hermitian block Hankel matrices, matrix polynomials, the Hamburger moment problem, interpolation and maximum entropy. Integral Equations and Operator Theory, 1989,12:757－812.

[17] Yu N Dyukarev. The matrix Stieltjes moment problem. Manuscript deposited at VINITI, No. 2628, 22 March 1981, Moscow, 1981.

[18] Yu N Dyukarev, V E Katsnel'son. Multiplicative and additive classes of Stieltjes analytic matrix valued functions, and interpolation problems associated with them. Amer. Math. Soc. Transl., 1986,131(2): 55－70.

[19] C Foias, A E Frazho. The Commutant Lifting Approach to Interpolation Problems. Vol. 44, Birkhäuser, Basel, OT, 1990.

[20] L B Golinskii. On one generalization of the Nevanlinna-Pick matrix problem. Izvest. Akad. Nauk. Arm. SSR., 1983,18: 187－205.

[21] T S lvanchenko, L A Sakhnovich. An operator approach to the Potapov scheme for the solution of interpolation problems. in: I. Gohberg, L. A. Sakhnovich (Eds.), Matrix and Operator Valued Functions, Oper. Theory Adv. Appl., 72, Birkhäuser Basel, 1994: 48－86.

[22] I V Kovalishina. Analytic theory of a class of interpolation problems. Math. USSR. Izv, 1984,22:419－463.

[23] I V Kovalishina, V P Potapov. An indefinite metric in the Nevanlinna-Pick problem. Akad. Nauk, Arm. SSR. Dokl., 1974,59: 17－22.

[24] M G Krein, A A Nudel'man. The Markov Moment Problem and Extremal Problems. Transl. Math. Monogr. 50, American Mathematical Society, Providence, 1977.

[25] M G Krein, A A Nudel'man. On an interpolation problem related to the Stieltjes moment problem. Doklady Akad Nauk Ukrainian SSR., 1977, 12A:1 068－1 072.

[26] M G Krein, A A Nudel'man. An interpolation problem in the class of Stieltjes functions and its connection with other problems. Integral Equations and Operator Theory, 1998, 30:251－278.

[27] F J Narcowich, R-operators II. On the approximation of certain operator-valued analytic functions and the Hermitian moment problem, Indiana U. Math. J., 1977,26:483－513.

[28] A A Nudel'man. Some generalizations of classical interpolation problems. Oper Theory Adv. Appl., 61, Birkhäuser Basel, 1993: 171－188.

[29] M Rosenblum, J Rovnyak. An operator theoretic approach to theorems of

the Pick-Nenanlinna and Loewner types, Ⅰ,Ⅱ. Integral Equations and Operator Theory,1980,3:408－436;1982,5:870－887.

[30] M Rosenblum, J Rovnyak. Hardy Classes and Operator Theory. Oxford University Press. Oxford, 1985.

[31] D Sarason. Generalized interpolation in H^{∞}. Trans. Amer. Math. Soc., 1967,127(2):180－203.

[32] D Sarason. Nevanlinna-Pick interpolation with boundary data. Integral Equations and Operator Theory, 1998,30:231－250.

[33] N Young. The Nevanlinna-Pick problem for the matrix valued function. J. Operator Theory,1986,15:239－265.

Linear Algebra and Its Applications
1998;277,199－236

非退化与退化情形截断 Hamburger 矩阵矩量问题与矩阵连分式[①]

The Truncated Hamburger Matrix Moment Problems in the Nondegenerate and Degenerate Cases, and Matrix Continued Fractions

Abstract The present paper deals simultaneously with the nondegenerate and degenerate truncated Hamburger matrix moment problems in a unified way based on the use of the Schur algorithm involving matrix continued fractions. A full analysis of them together with a relative matrix moment problem on the real axis is given. With the help of the correspondence between the moment problem on the real axis and the Nevanlinna-Pick(NP) interpolation, the solutions of the nontangential NP interpolation in the Nevanlinna class are derived as an application.

Keywords Hamburger matrix moment problem; Matrix continued fraction; Nevanlinna class; Nonnegatively extendable; Nevanlinna-Pick interpolation; Block-Hankel vector

§ 1. Introduction and preliminaries

The natural matrix version of the classical Hamburger moment

---

① Received 4 February 1997; accepted 24 September 1997. Partially supported by the National Natural Science Foundation of China. 本文与胡永建合作. Submitted by R. Brualdi.

problem (see, e.g., [2]) consists in finding a bounded Hermitian measure $\sigma(u)(-\infty<u<+\infty)$ such that

$$S_k = \int_{-\infty}^{+\infty} u^k \mathrm{d}\sigma(u), \quad k = 0,1,\cdots \tag{1.1}$$

Of particular importance are the questions of solvability, of the number of solutions, and of the construction of all solutions if they exist.

In the present paper we give a careful treatment of the truncated Hamburger matrix moment problem (TH problem), where one asks to describe the solutions $\sigma(u)(-\infty<u<+\infty)$ having precribed matrix moments $S_0, S_1, \cdots, S_{2n}$ only. A variation on the problem with a more natural mathematical solution is where one asks only for an inequality "$\leqslant$" in the moment condition for $k=2n$. Also very different is the theory for the case where one prescribes the moments up to an odd index $k=2n+1$. The present paper deals with the case of maximum k equal to $2n$ only and primarily with equality on the last moment condition.

The subject now has a long history and has given rise to important applications in many branches of analysis and others. In the scalar case: $p=1$, the reader may consult to the fundamental books [2,19,21] for details of numerous moment problems. In the matrix case, the TH problem for operators was first studied systematically by Ando [4], who gave the conditions for the problem to be solvable. In the nondegenerate case, Kovalishina [17] has presented a method for the solution of the Hamburger moment problem(1.1)based on the use of the fundamental matrix inequality and the analytic J-theory. Bolotnikov[6] treated the solution of the TH problem for the degenerate case by means of the Schur's stepwise algorithm. (Unfortunately, one of his basic results, Theorem 1.1 of [6], is untrue for the TH problem even for the scalar case.) Further on, the TH problem can be solved using a number of other approaches, e.g. reproducing kernels method [12,13], methods based on operator theory[2,18], or on realization theory of matrix-valued functions [5]. There has also been recent work of Kheifets [16], where the equality versus inequality issue is analyzed via a different approach (embedding the problem in the so-called abstract interpolation problem which identifies solutions of the TH problem with unitary colligation extensions of a given isometric colligation directly constructible from the data of the problem).

In the present paper we present a common method of solving simultaneously the nondegenerate and degenerate truncated Hamburger matrix moment problems based on the use of a matrix version of the Schur algorithm involving matrix continued fractions, suitably adapted to the present framework. Some new results and more explicit version of known results for the TH problem and for its variation mentioned above are given. The notion of the nonnegatively extendable Hermitian-block Hankel matrix plays a key role in our investigations. It turns out that the TH problem has a solution if and only if the corresponding block-Hankel matrix is nonnegatively extendable, i. e., if and only if the block-Hankel matrix can be enlarged to a block-Hankel matrix of larger size(one more block-row and-column) which is still nonnegative definite. As the main result, the general solution $\sigma(u)$ to the TH problem is formulated in both an algorithm form and a more closed form (Theorems 3.4 and 4.12). With the help of the one-to-one correspondence between the moment problem and the Nevanlinna-Pick (NP) interpolation problem [10] the solution of the NP problem in the class of the Nevanlinna matrix-valued functions is acheived as an application.

Although the concern of this paper is mainly the "truncated" moment problems, our descriptions of the solutions in terms of matrix continued fractions are complete to the degree that transition to the infinite moment problems becomes quite transparent.

This work could reveal certain interesting connections with other problems in analysis in addition to the interpolation problem of NP type, for example to the work of Adamjan-Arov-Krein [1] on the reduction of the Nehari problem to the contractive completion problem for infinite, contractive block-Hankel matrices, where the issue is to extend an infinite block-Hankel matrix by one more block-row while maintaining the property of being a contraction. In the Nehari problem case, contractivity of the original matrix guarantees the extendability property. This presents an interesting theory parallel to the theory of the TH problem. As was mentioned above, the corresponding phenomenon for the TH problem situation fails—nonnegativity of the original block-Hankel matrix in general does not guarantee that it admits an extension of a larger block-Hankel matrix which is still nonnegative. This illustrates the subtlely of the TH problem compared to a seemingly analogous Nehari problem.

To introduce the discussion, we devote Section 2 to establish the various auxiliary propositions useful for our investigations later on. The criteria of existence and uniqueness of the solutions to the TH problem, and a unified description of the solutions to the TH problems in the non-degenerate and degenerate cases are settled in Section 3. As a consequence, the solution of the matrix moment problem on the real axis is also derived with little additional effort. In Section 4 it is shown how the descriptions of the solutions to the moment problems in terms of matrix continued fractions are interlaced with other existing ones in terms of linear fractional transformations. Some properties of nonnegatively extendable Hermitian block-Hankel matrices are given as well. Section 5 traces out an application of our results on moment problems to the solution of the NP problem in the class of the Nevanlinna matrix-valued functions with the help of the correspondence between the moment problem on the real axis and that NP interpolation.

The following notation and conventions will be used throughout the paper. All matrices and vectors are assumed to be complex. The positive semidefiniteness (definiteness) of a Hermitian matrix $\boldsymbol{A}$ is denoted by $\boldsymbol{A}\geqslant\boldsymbol{O}(\boldsymbol{A}>\boldsymbol{O})$. By π^+ denote the open upper-half plane, its subset $\{z\,|\,\varepsilon\leqslant\arg z\leqslant\pi-\varepsilon,\varepsilon\in(0,\pi/2)\,|\}$ by π_ε. For a square matrix $\boldsymbol{B}$, the symbol $\boldsymbol{B}^{\mathrm{D}}$

denotes its Drazin inverse, i. e. , the unique solution of the equations: $\boldsymbol{XBX}=\boldsymbol{X}$, $\boldsymbol{BX}=\boldsymbol{XB}$, and $\boldsymbol{B}^k=\boldsymbol{XB}^{k+1}$, $k=\text{index}(\boldsymbol{B})$. For a polynomial matrix $\boldsymbol{A}(\lambda)=\sum_{k=0}^{m}\boldsymbol{A}_k\lambda^k$, $\boldsymbol{A}_k\in\mathbf{C}^{p\times p}$, the symbol $\hat{A}(\lambda)$ designates the polynomial matrix $\sum_{k=0}^{m}\boldsymbol{A}_k^*\lambda^k$. The notation $\dfrac{\boldsymbol{B}}{\boldsymbol{A}}$ stands for $\boldsymbol{BA}^{-1}$ if $\boldsymbol{A},\boldsymbol{B}$ are $p\times p$ matrices and $\boldsymbol{A}$ is nonsingular, and the following notation for the matrix continued fraction:

$$\frac{\boldsymbol{B}_1}{\boldsymbol{A}_1}\pm\frac{\boldsymbol{B}_2}{\boldsymbol{A}_2}\pm\frac{\boldsymbol{B}_3}{\boldsymbol{A}_3}\pm\cdots\pm\frac{\boldsymbol{B}_k}{\boldsymbol{A}_k}\pm\cdots,$$

in which all fractions are assumed to be meaningful. A $\mathbf{C}^{p\times p}$-valued function $\sigma(u)$ defined on the real axis in called a *Hermitian measure* if is nondecreasing, i. e. , $\sigma(\lambda)-\sigma(\mu)\geqslant\mathbf{O}$ for all $\lambda>\mu$. Without being pointed out explicitly, all Hermitian measures in integral formulas of this paper are assumed to be bounded: $\int_{-\infty}^{+\infty}\operatorname{tr}\,\mathrm{d}\sigma(u)<+\infty$. A$\mathbf{C}^{p\times p}$-valued function $F(\lambda)$ is of the *Nevanlinna class* $\mathbf{N}_p$ if it is analytic in π^+ and such that

$$\frac{F(\lambda)-F(\lambda)^*}{2\mathrm{i}}\geqslant\mathbf{O},\quad \lambda\in\pi^+.$$

Each $F(\lambda)\in\mathbf{N}_p$ can be continued onto the open lower-half plane $\operatorname{Im}\lambda<0$ by reflection: $F(\lambda)=F(\bar{\lambda})^*$, $\operatorname{Im}\lambda<0$, and it admits an integral representation

$$F(\lambda)=\boldsymbol{\alpha}\lambda+\boldsymbol{\beta}+\int_{-\infty}^{+\infty}\frac{1+u\lambda}{u-\lambda}\mathrm{d}\tau(u),\quad \operatorname{Im}\lambda\neq 0, \tag{1.2}$$

where $\boldsymbol{\alpha}\geqslant 0$, $\boldsymbol{\beta}=\boldsymbol{\beta}^*$, $\tau(u)$ is suitable Hermitian measure. The collection of functions $F(\lambda)\in\mathbf{N}_p$, represented in the from

$$F(\lambda)=\int_{-\infty}^{+\infty}\frac{1}{u-\lambda}\mathrm{d}\sigma(u) \tag{1.3}$$

forms a subclass $\mathbf{N}_p^0$ of the class $\mathbf{N}_p$, where $\sigma(u)$ is as before. This subclass. $\mathbf{N}_p^0$ can be characterized intrinsically: in order that $F(\lambda)\in\mathbf{N}_p$ belongs to N_p^0, it is necessary and sufficient that

$$\sup_{y>0}\|yF(iy)\|<+\infty, \tag{1.4}$$

where $\|\cdot\|$ is a certain matrix norm on $\mathbf{C}^{p\times p}$ (see [2] for the scalar case. The extension to the matrix-valued functions causes no difficulty).

§ 2. Some important lemmata

we shall establish some auxiliary propositions which are useful later on.

To begin with, we formulate the TH problem on the real axis. Given a sequence of $p\times p$ Hermitian matrices, $\boldsymbol{S}_0, \boldsymbol{S}_1, \cdots, \boldsymbol{S}_{2n}$, we seek a Hermitian measure $\sigma(u)(-\infty<u<+\infty)$ such that

$$\boldsymbol{S}_k = \int_{-\infty}^{+\infty} u^k \mathrm{d}\sigma(u), \quad k = 0,1,\cdots,2n. \tag{2.1}$$

It is required to find conditions for a solution and for a unique solution $\sigma(u)$ to exist, and to describe the solutions when these conditions are met.

The well-known Hamburger-Nevanlinna theorem shows the close relation of the TH problem (2.1) to the following problem: Given a sequence of $p\times p$ Hermitian matrices, $\boldsymbol{S}_0, \boldsymbol{S}_1, \cdots, \boldsymbol{S}_{2n}$, it is required to describe all $F(\lambda)\in\mathbf{N}_p$ with a given asymptotical expansion of the form

$$F(\lambda) = -\frac{\boldsymbol{S}_0}{\lambda} - \frac{\boldsymbol{S}_1}{\lambda^2} - \cdots - \frac{\boldsymbol{S}_{2n}}{\lambda^{2n+1}} - R_{2n+1}(\lambda), \tag{2.2}$$

where $R_{2n+1}(\lambda) = o(\lambda^{-2n-1})$ as $\lambda\to\infty$ uniformly in each sector π_ε.

In what follows, all other asymptotic expansions will be assumed to be applicable in the same range of sectors π_ε as also will the symbol $o(\lambda^{-m})$.

Lemma 2.1 (Hamburger-Nevanlinna)[2,17] *If $\sigma(u)(-\infty<u<+\infty)$ is a solution to the TH problem (2.1), then there exists $F(\lambda)\in\mathbf{N}_p$*

$$F(\lambda) = \int_{-\infty}^{+\infty} \frac{1}{u-\lambda} \mathrm{d}\sigma(u) \tag{2.3}$$

for which

$$\lim_{\lambda\to\infty} \lambda^{2n+1}\left[F(\lambda) + \frac{\boldsymbol{S}_0}{\lambda} + \frac{\boldsymbol{S}_1}{\lambda^2} + \cdots + \frac{\boldsymbol{S}_{2n-1}}{\lambda^{2n}}\right] = -\boldsymbol{S}_{2n} \tag{2.4}$$

uniformly in each π_ε. Conversely, if (2.4) holds, at least for $\lambda = \mathrm{i}y$ $(y\to+\infty)$, for some $F(\lambda)\in\mathbf{N}_p$, then $F(\lambda)$ has the representation (2.3), where $\sigma(u)$ has $2n+1$ moments $\boldsymbol{S}_0, \boldsymbol{S}_1, \cdots, \boldsymbol{S}_{2n}$.

Lemma 2.2 *Suppose that $F(\lambda)\in\mathbf{N}_p$ admits the representation* (1.2).

If either $\boldsymbol{\alpha}>\boldsymbol{O}$ *or* $\int_{-\infty}^{+\infty}\mathrm{d}\tau(u)>\boldsymbol{O}$, *then* $-F^{-1}(z)$ *exists in* π^{+} *and belongs to the class* $\mathbf{N}_p$.

Proof Observe that for each $\lambda=a+b\mathrm{i}\in\pi^{+}(a,b\in\mathbb{R},b>0)$,

$$\operatorname{Im}F(\lambda)=\boldsymbol{\alpha}b+\int_{-\infty}^{+\infty}\frac{b^2(1+u^2)}{(u-a)^2+b^2}\mathrm{d}\tau(u).$$

If $\boldsymbol{\alpha}>\boldsymbol{O}$, $\operatorname{Im}F(\lambda)>\boldsymbol{O}$, $\forall\lambda\in\pi^{+}$. If $\int_{-\infty}^{+\infty}\mathrm{d}\tau(u)>\boldsymbol{O}$, $\int_{-\delta}^{\delta}\mathrm{d}\tau(u)>\boldsymbol{O}$ for a certain $\delta>0$, and therefore

$$\operatorname{Im}F(\lambda)\geqslant\frac{b^2}{(\delta+|a|)^2+b^2}\int_{-\delta}^{\delta}\mathrm{d}\tau(u)>\boldsymbol{O}.$$

On the other hand, we have that $|\det F(\lambda)|=|\det(\mathrm{i}F(\lambda))|\geqslant\det\operatorname{Im}F(\lambda)>0$ if $\boldsymbol{\alpha}>\boldsymbol{O}$ or $\int_{-\infty}^{+\infty}\mathrm{d}\tau(u)>\boldsymbol{O}$, and therefore $-F^{-1}(\lambda)$ exists and is analytic in π^{+} under the hypothesis. Finally, the fact that $\operatorname{Im}(-F^{-1}(\lambda))\geqslant\boldsymbol{O}$, $\forall\lambda\in\pi^{+}$ follows from the relation

$$\operatorname{Im}(-F^{-1}(\lambda))=F^{-1}(\lambda)\operatorname{Im}F(\lambda)F^{-1}(\lambda)^{*}\geqslant\boldsymbol{O},\quad\lambda\in\pi^{+}.$$

Hence, $-F^{-1}(\lambda)\in\mathbf{N}_p$. ■

Lemma 2.3 *Suppose that the Hermitian block-Hankel matrix* $\boldsymbol{\Gamma}_k=(\boldsymbol{S}_{i+j})_{i,j=0}^{k}$ *is nonnegative Hermitian*: $\boldsymbol{\Gamma}_k\geqslant\boldsymbol{O}$, *where each* $\boldsymbol{S}_k$ *of order p is Hermitian. If* $\boldsymbol{S}_0=\boldsymbol{O}$, *then* $\boldsymbol{S}_1=\cdots=\boldsymbol{S}_{2k-1}=\boldsymbol{O}$; *if* $\boldsymbol{S}_0\neq\boldsymbol{O}$ *is singular, then there exists a certain nonsingular* $\boldsymbol{T}$ *such that*

$$\boldsymbol{T}^{*}\boldsymbol{S}_i\boldsymbol{T}=\begin{bmatrix}\tilde{\boldsymbol{S}}_i & \boldsymbol{O}\\ \boldsymbol{O} & \boldsymbol{O}\end{bmatrix},\ i=0,1,\cdots,2k-1,\quad \boldsymbol{T}^{*}\boldsymbol{S}_{2k}\boldsymbol{T}=\begin{bmatrix}\tilde{\boldsymbol{S}}_{2k} & \boldsymbol{O}\\ \boldsymbol{O} & \boldsymbol{D}\end{bmatrix},\tag{2.5}$$

where $\tilde{\boldsymbol{S}}_0>\boldsymbol{O},\boldsymbol{D}\geqslant\boldsymbol{O}$, *and all* $\tilde{\boldsymbol{S}}_i$ *are of the same order.*

We remark that there exists a misstatement in [6](p. 1255) in constructing Schur's stepwise algorithm. He asserted that if $\det\boldsymbol{S}_0=0$ then from $\boldsymbol{\Gamma}_k\geqslant\boldsymbol{O}$ there follows the existence of a unitary matrix $\boldsymbol{U}$ such that

$$\boldsymbol{U}^{*}\boldsymbol{S}_i\boldsymbol{U}=\begin{bmatrix}\tilde{\boldsymbol{S}}_i & \boldsymbol{O}\\ \boldsymbol{O} & \boldsymbol{O}\end{bmatrix},\quad i=0,1,\cdots,2k;\quad \tilde{\boldsymbol{S}}_0>\boldsymbol{O}.$$

This, however, is not correct, as an example.

$$\boldsymbol{\Gamma}_2=\begin{bmatrix}\boldsymbol{O} & \boldsymbol{O}\\ \boldsymbol{O} & \boldsymbol{I}_2\end{bmatrix}\geqslant\boldsymbol{O}.$$

The proof of Lemma 2.3 *If* $S_0=O$ then $S_1=\cdots=S_{2k-1}=O$ evidently. If $S_0\neq O$ is singular, then there exists a certain unitary matrix V such that

$$V^*S_0V=\begin{bmatrix}\tilde{S}_0 & O\\ O & O\end{bmatrix},\quad \tilde{S}_0>O.$$

From the fact $(I_p\otimes V)^*\Gamma_k(I_p\otimes V)\geqslant O$, where $A\otimes B$ stands for the tensor product of a pair of matrices A and B, we obtain that

$$V^*S_iV=\begin{bmatrix}\tilde{S}_i & O\\ O & O\end{bmatrix},\qquad i=0,1,\cdots,2k-1,$$

where all $\tilde{S}_i$ have the same order as that of $\tilde{S}_0$. Let us write $V^*S_{2k}V$ in the block from

$$V^*S_{2k}V=\begin{bmatrix}A_{11} & A_{12}\\ A_{12}^* & A_{22}\end{bmatrix}\geqslant O,$$

where order A_{11} = order $\tilde{S}_0$. We distinguish three cases. In the case $A_{22}>O$, we have

$$P=\begin{bmatrix}I & O\\ -A_{22}^{-1}A_{12}^* & I\end{bmatrix}$$

such that

$$P^*V^*S_{2k}VP=\begin{bmatrix}\tilde{S}_{2k} & O\\ O & A_{22}\end{bmatrix}$$

in which $\tilde{S}_{2k}=A_{11}-A_{12}A_{22}^{-1}A_{12}^*$. In the case $A_{22}=O$, then $A_{12}=O$, and thus

$$V^*S_iV=\begin{bmatrix}A_{11} & O\\ O & O\end{bmatrix}.$$

In the case when $A_{22}\neq O$ is singular, there exists a unitary matrix $\tilde{V}$ such that

$$\tilde{V}^*A_{22}\tilde{V}=\begin{bmatrix}\tilde{A}_{22} & O\\ O & O\end{bmatrix},\quad \tilde{A}_{22}>O,$$

and thus

$$\begin{bmatrix}I & O\\ O & \tilde{V}^*\end{bmatrix}V^*S_{2k}V\begin{bmatrix}I & O\\ O & \tilde{V}\end{bmatrix}=\begin{bmatrix}\tilde{A}_{11} & \tilde{A}_{12} & O\\ \tilde{A}_{12}^* & \tilde{A}_{22} & O\\ O & O & O\end{bmatrix},\quad \tilde{A}_{22}>O.$$

So, like the first case, there exists a nonsingular matrix $\boldsymbol{P}$ such that

$$\boldsymbol{P}^*\begin{bmatrix}\widetilde{\boldsymbol{A}}_{11} & \widetilde{\boldsymbol{A}}_{12} & \boldsymbol{O}\\ \widetilde{\boldsymbol{A}}_{12}^* & \widetilde{\boldsymbol{A}}_{22} & \boldsymbol{O}\\ \boldsymbol{O} & \boldsymbol{O} & \boldsymbol{O}\end{bmatrix}\boldsymbol{P}=\begin{bmatrix}\widetilde{\boldsymbol{S}}_{2k} & & \\ & \widetilde{\boldsymbol{A}}_{22} & \boldsymbol{O}\\ & \boldsymbol{O} & \boldsymbol{O}\end{bmatrix},$$

where order $\widetilde{\boldsymbol{S}}_{2k}$ = order $\widetilde{\boldsymbol{A}}_{11}$. Thus, we complete the proof by setting $\boldsymbol{T}=\boldsymbol{VP}$ in the first two cases and

$$\boldsymbol{T}=\boldsymbol{V}\begin{bmatrix}\boldsymbol{I} & \boldsymbol{O}\\ \boldsymbol{O} & \widetilde{\boldsymbol{V}}\end{bmatrix}\boldsymbol{P}$$

in the last case. ■

The key role of the TH Problem is played by the following notion.

Definition 2.4 A Hermitian block-Hankel matrix $\boldsymbol{\Gamma}_k=(\boldsymbol{S}_{i+j})_{i,j=0}^{k}$, is called nonnegatively extendable (n. e., for short) if there exist Hermitian matrices $\boldsymbol{S}_{2k+1},\boldsymbol{S}_{2k+2}$ of order p such that $\boldsymbol{\Gamma}_{k+1}=(\boldsymbol{S}_{i+j})_{i,j=0}^{k+1}\geqslant\boldsymbol{O}$.

It is clear that $\boldsymbol{\Gamma}_k>\boldsymbol{O}$ implies the nonnegative extendability of $\boldsymbol{\Gamma}_k$, and that the latter in turn implies $\boldsymbol{\Gamma}_k\geqslant\boldsymbol{O}$, but each of the converse propositions is not true, for instance,

$$\boldsymbol{\Gamma}_2=\begin{bmatrix}\boldsymbol{O} & \boldsymbol{O}\\ \boldsymbol{O} & \boldsymbol{O}\end{bmatrix}\quad and\quad \boldsymbol{\Gamma}_2=\begin{bmatrix}\boldsymbol{O} & \boldsymbol{O}\\ \boldsymbol{O} & \boldsymbol{I}_2\end{bmatrix}.$$

In the scalar case: $p=1$, $\boldsymbol{\Gamma}_k$ is n. e., if and only if either $\boldsymbol{\Gamma}_k>\boldsymbol{O}$ or $\boldsymbol{\Gamma}_k\geqslant\boldsymbol{O}$ is singular and proper, i. e., the leading principal submatrix of $\boldsymbol{\Gamma}_k$ of order $m=\operatorname{rank}\boldsymbol{\Gamma}_k$ is positive Hermitian (see [8] for details).

In the Sections 4 and 5 we shall characterize the nonnegative extendability of a Hermitian block-Hankel matrix (see Theorem 3.9 and Corollary 4.10).

In the case when $\boldsymbol{\Gamma}_k$ is n. e., from Lemma 2.3 we obtain

Corollary 2.5 *Suppose* $\boldsymbol{\Gamma}_k$ *is n. e. If* $\boldsymbol{S}_0=\boldsymbol{O}$, *then* $\boldsymbol{\Gamma}_k=\boldsymbol{O}$. *If* $\boldsymbol{S}_0\neq\boldsymbol{O}$, *there exists a unitary matrix* $\boldsymbol{U}$ such that

$$\boldsymbol{U}^*\boldsymbol{S}_k\boldsymbol{U}=\begin{bmatrix}\widetilde{\boldsymbol{S}}_i & \boldsymbol{O}\\ \boldsymbol{O} & \boldsymbol{O}\end{bmatrix},\quad i=0,1,\cdots,2k,\tag{2.6}$$

in which $\widetilde{\boldsymbol{S}}>\boldsymbol{O}$ and all $\widetilde{\boldsymbol{S}}_k$ *have the same order as that of* $\widetilde{\boldsymbol{S}}_0$.

Lemma 2.6 *Suppose that* $F(\lambda)\in\mathbf{N}_p$ *has the asymptotical expansion*

(2.2), $n \geqslant 1$. *Then either* $\boldsymbol{S}_0 = \boldsymbol{O}$ *and* $F(\lambda) \equiv \boldsymbol{O}$ *or* $\boldsymbol{S}_0 \neq \boldsymbol{O}$ *and*

$$F(\lambda) = -\frac{\boldsymbol{S}_0}{\lambda \boldsymbol{I}_P - \boldsymbol{S}_0^D \boldsymbol{S}_1 + \boldsymbol{S}_0^D F_1(\lambda)}, \tag{2.7}$$

where $F_1(\lambda) \in \mathbf{N}_p$, *and*

$$F_1(\lambda) = -\frac{\boldsymbol{S}_0^{(1)}}{\lambda} - \frac{\boldsymbol{S}_1^{(1)}}{\lambda^2} - \cdots - \frac{\boldsymbol{S}_{2n-2}^{(1)}}{\lambda^{2n-1}} + o(\lambda^{-2n+1}) \tag{2.8}$$

in which $\boldsymbol{S}_0^{(1)}, \boldsymbol{S}_1^{(1)}, \cdots, \boldsymbol{S}_{2n-2}^{(1)}$ *are defined by the formula*:

$$\boldsymbol{\Gamma}_{n-1}^{(1)} = [\boldsymbol{S}_{i+j}^{(1)}]_{i,j=0}^{n-1} = \begin{bmatrix} \boldsymbol{S}_0 & & \boldsymbol{O} \\ & \ddots & \\ \boldsymbol{O} & & \boldsymbol{S}_0 \end{bmatrix} \begin{bmatrix} \boldsymbol{S}_0 & & \boldsymbol{O} \\ \vdots & \ddots & \\ \boldsymbol{S}_{n-1} & \cdots & \boldsymbol{S}_0 \end{bmatrix}^D \times$$

$$\left\{ [\boldsymbol{S}_{i+j}]_{i,j=1}^{n} - \begin{bmatrix} \boldsymbol{S}_1 \\ \vdots \\ \boldsymbol{S}_n \end{bmatrix} \boldsymbol{S}_0^D [\boldsymbol{S}_1, \boldsymbol{S}_2, \cdots, \boldsymbol{S}_n] \right\} \times$$

$$\begin{bmatrix} \boldsymbol{S}_0 & \cdots & \boldsymbol{S}_{n-1} \\ & \ddots & \vdots \\ \boldsymbol{O} & & \boldsymbol{S}_0 \end{bmatrix}^D \begin{bmatrix} \boldsymbol{S}_0 & & \boldsymbol{O} \\ & \ddots & \\ \boldsymbol{O} & & \boldsymbol{S}_0 \end{bmatrix}. \tag{2.9}$$

Proof In the case of $\boldsymbol{S}_0 = \boldsymbol{O}, F(\lambda) \equiv \boldsymbol{O}$ follows from Lemma 2.1. In the case of $\boldsymbol{S}_0 \neq \boldsymbol{O}$, we may make the assumption that $\boldsymbol{S}_0 > \boldsymbol{O}$ throughout our proof (for if $\boldsymbol{S}_0 \geqslant \boldsymbol{O}$ is nonzero and singular, by Lemma 2.1, then there exist $\sigma(u)$ such that $\boldsymbol{S}_i = \int_{-\infty}^{+\infty} u^i \mathrm{d}\sigma(u) \, (i = 0, 1, \cdots, 2n)$ and a unitary matrix $\boldsymbol{U}$ such that

$$\boldsymbol{U}^* \boldsymbol{S}_0 \boldsymbol{U} = \begin{bmatrix} \tilde{\boldsymbol{S}}_0 & \boldsymbol{O} \\ \boldsymbol{O} & \boldsymbol{O} \end{bmatrix}, \quad \tilde{\boldsymbol{S}}_0 > \boldsymbol{O}.$$

Thus

$$\boldsymbol{U}^* \mathrm{d}\sigma(u) \boldsymbol{U} = \begin{bmatrix} \mathrm{d}\tilde{\sigma}(u) & \boldsymbol{O} \\ \boldsymbol{O} & \boldsymbol{O} \end{bmatrix},$$

so that

$$\boldsymbol{U}^* \boldsymbol{S}_i \boldsymbol{U} = \begin{bmatrix} \tilde{\boldsymbol{S}}_i & \boldsymbol{O} \\ \boldsymbol{O} & \boldsymbol{O} \end{bmatrix}, \quad i = 0, 1, \cdots, 2n,$$

where $\tilde{\boldsymbol{S}}_i = \int_{-\infty}^{+\infty} u^i \mathrm{d}\tilde{\sigma}(u)$ of order equal to rank $\tilde{\boldsymbol{S}}_0$). Then by Lemma 2.1

again, there exists a Hermitian measure $\sigma(u)(-\infty<u<\infty)$ such that

$$F(\lambda)=\int_{-\infty}^{+\infty}\frac{1}{u-\lambda}\mathrm{d}\sigma(u)$$

and

$$\boldsymbol{S}_0=\int_{-\infty}^{+\infty}\mathrm{d}\sigma(u)>\boldsymbol{O},$$

whence $-F^{-1}(\lambda)\in\mathbf{N}_p$ by Lemma 2. 2. Suppose now that $-F^{-1}(\lambda)$ permits the representation(1. 2). It is readily seen that (2. 7) holds with

$$F_1(\lambda)=-\boldsymbol{S}_0F^{-1}(\lambda)\boldsymbol{S}_0-\boldsymbol{S}_0\lambda+\boldsymbol{S}_1$$
$$=(\boldsymbol{S}_0\boldsymbol{\alpha}\boldsymbol{S}_0-\boldsymbol{S}_0)\lambda+(\boldsymbol{S}_0\boldsymbol{\beta}\boldsymbol{S}_0+\boldsymbol{S}_1)+\int_{-\infty}^{+\infty}\frac{1+u\lambda}{u-\lambda}\mathrm{d}(\boldsymbol{S}_0\tau(u)\boldsymbol{S}_0). \tag{2.10}$$

In view of the fact that

$$\boldsymbol{\alpha}=\lim_{y\to+\infty}\frac{-F^{-1}(\mathrm{i}y)}{\mathrm{i}y}=\lim_{y\to+\infty}\frac{-\boldsymbol{I}_p}{\mathrm{i}yF(\mathrm{i}y)}=\boldsymbol{S}_0^{-1},$$

$F_1(\lambda)$ has the from

$$F_1(\lambda)=(\boldsymbol{S}_0\boldsymbol{\beta}\boldsymbol{S}_0+\boldsymbol{S}_1)+\int_{-\infty}^{+\infty}\frac{1+u\lambda}{u-\lambda}\mathrm{d}(\boldsymbol{S}_0\tau(u)\boldsymbol{S}_0),$$

and therefore $F_1(\lambda)\in\mathbf{N}_p$, since $\boldsymbol{S}_0\boldsymbol{\beta}\boldsymbol{S}_0+\boldsymbol{S}_1$ is Hermitian.

On the other hand, from (2. 10), $F_1(z)\in\mathbf{N}_p$ can be rewritten in the form

$$F_1(\lambda)=-\boldsymbol{S}_0[\lambda^{-1}F^{-1}(\lambda)][\boldsymbol{S}_0\lambda+F(\lambda)\lambda^2-F(\lambda)\boldsymbol{S}_0^{-1}\boldsymbol{S}_1\lambda]$$
$$=-\boldsymbol{S}_0\left[\lambda\left(F(\lambda)+\frac{\boldsymbol{S}_0}{\lambda}+\cdots+\frac{\boldsymbol{S}_{2n}}{\lambda^{2n+1}}\right)-\boldsymbol{S}_0-\frac{\boldsymbol{S}_1}{\lambda}-\cdots-\frac{\boldsymbol{S}_{2n}}{\lambda^{2n}}\right]^{-1}\times$$
$$\left[\boldsymbol{S}_0\lambda+\lambda^2\left(F(\lambda)+\frac{\boldsymbol{S}_0}{\lambda}+\cdots+\frac{\boldsymbol{S}_{2n}}{\lambda^{2n+1}}\right)-\boldsymbol{S}_0\lambda-\boldsymbol{S}_1-\cdots-\right.$$
$$\frac{\boldsymbol{S}_{2n}}{\lambda^{2n+1}}-\left(F(\lambda)+\frac{\boldsymbol{S}_0}{\lambda}+\cdots+\frac{\boldsymbol{S}_{2n}}{\lambda^{2n+1}}\right)\times$$
$$\left.\boldsymbol{S}_0^{-1}\boldsymbol{S}_1\lambda+\boldsymbol{S}_1+\frac{\boldsymbol{S}_1\boldsymbol{S}_0^{-1}\boldsymbol{S}_1}{\lambda}+\cdots+\frac{\boldsymbol{S}_{2n}\boldsymbol{S}_0^{-1}\boldsymbol{S}_1}{\lambda^{2n}}\right].$$

Thus $F_1(\lambda)$ has an asymptotic expansion of the form (2. 8) in which $\boldsymbol{S}_0^{(1)},\boldsymbol{S}_1^{(1)},\cdots,\boldsymbol{S}_{2n-1}^{(1)}$ are determined by $\boldsymbol{S}_0,\boldsymbol{S}_1,\cdots,\boldsymbol{S}_{2n}$. To complete the proof, it remains only to verify the formula (2. 9). Let

$$\Phi(\lambda)=-\frac{\boldsymbol{S}_0}{\lambda}-\frac{\boldsymbol{S}_1}{\lambda^2}-\cdots-\frac{\boldsymbol{S}_{2n}}{\lambda^{2n+1}}. \tag{2.11}$$

We will find the first $2n-1$ coefficients of the Laurent expansion of $\Phi_1(\lambda)=-\boldsymbol{S}_0\Phi^{-1}(\lambda)\boldsymbol{S}_0-\boldsymbol{S}_0\lambda+\boldsymbol{S}_1$ at infinity. In the remaining part of the proof, we always assume $\boldsymbol{\Gamma}_n>\boldsymbol{O}$ (since in the case when $\boldsymbol{\Gamma}_n\geqslant\boldsymbol{O}$ is singular, one may choose a certain Hermitian block-Hankel matrix $\widetilde{\boldsymbol{\Gamma}}_n>\boldsymbol{O}$ and let $\boldsymbol{\Gamma}_n(\varepsilon)=\Gamma_n+\varepsilon\widetilde{\Gamma}_n$, $\forall\,\varepsilon>0$, and prove that (2.9) holds for $\boldsymbol{\Gamma}_n(\varepsilon)$ and $[\boldsymbol{S}_{i+j}^{(1)}(\varepsilon)]_{i,j=0}^{n-1}$. Then go to the limit as $\varepsilon\to0$). Then, it follows from [15] that there exists a pair of polynomial matrices $\boldsymbol{A}(\lambda)=\boldsymbol{I}_p\lambda^{n+1}+\sum\limits_{i=0}^{n}\boldsymbol{A}_i\lambda^i$ and $\boldsymbol{B}(\lambda)=\sum\limits_{i=0}^{n}\boldsymbol{B}_i\lambda^i$ such that

$$\boldsymbol{A}^{-1}(\lambda)\boldsymbol{B}(\lambda)=\hat{B}(\lambda)\hat{A}^{-1}(\lambda),\tag{2.12}$$

where for a polynomial matrix $\boldsymbol{D}(\lambda)=\sum\limits_{i=0}^{n}\boldsymbol{D}_i\lambda^i$, $\hat{\boldsymbol{D}}(\lambda)$ denotes the corresponding polynomial matrix $\sum\limits_{i=0}^{n}\boldsymbol{D}_i^*\lambda^i$, and the Laurent expansion of $\boldsymbol{A}^{-1}(\lambda)\boldsymbol{B}(\lambda)$ at infinity has the form

$$\boldsymbol{A}^{-1}(\lambda)\boldsymbol{B}(\lambda)=-\frac{\boldsymbol{S}_0}{\lambda}-\frac{\boldsymbol{S}_1}{\lambda^2}-\cdots-\frac{\boldsymbol{S}_{2n}}{\lambda^{2n+1}}+o(\lambda^{-2n-1}).\tag{2.13}$$

Moreover, the generalized Bezoutian of the quadruple $(\boldsymbol{A}(\lambda),\boldsymbol{B}(\lambda);\hat{B}(\lambda),\hat{A}(\lambda))$ $\mathrm{Bez}(\boldsymbol{A},\boldsymbol{B};\hat{B},\hat{A})$ can be written in the form [3]

$$\mathrm{Bez}(\boldsymbol{A},\boldsymbol{B};\hat{B},\hat{A})=\boldsymbol{S}(A)^*\Gamma_n\boldsymbol{S}(A),\tag{2.14}$$

in which

$$\boldsymbol{S}(A)=\begin{bmatrix}\boldsymbol{A}_1 & \cdots & \boldsymbol{A}_n & \boldsymbol{I}_p\\ \vdots & \iddots & \iddots & \\ \boldsymbol{A}_n & \iddots & & \\ \boldsymbol{I}_p & & & \end{bmatrix}\tag{2.15}$$

Let $\boldsymbol{C}(\lambda),\boldsymbol{R}(\lambda)$ be the polynomial matrices such that

$$\boldsymbol{A}(\lambda)=\boldsymbol{R}(\lambda)\boldsymbol{C}(\lambda)+\boldsymbol{R}(\lambda),\quad \deg\boldsymbol{R}(\lambda)<\deg\boldsymbol{B}(\lambda),\tag{2.16}$$

where $\deg\boldsymbol{R}(\lambda)$ stands for the formal degree of the polynomial matrix $\boldsymbol{R}(\lambda)$. Then $[\boldsymbol{A}^{-1}(\lambda)\boldsymbol{B}(\lambda)]^{-1}=\boldsymbol{B}^{-1}(\lambda)\boldsymbol{A}(\lambda)=\boldsymbol{C}(\lambda)+\boldsymbol{B}^{-1}(\lambda)\boldsymbol{R}(\lambda)$, so that, by the argument given above,

$$\boldsymbol{B}^{-1}(\lambda)\boldsymbol{R}(\lambda)=\hat{R}(\lambda)\hat{B}^{-1}(\lambda)$$

$$=-\frac{S_0^{-1}S_0^{(1)}S_0^{-1}}{\lambda}-\frac{S_0^{-1}S_1^{(1)}S_0^{-1}}{\lambda^2}-\cdots-\frac{S_0^{-1}S_{2n-2}^{(1)}S_0^{-1}}{\lambda^{2n-1}}+o(\lambda^{-2n-1}). \quad (2.17)$$

Thanks to the relation

$$\begin{aligned}\mathrm{Bez}(\boldsymbol{A},\boldsymbol{B};\hat{B},\hat{A})&=\mathrm{Bez}(\boldsymbol{BC}+\boldsymbol{R},\boldsymbol{B};\hat{B},\hat{C}\hat{B}+\hat{R})\\&=\mathrm{Bez}(\boldsymbol{BC},\boldsymbol{B};\hat{B},\hat{C}\hat{B})+\begin{bmatrix}\mathrm{Bez}(\boldsymbol{R},\boldsymbol{B};\hat{B},\hat{R}) & \boldsymbol{O}\\ \boldsymbol{O} & \boldsymbol{O}\end{bmatrix},\end{aligned}$$

we obtain from (2.14) that

$$\begin{bmatrix}\boldsymbol{I}_p & \boldsymbol{O} & \cdots & -\boldsymbol{B}_0\boldsymbol{S}_0^{-1}\\ & \ddots & \ddots & \vdots\\ & & \ddots & -\boldsymbol{B}_{n-1}\boldsymbol{S}_0^{-1}\\ \boldsymbol{O} & & & \boldsymbol{I}_p\end{bmatrix}\boldsymbol{S}(A)^*\boldsymbol{\Gamma}_n\boldsymbol{S}(A)\begin{bmatrix}\boldsymbol{I}_p & & & \boldsymbol{O}\\ \boldsymbol{O} & \ddots & & \\ \vdots & \ddots & \ddots & \\ -\boldsymbol{S}_0^{-1}\boldsymbol{B}_0^* & \cdots & -\boldsymbol{S}_0^{-1}\boldsymbol{B}_{n-1}^* & \boldsymbol{I}_p\end{bmatrix}$$

$$=\begin{bmatrix}\boldsymbol{S}(B)^*\begin{bmatrix}\boldsymbol{S}_0 & & \\ & \ddots & \\ & & \boldsymbol{S}_0\end{bmatrix}^{-1}\boldsymbol{\Gamma}_{n-1}^{(1)}\begin{bmatrix}\boldsymbol{S}_0 & & \\ & \ddots & \\ & & \boldsymbol{S}_0\end{bmatrix}^{-1}\boldsymbol{S}(\boldsymbol{B}) & \boldsymbol{O}\\ \boldsymbol{O} & \boldsymbol{S}_0\end{bmatrix}.$$

Multiplying by

$$\boldsymbol{T}=\mathrm{diag}\left[\begin{bmatrix}\boldsymbol{S}_0 & & \boldsymbol{O}\\ \vdots & \ddots & \\ \boldsymbol{S}_{n-1} & \cdots & \boldsymbol{S}_0\end{bmatrix},\boldsymbol{I}_p\right]\begin{bmatrix}\boldsymbol{S}(B)^* & \boldsymbol{O}\\ \boldsymbol{O} & \boldsymbol{I}_p\end{bmatrix}^{-1}$$

on the left, by $\boldsymbol{T}^*$ on the right, and then computing the existing result, we have

$$\begin{bmatrix}-\boldsymbol{S}_1\boldsymbol{S}_0^{-1} & \boldsymbol{I}_p & & \boldsymbol{O}\\ \vdots & & \ddots & \\ -\boldsymbol{S}_n\boldsymbol{S}_0^{-1} & & & \boldsymbol{I}_p\\ \boldsymbol{I}_p & \boldsymbol{O} & \cdots & \boldsymbol{O}\end{bmatrix}\boldsymbol{\Gamma}_n\begin{bmatrix}-\boldsymbol{S}_1\boldsymbol{S}_0^{-1} & \cdots & -\boldsymbol{S}_n\boldsymbol{S}_0^{-1} & \boldsymbol{I}_p\\ \boldsymbol{I}_p & & & \boldsymbol{O}\\ & \ddots & & \vdots\\ \boldsymbol{O} & & \boldsymbol{I}_p & \boldsymbol{O}\end{bmatrix}$$

$$=\begin{bmatrix}\begin{bmatrix}\boldsymbol{S}_0 & & \boldsymbol{O}\\ \vdots & \ddots & \\ \boldsymbol{S}_{n-1} & \cdots & \boldsymbol{S}_0\end{bmatrix}(\boldsymbol{I}_n\otimes\boldsymbol{S}_0)^{-1}\boldsymbol{\Gamma}_{n-1}^{(1)}(\boldsymbol{I}_n\otimes\boldsymbol{S}_0)^{-1}\begin{bmatrix}\boldsymbol{S}_0 & \cdots & \boldsymbol{S}_{n-1}\\ & \ddots & \vdots\\ & & \boldsymbol{S}_0\end{bmatrix} & \boldsymbol{O}\\ \boldsymbol{O} & \boldsymbol{S}_0\end{bmatrix},$$

whence

$$[S_{i+j}]_{i,j=1}^{n}-\begin{bmatrix}S_1\\ \vdots\\ S_n\end{bmatrix}S_0^{-1}[S_1,S_2,\cdots,S_n]=\begin{bmatrix}S_0 & & O\\ \vdots & \ddots & \\ S_{n-1} & \cdots & S_0\end{bmatrix}\begin{bmatrix}S_0 & & \\ & \ddots & \\ & & S_0\end{bmatrix}^{-1}\times$$

$$\boldsymbol{\Gamma}_{n-1}^{(1)}\begin{bmatrix}S_0 & & \\ & \ddots & \\ & & S_0\end{bmatrix}^{-1}\begin{bmatrix}S_0 & \cdots & S_{n-1}\\ & \ddots & \vdots\\ & & S_0\end{bmatrix}.$$

Hence, the formula (2.9) follows from the last equation. ■

Lemma 2.7 *Suppose that* $n\geqslant 1$ *and* $\boldsymbol{\Gamma}_n$ *is n. e. and that* $F_1(\lambda)\in \mathbf{N}_p$ *has the asymptotic expansion* (2.13), *where* $S_0^{(1)},\cdots,S_{2n-2}^{(1)}$ *are defined by Eq.* (2.9). *Let* $F(\lambda)$ *be of the form* (2.7). *Then* $\boldsymbol{\Gamma}_{n-1}^{(1)}=[S_{i+j}^{(1)}]_{i,j=0}^{n-1}$ *is also n. e., and* $F(\lambda)\in\mathbf{N}_p$, *which has the asymptotic expansion* (2.2).

Proof Suppose that $\boldsymbol{\Gamma}_n$ is n. e. ($n\geqslant 1$). By Corollary 2.5, we may assume that $S>O$. Then by definition, $(S_{i+j})_{i,j=0}^{n+1}\geqslant O$ for a suitable pair of Hermitian matrices S_{2n+1} and S_{2n+2}. Define $S_{2n-1}^{(1)}$ and $S_{2n}^{(1)}$ via the formula

$$[S_{i+j}^{(1)}]_{i,j=0}^{n}=(I_{n+1}\otimes S_0)\begin{bmatrix}S_0 & & & O\\ \vdots & \ddots & & \\ S_{n-1} & \cdots & S_0 & \\ S_n & \cdots & S_1 & S_0\end{bmatrix}^{-1}\times$$

$$\left([S_{i+j}]_{i,j=1}^{n}-\begin{bmatrix}S_1\\ \vdots\\ S_{n+1}\end{bmatrix}S_0^{-1}[S_1,S_2,\cdots,S_{n+1}]\right)\begin{bmatrix}S_0 & \cdots & S_{n-1} & S_n\\ & \ddots & \vdots & \vdots\\ & & S_0 & S_1\\ & & & S_0\end{bmatrix}^{-1}(I_{n+1}\otimes S_0).$$

Then obviously $[S_{i+j}^{(1)}]_{i,j=0}^{n-1}\geqslant O$, so that $\boldsymbol{\Gamma}_n^{(1)}$ is n. e. On the other hand, it is easy to prove in much the same way as that used in Lemma 2.6 that $F(\lambda)\in\mathbf{N}_p$ and it has an asymptotic expansion of the form

$$F(\lambda)=-\frac{S_0}{\lambda}-\frac{S_1}{\lambda^2}-\frac{S'_2}{\lambda^3}-\cdots-\frac{S'_{2n}}{\lambda^{2n+1}}+o(\lambda^{-2n-1}).$$

Then Eq. (2.9) holds with $S'_i\,(i\geqslant 2)$ in place of $S_i\,(i\geqslant 2)$ therein, and therefore $S_i=S'_i\,(i\geqslant 2)$, noting that if S_0 and S_1 are fixed, then $S_2,\cdots,S_{2n}$ are uniquely determined by $S_0^{(1)},S_1^{(1)},\cdots,S_{2n-2}^{(1)}$ via the formula (2.9). ■

§ 3. The solution of the TH problem

In this section, the nondegenerate and degenerate TH problems are treated in a unified way of constructing matrix continued fractions. Some results on the matrix moment problem on the real axis are given as well with little additional effort.

The following theorem gives a criterion of existence of the solutions to the TH problem in terms of the notion of n. e. , which is a natural generalization of the corresponding theorem in the scalar case (see, e. g. , [21], Theorem 2. 3, pp. 31～32; [8], Theorem 3. 6). see[4], Theorem 1, for the same result in the operator case.

Theorem 3. 1 *The TH problem* (2. 1) *is solvable if and only if* $\boldsymbol{\Gamma}_n=(\boldsymbol{S}_{i+j})_{i,j=0}^n$ *is n. e.*

Proof Assume that the problem (2. 1) has a solution $\sigma(u)$. Then the problem (2. 2) also has a solution $F(\lambda)$ by Lemma 2. 1. It is obvious that $\boldsymbol{\Gamma}_n=(\boldsymbol{S}_{i+j})_{i,j=0}^n=[\int_{-\infty}^{+\infty}u^{i+j}\mathrm{d}\sigma(u)]_{i,j=0}^n\geqslant \boldsymbol{O}$. Applying Lemma 2. 6 successively, one can derive that

$$F(\lambda)=-\frac{\boldsymbol{S}_0}{\lambda\boldsymbol{I}_p-\boldsymbol{S}_0^{\mathrm{D}}\boldsymbol{S}_1}-\frac{\boldsymbol{S}_0^{\mathrm{D}}\boldsymbol{S}_0^{(1)}}{\lambda\boldsymbol{I}_p-\boldsymbol{S}_0^{(1)\mathrm{D}}\boldsymbol{S}_1^{(1)}}-\cdots-\frac{\boldsymbol{S}_0^{(n-3)\mathrm{D}}\boldsymbol{S}_0^{(n-2)}}{\lambda\boldsymbol{I}_p-\boldsymbol{S}_0^{(n-2)\mathrm{D}}\boldsymbol{S}_1^{(n-2)}}$$

$$-\frac{\boldsymbol{S}_0^{(n-2)\mathrm{D}}\boldsymbol{S}_0^{n-1}}{\lambda\boldsymbol{I}_p-\boldsymbol{S}_0^{(n-1)\mathrm{D}}\boldsymbol{S}_1^{(n-1)}+\boldsymbol{S}_0^{(n-1)\mathrm{D}}\Phi(\lambda)} \tag{3. 1}$$

for a certain $\Phi(\lambda)=-\boldsymbol{S}_0^{(n)}/\lambda+o(\lambda^{-1})\in\mathbf{N}_p$, where $\boldsymbol{S}_0^{(1)},\boldsymbol{S}_0^{(1)},\cdots,\boldsymbol{S}_0^{(n-1)}$, $\boldsymbol{S}_0^{(n)}$ are uniquely determined by $\boldsymbol{\Gamma}_n$. Let $\widetilde{F}(\lambda)$ denote the matrix-valued function $F(\lambda)$ given in Eq. (3. 1) but with $\Phi(\lambda)$ replaced by $\widetilde{\Phi}(\lambda)=-\boldsymbol{S}_0^{(n)}/\lambda\in\mathbf{N}_p$. Obviously, by Lemma 2. 7, $\widetilde{F}(\lambda)$ is also a solution to the problem(2. 2), which is rational. Suppose now that the Laurent expansion of $\widetilde{F}(\lambda)$ at infinity has the form

$$\widetilde{F}(\lambda)=-\frac{\boldsymbol{S}_0}{\lambda}-\frac{\boldsymbol{S}_1}{\lambda^2}-\frac{\boldsymbol{S}_2}{\lambda^3}-\cdots-\frac{\boldsymbol{S}_{2n}}{\lambda^{2n+1}}-\frac{\boldsymbol{S}_{2n+1}}{\lambda^{2n+2}}-\frac{\boldsymbol{S}_{2n+2}}{\lambda^{2n+3}}+o(\lambda^{-2n-3}).$$

Then Lemma 2. 1 implies that there exists a Hermitian measure $\tau(u)$ such that

$$\boldsymbol{S}_k = \int_{-\infty}^{+\infty} u^k \mathrm{d}\tau(u), \quad k = 0,1,\cdots,2n+2,$$

so that

$$\boldsymbol{\Gamma}_{n+1} = [\boldsymbol{S}_{i+j}]_{i,j=0}^{n+1} = \left[\int_{-\infty}^{+\infty} u^{i+j} \mathrm{d}\tau(u)\right]_{i,j=0}^{n+1} \geqslant \boldsymbol{O}.$$

Hence, $\boldsymbol{\Gamma}_n$ is n. e.

Conversely, if $\boldsymbol{\Gamma}_n$ is n. e., we may assume $\boldsymbol{S}_O > \boldsymbol{O}$ by Corollary 2. 5. Let

$$F(\lambda) = -\frac{\boldsymbol{S}_0}{\lambda} - \frac{\boldsymbol{S}_1}{\lambda^2} - \frac{\boldsymbol{S}_2}{\lambda^3} - \cdots - \frac{\boldsymbol{S}_{2n}}{\lambda^{2n+1}} + o(\lambda^{-2n-1}).$$

Then we have

$$F(\lambda) = -\frac{\boldsymbol{S}_0}{\lambda \boldsymbol{I}_p - \boldsymbol{S}_0^{-1}\boldsymbol{S}_1 + \boldsymbol{S}_0^{-1} F_1(\lambda)} \tag{3. 2}$$

in which

$$F_1(\lambda) = -\frac{\boldsymbol{S}_0^{(1)}}{\lambda} - \frac{\boldsymbol{S}_1^{(1)}}{\lambda^2} - \cdots - \frac{\boldsymbol{S}_{2n-2}^{(1)}}{\lambda^{2n-1}} + o(\lambda^{-2n+1}) \tag{3. 3}$$

with coefficients $\boldsymbol{S}_0^{(1)}, \boldsymbol{S}_1^{(1)}, \cdots, \boldsymbol{S}_{2n-2}^{(1)}$ defined by (2. 9). By Lemma 2. 7, $\boldsymbol{\Gamma}_{n-1}^{(1)}$ is n. e. Also, $F(\lambda) \in \mathbf{N}_p$ is equivalent to the fact $F_1(\lambda) \in \mathbf{N}_p$ by the proof of Lemma 2. 6. For $F_1(\lambda)$, we may assume $\boldsymbol{S}_0^{(1)} > \boldsymbol{O}$, and put

$$F_1(\lambda) = -\frac{\boldsymbol{S}_0^{(1)}}{\lambda \boldsymbol{I}_p - \boldsymbol{S}_0^{(1)-1}\boldsymbol{S}_1^{(1)} + \boldsymbol{S}_0^{(1)-1} F_2(\lambda)}$$

in which

$$F_2(\lambda) = -\frac{\boldsymbol{S}_0^{(2)}}{\lambda} - \frac{\boldsymbol{S}_1^{(2)}}{\lambda^2} - \cdots - \frac{\boldsymbol{S}_{2n-4}^{(2)}}{\lambda^{2n-3}} + o(\lambda^{-2n+3})$$

with coefficient $\boldsymbol{S}_0^{(2)}, \boldsymbol{S}_1^{(2)}, \cdots, \boldsymbol{S}_{2n-4}^{(2)}$ defined by $\boldsymbol{\Gamma}_{n-1}^{(1)}$ via a formula similar to (2. 9). The further continuation of this procedure is evident. Hence, in order to find a solution $F(\lambda)$ to the problem (2. 2), it is only required to find a matrix-valued function $F_n(\lambda)$, which belongs to $\mathbf{N}_p$ and has the asymptotic expansion $F_n(\lambda) = -\boldsymbol{S}_0^{(n)}/\lambda + o(\lambda^{-1}), \boldsymbol{S}_0^{(n)} \geqslant \boldsymbol{O}$. It is clear that such a function $F_n(\lambda)$ exists, and can be taken as $F_n(\lambda) = -\boldsymbol{S}_0^{(n)}/\lambda$. Thus the problem (2. 2) (and therefore the TH problem) is solvable. ■

It is noteworthy that in [6] the author concluded that $\boldsymbol{\Gamma}_n \geqslant \boldsymbol{O}$ was a sufficient and necessary condition for a solution of the TH problem to exist. This, however, is not true even for the scalar case [8]. But it can be verified that $\Gamma_n \geqslant O$ is actually a criterion of solvability for the matrix moment problem on the real axis of finding a Hermitian measure $\sigma(u)(-\infty<u<+\infty)$ such that

$$S_k = \int_{-\infty}^{+\infty} u^k \mathrm{d}\sigma(u), \quad i = 0,1,\cdots,2n-1;\ S_{2n} \geqslant \int_{-\infty}^{+\infty} u^{2n} \mathrm{d}\sigma(u), \tag{3.4}$$

where $S_0, S_1, \cdots, S_{2n}$ are given $p \times p$ Hermitian matrices. (See [2] for scalar case.)

Theorem 3.2 *The moment problem* (3.4) *is solvable if and only if* $\Gamma_n = [S_{i+j}]_{i,j=0}^n \geqslant O$.

Proof Suppose that the problem (3.4) has a solution $\sigma(u)$. Let

$$\tilde{S}_{2n} = \int_{-\infty}^{+\infty} u^{2n} \mathrm{d}\sigma(u),$$

then $O \leqslant \tilde{S}_{2n} \leqslant S_{2n}$, Therefore

$$\Gamma_n = \left[\int_{-\infty}^{+\infty} u^{i+j} \mathrm{d}\sigma(u)\right]_{i,j=0}^n + \mathrm{diag}[O,\cdots,O,S_{2n}-\tilde{S}_{2n}] \geqslant O. \tag{3.5}$$

Suppose conversely that $\Gamma_n \geqslant O$. Now put

$$F(\lambda) = -\frac{S_0}{\lambda I_p - S_0^{\mathrm{D}} S_1} - \frac{S_0^{\mathrm{D}} S_0^{(1)}}{\lambda I_p - S_0^{(1)\mathrm{D}} S_1^{(1)}} - \cdots - \frac{S_0^{(n-3)\mathrm{D}} S_0^{(n-2)}}{\lambda I_p - S_0^{(n-2)\mathrm{D}} S_1^{(n-2)}}$$

$$- \frac{S_0^{(n-2)\mathrm{D}} S_0^{(n-1)}}{\lambda I_p S_0^{(n-1)\mathrm{D}} S_1^{(n-1)} + S_0^{(n-1)\mathrm{D}} \Phi(\lambda)},$$

in which $\Phi(\lambda) = -\alpha/\lambda \in \mathbf{N}_p (O \leqslant \alpha \leqslant S_0^{(n)})$ and $S_0^{(1)}, S_1^{(1)}, \cdots, S_0^{(n-1)}, S_1^{(n-1)}$, $S_0^{(n)}$ are defined by the formula (2.9) successively. By using Lemma 2.7 repeatedly, we have that $F(\lambda) \in \mathbf{N}_p$ which is rational, and has Laurent expansion at infinity

$$F(\lambda) = -\frac{S_0}{\lambda} - \frac{S_1}{\lambda^2} - \frac{S_2}{\lambda^3} - \cdots - \frac{S_{2n-1}}{\lambda^{2n}} - \frac{\tilde{S}_{2n}}{\lambda^{2n+1}} + o(\lambda^{-2n-1}).$$

Since $O \leqslant \alpha \leqslant S_0^{(n)}$, we conclude from Eq. (2.9) that $O \leqslant \tilde{S}_{2n} \leqslant S_{2n}$. Hence, by Lemma 2.1, there exists a certain Hermitian measure $\sigma(u)$ with moments $S_0, S_1, \cdots, S_{2n-1}, \tilde{S}_{2n}$, that is, $\sigma(u)$ is a solution to the problem (3.4). ■

As a consequence of Theorem 3.2 we have a useful conclusion

which is actually implied by Ando in [4], Corollary 3.

Corollary 3.3 (Ando [4]) *If $\boldsymbol{\Gamma}_n=[\boldsymbol{S}_{i+j}]_{i,j=0}^{n}\geqslant \boldsymbol{O}$, then there exists a $\boldsymbol{C}\geqslant\boldsymbol{O}$ of order p such that $\boldsymbol{\Gamma}_n$ can be written as a certain sum of two nonnegative Hermitian block-Hankel matrices:*

$$\boldsymbol{\Gamma}_n=\begin{bmatrix}\boldsymbol{S}_1 & \cdots & \boldsymbol{S}_n\\ \vdots & & \vdots\\ \boldsymbol{S}_n & \cdots & \boldsymbol{S}_{2n}-\boldsymbol{C}\end{bmatrix}+\begin{bmatrix}\boldsymbol{O} & & \boldsymbol{O}\\ & \ddots & \\ \boldsymbol{O} & & \boldsymbol{C}\end{bmatrix},\tag{3.6}$$

where the first matrix on the right side of the last equality is n. e.

Note that in the scalar case: $p=1$, if $\boldsymbol{\Gamma}_n\geqslant\boldsymbol{O}$ is singular, the form (3.6), alias its quasidirect decomposition, is uniquely defined by $\boldsymbol{\Gamma}_n$ itself (see [8], Lemma 2.7), and the singularity of $\boldsymbol{\Gamma}_n$ always leads to a unique solution to the problem (3.4), which has a finite number of points of increase.

Although, the singularity of $\boldsymbol{\Gamma}_n$ in the matrix case will not always lead to uniqueness, but to reduction of the dimension of the solutions (see Theorem 4.5). We will consider these questions in the following results. To begin with we present the explicit forms of the solutions to the TH problem and to the moment problem (3.4), which follow from Lemma 2.1 and Theorems 3.2 and 3.3 at once.

Theorem 3.4 *The general solution $\sigma(u)$ to the TH problem is representable as a matrix continued fraction*

$$\int_{-\infty}^{+\infty}\frac{\mathrm{d}\sigma(u)}{u-\lambda}=-\frac{\boldsymbol{S}_0}{\lambda\boldsymbol{I}_p-\boldsymbol{S}_0^{\mathrm{D}}\boldsymbol{S}_1}-\frac{\boldsymbol{S}_0^{\mathrm{D}}\boldsymbol{S}_0^{(1)}}{\lambda\boldsymbol{I}_p-\boldsymbol{S}_0^{(1)\mathrm{D}}\boldsymbol{S}_1^{(1)}}-\cdots-\frac{\boldsymbol{S}_0^{(n-3)\mathrm{D}}\boldsymbol{S}_0^{(n-2)}}{\lambda\boldsymbol{I}_p-\boldsymbol{S}_0^{(n-2)\mathrm{D}}\boldsymbol{S}_1^{(n-2)}}$$

$$-\frac{\boldsymbol{S}_0^{(n-2)\mathrm{D}}\boldsymbol{S}_0^{(n-1)}}{\lambda\boldsymbol{I}_p-\boldsymbol{S}_0^{(n-1)\mathrm{D}}\boldsymbol{S}_1^{(n-1)}+\boldsymbol{S}_0^{(n-1)\mathrm{D}}\Phi(\lambda)},\tag{3.7}$$

where $\Phi(\lambda)=\int_{-\infty}^{+\infty}1/(u-\lambda)\mathrm{d}\tau(u)$ and $\boldsymbol{S}_0^{(n)}=\int_{-\infty}^{+\infty}\mathrm{d}\tau(u)$ for a Hermitian measure $\tau(u)$.

(To determine the Hermitian measure $\sigma(u)$ from the function $\Phi(\lambda)$ one can use the Stieltjes-Perron inversion formula in the matrix case.)

It is easy to see that in the case when $\boldsymbol{\Gamma}_n$ is n. e., the TH problem has infinitely many solutions provided $\boldsymbol{S}_0^{(n)}\neq\boldsymbol{O}$ (see the proof of Theorem 4.1 for details).

Corollary 3.5 *The TH problem has a unique solution $\sigma(u)$ if and only if $\boldsymbol{\Gamma}_n$ is n.e., and $\boldsymbol{S}_0^n=\boldsymbol{O}$. In this case, the unique solution $\sigma(u)$ has a finite number of points of increase such that Eq.(3.7) holds.*

Proof Follows from Theorems 3.1 and 3.4 and the Stieltjes-Perron inversion formula. ■

Theorems 3.1 and 3.4, and Corollary 3.5 support that a great amount of all information about the TH problem is contained in the matrix $\boldsymbol{\Gamma}_n$ and the matrix continued fraction(3.7).

Theorem 3.6 *The general solution to the moment problem* (3.4) *is representable as a matrix continued fraction given as in* (3.7), *where* $\Phi(\lambda)=\int_{-\infty}^{+\infty} 1/(u-\lambda)\,\mathrm{d}\tau(u)$ *but only* $\boldsymbol{S}_0^{(n)}\geqslant\int_{-\infty}^{+\infty}\mathrm{d}\tau(u)$ *for a Hermitian measure* $\tau(u)$.

In the next section, we will generalize a theorem of Kovalishina [17] and use it to rewrite the aforementioned general solution to the problem (3.4) as a linear fractional transformation of an arbitrary Nevanlinna pair.

Corollary 3.7 *The moment problem* (3.4) *has a unique solution $\sigma(u)$ if and only if $\boldsymbol{\Gamma}_n\geqslant\boldsymbol{O}$ and $\boldsymbol{S}_0^{(n)}=\boldsymbol{O}$. In this case, the unique solution $\sigma(u)$ has a finite number of points of increase.*

Proof Follows from Theorems 3.2 and 3.5, and the Stieltjes-Perron inversion formula. ■

Corollary 3.8 *If $\boldsymbol{\Gamma}_n\geqslant\boldsymbol{O}$ and $\boldsymbol{S}_0^{(n)}=\boldsymbol{O}$, then $\boldsymbol{\Gamma}_n$ has only a trivial decomposition of the form* (3.6).

Proof Follows from Corollaries 3.3, 3.5 and 3.7. ■

Note that under the assumption of Corollary 3.8, $\boldsymbol{\Gamma}_n$ is always singular (see Theorem 3.9).

Theorem 3.9 *If $\boldsymbol{\Gamma}_n$ is n.e., then there exists a nonsingular matrix $\boldsymbol{P}$ of order $(n+1)p$ such that*

$$P\Gamma_n P^* = \begin{bmatrix} S_0^{(0)} & & & \\ & S_0^{(1)} & & \\ & & \ddots & \\ & & & S_0^{(n)} \end{bmatrix},$$

$$R(S_0^{(n)}) \subseteq R(S_0^{(n-1)}) \subseteq \cdots \subseteq R(S_0^{(0)}), \tag{3.8}$$

where $R(\mathbf{A})$ *stands for the column space of a matrix* $\mathbf{A}$.

Proof Set $\Gamma_n^{(0)} = \Gamma_n$. From (2.9) we obtain that

$$\begin{bmatrix} S^{(0)} & O \\ O & \Gamma_{n-1}^{(1)} \end{bmatrix} = \begin{bmatrix} I_P & & & O \\ & S_0^{(0)} & & \\ & & \ddots & \\ O & & & S_0^{(0)} \end{bmatrix} \begin{bmatrix} I_P & & & \\ & S_0^{(0)} & & \\ & \vdots & \ddots & \\ & S_{n-1}^{(0)} & \cdots & S_0^{(0)} \end{bmatrix}^D \times$$

$$\begin{bmatrix} I_P & & & \\ -S_1 S_0^D & I_P & & \\ \vdots & & \ddots & \\ -S_n S_0^D & O & \cdots & I_p \end{bmatrix} \Gamma_n^{(0)} \begin{bmatrix} I_p & -S_0^D S_1 & \cdots & -S_0^D S_n \\ & I_p & & O \\ & & \ddots & \vdots \\ & & & I_p \end{bmatrix} \times$$

$$\begin{bmatrix} I_p & & & \\ & S_0^{(0)} & \cdots & S_{n-1}^{(0)} \\ & & \ddots & \vdots \\ & & & S_0^{(0)} \end{bmatrix}^D \begin{bmatrix} I_p & & & O \\ & S_0^{(0)} & & \\ & & \ddots & \\ O & & & S_0^{(0)} \end{bmatrix}.$$

By using the operation similar to that given in the last equation on $\Gamma_{n-1}^{(1)}$, we obtain a formula similar to the last one with $S_0^{(0)}, \Gamma_{n-1}^{(1)}$ replaced by $S_0^{(1)}$, $\Gamma_{n-2}^{(2)}$ respectively, so that

$$\begin{bmatrix} S_0^{(0)} & & \\ & S_0^{(1)} & \\ & & \Gamma_{n-2}^{(2)} \end{bmatrix}$$

$$= \begin{bmatrix} I_p & & & & \\ & I_p & & & \\ & & S_0^{(1)} & & \\ & & & \ddots & \\ & & & & S_0^{(1)} \end{bmatrix} \begin{bmatrix} I_p & & & & O \\ & I_p & & & \\ & & S_0^{(1)} & & \\ & & \vdots & \ddots & \\ & & S_{n-2}^{(1)} & \cdots & S_0^{(1)} \end{bmatrix}^D \times$$

$$\begin{bmatrix} \boldsymbol{I}_p & & & & \\ & \boldsymbol{I}_p & & & \\ & -\boldsymbol{S}_1^{(1)}\boldsymbol{S}_0^{(1)\mathrm{D}} & \boldsymbol{I}_p & & \\ & \vdots & & \ddots & \\ & -\boldsymbol{S}_{n-1}^{(1)}\boldsymbol{S}_0^{(1)\mathrm{D}} & \boldsymbol{O} & \cdots & \boldsymbol{I}_p \end{bmatrix} \begin{bmatrix} \boldsymbol{S}_0^{(0)} & \\ & \boldsymbol{\Gamma}_{n-1}^{(1)} \end{bmatrix} \times$$

$$\begin{bmatrix} \boldsymbol{I}_p & & & & \\ & \boldsymbol{I}_p & & & \\ & -\boldsymbol{S}_1^{(1)}\boldsymbol{S}_0^{(1)\mathrm{D}} & \boldsymbol{I}_p & & \\ & \vdots & & \ddots & \\ & -\boldsymbol{S}_{n-1}^{(1)\mathrm{D}}\boldsymbol{S}_0^{(1)\mathrm{D}} & \boldsymbol{O} & \cdots & \boldsymbol{I}_p \end{bmatrix}^* \times$$

$$\begin{bmatrix} \boldsymbol{I}_p & & & & \\ & \boldsymbol{I}_p & & & \\ & & \boldsymbol{S}_0^{(1)} & \cdots & \boldsymbol{S}_{n-2}^{(1)} \\ & & & \ddots & \vdots \\ & & & & \boldsymbol{S}_0^{(1)} \end{bmatrix}^{\mathrm{D}} \begin{bmatrix} \boldsymbol{I}_p & & & & \\ & \boldsymbol{I}_p & & & \\ & & \boldsymbol{S}_0^{(1)} & & \\ & & & \ddots & \\ & & & & \boldsymbol{S}_0^{(1)} \end{bmatrix}.$$

The further continuation of the procedure is evident. After n such steps we get to a block diagonal matrix $\mathrm{diag}[\boldsymbol{S}_0^{(0)}, \boldsymbol{S}_0^{(1)}, \cdots, \boldsymbol{S}_0^{(n)}]$ which coincides with $\boldsymbol{P}\boldsymbol{\Gamma}_n\boldsymbol{P}^*$, where $\boldsymbol{P}$ is a certain nonsingular matrix of order $(n+1)p$. The second assertion follows directly from the nonnegative extendability of $\boldsymbol{\Gamma}_{n-k}^{(k)}(k=0,1,\cdots,n-1)$. ■

We point out that Theorem 3.9 is not true for the case of $\boldsymbol{\Gamma}_n \geqslant \boldsymbol{O}$. As an example,

$$\boldsymbol{\Gamma}_2 = \begin{bmatrix} \boldsymbol{O} & \boldsymbol{O} \\ \boldsymbol{O} & \boldsymbol{I}_2 \end{bmatrix}, \quad R(\boldsymbol{S}_0^{(1)}) = \mathbf{C}^2, \quad R(\boldsymbol{S}_0^{(0)}) = \{\mathbf{0}\}.$$

Also, by the proof of Theorem 3.9 we can take

$$\boldsymbol{P}_n = \prod_{i=0}^{\overrightarrow{n-1}} \begin{bmatrix} \boldsymbol{I}_p & & & & & \\ & \ddots & & & & \\ & & \boldsymbol{I}_p & & & \\ & & -\boldsymbol{S}_1^{(i)}\boldsymbol{S}_0^{(i)\mathrm{D}} & \boldsymbol{I}_p & & \\ & & \vdots & \ddots & \ddots & \\ & & -\boldsymbol{S}_{n-1-i}^{(i)}\boldsymbol{S}_0^{(i)\mathrm{D}} & \cdots & -\boldsymbol{S}_1^{(i)}\boldsymbol{S}_0^{(i)\mathrm{D}} & \boldsymbol{I}_p \end{bmatrix} \times$$

$$\begin{bmatrix} I_p & & & & & \\ & \ddots & & & & \\ & & I_p & & & \\ & & -S_1^{(i)}S_0^{(i)D} & I_p & & \\ & & \vdots & & \ddots & \\ & & -S_{n-i}^{(i)}S_0^{(i)D} & O & \cdots & I_p \end{bmatrix} \tag{3.9}$$

as the matrix $\boldsymbol{P}$ given is Theorem 3. 9, which can be considered as the product of a number of elementary operations on block rows without exchange of block rows.

Therefore, we have the following

Corollary 3. 10 *If $\boldsymbol{\Gamma}_n$ is n. e. , there exists a certain principal submatrix, of order equal to rank of $\boldsymbol{\Gamma}_n$, which is positive Hermitian, and is located at the left upper corner as it possibly can, that is, the ith row(column)of it is located at the jth block row (column) of $\boldsymbol{\Gamma}_n$, $j\leqslant i$.*

Note that in the scalar case: $p=1$, if $\boldsymbol{\Gamma}_n$ is n. e. , then the leading principal submatrix of $\boldsymbol{\Gamma}_n$, of order equal to rank of $\boldsymbol{\Gamma}_n$ is always positive Hermitian, and vice versa.

The following result characterizes the class of Hermitian block-Hankel matrices which are n. e.

Theorem 3. 11 *$\boldsymbol{\Gamma}_n$ is n. e. , if and only if $\boldsymbol{\Gamma}_n\geqslant \boldsymbol{O}$ and $R(\tilde{\boldsymbol{S}}_0^{(n)})\subseteq R(\boldsymbol{S}_0^{n-1})$, where $\tilde{\boldsymbol{S}}_0^{(n)}$ is defined via $\boldsymbol{P}_n\boldsymbol{\Gamma}_n\boldsymbol{P}_n^* =\mathrm{diag}[\boldsymbol{S}_0^{(0)},\cdots,\boldsymbol{S}_0^{(n-1)},\tilde{\boldsymbol{S}}_0^{(n)}]$ in which $\boldsymbol{P}_n$ is as in (3. 9).*

Proof The "only if" part follows immediately from Theorem 3. 9, the "if"part can be derived in much the same way as that used in Theorem 3. 1. ■

It is worth to point out that we deal only with the TH problem of the form (2. 1) in the present paper. As for another type of the TH problem of finding a Hermitian measure $\sigma(u)$ subject to

$$\boldsymbol{S}_i = \int_{-\infty}^{+\infty} u_i \mathrm{d}\sigma(u), \quad i = 0,1,\cdots, 2n-1, \tag{3.10}$$

one may also give its criteria of solvability and uniqueness in a similar way to that used in the TH problem (2. 1), and will be omitted here. (The change, however, from the TH problem(2. 1) to (3. 10) seems to be not a simple matter, and some details need to be discussed rather carefully.)

§ 4. The relation between the solution forms of matrix continued fractions and of linear fractional transformations

In this section we will consider an intrinsic relation between the descriptions of the solutions to the moment problems (2. 1) and (3. 4) in terms of matrix continued fractions and in terms of the general and popular way of linear fractional transformations.

A well-known theorem of Kovalishina ([17], Theorem 3) is modified and extended to some extent.

Starting from Theorem 3. 4, we can readily show the close relation of the solution form (3. 7) of matrix continued fractions of the TH problem to the solution form of linear fractional transformations [17]. Note that here a uniform description of the solutions is given in nondegenerate case($\boldsymbol{\Gamma}_n > \boldsymbol{O}$) or not ($\boldsymbol{\Gamma}_n \geqslant \boldsymbol{O}$ is singular and n. e.).

Theorem 4. 1 *The general solution to the TH problem* (2. 1) *is representable as a linear fractional transformation of an arbitrary* $G(\lambda) \in \mathbf{N}_p$ *such that* $\lim\limits_{\lambda \to \infty} G(\lambda)/\lambda = \boldsymbol{O}$ *uniformly in each* π_ε:

$$\int_{-\infty}^{+\infty} \frac{\mathrm{d}\sigma(u)}{u-\lambda} = \frac{\theta_{11}(\lambda)G(\lambda)+\theta_{12}(\lambda)}{\theta_{21}(\lambda)G(\lambda)+\theta_{22}(\lambda)}, \tag{4.1}$$

whose coefficient matrix is of the decomposable form

$$\boldsymbol{\Theta}(\lambda) = \begin{bmatrix} \theta_{11}(\lambda) & \theta_{12}(\lambda) \\ \theta_{21}(\lambda) & \theta_{22}(\lambda) \end{bmatrix} = \overrightarrow{\prod_{i=0}^{n}} \begin{bmatrix} \boldsymbol{O} & \boldsymbol{S}_0^{(i)} \\ \boldsymbol{S}_0^{(i)\mathrm{D}} & \lambda \boldsymbol{I}_p - \boldsymbol{S}_0^{(i)\mathrm{D}} \boldsymbol{S}_1^{(i)} \end{bmatrix}, \boldsymbol{S}_1^{(n)} = \boldsymbol{O}. \tag{4.2}$$

In the degenerate case, if

$$\boldsymbol{U}^* \boldsymbol{S}_0^{(n)} \boldsymbol{U} = \begin{bmatrix} \tilde{\boldsymbol{S}}_0^{(n)} & \boldsymbol{O} \\ \boldsymbol{O} & \boldsymbol{O}_r \end{bmatrix}, \quad \tilde{\boldsymbol{S}}_0^{(n)} > \boldsymbol{O},\ 0 < r \leqslant p,$$

for a certain unitary matrix $\boldsymbol{U}$, *then*

$$\boldsymbol{U}^* G(\lambda)\boldsymbol{U}=\begin{bmatrix}\widetilde{G}(\lambda) & \boldsymbol{O}\\ \boldsymbol{O} & \boldsymbol{O}_r\end{bmatrix}.$$

Further, the problem has only a solution $\sigma(u)$ *if and only if* $\boldsymbol{\Gamma}_n$ *is n.e. and* $\boldsymbol{S}_0^{(n)}=\boldsymbol{O}$. *In this case, the unique solution* $\sigma(u)$ *is such that* (4.1) *holds with* $G(\lambda)\equiv\boldsymbol{O}$.

Proof Observe that in order to prove the theorem, it is sufficient to show the following assertion: if $\Phi(\lambda)=\int_{-\infty}^{\infty}(u-\lambda)^{-1}\mathrm{d}\tau(u)$, where $\int_{-\infty}^{\infty}\mathrm{d}\tau(u)=\boldsymbol{S}_0^{(n)}$, it can be rewritten as the form

$$\Phi(\lambda)=-\frac{\boldsymbol{S}_0^{(n)}}{\lambda\boldsymbol{I}_p+\boldsymbol{S}_0^{(n)\mathrm{D}}G(\lambda)},\tag{4.3}$$

where $G(\lambda)$ is the parameter function described in the theorem, and vice versa.

As before, we may assume that $\boldsymbol{S}_0^{(n)}>\boldsymbol{O}$. Suppose that $\Phi(\lambda)$ admits an integral representation $\Phi(\lambda)=\int_{-\infty}^{+\infty}(u-\lambda)^{-1}\mathrm{d}\tau(u)$, $\int_{-\infty}^{+\infty}\mathrm{d}\tau(u)=\boldsymbol{S}_0^{(n)}$. Then, by Lemma 2.2, $\Phi^{-1}(\lambda)$ exists and $-\Phi^{-1}(\lambda)\in\mathbf{N}_p$. On the other hand,

$$\lim_{y\to\infty}\frac{-\Phi^{-1}(\mathrm{i}y)}{\mathrm{i}y}=\lim_{y\to\infty}\frac{-\boldsymbol{I}_p}{\mathrm{i}y\Phi(\mathrm{i}y)}=\boldsymbol{S}_0^{(n)-1},$$

so that $-\Phi^{-1}(\lambda)$ permits an integral representation of the form

$$-\Phi^{-1}(\lambda)=\boldsymbol{S}_0^{(n)-1}\lambda+\tilde{\boldsymbol{\beta}}+\int_{-\infty}^{+\infty}\frac{(1+u\lambda)}{(u-\lambda)}\mathrm{d}\tilde{\tau}(u),\quad \tilde{\boldsymbol{\beta}}=\tilde{\boldsymbol{\beta}}^*.$$

We have in turn

$$\Phi(\lambda)=\frac{-\boldsymbol{S}_0^{(n)}}{\lambda\boldsymbol{I}_p+\boldsymbol{S}_0^{(n)-1}\boldsymbol{S}_0^{(n)}\left(\tilde{\boldsymbol{\beta}}+\int_{-\infty}^{+\infty}(1+u\lambda)/(u-\lambda)\mathrm{d}\tilde{\tau}(u)\right)\boldsymbol{S}_0^{(n)}},$$

which coincides with Eq. (4.3) with

$$G(\lambda)=\boldsymbol{S}_0^{(n)}\left(\tilde{\boldsymbol{\beta}}+\int_{-\infty}^{+\infty}(1+u\lambda)/(u-\lambda)\mathrm{d}\tilde{\tau}(u)\right)\boldsymbol{S}_0^{(n)}.$$

Conversely, if $\Phi(\lambda)$ is of the form (4.3), then $\Phi(\lambda)\in\mathbf{N}_p$ and $\lim_{y\to\infty}-\mathrm{i}y\cdot\Phi(\mathrm{i}y)=\boldsymbol{S}_0^{(n)}$, so that $\Phi(\lambda)\in\mathbf{N}_p^0$ and $\Phi(\lambda)=\int_{-\infty}^{+\infty}(u-\lambda)^{-1}\mathrm{d}\tau(u)$ for a certain

$\tau(u)$ satisfying $\int_{-\infty}^{+\infty} \mathrm{d}\tau(u) = \boldsymbol{S}_0^{(n)}$, as desired. The rest is plain. ■

Note that in the case of $\boldsymbol{\Gamma}_n > \boldsymbol{O}$, let

$$\boldsymbol{J} = \begin{bmatrix} \boldsymbol{O} & \mathrm{i}\boldsymbol{I}_p \\ -\mathrm{i}\boldsymbol{I}_p & \boldsymbol{O} \end{bmatrix},$$

then, by direct verification, each factor on the right side of Eq. (4.2) (and therefore the matrix $\boldsymbol{\Theta}(\lambda)$) is a matrix-valued function which is $\boldsymbol{J}$-expanding in π^+, $\boldsymbol{J}$-unitary on the real axis, of full rank, and has a pole of order one at $\lambda = \infty$.

The correspondence defined by Eq. (4.1) between Hermitian measures $\sigma(u)$ and functions $G(\lambda)$ with the properties given in Theorem 4.1 is one-to-one.

Before parametrizing the general solution to the moment problem (3.4) in terms of a linear fractional transformation, we need a result which is an extension, to some extent, of a well-known theorem of [17] (Theorem 3).

Theorem 4.2 *If a matrix-valued function* $W(\lambda)$ *holomorphic in* π^+ *satisfying the fundamental matrix inequality*:

$$\begin{bmatrix} \boldsymbol{S}_0 & \boldsymbol{S}_1 & \cdots & \boldsymbol{S}_n & \vdots & W(\lambda) \\ \boldsymbol{S}_1 & \boldsymbol{S}_2 & \cdots & \boldsymbol{S}_{n+1} & \vdots & \lambda[W(\lambda)+\boldsymbol{S}_0/\lambda] \\ \vdots & \vdots & \vdots & \vdots & \vdots & \vdots \\ \boldsymbol{S}_n & \boldsymbol{S}_{n+1} & \cdots & \boldsymbol{S}_{2n} & \vdots & \lambda^n[W(\lambda)+\boldsymbol{S}_0/\lambda+\cdots+\boldsymbol{S}_{n-1}/\lambda^n] \\ \vdots & \vdots & \vdots & \vdots & \vdots & \vdots \\ & & * & & \vdots & \dfrac{W(\lambda)-W(\lambda)^*}{\lambda-\bar{\lambda}} \end{bmatrix} \geqslant \boldsymbol{O},$$

$$\lambda \in \pi^+, \qquad (\mathrm{FMI}(\mathscr{H}))$$

then $W(\lambda)$ *belongs to* $\mathbf{N}_p$ *and has the asymptotic representation*

$$\lim_{\lambda\to\infty} \lambda^{2n+1}\left[W(\lambda)+\frac{\boldsymbol{S}_0}{\lambda}+\cdots+\frac{\boldsymbol{S}_{2n-1}}{\lambda^{2n}}\right] = -\tilde{\boldsymbol{S}}_{2n},$$

where $\boldsymbol{O} \leqslant \tilde{\boldsymbol{S}}_{2n} \leqslant \boldsymbol{S}_{2n}$. *Conversely, if* $W(\lambda) \in \mathbf{N}_p$ *has the asymptotic representation*

$$\lim_{\lambda\to\infty} \lambda^{2n+1}\left[W(\lambda)+\frac{\boldsymbol{S}_0}{\lambda}+\cdots+\frac{\boldsymbol{S}_{2n-1}}{\lambda^{2n}}\right] = -\boldsymbol{S}_{2n},$$

then it satisfies FMI($\mathscr{H}$).

Proof The second part of the theorem has been proved by Kovalishina in [17]. To verify the first part, let us begin with the problem in the case of $n=0$. Then FMI($\mathscr{H}$) is reduced to

$$\begin{bmatrix} \boldsymbol{S}_0 & W(\lambda) \\ W(\lambda)^* & \dfrac{W(\lambda)-W(\lambda)^*}{\lambda-\bar{\lambda}} \end{bmatrix} \geqslant \boldsymbol{O}, \quad \lambda \in \pi^+. \tag{4.4}$$

Letting $\lambda = \mathrm{i}y\,(y>0)$ and multiplying by $\boldsymbol{T}=\mathrm{diag}[\boldsymbol{I}_p, -\mathrm{i}y\boldsymbol{I}_p]$ on the right and by $\boldsymbol{T}^*$ on the left, from the last matrix inequality we obtain

$$\begin{bmatrix} \boldsymbol{S}_0 & -\mathrm{i}yW(\mathrm{i}y) \\ (-\mathrm{i}yW(\mathrm{i}y))^* & \mathrm{Re}[-\mathrm{i}yW(\mathrm{i}y)] \end{bmatrix} \geqslant \boldsymbol{O}, \quad y>0, \tag{4.5}$$

so that

$$(\boldsymbol{S}_0)_{kk}\mathrm{Re}[-\mathrm{i}yW(\mathrm{i}y)]_{ll} - |-\mathrm{i}yW(\mathrm{i}y)_{kl}|^2 \geqslant 0, \quad 1 \leqslant k,l \leqslant p,$$

whence

$$|-\mathrm{i}yW(\mathrm{i}y)_{kl}| \leqslant (\boldsymbol{S}_0)_{kk}, \quad 1 \leqslant k,l \leqslant p,$$

where $(\boldsymbol{A})_{kl}$ denotes the (k,l) entry of a matrix $\boldsymbol{A}$. Thus, $\sup\limits_{y\geqslant 1} \| \mathrm{i}yW(\mathrm{i}y) \| < +\infty$ for a certain matrix norm $\| \cdot \|$ on $\mathbf{C}^{p\times p}$, and therefore, by (1.4), $W(\lambda) = \int_{-\infty}^{+\infty} (u-\lambda)\mathrm{d}\sigma(u)$ which has the asymptotic representation: $\lim\limits_{\lambda\to\infty} -\lambda W(\lambda) = \int_{-\infty}^{+\infty} \mathrm{d}\sigma(u)$ by Lemma 2.1. Going to the limit as $y \to +\infty$, from (4.5) we obtain

$$\begin{bmatrix} \boldsymbol{S}_0 & \int_{-\infty}^{+\infty} \mathrm{d}\sigma(u) \\ \int_{-\infty}^{+\infty} \mathrm{d}\sigma(u) & \int_{-\infty}^{+\infty} \mathrm{d}\sigma(u) \end{bmatrix} \geqslant \boldsymbol{O},$$

which is equivalent to

$$\boldsymbol{S}_0 - \int_{-\infty}^{+\infty} \mathrm{d}\sigma(u) \left[\int_{-\infty}^{+\infty} \mathrm{d}\sigma(u)\right]^{\mathrm{D}} \int_{-\infty}^{+\infty} \mathrm{d}\sigma(u) = \boldsymbol{S}_0 - \int_{-\infty}^{+\infty} \mathrm{d}\sigma(u) \geqslant \boldsymbol{O},$$

since index $\left[\int_{-\infty}^{+\infty} \mathrm{d}\sigma(u)\right] = 1$ obviously, that is, $\int_{-\infty}^{+\infty} \mathrm{d}\sigma(u) \leqslant \boldsymbol{S}_0$.

Next, we consider FMI($\mathscr{H}$) in the case of $n \geqslant 1$. If we multiply FMI($\mathscr{H}$), step by step, on the left by

$$T_i=\begin{bmatrix} I_{(n+1)p} & & & & \vdots & O \\ \cdots & \cdots & \cdots & \cdots & \cdots & \cdots \\ O,\cdots,O,I_p, & O, & \cdots, & O & \vdots & \bar{\lambda}I_p \end{bmatrix},\quad i=1,2,\cdots,n,$$

and on the left by T_i^*, and applying the similar argument to that given in [17], we obtain

$$\begin{bmatrix} S_0 & S_1 & \cdots & S_n & \vdots & \lambda^n[W(\lambda)+S_0/\lambda+\cdots+S_{n-1}/\lambda^n] \\ S_1 & S_2 & \cdots & S_{n+1} & \vdots & \lambda^{n+1}[W(\lambda)+S_0/\lambda+\cdots+S_n/\lambda^{n+1}] \\ \vdots & \vdots & \vdots & \vdots & \vdots & \vdots \\ S_n & S_{n+1} & \cdots & S_{2n} & \vdots & E(\lambda) \\ \vdots & \vdots & \vdots & \vdots & \vdots & \vdots \\ & & * & & \vdots & \dfrac{E(\lambda)-E(\lambda)^*}{\lambda-\bar{\lambda}} \end{bmatrix}\geqslant O,$$

$\lambda\in\pi^+$, in which

$$E(\lambda)=\lambda^{2n}\left[W(\lambda)+\frac{S_0}{\lambda}+\cdots+\frac{S_{2n-1}}{\lambda^{2n}}\right],$$

and therefore, in particular,

$$\begin{bmatrix} S_{2n} & E(\lambda) \\ E(\lambda)^* & \dfrac{E(\lambda)-E(\lambda)^*}{\lambda-\bar{\lambda}} \end{bmatrix}\geqslant O,\quad \lambda\in\pi^+.$$

From the result proved for the case of $n=0$, it follows that $E(\lambda)\in \mathbf{N}_p$ has the asymptotic representation: $-\lim\limits_{\lambda\to\infty}\lambda E(\lambda)=\tilde{S}_{2n}\leqslant S_{2n}$, where $\tilde{S}_{2n}\geqslant O$, that is,

$$-\lim_{\lambda\to\infty}\lambda^{2n+1}\left[W(\lambda)+\frac{S_0}{\lambda}+\cdots+\frac{S_{2n-1}}{\lambda^{2n}}\right]=\tilde{S}_{2n}\leqslant S_{2n}.$$

This completes the proof. ■

Thanks to Lemma 2. 1, it is easy to see that Theorem 4. 2 is equivalent to the following result, which is at first considered in [20] (Theorem 2. 10) in the nondegenerate case.

Theorem 4. 3 *The moment problem* (3. 4) *is equivalent to* FMI($\mathscr{H}$). *Further, the relation between their solutions can be formulated as*

$$W(\lambda)=\int_{-\infty}^{+\infty}\frac{\mathrm{d}\sigma(u)}{u-\lambda},\tag{4. 6}$$

where $\sigma(u)$, $W(\lambda)$ *are the solutions to the problem* (3.4) *and to* FMI($\mathscr{H}$), *respectively*.

It is known ([17], p. 448) that the general solution to FMI ($\mathscr{H}$) in the case when $n=0$ and $\boldsymbol{S}_0>\boldsymbol{O}$ can be described as a linear fractional transformation of an arbitrary Nevanlinna pair $\begin{bmatrix} p(\lambda) \\ q(\lambda) \end{bmatrix}$:

$$W(\lambda)=\frac{-q(\lambda)}{p(\lambda)+\lambda\boldsymbol{S}_0^{-1}q(\lambda)}, \tag{4.7}$$

which can be rewritten as

$$W(\lambda)=\frac{-\boldsymbol{S}_0[\boldsymbol{S}_0^{-1}q(\lambda)\boldsymbol{S}_0^{-1}]}{\boldsymbol{S}_0^{-1}[\boldsymbol{S}_0p(\lambda)\boldsymbol{S}_0^{-1}]+\boldsymbol{\lambda}[\boldsymbol{S}_0^{-1}q(\lambda)\boldsymbol{S}_0^{-1}]}.$$

Let $\tilde{p}(\lambda)=\boldsymbol{S}_0p(\lambda)\boldsymbol{S}_0^{-1}$ and $\tilde{q}(\lambda)=\boldsymbol{S}_0^{-1}q(\lambda)\boldsymbol{S}_0^{-1}$, then $\begin{bmatrix} \tilde{p}(\lambda) \\ \tilde{q}(\lambda) \end{bmatrix}$ is also an arbitrary Nevanlinna pair, if $\begin{bmatrix} p(\lambda) \\ q(\lambda) \end{bmatrix}$ is, and vice versa. Then we have an equivalent form of Eq. (4.7)

$$W(\lambda)=\frac{-\boldsymbol{S}_0\tilde{q}(\lambda)}{\boldsymbol{S}_0^{-1}\tilde{p}(\lambda)+\lambda\tilde{q}(\lambda)}. \tag{4.8}$$

Thus, we have proved the following lemma.

Lemma 4.4 *Let* $\boldsymbol{S}_0>\boldsymbol{O}$. *Then the general solution* $\sigma(u)$ *to the moment problem* (3.4) *for the case of* $n=0$ *is representable as a linear fractional transformation of an arbitrary Nevanlinna pair* $\begin{bmatrix} p(\lambda) \\ q(\lambda) \end{bmatrix}$:

$$\int_{-\infty}^{+\infty}\frac{\mathrm{d}\sigma(u)}{u-\lambda}=\frac{-\boldsymbol{S}_0q(\lambda)}{\boldsymbol{S}_0^{-1}p(\lambda)+\lambda q(\lambda)} \tag{4.9}$$

with coefficient matrix

$$\begin{bmatrix} \boldsymbol{O} & -\boldsymbol{S}_0 \\ \boldsymbol{S}_0^{-1} & \lambda\boldsymbol{I}_p \end{bmatrix}. \tag{4.10}$$

Note that let

$$\boldsymbol{J}=\begin{bmatrix} \boldsymbol{O} & i\boldsymbol{I}_p \\ -i\boldsymbol{I}_p & \boldsymbol{O} \end{bmatrix},$$

then, as before, the matrix valued function (4.10) is $\boldsymbol{J}$-expanding in

π^+, $\boldsymbol{J}$-unitary on the real axis, of full rank, and has a pole of order one at ∞.

Also, if $\boldsymbol{S}_0 \geqslant \boldsymbol{O}$ is singular, it is not difficult to prove that Lemma 4.4 is valid as well with $\boldsymbol{S}_0^{\mathrm{D}}$ in place of $\boldsymbol{S}_0^{-1}$ therein and that if

$$\boldsymbol{U}^* \boldsymbol{S}_0 \boldsymbol{U} = \begin{bmatrix} \tilde{\boldsymbol{S}}_0 & \boldsymbol{O} \\ \boldsymbol{O} & \boldsymbol{O}_r \end{bmatrix}, \quad \tilde{\boldsymbol{S}}_0 > \boldsymbol{O},$$

then

$$\boldsymbol{U}^* p(\lambda) \boldsymbol{U} = \begin{bmatrix} \tilde{p}(\lambda) & \boldsymbol{O} \\ \boldsymbol{O} & \boldsymbol{O}_r \end{bmatrix} \quad \text{and} \quad \boldsymbol{U}^* q(\lambda) \boldsymbol{U} = \begin{bmatrix} \tilde{q}(\lambda) & \boldsymbol{O} \\ \boldsymbol{O} & \boldsymbol{I}_r \end{bmatrix}, \quad 0 < r \leqslant p,$$

where $\begin{bmatrix} \tilde{p}(\lambda) \\ \tilde{q}(\lambda) \end{bmatrix}$ is also an arbitrary Nevanlinna pair with reduced dimension.

Combining Theorem 3.6 with Lemma 4.4, we have

Theorem 4.5 *The general solution to the moment problem* (3.4) *is representable as a linear fractional transformation of an arbitrary Nevanlinna pair* $\begin{bmatrix} p(\lambda) \\ q(\lambda) \end{bmatrix}$:

$$\int_{-\infty}^{+\infty} \frac{\mathrm{d}\sigma(u)}{u-\lambda} = \frac{\theta_{11}(\lambda) p(\lambda) + \theta_{12}(\lambda) q(\lambda)}{\theta_{21}(\lambda) p(\lambda) + \theta_{22}(\lambda) q(\lambda)}, \tag{4.11}$$

whose coefficient matrix is of the decomposable form

$$\boldsymbol{\Theta}(\lambda) = \begin{bmatrix} \theta_{11}(\lambda) & \theta_{12}(\lambda) \\ \theta_{21}(\lambda) & \theta_{22}(\lambda) \end{bmatrix} = \overrightarrow{\prod_{i=0}^{n}} \begin{bmatrix} \boldsymbol{O} & -\boldsymbol{S}_0^{(i)} \\ \boldsymbol{S}_0^{(i)\mathrm{D}} & \lambda \boldsymbol{I}_p - \boldsymbol{S}_0^{(i)\mathrm{D}} \boldsymbol{S}_1^{(i)} \end{bmatrix}, \quad \boldsymbol{S}_1^{(n)} = \boldsymbol{O}. \tag{4.12}$$

In the degenerate case ($\boldsymbol{\Gamma}_n \geqslant \boldsymbol{O}$ *is singular*), *if*

$$\boldsymbol{U}^* \boldsymbol{S}^{(n)} \boldsymbol{U} = \begin{bmatrix} \hat{\boldsymbol{S}}_0^{(n)} & \boldsymbol{O} \\ \boldsymbol{O} & \boldsymbol{O}_r \end{bmatrix}, \quad \hat{\boldsymbol{S}}_0^{(n)} > \boldsymbol{O},\ 0 < r \leqslant p,$$

then

$$\boldsymbol{U}^* p(\lambda) \boldsymbol{U} = \begin{bmatrix} \tilde{p}(\lambda) & \boldsymbol{O} \\ \boldsymbol{O} & \boldsymbol{O}_r \end{bmatrix} \quad \textit{and} \quad \boldsymbol{U}^* q(\lambda) \boldsymbol{U} = \begin{bmatrix} \tilde{q}(\lambda) & \boldsymbol{O} \\ \boldsymbol{O} & \boldsymbol{I}_r \end{bmatrix},$$

where $\begin{bmatrix} \tilde{p}(\lambda) \\ \tilde{q}(\lambda) \end{bmatrix}$ *is also an arbitrary Nevanlinna pair. Further. if and on-*

ly if $S_0^{(n)}=O$, *the problem* (3.4) *has only one solution* $\sigma(u)$, *which corresponds to* (4.11) *with* $p(\lambda)=O$ *and* $q(\lambda)=I_p$.

It is worth to point out that in the nondegenerate case: $\Gamma_n>O$, the matrix decomposition of the form (4.12) is essentially nothing other than a superposition of elementary linear fractional transformations which has been presented by Kovalishina [17] as a Blaschke-Potopov product in the so-called Schur's stepwise process. Thus this leads to the solution of the full moment problem consisting of finding a Hermitian measure $\sigma(u)$ subject to $S_k=\int_{-\infty}^{+\infty}\mathrm{d}\sigma(u), k=0,1,\cdots$, to an infinite product of the form

$$\overrightarrow{\prod_{i=0}^{\infty}}\begin{bmatrix} O & -S_0^{(i)} \\ S_0^{(i)-1} & \lambda I_p - S_0^{(i)-1}S_1^{(i)} \end{bmatrix}, \tag{4.13}$$

each factor of which is J-expanding in π^+, J-unitary on the real axis, of full rank and has only a pole of order one at $\lambda=\infty$.

In the degenerate case ($\Gamma_n\geqslant O$ is singular), Theorem 4.5 coincides essentially with Theorem 1.1 of [6](pp. 25～27) but with "the moment problem (3.4)" in place of "the problem H"therein. Consequently, this also leads to the solution of the full moment problem mentioned above under the condition $\Gamma_n\geqslant O, n=0,1,\cdots$ We will omit this investigation here.

As a rule, the independent parameters in Eq. (4.11) are the equivalence classes of Nevanlinna pairs: different pairs lead to the same measure $\sigma(u)$ if and only if they are equivalent to each other.

Our next object is to consider the close relation of Theorems 3.4 and 3.6 to the corresponding results given in [11,9] based on the use of the theory of orthogonal polynomial matrices in the nondegenerate case, which receive further development suitable to the degenerate case.

Defined the recurrence relations under the condition $\Gamma_n\geqslant O$:

$$\begin{aligned} M_{i+1}(\lambda) &= M_i(\lambda)[\lambda I_p - S_0^{(i)\mathrm{D}}S_1^{(i)}] - M_{i-1}(\lambda)S_0^{(i-1)\mathrm{D}}S_1^{(i)}, \\ N_{i+1}(\lambda) &= N_i(\lambda)[\lambda I_p - S_0^{(i)\mathrm{D}}S_1^{(i)}] - N_{i-1}(\lambda)S_0^{(i-1)\mathrm{D}}S_1^{(i)}, \end{aligned} \tag{4.14}$$

for $i=1,2,\cdots,n$, with initial conditions:

$$M_0(\lambda)=\boldsymbol{I}_p,\quad M_1(\lambda)=\lambda\boldsymbol{I}_p-\boldsymbol{S}_0^{(0)\mathrm{D}}\boldsymbol{S}_1^{(0)},N_0(\lambda)=\boldsymbol{O},\quad N_1(\lambda)=\boldsymbol{S}_0^{(0)},$$

where $\boldsymbol{S}_0^{(0)}$, $\boldsymbol{S}_1^{(0)}$, …, $\boldsymbol{S}_0^{(n-1)}$, $\boldsymbol{S}_1^{(n-1)}$, $\boldsymbol{S}_0^{(n)}$ are defined as in Section 3. Thus

$$F_1(\lambda)=-\frac{\boldsymbol{S}_0^{(0)}}{\lambda\boldsymbol{I}_p-\boldsymbol{S}_0^{(0)\mathrm{D}}\boldsymbol{S}_1^{(0)}}=-\frac{N_1(\lambda)}{M_1(\lambda)},$$

$$F_2(\lambda)=-\frac{\boldsymbol{S}_0^{(0)}}{\lambda\boldsymbol{I}_P-\boldsymbol{S}_0^{(0)\mathrm{D}}\boldsymbol{S}_1^{(0)}-\dfrac{\boldsymbol{S}_0^{(0)\mathrm{D}}\boldsymbol{S}_0^{(1)}}{\lambda\boldsymbol{I}_P-\boldsymbol{S}_0^{(1)\mathrm{D}}\boldsymbol{S}_1^{(1)}}}=-\frac{N_2(\lambda)}{M_2(\lambda)},\tag{4.15}$$

…

$$F_{n-1}(\lambda)=-\frac{N_{n-1}(\lambda)}{M_{n-1}(\lambda)},$$

which can be considered as the truncated parts of the matrix continued fraction given in Eq. (3.7). We now show that if $\boldsymbol{\Gamma}_n$ is n. e., the matrix sequence of $M_0(\lambda),M_1(\lambda),\cdots,M_n(\lambda)$ defined by Eq. (4.14)has a generalized orthogonal property with respect to the sequence $\boldsymbol{S}_0,\boldsymbol{S}_1,\cdots,\boldsymbol{S}_{2n}$ in the sense that

$$\int_{-\infty}^{+\infty}M_i^*(u)\mathrm{d}\sigma(u)M_j(u)$$

$$=\begin{cases}\boldsymbol{O}, & i\neq j,\\ \boldsymbol{S}_{2i}-[\boldsymbol{S}_i,\cdots,\boldsymbol{S}_{2i-1}]\boldsymbol{\Gamma}_{i-1}^{\mathrm{D}}[\boldsymbol{S}_i,\cdots,\boldsymbol{S}_{2i-1}]^*, & i=j\geqslant 1,\\ \boldsymbol{S}_0, & i=j=0,\end{cases}\tag{4.16}$$

where $\sigma(u)$ is a solution to the TH problem (2.1). (In the case of $\boldsymbol{\Gamma}_n>\boldsymbol{O}$, Eq. (4.16)means that $\{M_i(\lambda)\}$ is a sequence of right orthogonal matrices with respect to the Hermitian measure $\sigma(u)$(or the sequence of $\boldsymbol{S}_0,\boldsymbol{S}_1,\cdots,\boldsymbol{S}_{2n}$)[11].)

Indeed, assume $i\geqslant j$ and $i\geqslant 1$ (for if $i=j=0$, Eq. (4.16)holds obviously), and rewrite $M_i(\lambda)$ and $M_j(\lambda)$ in the form

$$M_i(\lambda)=\boldsymbol{M}_{i0}+\boldsymbol{M}_{i1}\lambda+\cdots+\boldsymbol{M}_{i,i-1}\lambda^{i-1}+\boldsymbol{I}_p\lambda^i,$$
$$M_j(\lambda)=\boldsymbol{M}_{j0}+\boldsymbol{M}_{j1}\lambda+\cdots+\boldsymbol{M}_{j,j-1}\lambda^{j-1}+\boldsymbol{I}_p\lambda^j.\tag{4.17}$$

Then

$$\int_{-\infty}^{+\infty}M_i^*(u)\mathrm{d}\sigma(u)M_j(u)$$

$$=[\boldsymbol{M}_{i0}^*,\cdots,\boldsymbol{M}_{i,i-1}^*,\boldsymbol{I}_p]\boldsymbol{\Gamma}_i[\boldsymbol{M}_{j0}^*,\cdots,\boldsymbol{M}_{j,j-1}^*,\boldsymbol{I}_p,\boldsymbol{O},\cdots,\boldsymbol{O}]^*.\tag{4.18}$$

On the other hand, it is not difficult to verify that the Laurent expansion of the rational matrix $N_i(\lambda)/M_i(\lambda)$ at $\lambda=\infty$ admits the form

$$\frac{N_i(\lambda)}{M_i(\lambda)} = \frac{\boldsymbol{S}_0}{\lambda} + \cdots + \frac{\boldsymbol{S}_{2i-1}}{\lambda^{2i}} + \frac{\tilde{\boldsymbol{S}}_{2i}}{\lambda^{2i+1}} + o(\lambda^{-2i-1}),$$

so that

$$[\boldsymbol{M}_{i0}^*, \cdots, \boldsymbol{M}_{i,i-1}^*, \boldsymbol{I}_p]\begin{bmatrix} \boldsymbol{S}_0 & \boldsymbol{S}_1 & \cdots & \boldsymbol{S}_i \\ \boldsymbol{S}_1 & \boldsymbol{S}_2 & \cdots & \boldsymbol{S}_{i+1} \\ \vdots & \vdots & \vdots & \vdots \\ \boldsymbol{S}_i & \boldsymbol{S}_{i+1} & \cdots & \tilde{\boldsymbol{S}}_{2i} \end{bmatrix} = [\boldsymbol{O}, \boldsymbol{O}, \cdots, \boldsymbol{O}].$$

or equivalently

$$[\boldsymbol{M}_{i0}^*, \cdots, \boldsymbol{M}_{i,i-1}^*]\boldsymbol{\Gamma}_{i-1} = -[\boldsymbol{S}_i, \cdots, \boldsymbol{S}_{2i-1}],$$

$$\tilde{\boldsymbol{S}}_{2i} = -\boldsymbol{M}_{i0}^* \boldsymbol{S}_i - \cdots - \boldsymbol{M}_{i,i-1}^* \boldsymbol{S}_{2i-1}.$$

Thus, since $\boldsymbol{\Gamma}_{i-1} \boldsymbol{\Gamma}_{i-1}^{\mathrm{D}} = \boldsymbol{\Gamma}_{i-1}^{\mathrm{D}} \boldsymbol{\Gamma}_{i-1}$ is the orthogonal projection onto $R(\boldsymbol{\Gamma}_{i-1})$ we have

$$-[\boldsymbol{S}_i, \cdots, \boldsymbol{S}_{2i-1}] = -[\boldsymbol{S}_i, \cdots, \boldsymbol{S}_{2i-1}]\boldsymbol{\Gamma}_{i-1}^{\mathrm{D}} \boldsymbol{\Gamma}_{i-1},$$

and therefore

$$\boldsymbol{S}_{2i} - \tilde{\boldsymbol{S}}_{2i} = \boldsymbol{S}_{2i} - [\boldsymbol{S}_i, \cdots, \boldsymbol{S}_{2i-1}]\boldsymbol{\Gamma}_{i-1}^{\mathrm{D}}[\boldsymbol{S}_1, \cdots, \boldsymbol{S}_{2i-1}]^*. \tag{4.19}$$

Then

$$\int_{-\infty}^{+\infty} M_i^*(u) \mathrm{d}\sigma(u) M_j(u) = [\boldsymbol{O}, \cdots, \boldsymbol{O}, \boldsymbol{S}_{2i} - \tilde{\boldsymbol{S}}_{2i}][\boldsymbol{M}_{j0}^*, \cdots, \boldsymbol{M}_{j,j-1}^*, \boldsymbol{I}_p, \boldsymbol{O}, \cdots, \boldsymbol{O}]^*$$

$$= \begin{cases} \boldsymbol{O}, & i \neq j, \\ \boldsymbol{S}_{2i} - \tilde{\boldsymbol{S}}_{2i}, & i = j \geqslant 1, \end{cases}$$

as needed thanks to (4.19).

Hence we have shown the following result.

Lemma 4.6 *Let $\boldsymbol{\Gamma}_n$ be n. e. and $M_i(\lambda)(i=0,1,\cdots,n)$ be defined by* (4.14). *Then $\{M_i(\lambda)\}_{i=0}^n$ has the generalized orthonogal property with respect to the sequence $\boldsymbol{S}_0, \boldsymbol{S}_1, \cdots, \boldsymbol{S}_{2n}$ in the sense given in* (4.16).

Note that the matrix $\boldsymbol{S}_{2i} - \tilde{\boldsymbol{S}}_{2i}$ given in Eq. (4.19) is as a general rule called the generalized Schur complement of $\boldsymbol{\Gamma}_{i-1}$ in $\boldsymbol{\Gamma}_i$, denoted by $\boldsymbol{\Gamma}_i / \boldsymbol{\Gamma}_{i-1}, i=1,2,\cdots,n$.

As a consequence of Lemma 4.6, we have

Corollary 4.7 *Let $\boldsymbol{\Gamma}_n$ be n. e. and $M_i(\lambda)(i=0,1,\cdots,n)$ be defined by* (4.14) *or* (4.17). *Then*

$$\begin{bmatrix} M_0(\lambda) \\ M_1(\lambda) \\ \vdots \\ M_n(\lambda) \end{bmatrix} = \boldsymbol{P}_n \begin{bmatrix} \boldsymbol{I}_p \\ \lambda \boldsymbol{I}_p \\ \vdots \\ \lambda^n \boldsymbol{I}_p \end{bmatrix}, \tag{4.20}$$

in which $\boldsymbol{P}_n$ *is as in* (3.9).

Proof Write the lower block triangular matrix $\boldsymbol{P}_n$ in the form

$$\boldsymbol{P}_n = \begin{bmatrix} \boldsymbol{I}_p & & & \boldsymbol{O} \\ \boldsymbol{P}_{10} & \boldsymbol{I}_p & & \\ \vdots & \ddots & \ddots & \\ \boldsymbol{P}_{n0} & \cdots & \boldsymbol{P}_{n,n-1} & \boldsymbol{I}_p \end{bmatrix}.$$

Then Eq. (4.20) is equivalent to

$$\boldsymbol{M}_n = \begin{bmatrix} \boldsymbol{I}_p & & & \boldsymbol{O} \\ \boldsymbol{M}_{10} & \boldsymbol{I}_p & & \\ \vdots & \ddots & \ddots & \\ \boldsymbol{M}_{n0} & \cdots & \boldsymbol{M}_{n,n-1} & \boldsymbol{I}_p \end{bmatrix} = \boldsymbol{P}_n. \tag{4.21}$$

But, by Lemma 4.6, $\int_{-\infty}^{+\infty} M_i^*(u)\mathrm{d}\sigma(u)M_j(u) = \hat{\boldsymbol{S}}_0^{(i)}\delta_{ij}$, where $\hat{\boldsymbol{S}}_0^{(0)} = \boldsymbol{S}_0$ and $\hat{\boldsymbol{S}}_0^{(i)} = \boldsymbol{\Gamma}_i/\boldsymbol{\Gamma}_{i-1}, i \geqslant 1$, and therefore

$$\boldsymbol{M}_n\boldsymbol{\Gamma}_n\boldsymbol{M}_n^* = \begin{bmatrix} \hat{\boldsymbol{S}}_0^{(0)} & & & \boldsymbol{O} \\ & \hat{\boldsymbol{S}}_0^{(1)} & & \\ & & \ddots & \\ \boldsymbol{O} & & & \hat{\boldsymbol{S}}_0^{(n)} \end{bmatrix}.$$

Now put

$$\boldsymbol{P}_n^{-1}\boldsymbol{M}_n = \begin{bmatrix} \boldsymbol{I}_p & & & \boldsymbol{O} \\ \boldsymbol{C}_{10} & \boldsymbol{I}_p & & \\ \vdots & \ddots & \ddots & \\ \boldsymbol{C}_{n0} & \cdots & \boldsymbol{C}_{n,n-1} & \boldsymbol{I}_p \end{bmatrix} = \boldsymbol{C}_n.$$

To prove Eq. (4.21), we need only to verify $\boldsymbol{C}_{ij} = \boldsymbol{O}$ $(0 \leqslant j < i \leqslant n)$. But this follows from Theorem 3.9 and the following equality.

$$C_n\begin{bmatrix} \boldsymbol{S}_0^{(0)} & & & \boldsymbol{O} \\ & \boldsymbol{S}_0^{(1)} & & \\ & & \ddots & \\ \boldsymbol{O} & & & \boldsymbol{S}_0^{(n)} \end{bmatrix} C_n^* = \begin{bmatrix} \hat{\boldsymbol{S}}_0^{(0)} & & & \boldsymbol{O} \\ & \hat{\boldsymbol{S}}_0^{(1)} & & \\ & & \ddots & \\ \boldsymbol{O} & & & \hat{\boldsymbol{S}}_0^{(n)} \end{bmatrix}.$$

This completes the proof. ■

As a useful by-product of the proof of Corollary 4.7, we obtain

$$\hat{\boldsymbol{S}}_0^{(0)} = \boldsymbol{S}_0^{(0)}, \quad \hat{\boldsymbol{S}}_0^{(i)} = \boldsymbol{\Gamma}_i/\boldsymbol{\Gamma}_{i-1} = \boldsymbol{S}^{(i)}, \quad i=1,2,\cdots,n. \tag{4.22}$$

Thus, Lemma 4.6 can be reformulated as follows.

Lemma 4.8 *Let $\boldsymbol{\Gamma}_n$ be n. e. and $M_i(\lambda)(i=0,1,\cdots,n)$ be defined by* (4.14). *Then $\{M_i(\lambda)\}_{i=0}^n$ has the generalized orthogonal property with respect to the sequence $\boldsymbol{S}_0, \boldsymbol{S}_1, \cdots, \boldsymbol{S}_{2n}$ in the sense that*

$$\int_{-\infty}^{+\infty} M_i^*(i)\mathrm{d}\sigma(u)M_j(u) = \boldsymbol{S}_0^{(i)}\delta_{ij}, \tag{4.16'}$$

where $\sigma(u)$ is a solution to the TH problem (2.1).

As for the moment problem (3.4), we have a result little short of Eq. (4.16′).

Lemma 4.9 *Let $\boldsymbol{\Gamma}_n \geqslant \boldsymbol{O}$ and $M_i(\lambda)(i=0,1,\cdots,n)$ be as before. Then*

$$\int_{-\infty}^{+\infty} M_i^*(u)\mathrm{d}\sigma(u)M_j(u)\begin{cases} \leqslant \tilde{\boldsymbol{S}}_0^{(n)}, & i=j=n, \\ = \boldsymbol{S}_0^{(i)}\delta_{ij}, & \text{otherwise}, \end{cases}$$

where $\sigma(u)$ is a solution to the moment problem (3.4), *and $\tilde{\boldsymbol{S}}_0^{(n)}$ is as in Theorem* 3.11.

Thanks to Theorems 3.4 and 3.6, and letting $\boldsymbol{\Gamma}_{n-1}/\boldsymbol{\Gamma}_{n-2} = \boldsymbol{S}_0$ if $n=1$; $\boldsymbol{\Gamma}_{n-1}/\boldsymbol{\Gamma}_{n-2} = \boldsymbol{I}_p$ if $n \leqslant 0$, Lemma 4.9 together with Theorem 3.11 leads to the following.

Corollary 4.10 *$\boldsymbol{\Gamma}_n$ is n. e. if and only if $\boldsymbol{\Gamma}_n \geqslant \boldsymbol{O}$ and*

$$R(\boldsymbol{\Gamma}_n/\boldsymbol{\Gamma}_{n-1}) \subset R(\boldsymbol{\Gamma}_{n-1}/\boldsymbol{\Gamma}_{n-2}). \tag{4.23}$$

Corollary 4.11 *The TH problem* (2.1) (*or the moment problem* (3.4)) *has at most one solution if and only if*

$$R(\boldsymbol{\Gamma}_n/\boldsymbol{\Gamma}_{n-1}) \perp R(\boldsymbol{\Gamma}_{n-1}/\boldsymbol{\Gamma}_{n-2}). \tag{4.24}$$

Proof Write

$$\boldsymbol{\Gamma}_n/\boldsymbol{\Gamma}_{n-1} = [\boldsymbol{I}_p - \boldsymbol{S}_0^{(n-1)}\boldsymbol{S}_0^{(n-1)\mathrm{D}}]\boldsymbol{\Gamma}_n/\boldsymbol{\Gamma}_{n-1} + \boldsymbol{S}_0^{(n-1)}\boldsymbol{S}_0^{(n-1)\mathrm{D}}\boldsymbol{\Gamma}_n/\boldsymbol{\Gamma}_{n-1}.$$

It is easy to check that $\boldsymbol{S}_0^{(n)}=\boldsymbol{S}_0^{(n-1)}\boldsymbol{S}_0^{(n-1)\mathrm{D}}\boldsymbol{\Gamma}_n/\boldsymbol{\Gamma}_{(n-1)}\boldsymbol{S}_0^{(n-1)\mathrm{D}}\boldsymbol{S}_0^{(n-1)}$. In view of the facts $(\boldsymbol{S}_0^{(n-1)\mathrm{D}})^{\mathrm{D}}=\boldsymbol{S}_0^{(n-1)}$ and $\boldsymbol{S}_0^{(n-1)}=\boldsymbol{\Gamma}_{n-1}/\boldsymbol{\Gamma}_{n-2}$, $R((\boldsymbol{I}_p-\boldsymbol{S}_0^{(n-1)}\boldsymbol{S}_0^{(n-1)\mathrm{D}})\boldsymbol{\Gamma}_n/\boldsymbol{\Gamma}_{n-1})\perp R(\boldsymbol{S}_0^{(n)})$. If now (4. 24) holds, then $(\boldsymbol{S}_0^{(n-1)})^2\boldsymbol{S}_0^{(n-1)\mathrm{D}}\boldsymbol{\Gamma}_n/\boldsymbol{\Gamma}_{n-1}=\boldsymbol{S}_0^{(n-1)}\boldsymbol{\Gamma}_n/\boldsymbol{\Gamma}_{n-1}=\boldsymbol{O}$, so that

$$\boldsymbol{S}_0^{(n-1)\mathrm{D}}\boldsymbol{S}_0^{(n-1)}\boldsymbol{\Gamma}_n/\boldsymbol{\Gamma}_{n-1}\boldsymbol{S}_0^{(n-1)\mathrm{D}}\boldsymbol{S}_0^{(n-1)}=\boldsymbol{S}_0^{(n-1)}\boldsymbol{S}_0^{(n-1)\mathrm{D}}\boldsymbol{\Gamma}_n/\boldsymbol{\Gamma}_{n-1}\boldsymbol{S}_0^{(n-1)\mathrm{D}}\boldsymbol{S}_0^{(n-1)}$$
$$=\boldsymbol{S}_0^{(n)}=\boldsymbol{O}.$$

Applying Corollary 3. 5 or 3. 7, we obtain that the problem (2. 1) or (3. 4) has at most one solution. Conversely, if

$$\boldsymbol{S}_0^{(n)}=\boldsymbol{S}_0^{(n-1)\mathrm{D}}\boldsymbol{S}_0^{(n-1)}\boldsymbol{\Gamma}_n/\boldsymbol{\Gamma}_{n-1}\boldsymbol{S}_0^{(n-1)\mathrm{D}}\boldsymbol{S}_0^{(n-1)}=\boldsymbol{O},$$

then $\boldsymbol{S}_0^{(n-1)\mathrm{D}}(\boldsymbol{\Gamma}_n/\boldsymbol{\Gamma}_{n-1})^{1/2}=\boldsymbol{O}$, and therefore $(\boldsymbol{S}_0^{(n-1)})^2\boldsymbol{S}_0^{(n-1)\mathrm{D}}\boldsymbol{\Gamma}_n/\boldsymbol{\Gamma}_{n-1}=\boldsymbol{O}$. Hence Eq. (4. 24) holds, as claimed. ■

Note that Ando ([4], Section 3) has settled the extension theorems for a bounded positive operator.

In conclusion, thanks to Theorems 3. 4 and 3. 6 together with Lemma 4. 8 and Lemma 4. 9, we may derive the general solution to the TH problem or to the moment problem (3. 4) in terms of a linear fractional transformation based on the use of the orthogonal polynomial matrices $M_0(\lambda), M_1(\lambda), \cdots, M_n(\lambda)$, and the corresponding polynomial matrices of the second kind, $N_0(\lambda), N_1(\lambda), \cdots, N_n(\lambda)$ (see [11,9] for the nondegenerate case).

Theorem 4. 12 *The general solution $\sigma(u)$ to the TH problem (2. 1) (The moment problem (3. 4) resp.) is representable as a lineal fractional transformation*

$$\int_{-\infty}^{+\infty}\frac{\mathrm{d}\sigma(u)}{u-\lambda}=-\frac{N_n(\lambda)+N_{n-1}(\lambda)\boldsymbol{S}_0^{(n-1)\mathrm{D}}\Phi(\lambda)}{M_n(\lambda)+M_{n-1}(\lambda)\boldsymbol{S}_0^{(n-1)\mathrm{D}}\Phi(\lambda)}, \tag{4. 25}$$

where $\Phi(\lambda)=\int_{-\infty}^{+\infty}\mathrm{d}\sigma(u)/(u-\lambda)$ for an arbitrary Hermitian measure $\sigma(u)$ satisfying $\int_{-\infty}^{+\infty}\mathrm{d}\tau(u)=\boldsymbol{S}_0^{(n)}$ $\left(\int_{-\infty}^{+\infty}\mathrm{d}\tau(u)\leqslant\boldsymbol{S}_0^{(n)}, resp.\right)$.

Note that the correspondence defined by Eq. (4. 25) between Hermitian measures $\sigma(u)$ and function $\Phi(\lambda)$ with the aforementioned properties is one-to-one.

In the case when $\boldsymbol{\Gamma}_n$ is n. e., the general solution $\sigma(u)$ to the TH problem of the form (4. 25) can be rewritten, by Eq. (4. 3), in the form

$$\int_{-\infty}^{+\infty}\frac{d\sigma(u)}{u-\lambda}=-\frac{N_n(\lambda)[\lambda I_p+S_0^{(n)D}G(\lambda)]-N_{n-1}(\lambda)S_0^{(n-1)D}S_0^{(n)}}{M_n(\lambda)[\lambda I_p+S_0^{(n)D}G(\lambda)]-M_{n-1}(\lambda)S_0^{(n-1)D}S_0^{(n)}},\tag{4.26}$$

where $G(\lambda)\in \mathbf{N}_p$ is an arbitrary matrix-valued function such that $\lim\limits_{\lambda\to\infty} G(\lambda)/\lambda=\boldsymbol{O}$. In particular, we may set $\boldsymbol{S}_1^{(n)}=\boldsymbol{O}$, so that $M_{n+1}(\lambda)$ and $N_{n+1}(\lambda)$ make sense by Eq. (4.14). Then Eq. (4.26) has the further form

$$\int_{-\infty}^{+\infty}\frac{d\sigma(u)}{u-\lambda}=-\frac{N_{n+1}(\lambda)+N_n(\lambda)S_0^{(n)D}G(\lambda)}{M_{n+1}(\lambda)+M_n(\lambda)S_0^{(n)D}G(\lambda)},\quad S_1^{(n)}=O.\tag{4.26'}$$

Also, in the case of $\boldsymbol{\Gamma}_n\geqslant\boldsymbol{O}$, the general solution $\sigma(u)$ to the moment problem (3.4) can be rewritten, by Lemma 4.3 and the remark about it, in the form

$$\int_{-\infty}^{+\infty}\frac{d\sigma(u)}{u-\lambda}=-\frac{N_n(\lambda)S_0^{(n)D}p(\lambda)+[\lambda N_n(\lambda)-N_{n-1}(\lambda)S_0^{(n-1)D}S_0^{(n)}]q(\lambda)}{M_n(\lambda)S_0^{(n)D}p(\lambda)+[\lambda M_n(\lambda)-M_{n-1}(\lambda)S_0^{(n-1)D}S_0^{(n)}]q(\lambda)},\tag{4.27}$$

where $\begin{bmatrix}p(\lambda)\\q(\lambda)\end{bmatrix}$ is an arbitrary Nevanlinna pair of the form formulated as in Theorem 4.5.

The correspondence defined by Eq. (4.27) between $\sigma(u)$ and all equivalence classes of Nevanlinna pairs $\begin{bmatrix}p(\lambda)\\q(\lambda)\end{bmatrix}$ is one-to-one. Analogous to (4.26′), (4.27) has the further form

$$\int_{-\infty}^{+\infty}\frac{d\sigma(u)}{u-\lambda}=-\frac{N_n(\lambda)S_0^{(n)D}p(\lambda)+N_{n+1}(\lambda)q(\lambda)}{M_n(\lambda)S_0^{(n)D}p(\lambda)+M_{n+1}(\lambda)q(\lambda)},\quad S_1^{(n)}=O.\tag{4.27'}$$

§ 5. The Nevanlinna-Pick interpolation with multiple nodes in the Nevanlinna class $\mathbf{N}_p$

As an application of the aforementioned results in this paper, in this section, we consider the nontangential Nevanlinna-Pick interpolation (NP) problem with multiple nodes in the class $\mathbf{N}_p$. (See [5,12,14] and references therein for more information.)

The problem is as follows. Given $\lambda_1,\lambda_2,\cdots,\lambda_\theta\in\pi^+$, which are distinct, with multiplicities $\tau_1,\tau_2,\cdots,\tau_\theta$, respectively, $n=\sum\limits_{j=1}^{\theta}\tau_j$, and n

matrices of order p, $\boldsymbol{C}_{ik}$ ($i=1, 2, \cdots, \theta, k=0,1,\cdots, \tau_i-1$), it is required to find the conditions for a $F(\lambda)\in\mathbf{N}_p$ (or for only one $F(\lambda)\in\mathbf{N}_p$) to exist subject to

$$\frac{1}{k!}F^{(k)}(\lambda_i)=\boldsymbol{C}_{ik}, \quad i=1,2,\cdots,\theta, \quad k=0,1,\cdots,\tau_i-1, \tag{5.1a}$$

and to describe all the solutions if these conditions are met.

It is known [8～10] that to the NP problem (5.1a) there corresponds a unique Hermitian block vector $(\boldsymbol{S}_0,\boldsymbol{S}_1,\cdots,\boldsymbol{S}_{2n-2})$, $\boldsymbol{S}_j=\boldsymbol{S}_j^*$, order $\boldsymbol{S}_j=p$ such that the NP problem (5.1a) is equivalent to a certain matrix moment problem on the real axis associated with that vector $(\boldsymbol{S}_0, \boldsymbol{S}_1,\cdots,\boldsymbol{S}_{2n-2})$:

$$\boldsymbol{S}_k=\int_{-\infty}^{+\infty}u^k\mathrm{d}\sigma(u), \quad k=0,1,\cdots,2n-3, \tag{5.2}$$

$$\boldsymbol{S}_{n-2}\geqslant\int_{-\infty}^{+\infty}u^{2n-2}\mathrm{d}\sigma(u).$$

Moreover, if $\sigma(u)$ is a solution to the moment problem (5.2), then

$$F(\lambda)=\Omega(\lambda)+\int_{-\infty}^{+\infty}\frac{\mathrm{d}\sigma(u)}{u-\lambda}A(\lambda) \tag{5.3}$$

is a solution to that NP problem, and vice versa, where $\Omega(\lambda)$ is the (unique) polynomial matrix of degree $2n-1$ at most, subject only to Eq. (5.1a) and

$$\frac{1}{k!}F^k(\bar{\lambda}_i)=\boldsymbol{C}_{ik}^*, \quad i=1,2,\cdots,\theta,\ k=0,1,\cdots,\tau_i-1, \tag{5.1b}$$

and

$$A(\lambda)=\prod_{i=1}^{\theta}(\lambda-\lambda_i)^{\tau_i}(\lambda-\bar{\lambda}_i)^{\tau_i}. \tag{5.4}$$

That Hermitian block vector $(\boldsymbol{S}_0,\boldsymbol{S}_1,\cdots,\boldsymbol{S}_{2n-2})$ is called the Hankel block-vector of the NP problem (5.1a) and plays a key role in the NP problem. It is also known [8,9] that the Hankel block-vector can be found from the Laurent expension of $\Omega(\lambda)/A(\lambda)$ at $\lambda=\infty$:

$$\frac{\Omega(\lambda)}{A(\lambda)}=\frac{\boldsymbol{S}_0}{\lambda}+\frac{\boldsymbol{S}_1}{\lambda^2}+\cdots+\frac{\boldsymbol{S}_{2n-2}}{\lambda^{2n-1}}+\cdots \tag{5.5}$$

Thanks to the one-to-one correspondence defined by Eq. (5.3) between $\sigma(u)$ and $F(\lambda)$, the moment problem enables one not only to find the criteria of existence and uniqueness for the solutions to that NP problem but also to describe all the solutions in the nondegenerate case or not, and so on.

By now we are in such a more advantageous position that we are able to tackle seriously that NP problem with little additional effort. The following assertions are all at hand.

Theorem 5.1 ([20]) *The NP problem (5.1a) has a solution if and only if* $\boldsymbol{\Gamma}_{n-1}=[\boldsymbol{S}_{i+j}]_{i,j=0}^{n-1}\geqslant\boldsymbol{O}$, *where* $(\boldsymbol{S}_0,\boldsymbol{S}_1,\cdots,\boldsymbol{S}_{2n-2})$ *is the Hankel block-vector of that NP problem.*

Theorem 5.2 *The general solution* $F(\lambda)$ *to the NP problem (5.1a) is representable as a linear fractional transformation of an arbitrary Nevanlinna pair* $\begin{bmatrix}p(\lambda)\\ q(\lambda)\end{bmatrix}$:

$$F(\lambda)=\frac{\alpha(\lambda)p(\lambda)+\beta(\lambda)q(\lambda)}{\gamma(\lambda)p(\lambda)+\delta(\lambda)q(\lambda)} \tag{5.6}$$

whose coefficient matrix is of the decomposable form

$$\begin{bmatrix}\alpha(\lambda) & \beta(\lambda)\\ \gamma(\lambda) & \delta(\lambda)\end{bmatrix}=\begin{bmatrix}A(\lambda)\boldsymbol{I}_p & \Omega(\lambda)\\ \boldsymbol{O} & \boldsymbol{I}_p\end{bmatrix}\overrightarrow{\prod_{i=0}^{n-1}}\begin{bmatrix}\boldsymbol{O} & -\boldsymbol{S}_0^{(i)}\\ \boldsymbol{S}_0^{(i)\mathrm{D}} & \lambda\boldsymbol{I}_p-\boldsymbol{S}_0^{(i)\mathrm{D}}\boldsymbol{S}_1^{(i)}\end{bmatrix},\quad \boldsymbol{S}_1^{(n-1)}=\boldsymbol{O},$$

where $\boldsymbol{S}_0^{(0)},\boldsymbol{S}_0^{(1)},\cdots,\boldsymbol{S}_0^{(n-1)}$ *are defined as before from* $\boldsymbol{\Gamma}_{n-1}$.

In the degenerate case ($\boldsymbol{\Gamma}_{n-1}\geqslant\boldsymbol{O}$ *is singular*), *if*

$$\boldsymbol{U}^*\boldsymbol{S}_0^{(n-1)}\boldsymbol{U}=\begin{bmatrix}\hat{\boldsymbol{S}}_0^{(n-1)} & \boldsymbol{O}\\ \boldsymbol{O} & \boldsymbol{O}_r\end{bmatrix},\quad \hat{\boldsymbol{S}}_0^{(n-1)}>\boldsymbol{O},\quad 0<r\leqslant p,$$

then

$$\boldsymbol{U}^*p(\lambda)\boldsymbol{U}=\begin{bmatrix}\tilde{p}(\lambda) & \boldsymbol{O}\\ \boldsymbol{O} & \boldsymbol{O}_r\end{bmatrix}\quad\text{and}\quad \boldsymbol{U}^*q(\lambda)\boldsymbol{U}=\begin{bmatrix}\tilde{q}(\lambda) & \boldsymbol{O}\\ \boldsymbol{O} & \boldsymbol{I}_r\end{bmatrix},$$

where $\begin{bmatrix}\tilde{p}(\lambda)\\ \tilde{q}(\lambda)\end{bmatrix}$ *is also an arbitrary Nevanlinna pair with reduced dimension. Further, the NP problem* (5.1a) *has only one solution, if and only if* $\boldsymbol{\Gamma}_{n-1}\geqslant\boldsymbol{O}$ *and* $\boldsymbol{S}_0^{(n-1)}=\boldsymbol{O}$. *In this case, the unique solution* $F(\lambda)$ *is rational, of the form*

$$F(\lambda)=\frac{\beta(\lambda)}{\delta(\lambda)}. \tag{5.7}$$

Corollary 5.3 *Let* $n\geqslant 3$. *Then the NP problem (5.1a) has only one solution* $F(\lambda)$, *if and only if* $\boldsymbol{\Gamma}_{n-1}\geqslant\boldsymbol{O}$ *and*

$$R(\boldsymbol{\Gamma}_{n-1}/\boldsymbol{\Gamma}_{n-2})\perp R(\boldsymbol{\Gamma}_{n-2}/\boldsymbol{\Gamma}_{n-3}),\ \text{where } \boldsymbol{\Gamma}_k=[\boldsymbol{S}_{i+j}]_{i,j=0}^{k}.$$

Starting from Theorem 4.12, we have

Theorem 5.4 *The general solution $F(\lambda)$ to the NP problem* (5.1a) *is representable as a linear fractional transformation*

$$F(\lambda)=\Omega(\lambda)-\frac{N_{n-1}(\lambda)+N_{n-2}(\lambda)\boldsymbol{S}_0^{(n-2)\mathrm{D}}\Phi(\lambda)}{M_{n-1}(\lambda)+M_{n-2}(\lambda)\boldsymbol{S}_0^{(n-2)\mathrm{D}}\Phi(\lambda)}A(\lambda), \tag{5.8}$$

where $\Phi(\lambda)=\int_{-\infty}^{+\infty}1/(u-z)\mathrm{d}\tau(u)$ *for an arbitrary Hermitian measure* $\tau(u)$ *satisfying*

$$\int_{-\infty}^{+\infty}\mathrm{d}\tau(u)\leqslant \boldsymbol{S}_0^{(n-1)}.$$

The NP problem has only a solution if and only if $\boldsymbol{\Gamma}_{n-1}\geqslant\boldsymbol{O}$ *and* $\boldsymbol{S}_0^{(n-1)}=\boldsymbol{O}$. *In this case, the unique solution has the form*

$$F(\lambda)=\Omega(\lambda)-\frac{N_{n-1}(\lambda)}{M_{n-1}(\lambda)}A(\lambda). \tag{5.9}$$

In the scalar case, $\boldsymbol{\Gamma}_{n-1}\geqslant\boldsymbol{O}$ and $\boldsymbol{S}_0^{(n-1)}=\boldsymbol{O}$, if and only if $\boldsymbol{\Gamma}_{n-1}\geqslant\boldsymbol{O}$ is singular. Thus, the singularity of $\boldsymbol{\Gamma}_{n-1}$ will always lead to uniqueness of solutions to the NP problem (5.1a).

Note that Eq. (5.8) be rewritten in another form by means of Eq. (4.27)(or Eq. (4.27′), if we set $\boldsymbol{S}_1^{(n-1)}=\boldsymbol{O}$).

Additional Note While this article was in the course of the final manuscript the authors discovered that there is a paper by Bolotnikov [7] where the degenerate Hamburger matrix moment problem (which coincides with the moment problem (3.4) here) and extensions of non-negative block-Hankel matrices are considered via a different approach following the Potapov's method of the fundamental matrix inequality [17]. That paper appears to have several points in common with our paper (see Lemma 2.10 and Theorem 4.6 of [7]), and by the way the misstatement in [6] referred to by us here is corrected there.

Acknowledgements

We wish to thank the referee for his extremely useful suggestions and comments on the introductory remarks.

References

[1]V M Adamian, D Z Arov, M G Krein. Infinite Hankel block matrices and related extensions. Amer. Math . Soc. Transl, 1978, 111:133—156.

[2]N I Akhiezer, The Classical Moment Problem and Some Related Questions in Analysis. Oliver and Boyd, London, 1965.

[3]B D O Anderson, E I Jury. Generalized Bezoutian and Sylvester matrices in multivariable linear control. IEEE Trans. Automat. Control, 1976, AC-21, 551—556.

[4]T Ando. Truncated moment problems for operations. Acta. Sci. Math. (Szeged), 1970,31:319-334.

[5]J A Ball, I Gohberg, L Rodman. Interpolation of Rational Matrix Functions. Birkhäuser, Basel, 1990.

[6]V Bolotnikov. Description of the solutions of a degenerate moment problem on the axis and the semiaxis. Teoriya Funktsii Funktsional. Anal. i Prilozhen. 1988, 50:25-31.

[7]V Bolotnikov. On degenerate Hamburger moment problem and extensions of non-negative Hankel block matrices. Int. Eq. Oper. Theory, 1997, 25:253—276.

[8]G-N Chen. The general rational interpolation problem and its connection with the Nevanlinna-Pick interpolation and power moment problem. Linear Algebra Appl. 1998,273:83—117.

[9]G -N Chen, X-Q Li. The Nevanlinna-Pick interpolation problems and the power moment problems for matrix-valued functions. Linear Algebra Appl. , 1999,288:123-148.

[10]G-N Chen, B Zhao, H-P Zhang. The general rational interpolation problem in the scalar case and its Hankel vector. Linear Algebra Appl, 1996, 244: 165-201.

[11]A Draux. Formal orthogonal polynomials and Padé approximants in a non-commutative algebra. in: Proceedings of the M T N S. 83, Lecture Notes in Control and Inform. Sci. , Vol. 58, Springer, Berlin, 1984:278-292.

[12]H Dym. J-contractive matrix functions, reproducing kernel Hilbert spaces and interpolation. in: Regional Conference Series in Math. , Vol. 71, Amer. Math. Soc. , Providence, R I, 1989.

[13]H Dym. On Hermitian block Hankel matrices, matrix polynomials, the Hamburger moment problem, interpolation and maximum entropy. Int. Eq. Oper. Theory,1989,12:757-812.

[14] C Foias, A E Frazo. The Commutant Lifting Approach to Interpolation Problems. Birkhäuser, Basel, 1990.

[15] M I Gekhtman, M Shmoish. On invertibility of nonsquare generalized Bezoutians. Linear Algebra Appl. 1995, 223-224:205-241.

[16] A Kheifets. Hamburger moment problem: Parseval equality and A-singularity. J Funct. Anal. 1996,141:374-420.

[17] I Kovalishina. Analytic theory of a class of interpolation problems. Math. USSR Izv. 1984,22(3):419-463.

[18] M Krein, H Langer, On some extension problems which are clearly connected with theory of Hermitian operators in a space Π_k, Ⅲ. Indefinite analogous of the Hamburger and Stieltjes moment problems, Beit. Anal. 1979,14:25-40; 1980,15:27-45.

[19] M G Krein, A A Nudelman. The Markov Moment Problem and Extremal Problems. Transl. Math. Monographs, Vol. 50, Amer. Math. Soc., Providence, R. I., 1977.

[20] X-Q Li. A Class of Interpolation Problems and Moment Problems. Thesis, Beijing Normal Univ., Beijing, April 1996.

[21] J A Shohat, J D Tamarkin. The Problem of Moments. Amer. Math. Soc., Providence, R. I, 1943.

Linear Algebra and Its Applications
1998,283:179－203

$\mathbf{C}_p$ 类中多重 Nevanlinna-Pick 矩阵插值与 Carathéodory 矩阵系数问题①

On the Multiple Nevanlinna-Pick Matrix Interpolation in the Class $\mathbf{C}_p$ and the Carathéodory Matrix Coefficient Problem

Abstract In the present paper an intrinsic one-to-one correspondence between the (finite) multiple Nevanlinna-Pick matrix interpolation (NP) problem in the Carathéodory class and the Carathéodory matrix coefficient (CC) problem (or, equivalently, the truncated trigonometric matrix moment (TM) problem) is established. Thanks to this correspondence, the NP problem is reduced to what amounts to solving the CC problem (or the TM problem) associated with the so-called Toeplitz block-vector of the former in both the nondegenerate and degenerate cases simultaneously.

Keywords Nevanlinna-Pick interpolation; Carathéodory class; Carathéodory coefficient problem; Trigonometric moment problem; Toeplitz block-vector; Generalized orthogonal polynomial matrix

① Received 2 November 1997; received in revised form 27 April 1998; accepted 12 May 1998. Project supported by NSFC and Doctoral Programme Foundation of Institution of Higher Education. Submitted by R. A. Brualdi. 本文与胡永建合作. *AMS classification*: 15A57; 30E05;47A57.

§ 1. Introduction

As is well known, it is frequently the case that the finite interpolation problems of Nevanlinna-Pick type are bound up with and can be reduced, in a suitable way, to the study of the truncated power or trigonometric moment problems, which in turn can be regarded as the limiting cases at a single node of the former. There exist several approaches to the solution of these problems via this line (see, e. g. ,[18,2,13,15,16,19]). The transition, however, from the Nevanlinna-Pick type problems to their limiting cases seems to be not a simple matter even in the scalar case or in the case of simple nodes, and some details need to be considered and verified rather carefully.

The multiple Nevanlinna-Pick interpolation (NP) problem in the Nevanlinna class $\mathbf{N}_p$ has been recently examined via the so-called Hankel vector approach in [7] for the scalar case and in [9] for the matrix-valued function case. The studies were concerned mainly with an explicit one-to-one correspondence between the solutions to that NP problem and the solutions to a certain power moment problem associated with the so-called Hankel (block-)vector of the former.

In the present paper we primarily study an interesting connection between the (finite) NP problem but in the class $\mathbf{C}_p$ (Carathéodory) and the Carathéodory matrix coefficient (CC) problem (or, equivalently, the truncated trigonometric matrix moment (TM) problem). This then leads to a reduction of the solution of the finite NP problem to the solution of the corresponding CC problem (or the TM problem) in both the nondegenerate and degenerate cases simultaneously.

The overall strategy used here is much the same as that for the NP problem in the class $\mathbf{N}_p$[7,9]. The main tool adopted in our analysis is the notion of what we shall call the *Toeplitz block-vector* of an NP problem in the class $\mathbf{C}_p$. In analogy to [7,9] we shall show that with a fixed

NP problem in the class $\mathbf{C}_p$ there corresponds a unique block-vector, referred to as the *Toeplitz block-vector* of the NP problem, such that the NP problem can be reduced to what amounts to a certain CC problem (or, equivalently, a certain TM problem) associated with that Toeplitz block-vector. This development leads to an intrinsic one-to-one correspondence between the solutions to these two problems, and thus permits to reduce the NP problem in $\mathbf{C}_p$ to solving the CC or the TM problem.

Further on, let $\mathbf{S}_p$ denote the Schur class of $\mathbf{C}^{p\times p}$-valued functions. Since functions of classes $\mathbf{C}_p$ and $\mathbf{S}_p$ can be transformed into each other via a homographic transformation, the NP problems in these two classes are treated in essentially the same way.

Explicit formulations of the afore-cited problems are as follows. Throughout this paper, for Hermitian matrices $\boldsymbol{A}$ and $\boldsymbol{B}$, we write $\boldsymbol{A}\geqslant\boldsymbol{B}$ [$\boldsymbol{A}>\boldsymbol{B}$, resp.] to indicate $\boldsymbol{A}-\boldsymbol{B}$ is nonnegative [positive] Hermitian. $\|\boldsymbol{E}\|$ stands for the spectral norm (i. e., the largest singular value) of any matrix $\boldsymbol{E}$. Notice that if $\|\boldsymbol{E}\|\leqslant 1$, $\boldsymbol{E}\in\mathbf{C}^{p\times q}$, then and only then $\boldsymbol{E}\boldsymbol{E}^*\leqslant\boldsymbol{I}_p$ (or $\boldsymbol{E}^*\boldsymbol{E}\leqslant\boldsymbol{I}_q$). We sometimes write the function $z\to F(z)$ as simply $F(z)$.

(1) *The NP problem in the class* $\mathbf{C}_p$ [$\mathbf{S}_p$ resp.]: Given ρ distinct points $z_1, z_2, \cdots, z_\rho$ in the open unit disc D of the complex plane with multiplicities $\tau_1, \tau_2, \cdots, \tau_\rho, \tau_i\geqslant 1$, respectively, the number of nodes $n=\sum_{i=1}^{\rho}\tau_i\geqslant 2$ and $\boldsymbol{C}_{ik}\in\mathbf{C}^{p\times p}, i=1,2,\cdots,\rho, k=0,1,\cdots,\tau_i-1$, it is required to find conditions for the existence of a $G(z)\in\mathbf{C}_p$ [$\mathbf{S}_p$ resp.], the Carathéodory [Schur, resp.] class of $\mathbf{C}^{p\times p}$-valued functions $G(z)$ holomorphic in D and with nonnegative real part [contractive: $\|G(z)\|\leqslant 1$] there, such that

$$\frac{1}{k!}\frac{\mathrm{d}^k}{\mathrm{d}\lambda^k}G(\lambda)\bigg|_{\lambda=z_i}=\boldsymbol{C}_{ik},\quad i=1,2,\cdots,\rho,\ k=0,1,\cdots,\tau_i-1 \tag{1.1}$$

and to describe the solutions if these conditions are met.

(2) *The CC problem*: Given a $p\times p$ matrix sequence $\{\boldsymbol{c}_k\}_{k=0}^{n-1}$ with $\boldsymbol{c}_0=\boldsymbol{c}_0^*$, find $\mathbf{C}^{p\times p}$-valued functions $F(z)\in\mathbf{C}_p$ whose power series expan-

sions at $z=0$ begin with the terms $\boldsymbol{c}_0+2\boldsymbol{c}_1 z+\cdots+2\boldsymbol{c}_{n-1}z^{n-1}$, that is,

$$F(z)=\boldsymbol{c}_0+2\boldsymbol{c}_1 z+\cdots+2\boldsymbol{c}_{n-1}z^{n-1}+O(z^n). \tag{1.2}$$

It is well known (see, e.g., [18, 2, 12]) that the CC problem (1.2) is equivalent to the following TM problem.

(3) *The TM problem*: Given $\{\boldsymbol{c}_k\}_{k=0}^{n-1}$ with $\boldsymbol{c}_0=\boldsymbol{c}_0^*$ as in Eq. (1.2), it is required to find Hermitian measures $\sigma(\theta)$ on $[0,2\pi]$ (i.e., $\sigma(\theta)$ are bounded and nondecreasing $\mathbf{C}^{p\times p}$-valued functions on $[0,2\pi]$) subject to

$$\boldsymbol{c}_k=\frac{1}{2\pi}\int_0^{2\pi}\mathrm{e}^{-\mathrm{i}k\theta}\,\mathrm{d}\sigma(\theta),\quad k=0,1,\cdots,n-1. \tag{1.3}$$

A explicit relation between the solutions $F(z)$ to the CC problem (1.2) and the solutions $\sigma(\theta)$ to the TM problem (1.3) is expressed by the matrix extension of the Riesz - Herglotz integral representation,

$$F(z)=\mathrm{i}\ \mathrm{Im}\ F(0)+\frac{1}{2\pi}\int_0^{2\pi}\frac{\mathrm{e}^{\mathrm{i}\theta}+z}{\mathrm{e}^{\mathrm{i}\theta}-z}\mathrm{d}\sigma(\theta), \tag{1.4}$$

by which the class $\mathbf{C}_p$ is characterised as well, and with its inverse formula (see, e.g., [12], p. 36),

$$\sigma(\theta+)+\sigma(\theta-)=\boldsymbol{C}+\lim_{r\to 1-}\int_0^{\theta}[F(r\mathrm{e}^{\mathrm{i}\phi})+F^*(r\mathrm{e}^{\mathrm{i}\phi})]\mathrm{d}\phi, \tag{1.5}$$

where $\boldsymbol{C}$ is a constan matrix, and an asterisk (*) is used for Hermitian conjugate of the indicated matrix.

A popular criterion for the solvability of the NP problem (1.1) is the nonnegativity of a generalized Pick-block matrix (see (2.12) for definition). While the solvability of the CC problem (1.2) or the TM problem (1.3) depends on the nonnegativity of the associated block-Toeplitz matrix $(\boldsymbol{c}_{i-j})_{i,j=0}^{n-1}$ with $\boldsymbol{c}_{-k}=\boldsymbol{c}_k^*$ (see, e.g., [3, 20, 24]). A deep connection between the generalized block-Pick and block-Toeplitz matrices, as well as the mentioning connection with the NP problem (1.1) (see Theorem 2.4 and Corollaries 2.5～2.7) enables us to show the equivalence of these two solvability criteria. It should be noted ([20], pp. 17～19) that the nonnegativity of a block-Toeplitz matrix $(\boldsymbol{c}_{i-j})_{i,j=0}^{n-1}$ with $\boldsymbol{c}_{-k}=\boldsymbol{c}_k^*$ guarantees, in addition to the solvability of the TM problem (1.3), the extendability property in the sense that there exists some $\boldsymbol{c}_n$ such that $(\boldsymbol{c}_{i-j})_{i,j=0}^{n}\geqslant 0$, $\boldsymbol{c}_{-n}=\boldsymbol{c}_n^*$. This presents an interesting parallel to the theory of the block-Hankel matrices in the power moment problems. The corresponding phenomenon fails for the block-Hankel matrices, as well as for the truncated Hamburger matrix moment problem situation—nonnegativity of the original block-Hankel matrix could guarantee neither the extendability in the sense that it can be enlarged to a block-Hankel matrix of larger size (one more block-row and block-column) which is still nonnegative definite, nor the solvability property of the associated truncated Hamburger matrix moment problem (with equality on all the moment conditions). An evident example is the matrix

$$\begin{bmatrix} \mathbf{0}_p & \mathbf{0}_p \\ \mathbf{0}_p & \boldsymbol{I}_p \end{bmatrix}.$$

It is this case, if and only if that block-Hankel matrix is "proper", roughly speaking, its rank is concentrated in the upper lefthand corner as far as possible [6~8]. These reasons make evident the fact that the NP problem in $\mathbf{C}_p$ can be reduced to a certain TM problem (or a certain CC problem), whereas the NP problem in $\mathbf{N}_p$ cannot always be reduced to a certain Hamburger matrix moment problem. This illustrates the subtlety of the trigonometric moment problem compared to a seemingly analogous power moment problem.

This paper is divided in four sections in addition to the introduction. In Section 2 the notion of the Toeplitz block-vector of the NP problem (1.1) in the class $\mathbf{C}_p$ is introduced and examined, and then we will associate the CC problem (1.2), the TM problem (1.3), and the block-Toeplitz matrix with this block-vector and study the connection of the NP problem (1.1) with each of these three objects. In Section 3, the issue of solving the CC problem (1.2) is revisited for both the nondegenerate and the degenerate cases simultaneously. The main technique is classical: a matrix version of the Nevanlinna - Schur algorithm for the CC problem involving linear fractional transformations. As a consequence a parametrization of solutions to the NP problem (1.1) is described in the algorithmic form in a new light of the so-called the Toeplitz block-vector approach. The generalized left-and right-orthogonal polynomial matrices on the unit circle associated with the specific three-term recurrence relations are given in Section 4. The complete solutions of both the CC problem (1.2) and the NP problem (1.1) can be expressed in a more closed form with the help of these polynomial matrices.

For more information and further generalizations see, for instance, [5,19,20,14,17,22] and references therein.

§ 2. Connection between the NP problem (1.1) and the CC problem (1.2) or the TM problem (1.3)

Let $F(z)\in \mathbf{C}_p$, then $F(z)$ can be analytically continued out of the closed unit disc $\overline{D}$ by setting $F(z)=-F^*\left(\frac{1}{\bar z}\right)$, $|z|>1$. For convenience we may always assume without loss of generality [18] that all $z_i\neq 0$. Thus each solution $G(z)\in \mathbf{C}_p$ to the problem (1.1) is also an interpolant of the conditions,

$$\frac{1}{k!}\frac{d^k}{d\lambda^k}G\left(\frac{1}{\lambda}\right)\bigg|_{\lambda=\bar z_i}=-\boldsymbol{C}_{ik}^*\ ,\ i=1,2,\cdots,\rho,\ k=0,1,\cdots,\tau_i-1. \tag{1.1'}$$

It is readily verified (see, e.g., [21]) that there exists only one quasi-polynomial matrix $W(z)$ of the form $W(z)=\sum_{i=-n}^{n-1}\boldsymbol{w}_i z^i$, $\boldsymbol{w}_i\in \mathbf{C}^{p\times p}$, subject only to Eqs. (1.1) and (1.1′). Put now

$$\Omega(z)=\frac{1}{2}\left(W(z)-W^*\left(\frac{1}{\bar z}\right)\right), \tag{2.1}$$

$$A(z)=\prod_{i=1}^{\rho}(z-z_i)^{\tau_i}\left(\frac{1}{z}-\bar z_i\right)^{\tau_i}. \tag{2.2}$$

Then $\Omega(z)$ is a symmetric (relative to the unit circle) quasi-polynomial matrix: $\Omega(z)=-\Omega^*(\bar z^{-1})$ for $z\neq 0$, and subject only to Eqs. (1.1) and (1.1′). Consider the power series expansion of the rational matrix function $\Omega(z)/A(z)$ at $z=0$:

$$\frac{\Omega(z)}{A(z)}=-\boldsymbol{c}_0-2\boldsymbol{c}_1 z-\cdots-2\boldsymbol{c}_{n-1}z^{n-1}-2\boldsymbol{c}_n z^n+\cdots, \tag{2.3}$$

in which $\boldsymbol{c}_0=\boldsymbol{c}_0^*$ obviously. With the $p\times p$ matrix sequence $\{\boldsymbol{c}_k\}_{k=0}^{n-1}$ being sought via Eq. (2.3), we will associate the following objects:

1. the CC problem (1.2) with $\{\boldsymbol{c}_k\}_{k=0}^{n-1}$ defined via Eq. (2.3),

2. the TM problem (1.3) with $\{\boldsymbol{c}_k\}_{k=0}^{n-1}$ defined viq Eq. (2.3),

3. the block-Toeplitz matrix $(\boldsymbol{c}_{j-i})_{i,j=0}^{n-1}$ in which $\boldsymbol{c}_{-k}=\boldsymbol{c}_k^*$.

In this section, we primarily study the connection of the NP problem (1.1) with each of these three objects (see Theorems 2.1, 2.2 and 2.4).

Theorem 2.1 *The NP problem* (1.1) *is solvable, if and only if the CC problem* (1.2) *with* $\{\boldsymbol{c}_k\}_{k=0}^{n-1}$ *defined via Eq.* (2.3) *is solvable. Furthermore, there exists an explicit one-to-one correspondence between the solutions* $G(z)$ *to the NP problem* (1.1) *and the solutions* $F(z)$ *to the CC problem* (1.2), *which can be formulated as*

$$G(z)=\Omega(z)+A(z)F(z). \tag{2.4}$$

Proof Let first $F(z)\in\mathbf{C}_p$ be a solution to the CC problem (1.2), which admits an integral representation with Im $F(0)=\boldsymbol{O}$,

$$F(z)=\frac{1}{2\pi}\int_0^{2\pi}\frac{e^{i\theta}+z}{e^{i\theta}-z}d\sigma(\theta). \tag{2.5}$$

Here the Hermitian measure $\sigma(\theta)$ on $[0,2\pi]$ has $\boldsymbol{c}_0,\boldsymbol{c}_1,\cdots,\boldsymbol{c}_{n-1}$ as the first n moments:

$$\boldsymbol{c}_k=\frac{1}{2\pi}\int_0^{2\pi}e^{-ik\theta}d\sigma(\theta),\ k=0,1,\cdots,n-1.$$

Define $G(z)$ via Eq. (2.4). We observe that $G(z)$ is clearly an interpolant of the conditions (1.1) and (1.1′). Further on, since $G(z)=A(z)\cdot[\Omega(z)/A(z)+F(z)]=A(z)O(z^n)$, $G(z)$ is analytic in D. It remains to be checked that the resulting $G(z)$ belongs to $\mathbf{C}_p$. Let now

$$\widetilde{G}(z)=\Omega(z)-\frac{1}{2\pi}\int_0^{2\pi}\frac{A(z)-A(e^{i\theta})}{z-e^{i\theta}}(e^{i\theta}+z)d\sigma(\theta), \tag{2.6}$$

$$\Gamma(\theta)=\int_0^{\theta}A(e^{i\phi})d\sigma(\phi)=\int_0^{\theta}\prod_{k=1}^{\rho}|e^{i\phi}-z_k|^{2\tau_k}d\sigma(\phi),\ 0\leqslant\theta\leqslant 2\pi,$$

a Hermitian measure on $[0,2\pi]$. Then

$$G(z) = \widetilde{G}(z) + \frac{1}{2\pi}\int_0^{2\pi} \frac{e^{i\theta} + z}{e^{i\theta} - z} d\Gamma(\theta) \tag{2.7}$$

and $\widetilde{G}(z)$ is analytic in D by Eqs. (2.6) and (2.1), and of the form

$$\widetilde{G}(z) = \sum_{i=-n}^{n} \widetilde{\boldsymbol{g}}_i z^i, \quad \widetilde{\boldsymbol{g}}_i \in \mathbf{C}^{p\times p}. \tag{2.8}$$

A straightforward evaluation of Eq. (2.6) yields that $-\widetilde{G}(1/\bar{z})^* = \widetilde{G}(z)$, so that $\widetilde{\boldsymbol{g}}_i = -\widetilde{\boldsymbol{g}}_i^* = \boldsymbol{O}$ for all $i \neq 0$ in Eq. (2.8), and thus $\widetilde{G}(z) = \widetilde{\boldsymbol{g}}_0 = \mathrm{i}\ \mathrm{Im}\ G(0)$. Hence, from Eqs. (2.7) and (1.4), we deduce that

$$G(z) = \mathrm{i}\ \mathrm{Im}\ G(0) + \frac{1}{2\pi}\int_0^{2\pi} \frac{e^{i\theta} + z}{e^{i\theta} - z} d\Gamma(\theta) \in \mathbf{C}_p. \tag{2.9}$$

Conversely, let $G(z) \in \mathbf{C}_p$ be a solution to the NP problem (1.1) with an integral representation of the form (2.9). Define $F(z)$ via Eq. (2.4),

$$F(z) = \frac{G(z) - \Omega(z)}{A(z)} = \frac{z^n G(z) - z^n \Omega(z)}{z^n A(z)},$$

which is analytic in D, since the polynomial $z^n A(z)$ vanishes only at $z = z_i$ and $z = 1/\bar{z}_i, i = 1, 2, \cdots, \rho$. $F(z)$ has the power series expansion at $z = 0$ of the form (1.2) by Eqs. (2.3) and (2.9). To complete the proof, we need only to verify $F(z) \in \mathbf{C}_p$. Let now

$$\widetilde{F}(z) = \mathrm{i}\ \mathrm{Im}\ G(0) - \frac{1}{2\pi}\int_0^{2\pi} \frac{e^{i\theta} + z}{e^{i\theta} - z} \frac{A(e^{i\theta}) - A(z)}{A(e^{i\theta})} d\Gamma(\theta),$$

$$\sigma(\theta) = \int_0^{\theta} \frac{1}{A(e^{i\phi})} d\Gamma(\phi) = \int_0^{\theta} \prod_{j=1}^{\rho} | e^{i\phi} - z_j |^{-2\tau_j} d\Gamma(\phi), \quad 0 \leqslant \theta \leqslant 2\pi.$$

Then $\sigma(\theta)$ is a Hermitian measure on $[0, 2\pi]$, and

$$F(z) = \frac{\widetilde{F}(z) - \Omega(z)}{A(z)} + \frac{1}{2\pi}\int_0^{2\pi} \frac{e^{i\theta} + z}{e^{i\theta} - z} d\sigma(\theta).$$

It readily follows from the last equality that $\widetilde{F}(z)$ is an interpolant of the conditions (1.1) and (1.1′), and thus $\widetilde{F}(z) - \Omega(z) = \mathrm{i}\boldsymbol{\alpha} A(z)$ for a suitable $\boldsymbol{\alpha} = \boldsymbol{\alpha}^* \in \mathbf{C}^{p\times p}$. Since $F(0) = F^*(0)$, $\boldsymbol{\alpha} = \boldsymbol{O}$, and therefore $\widetilde{F}(z) = \Omega(z)$ and

$$F(z) = \frac{1}{2\pi}\int_0^{2\pi} \frac{e^{i\theta} + z}{e^{i\theta} - z} d\sigma(\theta) \in \mathbf{C}_p,$$

as claimed. ■

As to the TM problem, in [13] (p. 51) a close relation with the NP problem (1.1) with all $\tau_i = 1$ has been brought to light. The following direct consequence of Theorem 2.1 brings that relation to completion and to the more general case of multiple nodes.

Theorem 2.2 *The NP problem* (1.1) *is solvable if and only if the TM problem* (1.3) *with* $\boldsymbol{c}_0=\boldsymbol{c}_0^*$, $\boldsymbol{c}_1,\cdots,\boldsymbol{c}_{n-1}$ *defined via Eq.* (2.3) *is solvable. Furthermore, there exists an explicit on-to-one correspondence between the solutions* $G(z)$ *to the NP problem* (1.1) *and the solutions* $\sigma(\theta)$ *to the TM problem* (1.3), *which can be formulated as*

$$G(z)=\Omega(z)+A(z)\frac{1}{2\pi}\int_0^{2\pi}\frac{\mathrm{e}^{\mathrm{i}\theta}+z}{\mathrm{e}^{\mathrm{i}\theta}-z}\mathrm{d}\sigma(\theta). \tag{2.10}$$

As a result of Theorem 2.1 (Theorem 2.2, resp.), the finite NP problem (1.1) can be reduced to a certain CC problem (a certain TM problem) associated with the $p\times p$ matrix sequence $\{\boldsymbol{c}_k\}_{k=0}^{n-1}$ with $\boldsymbol{c}_0=\boldsymbol{c}_0^*$ defined via Eq. (2.3). In the sequel, we refer to the block-vector $\boldsymbol{c}=(\boldsymbol{c}_0,\boldsymbol{c}_1,\cdots,\boldsymbol{c}_{n-1})$ with $\boldsymbol{c}_0=\boldsymbol{c}_0^*$ and $\boldsymbol{c}_k\in\mathbf{C}^{p\times p}$ defined via Eq. (2.3) as the *Toeplitz block-vector* of the NP problem (1.1) compared with the Hankel block-vector of the NP problem in the Nevanlinna class [7,9]. It plays an essential role in our analysis throughout.

Starting from the Toeplitz block-vector once more, we will deduce an interesting decomposition of a generalized block-Pick matrix of the NP problem (1.1). Some notations are needed. Let us define,

$$a(z)=\prod_{i=1}^{\rho}(z-z_i)^{\tau_i},\quad a_{ij}(z)=\frac{a(z)}{(z-z_i)^{j+1}},$$

$$i=1,2,\cdots,\rho,\quad j=0,1,\cdots,\tau_i-1,$$

$$\boldsymbol{W}_i=\left(\frac{a_{ik}^{(\mu)}(0)}{\mu!}\right)_{k=0,\mu=0}^{\tau_i-1,n-1},\quad i=1,2,\cdots,\rho, \tag{2.11}$$

$$\boldsymbol{W}=\begin{bmatrix}\boldsymbol{W}_1\\ \boldsymbol{W}_2\\ \vdots\\ \boldsymbol{W}_\rho\end{bmatrix}\in\mathbf{C}^{n\times n}.$$

It is known (see, e.g., [10]) that $\boldsymbol{W}$ is nonsingular. Further on, form the data of the NP problem (1.1) a Hermitian *generalized block-Pick matrix* $\boldsymbol{P}$ (of the first kind) of the NP problem (1.1) is defined as follows:

$$\boldsymbol{P}=(\boldsymbol{P}_{ij})_{i,j=1}^{\rho}\in\mathbf{C}^{np\times np}, \tag{2.12a}$$

where $\boldsymbol{P}_{ij} \in \mathbf{C}^{\tau_i p \times \tau_j p}$ with entries $\boldsymbol{p}_{ij}^{kl} \in \mathbf{C}^{p \times p}$ determined by

$$\boldsymbol{p}_{ij}^{kl} = \frac{1}{k!l!}\frac{\partial^{k+l}}{\partial\lambda^k\partial\mu^l}\left[\frac{G(\lambda)+G^*(\bar{\mu})}{1-\lambda\mu}\right]_{\lambda=z_i,\mu=\bar{z}_j} \tag{2.12b}$$

in which $G(\lambda)$ is an arbitrary $\mathbf{C}^{p\times p}$-valued function subject only to Eqs. (1.1) and (1.1′).

Further on, we introduce a linear operator $\mathscr{L}$ from $\mathscr{H}$ into $\mathbf{C}^{p\times p}$, where $\mathscr{H}$ stands for the space of all rational scalar-valued functions of the complex variable z. More precisely, if a $R(z)\in\mathscr{H}$ has the power series expansions at $z=0$ and at $z=\infty$, respectively:

$$R(z) = \sum_{s=-\infty}^{\infty} r_s z^s \quad (z\to 0) \quad \text{and} \quad R(z) = \sum_{s=-\infty}^{\infty} \tilde{r}_s z^s \quad (z\to\infty),$$

where the numbers of $r_s\neq 0$ for $s\leqslant 0$ and of $\tilde{r}_s\neq 0$ for $s\geqslant 0$ are finite, we define its imagine by putting

$$\mathscr{L}\{R(z)\} = \frac{1}{2}\boldsymbol{c}_0(r_0+\tilde{r}_0) + \sum_{s=1}^{n}(\tilde{r}_s\boldsymbol{c}_s^* + r_{-s}\boldsymbol{c}_s),$$

where the $\boldsymbol{c}_s$ are as in Eq. (2.3), and n is fixed in advance; in particular,

$$\mathscr{L}\{z^k\}=\boldsymbol{c}_k^*, \quad \mathscr{L}\{z^{-k}\}=\boldsymbol{c}_k, \quad k=0,1,2,\cdots$$

A direct manipulation leads to

$$\mathscr{L}\{R(z)\}=\frac{1}{2}\left[\left\langle R(z)\frac{\Omega(z)}{A(z)}\right\rangle_{\infty 0} - \left\langle R(z)\frac{\Omega(z)}{A(z)}\right\rangle_{00}\right], \tag{2.13}$$

where $\langle Q(z)\rangle_{\infty 0}$ and $\langle Q(z)\rangle_{00}$ stand for the constant terms of the power series expansions for $Q(z)\in\mathscr{H}$ at $z=0$ and at $z=\infty$, respectively.

Lemma 2.3 *Let $\mathscr{H}$, $\mathscr{L}$, $A(z)$ and $\Omega(z)$ be defined as above. Then*

$$\mathscr{L}\left\{\frac{z+\lambda}{z-\lambda}A(z)\right\}=\Omega(\lambda). \tag{2.14}$$

Proof From Eq. (2.13) we obtain

$$\mathscr{L}\left\{\frac{z+\lambda}{z-\lambda}A(z)\right\}=\frac{1}{2}\left[\left\langle\frac{z+\lambda}{z-\lambda}\Omega(z)\right\rangle_{\infty 0} - \left\langle\frac{z+\lambda}{z-\lambda}\Omega(z)\right\rangle_{00}\right],$$

which is defined as a function of the variable λ. But, since $\Omega(z) = \sum_{i=-n}^{n}\boldsymbol{\omega}_i z^i$ with $\boldsymbol{\omega}_{-i}=\boldsymbol{\omega}_i^*$,

$$\left\langle\frac{z+\lambda}{z-\lambda}\Omega(z)\right\rangle_{\infty 0} = \boldsymbol{\omega}_0 + 2\boldsymbol{\omega}_1\lambda+\cdots+2\boldsymbol{\omega}_n\lambda^n,$$

$$\left\langle \frac{z+\lambda}{z-\lambda}\Omega(z)\right\rangle_{00} = -\boldsymbol{\omega}_0 - \frac{2\boldsymbol{\omega}_{-1}}{\lambda} - \cdots - \frac{2\boldsymbol{\omega}_{-n}}{\lambda^n},$$

so that Eq. (2.14) holds. ■

Theorem 2.4 *Let* $\boldsymbol{P}$ *be a generalized block-Pick matrix of the NP problem* (1.1). *Then*

$$\boldsymbol{P} = 2(\boldsymbol{W}\otimes\boldsymbol{I}_p)\boldsymbol{T}(\boldsymbol{W}\otimes\boldsymbol{I}_p)^*, \tag{2.15}$$

where $\boldsymbol{T}=(\boldsymbol{c}_{j-i})_{i,j=0}^{n-1}$ *with* $\boldsymbol{c}_{-k}=\boldsymbol{c}_k^*$ *is the block-Toeplitz matrix built on the Toeplitz block-vector* $\mathbf{c}$ *of the NP problem* (1.1), *and the symbol "*$\otimes$*" denotes the tensor product of two matrices.*

Proof Observe that the coefficients of the power series expansions for $((z+\lambda/z)-\lambda)A(z)$ at $z=0$ and $z=\infty$ are all polynomials in λ. This fact and the linearity of $\mathscr{L}$ imply that we can interchange $\mathscr{L}$ with the operation of differenting. Thus by Lemma 2.3 we have

$$\begin{aligned}
\boldsymbol{p}_{ij}^{kl} &= \frac{1}{k!\,l!}\frac{\partial^{k+1}}{\partial\lambda^k\partial\mu^l}\left[\frac{G(\lambda)+G^*(\bar{\mu})}{1-\lambda\mu}\right]_{\lambda=z_i,\mu=\bar{z}_j} \\
&= \frac{1}{k!\,l!}\frac{\partial^{k+1}}{\partial\lambda^k\partial\mu^l}\left[\frac{\Omega(\lambda)-\Omega(1/\mu)}{1-\lambda\mu}\right]_{\lambda=z_i,\mu=\bar{z}_j} \\
&= \mathscr{L}\left\{\frac{1}{k!\,l!}\frac{\partial^{k+1}}{\partial\lambda^k\partial\mu^l}\left[\frac{\frac{z+\lambda}{z-\lambda}A(z)-\frac{z+\mu^{-1}}{z-\mu^{-1}}A(z)}{1-\lambda\mu}\right]_{\lambda=z_i,\mu=\bar{z}_j}\right\} \\
&= 2\mathscr{L}\left\{\frac{A(z)}{(z-z_i)^{k+1}\left(\frac{1}{z}-\bar{z}_j\right)^{l+1}}\right\} = 2\mathscr{L}\left\{a_{ik}(z)a_{jl}^*\left(\frac{1}{\bar{z}}\right)\right\} \\
&= 2\left\{\left[a_{ik}^{(0)}(0),\cdots,\frac{a_{ik}^{(m-1)}(0)}{(m-1)!}\right]\otimes\boldsymbol{I}_p\right\}\mathscr{L}\{z^{u-v}\}_{u,v=0}^{n-1}\left\{\begin{bmatrix}\overline{a_{jl}^{(0)}(0)}\\ \vdots \\ \overline{\frac{a_{jl}(0)}{(n-1)!}}\end{bmatrix}\otimes\boldsymbol{I}_p\right\}.
\end{aligned}$$

We deduce from the last formula,

$$\boldsymbol{P}_{ij} = (\boldsymbol{p}_{ij}^{kl})_{k,l=0}^{\tau_i-1,\tau_j-1} = 2(\boldsymbol{W}_i\otimes\boldsymbol{I}_p)(\boldsymbol{c}_{j-i})_{i,j=0}^{n-1}(\boldsymbol{W}_j\otimes\boldsymbol{I}_p)^*.$$

This leads to Eq. (2.15) immediately. ■

Note that a little more general statement than those which arise out of Theorem 2.4 can be found in Theorem 1 of [8].

We emphasize that the relation (2.15) holds with the way in which the nodes $z_1, z_2, \cdots, z_\rho$ are organized in an arbitrary ordering. Different orderings may give rise to different $\boldsymbol{P}$ and $\boldsymbol{W}$, but all will be with one and the same $\boldsymbol{T}=(\boldsymbol{c}_{j-i})_{i,j=0}^{n-1}$. Thus, we will associate only one block vector $(\boldsymbol{c}_0, \boldsymbol{c}_1, \cdots, \boldsymbol{c}_{n-1}), \boldsymbol{c}_0=\boldsymbol{c}_0^*$, with the NP problem (1.1) such that Eq. (2.15) holds independently of the particular ordering chosen in which the nodes $z_1, z_2, \cdots, z_\rho$ are organized in Eqs. (2.11) and (2.12). It is the Toeplitz block-vector of the NP problem (1.1) that plays this role. Conversely, given a set $\mathcal{Q}$ of interpolation nodes with length n, an arbitrary block-vector $\boldsymbol{c}=(\boldsymbol{c}_0, \boldsymbol{c}_1, \cdots, \boldsymbol{c}_{n-1})$ with $\boldsymbol{c}_i \in \mathbf{C}^{p\times p}$ and $\boldsymbol{c}_0=\boldsymbol{c}_0^*$ sets up only one NP problem of the form (1.1) via Eq. (2.15) within a constant skew-Hermitian term for $\boldsymbol{c}_{i0}$, which takes $\boldsymbol{c}$ as its Toeplitz block-vector ([8], Theorem 2).

Let $\mathcal{T}_n$ denote the class of all Hermitian block-Toeplitz matrices of order np, which take $p\times p$ submatrices as entries, and let $\mathcal{P}_n$ denote the class of all Hermitian block matrices of order np, which are of the form (2.12), corresponding to a fixed set of nodes (and thus to a fixed $\boldsymbol{W}$ defined as in Eq. (2.11)). Observe that since any block matrix of the form (2.12) can be viewed as a generalized block-Pick matrix (of the first kind) associated with a certain NP problem, Theorem 2.4 directly leads to the following result (see [13] for the case of simple nodes only).

Corollary 2.5 *Let $z_1, z_2, \cdots, z_\rho$ be distinct with multiplicities $\tau_1, \tau_2, \cdots, \tau_\rho$, respectively, $n=\sum_{j=1}^{\rho}\tau_j$. Then the mapping $\mathcal{I}$ defined on $\mathcal{T}_n$ by Eq.* (2.16) *is an isomorphism of the classes $\mathcal{T}_n$ and $\mathcal{P}_n$:*

$$\mathcal{I}(\boldsymbol{T}_0)=(\boldsymbol{W}\otimes \boldsymbol{I}_p)\boldsymbol{T}_0(\boldsymbol{W}\otimes \boldsymbol{I}_p)^* \quad \forall\, \boldsymbol{T}_0 \in \mathcal{T}_n, \tag{2.16}$$

where $\boldsymbol{W}$ is given by Eq. (2.11).

We remark that a generalized block-Pick matrix of the second kind, defined by interchanging the places of λ and μ in Eq. (2.12) and written as $\widetilde{\boldsymbol{P}}$, may also be naturally considered for the finite NP problem (1.1). Such a matrix $\widetilde{\boldsymbol{P}}$ coincides with a *block-Pick matrix* for the case of simple nodes in [13].

A similar congruent relation to Eq. (2.15) but between the block-Pick matrix $\widetilde{\boldsymbol{P}}$ and the block-Toeplitz matrix $\widetilde{\boldsymbol{T}}=(\boldsymbol{c}_{i-j})_{i,j=0}^{n-1}$ with $\boldsymbol{c}_{-i}=\boldsymbol{c}_i^*$ may be established as follows.

Corollary 2.6 *Let $\widetilde{\boldsymbol{P}}$ be a generalized block-Pick matrix of the second kind of the NP problem* (1.1). *Then*

$$\widetilde{\boldsymbol{P}}=2(\boldsymbol{W}^{\mathrm{T}}\otimes\boldsymbol{I}_p)^*\widetilde{\boldsymbol{T}}(\boldsymbol{W}^{\mathrm{T}}\otimes\boldsymbol{I}_p),\tag{2.17}$$

where $\widetilde{\boldsymbol{T}}=(\boldsymbol{c}_{i-j})_{i,j=0}^{n-1}$ with $\boldsymbol{c}_{-i}=\boldsymbol{c}_i^*$ being the same as that given in Theorem 2.4 *and the symbol* T *denotes the transpose of the indicated matrix.*

By means of Corollary 2.6 a similar result to that given in Corollary 2.5 holds corresponding to the generalized block-Pick matrices of the second kind.

Due to $\widetilde{\boldsymbol{T}}=(\boldsymbol{J}_n\otimes\boldsymbol{I}_p)\boldsymbol{T}(\boldsymbol{J}_n\otimes\boldsymbol{I}_p)$, where $\boldsymbol{T}=(\boldsymbol{c}_{j-i})_{i,j=0}^{n-1}$ with $\boldsymbol{c}_{-k}=\boldsymbol{c}_k^*$, and

$$\boldsymbol{J}_n=\begin{bmatrix}\boldsymbol{O} & & 1\\ & \cdot^{\cdot^{\cdot}} & \\ 1 & & \boldsymbol{O}\end{bmatrix}\in\mathbf{C}^{n\times n},$$

both $\boldsymbol{P}$ and $\widetilde{\boldsymbol{P}}$ are congruent to one and the same block-Toeplitz matrix $\boldsymbol{T}$ (or $\widetilde{\boldsymbol{T}}$), and thus the nonnegativity properties of $\boldsymbol{P}$ and $\widetilde{\boldsymbol{P}}$ are equivalent each other.

From Theorems 2.2 and 2.4 together with Theorem 4 of [3] one deduces the following.

Corollary 2.7 *With the notation mentioned above, the following are equivalent*:

(a) *The NP problem* (1.1) *is solvable.*

(b) *The block-Toeplitz matrix $\boldsymbol{T}=(\boldsymbol{c}_{j-i})_{i,j=0}^{n-1}$ with $\boldsymbol{c}_{-k}=\boldsymbol{c}_k^*$ built on the Toeplitz block-vector of the NP problem* (1.1) *is Hermitian nonnegative.*

(c) *A generalized block-Pick matrix* $\boldsymbol{P}$ (*of the first kind*) of the NP problem(1.1) *is Hermitian nonnegative.*

(d) *The block-Toeplitz matrix $\widetilde{\boldsymbol{T}}=(\boldsymbol{c}_{i-j})_{i,j=0}^{n-1}\geqslant\boldsymbol{O}$, where the $\boldsymbol{c}_i$ are* *as in* (b).

(e) *A generalized block-Pick matrix* $\widetilde{\mathbf{P}}$ *of the second kind of the* NP *problem*(1. 1) *is Hermitian nonnegative*.

In the most general setting, the set of interpolation nodes is not necessarily finite, nor even countable (but with finite multiplicity for each node). For the corresponding multiple NP problem, we call the NP problems associated with finite subsets of the data its NP sub-problems. Then the following generalized Pick criterion holds.

Corollary 2. 8 (Refs. [15,16]) *A necessary and sufficient condition for the solvability of the multiple NP problem (with finite set of nodes or not) is that the generalized block-Pick matrices of the first (or the second) kind built on the data of all its NP sub-problems are Hermitian nonnegative*.

Proof The necessity is a direct consequence of the Riesz-Herglotz representation (1. 4) by a straightforward calculation. For the sufficiency, first in the case of finite number of nodes it follows from Corollary 2. 7. When the interpolation node set is denumerable or not countable, the argument is much the same as that made in [13],(p. 51) the details of which are omitted. ■

§ 3. Solutions to the CC problem (1. 2) and to the NP problem (1. 1) in algorithmic form

This section concerns mainly the problem of parametrizing the solutions to the CC problem (1. 2) (or, equivalently, the TM problem (1. 3)) in both the nondegenerate (i. e. , $\boldsymbol{T}>\boldsymbol{O}$) and degenerate (i. e. , $\boldsymbol{T}\geqslant\boldsymbol{O}$ is singular) cases simultaneously, based on the use the algorithm of consecutive linear fractional transformations. As a direct consequence, a parametrization of solutions to the NP problem (1. 1) follows from Theorem 2. 1. In the scalar case $p=1$, the algorithm we are going to perform is essentially nothing other than the well known Nevanlinna - Schur algorithm (see e. g. [2]). In the matrix case $p>1$, it is a natural extension of Schur's theory [26], and the nondegenerate case has been treated in [12,19]. Our analysis here is an elaboration of the work in Section 3 of [12].

For the other relevant results, we refer to the following: [20] (Ch. Ⅱ); [14] (Ch. 6, Ch. 7. 2); [11](Section 6); [13,22,24] and the references therein.

Throughout the rest of this paper, denote by $\boldsymbol{U}$, $\boldsymbol{V}$ (or $\boldsymbol{U}_0$, $\boldsymbol{V}_0$, $\widetilde{\boldsymbol{U}}$, $\widetilde{\boldsymbol{V}}$, $\hat{\hat{\boldsymbol{U}}}$, $\hat{\hat{\boldsymbol{V}}}$, and so forth) constant unitary matrices of order p.

Let us first remark that under the assumption $\boldsymbol{T}=(\boldsymbol{c}_{j-i})_{i,j=0}^{n-1}\geqslant\boldsymbol{O}$, if $\boldsymbol{c}_0=\boldsymbol{0}$, then $\boldsymbol{c}_i=\boldsymbol{0}$, $i=1,2,\cdots,n-1$, so that $F(z)\equiv\boldsymbol{O}$ is only one solution to Eq. (1. 2) by the well known Carathéodory - Toeplitz theorem; if $\boldsymbol{c}_0\neq\boldsymbol{0}$ (singular or not), then there exists a unitary matrix $\boldsymbol{U}$ such that

$$\boldsymbol{c}_i=\boldsymbol{U}(\tilde{\boldsymbol{c}}_i\oplus\boldsymbol{0}_r)\boldsymbol{U}^*,\quad i=0,1,\cdots,n-1,\tag{3. 1}$$

in which $r=p-\text{rank}\ \boldsymbol{c}_0$ and $\tilde{\boldsymbol{c}}_0>\boldsymbol{0}$. Let now $F(z)\in\mathbf{C}_p$ be a solution to the CC problem (1. 2) with $F(0)=\boldsymbol{c}_0\neq\boldsymbol{0}$; then, by the Carathéodory - Toeplitz theorem we have that $\widetilde{\boldsymbol{T}}=(\boldsymbol{c}_{i-j})_{i,j=0}^{n-1}\geqslant\boldsymbol{O}$, or equivalently, $\boldsymbol{T}=(\boldsymbol{c}_{j-i})_{i,j=0}^{n-1}\geqslant\boldsymbol{O}$, so that $F(z)$ has the form

$$F(z)=\boldsymbol{U}(\widetilde{F}(z)\oplus\boldsymbol{0}_r)\boldsymbol{U}^*,\tag{3. 2a}$$

where $\widetilde{\mathrm{F}}(z)\in\mathbf{C}_{p-r}$ is defined by

$$\widetilde{F}(z)=\tilde{\boldsymbol{c}}_0+2\tilde{\boldsymbol{c}}_1z+\cdots+2\tilde{\boldsymbol{c}}_{n-1}z^{n-1}+\mathrm{O}(z^n).\tag{3. 2b}$$

Let

$$\hat{\boldsymbol{c}}_0=\boldsymbol{c}_0+\boldsymbol{U}(\boldsymbol{0}_{p-r}\oplus\boldsymbol{I}_r)\boldsymbol{U}^*=\boldsymbol{U}(\tilde{\boldsymbol{c}}_0\oplus\boldsymbol{I}_r)\boldsymbol{U}^*,\tag{3. 3}$$

which is Hermitian positive, and let

$$\Phi(z)=\frac{1}{z}\boldsymbol{c}_0^{1/2}(F(z)+\hat{\boldsymbol{c}}_0)^{-1}(F(z)-\boldsymbol{c}_0)\hat{\boldsymbol{c}}_0^{-1/2},\tag{3. 4a}$$

where the symbol $\boldsymbol{H}^{1/2}$ is used for the Hermitian square root of a matrix $\boldsymbol{H}\geqslant\boldsymbol{O}$. Then, by Eqs. (3. 1) and (3. 2a), $\Phi(z)$ is well defined and has the form

$$\Phi(z)=\boldsymbol{U}(\widetilde{\Phi}(z)\oplus\boldsymbol{0}_r)\boldsymbol{U}^*,\tag{3. 4b}$$

in which $\widetilde{\Phi}(z)$ is defined by

$$\widetilde{\Phi}(z)=\frac{1}{z}\tilde{\boldsymbol{c}}_0^{1/2}(\widetilde{F}(z)+\tilde{\boldsymbol{c}}_0)^{-1}(\widetilde{F}(z)-\tilde{\boldsymbol{c}}_0)\tilde{\boldsymbol{c}}_0^{-1/2}.\tag{3. 4c}$$

It follows from Lemma 2 of [4] and by the use of the maximum modulus principle theorem (see e. g. [23]) that $\widetilde{\Phi}(z)\in\mathbf{S}_{p-r}$, and thus $\Phi(z)\in\mathbf{S}_p$. The inverse forms of Eqs. (3.4a) and (3.4c) are, respectively,

$$F(z)=\boldsymbol{c}_0^{1/2}(\boldsymbol{I}_p+z\Phi(z))(\boldsymbol{I}_p-z\Phi(z))^{-1}\hat{\boldsymbol{c}}_0^{1/2},\tag{3.5a}$$

$$\widetilde{F}(z)=\tilde{\boldsymbol{c}}_0^{1/2}(\boldsymbol{I}_{p-r}+z\widetilde{\Phi}(z))(\boldsymbol{I}_{p-r}-z\widetilde{\Phi}(z))^{-1}\tilde{\boldsymbol{c}}_0^{1/2}.\tag{3.5b}$$

It is easily verified that if $\Phi(z)\in\mathbf{S}_p$[$\widetilde{\Phi}(z)\in\mathbf{S}_{p-r}$, resp.] then the matrix function Eq. (3.5a) [Eq. (3.5b)] belongs to the class $\mathbf{C}_p$[$\mathbf{C}_{p-r}$].

Our basic strategy used here is to convert the solution of the CC problem (1.2) into the solution of a certain Schur matrix coefficient problem. Here, given $\boldsymbol{\alpha}_0,\boldsymbol{\alpha}_1,\cdots,\boldsymbol{\alpha}_k\in\mathbf{C}^{p\times p}$, the *Schur matrix coefficient (SC) problem* consists in finding $\Phi(z)\in\mathbf{S}_p$ such that

$$\Phi(z)=\boldsymbol{\alpha}_0+\boldsymbol{\alpha}_1 z+\cdots+\boldsymbol{\alpha}_k z^k+\mathrm{O}(z^{k+1}).\tag{3.6}$$

The matrix version of the Carathéodory－Fejér theorem says that the SC problem (3.6) is solvable if and only if the matrix of order $(k+1)p$

$$\boldsymbol{\Gamma}_k=\begin{bmatrix}\boldsymbol{\alpha}_0 & & \boldsymbol{O}\\ \vdots & \ddots & \\ \boldsymbol{\alpha}_k & \cdots & \boldsymbol{\alpha}_0\end{bmatrix}\tag{3.7}$$

is contractive: $\|\boldsymbol{\Gamma}_k\|\leqslant 1$.

To begin with, if $F(z)\in\mathbf{C}_p$ is a solution to the CC problem (1.2) then it is readily verified that the function $\Phi(z)\in\mathbf{S}_p$ defined via (3.4) is actually a solution to the SC problem (3.6) with $k=n-2$, and with $\boldsymbol{\alpha}_0$, $\boldsymbol{\alpha}_1,\cdots,\boldsymbol{\alpha}_{n-2}$ defined uniquely, in terms of the initial coefficient $\boldsymbol{c}_0,\boldsymbol{c}_1,\cdots$, $\boldsymbol{c}_{n-1}$, by

$$\begin{bmatrix}\boldsymbol{\alpha}_0\\ \boldsymbol{\alpha}_1\\ \vdots\\ \boldsymbol{\alpha}_{n-2}\end{bmatrix}=(\boldsymbol{I}_{n-1}\otimes\boldsymbol{c}_0^{1/2})\begin{bmatrix}\hat{\boldsymbol{c}}_0 & & & 0\\ \boldsymbol{c}_1 & \hat{\boldsymbol{c}}_0 & & \\ \vdots & \ddots & \ddots & \\ \boldsymbol{c}_{n-2} & \cdots & \boldsymbol{c}_1 & \hat{\boldsymbol{c}}_0\end{bmatrix}^{-1}\begin{bmatrix}\boldsymbol{c}_1\\ \boldsymbol{c}_2\\ \vdots\\ \boldsymbol{c}_{n-1}\end{bmatrix}\hat{\boldsymbol{c}}_0^{-1/2}.\tag{3.8}$$

Conversely, if $\Phi(z)\in\mathbf{S}_p$ is a solution to the SC problem (3.6) with $k=n-2$ and $\boldsymbol{\alpha}_0,\boldsymbol{\alpha}_1,\cdots,\boldsymbol{\alpha}_{n-2}$ defined via Eq. (3.8), we insert $\Phi(z)$ into Eq. (3.5a) and obtain a $\mathbf{C}^{p\times p}$-valued function $F(z)$. It is readily verified that this $F(z)$ belongs to $\mathbf{C}_p$ and satisfies Eq. (1.2).

Therefore, the functions $F(z) \in \mathbf{C}_p$ satisfying Eq. (1.2) can be characterized as follows.

Lemma 3.1 *If* $\boldsymbol{T}=(\boldsymbol{c}_{j-i})_{i,j=0}^{n-1} \geqslant \boldsymbol{O}$, *then the solutions* $F(z)$ *to the CC problem* (1.2) *can be parametrized by a linear fractional transformation*,

$$F(z)=\boldsymbol{c}_0^{1/2}(\boldsymbol{I}_p+z\Phi(z))(\boldsymbol{I}_p-z\Phi(z))^{-1}\hat{\boldsymbol{c}}_0^{1/2}, \qquad (3.9)$$

with the coefficient matrix

$$\begin{bmatrix} \boldsymbol{c}_0^{1/2} & \boldsymbol{O} \\ \boldsymbol{O} & \hat{\boldsymbol{c}}_0^{-1/2} \end{bmatrix}\begin{bmatrix} z\boldsymbol{I}_p & \boldsymbol{I}_p \\ -z\boldsymbol{I}_p & \boldsymbol{I}_p \end{bmatrix}, \qquad (3.10)$$

where the parameter $\Phi(z)$ *varies in the set of solutions of the SC problem* (3.6) *with* $k=n-2$ *and with* $\boldsymbol{\alpha}_0, \boldsymbol{\alpha}_1, \cdots, \boldsymbol{\alpha}_{n-2}$ *defined via Eq.* (3.8); *in particular, if Eq.* (3.1) *holds with* $\tilde{\boldsymbol{c}}_0 > \boldsymbol{0}$, *then* $\Phi(z)$ *has the form* (3.4b).

We remark that Lemma 3.1 implies incidentally that if $\boldsymbol{T}=(\boldsymbol{c}_{j-i})_{i,j=0}^{n-1} \geqslant \boldsymbol{O}$, then and only then the matrix $\boldsymbol{\Gamma}_{n-2}$ (see Eq. (3.7) with $k=n-2$) with $\boldsymbol{\alpha}_0, \boldsymbol{\alpha}_1, \cdots, \boldsymbol{\alpha}_{n-2}$ defined via Eq. (3.8) is contractive: $\|\boldsymbol{\Gamma}_{n-2}\| \leqslant 1$.

Further on, for an arbitrary contractive matrix $\boldsymbol{\Gamma}_k$ defined as in Eq. (3.7), if $\boldsymbol{\alpha}_0$ is unitary, $\boldsymbol{\alpha}_0^*\boldsymbol{\alpha}_0=\boldsymbol{\alpha}_0\boldsymbol{\alpha}_0^*=\boldsymbol{I}_p$, then $\boldsymbol{\alpha}_i=\boldsymbol{0}, i=1,2,\cdots,k$, obviously, so that the problem (3.6) has only one solution $\Phi(z)\equiv\boldsymbol{\alpha}_0$; if $\boldsymbol{\alpha}_0$ is not unitary with the defect index: $\operatorname{rank}(\boldsymbol{I}_p-\boldsymbol{\alpha}_0^*\boldsymbol{\alpha}_0)=p-r_0$ and if there exist unitary matrices $\boldsymbol{U}_0$ and $\boldsymbol{V}_0$ such that $\boldsymbol{\alpha}_0=\boldsymbol{U}_0(\tilde{\boldsymbol{\alpha}}_0\oplus\boldsymbol{I}_{r_0})\boldsymbol{V}_0^*$, where $\tilde{\boldsymbol{\alpha}}_0$ is real diagonal of order $p-r_0$, $\boldsymbol{O}\leqslant\tilde{\boldsymbol{\alpha}}_0<\boldsymbol{I}_{p-r_0}$, then, applying a theorem of [23](p. 147) to a solution $\Phi(z)$ of the SC problem (3.6), we have

$$\Phi(z)=\boldsymbol{U}_0(\tilde{\Phi}(z)\oplus\boldsymbol{I}_{r_0})\boldsymbol{V}_0^*, \qquad (3.11)$$

where $\tilde{\Phi}(z)\in\mathbf{S}_{p-r_0}$, and thus, by Eq. (3.6), the $\boldsymbol{\alpha}_i\,(i\geqslant1)$ are of the form

$$\boldsymbol{\alpha}_i=\boldsymbol{U}_0(\tilde{\boldsymbol{\alpha}}_i\oplus\boldsymbol{0}_{r_0})\boldsymbol{V}_0^*, \quad i=1,2,\cdots,k. \qquad (3.12)$$

Throughout the rest of this paper, if $\boldsymbol{v}$ is a contracive matrix of order p and if $\boldsymbol{v}=\boldsymbol{U}_v(\tilde{\boldsymbol{v}}\oplus\boldsymbol{I}_q)\boldsymbol{V}_v^*$ where $\|\tilde{\boldsymbol{v}}\|<1$ and $q=p-\operatorname{rank}(\boldsymbol{I}_p-$

$\boldsymbol{v}\boldsymbol{v}^*$), then we denote by $\hat{\boldsymbol{v}}$ the *strictly contractive part* of $\boldsymbol{v}$:

$$\hat{\boldsymbol{v}}=\boldsymbol{U}_v(\tilde{\boldsymbol{v}}\oplus\boldsymbol{0}_q)\boldsymbol{V}_v^*, \tag{3.13}$$

$\|\tilde{\boldsymbol{v}}\|<1$. Note that if $\boldsymbol{v}$ is such that $\|\boldsymbol{v}\|<1$, then $\boldsymbol{v}=\hat{\boldsymbol{v}}$.

For the convenience, we introduce a mapping Δ_k defined on the set of all matrix sequences $\{\boldsymbol{\alpha}_i\}_{i=0}^k$ such that $\|\boldsymbol{\Gamma}_k\|\leqslant 1$ by putting

$$(\boldsymbol{\beta}_0,\boldsymbol{\beta}_1,\cdots,\boldsymbol{\beta}_{k-1})=\Delta_k(\boldsymbol{\alpha}_0,\boldsymbol{\alpha}_1,\cdots,\boldsymbol{\alpha}_k), \tag{3.14a}$$

where $\boldsymbol{\beta}_0,\boldsymbol{\beta}_1,\cdots,\boldsymbol{\beta}_{k-1}\in\mathbf{C}^{p\times p}$ are connected with $\boldsymbol{\alpha}_0,\boldsymbol{\alpha}_1,\cdots,\boldsymbol{\alpha}_k$ by the relation

$$\begin{bmatrix}\boldsymbol{\beta}_0\\ \boldsymbol{\beta}_1\\ \vdots\\ \boldsymbol{\beta}_{k-1}\end{bmatrix}=(\boldsymbol{I}_k\otimes\boldsymbol{D}_{\hat{a}_0})\begin{bmatrix}\boldsymbol{D}_{\hat{a}_0}^2 & & & \boldsymbol{O}\\ -\boldsymbol{\alpha}_1\boldsymbol{\alpha}_0^* & \ddots & & \\ \vdots & \ddots & \ddots & \\ -\boldsymbol{\alpha}_{k-1}\boldsymbol{\alpha}_0^* & \cdots & -\boldsymbol{\alpha}_1\boldsymbol{\alpha}_0^* & \boldsymbol{D}_{\hat{a}_0}^2\end{bmatrix}^{-1}\begin{bmatrix}\boldsymbol{\alpha}_1\\ \boldsymbol{\alpha}_2\\ \vdots\\ \boldsymbol{\alpha}_k\end{bmatrix}\boldsymbol{D}_{\hat{a}_0^*}^{-1}. \tag{3.14b}$$

in which $\hat{\boldsymbol{a}}_0$ is the strictly contractive part of $\boldsymbol{\alpha}_0$, whereas

$$\boldsymbol{D}_{\hat{a}_0}=(\boldsymbol{I}_p-\hat{\boldsymbol{a}}_0\hat{\boldsymbol{a}}_0^*)^{1/2}\quad\text{and}\quad\boldsymbol{D}_{\hat{a}_0^*}=(\boldsymbol{I}_p-\hat{\boldsymbol{a}}_0^*\hat{\boldsymbol{a}}_0)^{1/2}. \tag{3.14c}$$

Let us now give a parametrization of solutions to the SC problem (3.6) in terms of the generalized Schur decomposition of a function matrix $\Phi(z)\in\mathbf{S}_p$. We first prove a simple lemma.

Lemma 3.2 *Let $\boldsymbol{\Gamma}_k$ defined via Eq.* (3.7) *be contractive*, $k\geqslant 1$, *and let $\boldsymbol{D}_{\hat{a}_0}$ and $\boldsymbol{D}_{\hat{a}_0^*}$ be as in Eq.* (3.14c). *Then the solutions $\Phi(z)$ to the problem* (3.6) *can be parametrized by a linear fractional transformation*,

$$\Phi(z)=\boldsymbol{D}_{\hat{a}_0}^{-1}(\boldsymbol{\alpha}_0+z\Phi_1(z))(\boldsymbol{I}_p+\hat{\boldsymbol{a}}_0^*z\Phi_1(z))^{-1}\boldsymbol{D}_{\hat{a}_0^*} \tag{3.15}$$

with the coefficient matrix

$$\begin{bmatrix}\boldsymbol{D}_{\hat{a}_0}^{-1} & \boldsymbol{O}\\ \boldsymbol{O} & \boldsymbol{D}_{\hat{a}_0^*}^{-1}\end{bmatrix}\begin{bmatrix}z\boldsymbol{I}_p & \boldsymbol{\alpha}_0\\ z\boldsymbol{\alpha}_0^* & \boldsymbol{I}_p\end{bmatrix}, \tag{3.16}$$

where the parameter $\Phi_1(z)\in\mathbf{S}_p$ runs over the solutions of the SC problem (3.6) *with $k-1$ and $\boldsymbol{\alpha}_0^{(1)},\boldsymbol{\alpha}_1^{(1)},\cdots,\boldsymbol{\alpha}_{k-1}^{(1)}$ in place of k and $\boldsymbol{\alpha}_0,\boldsymbol{\alpha}_1,\cdots,\boldsymbol{\alpha}_k$, respectively, where $\boldsymbol{\alpha}_0^{(1)},\boldsymbol{\alpha}_1^{(1)},\cdots,\boldsymbol{\alpha}_{k-1}^{(1)}$ are defined uniquely by*

$$(\boldsymbol{\alpha}_0^{(1)},\boldsymbol{\alpha}_1^{(1)},\cdots,\boldsymbol{\alpha}_{k-1}^{(1)})=\Delta_k(\boldsymbol{\alpha}_0,\boldsymbol{\alpha}_1,\cdots,\boldsymbol{\alpha}_k). \tag{3.17}$$

In particular, if there exist unitary matrices $\boldsymbol{U}_0$ and $\boldsymbol{V}_0$ such that $\boldsymbol{\alpha}_0=$

$U_0(\tilde{\alpha}_0 \oplus I_{r_0})V_0^*$ *with* $\|\tilde{\alpha}_0\| < 1$, *then* $\Phi_1(z)$ *has the form*

$$\Phi_1(z) = U_0(\widetilde{\Phi}_1(z) \oplus 0_{r_0})V_0^*. \tag{3.18}$$

Proof Let $\Phi(z) \in \mathbf{S}_p$ be a solution to the SC problem (3.6); then the matrix Γ_k defined via Eq. (3.7) is contractive. We may assume throughout this proof that $\|\alpha\| < 1$ (and thus $\alpha_0 = \hat{\alpha}_0$), which is not really restrictive by the reason mentioned above. Put now

$$\Phi_1(z) = \frac{1}{z}(I_p - \alpha_0\alpha_0^*)^{1/2}(I_p - \Phi(z)\alpha_0^*)^{-1}(\Phi(z) - \alpha_0)(I_p - \alpha_0^*\alpha_0)^{-1/2}. \tag{3.19}$$

It follows from Theorem 1 of [23] (p. 145) and the maximum modulus theorem that $\Phi_1(z)$ also belongs to the class $\mathbf{S}_p$. Thus Eq. (3.15) holds with α_0 in place of $\hat{\alpha}_0$ therein. Thus,

$$(I_p - \Phi(z)\alpha_0^*)(I_p - \alpha_0\alpha_0^*)^{-1/2} z\Phi_1(z) =$$
$$(\alpha_1 z + \cdots + \alpha_k z^k + O(z^{k+1}))(I_p - \alpha_0^*\alpha_0)^{-1/2}.$$

From the last we deduce the formula (3.17) for the case of $\|\alpha_0\| < 1$.

Conversely, if $\Phi_1(z) = \alpha_0^{(1)} + \alpha_1^{(1)} z + \cdots + \alpha_{k-1}^{(1)} z^{k-1} + O(z^k) \in \mathbf{S}_p$ with $\alpha_0^{(1)}, \alpha_1^{(1)}, \cdots, \alpha_{k-1}^{(1)}$ defined via Eq. (3.17), then it is seen by reading backwards the proceding argument that the $\mathbf{C}^{p\times p}$-valued function $\Phi(z)$ defined by Eq. (3.15) belongs to $\mathbf{S}_p$ and is a solution to the SC problem (3.6). ■

We now start with an arbitrary solution $\Phi(z) \in \mathbf{S}_p$ to the SC problem (3.6) with $k = n-2$. Then the matrix Γ_{n-2} defined by Eq. (3.7) with $k = n-2$ is contractive. Let us construct a sequence of $\Phi_k(z) \in \mathbf{S}_p$ by means of the following recurrence relation:

$$\Phi_{j+1}(z) =$$

$$\frac{1}{z}(I_p - \hat{\alpha}_0^{(j)}\hat{\alpha}_0^{(j)*})^{1/2}(I_p - \Phi_j(z)\hat{\alpha}_0^{(j)*})^{-1}(\Phi_j(z) - \alpha_0^{(j)})(I_p - \hat{\alpha}_0^{(j)*}\hat{\alpha}_0^{(j)})^{-1/2} \tag{3.20}$$

for $j = 0, 1, \cdots$, with the initialization $\Phi_0(z) = \Phi(z)$. If $\alpha_0 = \Phi(0)$ is unitary, then $\Phi(z) \equiv \alpha_0$ as was mentioned above. Assuming that α_0 is not unitary, we have from Lemma 3.2 with $k = n-2$ that $\Phi_1(z)$ defined uniquely via Eq. (3.15) belongs to $\mathbf{S}_p$ and has the form,

$$\Phi_1(z) = \alpha_0^{(1)} + \alpha_1^{(1)} z + \cdots + \alpha_{n-3}^{(1)} z^{n-3} + O(z^{n-2}),$$

where the coefficients $\boldsymbol{\alpha}_0^{(1)}, \boldsymbol{\alpha}_1^{(1)}, \cdots, \boldsymbol{\alpha}_{n-3}^{(1)}$ are defined by Eq. (3.17) with $k=n-2$. In particular, if $\boldsymbol{\alpha}_0=\boldsymbol{U}_0(\tilde{\boldsymbol{\alpha}}_0\oplus\boldsymbol{I}_{r_0})\boldsymbol{V}_0^*$ with $\mathbf{0}\leqslant\tilde{\boldsymbol{\alpha}}_0<\boldsymbol{I}_{p-r_0}$, then $\Phi_1(0)=\boldsymbol{\alpha}_0^{(1)}=\boldsymbol{U}_0(\boldsymbol{\beta}_0^{(1)}\oplus\mathbf{0}_{r_0})\boldsymbol{V}_0^*$ for a suitable $\boldsymbol{\beta}_0^{(1)}$ with $\|\boldsymbol{\beta}_0^{(1)}\|\leqslant 1$, so that there exist unitary matrices $\boldsymbol{U}_1$ and $\boldsymbol{V}_1$, which are of the form diag $(*, \boldsymbol{I}_{r_0})$, such that $\boldsymbol{\alpha}_0^{(1)}=\boldsymbol{U}_0\boldsymbol{U}_1(\tilde{\boldsymbol{\beta}}_0^{(1)}\oplus\boldsymbol{I}_{r_1}\oplus\mathbf{0}_{r_0})\boldsymbol{V}_1^*\boldsymbol{V}_0^*$ with $\|\tilde{\boldsymbol{\beta}}_0^{(1)}\|<1$, where $r_1=p-\mathrm{rank}(\boldsymbol{I}_p-\boldsymbol{\alpha}_0^{(1)}\boldsymbol{\alpha}_0^{(1)*})$, and therefore $\boldsymbol{\alpha}_0=\boldsymbol{U}_0\boldsymbol{U}_1\boldsymbol{U}_1^*(\tilde{\boldsymbol{\alpha}}_0^*\oplus\boldsymbol{I}_{r_0})\boldsymbol{V}_1\boldsymbol{V}_1^*\boldsymbol{V}_0^*=\boldsymbol{U}_0\boldsymbol{U}_1(\tilde{\boldsymbol{\beta}}_0\oplus\boldsymbol{I}_{r_0})\boldsymbol{V}_1^*\boldsymbol{V}_0^*$ for a suitable $\tilde{\boldsymbol{\beta}}_0$ with $\|\tilde{\boldsymbol{\beta}}_0\|<1$. Thus, $\Phi_1(z)$ has the form $\Phi_1(z)=\boldsymbol{U}_0\boldsymbol{U}_1(\tilde{\Phi}_1(z)\oplus\boldsymbol{I}_{r_1}\oplus\mathbf{0}_{r_0})\boldsymbol{V}_1^*\boldsymbol{V}_0^*$ with $\tilde{\Phi}_1(z)\in\mathbf{S}_{p-r_0-r_1}$. Let

$$\boldsymbol{\Gamma}_{n-3}^{(1)}=\begin{bmatrix}\boldsymbol{\alpha}_0^{(1)} & & \boldsymbol{O}\\ \vdots & \ddots & \\ \boldsymbol{\alpha}_{n-3}^{(1)} & \cdots & \boldsymbol{\alpha}_0^{(1)}\end{bmatrix},$$

then it is contractive. Here the inequality $\|\boldsymbol{\alpha}_0^{(1)}\|\leqslant 1$ is a condition imposed on $\boldsymbol{\alpha}_0$ and $\boldsymbol{\alpha}_1$. The condition that $\boldsymbol{\alpha}_0^{(1)}$ is unitary is possible only for $\Phi_1(z)\equiv\boldsymbol{\alpha}_0^{(1)}$. Further continuation of the recurrence procedure is evident. The algorithm terminates with a unitary $\boldsymbol{\alpha}_0^{(j)}$ for a certain j, then $\Phi_j(z)\equiv\boldsymbol{\alpha}_0^{(j)}$. Continuing the process $n-1$ times otherwise, we obtain the parametrized solutions to the SC problem (3.6) in both the nondegenerate ($\boldsymbol{\Gamma}_{n-2}>\boldsymbol{O}$) and degenerate ($\boldsymbol{\Gamma}_{n-2}\geqslant\boldsymbol{O}$ is singular) cases simultaneously.

For convenience, we take $\boldsymbol{\alpha}_i^{(0)}=\boldsymbol{\alpha}_i, i=0,1,\cdots,n-2$, and,

$$(\boldsymbol{\alpha}_0^{(i+1)},\boldsymbol{\alpha}_1^{(i+1)},\cdots,\boldsymbol{\alpha}_{n-i-3}^{(i+1)})=\Delta_{n-i-2}(\boldsymbol{\alpha}_0^{(i)},\boldsymbol{\alpha}_1^{(i)},\cdots,\boldsymbol{\alpha}_{n-i-2}^{(i)}),\quad i=0,1,\cdots,n-3. \tag{3.21}$$

From Eq. (3.21) and the result of the proceding paragraph we deduce that there exist unitary matrices $\tilde{\boldsymbol{U}}$ and $\tilde{\boldsymbol{V}}$ such that

$$\begin{aligned}&\boldsymbol{\alpha}_0^{(0)}=\tilde{\boldsymbol{U}}(\boldsymbol{\beta}_0^{(0)}\oplus\boldsymbol{I}_{r_0})\tilde{\boldsymbol{V}}^*,\quad \|\boldsymbol{\beta}_0^{(0)}\|<1,\\ &\boldsymbol{\alpha}_0^{(i)}=\tilde{\boldsymbol{U}}(\boldsymbol{\beta}_0^{(i)}\oplus\boldsymbol{I}_{r_i}\oplus\mathbf{0}_{r_{i-1}}\cdots\oplus\mathbf{0}_{r_0})\tilde{\boldsymbol{V}}^*,\quad \|\boldsymbol{\beta}_0^{(i)}\|<1,\\ &\qquad i=1,2,\cdots,n-2,\end{aligned} \tag{3.22}$$

where $r_i=p-\mathrm{rank}(\boldsymbol{I}_p-\boldsymbol{\alpha}_0^{(i)}\boldsymbol{\alpha}_0^{(i)*}), i=0,1,\cdots,n-2$, and $\boldsymbol{\beta}_0^{(i)}$ does not appear if $r_0+r_1+\cdots+r_i=p$. The matrices $\boldsymbol{\alpha}_0^{(0)},\boldsymbol{\alpha}_0^{(1)},\cdots,\boldsymbol{\alpha}_0^{(n-2)}$ produced by the generalized Schur decomposition of $\Phi(z)\in\mathbf{S}_p$ and found successively by the formulas (3.8) and (3.21) will be referred to as the so-called *generalized Schur parameters* of $\Phi(z)$.

These lead to the next result in the framework of the Schur's theory.

Theorem 3.3 *If* $\boldsymbol{\Gamma}_{n-2}$ *defined by Eq.* (3.7) *with* $k=n-2\geqslant 1$ *is contractive, then and only then the SC problem* (3.6) *with* $k=n-2$ *is solvable; its solutions* $\Phi(z)$ *can be parametrized by a linear fractional transformation*

$$\Phi(z)=(\tilde{\theta}_{11}(z)\phi(z)+\tilde{\theta}_{12}(z))(\tilde{\theta}_{21}(z)\phi(z)+\tilde{\theta}_{22}(z))^{-1} \qquad (3.23)$$

with the coefficient matrix

$$\begin{bmatrix}\tilde{\theta}_{11} & \tilde{\theta}_{12}\\ \tilde{\theta}_{21} & \tilde{\theta}_{22}\end{bmatrix}=\overset{\curvearrowright}{\prod_{i=0}^{n-2}}\begin{bmatrix}(\boldsymbol{I}_p-\hat{\boldsymbol{a}}_0^{(i)}\hat{\boldsymbol{a}}_0^{(i)*})^{-1/2} & \boldsymbol{O}\\ \boldsymbol{O} & (\boldsymbol{I}_p-\hat{\boldsymbol{a}}_0^{(i)*}\hat{\boldsymbol{a}}_0^{(i)})^{-1/2}\end{bmatrix}\begin{bmatrix}z\boldsymbol{I}_p & \boldsymbol{\alpha}_0^{(i)}\\ z\boldsymbol{\alpha}_0^{(i)*} & \boldsymbol{I}_p\end{bmatrix}, \qquad (3.24)$$

where $\phi(z)$ *is an arbitrary* $\mathbf{C}^{p\times p}$*-valued functions of the class* $\mathbf{S}_p$, *and of the form* $\phi(z)=\tilde{\boldsymbol{U}}(\tilde{\phi}(z)\oplus\mathbf{0}_{s'})\tilde{\mathbf{V}}^*$ *with* $s'=\sum_{i=0}^{n-2}r_i$ *and* $\tilde{\boldsymbol{U}},\tilde{\mathbf{V}}$ *as in Eq.* (3.22), *and the symbol* $\hat{\boldsymbol{a}}_0^{(i)}$ *stands for the strictly contractive part of* $\boldsymbol{\alpha}_0^{(i)}$ *defined as in Eq.* (3.13), *while* $\boldsymbol{\alpha}_0^{(1)},\boldsymbol{\alpha}_0^{(2)},\cdots,\boldsymbol{\alpha}_0^{(n-2)}$ *are the generalized Schur parameters of* $\phi(z)$ *defined via Eq.* (3.21) *iteratively.*

Corollary 3.4 *If* $\boldsymbol{\Gamma}_{n-2}$ *is as in Theorem* 3.3, *then the SC problem* (3.6) *with* $k=n-2$ *has a unique solution* $\Phi(z)$ *if and only if* $p=\sum_{i=0}^{n-2}r_i$. *In this case,* $\Phi(z)=\tilde{\theta}_{12}(z)\tilde{\theta}_{22}^{-1}(z)$. *In particular, if* $\boldsymbol{\Gamma}_{n-2}$ *is contractive, and if* $p=1$, *then the SC problem* (3.6) *with* $k=n-2$ *has a unique solution if and only if* $\boldsymbol{I}_{n-1}-\boldsymbol{\Gamma}_{n-2}\boldsymbol{\Gamma}_{n-2}^*$ (*or* $\boldsymbol{I}_{n-1}-\boldsymbol{\Gamma}_{n-2}^*\boldsymbol{\Gamma}_{n-2}$) *is singular.*

We now return to the solution of the CC problem (1.2). Using the observation made before Theorem 3.3, one can see from Eqs. (3.8) and (3.21) that an analogous result holds for $\boldsymbol{c}_0,\boldsymbol{\alpha}_0^{(0)},\cdots,\boldsymbol{\alpha}_{n-2}^{(0)}$: if $\boldsymbol{c}_0=\boldsymbol{U}(\tilde{\boldsymbol{c}}_0\oplus\mathbf{0}_r)\boldsymbol{U}^*$ with $\tilde{\boldsymbol{c}}_0>\mathbf{0}$, there exist unitary matrices $\breve{\boldsymbol{U}}$ and $\breve{\mathbf{V}}$ such that

$$\boldsymbol{c}_0=\breve{\boldsymbol{U}}(\breve{\boldsymbol{\beta}}_0\oplus\mathbf{0}_r)\breve{\mathbf{V}}^*,\quad \boldsymbol{\alpha}_0^{(0)}=\breve{\boldsymbol{U}}(\breve{\boldsymbol{\alpha}}_0^{(0)}\oplus\boldsymbol{I}_{r_0}\oplus\mathbf{0}_r)\breve{\mathbf{V}}^*,$$

$$\boldsymbol{\alpha}_0^{(i)}=\breve{\boldsymbol{U}}(\breve{\boldsymbol{\alpha}}_0^{(i)}\oplus\boldsymbol{I}_{r_i}\oplus\mathbf{0}_{r_{i-1}}\oplus\cdots\oplus\mathbf{0}_{r_0}\oplus\mathbf{0}_r)\breve{\mathbf{V}}^*,\quad i=1,2,\cdots,n-2, \qquad (3.25)$$

where det $\check{\boldsymbol{\beta}}_0 \neq 0$, $\| \check{\boldsymbol{\alpha}}_0^{(i)} \| < 1, i=0,1,\cdots,n-2$, and $\check{\boldsymbol{\alpha}}_0^{(i)}$ does not appear if

$$r+r_0+\cdots+r_i=p.$$

The proof of the next theorem is the combination of Lemma 3. 1 and Theorem 3. 3.

Theorem 3. 5 *If $\mathbf{T}=(\mathbf{c}_{j-i})_{i,j=0}^{n-1} \geqslant \mathbf{O}$ and the $\mathbf{c}_i$ are of the form* (3. 1), *then the solutions $F(z)$ to the CC problem* (1. 2) *can be parametrized by a linear fractional transformation*

$$F(z)=(\theta_{11}(z)H(z)+\theta_{12}(z))(\theta_{21}(z)H(z)+\theta_{22}(z))^{-1} \quad (3.26)$$

with the coefficient matrix

$$\begin{bmatrix} \theta_{11}(z) & \theta_{12}(z) \\ \theta_{21}(z) & \theta_{22}(z) \end{bmatrix} = \begin{bmatrix} z\mathbf{c}_0^{1/2} & \mathbf{c}_0^{1/2} \\ -z\hat{\mathbf{c}}_0^{-1/2} & \hat{\mathbf{c}}_0^{-1/2} \end{bmatrix} \begin{bmatrix} \tilde{\theta}_{11}(z) & \tilde{\theta}_{12}(z) \\ \tilde{\theta}_{21}(z) & \tilde{\theta}_{22}(z) \end{bmatrix}, \quad (3.27)$$

where the matrix $(\tilde{\theta}_{ij}(z))_{i,j=1}^{2}$ is as in Eq. (3. 24), *$H(z)$ is an arbitrary $\mathbb{C}^{p\times p}$-valued function of the class $\mathbf{S}_p$, and of the form $H(z)=\check{\mathbf{U}}(\widetilde{H}(z)\oplus \mathbf{0}_s)\check{\mathbf{V}}^*$ in which $s=r+\sum_{i=0}^{n-2} r_i$, and $\check{\mathbf{U}},\check{\mathbf{V}}$ are as in Eq.* (3. 25).

Corollary 3. 6 *Let $\mathbf{T}=(\mathbf{c}_{j-i})_{i,j=0}^{n-1} \geqslant \mathbf{O}$. Then the CC problem* (1. 2) *has a unique solution $F(z)$ if and only if $p=r+\sum_{i=0}^{n-2} r_i$. In this case, $F(z)=\theta_{12}(z)\theta_{22}^{-1}(z)$. In particular, if $\mathbf{T}\geqslant \mathbf{O}$ and $p=1$, then the CC problem* (1. 2) *has a unique solution if and only if* **T** *is singular.*

We conclude this section with a description of solutions to the NP problem (1. 1) in the class $\mathbf{C}_p$ in both the nondegenerate and degenerate cases simultaneously. On the basis of Theorems 2. 1 and 3. 5, we can draw the following.

Theorem 3. 7 *If the NP problem* (1. 1) *is solvable, then its solutions $G(z)$ can be parametrized by a linear fractional transformation:*

$$G(z)=(a_{11}(z)H(z)+a_{12}(z))(a_{21}(z)H(z)+a_{22}(z))^{-1} \quad (3.28)$$

with the coefficient matrix

$$\begin{bmatrix} a_{11}(z) & a_{12}(z) \\ a_{21}(z) & a_{22}(z) \end{bmatrix} = \begin{bmatrix} A(z)\boldsymbol{I}_p & \Omega(z) \\ \boldsymbol{O} & \boldsymbol{I}_p \end{bmatrix} \begin{bmatrix} \theta_{11}(z) & \theta_{12}(z) \\ \theta_{21}(z) & \theta_{22}(z) \end{bmatrix}, \quad (3.29)$$

where $A(z)$, $\Omega(z)$ are as in Eqs. (2. 2) *and* (2. 1), *and $(\theta_{ij}(z))$ and*

$H(z)$ are given as in Eq. (3. 27). *In particular, if* $p = r + \sum_{i=0}^{n-1} r_i$, *then and only then the NP problem* (1. 1) *has exactly one solution. In this case*, $G(z) = a_{12}(z)a_{22}^{-1}(z)$. *In particular, for the scalar case*: $p=1$, *the NP problem* (1. 1) *has exactly one solution if and only if* $\boldsymbol{T}$ (*or* $\boldsymbol{P}$) *is nonnegative definite and singular.*

§ 4. Solutions to the CC problem (1. 2) and to the NP problem (1. 1) in a more closed form

This short section contains an alternative approach to the solution of the CC problem (1. 2), based on the use of the generalized left-and right-orthogonal polynomial matrices, of the first and second type, associated with the block-Toeplitz matrix $\boldsymbol{T} \geqslant \boldsymbol{O}$ (see, e. g. [25] for definition). As we shall see below, it allows one to give an explicit parametrization of the solutions to the CC problem (1. 2), and further to the NP problem (1. 1) in a more closed form. The same approach has been presented in of Ch. 5. 2 of[2] for the scalar situation, and in [12] for the nondegenerate case only.

Let $\hat{B}(z)$ denote the reciprocal of a polynomial matrix $B(z) = \sum_{i=0}^{s} \boldsymbol{B}_i z^i (\boldsymbol{B}_s \neq \boldsymbol{O})$, i. e., $\hat{B}(z) = z^s B^*(1/\bar{z})$. For $\boldsymbol{T} = (\boldsymbol{c}_{j-i})_{i,j=0}^{n-1} \geqslant \boldsymbol{O}$, motivated by [2,12], we consider the Szegö recurrence relations with coefficients entirely determined from the generalized Schur parameters $\boldsymbol{\alpha}_0^{(k)}$ $(0 \leqslant k \leqslant n-2)$ of $\Phi(z) \in \mathbf{S}_p$ defined via (3. 4):

$$\begin{aligned}
&M_{k+1}(z)(\boldsymbol{I}_p - \hat{\boldsymbol{a}}_0^{(k)} \hat{\boldsymbol{a}}_0^{(k)*})^{1/2} = zM_k(z) - \hat{P}_k(z)\boldsymbol{\alpha}_0^{(k)*},\\
&N_{k+1}(z)(\boldsymbol{I}_p - \hat{\boldsymbol{a}}_0^{(k)} \hat{\boldsymbol{a}}_0^{(k)*})^{1/2} = zN_k(z) + \hat{Q}_k(z)\boldsymbol{\alpha}_0^{(k)*},\\
&(\boldsymbol{I}_p - \hat{\boldsymbol{a}}_0^{(k)*} \hat{\boldsymbol{a}}_0^{(k)})^{1/2} P_{k+1}(z) = zP_k(z) - \boldsymbol{\alpha}_0^{(k)*} \hat{M}_k(z),\\
&(\boldsymbol{I}_p - \hat{\boldsymbol{a}}_0^{(k)*} \hat{\boldsymbol{a}}_0^{(k)})^{1/2} Q_{k+1}(z) = zQ_k(z) + \boldsymbol{\alpha}_0^{(k)*} \hat{N}_k(z),\\
&\qquad k = 0, 1, \cdots, n-2,
\end{aligned} \tag{4. 1}$$

with the initializations

$$M_0(z) = P_0(z) = \hat{\boldsymbol{c}}_0^{-1/2}, \quad N_0(z) = Q_0(z) = \boldsymbol{c}_0^{1/2}. \tag{4. 2}$$

Here, notation is as in the Section 3; in particular, $\hat{\boldsymbol{a}}_0^{(k)}$ denotes the strictly contractive part of the contractive matrix $\boldsymbol{\alpha}_0^{(k)}$, and $\hat{\boldsymbol{c}}_0$ is defined as in Eq. (3.3). On account of the identity

$$(\boldsymbol{I}_p-\hat{\boldsymbol{a}}_0^{(k)*}\hat{\boldsymbol{a}}_0^{(k)})^{-1/2}\boldsymbol{\alpha}_0^{(k)*}=\boldsymbol{\alpha}_0^{(k)*}(\boldsymbol{I}_p-\hat{\boldsymbol{a}}_0^{(k)}\boldsymbol{\alpha}_0^{(k)*})^{-1/2},$$

we can establish the connection between the representation (3.26) and the following representation for the solutions $F(z)$ to the CC problem (1.2):

$$F(z)=(zN_{n-1}(z)H(z)+\hat{Q}_{n-1}(z))(-zM_{n-1}(z)H(z)+\hat{P}_{n-1}(z))^{-1}, \tag{4.3}$$

where the parameter $H(z)\in\mathbf{S}_p$ is as in Theorem 3.5. Thus, we have the following theorem.

Theorem 4.1 *If the CC problem* (1.2) *is solvable, then its solutions $F(z)$ can be parametrized by Eq.* (4.3), *where the parameter $H(z)\in\mathbf{S}_p$ is as in Theorem* 3.5.

As for the NP problem (1.1), from Theorems 2.1 and 4.1 we deduce its solutions in yet another form.

Theorem 4.2 *If the NP problem* (1.1) *is solvable, then its solutions $G(z)$ can be parametrized by Eq.* (3.28) *but with the coefficient matrix*

$$\begin{bmatrix} a_{11}(z) & a_{12}(z) \\ a_{21}(z) & a_{22}(z) \end{bmatrix}=\begin{bmatrix} A(z)\boldsymbol{I}_p & \Omega(z) \\ \boldsymbol{O} & \boldsymbol{I}_p \end{bmatrix}\begin{bmatrix} zN_{n-1}(z) & \hat{Q}_{n-1}(z) \\ -zM_{n-1}(z) & \hat{P}_{n-1}(z) \end{bmatrix}, \tag{4.4}$$

where the parameter $H(z)\in\mathbf{S}_p$ is as in Theorem 3.7.

For $\boldsymbol{T}=(\boldsymbol{C}_{j-i})_{i,j=0}^{n-1}\geqslant\boldsymbol{O}$, the polynomial matrices $P_k(z)$ and $Q_k(z)$, defined by Eqs. (4.1) and (4.2), obey a certain generalized left-orthogonal property. To prove it, a few more notations are needed. In the sequel, by $\boldsymbol{B}^{\mathrm{D}}$ denote the Drazin inverse of a square matrix $\boldsymbol{B}$, i.e., the unique solution of matrix equations: $\boldsymbol{XBX}=\boldsymbol{X}$, $\boldsymbol{BX}=\boldsymbol{XB}$, $\boldsymbol{B}^k=\boldsymbol{XB}^{k+1}$, where $k=\mathrm{index}(\boldsymbol{B})$. Let

$$\boldsymbol{c}_0^{(k)}=\boldsymbol{c}_0-(\boldsymbol{c}_k^*,\cdots,\boldsymbol{c}_1^*)[(\boldsymbol{c}_{j-i})_{i,j=0}^{k-1}]^{\mathrm{D}}\begin{bmatrix}\boldsymbol{c}_k\\ \vdots \\ \boldsymbol{c}_1\end{bmatrix}, \quad k=1,2,\cdots,n-1. \tag{4.5}$$

We call $\boldsymbol{c}_0^{(0)}=\boldsymbol{c}_0, \boldsymbol{c}_0^{(1)}, \cdots, \boldsymbol{c}_0^{(n-1)}$ the *generalized Schur complements* of the block-Toeplitz matrix $\boldsymbol{T}=(\boldsymbol{c}_{j-i})_{i,j=0}^{n-1}\geqslant \boldsymbol{O}$. The symbol δ_{lk} denotes the Kronecker δ.

Theorem 4.3 *Let* $\boldsymbol{T}=(\boldsymbol{c}_{j-i})_{i,j=0}^{n-1}\geqslant \boldsymbol{O}$, *and let* $P_k(z)=\sum_{i=0}^{k}\boldsymbol{p}_{ki}z^i$, $k=0,1,\cdots,n-1$, *be defined by Eqs.* (4.1) *and* (4.2). *Then, for any solution* $\sigma(\theta)$ *to the TM problem* (1.3), *the polynomial matrices* $P_k(z)$ *satisfy the generalized left-orthogonal relations*:

$$\frac{1}{2\pi}\int_0^{2\pi}P_l(\mathrm{e}^{\mathrm{i}\theta})\mathrm{d}\sigma(\theta)P_k^*(\mathrm{e}^{\mathrm{i}\theta})=\delta_{lk}\boldsymbol{p}_{kk}\boldsymbol{c}_0^{(k)}\boldsymbol{p}_{kk}^*,\quad k,l=0,1,\cdots,n-1. \tag{4.6}$$

Proof Let $Q_k(z)=\sum_{i=0}^{k}\boldsymbol{q}_{ki}z^i$, $k=0,1,\cdots,n-1$, be defined via Eqs. (4.1) and (4.2). It follows from Eq. (4.3) with k in place of $n-1$ that

$$\widetilde{F}(z)\equiv\hat{Q}_k(z)\hat{P}_k^{-1}(z)=\boldsymbol{c}_0+2\boldsymbol{c}_1z+\cdots+2\boldsymbol{c}_kz^k+\mathrm{O}(z^{k+1}),\qquad k=0,1,\cdots,n-1, \tag{4.7}$$

so that $\widetilde{F}(z)$ is a certain solution to the CC problem (1.2) with $n=k+1$ and

$$\begin{bmatrix}\boldsymbol{c}_0 & 2\boldsymbol{c}_1 & \cdots & 2\boldsymbol{c}_{n-1}\\ & \boldsymbol{c}_0 & \cdots & 2\boldsymbol{c}_{n-2}\\ & & \ddots & 2\boldsymbol{c}_1\\ \boldsymbol{O} & & & \boldsymbol{c}_0\end{bmatrix}\begin{bmatrix}\boldsymbol{p}_{k0}^*\\ \vdots\\ \boldsymbol{p}_{kk}^*\\ \boldsymbol{O}\\ \vdots\\ \boldsymbol{O}\end{bmatrix}=\begin{bmatrix}\boldsymbol{q}_{k0}^*\\ \vdots\\ \boldsymbol{q}_{kk}^*\\ \boldsymbol{O}\\ \vdots\\ \boldsymbol{O}\end{bmatrix},\quad k=0,1,\cdots,n-1. \tag{4.8}$$

Further on, $F_k(z)\in\mathbf{C}_p$ implies the relation $F_k(z)=-F_k^*(1/\bar{z})$, so we obtain that

$$\hat{Q}_k(z)\hat{P}_k^{-1}(z)=-P_k^{-1}(z)Q_k(z),\quad k=0,1,\cdots,n-1.$$

The last equation and Eq. (4.7) yield

$$[\boldsymbol{p}_{k0},\cdots,\boldsymbol{p}_{kk},\boldsymbol{O},\cdots,\boldsymbol{O}]\times\begin{bmatrix}\boldsymbol{c}_0 & 2\boldsymbol{c}_1 & \cdots & 2\boldsymbol{c}_{n-1}\\ & \ddots & \ddots & \vdots\\ & & \ddots & 2\boldsymbol{c}_1\\ \boldsymbol{O} & & & \boldsymbol{c}_0\end{bmatrix}=-[\boldsymbol{q}_{k0},\cdots,\boldsymbol{q}_{k,k-1},\boldsymbol{q}_{kk}-\boldsymbol{r}_{kk},\boldsymbol{O},\cdots,\boldsymbol{O}]. \tag{4.9}$$

Combining Eqs. (4.8) and (4.9) we come to

$$\begin{bmatrix} \boldsymbol{c}_0 & \boldsymbol{c}_1 & \cdots & \boldsymbol{c}_{n-1} \\ \boldsymbol{c}_1^* & \boldsymbol{c}_0 & \ddots & \vdots \\ \vdots & \ddots & \ddots & \boldsymbol{c}_1 \\ \boldsymbol{c}_{n-1}^* & \cdots & \boldsymbol{c}_1^* & \boldsymbol{c}_0 \end{bmatrix} \begin{bmatrix} \boldsymbol{p}_{k0}^* \\ \vdots \\ \boldsymbol{p}_{kk}^* \\ \boldsymbol{0} \\ \vdots \\ \boldsymbol{0} \end{bmatrix} = \begin{bmatrix} \boldsymbol{0} \\ \vdots \\ \boldsymbol{0} \\ \frac{1}{2}\boldsymbol{r}_{kk}^* \\ \boldsymbol{0} \\ \vdots \\ \boldsymbol{0} \end{bmatrix}, \tag{4.10}$$

so that for $k \geqslant l$ and for any solution $\sigma(\theta)$ to the TM problem (1.3),

$$\begin{aligned} \frac{1}{2\pi}\int_0^{2\pi} P_l(\mathrm{e}^{\mathrm{i}\theta})\mathrm{d}\sigma(\theta)P_k^*(\mathrm{e}^{\mathrm{i}\theta}) &= [\boldsymbol{p}_{l0},\cdots,\boldsymbol{p}_{ll},\boldsymbol{0},\cdots,\boldsymbol{0}]\left[\boldsymbol{0},\cdots,\boldsymbol{0},\frac{1}{2}\boldsymbol{r}_{kk},\boldsymbol{0},\cdots,\boldsymbol{0}\right]^* \\ &= \frac{1}{2}\delta_{lk}\boldsymbol{p}_{kk}\boldsymbol{r}_{kk}^*. \end{aligned} \tag{4.11}$$

On the other hand Eq. (4.10) implies

$$\begin{bmatrix} \boldsymbol{c}_0 & \cdots & \boldsymbol{c}_{k-1} \\ \vdots & \ddots & \vdots \\ \boldsymbol{c}_{k-1}^* & \cdots & \boldsymbol{c}_0 \end{bmatrix} \begin{bmatrix} \boldsymbol{p}_{k0}^* \\ \vdots \\ \boldsymbol{p}_{k,k-1}^* \end{bmatrix} + \begin{bmatrix} \boldsymbol{c}_k\boldsymbol{p}_{kk}^* \\ \vdots \\ \boldsymbol{c}_1\boldsymbol{p}_{kk}^* \end{bmatrix} = \boldsymbol{O},$$

$$[\boldsymbol{c}_k^*,\cdots,\boldsymbol{c}_1^*][\boldsymbol{p}_{k0},\cdots,\boldsymbol{p}_{k,k-1}]^* + \boldsymbol{c}_0\boldsymbol{p}_{kk}^* = \frac{1}{2}\boldsymbol{r}_{kk}^*.$$

Since $\boldsymbol{T} \geqslant \boldsymbol{O}$ and $\boldsymbol{p}_{kk}^* = \hat{P}_k(0)$ is nonsingular, it follows from the last that

$$\frac{1}{2}\boldsymbol{r}_{kk}^*(\boldsymbol{p}_{kk}^*)^{-1} = \boldsymbol{c}_0 - [\boldsymbol{c}_k^*,\cdots,\boldsymbol{c}_1^*]\begin{bmatrix} \boldsymbol{c}_0 & \cdots & \boldsymbol{c}_{k-1} \\ \vdots & \ddots & \vdots \\ \boldsymbol{c}_{k-1}^* & \cdots & \boldsymbol{c}_0 \end{bmatrix}^{\mathrm{D}}\begin{bmatrix} \boldsymbol{c}_k \\ \vdots \\ \boldsymbol{c}_1 \end{bmatrix} = \boldsymbol{c}_0^{(k)}, \tag{4.12}$$

and thus the equality holds in Eq. (4.6). ■

We remark that in the nondegenerate situation: $\boldsymbol{T} > \boldsymbol{O}$, all $\boldsymbol{c}_0^{(k)} > \boldsymbol{0}$ and $\boldsymbol{T}_k = (\boldsymbol{c}_{j-i})_{i,j=0}^{k-1} > \boldsymbol{O}$, and therefore all the matrices $\boldsymbol{r}_{kk}$ are nonsingular. Then Eq. (4.10) yields

$$(\boldsymbol{c}_0^{(k)})^{-1}\boldsymbol{p}_{kk}^{-1}[\boldsymbol{p}_{k0},\boldsymbol{p}_{k1},\cdots,\boldsymbol{p}_{kk}] = [\boldsymbol{0},\cdots,\boldsymbol{0},\boldsymbol{I}_p]\boldsymbol{T}_k^{-1},\ k=0,1,\cdots,n-1, \tag{4.13}$$

so that the left side block-vector of the last equality coincides with the last block-row of $\boldsymbol{T}_k^{-1}$. Consequently, the polynomial matrices $P_k(z)$ defined via Eqs. (4.1) and (4.2) coincide with those given in ([1], p. 168) and ([12], p. 36) within a certain left normalization matrix factor. They form the family of *left-orthogonal polynomial matrices* associated with the measure $\sigma(\theta)$ whose first n trigonometric moments are $\boldsymbol{c}_0, \boldsymbol{c}_1, \cdots, \boldsymbol{c}_{n-1}$.

Next, we shall show that the polynomial matrices $Q_k(z)$, defined via Eqs. (4.1) and (4.2) form the family of generalized left-orthogonal polynomial matrices of the *second type*.

Let $F(z) \in \mathbf{C}_p$ be a solution to the CC problem (1.2). It is easy to prove that $F^{\mathrm{D}}(z)$ belongs also to $\mathbf{C}_p$. Let $F^{\mathrm{D}}(z) = \boldsymbol{c}'_0 + 2\boldsymbol{c}'_1 z + \cdots + 2\boldsymbol{c}'_{n-1} z^{n-1} + \mathrm{O}(z^{n-1})$, and $\boldsymbol{T}' = (\boldsymbol{c}'_{j-i})_{i,j=0}^{n-1} (\geqslant \boldsymbol{O})$ with $\boldsymbol{c}'_{-i} = \boldsymbol{c}'^{*}_i$, and Let $\boldsymbol{c}'^{(0)}_0, \cdots, \boldsymbol{c}'^{(n-1)}_0$ denote the generalized Schur complements of $\boldsymbol{T}'$ in the similar manner to that defined for $\boldsymbol{T} \geqslant \boldsymbol{O}$. Then we have the following Corollary.

Corollary 4.4 *Let* $\mathbf{T}' = (\boldsymbol{c}'_{j-i})_{i,j=0}^{n-1}$ *be the block-Toeplitz matrix built on the coefficients of* $F^{D}(z)$. *Let* $\boldsymbol{c}'^{(0)}_0, \cdots, \boldsymbol{c}'^{(n-1)}_0$ *denote the generalized Schur complements of* $\mathbf{T}'$. *Then, for any measure* $\sigma'(\theta)$ *whose first n trigonometric moments are* $\boldsymbol{c}'_0, \cdots, \boldsymbol{c}'_{n-1}$, *the polynomial matrices* $Q_k(z)$ *satisfy the generalized left-orthogonal relations*,

$$\frac{1}{2\pi}\int_0^{2\pi} Q_l(\mathrm{e}^{\mathrm{i}\theta})\mathrm{d}\sigma'(\theta) Q_k^*(\mathrm{e}^{\mathrm{i}\theta}) = \delta_{lk} \boldsymbol{q}_{kk} \boldsymbol{c}'^{(k)}_0 \boldsymbol{q}_{kk}^*, \quad k, l = 0, 1, \cdots, n-1. \tag{4.14}$$

On the other hand, starting from $\widetilde{\boldsymbol{T}} = (\boldsymbol{c}_{i-j})_{i,j=0}^{n-1} \geqslant \boldsymbol{O}$ in the CC problem (1.2) and the TM problem (1.3) (that is, the $\boldsymbol{c}_i^*$ are substituted for the $\boldsymbol{c}_i$, the $F^*(\bar{z})$ for the $F(z)$, and the $-\sigma(2\pi-\theta)$ for $\sigma(\theta)$, and thus the $\boldsymbol{\alpha}_0^{(k)*}$ are substituted for the $\boldsymbol{\alpha}_0^{(k)}$ as the generalized Schur parameters of $F^*(\bar{z})$), one deduces from Eqs. (4.1) and (4.2),

$$P_{k+1}^*(\bar{z})(\boldsymbol{I}_p - \hat{\boldsymbol{a}}_0^{(k)*}\hat{\boldsymbol{a}}_0^{(k)})^{1/2} = zP_k^*(\bar{z}) - \hat{M}_k^*(\bar{z})\boldsymbol{\alpha}_0^{(k)},$$

$$(\boldsymbol{I}_p - \hat{\boldsymbol{a}}_0^{(k)}\hat{\boldsymbol{a}}_0^{(k)*})^{1/2} M_{k+1}^*(\bar{z}) = zM_k^*(\bar{z}) - \boldsymbol{\alpha}_0^{(k)}\hat{P}_k^*(\bar{z}).$$

According to Theorem 4.3, the $M_k^*(\bar{z})\ (k=0,1,\cdots,n-1)$ form the family of generalized left-orthogonal polynomial matrices associated with the measure $-\sigma(2\pi-\theta)$ whose first n trigonometric moments are $\boldsymbol{c}_0, \boldsymbol{c}_1^*, \cdots, \boldsymbol{c}_{n-1}^*$, where the $\sigma(\theta)$ are defined as in Theorem 4.3. Then it is easily seen that the $M_k(z)$ form the family of *generalized right-orthogonal polynomial matrices* associated with $\sigma(\theta)$. In particular, if $\widetilde{\boldsymbol{T}}>\boldsymbol{O}$ (or, equivalently, $\boldsymbol{T}>\boldsymbol{O}$), then

$$[\boldsymbol{m}_{k0}^{\mathrm{T}}, \boldsymbol{m}_{k1}^{\mathrm{T}}, \cdots, \boldsymbol{m}_{kk}^{\mathrm{T}}]^{\mathrm{T}} \boldsymbol{m}_{kk}^{-1} (\tilde{\boldsymbol{c}}_0^{(k)})^{-1} = \widetilde{\boldsymbol{T}}_k^{-1} [\boldsymbol{0}, \cdots, \boldsymbol{0}, \boldsymbol{I}_p]^{\mathrm{T}},$$
$$k=0,1,\cdots,n-1, \tag{4.15}$$

where $\tilde{\boldsymbol{c}}_0^{(0)} = \boldsymbol{c}_0, \tilde{\boldsymbol{c}}_0^{(1)}, \cdots, \tilde{\boldsymbol{c}}_0^{(n-1)}$ stand for the generalized Schur complements of block-Toeplitz matrix $\widetilde{\boldsymbol{T}} = (\boldsymbol{c}_{i-j})_{i,j=0}^{n-1}$ defined as in Eq. (4.5) (but with $\boldsymbol{T}$ replaced by $\widetilde{\boldsymbol{T}}$), and they form the family of *right-orthogonal polynomial matrices* associated with $\sigma(\theta)$ ([12], p. 37).

As an analogous result, we can prove by using the observation made above that the $N_k(z)$ form the family of generalized right-orthogonal polynomial matrices associated with $\sigma'(\theta)$, defined as in Corollary 4.4, of the *second type*.

Acknowledgements

The authors acknowledge with many thanks the helpful comments and suggestions of the referee.

References

[1] D Alpay, V Bolotnikov, Ph Loubaton. On tangential H_2 interpolation with second order norm constraints. Int. Eq. and Oper. Theory, 1996, 24: 156—178.

[2] N I Akhiezer. The Classical Moment Problem and Some Related Questions in Analysis. Oliver and Boyd, London, 1965.

[3] T Ando. Truncated moment problems for operators. Acta. Sci. Math. (Szeged), 1970, 31: 319—334.

[4] T Ando, K Fan. Pick-Julia theorems for operators. Math. Z., 1979, 168: 23—34.

[5] J A Ball, I Gohberg, L Rodman. Interpolation of Rational Matrix Functions. Birkhäuser, Basel, 1990.

[6] V Bolotnikov. On degenerate Hamburger moment problem and extensions of nonnegative Hankel matrices. Int. Eq. and Oper. Theory, 1996, 25: 253—276.

[7] G-N Chen. The general rational interpolation problem and its connection with the Nevanlinna-Pick interpolation and power moment problem. Linear Algebra Appl, 1998, 273: 83—117.

[8] G-N Chen, Y-J Hu. The generalized block-Pick type matrices and multiple Nevanlinna-Pick matrix interpolation. J. Beijing Normal University (Natur. Sci.), 1998, 34(3): 305—312.

[9] G-N Chen, X-Q Li. The Nevanlinna-Pick interpolation problems and the power moment problems for matrix-valued functions. Linear Algebra Appl., 1999, 288: 123—148.

[10] G-N Chen, H-P Zhang. More on Loewner matrices. Linear Algebra Appl., 1994, 203—204: 265—300.

[11] Ph Delsarte, Y Genin. The tridiagonal approach to Szegö's orthogonal polynomials, Toeplitz linear systems, and related interpolation problems. SIAM J. Math. Anal., 1988, 19: 718—735.

[12] Ph Delsarte, Y Genin, Y Kamp. Schur parametrization of positive definite block-Toeplitz systems. SIAM J. Appl. Math., 1979, 36: 34—46.

[13] Ph Delsarte, Y Genin, Y Kamp. The Nevanlinna-Pick problem for matrix-valued functions. SIAM J. Appl. Math. 1979, 36: 47—61.

[14] H Dym. J-contractive Matrix Functions, Reproducing Kernel Hilbert Spaces and Interpolation. Series in Math. 71, Amer. Math. Soc., Providence, RI, 1989.

[15] I P Fedchina. A criterion for the solvability of the Nevanlinna-Pick tangent problem. Mat. Issled., 1972, 71(4/26): 213—227.

[16] I P Fedchina. Tangential Nevanlinna-Pick problem with multiple points. Akad. Nauk Arm. SSR Dokl., 1975, 60: 37—42.

[17] C Foias, A E Frazo. The Commutant Lifting Approach to Interpolation Problems. Birkhäuser, Basel, 1990.

[18] M G Krein, A A Nudelman. The Markov moment problem and extremal problems. Transl. Math. Monographs 50, Amer. Math. Soc., Providence, RI, 1977.

[19] I V Kovalishina. Analytic theory of a class of interpolation problems. Math. USSR Izv., 1984, 22: 419—463.

[20] I V Kovalishina, V P Potapov. Integral representation of Hermitian positive functions. VINITI, 1982; Private translation by T. Ando, Sapporo, Ja-

pan, 1982.

[21] G Mühlbach. On Hermite interpolation by Cauchy-Vandermonde systems: The Lagrange formula, the adjoint and the inverse of a Cauchy-Vandermonde matrix. J. Comput. Appl. Math., 1996, 67: 147—159.

[22] A A Nudelman. Some generalizations of classical interpolation problems. in: Operator Theory: Advances and Applications. Vol. 61, Birkhäuser, Basel, 1993: 171—187.

[23] V P Potapov. The multiplicative structure of J-contractive matrix functions. Amer. Math. Soc. Transl., 1960, 15(2): 131—243.

[24] D Sarason. Generalized interpolation in H^∞. Trans. Amer. Math. Soc., 1967, 127: 179—203.

[25] A Sinap, W Van Assche. Orthogonal matrix polynomials and applications. J. Comput. Appl. Math., 1996, 66: 27—52.

[26] I Schur. Über potenzreihen, die im Innern des Einheitskreises beschränkt sind. Z. Reine Angew, Math., 1917, 147: 205—232; 1918, 148: 122—145.

Journal of Mathematical Analysis and Applications
2001,254:23－34

在非退化与退化两种情形矩阵Stieltjes矩量问题的统一求解[①]

A Unified Treatment for the Matrix Stieltjes Moment Problem in Both Nondegenerate and Degenerate Cases

Abstract This paper presents a unified method for the solution of the matrix Stieltjes moment problem in both the nondegenerate and degenerate cases, based on the use of our recent investigation of the matrix Hamburger moment problem. As a by-product, a complete solution of the matrix Nevanlinna-Pick interpolation problem in the class of the matrix-valued Stieltjes functions is derived as well.

§ 1. Introduction and Preliminaries

The matrix Stieltjes moment problem (S) is formulated in the following manner. Given $m+1$ Hermitian matrices $\boldsymbol{s}_0, \boldsymbol{s}_1, \cdots, \boldsymbol{s}_m$ of order p, it is required to find conditions for the existence of a Hermitian measure $\sigma(u)$ $(0 \leqslant u < +\infty)$ (i. e., a right continuous and nondecreasing $\mathbf{C}^{p\times p}$-valued function on $[0, +\infty)$) subject to

① Received March 25, 1997. Partially supported by the National Natural Science Foundation of China.

本文与胡永建合作. *Submitted by* J. W. Helton.

$$\begin{aligned} s_k &= \int_0^{+\infty} u^k \mathrm{d}\sigma(u), \quad k = 0,1,\cdots,m-1, \\ s_m &\geqslant \int_0^{+\infty} u^m \mathrm{d}\sigma(u), \end{aligned} \tag{S}$$

and to describe the solutions if these conditions are met.

This problem with $m = 2n - 1$ was first considered by Yu. N. Dyukarev [2] and by V. A. Bolotnikov [3], who gave the description of the set of all solutions for the nondegenerate and degenerate cases, respectively, in terms of a linear fractional transformation of a Stieltjes pair $(p(z), q(z))$. Recently, Chen and Li [4] also gave a complete solution to problem S with $m=2n-1$ for the nondegenerate case based on the theory of orthogonal polynomial matrices.

Following [1], we treat in the present paper the problem S in a unified method, not only for both the nondegenerate and degenerate cases, but also for the cases of $m=2n$ and $m=2n-1$. As an application, we shall give a complete solution of the matrix Nevanlinna-Pick interpolation problem as well.

Recall that a $\mathbf{C}^{p\times p}$-valued function $F(z)$ analytic on the open upper half-plane $\mathscr{C}^+$ is said to be a Nevanlinna function if Im $F(z)\geqslant \boldsymbol{O}$ for $z\in \mathscr{C}^+$. The set of such Nevanlinna functions will be denoted by $\mathbf{N}_p$. Each $F(z)\in \mathbf{N}_p$ admits a continuation into the lower half-plane by reflection: $f(z)=f(\bar{z})^*$ for Im $z<0$. A $F(z)\in \mathbf{N}_p$ will be called a Stieltjes function if in addition $F(z)$ is analytic outside the semiaxis $[0,+\infty)$ and $F(z)\geqslant \boldsymbol{O}$ for real negative z. The set of such Stieltjes functions will be denoted by $\tilde{\mathbf{S}}_p$. Each $F(z)\in \tilde{\mathbf{S}}_p$ permits an integral representation [5],

$$F(z) = \boldsymbol{\alpha} + \int_0^{+\infty} \frac{\mathrm{d}\sigma(u)}{u-z}, \tag{1}$$

where $\boldsymbol{\alpha}=\boldsymbol{\alpha}^* \geqslant \mathbf{0}$ and $\sigma(u)(0\leqslant u<+\infty)$ is a certain Hermitian measure such that tr $\int_0^{\infty} (1+u)^{-1}\mathrm{d}\sigma(u) <+\infty$.

Lemma 1.1[5] *$F(z)\in\tilde{\mathrm{S}}_p$ if and only if $F(z)\in \mathbf{N}_p$ and $zF(z)\in \mathbf{N}_p$.*

A pair of $\mathbf{C}^{p\times p}$-valued functions $p(z), q(z)$, meromorphic in $\mathscr{C}^+$, is said to be a Nevanlinna pair if

$$\det(p^*(z)p(z)+q^*(z)q(z))\not\equiv 0, \tag{2}$$

$$-\mathrm{i}(q^*(z)p(z)-p^*(z)q(z))\geqslant \boldsymbol{O}. \tag{3}$$

If, furthermore, $p(z), q(z)$ are meromorphic outside the semiaxis $[0, +\infty)$, and, parallel with the conditions (2)(3), satisfy

$$-\mathrm{i}(zq^*(z)p(z)-\bar{z}p^*(z)q(z))\geqslant \boldsymbol{O}, \tag{4}$$

then the pair $(p(z), q(z))$ is said to be a Stieltjes pair. Two pairs $(p(z), q(z))$ and $(p_1(z), q_1(z))$ are said to be equivalent if there exists a meromorphic matrix-valued function $G(z)$ with $\det G(z)\not\equiv 0$ such that $p_1(z)=p(z)G(z)$ and $q_1(z)=q(z)G(z)$.

Denote by $[h]$ the integer part of real h, and put now

$$\boldsymbol{H}_0=(\boldsymbol{s}_{i+j})_{i,j=0}^{[m/2]} \quad \text{and} \quad \boldsymbol{H}_1=(\boldsymbol{s}_{i+j+1})_{i,j=0}^{[(m-1)/2]},$$

which are Hermitian block-Hankel matrices associated with the initial data $\boldsymbol{s}_0, \boldsymbol{s}_1, \cdots, \boldsymbol{s}_m$. It follows from [6] that the problem S is actually equivalent to a certain matrix moment problem, which consists in finding all of the Hermitian measures $\tau(u)$ $(-\infty<u<+\infty)$ such that

$$\begin{aligned}
\boldsymbol{s}_k &= \int_{-\infty}^{+\infty} u^{2k} d\tau(u), \quad k=0,1,\cdots,m-1 \\
\boldsymbol{0} &= \int_{-\infty}^{+\infty} u^{2k-1} d\tau(u), \quad k=1,2,\cdots,m \\
\boldsymbol{s}_m &\geqslant \int_{-\infty}^{+\infty} u^{2m} d\tau(u).
\end{aligned} \tag{5}$$

Thus, a criterion for the solvability of the problem S is established as follows. (See [6] for the sufficiency and [7] for the case of $m=2n-1$ only.)

Lemma 1.2 *The problem S is solvable if and only if* $\boldsymbol{H}_0\geqslant\boldsymbol{O}$ *and* $\boldsymbol{H}_1\geqslant\boldsymbol{O}$.

Proof Observe [1] that the problem (5) is solvable if and only if

$$\boldsymbol{\Gamma}(m)=\begin{pmatrix}
\boldsymbol{s}_0 & \boldsymbol{0} & \boldsymbol{s}_1 & \cdot \\
 & & & \cdot \\
\boldsymbol{0} & \boldsymbol{s}_1 & \cdot & \boldsymbol{s}_{m-1} \\
 & & \cdot & \\
\boldsymbol{s}_1 & \cdot & \boldsymbol{s}_{m-1} & \boldsymbol{0} \\
 & \cdot & & \\
\cdot & \boldsymbol{s}_{m-1} & \boldsymbol{0} & \boldsymbol{s}_m
\end{pmatrix}\geqslant\boldsymbol{O}.$$

It is easy to check that $\boldsymbol{\Gamma}(m)$ is congruent to

$$\begin{pmatrix} \boldsymbol{H}_0 & \boldsymbol{O} \\ \boldsymbol{O} & \boldsymbol{H}_1 \end{pmatrix},$$

so that $\boldsymbol{\Gamma}(m)\geqslant\boldsymbol{O}$ is equivalent to $\boldsymbol{H}_0\geqslant\boldsymbol{O}$ and $\boldsymbol{H}_1\geqslant\boldsymbol{O}$. ■

§ 2. Description of the solutions

To describe the solutions to the problem S, we begin by considering that problem in the simplest case: $m=0$ and $\boldsymbol{s}_0>\boldsymbol{0}$. In what follows, for matrices $\boldsymbol{A}$ and $\boldsymbol{B}$ (nonsingular) of order p, denote $\boldsymbol{A}\boldsymbol{B}^{-1}$ by $\dfrac{\boldsymbol{A}}{\boldsymbol{B}}$ and the Drazin inverse of $\boldsymbol{A}$ by $\boldsymbol{A}^{\mathrm{D}}$.

Lemma 2. 0[1] *Let $\boldsymbol{s}_0>\boldsymbol{0}$. Then the set of the Hermitian measures $\sigma(u)$ $(-\infty<u<+\infty)$ subject to $\int_{-\infty}^{+\infty}\mathrm{d}\sigma(u)\leqslant\boldsymbol{s}_0$ can be parameterized by a linear fractional transformation of an arbitrary Nevanlinna pair $(p(z),q(z))$:*

$$\int_{-\infty}^{+\infty}\frac{\mathrm{d}\sigma(u)}{u-z}=\frac{-\boldsymbol{s}_0q(z)}{\boldsymbol{s}_0^{-1}p(z)+zq(z)}. \tag{6}$$

Lemma 2. 1 *Let $\boldsymbol{s}_0>\boldsymbol{0}$. Then the set of the solutions $\sigma(u)$ to the problem S with $m=0$, namely, $\int_0^{+\infty}\mathrm{d}\sigma(u)\leqslant\boldsymbol{s}_0$, can be parameterized by a linear fractional transformation of an arbitrary Stieltjes pair $(p(z),q(z))$,*

$$\int_0^{+\infty}\frac{\mathrm{d}\sigma(u)}{u-z}=\frac{-\boldsymbol{s}_0q(z)}{\boldsymbol{s}_0^{-1}zp(z)+zq(z)}, \tag{7}$$

where the matrix

$$\boldsymbol{\Theta}_0(z)=\begin{pmatrix} \boldsymbol{0} & -\boldsymbol{s}_0 \\ \boldsymbol{s}_0^{-1} & z\boldsymbol{I}_p \end{pmatrix}$$

is $\boldsymbol{J}$-expanding in $\mathscr{C}^+$ and $\boldsymbol{J}$-unitary on the real axis,

$$\boldsymbol{J}=\begin{pmatrix} \boldsymbol{0} & \mathrm{i}\boldsymbol{I}_p \\ -\mathrm{i}\boldsymbol{I}_p & \boldsymbol{0} \end{pmatrix},$$

i. e. , $\boldsymbol{\Theta}_0(z)\boldsymbol{J}\boldsymbol{\Theta}_0(z)^*\geqslant\boldsymbol{J}$ *for* $z\in\mathscr{C}^+$, *and* $\boldsymbol{\Theta}_0(z)\boldsymbol{J}\boldsymbol{\Theta}_0(z)^*=\boldsymbol{J}$ *for* $z\in\mathbf{R}$.

Proof Let $\sigma(u)$ be such that $\int_0^{+\infty}\mathrm{d}\sigma(u)\leqslant\boldsymbol{s}_0$. Define $\mathrm{d}\sigma(u)=\boldsymbol{0}$,

a. e. for $u<0$, so that $\int_{-\infty}^{+\infty} d\sigma(u) \leqslant \mathbf{s}_0$. Then Lemma 2. 0 implies that

$$F(z)=\int_0^{+\infty}\frac{d\sigma(u)}{u-z}=\int_{-\infty}^{+\infty}\frac{d\sigma(u)}{u-z}=\frac{\mathbf{s}_0\tilde{q}(z)}{\mathbf{s}_0^{-1}\tilde{p}(z)+z\tilde{q}(q)}$$

for a certain Nevanlinna pair $(\tilde{p}(z),\tilde{q}(z))$. But, by (1), the last matrix function $F(z)$ belongs to $\tilde{\mathbf{S}}_p$, so that $zF(z)\in\mathbf{N}_p$, and therefore

$$\begin{aligned}&-\mathrm{i}(zF(z)-\bar{z}F(z)^*)\\&=(\mathbf{s}_0^{-1}\tilde{p}(z)+z\tilde{q}(z))^{-*}(-\mathrm{i}(\bar{z}\tilde{q}(z)^*\tilde{p}(z)-z\tilde{p}(z)^*\tilde{q}(z)))\times\\&\quad(\mathbf{s}_0^{-1}\tilde{p}(z)+z\tilde{q}(z))^{-1}\geqslant\mathbf{0}.\end{aligned}$$

Thus, by definition, $(\tilde{p}(\lambda),z\tilde{q}(\lambda))$ is also a Nevanlinna pair. Set now $p(z)=\tilde{p}(z)$ and $q(z)=z\tilde{q}(z)$; then $(p(z),q(z))$ is actually a Stieltjes pair, and (7) holds.

Conversely, for an arbitrary Stieltjes pair $(p(z),q(z))$ (and therefore $p(z),q(z))$ and $(zp(z),q(z))$ are Nevanlinna pairs), by Lemma 2. 0, the matrix function defined on the right side of (6) corresponds to a Hermitian measure $\sigma(u)$ $(-\infty<u<+\infty)$ such that (7) holds with $zp(z)$ in place of $p(z)$ and $\int_{-\infty}^{+\infty} d\sigma(u)\leqslant\mathbf{s}_0$.

$(p(z),q(z))$ being also a Nevanlinna pair, $\int_{-\infty}^{+\infty} z d\sigma(u)/(u-z)\in\mathbf{N}_p$ by Lemma 2. 0 again, so that it admits an integral representation of the matrix class $\mathbf{N}_p$,

$$\int_{-\infty}^{+\infty}\frac{z d\sigma(u)}{u-z}=\boldsymbol{\beta}+\int_{-\infty}^{+\infty}\frac{1+uz}{u-z}d\tau(u),$$

in which $\boldsymbol{\beta}=\boldsymbol{\beta}^*$ and $\tau(u)(-\infty<u<+\infty)$ is a certain bounded Hermitian measure. Letting $z=\xi+\mathrm{i}\eta$ and comparing imaginary parts of both sides of the last equation, we have

$$\int_{-\infty}^{+\infty}\frac{u\eta}{(u-\xi)^2+\eta^2}d\sigma(u)=\int_{-\infty}^{+\infty}\frac{\eta(1+u^2)}{(u-\xi)^2+\eta^2}d\tau(u),$$

whence

$$\begin{aligned}&\int_{-\infty}^{+\infty}\left[\frac{\eta}{(u-\xi)^2+\eta^2}-\frac{\eta}{1+u^2}\right](1+u^2)\left[d\tau(u)-\frac{u}{1+u^2}d\sigma(u)\right]\\&\quad=\eta\left[\int_{-\infty}^{+\infty}\frac{u}{1+u^2}d\sigma(u)-\int_{-\infty}^{+\infty}d\tau(u)\right].\end{aligned}$$

Putting

$$\psi(u)=\int_0^u(1+\lambda)^2\left[\mathrm{d}\tau(\lambda)-\frac{\lambda}{1+\lambda^2}\mathrm{d}\sigma(\lambda)\right],$$

from the Stieltjes - Perron inverse formulas [8, p. 124], we conclude that

$$\frac{\psi(u+0)+\psi(u-0)}{2}=\text{const.}$$

Therefore $(1+u^2)\mathrm{d}\tau(u)=u\mathrm{d}\sigma(u)$ and $\mathrm{d}\sigma(u)=\mathbf{0}$ a. e. for $u<0$. Thus (7) holds and $\int_0^{+\infty}\mathrm{d}\sigma(u)\leqslant \boldsymbol{s}_0$ for the aforementioned $\sigma(u)$. The rest follows from the fact that

$$\boldsymbol{\Theta}_0(z)\boldsymbol{J}\boldsymbol{\Theta}_0(z)^*=\begin{pmatrix}\mathbf{0} & \mathrm{i}\boldsymbol{I}_p\\ -\mathrm{i}\boldsymbol{I}_p & \mathrm{i}(\bar{z}-z)\boldsymbol{s}_0^{-1}\end{pmatrix}. \qquad \blacksquare$$

Given an arbitrary block vector $(\boldsymbol{h}_0,\boldsymbol{h}_1,\cdots,\boldsymbol{h}_k)$, where each $\boldsymbol{h}_j\in\mathbf{C}^{p\times p}$, we define a transformation $\boldsymbol{\Gamma}_k$,

$$\boldsymbol{\Gamma}_k(\boldsymbol{h}_0,\boldsymbol{h}_1,\cdots,\boldsymbol{h}_k)=(\boldsymbol{l}_0,\boldsymbol{l}_1,\cdots,\boldsymbol{l}_{k-1}), \tag{8}$$

throughout, where $\boldsymbol{l}_0=\boldsymbol{h}_1$ and $\boldsymbol{l}_1,\boldsymbol{l}_2,\cdots,\boldsymbol{l}_{k-1}$ are connected with $\boldsymbol{h}_0,\boldsymbol{h}_1,\cdots,\boldsymbol{h}_k$ by the relation

$$\begin{aligned}&(\boldsymbol{l}_{i+j+1})_{i,j=0}^{[k/2]-1,[(k+1)/2]-1}\\ &=(\boldsymbol{I}_{[k/2]}\otimes\boldsymbol{h}_0)\begin{bmatrix}\boldsymbol{h}_0 & & \mathbf{0}\\ \vdots & \ddots & \\ \boldsymbol{h}_{[k/2]-1} & \cdots & \boldsymbol{h}_0\end{bmatrix}^{\mathrm{D}}\times\\ &\left[(\boldsymbol{h}_{i+j})_{i,j=1}^{[k/2],[(k+1)/2]}-\begin{bmatrix}\boldsymbol{h}_1\\ \vdots\\ \boldsymbol{h}_{[k/2]}\end{bmatrix}\boldsymbol{h}_0^{\mathrm{D}}(\boldsymbol{h}_1,\cdots,\boldsymbol{h}_{[(k+1)/2]})\right]\times\\ &\begin{bmatrix}\boldsymbol{h}_0 & \cdots & \boldsymbol{h}_{[(k+1)/2]-1}\\ & \ddots & \vdots\\ \mathbf{0} & & \boldsymbol{h}_0\end{bmatrix}(\boldsymbol{I}_{[(k+1)/2]}\otimes\boldsymbol{h}_0).\end{aligned} \tag{9}$$

Lemma 2.2 *Let $\sigma(u)$ be a solution to the problem S with $m\geqslant 2$. Then*

$$\int_0^{+\infty}\frac{\mathrm{d}\sigma(u)}{u-z}=\frac{-\boldsymbol{s}_0-\boldsymbol{s}_0\boldsymbol{s}_1^{\mathrm{D}}F_1(z)}{(z\boldsymbol{I}_p-\boldsymbol{s}_0^{\mathrm{D}}\boldsymbol{s}_1)+z\boldsymbol{s}_1^{\mathrm{D}}F_1(z)}, \tag{10}$$

where $F_1(z)=\int_0^{+\infty}\frac{d\sigma^{(1)}(u)}{u-z}$ *for a certain* $\sigma^{(1)}(u)$ *resolving the moment problem*

$$\mathbf{s}_k^{(1)}=\int_0^{+\infty}u^k\mathrm{d}\tau(u),\quad k=0,1,\cdots,m-3;\quad \mathbf{s}_{m-2}^{(1)}\geqslant\int_0^{+\infty}u^{m-2}\mathrm{d}\tau(u),\tag{11}$$

in which $\mathbf{s}_0^{(1)},\mathbf{s}_1^{(1)},\cdots,\mathbf{s}_{m-2}^{(1)}$ *are defined via the formula*

$$(\mathbf{s}_0^{(1)},\mathbf{s}_1^{(1)},\cdots,\mathbf{s}_{m-2}^{(1)})=\boldsymbol{\Gamma}_{m-1}\boldsymbol{\Gamma}_m(\mathbf{s}_0,\mathbf{s}_1,\cdots,\mathbf{s}_m).\tag{12}$$

Proof Without loss of generality, we may assume $\mathbf{s}_0>\boldsymbol{O}$. Put now $F(z)=\int_0^{+\infty}d\sigma(u)/(u-z)$, then $F(z)\in\widetilde{\mathbf{S}}_p$, and this permits an asymptotic expansion in $\mathscr{C}^+$:

$$\lim_{z\to\infty}z^{m+1}\left(F(z)+\frac{\mathbf{s}_0}{z}+\cdots+\frac{\mathbf{s}_{m-1}}{z^m}\right)=-\int_0^{+\infty}u^m\mathrm{d}\sigma(u)\geqslant-\mathbf{s}_m.$$

It follows from [1, Theorem 4.1] that $F(z)$ can be rewritten as

$$F(z)=-\frac{\mathbf{s}_0}{z\boldsymbol{I}_p+\mathbf{s}_0^{-1}\widetilde{F}(z)},\tag{13}$$

where $\widetilde{F}(z)\in\mathbf{N}_p$ and

$$\lim_{z\to\infty}z^{m-1}\left(\widetilde{F}(z)+\mathbf{s}_1+\frac{\tilde{\mathbf{s}}_0^{(1)}}{z}+\cdots+\frac{\tilde{\mathbf{s}}_{m-3}^{(1)}}{z^{m-2}}\right)\geqslant-\tilde{\mathbf{s}}_{m-2}^{(1)},$$

in which $\tilde{\mathbf{s}}_0^{(1)},\mathbf{s}_1^{(1)},\cdots,\tilde{\mathbf{s}}_{m-2}^{(1)}$ are defined via

$$(\mathbf{s}_1,\tilde{\mathbf{s}}_0^{(1)},\cdots,\tilde{\mathbf{s}}_{m-3}^{(1)},\tilde{\mathbf{s}}_{m-2}^{(1)})=\boldsymbol{\Gamma}_m(\mathbf{s}_0,\mathbf{s}_1,\cdots,\mathbf{s}_m).$$

Since $zF(z)\in\mathbf{N}_p$, $\widetilde{F}(z)/z\in\mathbf{N}_p$, which shows that $\widetilde{F}(z)/z\in\widetilde{\mathbf{S}}_p$ by Lemma 1.1. Also, from

$$\lim_{z\to\infty}z^m\left(\frac{\widetilde{F}(z)}{z}+\frac{\mathbf{s}_1}{z}+\frac{\tilde{\mathbf{s}}_0^{(1)}}{z^2}+\cdots+\frac{\tilde{\mathbf{s}}_{m-3}^{(1)}}{z^{m-1}}\right)\geqslant-\mathbf{s}_{m-2}^{(1)},$$

we find that

$$\frac{\widetilde{F}(z)}{z}=-\frac{\mathbf{s}_1}{z\boldsymbol{I}_p+\mathbf{s}_1^{\mathrm{D}}\widetilde{F}_1(z)},\tag{14}$$

where $\widetilde{F}_1(z)\in\mathbf{N}_p$ and

$$\lim_{z\to\infty}z^{m-2}\left(\widetilde{F}_1(z)+\mathbf{s}_0^{(1)}+\frac{\mathbf{s}_1^{(1)}}{z}+\cdots+\frac{\mathbf{s}_{m-3}^{(1)}}{z^{m-3}}\right)\geqslant-\mathbf{s}_{m-2}^{(1)},\tag{15}$$

in which $\mathbf{s}_0^{(1)},\mathbf{s}_1^{(1)},\cdots,\mathbf{s}_{m-2}^{(1)}$ are defined by (12). Let $F_1(z)=\widetilde{F}_1(z)/z$. From (14) we obtain

$$\widetilde{F}(z)=-\frac{\mathbf{s}_1}{I_p+\mathbf{s}_1^{\mathrm{D}}F_1(z)}\in\mathbf{N}p, \tag{16}$$

whence $F_1(z)\in\mathbf{N}_p$, which shows in turn that $F_1(z)\in\widetilde{\mathbf{S}}_p$ and, by (15),

$$\lim_{z\to\infty}z^{m-1}\left(\widetilde{F}_1(z)+\frac{\mathbf{s}_0^{(1)}}{z}+\cdots+\frac{\mathbf{s}_{m-3}^{(1)}}{z^{m-2}}\right)\geqslant-\mathbf{s}_{m-2}^{(1)}.$$

Applying arguments similar to that given in the proofs of the Nevanlinna - Hamburger theorem [8, pp. 95~97, 9] and Lemma 2.1 to $F_1(z)\in\widetilde{\mathbf{S}}_p$, one obtains that there exists a Hermitian measure $\sigma^{(1)}(u)(0\leqslant u<+\infty)$ such that

$$\mathbf{s}_i^{(1)}=\int_0^{+\infty}u^i\mathrm{d}\sigma^{(1)}(u),\ i=0,1,\cdots,m-3,\ \mathbf{s}_{m-2}^{(1)}\geqslant\int_0^{+\infty}u^{m-2}\mathrm{d}\sigma^{(1)}(u),$$

and $F_1(z)=\int_0^{+\infty}\mathrm{d}\sigma^{(1)}(u)/(u-z)$. Thus, $\sigma^{(1)}(u)$ is the Hermitian measure we expect. Inserting (16) into (13), we obtain (10). ■

Lemma 2.3 *Let $\mathbf{s}_0,\mathbf{s}_1,\cdots,\mathbf{s}_m$ be given such that $\boldsymbol{H}_0\geqslant\boldsymbol{O}$ and $\boldsymbol{H}_1\geqslant\boldsymbol{O}$, and let $\mathbf{s}_0^{(1)},\mathbf{s}_1^{(1)},\cdots,\mathbf{s}_{m-2}^{(1)}$ be defined via (12). If $\sigma^{(1)}(u)$ is a solution to the problem (11), and $F_1(z)=\int_0^{+\infty}\mathrm{d}\sigma^{(1)}(u)/(u-z)$, then*

$$\int_0^{+\infty}\frac{\mathrm{d}\sigma(u)}{u-z}=\frac{-\mathbf{s}_0-\mathbf{s}_0\mathbf{s}_1^{\mathrm{D}}F_1(z)}{(z\boldsymbol{I}_p-\mathbf{s}_0^{\mathrm{D}}\mathbf{s}_1)+z\mathbf{s}_1^{\mathrm{D}}F_1(z)}, \tag{17}$$

where $\sigma(u)$ is a certain solution to the problem S.

Proof Let

$$F(z)=\frac{-\mathbf{s}_0-\mathbf{s}_0\mathbf{s}_1^{\mathrm{D}}F_1(z)}{(z\boldsymbol{I}_p-\mathbf{s}_0^{\mathrm{D}}\mathbf{s}_1)+z\mathbf{s}_1^{\mathrm{D}}F_1(z)},$$

which makes sense by [1, Lemma 2.6]. According to the formula (12), it is not difficult to verify that $F(z)$ belongs to $\widetilde{\mathbf{S}}_p$ and admits an asymptotic expansion in $\mathscr{C}^+$:

$$\lim_{z\to\infty}z^{m+1}\left(F(z)+\frac{\mathbf{s}_0}{z}+\cdots+\frac{\mathbf{s}_{m-1}}{z^m}\right)=-\hat{\mathbf{s}}_m\geqslant-\mathbf{s}_m.$$

By reason of arguments appearing in the proof of Lemma 2.2, there exists a Hermitian measure $\sigma(u)$ $(0<u<+\infty)$ such that $F(z)=\int_0^{+\infty}\mathrm{d}\sigma(u)/(u-z)$ and

$$\int_0^{+\infty}u^k\mathrm{d}\sigma(u)=\mathbf{s}_k,\ k=0,1,\cdots,m-1;\quad\int_0^{+\infty}u^m\mathrm{d}\sigma(u)=\hat{\mathbf{s}}_m\leqslant\mathbf{s}_m.$$

Thus, $\sigma(u)$ is a certain solution to the problem S. ■

Now put $\boldsymbol{s}_i^{(0)}=\boldsymbol{s}_i$, $i=0,1,\cdots,m$. Formula (12) gives rise to $\boldsymbol{s}_0^{(1)}$, $\boldsymbol{s}_1^{(1)},\cdots,s_{m-2}^{(1)}$. The further continuation of the procedure is evident. In the case of $m=2n$ [$m=2n-1$, resp.] after n [$n-1$] such steps, we construct $\boldsymbol{s}_0^{(1)},\boldsymbol{s}_1^{(1)},\boldsymbol{s}_0^{(2)},\boldsymbol{s}_1^{(2)},\cdots,\boldsymbol{s}_0^{(n-1)},\boldsymbol{s}_1^{(n-1)},\boldsymbol{s}_0^{(n)}$ [$\boldsymbol{s}_0^{(1)},\boldsymbol{s}_1^{(1)},\boldsymbol{s}_0^{(2)},\boldsymbol{s}_1^{(2)},\cdots,\boldsymbol{s}_0^{(n-1)},\boldsymbol{s}_1^{(n-1)}$]. By using Lemmas 2.2 and 2.4 recursively, we get to the general form of all solutions $\sigma(u)$ to the problem S with $m=2n$ or $m=2n-1$. Thus we have the following.

Theorem 2.4 *If the problem S is solvable, then the set of all solutions $\sigma(u)$ can be parameterized by a linear fractional transformation,*

$$\int_0^{+\infty}\frac{\mathrm{d}\sigma(u)}{u-z}=\begin{cases}\dfrac{\theta_{11}(z)zp(z)+\theta_{12}(z)q(z)}{\theta_{21}(z)zp(z)+\theta_{22}(z)q(z)}, & \text{if } m=2n,\\[2ex] \dfrac{\theta_{11}(z)p(z)+\theta_{12}(z)q(z)}{\theta_{21}(z)p(z)+\theta_{22}(z)q(z)}, & \text{if } m=2n-1,\end{cases}\tag{18}$$

where

$$\boldsymbol{\Theta}(z)=(\boldsymbol{\theta}_{ij}(z))_{i,j=1}^2=\begin{cases}\overset{\frown}{\prod_{i=0}^{n-1}}\begin{pmatrix}-\boldsymbol{s}_0^{(i)}\boldsymbol{s}_1^{(i)\mathrm{D}} & -\boldsymbol{s}_0^{(i)}\\ z\boldsymbol{s}_1^{(i)\mathrm{D}} & z\boldsymbol{I}_p-\boldsymbol{s}_0^{(i)\mathrm{D}}\boldsymbol{s}_1^{(i)}\end{pmatrix}\cdot\begin{pmatrix}\boldsymbol{0} & -\boldsymbol{s}_0^{(n)}\\ \boldsymbol{s}_0^{(n)\mathrm{D}} & z\boldsymbol{I}_p\end{pmatrix}, \\ \quad \text{if } m=2n;\\ \overset{\frown}{\prod_{i=0}^{n-1}}\begin{pmatrix}-\boldsymbol{s}_0^{(i)}\boldsymbol{s}_1^{(i)\mathrm{D}} & -\boldsymbol{s}_0^{(i)}\\ z\boldsymbol{s}_1^{(i)\mathrm{D}} & z\boldsymbol{I}_p-\boldsymbol{s}_0^{(i)\mathrm{D}}\boldsymbol{s}_1^{(i)}\end{pmatrix}, \\ \quad \text{if } m=2n-1\end{cases}\tag{19}$$

and $(p(z),q(z))$ is an arbitrary Stieltjes pair subject to the condition that if one at least of $\boldsymbol{H}_0$ and $\boldsymbol{H}_1$ is singular and if there exists a unitary matrix $\boldsymbol{U}$ such that

$$\boldsymbol{U}^*\boldsymbol{s}_0^{(m/2)}\boldsymbol{U}=\begin{pmatrix}\hat{\boldsymbol{s}}_0^{(m/2)} & \boldsymbol{0}\\ \boldsymbol{0} & \boldsymbol{0}_r\end{pmatrix},\quad \hat{\boldsymbol{s}}_0^{(m/2)}>\boldsymbol{0}$$

with $m=2n$ and

$$\boldsymbol{U}^*\boldsymbol{s}_1^{((m-1)/2)}\boldsymbol{U}=\begin{pmatrix}\hat{\boldsymbol{s}}_1^{((m-1)/2)} & \boldsymbol{0}\\ \boldsymbol{0} & \boldsymbol{0}_r\end{pmatrix},\quad \hat{\boldsymbol{s}}_1^{((m-1)/2)}>\boldsymbol{0}$$

with $m=2n-1$, then

$$U^{*} p(z) U=\begin{pmatrix} \hat{p}(z) & \mathbf{0} \\ \mathbf{0} & \mathbf{0}_{r} \end{pmatrix} \quad and$$

$$U^{*} q(z) U=\begin{pmatrix} \hat{q}(z) & \mathbf{0} \\ \mathbf{0} & I_{r} \end{pmatrix} \quad (0<r \leqslant p). \tag{20}$$

In particular, if $s_0^{(m/2)}=\mathbf{0}[s_1^{((m-1)/2)}=\mathbf{0}$, *resp.*$](r=p)$, *the problem S has a unique solution.*

The independent parameters in (18) are the equivalent classes of the Stieltjes pairs: different pairs lead to the same measure $\sigma(u)$ if and only if they are equivalent.

We note also that the process described above can be applied also to the full matrix Stieltjes moment problem and admits a precise algorithmization to compute the resolvent matrix $\boldsymbol{\Theta}(z)$ *of the problem S. Furthermore, the matrix* $\boldsymbol{\Theta}(z)$ *given in* (19) *is in fact* $\boldsymbol{J}$*-expanding in* $\mathscr{C}^{+}$ and $\boldsymbol{J}$-unitary on the real axis if $\boldsymbol{H}_0>\boldsymbol{O}$ and $\boldsymbol{H}_1>\boldsymbol{O}$.

§ 3. Nevanlinna-Pick Interpolation Problem in the Class $\widetilde{\mathbf{S}}_p$

As an application of the previous results, we shall consider the truncated matrix Nevanlinna-Pick interpolation (NP) problem in the class $\widetilde{\mathbf{S}}_p$. Given $z_1, z_2, \cdots, z_d \in \mathscr{C}^{+}$ and $z_{d+1}, \cdots, z_s \in(-\infty, 0)$, and a $p \times p$ matrix sequence $\boldsymbol{c}_{ik}$, $i=1,2, \cdots, s, k=0,1, \cdots, \tau_i-1$, where $\boldsymbol{c}_{ik}=\boldsymbol{c}_{ik}^{*}$, $i \geqslant d+1$, it is required to establish the conditions for a $G(z) \in \widetilde{\mathbf{S}}_p$ to exist such that

$$\frac{1}{k!} G^{(k)}(z_i)=\boldsymbol{c}_{ik}, \quad i=1,2, \cdots, s, k=0,1, \cdots, \tau_i-1, \tag{21}$$

and to give a description of all of its solutions.

Let $\Omega(z)$ be the polynomial matrix of formal degree at most $N-1$, where $N=2(\tau_1+\tau_2+\cdots+\tau_d)+\tau_{d+1}+\cdots+\tau_s$ is the number of all interpolation nodes with multiplicity, such that

$$\frac{1}{k!} \Omega^{(k)}(z_i)=\boldsymbol{c}_{ik} \text{ and } \frac{1}{k!} \Omega^{(k)}(\bar{z}_i)=\boldsymbol{c}_{ik}^{*} \ \forall i, k.$$

Now put

$$A(z) = \prod_{i=1}^{d}[(z-z_i)(z-\bar{z}_i)]^{\tau_i}\prod_{i=d+1}^{s}(z-z_i)^{\tau_i} \tag{22}$$

If $\Omega(z)/A(z)$ has the Laurent expansion at infinity,

$$\frac{\boldsymbol{s}_0}{z}+\frac{\boldsymbol{s}_1}{z^2}+\cdots+\frac{\boldsymbol{s}_{m-1}}{z^m}+\frac{\boldsymbol{s}_m}{z^{m+1}}+\cdots,$$

where $m=N-1$, then it is known [4] that $\boldsymbol{s}_k=\boldsymbol{s}_k^*$ $(k=0,1,\cdots,m)$. We call $(\boldsymbol{s}_0,\boldsymbol{s}_1,\cdots,\boldsymbol{s}_m)$ the block-Hankel vector of that NP problem, which corresponds to a certain problem S mentioned in Section 1.

Lemma 3.1[4] *The NP problem is solvable if and only if the corresponding problem S is solvable, and there exists a one-to-one correspondence between their solutions $G(z)$ and $\sigma(u)$ $(u\geqslant 0)$ defined by*

$$G(z)=\Omega(z)+A(z)\int_0^{+\infty}\frac{\mathrm{d}\sigma(u)}{u-z}. \tag{23}$$

Thanks to this correspondence, the problem S enables one not only to find existence conditions for the solutions to the problem NP, but also to describe all solutions, and so on. (See also [4,7] for the case where $\boldsymbol{H}_0>\boldsymbol{O}$, $\boldsymbol{H}_1>\boldsymbol{O}$, and $m=2n-1$; see [9~12] for more information.)

Theorem 3.2 *The problem NP is solvable if and only if $\boldsymbol{H}_0\geqslant\boldsymbol{O}$ and $\boldsymbol{H}_1\geqslant\boldsymbol{O}$, in which $(\boldsymbol{s}_0,\boldsymbol{s}_1,\cdots,\boldsymbol{s}_m)$ stands for its block-Hankel vector.*

Theorem 3.3 *If the problem NP is solvable, then its general solution $G(z)$ can be represented in the form of the linear fractional transformation,*

$$G(z)=\begin{cases}\dfrac{a_{11}(z)zp(z)+a_{12}(z)q(z)}{a_{21}(z)zp(z)+a_{22}(z)q(z)}, & if\ m=2n,\\[2ex] \dfrac{a_{11}(z)p(z)+a_{12}(z)q(z)}{a_{21}(z)p(z)+a_{22}(z)q(z)}, & if\ m=2n-1,\end{cases} \tag{24}$$

where

$$\begin{pmatrix}a_{11}(z) & a_{12}(z)\\ a_{21}(z) & a_{22}(z)\end{pmatrix}=\begin{pmatrix}A(z)\boldsymbol{I}_p & \Omega(z)\\ \boldsymbol{0} & \boldsymbol{I}_p\end{pmatrix}\boldsymbol{\Theta}(z), \tag{25}$$

in which $\boldsymbol{\Theta}(z)$ is as in (19) and $(p(z),q(z))$ is an arbitrary Stieltjes pair of the form formulated as in Theorem 2.4. In particular, if $\boldsymbol{s}_0^{(m/2)}=\boldsymbol{0}$ with $m=2n$ or $\boldsymbol{s}_1^{((m-1)/2)}=\boldsymbol{0}$ with $m=2n-1$, then the problem NP has only one solution.

References

[1] G-N Chen and Y-J Hu. The truncated Hamburger matrix moment problem in the nondegenerate and degenerate cases, and matrix continued fractions. Linear Algebra Appl. ,1998,277:199—236.

[2] Y Dyukarev. The matrix Stieltjes moment problem. Manuscript deposited at VINITI, No. 2628, March 22, 1981, Moscow 1981.

[3] V A Bolotnikov. Description of the solutions of a degenerate moment problem on the axis and the semiaxis, Teor. Funktsiĭ, Funktsional. Anal. Prilozhen,1988,(50):25—31[translated from Russian].

[4] G-N Chen and X-Q Li. The Nevanlinna-Pick interpolation problems and the power moment problems for the matrix-valued functions. Linear Algebra Appl. ,1999,288:123—148.

[5] M G Krein and A A Nudelman. The Markov Moment and Extremal Problems. Transl. Math. Monogr. ,1977,50.

[6] T Ando. Truncated moment problems for operators. Acta Sci. Math. (Szeged),1970,31:319—334.

[7] Y Dyukarev and V Katsnelson. Multiplicative and additive classes of Stieltjes analytic matrix valued functions, and interpolation problems associated with them. Amer. Math. Soc. Transl. Ser. (2),1986,131:55—70.

[8] N I Akhiezer. The Classical Moment Problem and Some Related Questions in Analysis. Oliver and Boyd, London, 1965.

[9] I Kovalishina. Analytic theory of a class of interpolation problems. Math. USSR Izv. ,1984,22(3):449—463.

[10] V A Bolotnikov. The two-sided Nevanlinna-Pick problem in the Stieltjes class. Oper. Theory Adv. Appl. ,1993,61:15—37.

[11] D Alpay and V A Bolotnikov. Classes of pairs of meromorphic matrix-valued functions generated by non-negative kernels and associated Nevanlinna-Pick problems. Integral Equations Operator Theory,1995,23:245—267.

[12] D Alpay, J A Ball, I Gohberg, and L Rodman. The two-side residue interpolation in the Stieltjes class for matrix functions. Linear Algebra Appl. , 1994,208—209:485—521.

Acta Mathematicae Applicata Sinica, English Series
2002, 18(2): 255－262

对亚纯函数的边界 Nevanlinna-Pick 插值[①]

Boundary Nevanlinna-Pick Interpolation for Meromorphic Functions

Abstract Versions of the multiple Nevanlinna-Pick interpolation problem in the class $\hat{\mathrm{N}}$ involving both interior and boundary data are investigated. This leads to solvability criteria for the indicated problems and a description of their solutions.

Keywords Nevanlinna-Pick interpolation; generalized Pick matrix; Hankel vector; power moment problem; minimally positive; Hankel matrix

§ 1. Introduction and Preliminaries

The present note is devoted to the solution of versions the multiple Nevanlinna-Pick interpolation (NP) problem in the class $\hat{\mathrm{N}}$, consisting of meromorphic functions $f(z)$ such that $\mathrm{Im}\, f(z)/\mathrm{Im}\, z \geqslant 0$, for $\mathrm{Im}\, z \neq 0$, which involve both interior and boundary data. The method used here is based on a transition of the indicated boundary NP problems into a certain truncated power moment problem associated with the so-called *Hankel vector* of the former. We are given θ distinct points $\lambda_1, \lambda_2, \cdots, \lambda_\theta$

① Manuscript received Augest 26, 1999; Revised November 27, 2000. 2000 MR Subject Classification 30E05, 47A57. Supported by the National Natural Sciences Foundation of China (No. 19971009). 本文与贾利新合作.

in π^+ (the open upper half plane) with respective multiplicities $\tau_1, \tau_2, \cdots \tau_\theta (\geqslant 1)$ and $c_{ik} \in \mathbf{C}$ for $i=1,2,\cdots,\theta, k=0,1,\cdots,\tau_i-1$, and s distinct real points $\alpha_1, \alpha_2, \cdots, \alpha_s$ with respective multiplicities $2\eta_1, 2\eta_2, \cdots, 2\eta_s (\geqslant 2)$ and $d_{jl} \in \mathbf{R}$ for $j=1,2,\cdots,s, l=0,1,\cdots,2\eta_j-1$. The IBNP($\hat{\mathrm{N}}$) problem requires finding the conditions which assure the existence of a function $f(z) \in \hat{\mathrm{N}}$ subject to

$$\begin{aligned} &\frac{1}{k!} f^{(k)}(\lambda_i) = c_{ik}, \quad i=1,2,\cdots,\theta, k=0,1,\cdots,\tau_i-1, \\ &\frac{1}{l!} f^{(l)}(\alpha_j) = d_{jl}, \quad j=1,2,\cdots,s, l=0,1,\cdots,2\eta_j-1, \end{aligned} \tag{1.1}$$

and to describe the family of such functions $f(z)$ whenever these conditions are met.

Let $m = \sum_{i=1}^{\theta} \tau_i, r = \sum_{j=1}^{s} \eta_j$, and $N=m+r>1$. Let $w(z)$ denote the unique interpolation polynomial, of degree at most $2N-1$, subject to both (1.1) and $\frac{1}{k!} f^{(k)}(\bar{\lambda}_i) = \bar{c}_{ik}, \forall i, k$. It is easily seen that $w(z)$ has all the real coefficients. Denote the collection of interpolation nodes, with multiplicities counted, by

$$\boldsymbol{x} = (\overbrace{x_1, \cdots, x_1}^{\delta_1 \text{ times}}, \cdots, \overbrace{x_{\theta+s}, \cdots, x_{\theta+s}}^{\delta_{\theta+s} \text{ times}}) \tag{1.2}$$

where $x_i = \lambda_i, \delta_i = \tau_i$ if $1 \leqslant i \leqslant \theta$ and $x_i = \alpha_i, \delta_i = \eta_i$ *if* $\theta+1 \leqslant i \leqslant \theta+s$; $\sum_{i=1}^{\theta+s} \delta_i = N$. The *generalized Pick matrix* of the IBNP ($\hat{\mathrm{N}}$) problem (1.1), associated with $\boldsymbol{x}$, is the block matrix $\boldsymbol{P}_{\boldsymbol{x}} = (\boldsymbol{P}_{ij})_{i,j=1}^{\theta+s} \in \mathbf{C}^{N \times N}$, in which $\boldsymbol{P}_{ij} = (p_{ij}^{kl})_{k,l=0}^{\delta_i-1,\delta_j-1} \in \mathbf{C}^{\delta_i \times \delta_j}$ with entries p_{ij}^{kl} determined by

$$p_{ij}^{kl} = \begin{cases} \dfrac{1}{k!} \dfrac{1}{l!} \dfrac{\partial^{k+l}}{\partial \lambda^k \lambda \mu^l} \left[\dfrac{w(\lambda) - w(\mu)}{\lambda - \mu} \right]_{\lambda = x_i, \mu = \bar{x}_j}, & \text{if } x_i \neq \bar{x}_j, \\ \dfrac{1}{(k+l+1)!} \dfrac{\partial^{k+l+1}}{\partial \lambda^{k+l+1}} w(\lambda) \Big|_{\lambda = x_i}, & \text{if } x_i = \bar{x}_j. \end{cases} \tag{1.3}$$

Let $A(z)$ be the annihilator polynomial of interpolation nodes, of degree $2N$:

$$A(z) = \prod_{i=1}^{\theta} (z - \lambda_i)^{\tau_i} (z - \bar{\lambda}_i)^{\tau_i} \prod_{j=1}^{s} (z - \alpha_j)^{2\eta_j}. \tag{1.4}$$

It is clear that $A(z)$ has also real coefficients and $A(u) > 0$ for real $u \neq$

$\alpha_1, \alpha_2, \cdots, \alpha_s$. Let $h_0, h_1, \cdots$ be the Markov parameters of the rational function $w(z)/A(z)$, namely, for sufficiently large values of $|z|$,

$$\frac{w(z)}{A(z)} = \sum_{j=0}^{+\infty} \frac{h_j}{z^{j+1}}. \tag{1.5}$$

We refer to $\boldsymbol{b} = (h_0, h_1, \cdots, h_{2N-2})^{\mathrm{T}} \in \mathbf{R}^{2N-1}$ as *the Hankel vector* of the IBNP($\hat{\mathrm{N}}$) problem (1.1). With the real vector $\boldsymbol{b}$ there is associated a certain truncated power moment problem: Find out a nondecreasing function $\tau(u)(-\infty<u<+\infty)$ such that

$$\Phi(z) = \int_{-\infty}^{+\infty} \frac{\mathrm{d}\tau(u)}{u-z} \in \hat{\mathrm{N}}$$

and

$$h_j = \int_{-\infty}^{+\infty} u^j \mathrm{d}\tau(u), j = 0, 1, \cdots, 2N-3; h_{2N-2} \geqslant \int_{-\infty}^{+\infty} u^{2N-2} \mathrm{d}\tau(u). \tag{1.6}$$

With notations as above, a variant of the IBNP($\hat{\mathrm{N}}$) problem, referred to as the IBNP($\hat{\mathrm{N}}$)$'$ *problem*, is noteworthy, which is the same as the IBNP($\hat{\mathrm{N}}$) problem except for the inequalities

$$\frac{1}{(2\eta_l-1)!} f^{(2\eta_l-1)}(\alpha_l) \leqslant d_{l,2\eta_l-1}, \quad l=1,2,\cdots,s \tag{1.7}$$

in place of the corresponding equalities in (1.1). It turns out that the IBNP($\hat{\mathrm{N}}$)$'$ problem rather than the IBNP($\hat{\mathrm{N}}$) problem is the closest boundary analogues of the origianal NP interpolation in $\hat{\mathrm{N}}$.

The functions of the class $\hat{\mathrm{N}}$ are characterized by

Lemma 1.1 (Chebotarev's Theorem) *For a meromorphic function $f(z)$ to belong to $\hat{\mathrm{N}}$, it is necessary and sufficient for it to admit the absolutely convergent expansion*:

$$f(z) = \left(c_0 - \sum_j \frac{\sigma_j}{v_j}\right) + c_1 z - \frac{c_{-1}}{z} - \sum_j \frac{\sigma_j}{z - v_j}, \tag{1.8}$$

where $c_0, v_j (\neq 0)$ are real, $c_1, c_{-1} \geqslant 0$, and $\sigma_j \geqslant 0$ for $j=1,2,\cdots$.

Two remarks are in order. First of all, $f(z) \in \hat{\mathrm{N}}$ if and only if $f(z) \in \mathrm{N}$ (the class of functions holomorphic in τ^+ with $\mathrm{Im}\, f(z) \geqslant 0$ there) and $f(u)$ is real for all real u but the poles of $f(u)$. Secondly, denote by $\sigma_f(u)$ the nondecreasing function $(-\infty<u<+\infty)$ which possibly has as its only mass points the v_j and 0 with $\sigma_f(\{v_j\})=\sigma_j\,(j=1,2,\cdots)$ and $\sigma(\{0\})=c_{-1}$. Then (1.8) takes on the form

$$f(z)=\tilde{c}_0+c_1z+\int_{-\infty}^{+\infty}\left(\frac{1}{u-z}-\frac{u}{1+u^2}\right)\mathrm{d}\sigma_f(u) \tag{1.9}$$

in which $\int_{-\infty}^{+\infty}u(1+u^2)^{-1}\mathrm{d}\sigma_f(u)<+\infty$, since $\sum_j\sigma_j/v_j$ converges, and $\tilde{c}_0=c_0-\sum_j\sigma_j/v_j+\int_{-\infty}^{+\infty}u(1+u^2)^{-1}\mathrm{d}\sigma_f(u)$ is real.

Let $b(z)=\prod_{j=1}^{\theta+s}(z-x_j)^{\delta_j}\,(A(z)=b(z)\overline{b(\bar{z})})$ and $b_{ik}(z)=b(z)/(z-x_i)^{k+1}$, $i=1,2,\cdots,\theta+s$, $k=0,1,\cdots,\delta_i-1$. Denote by W_x the transition matrix of order N from the standard basis $\{1,z,\cdots,z^{N-1}\}$ of the space of all polynomials of degree $\leqslant N-1$ to the interpolation basis $\{b_{ik}(z)\}$. W_x is invertible obviously. Letting $\boldsymbol{H}_{NN}[\boldsymbol{b}]=(h_{i+j})_{i,j=0}^{N-1}$ built on $\boldsymbol{b}=(h_0,h_1,\cdots,h_{2N-2})$, we have[2]

$$\boldsymbol{P}_x=\boldsymbol{W}_x\boldsymbol{H}_{NN}[\boldsymbol{b}]\boldsymbol{W}_x^*, \tag{1.10}$$

whence

$$\boldsymbol{P}_x\geqslant\boldsymbol{O}\ (>\boldsymbol{O})\text{ if and only if }\boldsymbol{H}_{NN}[\boldsymbol{b}]\geqslant\boldsymbol{O}\ (>\boldsymbol{O}). \tag{1.11}$$

In the case when $\boldsymbol{H}_{NN}[\boldsymbol{b}]\geqslant\boldsymbol{O}$ is singular (or what is the same, $\boldsymbol{P}_x\geqslant\boldsymbol{O}$ is singular), there exist a unique real ξ with $0\leqslant\xi\leqslant h_{2N-2}$ such that $\boldsymbol{b}=\boldsymbol{b}_p+\boldsymbol{b}_d$, $\boldsymbol{b}_p=(h_0,h_1,\cdots,h_{2N-3},h_{2N-2}-\xi)^{\mathrm{T}}$, $\boldsymbol{b}_d=(0,\cdots,0,\xi)^{\mathrm{T}}$, and $\mathrm{H}_{NN}[\boldsymbol{b}]=\boldsymbol{H}_{NN}[\boldsymbol{b}_p]+\boldsymbol{H}_{NN}[\boldsymbol{b}_d]$, where $\boldsymbol{H}_{NN}[\boldsymbol{b}_p]\geqslant\boldsymbol{0}$ and $\boldsymbol{H}_{NN}[\boldsymbol{b}_d]\geqslant\boldsymbol{0}$ are the proper and degenerate parts of $\boldsymbol{H}_{NN}[\boldsymbol{b}]$, respectively, and

$$\operatorname{rank}\boldsymbol{H}_{NN}[\boldsymbol{b}]=\operatorname{rank}\boldsymbol{H}_{NN}[\boldsymbol{b}_p]+\operatorname{rank}\boldsymbol{H}_{NN}[\boldsymbol{b}_d]$$

(see, [2, Lemma 2.7]). From (1.10) we deduce that $\boldsymbol{P}_x\geqslant\boldsymbol{O}$ has the following (unique) quasidirect composition:

$$\boldsymbol{P}_x=\boldsymbol{P}_{xp}+\boldsymbol{P}_{xd}\overset{\text{def}}{=}\boldsymbol{W}_x\boldsymbol{H}_{NN}[\boldsymbol{b}_p]\boldsymbol{W}_x^*+\boldsymbol{W}_x\boldsymbol{H}_{NN}[\boldsymbol{b}_d]\boldsymbol{W}_x^*, \tag{1.12}$$

where $\boldsymbol{P}_{xp}\geqslant\boldsymbol{O}$, $\boldsymbol{P}_{xd}\geqslant\boldsymbol{O}$, and $\operatorname{rank}\boldsymbol{P}_x=\operatorname{rank}\boldsymbol{P}_{xp}+\operatorname{rank}\boldsymbol{P}_{xd}$.

It is well known that if $\boldsymbol{H}_{NN}[\boldsymbol{b}]\geqslant\boldsymbol{O}$ is singular, then and only then the problem(1.6) has only one solution $\tau(u)$ such that $\Phi(z)=\int_{-\infty}^{+\infty}\frac{\mathrm{d}\tau(u)}{u-z}$ coincides with $w(z)$ if $\boldsymbol{H}_{NN}[\boldsymbol{b}]=\boldsymbol{H}_{NN}[\boldsymbol{b}_d]$ and with $-\frac{\gamma_{n1}(z)}{p_{n1}(z)}$, other-

wise. Here $n_1 = \text{rank } \boldsymbol{H}_{NN}[\boldsymbol{b}]$, $p_{n_1}(z)$ is the Rth orthogonal polynomial with respect to the sequence $\{h_j\}_0^{2R-2}$, $R = \text{rank } \boldsymbol{H}_{NN}[\boldsymbol{b}_p]$($= n_1$ or $n_1 - 1$), whereas $\gamma_{n_1}(z)$ is the corresponding polynomial of the second kind (cf. [1, Section 1 of Ch. 1]). When $\Phi(z) = -\gamma_{n_1}(z)/p_{n_1}(z)$, $\tau(\{\alpha_j\}) = 0$ for some j amounts to saying that $p_{n_1}(\alpha_j) \neq 0$ for the same j.

A generalized Pick matrix $\boldsymbol{P}_x$ is called *minimally positive* if $\boldsymbol{P}_x \geqslant \boldsymbol{O}$ and it majorizes no nonzero diagonal matrix of the form diag$[\boldsymbol{O}_m$, diag$[\overbrace{0,\cdots,0}^{\eta_i - 1 \text{ times}}, a_i]_{i=1}^{s}]$ with all $a_i \geqslant 0$, relative to the usual ordering on Hermitian matrices. (Each minimally positive matrix is singular obviously.) Note that this definition is a slight extension of Sarason's[6].

§ 2. Theorems and Proofs

One of our basic assertions concerns the connection between the IBNP$(\hat{\mathrm{N}})'$ problem and the moment problem (1.6).

Theorem 2.1 *With the notations as above, the following are equivalent*: (a) *The IBNP* $(\hat{\mathrm{N}})'$ *problem is solvable*; (b) $\boldsymbol{P}_x \geqslant \boldsymbol{O}$; (c) $\boldsymbol{H}_{NN}[\boldsymbol{b}] \geqslant \boldsymbol{O}$; (d) *The moment problem* (1.6) *is solvable. The IBNP*$(\hat{\mathrm{N}})'$ *problem is determinate* (*namely, it has only one solution*) *if and only if the problem* (1.6) *is; also, if and only if* $\boldsymbol{P}_x \geqslant \boldsymbol{O}$ *is singular* (*or what is the same,* $\boldsymbol{H}_{NN}[\boldsymbol{b}] \geqslant \boldsymbol{O}$ *is singular*). *Moreover, if* $\tau(u)$ $(-\infty < u < +\infty)$ *is a solution to the problem* (1.6), *then* $f(z)$ *defined via the formula*

$$f(z) = w(z) + A(z)\int_{-\infty}^{+\infty} \frac{\mathrm{d}\tau(u)}{u - z} \tag{2.1}$$

is a solution to the IBNP$(\hat{\mathrm{N}})'$ *problem, which admits the representation* (1.9) *with*

$$\begin{aligned} \tilde{c}_0 &= h_{2N-1} + c_1 \alpha_{2N-1} - \int_{-\infty}^{+\infty} \left(u^{2N-1} - \frac{uA(u)}{1+u^2} \right) \mathrm{d}\tau(u), \\ c_1 &= h_{2N-2} - \int_{-\infty}^{+\infty} u^{2N-2}\,\mathrm{d}\tau(u), \quad \mathrm{d}\sigma_f(u) = A(u)\,\mathrm{d}\tau(u). \end{aligned} \tag{2.2}$$

In particular, if $\boldsymbol{P}_x \geqslant \boldsymbol{O}$ (*or* $\boldsymbol{H}_{NN}[\boldsymbol{b}] \geqslant \boldsymbol{O}$) *is singular, then and only then the unique solution* $f(z)$ *to the IBNP*$(\hat{\mathrm{N}})'$ *problem coincides with* $w(z)$

whenever $\boldsymbol{P}_x = \boldsymbol{P}_{xd}$ (*or* $\boldsymbol{H}_{NN}[\boldsymbol{b}] = \boldsymbol{H}_{NN}[\boldsymbol{b}_d]$), *and with* $w(z) - \dfrac{A(z)\gamma_{n_1}(z)}{p_{n_1}(z)}$ *otherwise, where* $p_{n_1}(z)$ *and* $\gamma_{n_1}(z)$ *are defined as in Section* 1.

Proof (a)⇒(b): Assume that $f(z) \in \hat{\mathrm{N}}$ is a solution to the IBNP$(\hat{\mathrm{N}})'$ problem which is of the form (1.8). Define the nondecreasing function $\tau(u)(-\infty < u < +\infty)$ by

$$\tau(u) = \frac{\sigma_f(u)}{A(u)} \quad \text{for } u \neq \alpha_1, \alpha_2, \cdots, \alpha_s; \quad \tau(\{\alpha_j\}) = 0, j = 1, 2, \cdots, s \tag{2.3}$$

so that $\tau_j = \tau(\{v_j\}) = \sigma_j / A(v_j), j = 1, 2, \cdots$, and $\tau_{-1} = \tau(\{0\}) = c_{-1}/A(0)$ (if $A(0) \neq 0$). Then the matrix $\boldsymbol{P}$, which is the same as $\boldsymbol{P}_x$ but with $f(z)$ in place of $w(z)$ in (1.3), has the

$$\begin{aligned}\tilde{p}_{ij}^{kl} &= c_1 \delta_{k0} \delta_{l0} + \frac{(-1)^{k+l} c_{-1}}{x_i^{k+1} \bar{x}_j^{l+1}} + \sum_d \frac{(-1)^{k+l} \sigma_d}{(x_i - v_d)^{k+1} (\bar{x}_j - v_d)^{l+1}} \\ &= c_1 \delta_{k0} \delta_{l0} + \int_{\infty}^{+\infty} b_{ik}(u) \overline{b_{jl}(u)} \mathrm{d}\tau(u), \qquad \forall i, j, k, l.\end{aligned}$$

This means that $\boldsymbol{P}$ is twice the Grammian matrix of the functions $\{b_{ik}(z)\}$ in $L^2(\tau)$ plus the matrix $((c_1 \delta_{k0} \delta_{l0})_{k,l=0}^{\delta_i - 1, \delta_j - 1})_{i,j=1}^{\theta+s} \geqslant \boldsymbol{O}$, whence $\boldsymbol{P}_x \geqslant \boldsymbol{P} \geqslant \boldsymbol{O}$, and (b) follows.

(b)⇒(c): Follows from (1.11).

(c)⇒(d): Assume (c). Then it is known[2] that there exists $\tau(u)(-\infty < u < +\infty)$ such that (1.6) holds true with $\Phi(z) = \int_{-\infty}^{+\infty} \mathrm{d}\tau(u)/(u - z)$ rational.

(d)⇒(a): Let $\tau(u)$ be a solution to the problem (1.6), and let $c_1 = h_{2N-2} - \int_{-\infty}^{+\infty} u^{2N-2} \mathrm{d}\tau(u) \geqslant 0$. Define $f(z)$ by the formula (2.1). Then $f(z)$ is meromorphic, real for each $z \in \mathbf{R}$ but the poles of $f(z)$, and satisfies (1.1) with $2\eta_j - 2$ in place of $2\eta_j - 1$ therein and (1.7) even if some α_j is a mass point of $\tau(u)$. It remains to verify $f(z) \in \mathrm{N}$. Let $\mathrm{d}\sigma_f(u) = A(u)\mathrm{d}\tau(u)(-\infty < u < +\infty)$. Then $\int_{-\infty}^{+\infty} (1+u^2)^{-1} \mathrm{d}\sigma_f(u) < +\infty$.

Rewrite $f(z)$ in the form:

$$f(z)=Q(z)+\int_{-\infty}^{+\infty}\left(\frac{1}{u-z}-\frac{u}{1+u^2}\right)\mathrm{d}\sigma_f(u), \tag{2.4}$$

where

$$Q(z)=w(z)+\int_{-\infty}^{+\infty}\left(\frac{A(z)-A(u)}{u-z}+\frac{uA(u)}{1+u^2}\right)\mathrm{d}\tau(u)$$

is a certain polynomial of z, of degree at most $2N-1$, and the integral on the right side of (2.4) stands for some function of the class N. Thus, in oredr to prove $f(z)\in$ N one only has to check $Q(z)=c_1z+\tilde{c}_0$ ($\tilde{c}_0$ real). To do this, we put $A(z)=\sum_{j=0}^{2N}a_jz^j\,(a_{2N}=1)$, so that

$$\frac{A(z)-A(u)}{u-z}=\sum_{j,k=0;j+k\leqslant 2N-1}^{2N-1}a_{j+k+1}u^jz^k, \tag{2.5}$$

and, by (1.5),

$$w(z)=\sum_{j,k=0;j+k\leqslant 2N-1}^{2N-1}a_{j+k+1}h_jz^k. \tag{2.6}$$

The polynomial $Q(z)$ will now be expressed as

$$Q(z)=h_{2N-1}+c_1a_{2N-1}+c_1z-\int_{-\infty}^{+\infty}\left(u^{2N-1}-\frac{uA(u)}{1+u^2}\right)\mathrm{d}\tau(u)=c_1z+\tilde{c}_0, \tag{2.7}$$

where

$$\tilde{c}_0=h_{2N-1}+c_1a_{2N-1}-\int_{-\infty}^{+\infty}\left(u^{2N-1}-\frac{uA(u)}{1+u^2}\right)\mathrm{d}\tau(u) \tag{2.8}$$

is real, noting that the last integral makes sense, since $(1+u^2)u^{2N-1}-uA(u)$ is a polynomial of degree at most $2N$. Thus, the equivalence of (a)~(d) is proved.

Furthermore, from a similar argument to that used in the proof of Corollary 3.5 of [2] we deduce that if IBNP $(\hat{\mathrm{N}})'$ problem is determinate, then and only then $\boldsymbol{P}_x\geqslant\boldsymbol{O}$ is singular, and that its unique solution is rational as asserted in the statement of the theorem. The proof of Theorem 2.1 is complete. ■

Let us turn to another basic assertion which concerns the connection between the IBNP $(\hat{\mathrm{N}})$ problem and the moment problem (1.6) with one additional constraint:

$$\tau(\{\alpha_j\})=0,\quad j=1,2,\cdots,s, \tag{2.9}$$

noting that (2.9) amounts to that $\varphi(z)=\int_{-\infty}^{+\infty}\frac{\mathrm{d}\tau(u)}{u-z}$ has no pole at each α_j.

Theorem 2.2 *The following are equivalent*: (a) *The* IBNP$(\hat{\mathrm{N}})$ *problem is solvable*; (b) $\boldsymbol{P}_x>\boldsymbol{O}$ *or* $\boldsymbol{P}_x$ *is minimally positive*; (c) $\boldsymbol{H}_{NN}[\boldsymbol{b}]>\boldsymbol{O}$ *or* $\boldsymbol{H}_{NN}[\boldsymbol{b}]\geqslant\boldsymbol{O}$ *is singular and the unique solution* $\tau(u)$ *to the problem* (1.6) *is such that* $\varphi(z)=\int_{-\infty}^{+\infty}\mathrm{d}\tau(u)/(u-z)$ *has no pole at* $\alpha_1,\alpha_2,\cdots,\alpha_s$; (d) *The moment problem* (1.6)(2.9) *is solvable. Moreover*,

(a′) *If* $\boldsymbol{P}_x>\boldsymbol{O}$ ($\boldsymbol{H}_{NN}[\boldsymbol{b}]>\boldsymbol{O}$), *then and only then the* IBNP$(\hat{\mathrm{N}})$ *problem is indeterminate* (*namely, it has infinitely many solutions*); *in that case, its solutions* $f(z)$ *can be parameterized by the linear fractional transformation*:

$$f(z)=w(z)-A(z)\frac{\gamma(z)v(z)-\delta(z)}{p(z)v(z)-q(z)}, \tag{2.10}$$

where $p(z),q(z)$ *are respectively the Nth and the* $(N-1)$*th orthogonal polynomials with respect to the sequence* $\{h_j\}_0^{2N-2}$, $\gamma(z)$ *and* $\delta(z)$ *the corresponding polynomials of the second kind, and* $v(z)$, *as the parameter, runs over the class* $\hat{\mathrm{N}}\cup\{\infty\}$ *such that*

$$p(\alpha_j)v(\alpha_j)-q(\alpha_j)\neq 0,\ j=1,2,\cdots,s. \tag{2.11}$$

(b′) *if* $\boldsymbol{H}_{NN}[\boldsymbol{b}]\geqslant\boldsymbol{O}$ *is singular, the unique solution to the* IBNP$(\hat{\mathrm{N}})$ *problem coincides with* $w(z)$ *of degree at most one if* $\boldsymbol{H}_{NN}[\boldsymbol{b}]$ *is degenerate, and with* $f(z)=w(z)-A(z)\gamma_{n1}(z)/p_{n1}(z)$ *if* $\boldsymbol{H}_{NN}[\boldsymbol{b}]$ *is not degenerate and* $p(\alpha_j)\neq 0$ *for* $j=1,2,\cdots,s$, *where* $p_{n_1}(z)$ *and* $\gamma_{n_1}(z)$ *are as in Section* 1.

Proof (a)⇒(b): Assume that $f(z)$ is a solution to the IBNP$(\hat{\mathrm{N}})$ problem. Then the IBNP$(\hat{\mathrm{N}})'$ problem is also solvable, and $\boldsymbol{P}_x\geqslant\boldsymbol{O}$ by Theorem 2.1. If, further, $\boldsymbol{P}_x\geqslant\boldsymbol{O}$ is singular, $\boldsymbol{P}_x$ has to be minimally positive. Indeed, if not, by definition, there exists a nonzero diagonal matrix $\boldsymbol{D}$ of the form: $\boldsymbol{D}=\mathrm{diag}[\boldsymbol{0}_m,\mathrm{diag}[0,\cdots,0,a_i]_{i=1}^s]$ with all $a_i\geqslant 0$ such that $\boldsymbol{P}_x-\boldsymbol{D}\overset{\text{def}}{=}\boldsymbol{P}_D\geqslant\boldsymbol{O}$. Thus, $\boldsymbol{P}_x\neq\boldsymbol{P}_D$ and $\boldsymbol{P}_x\geqslant\boldsymbol{P}_D\geqslant\boldsymbol{O}$, so that $\boldsymbol{P}_D\geqslant\boldsymbol{O}$ must be singular with the same off-diagonal entries as that of $\boldsymbol{P}_x$. But the IBNP($\hat{\mathrm{N}}$) problem whose generalized Pick matrix is $\boldsymbol{P}_D$ then has only one solution, $\tilde{f}(z)$, by Theorem 2.1, which is also a solution to the IBNP($\hat{\mathrm{N}}$)$'$ problem whose generalized Pick matrix is $\boldsymbol{P}_x$, whence $f(z)=\tilde{f}(z)$ by Theorem 2.1 again, and thus $\boldsymbol{P}_x=\boldsymbol{P}_D$, a contradiction.

(b)⇒(c): Assume (b). Then $\boldsymbol{H}_{NN}[\boldsymbol{b}]\geqslant \boldsymbol{O}$. If, further, $\boldsymbol{H}_{NN}[\boldsymbol{b}]\geqslant \boldsymbol{O}$ is singular, then the problem (1.6) has only one solution $\tau(u)$ such that $\tau(u)$ vanishes on real axis if $\boldsymbol{H}_{NN}[\boldsymbol{b}]$ is degenerate, and $\phi(z)=\int_{-\infty}^{+\infty}\frac{\mathrm{d}\tau(u)}{u-z}=-\gamma_{n_1}(z)/P_{n_1}(z)$ otherwise. For the latter case, we have that with the unique solution to the IBNP($\hat{\mathrm{N}}$)$'$ problem: $f(z)=w(z)-A(z)\gamma_{n_1}(z)/p_{n_1}(z)$ there corresponds a generalized Pick matrix $\boldsymbol{P}$, which is the same as $\boldsymbol{P}_x$ but with $f(z)$ in place of $w(z)$ therein. Then that $\boldsymbol{P}_x$ is minimally positive amounts to $\boldsymbol{P}=\boldsymbol{P}_x$, and thus $f(z)$ satisfies the IBNP($\hat{\mathrm{N}}$) problem as well. Thus, $p_{n_1}(\alpha_j)\neq 0$ for $j=1,2\cdots,s$, since $p_{n_1}(z)$ and $\gamma_{n_1}(z)$ are coprime, and $p_{n_1}(\lambda_i)\neq 0$ for $i=1,2,\cdots,\theta$, as asserted.

(c)⇒(d); if (c) holds true, the only thing that requires a proof is that the condition $\boldsymbol{H}_{NN}[\boldsymbol{b}]>\boldsymbol{O}$ is sufficient for th solvability of the problem (1.6)(2.9). In case $\boldsymbol{H}_{NN}[\boldsymbol{b}]>\boldsymbol{O}$, denote by $p(z)$, $q(z)$ the Nth and the $(N-1)$th orthogonal polynomials with respect to the sequence $\{h_j\}_{j=0}^{2N-2}$, respectively, and by $\gamma(z)$, $\delta(z)$ the corresponding polynomials of the second kind (cf. [1, Ch, 1]). Then there exists real c such that $p(\alpha_j)c-q(\alpha_j)\neq 0$ for $j=1,2,\cdots,s$ for which the function $\tau(u)$ defined by

$$\int_{-\infty}^{+\infty}\frac{\mathrm{d}\tau(u)}{u-z}=-\frac{\gamma(z)c-\delta(z)}{p(z)c-q(z)}$$

is a solution to the problem (1.6)(2.9), noting that $\tau(\{\alpha_j\})=0$ for some j is equivalent to $p(\alpha_j)c-q(\alpha_j)\neq 0$ for the same j, whence (d) follows.

(d)⇒(a): Assume now that $\tau(u)$ satisfies (1.6) and (2.9). Then $\boldsymbol{H}_{NN}[\boldsymbol{b}]\geqslant \boldsymbol{O}$. If, further, $\boldsymbol{H}_{NN}[\boldsymbol{b}]>\boldsymbol{O}$, then $\phi(z)=\int_{-\infty}^{+\infty}\mathrm{d}\tau(u)/(u-z)\in\hat{\mathrm{N}}$ has the form[2]:

$$\phi(z)=-\frac{\gamma(z)v(z)-\delta(z)}{p(z)v(z)-q(z)}$$

for some $v(z)\in\hat{\mathrm{N}}\cup\{\infty\}$ subject to (2. 11). Thus the function $f(z)$ defined by

$$f(z)=w(z)+A(z)\phi(z) \tag{2. 12}$$

is obviously a solution to the IBNP($\hat{\mathrm{N}}$) problem. On the other hand, when $\boldsymbol{H}_{NN}[\boldsymbol{b}]\geqslant\boldsymbol{O}$ is singular and degenerate, $w(z)$ is the unique solution to the IBNP($\hat{\mathrm{N}}$) problem; when $\boldsymbol{H}_{NN}[\boldsymbol{b}]\geqslant\boldsymbol{O}$ is singular and not degenerate, $f(z)=w(z)-A(z)\gamma_{n_1}(z)/p_{n_1}(z)$ satisfies uniquely the same problem, since the condition (2. 9) amounts to $p_{n_1}(\alpha_j)\neq 0$ for $j=1,2,\cdots,s$.

It remains to verify that if $\boldsymbol{H}_{NN}[\boldsymbol{b}]\geqslant\boldsymbol{O}$, each solution $f(z)$, of the form (1. 9), to the IBNP($\hat{\mathrm{N}}$) problem can be expressed in the form (2. 10)(2. 11). Let $\phi(z)$ be defined as in (2. 12). We need only to check that $\phi(z)=\int_{-\infty}^{+\infty}\mathrm{d}\tau(u)/(u-z)$ for some $\tau(u)$ satisfying the problem (1. 6)(2. 9). Let $\tau(u)$ be defined as in (2. 3). We have $d\tau(u)\geqslant 0(-\infty<u<+\infty)$ and

$$\begin{aligned}&\tilde{c}_0+c_1z+\int_{-\infty}^{+\infty}\left(\frac{A(u)-A(z)}{u-z}-\frac{ua(u)}{1+u^2}\right)\mathrm{d}\tau(u)-w(z)\\&=A(z)\left[\phi(z)-\int_{-\infty}^{+\infty}\frac{\mathrm{d}\tau(u)}{u-z}\right].\end{aligned}$$

From a similar argument to that used for results (2. 4)～(2. 8), we deduce that the left hand side of the last equations happens to be a polynomial of degree at most $2N-1$, so that $\phi(z)=\int_{-\infty}^{+\infty}\mathrm{d}\tau(u)/(u-z)$, since $\phi(z)-\int_{-\infty}^{+\infty}\mathrm{d}\tau(u)/(u-z)$ has no pole at each α_j, and further $h_j=\int_{-\infty}^{+\infty}u^j\,\mathrm{d}\tau(u)-\delta_{0,2N-2}c_1$, $j=0,1,\cdots,2N-2$. Thus, is follows from above discussions that $f(z)$ has the form (2. 10)(2. 11) for some $v(z)\in\hat{\mathrm{N}}\cup\{\infty\}$. The proof of Theorem 2. 2 is complete. ■

The statement of the proceeding theorem actually provides us with a simple one -to -one correspondence between solutions $f(z)$ to IBNP($\hat{\mathrm{N}}$) problem and solutions $\tau(u)$ to the problem (1.6) and (2.9):

$$f(z) = w(z) + A(z)\int_{-\infty}^{+\infty} \frac{\mathrm{d}\tau(u)}{u-z}.$$

This key observation thereby allows us to reduce the IBNP($\hat{\mathrm{N}}$) problem to what amounts to the more-studied moment problem (1.6)(2.9) in a comparatively understandable way.

For more information on the solution of the NP interpolation via the Hankel vector approach, see, e.g., [3~5] and the references therein.

§ 3. Examples

To conclude this note we illustrate the results of Theorems 2.1 and 2.2 in terms of numerical examples.

Example 3.1 Let $\alpha_1 = c = \frac{-1+\sqrt{5}}{2}$, $\eta_1 = 3$, $d_{10}=0$, $d_{11}=25-10\sqrt{5}$, $d_{12}=10-5\sqrt{5}$, $d_{13}=5-2\sqrt{5}$, $d_{14}=2-\sqrt{5}$, $d_{15}=2$. Consider the solutions of the corresponding IBNP($\hat{\mathrm{N}}$) and IBNP($\hat{\mathrm{N}}$)$'$ problems. Then $A(z)=(z-c)^6$, $\boldsymbol{x}=(c,c,c)$, $\boldsymbol{P}_x=(d_{1,i+j+1})_{i,j=0}^2 \geqslant \boldsymbol{O}$,

$$\boldsymbol{H}_{33}[\boldsymbol{b}] = \begin{bmatrix} 2 & 1 & 1 \\ 1 & 1 & 0 \\ 1 & 0 & 1 \end{bmatrix}$$

(singular and proper), $n_1 = \operatorname{rank} \boldsymbol{P}_x = \operatorname{rank} \boldsymbol{H}_{33}[\boldsymbol{b}] = 2$, $p_2(z) = z^2+z+1 = (z-c)\left(z+\frac{1+\sqrt{5}}{2}\right)$, $\gamma_2(z)=2z+3$. Since $p_2(c)=0$, or what is the same, $\boldsymbol{P}_x$ is not minimally positive, the IBNP($\hat{\mathrm{N}}$) problem then has no sulution. On the other hand, the problem (1.6) has only one solution $\tau(u)$ such that $\phi(z) = -\frac{2z+3}{z^2+z+1}$, so that the unique solution to the IBNP($\hat{\mathrm{N}}$)$'$ then coincides with $f(z)=w(z)-(z-c)^6(2z+3)/(z^2+z+1)$. Indeed, $f^{(j)}(c)=w^{(j)}(c)$ for $j=0,1,\cdots,4$, and

$$f^{(5)}(c) = w^{(5)}(c) - \frac{1}{5!}\frac{2+\sqrt{5}}{\sqrt{5}} < w^{(5)}(c).$$

Example 3.2 Let $\lambda_1 = i, c_{10} = i; \alpha_1 = 0, d_{10} = 0, d_{11} = 2$. Then $\boldsymbol{x} = (i, 0), \boldsymbol{P}_x = \begin{bmatrix} 1 & 1 \\ 1 & 2 \end{bmatrix} > \boldsymbol{O}, \boldsymbol{H}_{22}[\boldsymbol{b}] = \begin{bmatrix} 1 & 0 \\ 0 & 1 \end{bmatrix} > \boldsymbol{O}, w(z) = z^3 + 2z, A(z) = z^2(z^2+1), p(z) = z^2 - 1, q(z) = z, \gamma(z) = z, \delta(z) = 1, \alpha(z) = -2z, \beta(z) = z^2$. Thus all the solutions of the corresponding IBNP($\hat{\mathrm{N}}$) problem are parametrized by

$$f(z) = -\frac{2zv(z) + z^2}{(z^2-1)v(z) - z},$$

where $v(z)$ ranges over the class $\hat{\mathrm{N}} \cup \{\infty\}$ whenever $v(0) \neq 0$.

References

[1] N I Akhiezer. The classical moment problem. Oliver & Boyd, London, 1965.

[2] G-N Chen. The general rational interpolation problem and its connection with the Nevanlinna-Pick interpolation and power moment problem. Linear Algebra Appl., 1998, 273: 83−117.

[3] G-N. Hu Y-J Chen. The Nevanlinna-Pick interpolation problems and power moment problems for matrix-valued functions Ⅲ: the infinitely many data case. Linear Algebra Appl., 2000, 306: 59−86.

[4] G-N Chen Y-J Hu. The truncated Hamburger matrix moment problem in the nondegenerate and degenerate cases, and matrix continued fractions. Linear Algebra Appl., 1998, 277: 199−236.

[5] G-N Chen, X-Q Li. The Nevanlinna-Pick interpolation problems and power moment problems for matrix-valued functions. Linear Algebra Appl., 1999, 288: 123−148.

[6] D Sarason. Nevanlinna-Pick interpolation with boundary data. Integr. Equ. Oper. Theory, 1998, 30: 231−250.

Linear Algebra and Its Applications
2011,434:1 851－1 878

矩阵 Carathéodory 函数的极小 w-熵内插式与对应块 Pick 矩阵的极大行列式填充①

Minimum w-Entropy Interpolants for Matricial Carathéodory Functions and Maximum Determinant Completions of Associated Block Pick Matrix

Abstract The so-called modified block Toeplitz vector approach is used to connect a class of particular solutions G_w for $w \in D$ of a nondegenerate interpolation problem of the Nevanlinna-Pick type with a class of particular solutions F_w of a certain matricial Carathéodory coefficient problem in a transparent way. This will suggest a simple approach to the minimum w-entropy interpolants and the maximum determinant completions of the associated block Pick matrix within the framework of that Nevanlinna-Pick type interpolation problem by using the known assertions corresponding to F_w. It turns out that $G_w (w \in D)$ is exactly or provides us with the unique solution to these two extremal problems in a manner.

Keywords Matricial Carathéodory function; Nevanlinna-Pick type

---

① Received 24 February 2010; Accepted 22 November 2010. This work was supported by grants from the National Natural Science Foundation of China (Nos. 10701009 and 11071017).
本文与胡永建合作.

Submitted by V. Mehrmann. AMS Classification: 30E05. 47A56.

interpolation; Carathéodory coefficient problem; w-Entropy integral; Maximum determinant completions

§ 1. Introduction

The Nevanlinna-Pick type interpolation problem (Problem(NP)) together with the Carathéodory coefficient problem (Problem (C)) stands at the source of several major streams in analysis. In the present paper, we connect them with both the minimum w-entropy interpolants and maximum determinant completions of the associated generalized block Pick matrix $\boldsymbol{P}_{\triangle}$ (see(5)(6) for definition) of Problem (NP) in the nondegenerate case. Here Problem (NP) denotes specially a multiple Nevanlinna-Pick interpolation with the data Δ given in(3), involving zero as one of multiple nodes for the class $C_q(D)$ of $q \times q$ Carathéodory functions in the open unit disk D(see Section 2. 1 for the formulation), while the nondegenerate case means the particular situation that the generalized block Pick matrix $\boldsymbol{P}_{\triangle}$ mentioned is positive Hermitian. It is well known [1,3,5,7~10,16,18~20,22,23] that the nondegenerate Problem (NP) and Problem(C) have infinitely many (in a strong sense) solutions, each of which may be described in terms of special linear fractional transformations which are obtained using quite different approaches. Problem(NP) takes Problem (C) as a special case, which in turn is, in essence, a certain truncated trigonometric matrix moment problem (Problem(TM)) on one hand, and on the other hand, by using the so-called modified block Toeplitz vector approach, Problem(NP) may be reduced to a certain problem (C) in the class $C_q(D)$ (or equivalently, a certain Problem (TM)). In fact, in Section 2, we will introduce the notion of the block Toeplitz vector $\boldsymbol{c}=(\boldsymbol{C}_0, \boldsymbol{C}_1, \cdots, \boldsymbol{C}_{m-1})$ of Problem (NP), where m is the number of interpolation nodes (accounting for multiplicities) in the data Δ, to unify Problem (NP) and the other two previously mentioned problems based on that block vector $\boldsymbol{c}$. Furthermore, we will recall three essential results form [7,8] as follows: (1) the information block matrices of Problem(NP) and Problem (C) are congruent (Theorem 1); (2) there exists an explicit one-to-one correspondence between the solutions (if any) of these three problems mentioned above (Theorem 2);and thus (3) these three problems have the same solvability and the same uniqueness of solution, and the same nondegenerate property (Theorem 3). (The reader might be referred to [7,8] for the detailed statements and proofs.) Also, we connect a class of the particular solutions G_w of (nondegenerate) Problem (NP) with a corresponding class of the particular solutions F_w of (nondegenerate) Problem (C) in a transparent way (see(19) and (20)). Both G_w and F_w for each $w \in D$ play a key role in our analysis. In turn , this will suggest a simple approach , by using the known results corresponding to F_w and the information matrix in the setting of (nondegenerate) Problem (C), to the minimum w-entropy interpolants and maximum determinant completions of the block Pick matrix $\boldsymbol{P}_\triangle$ in the setting of (nondegenerate) Problem (NP), and vice versa.

Section 3 is devoted to the minimum w-entropy interpolants for (nondegenerate)Problem(NP) and related problems. Let $w \in D$, and let the w-entropy functional $\eta_w(H)$ be defined in the class $C_q(D)$ by the formula

$$\eta_w(H) = -\frac{1}{4\pi}\int_T \frac{1-|w|^2}{|u-w|^2}\log(\det(\mathrm{Re}(\underline{H}(u))))\underline{\lambda}du, \quad H \in C_q(D) ,$$

where $\underline{\lambda}$ is the linear Lebesgue-Borel measure on the unit circle T, and $\underline{H}$ stands for a boundary function of H. Following the studies of [3,4]and [19, Section 8] together with a basic relation between G_w and F_w (see (20)), we obtain mainly that for every solution G of either (nondegenerate) Problem (NP) or related subproblems, the inequality

$$\eta_w(G) \geqslant \eta_w(G_w)$$

holds, with equality if and only if $G = G_w$, and that for every solution F of either (nondegenerate)Problem (C) or its related problems, the inequality

$$\eta_w(F) \geqslant \eta_w(\widetilde{F}_w)$$

holds, with equality if and only if $F=\widetilde{F}_w$, where $\widetilde{F}_w$ is either F_w or the so-called central solution of some related matricial Carathéodory coefficient problem (see Theorem 5 and Corollaries 1~3). An explicit formula is also given for the value of this minimum w-entropy $\eta_w(G_w)$ or $\eta_w(\widetilde{F}_w)$. Furthermore, matricial generalizations of these entropy inequalities can be presented and proved to be valid as well (Theorem 7 and Corollaries 4~6). The strategy we adopt here seems to fit particularly well with the solution of these extremal problems given in Section 3.

Finally, in Section 4 we consider the maximum determinant completions of the $mq\times mq$ block Pick matrix $\boldsymbol{P}_\Delta$ of (nondegenerate) Problem (NP) with the data Δ (see(3)). This subject matter was first dealt with in [13] by Dym and [14] by Dym and Gohberg. Let $\Delta'(\boldsymbol{D}_{0,\tau_0})$, $\hat{\Delta}(\boldsymbol{D}_{\rho+1,0})$ and $\check{\Delta}(\boldsymbol{D}_{k,\tau_k})(1\leqslant k\leqslant\rho)$ be one-step extensions of the data Δ each with a single parameter indicated (see (22)(23) and (24)), which are the only possible one-step extensions of the date Δ with respect to the zero node z_0, a new simple node $z_{\rho+1}$ different from $z_0, z_1, \cdots, z_\rho$, and the nonzero node $z_k(1\leqslant k\leqslant\rho)$, respectively, and let $\boldsymbol{P}_{\Delta'(\boldsymbol{D}_{0,\tau_0})}$, $\boldsymbol{P}_{\hat{\Delta}(D_{\rho+1,0})}$ *and* $\boldsymbol{P}_{\check{\Delta}(\boldsymbol{D}_{k,\tau_k})}$ $(1\leqslant k\leqslant\rho)$ be generalized block Pick matrices associated with the data $\Delta'(\boldsymbol{D}_{0,\tau_0})$, $\hat{\Delta}(\boldsymbol{P}_{\rho+1,0})$ and $\check{\Delta}(\boldsymbol{D}_{k,\tau_k})$ $(1\leqslant k\leqslant\rho)$, respectively. Let now

$$A'_\Delta:=\{\boldsymbol{D}_{0,\tau_0}\in\mathbf{C}^{q\times q}:\boldsymbol{P}_{\Delta'(\boldsymbol{D}_{0,\tau_0})}\geqslant\boldsymbol{O}\},$$

$$\hat{\mathcal{A}}_\Delta=\{\boldsymbol{D}_{\rho+1,0}\in\mathbf{C}^{q\times q}:\boldsymbol{P}_{\hat{\Delta}(\boldsymbol{D}_{\rho+1,0})}\geqslant\boldsymbol{O}\},$$

$$\check{\mathcal{A}}_{\Delta,\mathrm{k}}:=\{\boldsymbol{D}_{k,\tau_k}\in\mathbf{C}^{q\times q}:\boldsymbol{P}_{\check{\Delta}(D_{k,\tau_k})}\geqslant\boldsymbol{O}\},1\leqslant k\leqslant\rho.$$

Based on the modified block Toeplitz vector approach, and the theory of the fundamental matrix inequalities in [20], we will indicate that the inequalities

$$\det\boldsymbol{P}_{\Delta'(D_{0,\tau_0})}\leqslant\det\boldsymbol{P}_{\Delta'(\frac{1}{\tau_0!}G_0^{(\tau_0)}(0))},\ \forall\,\boldsymbol{D}_{0,\tau_0}\in\mathcal{A}'_\Delta,$$

$$\det\boldsymbol{P}_{\hat{\Delta}(D_{\rho+1,0})}\leqslant\det\boldsymbol{P}_{\hat{\Delta}(G_{z_{\rho+1}}(z_{\rho+1}))},\ \forall\,\boldsymbol{D}_{\rho+1,0}\in\mathcal{A}_{\wedge\Delta},$$

$$\det\boldsymbol{P}_{\check{\Delta}(D_{k,\tau_k})}\leqslant\det\boldsymbol{P}_{\check{\Delta}(\frac{1}{\tau_k!}G_{z_k}^{(\tau_k)}(z_k))},\ \forall\,\boldsymbol{D}_{k,\tau_k}\in\mathcal{A}_{\smile\Delta,k}$$

hold each with equality if and only if

$$\boldsymbol{D}_{0,\tau_0}=\frac{1}{\tau_0\,!}G_0^{(\tau_0)}(0),\quad \boldsymbol{D}_{\rho+1,0}=G_{z_{\rho+1}}(z_{\rho+1}),$$

$$\boldsymbol{D}_{k,\tau_k}=\frac{1}{\tau_k\,!}G_{z_K}^{(\tau_k)}(z_k)\quad(1\leqslant k\leqslant\rho),$$

respectively (see Theorems 9, 11 and 12). Here and later, the symbol $G_0^{(\tau_0)}(0)$ stands of $\frac{d^{\tau_0}}{dz^{\tau_0}}G_0(z)|_{z=0}$, and so on.

A few words about notation: for Hermitian matrices $\boldsymbol{A}$ and $\boldsymbol{B}$ of the same order, write $\boldsymbol{A}\geqslant\boldsymbol{B}$ or $\boldsymbol{B}\leqslant\boldsymbol{A}$ to indicate the matrix $\boldsymbol{A}-\boldsymbol{B}$ is nonnegative Hermitian. Similarly, $\boldsymbol{A}>\boldsymbol{B}$ or $\boldsymbol{B}<\boldsymbol{A}$ means that $\boldsymbol{A}-\boldsymbol{B}$ is positive Hermitian. If $\boldsymbol{A}\in\mathbf{C}^{q\times q}$, then $\det\boldsymbol{A}$ stands for the determinant of $\boldsymbol{A}$, $\boldsymbol{A}^*$ will be the Hermitian transpose of $\boldsymbol{A}$, Re $\boldsymbol{A}$ and Im $\boldsymbol{A}$ are the real part and the imaginary part of $\boldsymbol{A}$, respectively. The symbol $\boldsymbol{I}_q$ denotes the identity matrix of order q. The symbol T stands for the unit circle and D for its interior and $\overline{D}$ for the unit close disk, i. e., $T:=\{z\in\boldsymbol{C}:|z|=1\}$, $D:=\{z\in\mathbf{C}:|z|<1\}$ *and* $\overline{D}:=\{z\in\mathbf{C}:|z|=1\}$.

§ 2. The unity of Problem (NP) and Its Related Problem (C)

This section is devoted to the investigation on the unity of Problem (NP) and Problem (C) associated with the block Toeplitz vector of Problem (NP). We will summarize several intrinsic relations between these two interpolation problems, which are useful later. The contents are taken largely from [7,8] for Sections 2. 1 and 2. 2 here and [19, Section 2] for Section 2. 3.

A $\mathbf{C}^{q\times q}$-valued function $\Omega(z)$ which is holomorphic in D with the nonnegative Hermitian real part: $\mathrm{Re}\,\Omega(z):=\frac{1}{2}(\Omega(z)+\Omega(z)^*)\geqslant\boldsymbol{O}$ for each $z\in D$ is called $q\times q$ Carathéodory function (in D). The set of all $q\times q$ Carathéodory functions in D will be denoted by $C_q(D)$. Each $\Omega\in C_q(D)$ permits an integral representation of the form

$$\Omega(z) = \int_T \frac{u+z}{u-z}\mu(\mathrm{d}u) + \mathrm{i}\,\mathrm{Im}\ \Omega(0), z \in D, \tag{1}$$

where μ is a nonnegative Hermitian $q\times q$ Borel measure on T, which is uniquely determined by Ω, and is called the *Riesz-Herglotz measure* of Ω, By relection, we define for $\Omega\in C_q(D)$,

$$\Omega(z) = -\Omega^*(\bar{z}^{-1}), z\in \mathbf{C}/\overline{D} \tag{2}$$

§ 2.1 Problem (NP) and Problem (C) Associated with the Block Toeplitz Vector

Problem (NP) under consideration is the following multiple Nevanlinna-Pick type interpolation for the class $C_q(D)$:

Problem (NP) Let ρ be a fixed nonnegative integer, let $z_0, z_1, \cdots, z_\rho$ be pairwise different points in D, let $\tau_0, \tau_1, \cdots, \tau_\rho$ be positive integers, and let $\boldsymbol{D}_{ik}$ be given in $\mathbf{C}^{q\times q}$ for $k=0,\cdots,\tau_i-1$ and $i=0,1,\cdots,\rho$. Denote by Δ such interpolation data:

$$\Delta := \{(z_k)_{k=0}^{\rho};\ (\tau_k)_{k=0}^{\rho};\boldsymbol{D}_{ik}, k=0,1,\cdots,\tau_i-1, i=0,1,\cdots,\rho\}. \tag{3}$$

Find necessary and sufficient conditions for the existence of a $G\in C_q(D)$ *satisfying the conditions*

$$\frac{1}{k!}\frac{\mathrm{d}^k}{\mathrm{d}z^k}G(z)\Big|_{z=z_i} = \boldsymbol{D}_{ik}, k=0,1,\cdots,\tau_i-1, \quad i=0,1,\cdots,\rho. \tag{4}$$

Furthermore, find a description of the solution set $\boldsymbol{C}_\Delta$ of Problem (NP) when the solvability conditions are met.

Note that Problem(NP) can be regarded as a generalization of the same problem in the case of $0\notin\{z_0, z_1, \cdots, z_\rho\}$ of which a complete solution in both nondegenerate and degenerate case has been given via the block Toeplitz vector approach (see, e. g., [7]). For the general case (that is, $0\in\{z_0, z_1, \cdots, z_\rho\}$ or not) we refer to [5,22] and the references therein for the solution of this kind of interpolation problems for the class $C_q(D)$ or for similar classes of holomorphic matrix functions.

The so-called generalized block Pick matrix $\boldsymbol{P}_\Delta$ will play an important role by solving Problem (NP). Let $m := \sum_{j=0}^{\rho}\tau_j$ and let G be a $\mathbf{C}^{q\times q}$-*valued function holomorphic in* D and subject to (4) (such a function exists always). We define

$$P_{\Delta}:=(P_{ij})_{i,j=0}^{\rho}\in \mathbf{C}^{mq\times mq},\tag{5}$$

where $P_{ij}:=(P_{ij}^{kl})_{k=0,l=0}^{\tau_i-1,\tau_j-1}\in \mathbf{C}^{\tau_i q\times \tau_j q}$ with entries $P_{ij}^{kl}\in \mathbf{C}^{q\times q}$ determined by

$$P_{ij}^{kl}:=\frac{1}{k!\ l!}\frac{\partial^{k+l}}{\partial\lambda^k\partial\bar{\mu}^l}\left(\frac{G(\lambda)+G(\bar{\mu})^*}{1-\lambda\bar{\mu}}\right)\Bigg|_{\lambda=z_i,\mu=\bar{z}_j}\tag{6}$$

Note that the entries P_{ij}^{kl} (and therefore P_{Δ} itself) can be expressed explicitly by the given data Δ of Problem (NP). It is well known that Problem (NP) with the data Δ is solvable if and only if $P_{\Delta}\geqslant O$. (For an alternative proof see also Theorem 3 below.)

Throughout this paper, we presume that $0\in\{z_0,z_1,\cdots,z_p\}$ and $z_0=0$. Write $D_i=D_{0i}, i=0,1,\cdots,\tau_0-1$ for short. Similar to in [7], let

$$A(z):=\prod_{i=1}^{\rho}(z-z_i)^{\tau_i}\left(\frac{1}{z}-\bar{z}_i\right)^{\tau_i}=:\sum_{i=-n}^{n}A_i z^i,\tag{7}$$

where $n:=\tau_1+\tau_2+\cdots+\tau_\rho=m-\tau_0$, which will be called the annihilating quasi-polynomial of nonzero interpolation nodes (with their symmetric points with respect to T). Then $A_{-n}=\bar{A}_n=(-1)^n\prod_{i=1}^{\rho}z_i^{\tau_i}\neq 0, A(z)=A(\bar{z}^{-1})^*$, $\forall z\neq 0$ and $A(z)>0$. $\forall z\in T$. It is easily verified that there exists a unique quasi-$q\times q$ matrix polynomial of the form

$$\widetilde{W}(z):=\sum_{i=-n}^{n-1}\widetilde{W}_i z^i, \widetilde{W}_i\in \mathbf{C}^{q\times q}$$

satisfying both (4) with $i\neq 0$ and the conditions

$$\frac{1}{k!}\frac{\mathrm{d}^k}{\mathrm{d}z^k}\widetilde{W}\left(\frac{1}{z}\right)\Bigg|_{z=\bar{z}_i}=-D_{ik}^*, k=0,1,\cdots,\tau_i-1,\quad i=1,2,\cdots,\rho.\tag{8}$$

Let further

$$W(z):=\frac{1}{2}\left[\widetilde{W}(z)-\widetilde{W}\left(\frac{1}{z}\right)\right]=:\sum_{i=-n}^{n}W_i z^i, W_i\in \mathbf{C}^{q\times q}\tag{9}$$

Then $W(z)$ is also subject to the conditions (4) with $i\neq 0$ and (8), $W(z)=-W(\bar{z}^{-1})^*$, $\forall z\neq 0$, and thus its real part vanishes on T: $\operatorname{Re} W(z):=\frac{1}{2}(W(z)+W(z)^*)=O$, $\forall z\in T$. We will refer such $W(z)$ to as the interpolation quasi-matrix polynomial of Problem (NP). Consider the Taylor series representation at $z=0$ of $\widetilde{W}(z)/A(z)$:

$$\frac{\widetilde{W}(z)}{A(z)}=-2\widetilde{C}_0-2\widetilde{C}_1 z-\cdots-2\widetilde{C}_n z^n+\cdots,(z\to 0)$$

and further

$$\frac{W(z)}{A(z)}=-\widetilde{C}_0-2\widetilde{C}_1 z-\cdots-2\widetilde{C}_n z^n+\cdots,(z\to 0)$$

where $\widetilde{C}_j\in\mathbf{C}^{q\times q}$. Then it is not hard to accept that

$$\widetilde{W}(z)-W(z)=\frac{1}{2}\left[\widetilde{W}(z)+\widetilde{W}\left(\frac{1}{z}\right)\right]=-\widetilde{C}_0 A(z),$$

in which $\widetilde{C}_0=\widetilde{C}_0^*$. Let $C_0,C_1,\cdots,C_{m-1}$ be defined recursively via

$$C_i:=\widetilde{C}_i,i=0,1,\cdots,n-1 \tag{10}$$

$$C_n:=\frac{1}{2}A_{-n}^{-1}D_0+\widetilde{C}_n,$$

$$C_{n+k}:=A_{-n}^{-1}\left(\frac{1}{2}D_k-\sum_{l=-n+1}^{n}A_l C_{k-l}\right),k=1,2,\cdots,\tau_0-1,$$

where $C_{-k}=C_k^*$ for $k\geqslant 0$. We refer to the block vector $c:=(C_0,C_1,\cdots,C_{m-1})$ found as the *block Toeplitz vector* of Problem (NP), which is uniquely defined by the data Δ of Problem (NP) given in (3). With such $c=(C_0,C_1,\cdots,C_{m-1})$ we can associate a matricial Carathéodory coefficient problem.

Problem(C) Let $c=(C_0,C_1,\cdots,C_{m-1})$ be the block Toeplitz vector of Problem (NP). Find necessary and sufficient conditions for the existence of an $F\in C_q(D)$ satisfying

$$F(0)=C_0,\quad \frac{1}{k!}\frac{\mathrm{d}^k}{\mathrm{d}z^k}F(z)\bigg|_{z=0}=2C_k,\quad k=1,2,\cdots,m-1, \tag{11}$$

and describe the solution set $C_q[D,c]$ of Problem (C) when these conditions are met.

Problem (C) can be conceived as an extreme case of Problem (NP) when $\tau_0=m>0$ and $\rho=0$ on one hand, and on the other hand it, in essence, is nothing other than a certain truncated trigonometric matrix moment problem. The statement of this moment problem is as follows.

Problem(TM) Let $c=(C_0,C_1,\cdots,C_{m-1})$ be the block Toeplitz vector of Problem (NP). It is required to find the nonnegative Hermitian Borel measures μ on T such that

$$C_k = \int_T u^{-k}\mu(\mathrm{d}u), \quad k = 0,1,\cdots,m-1. \tag{12}$$

The solution set $C_q[D,c]$ of Problem (C) (or equivalently, the solution set M_c of Problem (TM)) is nonempty if and only if

$$T_c := (C_{j-i})_{i,j=0}^{m-1} \geqslant O, \tag{13}$$

where $C_{-k}=C_k^*$ for $k\geqslant 0$ (see, e. g., [1,5,10,20]).

§ 2.2 The Unity of Problem (NP) and Problem (C)

We start with a congruence relation of P_Δ *with* T_c. Define $W_\Delta \in \mathbf{C}^{m\times m}$ via

$$\left(\mathrm{row}\left(\frac{1}{z^i}B(z)\right)_{i=\tau_0}^{1}, \mathrm{row}(\mathrm{row}(B_{ik}(z))_{i=1}^{\rho})_{k=0}^{\tau_i-1}\right)^{\mathrm{T}} = W_\Delta(\mathrm{row}(z^i)_{i=-\tau_0}^{n-1})^{\mathrm{T}}, \tag{14}$$

where

$$B(z) = \prod_{i=1}^{\rho}(z-z_i)^{\tau_i}, B_{ik}(z) = \frac{B(z)}{(z-z_i)^{k+1}}, 0 \leqslant k \leqslant \tau_i - 1, 1 \leqslant i \leqslant \rho.$$

Then $W_\Delta \in \mathbf{C}^{m\times m}$ is nonsingular, since it is a transition matrix of a pair of bases of the space of all scalar quasi-polynomials of the form $\sum_{j=-\tau_0}^{n-1} b_j z^j$.

Theorem 1[7,8] *Let P_Δ be as in (5)(6), and $T_c=(C_{j-i})_{i,j=0}^{m-1}$. Then*

$$P_\Delta = 2(W_\Delta \otimes I_q)T_c(W_\Delta \otimes I_q)^*, \tag{15}$$

where the symbol $\otimes$ stands for the tensor product of two matrices.

The next step consists of establishing a particular one-to-one correspondence between the solutions C_Δ of Problem (NP) and the solutions $C_q[\mathbf{D},\mathbf{c}]$ of Problem(C).

Theorem 2 [7,8] *Problem (NP) is solvable if and only if Problem (C) is solvable. In that case, if G is a solution to problem (NP) and permits an integral representation of the form*

$$G(z) = \mathrm{i}\,\mathrm{Im}\,G(0) + \int_T \frac{u+z}{u-z}\sigma(\mathrm{d}u), \forall z \in D,$$

then F defined via

$$G(z) = W(z) + A(z)F(z) \tag{16}$$

belongs to $C_q(D)$ and is a solution to Problem(C), which has an integral *representation of the form*

$$F(z) = \int_T \frac{u+z}{u-z}\mu(\mathrm{d}u) \tag{17}$$

where for a Borel subset B of the unit circle T,

$$\mu(B):=\int_B A(u)^{-1}\sigma(\mathrm{d}u).$$

Conversely, if $F \in C_q(D)$ is a solution to Problem (C) with the integral representation (17), *then G defined by* (16) *belongs to $C_q(D)$ and is a solution to Problem (NP), which has an integral representation of the form*

$$G(z) = \mathrm{i}\ \mathrm{Im}\ \boldsymbol{D}_0 + \int_T \frac{u+z}{u-z}\sigma(\mathrm{d}u),$$

where $\boldsymbol{D}_0 = \boldsymbol{D}_{00}$ and for a Borel subset B of the unit circle T,

$$\sigma(B):=\int_B A(u)\mu(\mathrm{d}u).$$

As a consequence of Theorems 1 and 2, we have

Theorem 3[7,8] *The following are all equivalent:*

(1) *Problem (NP) is solvable.*

(2) *Problem (C) associated with $\boldsymbol{c}=(\boldsymbol{C}_0,\boldsymbol{C}_1,\cdots,\boldsymbol{C}_{m-1})$ is solvable.*

(3) *Problem (TM) is solvable.*

(4) $\boldsymbol{P}_\Delta \geqslant \boldsymbol{O}$.

(5) $\boldsymbol{T}_c \geqslant \boldsymbol{O}$.

Moreover, $\boldsymbol{P}_\Delta > \boldsymbol{O}$ if and only if $\boldsymbol{T}_c > \boldsymbol{O}$.

Note that $\boldsymbol{P}_\Delta$ and $\boldsymbol{T}_c$ are often referred to as the *information matrices* of Problem (NP) with the data Δ and Problem (C) with the moment vector $\boldsymbol{c}$, respectively.

§ 2.3 The Particular Solutions to Problem (C) and to Problem (NP) in the Nondegenerate Case

Hereafter, we always assume that the block Toeplitz vector $\boldsymbol{c}=(\boldsymbol{C}_0, \boldsymbol{C}_1,\cdots,\boldsymbol{C}_{m-1})$ is such that $\boldsymbol{T}_c=(\boldsymbol{C}_{j-i})_{i,j=0}^{m-1}>\boldsymbol{O}$, where $\boldsymbol{C}_{-k}=\boldsymbol{C}_k^*$ for $k \geqslant 0$. This means that Problem (C) (and therefore Problem (NP)) is *nondegenerate*. Then $\hat{\boldsymbol{T}}_c:=(\boldsymbol{C}_{i-j})_{i,j=0}^{m-1}>\boldsymbol{O}$ as well, since $\hat{\boldsymbol{T}}_c=\boldsymbol{Q}_m\boldsymbol{T}_c\boldsymbol{Q}_m$, where $\boldsymbol{Q}_m:=(\delta_{i,m-1-j}\boldsymbol{I}_q)_{i,j=0}^{m-1}$ in which δ_{ij} is the Kronecker symbol.

Let $T_c^{-1} := (v_{ij})_{i,j=0}^{m-1}$, where $v_{ij} \in \mathbf{C}^{q\times q}$. Then $T_c^{-1} > O$ as well. For $w \in D$, the particular solutions of special interest here to Problem (C) in the nondegenerate case are those F given in (17) of which the Riesz-Herglotz measures μ are chosen to be of the form

$$\mu_w(B) := \frac{1}{2\pi}\int_B \frac{1-|w|^2}{|u-w|^2} A_w(u)^{-*} A_w(w) A_w(u)^{-1} \underline{\lambda}(du), \quad (18)$$

where B is a Borel subset of T, and for $z \in \overline{D}$,

$$A_w(z) := \sum_{i,j=0}^{m-1} z^{m-1-i} v_{ij} \overline{w}^{m-1-j},$$

which is invertible in $\overline{D}$. Let now for every $w \in D$,

$$F_w(z) := \int_T \frac{u+z}{u-z} \mu_w(du), \quad (19)$$

Then F_w belongs to $C_q[D, \mathbf{c}]$ (see [19, Proposition 2.4]), and therefore from Theorem 2 we deduce that G_w defined, for each $z \in D$, *by*

$$G_w(z) := W(z) + A(z) F_w(z) \quad (20)$$

is a particular solution to (nondegenerate) Problem (NP): $G_w \in \mathbf{C}_\Delta$, where F_w is as in (19), and admits, for each $z \in d$, the representation

$$G_w(z) = \mathrm{i}\,\mathrm{Im}\,\mathbf{D}_0 + \int_T \frac{u+z}{u-z} A(u) \mu_w(du). \quad (21)$$

Both F_w and G_w mentioned above will play a key role in our analysis later on. It is known [16, Theorem 16 in Part Ⅲ] that F_0 defined by (18), (19) with $w=0$ is exactly the so-called central $q\times q$ Carathéodory function corresponding to $\mathbf{c} = (\mathbf{C}_0, \mathbf{C}_1, \cdots, \mathbf{C}_{m-1})$, and therefore G_0 defined by (18)~(21) with $w=0$ is, in a sense, the central solution of (nondegenerate) Problem (NP) with the data Δ.

§ 3. Minimum *w*-entropy

In this section, as usual, we presume that the original Problem (NP) is nondegenerat (i. e., $\mathbf{P}_\Delta > \mathbf{O}$). Here we need some further notation, which will later be useful, and are indicated in Section 3.1. In the second part of this section, an explicit formula of w-entropy functional for each solution of (nondegenerate) Problem (NP) considered in Section 2 will be formulated (see(29)(26) below) in a slightly different setting form others (see e. g. ,[12～14,2]). The extremal value of the w-entropy over the solution set of each of Problem (NP) and related subproblems ,as well as the related problems of Problem (C) under consideration will be obtained in a simpler way by using the block Toeplitz vector approach . Note that in this setting ,the w-entropy of a matrix-valued function in $C_q(D)$ is defined to be in the interval $(-\infty,+\infty]$,so that the extremal value will be a minimum. Consequently ,the extremal value will become a maximum whenever the sign in the definition of w-entropy is reversed. In the third part,we deal with the matricial generalizations of the w-entropy inequalities given in Section 3. 2. More information of this subject matter can be found,e. g. ,in [2～4,6,11～15].

§ 3. 1　Some Further Notation

Besides the underlying data Δ (see(3)) of Problem (NP),let further

$$\Delta':=\Delta'(\boldsymbol{D}_{0,\tau_0})=\{(z_k)_{k=0}^{\rho};(\tau'_k)_{k=0}^{\rho};\ \boldsymbol{D}_{ik},0\leqslant i\leqslant\rho,o\leqslant k\leqslant\tau'_i-1\},\tag{22}$$

$$\hat{\Delta}:=\hat{\Delta}(\boldsymbol{D}_{\rho+1,0})=\{(z_k)_{k=0}^{\rho+1};(\tau_k)_{k=0}^{\rho+1};\ \boldsymbol{D}_{ik},0\leqslant i\leqslant\rho+1,0\leqslant k\leqslant\tau_i-1\},\tag{23}$$

$$\check{\Delta}_s:=\check{\Delta}(\boldsymbol{D}_{s,\tau_s})=\{(z_k)_{k=0}^{\rho};(\tau'_k)_{k=0}^{\rho};\boldsymbol{D}_{ik},0\leqslant i\leqslant\rho,0\leqslant k\leqslant\tau_i'-1\},1\leqslant s\leqslant\rho\tag{24}$$

where,in Δ':$\tau'_k=\tau_k$ for $1\leqslant k\leqslant\rho$ and $\tau'_0:=\tau_0+1$,and $\boldsymbol{D}_{0,\tau_0}$ is an arbitrary $q\times q$ matrix; in $\hat{\Delta}$:$z_{\rho+1}\in D$ is given and different from the original $z_0,z_1,\cdots,z_\rho,\tau_{\rho+1}=1$,and $\boldsymbol{D}_{\rho+1,0}$ is an arbitrary $q\times q$ matrix ; in $\check{\Delta}_s(1\leqslant s\leqslant\rho)$:$\tau'_k=\tau_k$ for $k\neq s$,and $\tau'_s=\tau_s+1$,and $\boldsymbol{D}_{s,\tau_\rho}$ is an arbitrary $q\times q$ matrix. Observe that Δ',$\hat{\Delta}$,and $\check{\Delta}_s(1\leqslant s\leqslant\rho)$ form all the possible one-step extensions of the data Δ. For each $w\in D$ and G_w defined by (18)～(21), let specially

$$\Delta'_w:=\Delta'\left(\frac{1}{\tau_0!}G_w^{(\tau_0)}(0)\right),\qquad \hat{\Delta}_w:=\hat{\Delta}(G_w(z_{\rho+1})),$$

$$\check{\Delta}_{s,w}:=\check{\Delta}_s\left(\frac{1}{\tau_s!}G_w^{(\tau_s)}(z_s)\right).\tag{25}$$

The following abbreviations will be used in order to keep the notation simple: Problem $(NP)_{\Delta'_w}$ = Problem (NP) with the data Δ'_w, Problem $(C)_{c'_w}$ = Problem (C) with the moment vector $\boldsymbol{c}'_w$, where $\boldsymbol{c}'_w$ is the block Toeplitz vector of Problem $(NP)_{\Delta'_w}$, similarly for the others, and therefore, naturally, Problem $(NP)_{\Delta}$ = the original Problem (NP) and Problem $(C)_c$ = the original Problem (C). It is obvious that $\boldsymbol{C}_{\Delta'_w} \subseteq \boldsymbol{C}_{\Delta}$, where $\boldsymbol{C}_{\Delta'_w}$ stands for the solution set of Problem $(NP)_{\Delta'_w}$, similarly for the others, so that all three cases of Problem (NP) type can be viewed as the subproblems of Problem $(NP)_{\Delta}$.

Later, we shall indicate that all three cases mentioned and their associated Problem (C) (with the moment vectors $\boldsymbol{c}'_w$, $\hat{\boldsymbol{c}}_w$ *and* $\check{\boldsymbol{c}}_{s,w}$ $(1 \leqslant s \leqslant \rho)$, respectively) are nondegenerate under the assumption

Table 1

Notation of Problem (NP) and the related problems.

Problem	Data	Annihilating quasipolynomial	Interpolation quasi-matrix polynomial	Information matrix	Solution set	Central solution
(NP)	Δ	$A(z)$	$W(z)$	$\boldsymbol{P}_{\Delta}$	$\boldsymbol{C}_{\Delta}$	$G_0(z)$
(C)	$\boldsymbol{c}$	1	0	$\boldsymbol{T}_c$	$C_q[D, \boldsymbol{c}]$	$F_0(z)$
$(NP)_{\Delta'_w}$	Δ'_w	$A(z)$	$W(z)$	$\boldsymbol{P}_{\Delta'_w}$	$\boldsymbol{C}_{\Delta'_w}$	$G_w(z)$
$(C)_{c'_w}$	$\boldsymbol{c}'_w$	1	0	$\boldsymbol{T}_{c'_w}$	$C_q[D, \boldsymbol{c}'_w]$	$F_w(z)$
$(NP)_{\hat{\Delta}_w}$	$\hat{\Delta}_w$	$A(z)(z - z_{\rho+1}) \cdot (z^{-1} - \bar{z}_{\rho+1})$	$W_{\hat{\Delta}_w}(z)$	$\boldsymbol{P}_{\hat{\Delta}_w}$	$\boldsymbol{C}_{\hat{\Delta}_w}$	$G_w(z)$
$(C)_{\hat{c}_w}$	$\hat{\boldsymbol{c}}_w$	1	0	$\boldsymbol{T}_{\hat{c}_w}$	$C_q[D, \hat{\boldsymbol{c}}_w]$	$\hat{F}_w(z)$
$(NP)_{\check{\Delta}_{s,w}}$	$\check{\Delta}_{s,w}$	$A(z)(z - z_s) \cdot (z^{-1} - \bar{z}_s)$	$W_{\check{\Delta}_{s,w}}(z)$	$\boldsymbol{P}_{\check{\Delta}_{s,w}}$	$\boldsymbol{C}_{\check{\Delta}_{s,w}}$	$G_w(z)$
$(C)_{\check{c}_{s,w}}$	$\check{\boldsymbol{c}}_{s,w}$	1	0	$\boldsymbol{T}_{\check{c}_{s,w}}$	$C_q[D, \check{\boldsymbol{c}}_{s,w}]$	$\check{F}_{s,w}(z)$

that the original Problem (NP) is nondegenerate, and can treat uniformly their minimum w-entropy interpolants by using the modified block Toeplitz approach. For the reader's convenience, we enumerate the notation which appear in pairs in these problems in Table 1.

§ 3.2 On minimum w-entropy interpolants for (nondegenerate) Problem (NP) and related problems

For a given $F \in C_q(D)$ with the Riesz-Herglotz measure μ, following Arov and Kreĭn [3] we call

$$\eta_w(F) := -\frac{1}{4\pi}\int_T \frac{1-|w|^2}{|u-w|^2}\log(\det \Lambda(u))\underline{\lambda}(du) \in (-\infty, +\infty]$$

the w-entropy of F with respect to the point $w \in D$. Here $\underline{\lambda}$ is the linear Lebesgue-Borel measure defined on the σ-algebra of all Borel subsets of the unit circle T, and Λ is the Radon-Nikodym derivative of the locally absolutely continuous part in the Lebesgue decomposition of $\mu(du)$ with respect to the normalized $\frac{1}{2\pi}\underline{\lambda}$ on T. It is known ([3, Theorem 1] and [17, Lemma 4]) that $\Lambda(u) = \mathrm{Re}\,\underline{F}(u)$ λ-a. e. on T, and therefore

$$\eta_w(F) = -\frac{1}{4\pi}\int_T \frac{1-|w|^2}{|u-w|^2}\log(\det(\mathrm{Re}\,\underline{F}(u)))\underline{\lambda}(du). \qquad (26)$$

Here for $H \in C_q(D)$ the symbol $\underline{H}$ stands for a boundary function of H such that

$$\lim_{s\to 1^-} H(su) = \underline{H}(u) \quad \lambda\text{-a,e. on } T$$

(such a $\underline{H}$ exists always for each $H \in C_q(D)$ in view of a well-known theorem of Fatou).

To the eight solution sets stated in Table 1 there correspond eight cases for the extremal problem being considered. To present and prove the main conclusion mentioned in the setting of Problem $(\mathrm{NP})_\Delta$/Problem $(\mathrm{C})_c$ and the pairs of related problems, we begin with two results of Problem $(\mathrm{C})_c$ and Problem $(\mathrm{C})_{c'_w}$, proofs of which are, in essence, given in [19, Section 8]

Let $w \in D$ and let F_w defined by (19) have the Taylor series representation of the form

$$F_w(z) = \boldsymbol{C}_{0,w} + 2\sum_{j=1}^{+\infty} \boldsymbol{C}_{j,w} z^j,\ z \in D, \qquad (27)$$

where $\boldsymbol{C}_{j,w} = \boldsymbol{C}_j$, $j = 0, 1, \cdots, m-1$. Let now $\boldsymbol{c}'_w := (\boldsymbol{C}_{0w}, \boldsymbol{C}_{1w}, \cdots, \boldsymbol{C}_{m-1,w}, C_{mw})$ based on (27) and $\boldsymbol{T}_{c'_w} := (\boldsymbol{C}_{j-i,w})_{i,j=0}^{m}$ in which $\boldsymbol{C}_{-j,w} = \boldsymbol{C}_{jw}^*$, $j = 0$,

$1,\cdots,m$. It follows form [19, Remark 2.9] that $\boldsymbol{T}_{c'_w} > \boldsymbol{O}$. Denote by $C_q[D,\boldsymbol{c}'_w]$ the solution set of the associated Problem $(\mathrm{C})_{c'_w}$. *Hence* $F_w \in C_q[D,\boldsymbol{c}'_w]$ as well, and it coincides with the central $q\times q$ Carathéodory function corresponding to $\boldsymbol{c}'_w$ (see [19, Corollary 3.5]).

Lemma 4 *Let $w\in D$, and let $\boldsymbol{c}=(\boldsymbol{C}_0,\boldsymbol{C}_1,\cdots,\boldsymbol{C}_{m-1})$ be the block Toeplitz vector of Problem $(NP)_\Delta$ such that $\boldsymbol{T}_c=(\boldsymbol{C}_{j-i})_{i,j=0}^{m-1}>\boldsymbol{O}$. Let $\boldsymbol{T}_c^{-1}=(\boldsymbol{v}_{ij})_{i,j=0}^{m-1}$, where $\boldsymbol{v}_{ij}\in\mathbf{C}^{q\times q}$. Furthermore, let $\boldsymbol{c}'_w=(\boldsymbol{C}_{0w},\cdots,\boldsymbol{C}_{m-1,w},\boldsymbol{C}_{mw})$ be defined via (27). Then Problem $(\mathrm{C})_{c'_w}$ is nondegenerate. Moreover,*

(1) *For each $F\in C_q[D,\boldsymbol{c}]$, the inequality*

$$\eta_w(F)\geqslant -\frac{1}{2}\log\left[\det\left((1-|w|^2)\sum_{i,j=0}^{m-1}\bar{w}^i\boldsymbol{v}_{ij}w^j\right)^{-1}\right] \tag{28}$$

holds with equality if and only if F coincides with the matrix-valued function $\mathrm{F_w}: D\to\mathbf{C}^{q\times q}$ defined by (19).

(2) *For each $F\in \boldsymbol{C}_q[D,\boldsymbol{c}'_w]$, the inequality (28) holds as well with equality if and only if $F=F_w$. Moreover, for each $F\in \boldsymbol{C}_q[D,\boldsymbol{c}'_w]$, the inequality*

$$\eta_0(F)\geqslant\eta_0(F_w)$$

holds with equality if and only if $F=F_w$.

Proof Observe that $\hat{\boldsymbol{T}}_c=(\boldsymbol{C}_{i-j})_{i,j=0}^{m-1}>\boldsymbol{O}$ as well, and

$$(\bar{w}^{m-1}\boldsymbol{I},\cdots,\bar{w}\boldsymbol{I},\boldsymbol{I})\hat{\boldsymbol{T}}_c^{-1}\operatorname{col}(w_{m-1}\boldsymbol{I},\cdots,w\boldsymbol{I},\boldsymbol{I})$$

$$=(\boldsymbol{I},\bar{w}\boldsymbol{I},\cdots,\bar{w}^{m-1}\boldsymbol{I})\boldsymbol{T}_c^{-1}\operatorname{col}(\boldsymbol{I},w\boldsymbol{I},\cdots,W^{m-1}\boldsymbol{I})=\sum_{i,j=0}^{m-1}\bar{w}^i\boldsymbol{v}_{ij}w^j.$$

The required assertions are an easy consequence of [19, Theorem 7.4 and Corollary 7.5] in combination with [16, Corollary 7 in Part Ⅳ], which shows that the expression of the right hand side of (28) is also equal to

$$-\frac{1}{2}\log\left[\det\left((1-|w|^2)\sum_{i,j=0}^{m-1}w^{m-1-i}\boldsymbol{v}_{ij}\bar{w}^{m-1-j}\right)^{-1}\right]. \qquad \blacksquare$$

§ 3.2.1 The Case of Problem (NP)

Now we shall be particularly interested in the extremal problem over the solution set $\boldsymbol{C}_\Delta$, which is applicable to each of the remaining cases. Recall that the $\mathbf{C}^{q\times q}$-valued quasi-polynomial $W(z)$ defined in (9) is subject to Re $W(u)=\boldsymbol{O}$, $\forall u\in\mathbf{T}$. From Theorem 2 it follows that for each $G\in\boldsymbol{C}_\Delta$ there is a unique $F\in\boldsymbol{C}_q[D,\boldsymbol{c}]$ such that $G(z)=W(z)+A(z)F(z)$, where $A(z)$ is as in (7) such that $A(u)=\prod_{i=1}^{\rho}|1-\bar{z}_i u|^{2\tau_i}>0$, $\forall u\in T$, so that

Re $\underline{G}(u)=$ Re $W(u)+A(u)$ Re $\underline{F}(u)=A(u)$ Re $\underline{F}(u)$ $\quad\underline{\lambda}$-a. e. on T.

Thus, for $w\in D$, from (26) we derive that

$$\begin{aligned}\eta_w(G) &=-\frac{1}{4\pi}\int_T\frac{1-|w|^2}{|u-w|^2}\log[A(u)]^q\det(\mathrm{Re}\,\underline{F}(u))]\underline{\lambda}(du)\\&=-\frac{1}{4\pi}\int_T\frac{1-|w|^2}{|u-w|^2}[q\log A(u)+\log\det(\mathrm{Re}\,\underline{F}(u))]\underline{\lambda}(du)\\&=-\frac{q}{2\pi}\sum_{i=1}^{\rho}\tau_i\int_T\frac{1-|w|^2}{|u-w|^2}\log|1-\bar{z}_i u|\underline{\lambda}(du)+\eta_w(F)\\&=-\log\prod_{i=1}^{\rho}|1-\bar{z}_i w|^{q\tau_i}+\eta_w(F),\end{aligned}$$

where the last equality rests on the formula of Poisson -Jensen (see, e. g. ,[24]): for $w\in D$,

$$\log|1-\bar{z}_i w|=\frac{1}{2\pi}\int_T\frac{1-|w|^2}{|u-w|^2}\log|1-\bar{z}_i u|\underline{\lambda}(du).$$

Hence

$$\eta_w(G)=-\log\prod_{i=1}^{\rho}|1-\bar{z}_i w|^{q\tau_i}+\eta_w(F).\tag{29}$$

Theorem 5 Let $w\in D$, *and let* $\boldsymbol{c}=(C_0,C_1,\cdots,C_{m-1})$ *and* $\mathbf{P}_\Delta$ *be the block Toeplitz vector and the generalized block Pick matrix of (nondegenerate) Problem* $(NP)_\Delta$, *respectively. Then for each* $G\in\boldsymbol{C}_\Delta$, *the inequality*

$$\eta_w(G)\geqslant-\log\prod_{i=1}^{\rho}|1-\bar{z}_i w|^{q\tau_i}-\frac{1}{2}\log\Big[\det\Big((1-|w|^2)\sum_{i,j=0}^{m-1}\bar{w}^i\boldsymbol{v}_{ij}w^j\Big)^{-1}\Big]$$

is fulfilled, where equality holds if and only if G coincides with the matrix -valued function $G_w:D\to\mathbf{C}^{q\times q}$ *defined by* (18)~(21).

Proof Observe that the required inequality is valid by virtue of (28) and (29) for each pair of $F\in\boldsymbol{C}_q[D,\boldsymbol{c}]$ and $G\in\boldsymbol{C}_\Delta$ connected by

$$G(z)=W(z)+A(z)F(z),\ \forall z\in D,$$

where equality holds if and only if F coincides with F_w defined by (19), i. e. , if and only if G coincides with G_w *defined by* (18)~(21). ■

§3.2.2 The Cases of Problem $(NP)_{\Delta'_w}$, Problem $(NP)_{\hat{\Delta}_w}$ and Problem $(NP)_{\check{\Delta}_{s,w}}$

The next objective is to discuss the minimum w-entropy intrepolants over the solution set of either Problem $(NP)_{\Delta'_w}$, Problem $(NP)_{\hat{\Delta}_w}$ or Problem $(NP)_{\check{\Delta}_{s,w}}$ $(1\leqslant s\leqslant \rho)$, which will be indicated in Corollaries 1 and 2, respectively.

For each $w\in D$, let Δ'_w be as in (25). Then since $F_w(z)=\int_{\mathrm{T}}\frac{u+z}{u-z}\mu_w(du)$ and $G_w(z)=\mathrm{i}\ \mathrm{Im}\ \boldsymbol{D}_0+\int_{\mathrm{T}}\frac{u+z}{u-z}A(u)\mu_w(du), z\in D$,

$$\boldsymbol{D}_{0,\tau_0}=\frac{1}{\tau_0\,!}G_w^{(\tau_0)}(0)=2\int_{\mathrm{T}}u^{-\tau_0}A(u)\mu_w(du)$$

$$=2\sum_{l=-n}^{n}A_l\int_{\mathrm{T}}u^{-\tau_0+l}\mu_w(du)=2\sum_{l=-n}^{n}A_l\boldsymbol{C}_{\tau_0-l,w},$$

and therefore

$$\boldsymbol{C}_{mw}=A_{-n}^{-1}\left(\frac{1}{2}\boldsymbol{D}_{0,\tau_0}-\sum_{l=-n+1}^{n}A_l\boldsymbol{C}_{\tau_0-1,w}\right).$$

By definition (see(10) and (27)), $\boldsymbol{c}'_w$ is exactly the block Toeplitz vector of Problem $(NP)_{\Delta'_w}$. The generalized block Pick matrix $\boldsymbol{P}_{\Delta'_w}$ of Problem $(NP)_{\Delta'_w}$ is also positive Hermitian: $\boldsymbol{P}_{\Delta'_w}>\boldsymbol{O}$ by using the fact that $\boldsymbol{T}_{c'_w}>\boldsymbol{O}$ and Theorem 3 with Δ and $\boldsymbol{c}$ replaced by Δ'_w and $\boldsymbol{c}'_w$, respectively. Denote by $\boldsymbol{C}_{\Delta'_w}$ the solution set of Problem $(NP)_{\Delta'_w}$. Then $\boldsymbol{C}_{\Delta'_w}\subseteq\boldsymbol{C}_{\Delta}$ and $G_w\in\boldsymbol{C}_{\Delta'_w}$, obviously.

Corollary 1 *Let* $\boldsymbol{c}=(\boldsymbol{C}_0,\boldsymbol{C}_1,\cdots,\boldsymbol{C}_{m-1})$ *be the block Toeplitz vector of Problem* $(NP)_\Delta$ *such that* $\boldsymbol{T}_c=(\boldsymbol{C}_{j-i})_{i,j=0}^{m-1}>\boldsymbol{O}$, and let $\boldsymbol{T}_c^{-1}=(\boldsymbol{v}_{ij})_{i,j=0}^{m-1}$, *where* $\boldsymbol{v}_{ij}\in\mathbf{C}^{q\times q}$. Furthermore, let $w\in D$, *and let* $G\in\boldsymbol{C}_{\Delta'_w}$. Then the inequality

$$\eta_w(G)\geqslant-\log\prod_{i=1}^{\rho}|1-\bar{z}_iw|^{q\tau_i}-\frac{1}{2}\log\Big[\det\Big((1-|w|)^2-\sum_{i,j=0}^{m-1}\bar{w}^i\boldsymbol{v}_{ij}w^j\Big)^{-1}\Big]$$

holds with equality if and only if G *coincides with* G_w *defined by* (18)~(21). *Moreover, for each* $G\in\boldsymbol{C}_{\Delta'_w}$, *the inequality*

$$\eta_0(G)\geqslant\eta_0(G_w)=\frac{1}{2}\log(\det\boldsymbol{v}_{00}^{-1})$$

holds with equality if and only if $G=G_w$.

Proof It is clear that the first result of this corollary is a direct consequence of Theorem 5. For the assertion on the entropy of G with respect to $w=0$, observe that Problem $(\mathrm{NP})_{\Delta'_w}$ is also nondegenerate, since $\boldsymbol{P}_{\Delta'_w}>\boldsymbol{O}$ as indicated above. Furthermore, both Problem$(\mathrm{NP})_{\Delta}$ and Problem $(\mathrm{NP})_{\Delta'_w}$ enjoy the same $A(z)$ and $W(z)$ defined by (7) and (9). By(29), for $G\in\boldsymbol{C}_{\Delta'_w}$, $\eta_0(G)=\eta_0(F)$, where $F\in\boldsymbol{C}_q[D,\boldsymbol{c}'_w]$ is connected with G the formula $G(z)=W(z)+A(z)F(z)$, $\forall\, z\in D$. From Part (2) of Lemma 4 it follows that

$$\eta_0(G)=\eta_0(F)\geqslant\eta_0(F_w)=\eta_0(G_w)=-\frac{1}{2}\log(\det v_{00}^{-1}),$$

where the equality $\eta_0(G)=\eta_0(G_w)$ holds if and only if $F=F_w$, i. e., if and only if $G=G_w$. ■

The following results will exhibit the connection of G_w with the minimum w-entropy interpolants of either $\mathbf{C}_{\hat{\Delta}_w}$ or $\mathbf{C}_{\check{\Delta}_{s,w}}$ $(1\leqslant s\leqslant\rho)$. It should be noted that Problem $(\mathrm{NP})_{\hat{\Delta}_w}$ and Problem $(\mathrm{NP})_{\check{\Delta}_{s,w}}$ $(1\leqslant s\leqslant\rho)$ are also nondegenerate. Indeed, It suffices to observe that $\boldsymbol{P}_{\hat{\Delta}_w}\geqslant\boldsymbol{O}$ and $\boldsymbol{P}_{\check{\Delta}_{s,w}}\geqslant\boldsymbol{O}(1\leqslant s\leqslant\rho)$ obviously, and

$$\operatorname{rank}\boldsymbol{P}_{\hat{\Delta}_w}=\operatorname{rank}\boldsymbol{P}_{\check{\Delta}_{s,w}}=\operatorname{rank}\boldsymbol{P}_{\Delta'_w}=(m+1)q,\quad 1\leqslant s\leqslant\rho,$$

in view of the rank invariance of block Pick matrices of functions in $\boldsymbol{C}_q(D)$ (see,. e. g., [21, Theorem 2]), so that

$$\boldsymbol{P}_{\hat{\Delta}_w}>\boldsymbol{O}\text{ and }\boldsymbol{P}_{\check{\Delta}_{s,w}}>\boldsymbol{O}\quad(1\leqslant s\leqslant\rho)$$

Furthermore, it is clear that, for each $w\in D$,

$$\mathbf{C}_{\hat{\Delta}_w}\subseteq\mathbf{C}_{\Delta},\quad\mathbf{C}_{\check{\Delta}_{s,w}}\subseteq\mathbf{C}_{\Delta}(1\leqslant s\leqslant\rho),$$

and

$$G_w\in\mathbf{C}_{\hat{\Delta}_w}\cap\mathbf{C}_{\check{\Delta}_{s,w}}(1\leqslant s\leqslant\rho).$$ ■

Corollary 2 *Let $w\in D$, and let $\mathbf{C}_{\hat{\Delta}_w}$ and $\mathbf{C}_{\check{\Delta}_{s,w}}$ $(1\leqslant s\leqslant\rho)$ be defined as in Table 1. Then the statement of Corollary 1 remains valid when $\boldsymbol{C}_{\Delta'_w}$ is replaced by $\mathbf{C}_{\hat{\Delta}_w}$ and $\mathbf{C}_{\check{\Delta}_{s,w}}$ $(1\leqslant s\leqslant\rho)$, respectively.*

Proof The first assertion of this corollary follows directly form Theorem 5. Taking on $w=0$, we in turn see that the inequality

$$\eta_0(G)\geqslant-\frac{1}{2}\log(\det \boldsymbol{v}_{00}^{-1})$$

holds for each $G\in\mathbf{C}_{\hat{\Delta}_w}$ and each $G\in\mathbf{C}_{\check{\Delta}_{s,w}}$ $(1\leqslant s\leqslant\rho)$, respectively, and that there exists only one member of $\mathbf{C}_{\hat{\Delta}_w}$ *and* $\mathbf{C}_{\check{\Delta}_{s,w}}$ $(1\leqslant s\leqslant\rho)$, respectively, such that its 0-entropy is equal to $-\frac{1}{2}\log(\det \boldsymbol{v}_{00}^{-1})$, which is exactly G_w defined by (21), since by Corollary 1, $\eta_0(G_w)=-\frac{1}{2}\log(\det \boldsymbol{v}_{00}^{-1})$, moreover,

$$G_w\in \mathbf{C}_{\hat{\Delta}_w}\cap \mathbf{C}_{\check{\Delta}_{s,w}}\ (1\leqslant s\leqslant\rho).$$ ■

§ 3. 2. 3 The Cases of Problem $(\mathrm{C})_{\hat{\mathrm{c}}_w}$ and Problem $(\mathrm{C})_{\check{\mathrm{c}}_{s,w}}$

Now we turn our attention back to (nondegenerate) Problem $(\mathrm{C})_{\hat{\mathrm{c}}_w}$ and Problem $(\mathrm{C})_{\check{\mathrm{c}}_{s,w}}$ $(1\leqslant s\leqslant\rho)$. We shall deduce the minimum w-entropy interpolants for each of these two problems.

Corollary 3 *Let $w\in D$, and let $\hat{\mathbf{c}}_w$ and $\check{\mathbf{c}}_{s,w}$ $(1\leqslant s\leqslant\rho)$ be the block Toeplitz vectors of (nondegenerate) Problem $(NP)_{\hat{\Delta}_w}$ and Problem $(NP)_{\check{\Delta}_{s,w}}$ $(1\leqslant s\leqslant\rho)$, respectively. Then Problem $(C)_{\hat{c}_w}$ and Problem $(C)_{\check{c}_{s,w}}$ $(1\leqslant s\leqslant\rho)$ are nondegenerate. Moreover,*

(1) *For each $F\in C_q[D,\hat{c}_w]$, the inequality*

$$\eta_w(F)\geqslant \log|1-\bar{z}_{\rho+1}w|^q-\frac{1}{2}\log\Big[\det\Big((1-|w|^2)\sum_{i,j=0}^{m-1}\bar{w}^i\boldsymbol{v}_{ij}w^j\Big)^{-1}\Big]$$

holds, with equality if and only if F coincides with the matrix-valued function $\hat{F}_w: D\to \mathbf{C}^{q\times q}$ defined, for $z\in D$, by

$$\hat{F}_w(z):=\int_{\mathrm{T}}\frac{u+z}{u-z}\frac{\mu_w(\mathrm{d}u)}{|u-z_{\rho+1}|^2},\tag{19'}$$

where μ_w is defined by (18) for $w\in D$.

For each $F\in \mathbf{C}_q[D,\check{\boldsymbol{c}}_{s,w}]$ $(1\leqslant s\leqslant\rho)$, the inequality

$$\eta_w(F)\geqslant \log|1-\bar{z}_s w|^q-\frac{1}{2}\log\Big[\det\Big((1-|w|^2)\sum_{i,j=0}^{m-1}\bar{w}^i\boldsymbol{v}_{ij}w^j\Big)^{-1}\Big]$$

holds, with equality if and only if F coincides with the matrix-valued function $\check{F}_{s,w}: D\to \mathbf{C}^{q\times q}$ $(1\leqslant s\leqslant\rho)$ defined, for $z\in D$, by

$$\check{F}_{s,w}(z):=\int_{\mathrm{T}}\frac{u+z}{u-z}\frac{\mu_w(\mathrm{d}u)}{|u-z_s|^2}\quad(1\leqslant s\leqslant\rho),\tag{19''}$$

where μ_w is defined by (18) for $w\in D$.

(2) *For each $F\in \mathbf{C}_q[D,\hat{\boldsymbol{c}}_w]$, the inequality*

$$\eta_0(F)\geqslant\eta_0(\hat{F}_w)$$

holds with equality if and only if $F=\hat{F}_w$.

For each $F\in C_q[D,\check{\boldsymbol{c}}_{s,w}(1\leqslant s\leqslant\rho)]$, *the inequality*

$$\eta_0(F)\geqslant\eta_0(\check{F}_{s,w})(1\leqslant s\leqslant\rho)$$

holds with equality if and only if $F=\check{F}_{s,w}(1\leqslant s\leqslant\rho)$.

Proof Observe that, by using Theorem 2, to each $F\in\boldsymbol{C}_q[D,\hat{\boldsymbol{c}}_w]$ there corresponds a unique $G\in\boldsymbol{C}_{\hat{\Delta}_w}$ such that

$$G(z)=W_{\hat{\Delta}_w}(z)+A(z)(z-z_{\rho+1})(z^{-1}-\bar{z}_{\rho+1})F(z),$$

where $W_{\hat{\Delta}_w}(z)$ is the counterpart of $W(z)$ in Problem $(\mathrm{NP})_{\hat{\Delta}_w}$, and thus, by (29) with $\rho+1$ in place of ρ,

$$\eta_w(G)=-\log\prod_{i=1}^{\rho+1}|1-\bar{z}_iw|^{q\tau_i}+\eta_w(F),$$

where $\tau_{\rho+1}=1$, and $\eta_0(G)=\eta_0(F)$. In particular, $\hat{F}_w$, as indicated in (19′), and $G_w\in C_{\hat{\Delta}_w}$ form a dual pair of the solution sets $\boldsymbol{C}_q[D,\hat{\boldsymbol{c}}_w]$ and $\boldsymbol{C}_{\hat{\Delta}_w}$, so that

$$\eta_w(G_w)=-\log\prod_{i=1}^{\rho+1}|1-\bar{z}_iw|^{q\tau_i}+\eta_w(\hat{F}_w)\text{ and }\eta_0(G_w)=\eta_0(\hat{F}_w).$$

In turn, for each $w\in D$ and each $F\in\boldsymbol{C}_q[D,\hat{\boldsymbol{c}}_w]$ and corresponding $G\in\boldsymbol{C}_{\hat{\Delta}_w}\subseteq\boldsymbol{C}_\Delta$, it follows from Theorem 5 that

$$\begin{aligned}\eta_w(F)&=\log\prod_{i=1}^{\rho+1}|1-\bar{z}_iw|^{q\tau_i}+\eta_w(G)\\&\geqslant\log\prod_{i=1}^{\rho+1}|1-\bar{z}_iw|^{q\tau_i}-\log\prod_{i=1}^{\rho}|1-\bar{z}_iw|^{q\tau_i}-\\&\quad\frac{1}{2}\log\Big[\det\Big((1-|w|^2)\sum_{i,j=0}^{m-1}\bar{w}^i\boldsymbol{v}_{ij}w^j\Big)^{-1}\Big]\\&=\log|1-\bar{z}_{\rho+1}w|^q-\frac{1}{2}\log\Big[\det\Big((1-|w|^2)\sum_{i,j=0}^{m-1}\bar{w}^i\boldsymbol{v}_{ij}w^j\Big)^{-1}\Big].\end{aligned}$$

Hence, the first inequality of Part (1) holds with equality if and only if $G=G_w$, or equivalently, if and only if $F=\hat{F}_w$, Similarly, we can prove that the other inequality holds true for $F\in\boldsymbol{C}_q[D,\check{\boldsymbol{c}}_{s,w}](1\leqslant s\leqslant\rho)$ with equality if and only if $F=\check{F}_{s,w}(1\leqslant s\leqslant\rho)$.

As for the statement of Part (2), from the result of Part (1) of this corollary if follows that for each $F \in \boldsymbol{C}_q[D, \hat{\boldsymbol{c}}_w]$ the inequality

$$\eta_0(F) \geqslant -\frac{1}{2}\log(\det \boldsymbol{v}_{00}^{-1})$$

holds, and there exists only one member of $\boldsymbol{C}_q[D, \hat{\boldsymbol{c}}_w]$ such that its 0-entropy is equal to $-\frac{1}{2}\log(\det \boldsymbol{v}_{00}^{-1})$, which is exactly $\hat{F}_w$, since by Corollary 1, $\eta_0(G_w) = -\frac{1}{2}\log(\det \boldsymbol{v}_{00}^{-1})$, and therefore

$$\eta_0(\hat{F}_w) = \eta_0(G_w) = -\frac{1}{2}\log(\det \boldsymbol{v}_{00}^{-1}).$$

Similarly, we can prove that the other result for $F \in \boldsymbol{C}_q[D, \check{\boldsymbol{c}}_{s,w}]$ $(1 \leqslant s \leqslant \rho)$ holds true as well. ■

It is worth noting that Corollary 3 says actually, for each $w \in D$, $\hat{F}_w$ and $\check{F}_{s,w}$ $(1 \leqslant s \leqslant \rho)$ coincide with the so-called central $q \times q$ Carathéodory functions corresponding to $\hat{\boldsymbol{c}}_w$ and $\check{\boldsymbol{c}}_{s,w}$ $(1 \leqslant s \leqslant \rho)$, respectively, whereas G_w coincides with the central solution of either (nondegenerate) Problem $(\mathrm{NP})_{\check{\Delta}_w}$ or Problem $(\mathrm{NP})\check{\Delta}_{s,w}$ $(1 \leqslant s \leqslant \rho)$ in a sense. We recall in passing also that for each $w \in D$, F_w defined by (19) coincides with the central $q \times q$ Carathéodory function corresponding to $\boldsymbol{c}'_w$ based on (27), whereas G_w coincides also with the central solution of (nondegenerate) Problem $(\mathrm{NP})_{\Delta'_m}$ in a sense.

§ 3.3 Matricial Generalizations of the w-entropy Inequalities

We now consider the matricial generalizations of the assertions given is Section 3. 2 in the setting of (nondegeerate) Problem $(\mathrm{NP})_{\Delta}$ and its related problems. It turns out that for each $w \in D$, either the $\mathbf{C}^{q \times q}$-valued function G_w defined in D or the central solutions of the associated Problem $(\mathrm{C})_c$ and its related problems is also the unique solution to these matricial generalizations.

Let $F \in \boldsymbol{C}_q(D)$ be of finite 0-entropy: $\eta_0(F) < +\infty$. Then by [19, Section 8], $\eta_w(F) < +\infty$ for each $w \in D$, and Re $\underline{F}$ is Lebesgue integrable on T, Re $\underline{F} \geqslant \boldsymbol{O}$ λ-a. e. on T, and thus there exist outer functions Φ_F and Ψ_F, which belong to $[H^2(D)]^{q \times q}$ such that

$$\operatorname{Re} \underline{F}=\underline{\Phi_F}\,\underline{\Phi_F}^{*} \text{ and } \operatorname{Re} \underline{F}=\underline{\Psi_F}^{*}\,\underline{\Psi_F} \quad \lambda-\text{a. e. on } T.$$

Here $[H^2(D)]^{q\times q}$ denotes the set of all $\mathbf{C}^{q\times q}$-valued functions each entry of which belongs to the scalar Hardy space $H^2(D)$. Moreover, the outer functions Φ_F and Ψ_F are uniquely defined up to constant $q\times q$ unitary right and left factors, respectively.

The following assertion is, in essence, established in [19, Theorem 7.6], which is related to the studies of [3,4], and is formulated in terms of the Loewner semi-ordering for Hermitian matrices.

Lemma 6 *Let* $w\in D$, *Let* $\boldsymbol{c}=(\boldsymbol{C}_0,\boldsymbol{C}_1,\cdots,\boldsymbol{C}_{m-1})$ *be the block Toeplitz vector of Problem* $(NP)_\Delta$ *such that* $\boldsymbol{T}_c=(\boldsymbol{C}_{j-i})_{i,j=0}^{m-1}>\boldsymbol{O}$. *Let* $\boldsymbol{T}_c^{-1}=(\boldsymbol{v}_{ij})_{i,j=0}^{m-1}$, *where* $\boldsymbol{v}_{ij}\in\mathbf{C}^{q\times q}$. *Furthermore, let* $\boldsymbol{c}'_w=(\boldsymbol{C}_{0w},\boldsymbol{C}_{1w},\cdots,\boldsymbol{C}_{mw})$ *be the block Toeplitz vector of Problem* $(NP)_{\Delta'_w}$.

(1) *Let* $F\in\boldsymbol{C}_q[D,\boldsymbol{c}]$ be of finite 0-entropy $\eta_0(F)$. Then the inequalities

$$\Psi_F^*(w)\Psi_F(w)\leqslant\Big[(1-|w|^2)\sum_{i,j=0}^{m-1}w^{m-1-j}\boldsymbol{v}_{ij}\overline{w}^{m-1-j}\Big]^{-1}$$

and

$$\Phi_F(w)\Phi_F^*(w)\leqslant\Big[(1-|w|^2)\sum_{i,j=0}^{m-1}\overline{w}^i\boldsymbol{v}_{ij}w^j\Big]^{-1}$$

hold each with equality if and only if $F=F_w$.

(2) *Let* $F\in\boldsymbol{C}_q[D,\boldsymbol{c}'_w]$ *be of finite 0-entropy* $\eta_0(F)$. *Then the inequalities indicated in Part (1) of this lemma hold each with equality if and only if* $F=F_w$. *Furthermore, the inequalities*

$$\Psi_F^*(0)\Psi_F(0)\leqslant\Psi_{F_w}^*(0)\Psi_{F_w}(0) \text{ and } \Phi_F(0)\Phi_F^*(0)\leqslant\Phi_{F_w}(0)\Phi_{F_w}^*(0)$$

hold as well each with equality if and only if $F=F_w$.

§ 3.3.1 The Matricial Generalizations of Theorem 5 and Corollary 1

Let now $G\in\boldsymbol{C}_\Delta$ be of finite 0-entropy: $\eta_0(G)<+\infty$. Then Theorem 2 says that there exists a unique $F\in\boldsymbol{C}_q[D,\boldsymbol{c}]$ such that $G(z)=W(z)+A(z)F(z)$, $z\in D$, and therefore, by (26) and (29),

$$\eta_0(F)=-\frac{1}{4\pi}\int_T\log(\det(\operatorname{Re}\underline{F}(u)))\underline{\lambda}(\mathrm{d}u)=\eta_0(G)<+\infty,$$

which means the result given in Part (1) of Lemma 6 is applicable. For such

$G \in \mathbf{C}_\Delta$, there also exist two outer functions Φ_G and Φ_G in $[H^2(D)]^{q\times q}$ satisfying

$$\operatorname{Re} \underline{G} = \underline{\Phi_G}\underline{\Phi_G}^* \text{ and } \operatorname{Re} \underline{G} = \underline{\Psi_G}^* \underline{\Psi_G} \quad \lambda\text{-a. e. on } T.$$

In this connection, the outer functions Φ_G and Ψ_G are uniquely defined by G up to constant $q\times q$ unitary right and left factors, respectively. Let now

$$\alpha(w) := \prod_{j=1}^{\rho} (1-\bar{z}_j w)^{\tau_j}, \forall w \in \mathbf{C},$$

so that

$$A(u) = \alpha(u)\alpha^*(u), \forall u \in T.$$

Furthermore, since $|\bar{z}_j w| < 1$ for $w \in D$, the scalar function α is outer in $H^2(D)$, and in turn, $\alpha\Phi_F$ and $\alpha\Psi_F$ are outer in $[H^2(D)]^{q\times q}$. Consequently, from the fact that $\operatorname{Re} \underline{G}(u) = A(u) \operatorname{Re} \underline{F}(u)$ λ-a. e. on T, it follows that $\Phi_G = \alpha\Phi_F U_G$ and $\Psi_G = V_G \alpha\Psi_F$ for some pair of constant unitary U_G and V_G in $\mathrm{C}^{q\times q}$. Thus

$$\underline{\Phi_G}(u)\underline{\Phi_G}^*(u) = \alpha(u)\underline{\Phi_F}(u)\underline{\Phi_F}^*(u)\alpha^*(u), \lambda\text{-a. e. on } T,$$

$$\underline{\Psi_G}(u)^* \underline{\Psi_G}(u) = \alpha^*(u)\underline{\Psi_F}(u)^* \underline{\Psi_G}(u)\alpha(u), \lambda\text{-a. e. on } T.$$

By using the results given above together with Lemma 6 we arrive immediately at the matricial generalizations of Theorem 5 and Corollary 1, respectively.

Theorem 7 *Let $\mathbf{c} = (\mathbf{C}_0, \mathbf{C}_1, \cdots, \mathbf{C}_{m-1})$ be the block Toeplitz vector of Problem $(NP)_\Delta$ such that $\mathbf{T}_c = (\mathbf{C}_{j-i})_{i,j=0}^{m-1} > \mathbf{O}$, and let $\mathbf{T}_c^{-1} = (\mathbf{v}_{ij})_{i,j=0}^{m-1}$, where $\mathbf{v}_{ij} \in \mathbf{C}^{q\times q}$. Furthermore, let $w \in D$, and let $G \in \mathbf{C}_\Delta$ be of finite o-entropy $\eta_0(G)$.*

(1) *The inequality*

$$\Psi_G^*(w)\Psi_G(w) \leqslant \prod_{j=1}^{\rho} |1-\bar{z}_j w|^{2\tau_j} \left[(1-|w|)^2 \sum_{i,j=0}^{m-1} w^{m-1-i}\mathbf{v}_{ij}\bar{w}^{m-1-j}\right]^{-1}$$

holds with equality if and only if $G = G_w$.

(2) *The inequality*

$$\Phi_G(w)\Phi_G^*(w) \leqslant \prod_{j=1}^{\rho} |1-\bar{z}_j w|^{2\tau_j} \left[(1-|w|)^2 \sum_{i,j=0}^{m-1} \bar{w}^i\mathbf{v}_{ij}w^j\right]^{-1}$$

holds with equality if and only if $G = G_w$.

Corollary 4 Let $\mathbf{c}=(\mathbf{C}_0,\mathbf{C}_1,\cdots,\mathbf{C}_{m-1})$ *be the block Toeplitz vector of Problem* $(NP)_\Delta$ *such that* $\mathbf{T}_c=(\mathbf{C}_{j-i})_{i,j=0}^{m-1}>\mathbf{O}$, *and let* $\mathbf{T}_c^{-1}=(\mathbf{v}_{ij})_{i,j=0}^{m-1}$, *where* $\mathbf{v}_{ij}\in\mathbf{C}^{q\times q}$. *Furthermore, let* $w\in D$ *and let* $G\in\mathbf{C}_{\Delta'_w}$ *be of finite* 0-*entropy* $\eta_0(G)$, *where* Δ'_w *is as in* (25).

(1) *The inequality*

$$\Psi_G^*(w)\Psi_G(w)\leqslant\prod_{j=1}^{\rho}|1-\bar{z}_j w|^{2\tau_j}\left[(1-|w|)^2\sum_{i,j=0}^{m-1}w^{m-1-i}\mathbf{v}_{ij}\bar{w}^{m-1-j}\right]^{-1}$$

holds with equality if and only if $G=G_w$. *Moreover, the inequality*

$$\Psi_G^*(0)\Psi_G(0)\leqslant\Psi_{G_w}^*(0)\Psi_{G_w}(0)=\mathbf{v}_{m-1,m-1}^{-1}$$

holds as well, where equality is fulfilled if and only if $\mathrm{G}=\mathrm{G}_w$.

(2) *The inequality*

$$\Phi_G(w)\Phi_G^*(w)\leqslant\prod_{j=1}^{\rho}|1-\bar{z}_j w|^{2\tau_j}\left[(1-|w|)^2\sum_{i,j=0}^{m-1}\bar{w}^i\mathbf{v}_{ij}w^j\right]^{-1}$$

holds with equality if and only if $G=G_w$. *Moreover, the inequality*

$$\Phi_G(0)\Phi_G^*(0)\leqslant\Phi_{G_w}(0)\Phi_{G_w}^*(0)=\mathbf{v}_{00}^{-1}$$

holds aw well, where equality is fulfilled if and only if $G=G_w$.

Proof The first assertion in either (1) or (2) here is obvious, since $\mathbf{C}_{\Delta'_w}\subseteq\mathbf{C}_\Delta$ and $G_w\in\mathbf{C}_{\Delta'_w}$. The second is a direct consequence of Part (2) of Lemma 6, since $\eta_0(F)=\eta_0(G)<+\infty$, where $F\in\mathbf{C}_q[D,\mathbf{c}'_w]$ and $\mathrm{G}(z)=W(z)+A(z)F(z)$, $\Psi_{G_w}^*(0)\Psi_{G_w}(0)=\Psi_{F_w}^*(0)\Psi_{F_w}(0)=\mathbf{v}_{m-1,m-1}^{-1}$ and $\Phi_{G_w}(0)\Phi_{G_w}^*(0)=\Phi_{F_w}(0)\Phi_{F_w}^*(0)=\mathbf{v}_{00}^{-1}$. ■

§ 3.3.2 The matricial generalizations of Corollaries 2 and 3

The next two results are to construct the matricial generalizations of the statements of Corollaries 2 and 3, respectively, as the integral complements of the assertions of Corollary 4 and Part (2) of Lemma 6.

Corollary 5 *Let* $w\in D$. *The statement of Corollary* 4 *remains also valid when the solution set* $\mathbf{C}_{\Delta'_w}$ *is replaced by* $\mathbf{C}_{\hat{\Delta}_w}$ *and* $\mathbf{C}_{\check{\Delta}_{s,w}}$ $(1\leqslant s\leqslant\rho)$, *respectively.*

Proof Use Theorem 7 in combination with the second assertions of Parts (1) and (2) of Corollary 4, which show that

$$\Psi_{G_w}^*(0)\Psi_{G_w}(0)=\mathbf{v}_{m-1,m-1}^{-1}\text{ and }\Phi_{G_w}(0)\Phi_{G_w}^*(0)=\mathbf{v}_{00}^{-1}. \quad ■$$

Corollary 6 *Let* $w \in D$ *and let* $\mathbf{c}=(\mathbf{C}_0,\mathbf{C}_1,\cdots,\mathbf{C}_{m-1})$ *be the block Toeplitz vector of Problem* $(NP)_\Delta$ *such that* $\mathbf{T}_c=(\mathbf{C}_{j-i})_{i,j=0}^{m-1}>\mathbf{O}$. *Let* $\mathbf{T}_c^{-1}=\mathbf{v}_{ij}{}_{i,j=0}^{m-1}$, *where* $\mathbf{v}_{ij}\in\mathbf{C}^{q\times q}$.

(1)*Let* $F\in\mathbf{C}_q[D,\hat{c}_w]$ *be of finite* 0-*entropy* $\eta_0(F)$. *Then the inequalities*

$$\Psi_F^*(w)\Psi_F(w)\leqslant|1-\bar{z}_{\rho+1}w|^{-2}\left[(1-|w|)^2\sum_{i,j=0}^{m-1}w^{m-1-i}\mathbf{v}_{ij}\bar{w}^{m-1-j}\right]^{-1}$$

and

$$\Phi_F(w)\Phi_F^*(w)\leqslant|1-\bar{z}_{\rho+1}w|^{-2}\left[(1-|w|)^2\sum_{i,j=0}^{m-1}\bar{w}^i\mathbf{v}_{ij}w^j\right]^{-1}$$

hold each with equality if and only if $F=\hat{F}_w$ *defined by* (19′). *Moreover, the inequalities*

$$\Psi_F^*(0)\Psi_F(0)\leqslant\Psi_{\hat{F}_w}^*(0)\Psi_{\hat{F}_w}(0)=\mathbf{v}_{m-1,m-1}^{-1}$$

and

$$\Phi_F(0)\Phi_F^*(0)\leqslant\Phi_{\hat{F}_w}(0)\Phi_{\hat{F}_w}^*(0)=\mathbf{v}_{00}^{-1}$$

holds as well each with equality if and only if $F=\hat{F}_w$

(2)*Let* $F\in\mathbf{C}_q[D,\check{c}_{s,w}](1\leqslant s\leqslant\rho)$ *be of finite* 0-*entropy* $\eta_0(F)$. *Then the inequalities*

$$\Psi_F^*(w)\Psi_F(w)\leqslant|1-\bar{z}_sw|^{-2}\left[(1-|w|^2)\sum_{i,j=0}^{m-1}w^{m-1-i}\mathbf{v}_{ij}\bar{w}^{m-1-j}\right]^{-1}(1\leqslant s\leqslant\rho)$$

and

$$\Phi_F(w)\Phi_F^*(w)\leqslant|1-\bar{z}_sw|^{-2}\left[(1-|w|)^2\sum_{i,j=0}^{m-1}\bar{w}^i\mathbf{v}_{ij}w^j\right]^{-1}(1\leqslant s\leqslant\rho)$$

hold each with equality if and only if $F=\check{F}_{s,w}(1\leqslant s\leqslant\rho)$ *defined as in* (19″). *Moreover, the inequalities*

$$\Psi_F^*(0)\Psi_F(0)\leqslant\Psi_{\check{F}_{s,w}}^*(0)\Psi_{\check{F}_{s,w}}(0)=\mathbf{v}_{m-1,m-1}^{-1}$$

and

$$\Phi_F(0)\Phi_F^*(0)\leqslant\Phi_{\check{F}_{s,w}}(0)\Phi_{\check{F}_{s,w}}^*(0)=\mathbf{v}_{00}^{-1}$$

hold as well each with equality if and only if $F=\check{F}_{s,w}(1\leqslant s\leqslant\rho)$.

Proof In view of Theorem 2, for each $F\in\mathbf{C}_q[D,\hat{\mathbf{c}}_w]$ with finite 0-entropy $\eta_0(F)$ there exists a unique $G\in\mathbf{C}_{\hat{\Delta}_w}\subseteq\mathbf{C}_\Delta$ such that

$$G(z)=W_{\hat{\Delta}_w}(z)+A(z)(z-z_{\rho+1})(z^{-1}-\bar{z}_{\rho+1})F(z),$$

where $W_{\hat{\Delta}_w}$ is the interpolation quasi-matrix polynomial of Problem $(\mathrm{NP})_{\hat{\Delta}_w}$, in particular,

$$G_w(z) = W_{\hat{\Delta}_w}(z) + A(z)(z - z_{\rho+1})(z^{-1} - \bar{z}_{\rho+1})\hat{F}_w(z),$$

where $\hat{F}_w$ is defined by $(19')$. Thus, for each $w \in D$,

$$\eta_0(\hat{F}_w) = \eta_0(G_w) \leqslant \eta_0(G) = \eta_0(F) < +\infty$$

by the argument of the proof of Corollary 3. In turn, it follows that there exist outer functions Φ_F, Ψ_F, Φ_G and Ψ_G in $[H^2(D)]^{q\times q}$ such that

$$\operatorname{Re} \underline{F}(u) = \underline{\Phi_F}(u)\underline{\Phi_F}^*(u), \quad \operatorname{Re} \underline{F}(u) = \underline{\Psi_F^*}(u)\underline{\Psi_F}(u), \ \lambda\text{-a. e. on } T,$$

and

$$\operatorname{Re} \underline{G}(u) = \underline{\Phi_G}(u)\underline{\Phi_G}^*(u), \quad \operatorname{Re} \underline{G}(u) = \underline{\Psi_G^*}(u)\underline{\Psi_G}(u), \ \lambda\text{-a. e. on } T,$$

In this connection, Φ_F, Ψ_F and Φ_G, Ψ_G are uniquely determined in some sense as indicated above. We will use the notation $\hat{\alpha}$ for the numerical polynomial given, for each $v \in D$, by $\hat{\alpha}(v) := \prod_{j=1}^{\rho+1}(1 - \bar{z}_j v)^{\tau_j}$. Consequently, since both $\hat{\alpha}\Phi_F$ and $\hat{\alpha}\Psi_F$ are outer in $[H^2(D)]^{q\times q}$, it follows that

$$\Phi_G = \hat{\alpha}\Phi_F U_G, \quad \Psi_G = \hat{\alpha} V_G \Psi_F$$

for some pair of constant unitary matrices U_G and V_G in $\mathbf{C}^{q\times q}$. Thus, Theorem 7 guarantees that

$$\begin{aligned}\Phi_G(w)\Psi_G^*(w) &= \prod_{j=1}^{\rho+1} |1 - \bar{z}_j w|^{2\tau_j}\Phi_F(w)\Phi_F^*(w)\\ &\leqslant \prod_{j=1}^{\rho} |1 - \bar{z}_j w|^{2\tau_j}\Big[(1 - |w|^2)\sum_{i,j=0}^{m-1}\bar{w}^i \boldsymbol{v}_{ij} w^j\Big]^{-1},\end{aligned}$$

where equality holds if and only if $G = G_w$, i. e., if and only if $F = \hat{F}_w$, and that

$$\begin{aligned}\Psi_G^*(w)\Psi_G(w) &= \prod_{j=1}^{\rho+1} |1 - \bar{z}_j w|^{2\tau_j}\Psi_F^*(w)\Psi_F(w)\\ &\leqslant \prod_{j=1}^{\rho} |1 - \bar{z}_j w|^{2\tau_j}\Big[(1 - |w|^2)\sum_{i,j=0}^{m-1} w^{m-1-i}\boldsymbol{v}_{ij}\bar{w}^{m-1-j}\Big]^{-1}\end{aligned}$$

where equality holds if and only if $G = G_w$, i. e., if and only if $F = \hat{F}_w$. Then, since $\tau_{\rho+1} = 1$, it is readily seen that the first statement of Part (1) follows from the last two inequalities. In turn, we have

$$\Phi_F(0)\Phi_F^*(0) \leqslant \boldsymbol{v}_{00}^{-1} \ \textit{and} \ \Psi_F^*(0)\Psi_F(0) \leqslant \boldsymbol{v}_{m-1,m-1}^{-1},$$

and there exists a unique member $\tilde{F}$ of $\boldsymbol{C}_q[D, \hat{\boldsymbol{c}}_w]$ with $\eta_0(\tilde{F}) < +\infty$

such that

$$\Phi_{\widetilde{F}}(0)\Phi_{\widetilde{F}}^{*}(0)=v_{00}^{-1} \text{ and } \Psi_{\widetilde{F}}^{*}(0)\Psi_{\widetilde{F}}(0)=v_{m-1,m-1}^{-1},$$

which is exactly $\hat{F}_w$, since, by Corollary 5,

$$\Phi_{\hat{F}_w}(0)\Psi_{\hat{F}_w}^{*}(0)=\Phi_{G_w}(0)\Phi_{G_w}^{*}(0)=v_{00}^{-1}$$

and

$$\Psi_{\hat{F}_w}^{*}(0)\Psi_{\hat{F}_w}(0)=\Psi_{G_w}^{*}(0)\Psi_{G_w}(0)=v_{m-1,m-1}^{-1}.$$

Part (2) of this corollary can be verified analogously. ■

§ 4. Maximum Determinant Completions of P_Δ

The problem under consideration in this section is the one -step nonnegative Hermitian extension problem for the information matrix P_Δ of (nondegenerate)Problem $(NP)_\Delta$ in such a way as to maximize the determinants of the specified one-step extensions of the matrix P_Δ, over the set of all associated one-step nonnegative Hermitian completions of Pick type.

§ 4. 1 The extension Problem from P_Δ to $P_{\Delta'}$

We shall begin with the maximum determinant completions of T_c. Assume that $T_c=(C_{j-i})_{i,j=0}^{m-1}>O$ and therefore $\hat{T}_c=(C_{i-j})_{i,j=0}^{m-1}>O$ as well. For any $C_m\in \mathbf{C}^{q\times q}$, let $T_c(C_m):=(C_{j-i})_{i,j=0}^{m}$ and $\hat{T}_c(C_m):=(C_{i-j})_{i,j=0}^{m}$, *where* $C_{-m}=C_m^*$. Find $C_m\in \mathbf{C}^{q\times q}$ such that $\det T_c(C_m)$ reaches the maximum when C_m takes the values in the set

$$\mathcal{A}:=\{C_m\in \mathbf{C}^{q\times q}: T_c(C_m)\geqslant O\}.$$

Note that $\det T_c(C_m)$ coincides with $\det \hat{T}_c(C_m)$ obviously. To this end. we need some further notation, which can only depend on $c=(C_0,C_1,\cdots,C_{m-1})$. Let $\hat{T}_{m-1}:=(C_{i-j})_{i,j=0}^{m-2}$ if $m\geqslant 2$, and $\hat{T}_0:=I_q$, and

$$b:=\begin{cases} C_0, & \text{if } m=1,\\ \begin{bmatrix} C_1\\ \vdots\\ C_{m-1}\end{bmatrix}, & \text{if } m\geqslant 2,\end{cases}\qquad c^*:=\begin{cases} C_0, & \text{if } m=1,\\ [C_{m-1},\cdots,C_1], & \text{if } m\geqslant 2,\end{cases}$$

$$R:=\begin{cases} C_0, & \text{if } m=1,\\ C_0-b^*\hat{T}_{m-1}^{-1}b, & \text{if } m\geqslant 2,\end{cases}\qquad L:=\begin{cases} C_0, & \text{if } m=1.\\ C_0-c^*\hat{T}_{m-1}^{-1}c, & \text{if } m\geqslant 2,\end{cases}$$

$$M:=\begin{cases}\boldsymbol{O}, & \text{if } m=1,\\ c^*\hat{\boldsymbol{T}}_{m-1}^{-1}b, & \text{if } m\geqslant 2,\end{cases}\quad E:=\begin{bmatrix}\boldsymbol{I}_q & \boldsymbol{O} & \boldsymbol{O}\\ -\hat{\boldsymbol{T}}_{m-1}^{-1}b & \boldsymbol{I}_{(m-1)q} & -\hat{\boldsymbol{T}}_{m-1}^{-1}c\\ \boldsymbol{O} & \boldsymbol{O} & \boldsymbol{I}_q\end{bmatrix}.$$

Then R [resp. ,L] is the Schur complement of $\hat{\boldsymbol{T}}_{m-1}$ [resp. ,$\boldsymbol{C}_0$] with respect to $\hat{\boldsymbol{T}}_c$ if $m\geqslant 2$, and $R=L=\boldsymbol{C}_0$ if $m=1$ so that $R>\boldsymbol{O}$ and $L>\boldsymbol{O}$ for $m\geqslant 1$.

Lemma 8 *Let* $\boldsymbol{T}_c=(\boldsymbol{C}_{j-i})_{i,j=0}^{m-1}>\boldsymbol{O}, \boldsymbol{T}_c(\boldsymbol{C}_m)=(\boldsymbol{C}_{j-i})_{i,j=0}^{m}$ *and* $\hat{\boldsymbol{T}}_c(\boldsymbol{C}_m)=(\boldsymbol{C}_{i-j})_{i,j=0}^{m}\ \forall\, \boldsymbol{C}_m\in\mathbf{C}^{q\times q}$. *Then for each* $\boldsymbol{C}_m\in\mathcal{A}$,

$$\det \boldsymbol{T}_c(\boldsymbol{C}_m)=\det\hat{\boldsymbol{T}}_c(\boldsymbol{C}_m)\leqslant \det R\cdot\det\hat{\boldsymbol{T}}_{m-1}\cdot\det L$$

holds, with equality if and only if $\boldsymbol{C}_m=\boldsymbol{O}$ *if* $m=1$ *and*

$$\boldsymbol{C}_m=M=\frac{1}{2(m!)}\frac{\mathrm{d}^m}{\mathrm{d}z^m}F_0(z)\bigg|_{z=0}$$

if $m\geqslant 2$, *in that case*,

$$\det\boldsymbol{T}_c(M)=\det\hat{\boldsymbol{T}}_c(M)=\max_{\boldsymbol{C}_m\in\mathcal{A}}\det\hat{\boldsymbol{T}}_c(\boldsymbol{C}_m)=\det R\cdot\det\hat{\boldsymbol{T}}_{m-1}\cdot\det L.$$

Proof Observe that if $m=1$, then $\det\hat{\boldsymbol{T}}_c(\boldsymbol{C}_1)\leqslant(\det\boldsymbol{C}_0)^2$, and $\det\hat{\boldsymbol{T}}_c(\boldsymbol{C}_1)=(\det\boldsymbol{C}_0)^2$ if and only if $\boldsymbol{C}_1=\boldsymbol{O}$. Let now $m\geqslant 2$. From [10, Section 3. 2] it follows that

$$E^*\hat{\boldsymbol{T}}_c(\boldsymbol{C}_m)E=\begin{bmatrix}R & 0 & \boldsymbol{C}_m^*-M^*\\ 0 & \hat{\boldsymbol{T}}_{m-1} & 0\\ \boldsymbol{C}_m-M & 0 & L\end{bmatrix},$$

so that the inequality

$$\det\hat{\boldsymbol{T}}_c(\boldsymbol{C}_m)=\det R\cdot\det\hat{\boldsymbol{T}}_{m-1}\cdot\det(\mathrm{L}-(\boldsymbol{C}_m-M)R^{-1}(\boldsymbol{C}_m^*-\mathrm{M}^*))$$
$$\leqslant\det R\cdot\det\hat{\boldsymbol{T}}_{\mathrm{m}-1}\cdot\det L$$

holds, with equality if and only if $\boldsymbol{C}_m=M$. From [10, Theorem 3. 4. 3 and Remark 3. 4. 5], we deduce that

$$M=\frac{1}{2(m!)}\frac{\mathrm{d}^m}{\mathrm{d}z^m}F_0(z)\bigg|_{z=0},$$

and $\hat{\boldsymbol{T}}_c(M)>\boldsymbol{O}$, since F_0 is the central solution of (nondegenerate) Problem $(\mathrm{C})_c$. ■

As for maximum determinant completions corresponding to the specific one-step block Pick-type extension problem form $\boldsymbol{P}_\Delta$ *to* $\boldsymbol{P}_{\Delta'}$, where

Δ' is as in (22) with D_{0,τ_0} as a single parameter, based on Theorems 1 and 2 in combination with Lemma 8, we have

Theorem 9 *Let* $m \geqslant 2$ *and* $P_{\Delta} > O$, *let* $\Delta' = \Delta'(D_{0,\tau_0})$ *be as in* (22) *with any* $D_{0,\tau_0} \in \mathbf{C}^{q\times q}$, *and let* $P_{\Delta'} \in \mathbf{C}^{(m+1)q\times(m+1)q}$ *be the generalized block Pick matrix associated with the data* Δ', *and let the set* $\mathcal{A}'_{\Delta}$, *be defined by*

$$\mathcal{A}'_{\Delta} := \{D_{0,\tau_0} \in \mathbf{C}^{q\times q} : P_{\Delta'} \geqslant O\}.$$

Then

$$\max\{\det P_{\Delta'} : D_{0,\tau_0} \in A'_{\Delta}\} > 0$$

and this maximum value is attained if and only if

$$D_{0,\tau_0} = \frac{1}{\tau_0 !}\frac{\mathrm{d}^{\tau_0}}{\mathrm{d}z^{\tau_0}} G_0(z)\bigg|_{z=0},$$

where G_0 *is defined by* (18)~(21) *with* $w=0$.

Proof For any $D_{0,\tau_0} \in A'_{\Delta}$, let

$$C_m := A_{-n}^{-1}\left(\frac{1}{2}D_{0,\tau_0} - \sum_{l=-n+1}^{n} A_l C_{\tau_0 - l}\right),$$

where $m = n + \tau_0$, $A_l(-n \leqslant l \leqslant n)$ and $C_k(0 \leqslant k \leqslant m-1)$ are as in (10). The formula(15) then is easily verified to be valid if P_{Δ}, T_c and W_{Δ} are replaced with $P_{\Delta'}$, $T_c(C_m)$ and $W_{\Delta'}$, respectively, where $W_{\Delta'}$ of order $m+1$ is nonsingular and of the form similar to W_{Δ} in (14) but with $\tau_0 + 1$ in place of τ_0, and $T_c(C_m) = (C_{j-i})_{i,j=0}^{m}$. This means that $P_{\Delta'} = 2(W_{\Delta'} \otimes I_q) T_c(C_m)(W_{\Delta'} \otimes I_q)^*$. Thus, by Lemma 8, the inequality

$$\begin{aligned}\det P_{\Delta'} &= 2^{(m+1)q}\,|\det(W_{\Delta'} \otimes I_q)|^2 \cdot \det T_c(C_m)\\ &\leqslant 2^{(m+1)q}\,|\det W_{\Delta'}|^2 \cdot \det R \cdot \det L \cdot \det \hat{T}_{m-1}\end{aligned}$$

holds with equality if and only if $C_m = M = \frac{1}{2(m!)}\frac{\mathrm{d}^m}{\mathrm{d}z^m} F_0(z)\Big|_{z=0}$. In that case, $T_c(M) > O$ (since $\hat{T}(M) > O$).

On the other hand, if $F_0(z)$ permits the integral representation of the form (19) with $w=0$, then form Theorem 2 we infer that $G_0(z) = W(z) + A(z)F_0(z)$ admits the integral representation of the form (21) with $w = 0$ and vice versa. Accordingly, if $C_m = M = \frac{1}{2(m!)}\frac{\mathrm{d}^m}{\mathrm{d}z^m} F_0(z)\Big|_{z=0}$, then

$$\frac{1}{\tau_0 !}\frac{\mathrm{d}^{\tau_0}}{\mathrm{d}z^{\tau_0}}G_0(z)\bigg|_{z=0} = 2\int_T u^{-\tau_0}A(u)\mu_0(\mathrm{d}u) = 2\sum_{j=-n}^{n}A_j\int_T u^{-\tau_0+j}\mu_0(du)$$

$$= 2\sum_{j=-n}^{n}A_j\boldsymbol{C}_{\tau_0-j} = \boldsymbol{D}_{0,\tau_0},$$

and vice versa. Hence $\boldsymbol{P}_{\Delta'}(\frac{1}{\tau_0 !}G_0^{(\tau_0)}(0))>\boldsymbol{O}$ (since $\boldsymbol{T}_c(M)>\boldsymbol{O}$) and

$$0<\max\{\det \boldsymbol{P}_{\Delta'}:\boldsymbol{D}_{0,\tau_0}\in \mathcal{A}'_\Delta\}$$
$$=2^{(m+1)q}\mid \det W_{\Delta'}\mid^2\cdot \det R\cdot \det \mathrm{L}\cdot \det \hat{\boldsymbol{T}}_{m-1},$$

and this maximum value is attained if and only if

$$\boldsymbol{D}_{0,\tau_0}=\frac{1}{\tau_0 !}\frac{\mathrm{d}^{\tau_0}}{\mathrm{d}z^{\tau_0}}G_0(z)\bigg|_{z=0}. \qquad \blacksquare$$

Note that the set $\mathcal{A}'_\Delta$ defined in the statement of Theorem 9 coincides with the set

$$\left\{\frac{1}{\tau_0 !}\frac{\mathrm{d}^{\tau_0}}{\mathrm{d}z\,\tau_0}G(z)\bigg|_{z=0}:G\in \boldsymbol{C}_\Delta\right\}.$$

Indeed, $\boldsymbol{C}_{\Delta'}\subseteq\boldsymbol{C}_\Delta$ if $\boldsymbol{C}_\Delta\neq\varnothing$, where $\boldsymbol{C}_{\Delta'}$ is the solution set of Problem $(\mathrm{NP})_{\Delta'}$.

§ 4.2 The Extension Problems from $\boldsymbol{P}_\Delta$ *to* $\boldsymbol{P}_{\hat{\Delta}}$ *and from* $\boldsymbol{P}_\Delta$ *to* $\boldsymbol{P}_{\check{\Delta}_s}$

In the remaining part of this section , we consider the other two maximum determinant completions corresponding to one-step nonnegative Hermitian extension problems of $\boldsymbol{P}_\Delta$ to $\boldsymbol{P}_{\hat{\Delta}}$, and $\boldsymbol{P}_\Delta$ to $\boldsymbol{P}_{\check{\Delta}}$. Here $\boldsymbol{P}_{\hat{\Delta}}$ and $\boldsymbol{P}_{\check{\Delta}}$of order $(m+1)q$ are the generalized block Pick matrices associated with the date $\hat{\Delta}$ and the date $\check{\Delta}$ defined by (23) and (24), respectively. Specially, let $\check{\Delta}:=\check{\Delta}_\rho$ for short. The problem under consideration is to find a $\boldsymbol{D}_{\rho+1,0}\in\mathbf{C}^{q\times q}$ and a $\boldsymbol{D}_{\rho,\tau_\rho}\in\mathbf{C}^{q\times q}$ such that det $\boldsymbol{P}_{\hat{\Delta}}$ and det $\boldsymbol{P}_{\check{\Delta}}$ associated, respectively, with such $\boldsymbol{D}_{\rho+1,0}$ and $\boldsymbol{D}_{\rho,\tau_\rho}$ attain to maximum values when $\boldsymbol{D}_{\rho+1,0}$ and $\boldsymbol{D}_{\rho,\tau_\rho}$ take values in the sets

$$\hat{\mathcal{A}}_\Delta:=\{\boldsymbol{D}_{\rho+1,0}\in\mathbf{C}^{q\times q}:\boldsymbol{P}_{\hat{\Delta}}\geqslant\boldsymbol{O}\}$$

and

$$\check{\mathcal{A}}_\Delta:=\{\boldsymbol{D}_{\rho,\tau_\rho}\in\mathbf{C}^{q\times q}:\boldsymbol{P}_{\check{\Delta}}\geqslant\boldsymbol{O}\},$$

respectively. Here both the fundamental matrix inequality FMI(C) in [20] with its transformed matrix inequality, and the Weyl matrix ball shall play a role in our analysis later on . It is convenient to first prove a preliminary result. This result is ,even by itself, very interesting.

Lemma 10 (1) *Let $\hat{\Delta}$ be as in* (23), *and let $\boldsymbol{P}_{\hat{\Delta}}$ be the associated generalized block Pick matrix with the data $\hat{\Delta}$. Let further F belong to $\mathbf{C}_q[D,\boldsymbol{c}]$. Then there exists invertible $\sum_{\hat{\Delta}}$ of order $m+1$, independent only of $A(z)(z-z_{\rho+1})(z^{-1}-\bar{z}_{\rho+1})$ such that*

$$\boldsymbol{P}_{\hat{\Delta}} = 2(\textstyle\sum_{\hat{\Delta}}\sum_{B}^{-1}\otimes\boldsymbol{I}_q)\boldsymbol{T}_{\hat{\Delta}}(\sum_{\hat{\Delta}}\sum_{B}^{-1}\otimes\boldsymbol{I}_q)^* \tag{30}$$

where

$$\boldsymbol{T}_{\hat{\Delta}} := \begin{bmatrix} \boldsymbol{T}_c & \frac{1}{2}(b_T(\bar{z}_{\rho+1})(F(z_{\rho+1}))^* - c_T(\bar{z}_{\rho+1})) \\ * & \frac{1}{2}\frac{F(z_{\rho+1})+(F(z_{\rho+1}))^*}{1-|z_{\rho+1}|^2} \end{bmatrix} \tag{31}$$

in which "$*$" at the left lower corner stands for the conjugate transpose of the block matrix located at the top right corner of $\boldsymbol{T}_{\hat{\Delta}}$,

$$b_T(z)=\begin{bmatrix} z^{-1}\boldsymbol{I}_q \\ z^{-2}\boldsymbol{I}_q \\ \vdots \\ z^{-m}\boldsymbol{I}_q \end{bmatrix},\quad C_T(z)=\begin{bmatrix} \boldsymbol{C}_0 & 0 & \cdots & 0 \\ 2\boldsymbol{C}_1^* & \boldsymbol{C}_0 & \cdots & 0 \\ \vdots & \vdots & \vdots & \vdots \\ 2\boldsymbol{C}_{m-1}^* & 2\boldsymbol{C}_{m-2}^* & \cdots & \boldsymbol{C}_0 \end{bmatrix} b_T(z),$$

and

$$\textstyle\sum_B := \begin{bmatrix} -z_{\rho+1} & 1 & 0 & \cdots & 0 & 0 \\ 0 & -z_{\rho+1} & 1 & \cdots & 0 & 0 \\ \vdots & \vdots & \vdots & \vdots & \vdots & \vdots \\ 0 & 0 & 0 & \cdots & -z_{\rho+1} & 1 \\ 1 & 0 & 0 & \cdots & 0 & 0 \end{bmatrix} \in \mathbf{C}^{(m+1)\times(m+1)}. \tag{32}$$

(2) *Let $\check{\Delta}=\check{\Delta}_\rho=\check{\Delta}(\boldsymbol{D}_{\rho,\tau_\rho})$ be as in* (24). *and let $\boldsymbol{P}_{\check{\Delta}}$ be the associated generalized block Pick matrix with the date $\check{\Delta}$. Let F belong to $\mathbf{C}_q[D,\boldsymbol{c}]$. Then there exists invertible $\sum_{\check{\Delta}}$ of order $m+1$, independent only of $A(z)(z-z_\rho)(z^{-1}-\bar{z}_\rho)$, such that*

$$\boldsymbol{P}_{\check{\Delta}} = 2(\textstyle\sum_{\check{\Delta}}\sum_{D}^{-1}\otimes\boldsymbol{I}_q)\boldsymbol{T}_{\check{\Delta}}(\sum_{\check{\Delta}}\sum_{D}^{-1}\otimes\boldsymbol{I}_q)^*, \tag{33}$$

where

$$\boldsymbol{T}_{\check{\Delta}} = \begin{bmatrix} \boldsymbol{T}_c & \frac{1}{2}(b_T(\bar{z}_\rho)(F(z_\rho))^* - c_T(\bar{z}_\rho)) \\ * & \frac{1}{2}\frac{F(z_\rho)+(F(z_\rho))^*}{1-|z_\rho|^2} \end{bmatrix}, \tag{34}$$

in which $b_T(z)$ *and* $c_T(z)$ *are as in* (31), *and*

$$\sum\nolimits_D := \begin{bmatrix} -z_\rho & 1 & 0 & \cdots & 0 & 0 \\ 0 & -z_\rho & 1 & \cdots & 0 & 0 \\ \vdots & \vdots & \vdots & & \vdots & \vdots \\ 0 & 0 & 0 & \cdots & -z_\rho & 1 \\ 1 & 0 & 0 & \cdots & 0 & 0 \end{bmatrix} \in \mathbf{C}^{(m+1)\times(m+1)}. \tag{35}$$

Proof Consider Problem $(\mathrm{NP})_{\hat{\Delta}}$ and Problem $(\mathrm{NP})_{\check{\Delta}}$. Then there exist $\hat{\boldsymbol{c}} := (\hat{\boldsymbol{C}}_0, \hat{\boldsymbol{C}}_1, \cdots, \hat{\boldsymbol{C}}_m)$, where each $\hat{\boldsymbol{C}}_k \in \mathbf{C}^{q\times q}$ and $\hat{\boldsymbol{C}}_0 = \hat{\boldsymbol{C}}_0^*$, *and* $\check{\boldsymbol{c}} := (\check{\boldsymbol{C}}_0, \check{\boldsymbol{C}}_1, \cdots, \check{\boldsymbol{C}}_m)$, where each $\check{\boldsymbol{C}}_k \in \mathbf{C}^{q\times q}$ and $\check{\boldsymbol{C}}_0 = \check{\boldsymbol{C}}_0^*$, as the block Toeplitz vectors of these two problems, respectively. From Theorem 1 we deduce that

$$\begin{aligned} \boldsymbol{P}_{\hat{\Delta}} &= 2\left(\sum\nolimits_{\hat{\Delta}} \otimes \boldsymbol{I}_q\right)\boldsymbol{T}_{\hat{c}}\left(\sum\nolimits_{\hat{\Delta}} \otimes \boldsymbol{I}_q\right)^*, \\ \boldsymbol{P}_{\check{\Delta}} &= 2\left(\sum\nolimits_{\check{\Delta}} \otimes \boldsymbol{I}_q\right)\boldsymbol{T}_{\check{c}}\left(\sum\nolimits_{\check{\Delta}} \otimes \boldsymbol{I}_q\right)^*, \end{aligned} \tag{36}$$

where

$$\boldsymbol{T}_{\hat{c}} := (\hat{\boldsymbol{C}}_{j-i})_{i,j=0}^m \; and \; \boldsymbol{T}_{\hat{c}} := (\hat{\boldsymbol{C}}_{j-i})_{i,j=0}^m, \tag{37}$$

and $\sum_{\hat{\Delta}}$ *and* $\sum_{\check{\Delta}}$ are of the form similar to $\sum_\Delta$ in (14) with B(z), $n-1$ and ρ replaced by $\mathrm{B}(z) := \prod_{i=1}^{\rho+1}(z-z_i)^{\tau_i}$, n and $\rho+1$ and with τ_ρ and $n-1$ replaced by $\tau_\rho+1$ and n, respectively. Let now $\boldsymbol{T}_{\hat{\Delta}}$ be as in (31). To prove (30), we need only to verify that

$$\boldsymbol{T}_{\hat{\Delta}} = \left(\sum\nolimits_B \otimes \boldsymbol{I}_q\right)\boldsymbol{T}_{\hat{c}}\left(\sum\nolimits_B \otimes \boldsymbol{I}_q\right)^*, \tag{38}$$

in which $\sum_B$ is as in (32).

Assume that $F \in \boldsymbol{C}_q[D, \boldsymbol{c}]$ has an integral representation of the form

$$F(z) := \int_T \frac{u+z}{u-z}\frac{\sigma(\mathrm{d}u)}{A(u)},$$

where σ is the Riesz-Herglotz measure of $G \in \boldsymbol{C}_\Delta$, which is connected with F by $G(z) = W(z) + A(z)F(z)$ (see Theorem 2). Let further $\hat{A}(z) := A(z)p(z)p(\bar{z}^{-1})$, $\forall z \neq 0$, where $p(z) = z - z_{\rho+1}$. Then a direct calculation leads to

$$\boldsymbol{T}_{\hat{\Delta}} = \begin{bmatrix} \left(\int_T u^{i-j} \frac{\sigma(\mathrm{d}u)}{A(u)}\right)_{i,j=0}^{m-1} & \left(\int_T u^i \frac{\sigma(\mathrm{d}u)}{(\bar{z} - \bar{z}_{\rho+1})A(u)}\right)_{i=0}^{m-1} \\ * & \frac{1}{2}\frac{F(z_{\rho+1}) + (F(z_{\rho+1}))^*}{1 - |z_{\rho+1}|^2} \end{bmatrix}$$

$$= \begin{bmatrix} \left(\int_T u^i p(u) \frac{\sigma(\mathrm{d}u)}{\hat{A}(u)} \overline{u^j p(u)}\right)_{i,j=0}^{m-1} & \left(\int_T u^i p(u) \frac{\sigma(\mathrm{d}u)}{\hat{A}(u)}\right)_{i=0}^{m-1} \\ * & \int_T \frac{\sigma(\mathrm{d}u)}{\hat{A}(u)} \end{bmatrix}$$

$$= \int_T \begin{bmatrix} p(u) \\ up(u) \\ \vdots \\ u^{m-1}p(u) \\ 1 \end{bmatrix} \frac{\sigma(\mathrm{d}u)}{\hat{A}(u)} [\overline{p(u)}, \overline{up(u)}, \cdots, \overline{u^{m-1}p(u)}, 1]$$

$$= (\textstyle\sum_B \otimes \boldsymbol{I}_q)\boldsymbol{T}_c^{\wedge}(\textstyle\sum_B \otimes \boldsymbol{I}_q)^*\ ,$$

where $\sum_B$ of order $m+1$ is the transition matrix of the basis of the space $\mathscr{B}_m$(of the numerical polynomials of degree at most m):

$$\mathcal{B} := \{p(z), zp(z), \cdots, z^{m-1}p(z), 1\}$$

to the standard basis $\{1, z, \cdots, z^m\}$, which is obviously invertible ,and of the form (32). Thus, from (36) and (38) if follows that

$$\boldsymbol{P}_{\hat{\Delta}} = 2(\textstyle\sum_{\hat{\Delta}} \sum_B^{-1} \otimes \boldsymbol{I}_q)\boldsymbol{T}_{\hat{\Delta}}(\sum_{\hat{\Delta}} \sum_B^{-1} \otimes \boldsymbol{I}_q)^*.$$

As for the proof of the second part of the lemma, similarly, one infers that $\boldsymbol{T}_{\hat{\Delta}}$ defined by (34) is of the form

$$\boldsymbol{T}_{\hat{\Delta}} = (\textstyle\sum_D \otimes \boldsymbol{I}_q)\boldsymbol{T}_c^{\wedge}(\sum_D \otimes \boldsymbol{I}_q)^*, \tag{39}$$

where $\sum_D$ is as in (35), so that form the last result in combination with (36) we deduce the formula (33). ■

Theorem 11 *Let* $m = \sum_{i=0}^{\rho} \tau_i \geqslant 2$. *Let* $\hat{\Delta} = \hat{\Delta}(\boldsymbol{D}_{\rho+1,0})$ *be as in* (23)*and let* $\boldsymbol{P}_{\hat{\Delta}}$ *be the associated generalized block Pick matrix with the data* $\hat{\Delta}$. *Then*

$$0 < \max\{\det \boldsymbol{P}_{\hat{\Delta}} : \boldsymbol{D}_{\rho+1,0} \in \hat{\mathcal{A}}_\Delta\} = 2^{mq} \mid \det \textstyle\sum_{\hat{\Delta}} \mid^2 \cdot \det T_c \cdot \det \hat{E},$$

where

$$\hat{E}:=-(c_T(\bar{z}_{\rho+1}))^*(2\boldsymbol{T}_c)^{-1}c_T(\bar{z}_{\rho+1})+\hat{L}\hat{K}^{-1}\hat{L}^*,$$
$$\hat{K}:=(b_T(\bar{z}_{\rho+1}))^*(2\boldsymbol{T}_c)^{-1}b_T(\bar{z}_{\rho+1}), \tag{40}$$
$$\hat{L}:=\frac{\boldsymbol{I}_q}{1-|z_{\rho+1}|^2}+(c_T(\bar{z}_{\rho+1}))^*(2\boldsymbol{T}_c)^{-1}b_T(\bar{z}_{\rho+1}),$$

and this maximum value is attained if and only if

$$\boldsymbol{D}_{\rho+1,0}=G_{z_{\rho+1}}(z_{\rho+1}).$$

Proof Observe that the matrix $\boldsymbol{T}_{\hat{\Delta}}$ defined by (31) is such that the inequality $\boldsymbol{T}_{\hat{\Delta}}\geqslant\boldsymbol{O}$ coincides with the fundamental matrix inequality FMI(C) of Problem $(C)_c$ at $z=\bar{z}_{\rho+1}$ of Potapov's interpolation theory (see, e. g., [20]) with $\boldsymbol{C}_0,\boldsymbol{C}_1,\cdots,\boldsymbol{C}_{m-1}$, and ζ replaced by $\boldsymbol{C}_0, 2\boldsymbol{C}_1^*,\cdots,2\boldsymbol{C}_{m-1}^*$, and $\bar{z}_{\rho+1}$, respectively. It is well known [20, Theorem 2] that in the nondegenerate case (that is, $\boldsymbol{T}_c>\boldsymbol{O}$), a $\mathbf{C}^{q\times q}$-valued function F holomorphic in D is a solution of Problem $(C)_c$, i. e., $F\in\boldsymbol{C}_q[D,\boldsymbol{c}]$, if and only if it satisfies FMI(C). Thus, for any $F\in\boldsymbol{C}_q[D,\boldsymbol{c}]$, $\boldsymbol{T}_{\hat{\Delta}}\geqslant\boldsymbol{O}$, and therefore, in particular, we have

$$\frac{\mathrm{F}(\mathrm{z}_{\rho+1})+(\mathrm{F}(\mathrm{z}_{\rho+1}))^*}{1-|\mathrm{z}_{\rho+1}|^2}-[\mathrm{F}(\mathrm{z}_{\rho+1})(\mathrm{b}_{\mathrm{T}}(\bar{\mathrm{z}}_{\rho+1}))^*-(\mathrm{c}_{\mathrm{T}}(\bar{\mathrm{z}}_{\rho+1}))^*]\times$$
$$(2\boldsymbol{T}_c)^{-1}[\mathrm{b}_{\mathrm{T}}(\bar{\mathrm{z}}_{\rho+1})(\mathrm{F}(\mathrm{z}_{\rho+1}))^*-c_{\mathrm{T}}(\bar{\mathrm{z}}_{\rho+1})]\geqslant\boldsymbol{O},$$

where $\mathrm{b_T(z)}$ and $\mathrm{c_T(z)}$ are as in (31). Let now $x:=(\mathrm{F}(\mathrm{z}_{\rho+1}))^*$ and $\hat{\mathrm{K}},\hat{L}$ and $\hat{E}$ as in (40). We have that $\hat{K}>0$ obviously, and that by (32)(38) and (31)

$$\det(2\boldsymbol{T}_{\hat{c}})=\det(2\boldsymbol{T}_{\hat{\Delta}})$$
$$=\det(2\boldsymbol{T}_c)\cdot\det\left[\frac{x+x^*}{1-|z_{\rho+1}|^2}-(b_T(\bar{z}_{\rho+1})x-c_T(\bar{z}_{\rho+1}))^*(2\boldsymbol{T}_c)^{-1}\cdot\right.$$
$$\left.(b_T(\bar{z}_{\rho+1})x-c_T(\bar{z}_{\rho+1}))\right]$$
$$=\det(2\boldsymbol{T}_c)\cdot\det[-(c_T(\bar{z}_{\rho+1}))^*(2\boldsymbol{T}_c)^{-1}c_T(\bar{z}_{\rho+1})-x^*\hat{K}x+$$
$$x^*\hat{L}^*+\hat{L}x],$$

where $\hat{\boldsymbol{c}}$ is the block Toeplitz vector of Problem $(\mathrm{NP})_{\hat{\Delta}}$. Thus, upon completing the square, we see that

$$-(c_T(\bar{z}_{\rho+1}))^*(2\boldsymbol{T}_c)^{-1}c_T(\bar{z}_{\rho+1})-x^*\hat{K}x+x^*\hat{L}^*+\hat{L}x$$
$$=\hat{E}-(x-\hat{K}^{-1}\hat{L}^*)^*\hat{K}(x-\hat{K}^{-1}\hat{L}^*),$$

in which $\hat{E}$ is as in (40). It is obvious that

$$\hat{E} \geqslant \hat{E} - (x - \hat{K}^{-1}\hat{L}^*)^* \hat{K} (x - \hat{K}^{-1}\hat{L}^*),$$

with equality if and only if

$$\begin{aligned} x &= \hat{K}^{-1}\hat{L}^* \\ &= [(b_T(\bar{z}_{\rho+1}))^* (2\boldsymbol{T}_c)^{-1} b_T(\bar{z}_{\rho+1})]^{-1} \left[\frac{\boldsymbol{I}_q}{1-|z_{\rho+1}|} + (b_T(\bar{z}_{\rho+1}))^* (2\boldsymbol{T}_c)^{-1} c_T(\bar{z}_{\rho+1}) \right]. \end{aligned}$$

Thus,

$$\det(2\boldsymbol{T}_{\hat{c}}) \leqslant 2^{mq} \det T_c \det \hat{E}$$

holds with equality is and only if $F(Z_{p+1}) = \hat{L}K^{-1}$. On the other hand, for an arbitrary fixed $Z_{\rho+1} \in D$, different form the original $z_0, z_1, \cdots, z_\rho$, the set

$$\{\Omega(z_{\rho+1})^* : \Omega \in \boldsymbol{C}_q[D, \boldsymbol{c}]\}$$

fills the Weyl matrix ball $\mathscr{B}$ with center $\hat{K}^{-1}\hat{L}^*$, some suitable right radius ρ_d and some suitable left radius ρ_g:

$$w = \hat{K}^{-1}\hat{L}^* + \rho_g^{\frac{1}{2}} u \rho_d^{\frac{1}{2}}, \ \forall\, u^* u \leqslant \boldsymbol{I}_q$$

(see [20, Section 2, the formulas (52) and (53)]). Moreover, it follows form [19, Section 6] that

$$\hat{K}^{-1}\hat{L}^* = (F_{z_{\rho+1}}(z_{\rho+1}))^*.$$

Due to the rank invariance of block Pick matrices of functions in $C_q[D]$ given in [21, Theorem 2] and Theorem 9, we see that

$$\operatorname{rank} \boldsymbol{P}_{\hat{\Delta}(\frac{1}{\tau_0!}G_0^{(\tau_0)}(0))} = \operatorname{rank} \boldsymbol{P}_{\hat{\Delta}(G_0(z_{\rho+1}))} = (m+1)q,$$

so that $\boldsymbol{P}_{\hat{\Delta}(G_0(z_{\rho+1}))} > \boldsymbol{O}$ as well.

Summing up what have proved and (30), we derive that

$$0 < \max\{\det \boldsymbol{P}_{\hat{\Delta}} : \mathbf{D}_{\rho+1,0} \in \hat{\mathcal{A}}_\Delta\} = 2^{mq} \mid \det \textstyle\sum_{\hat{\Delta}} \mid^2 \cdot \det \boldsymbol{T}_c \cdot \det \hat{E}$$

and this maximum value is attained if and only if $F(z_{\rho+1}) := F_{z_{\rho+1}}(z_{\rho+1})$, or equivalently, if and only if

$$\begin{aligned} \boldsymbol{D}_{\rho+1,0} &= W(z_{\rho+1}) + A(z_{\rho+1}) F(z_{\rho+1}) \\ &= W(z_{\rho+1}) + A(z_{\rho+1}) F_{z_{\rho+1}}(z_{\rho+1}) = G_{z_{\rho+1}}(z_{\rho+1}). \end{aligned}$$ ■

Nevertheless, on the basis of the formula (36), the problem of Theorem 11 can be formulated equivalently in terms of finding the solution of a particular specified extremal problem to maximize the determinant of the block Toeplitz matrix $\boldsymbol{T}_{\hat{c}}$ of order $(m+1)q$ with a single parameter $\hat{\boldsymbol{c}}$ which takes the values in the set $\hat{\mathcal{B}}_\Delta$ defined by

$\hat{\mathcal{B}}_\Delta := \{\hat{\boldsymbol{c}} : \hat{\boldsymbol{c}}$ is the block Toeplitz vector of Problem $(\mathrm{NP})_{\hat{\Delta}}$ such that $\boldsymbol{T}_{\hat{c}} \geqslant \boldsymbol{O}\}$.

Corollary 7 *Let $m = \sum_{i=0}^{\rho} \tau_i \geqslant 2$, and let $\mathbf{c} = (\boldsymbol{C}_0, \boldsymbol{C}_1, \cdots, \boldsymbol{C}_{m-1})$ be the block Toeplitz vector of Problem $(\mathrm{NP})_\Delta$ such that $\boldsymbol{T}_c = (\boldsymbol{C}_{j-i})_{i,j=0}^{m-1} > \boldsymbol{O}$. Furthermore, let $\mathbf{T}_{\hat{c}}$ be the information matrix of Problem $(\mathrm{C})_{\hat{c}}$. Then*

$$0 < \max\{\det \boldsymbol{T}_{\hat{c}} : \hat{\boldsymbol{c}} \in \hat{\mathcal{B}}_\Delta\} = 2^{-q} \det \boldsymbol{T}_c \cdot \det \hat{E},$$

where $\hat{E}$ is as in (40), and this maxium value is attained if and only if

$$\hat{\boldsymbol{c}} = \hat{\boldsymbol{c}}_{z_{\rho+1}},$$

where $\hat{\boldsymbol{c}}_{z_{\rho+1}}$ is the block Toeplitz vector of Problem $(\mathrm{NP})_{\hat{\Delta}_{z_{\rho+1}}}$, in which $\hat{\Delta}_{z_{\rho+1}}$ is defined by (25) with $w = z_{\rho+1}$, generated by the $\mathbf{C}^{q\times q}$-valued function $\hat{F}_{z_{\rho+1}}$ defined by (19′) with $w = z_{\rho+1}$, i. e.,

$$\hat{\boldsymbol{c}}_{z_{\rho+1}} = \left(\hat{F}_{z_{\rho+1}}(0), \frac{1}{2}\hat{F}'_{z_{\rho+1}}(0), \cdots, \frac{1}{2(m!)}\hat{F}^{m}_{z_{\rho+1}}(0)\right).$$

Proof Using the formula (36), and noting that the matrix $\sum_{\hat{\Delta}}$ of order $m+1$ is independent of the choice of $\boldsymbol{D}_{\rho+1,0}$, we know that

$$P_{\hat{\Delta}_{z_{\rho+1}}} = \left(\sum\nolimits_{\hat{\Delta}} \otimes \boldsymbol{I}_q\right) \boldsymbol{T}_{\hat{c}_{z_{\rho+1}}} \left(\sum\nolimits_{\hat{\Delta}} \otimes \boldsymbol{I}_q\right)^*.$$

Thus, referring to Theorem 11, we can conclude the required result. ■

Theorem 12 *Let $m = \sum\limits_{i=0}^{\rho} \tau_i \geqslant 2$. Let $\check{\Delta} = \check{\Delta}(\boldsymbol{D}_{\rho,\tau_\rho})$* be as in (24) with $s=\rho$, and let $\boldsymbol{P}_{\check{\Delta}}$ *be the associated generalized block Pick matrix with the data $\check{\Delta}$. Then*

$$0 < \max\{\det \boldsymbol{P}_{\check{\Delta}} : \mathbf{D}_{\rho,\tau_\rho} \in \check{\mathcal{A}}_\Delta\} = 2^{mq} \left|\det \sum\nolimits_{\check{\Delta}}\right|^2 \cdot \det \boldsymbol{T}_c \cdot \det \check{E},$$

where

$$\begin{aligned} \check{E} &:= -(c_T(\bar{z}_\rho))^* (2\boldsymbol{T}_c)^{-1} c_T(\bar{z}_\rho) + \check{L}\check{K}^{-1}\check{L}^*, \\ \check{K} &:= (b_T(\bar{z}_\rho))^* (2\boldsymbol{T}_c)^{-1} b_T(\bar{z}_\rho), \\ \check{L} &:= \frac{\boldsymbol{I}_q}{1-|z_\rho|^2} + (c_T(\bar{z}_\rho))^* (2\boldsymbol{T}_c)^{-1} b_T(\bar{z}_\rho), \end{aligned} \tag{41}$$

and this maximum value is attained if and only if

$$D_{\rho,\tau_\rho}=\frac{1}{\tau_\rho!}\frac{d^{\tau_\rho}}{dz^{\tau_\rho}}G_{z_\rho}(z)\Big|_{z=z_\rho},$$

where G_{z_ρ} *is as in* (18)～(21) *with* $w=z_\rho$.

Proof Observe that, as indicated in the proof on Theorem 11, the matrix inequality $\boldsymbol{T}_{\check{\Delta}}\geqslant\boldsymbol{O}$ coincides with the value of the fundamental matrix inequality FMI(C) of Problem $(C)_c$ with $\boldsymbol{C}_0,\boldsymbol{C}_1,\cdots,\boldsymbol{C}_{m-1}$, and ζ replaced by $\boldsymbol{C}_0, 2\boldsymbol{C}_1^*,\cdots,2\boldsymbol{C}_{m-1}^*$ and $\bar{z}_\rho$, respectively. Let $\check{E},\check{K}$ and $\check{L}$ be as in (41), and for $F\in C_q[\mathrm{D},c]$ let $x:=(F(z_\rho))^*$. Then, for the block Toeplitz vector $\check{\boldsymbol{c}}$ of Problem $(\mathrm{NP})_{\check{\Delta}}$, by (39)(34) and (35),

$$\begin{aligned}\det(2\boldsymbol{T}_{\check{c}})&=\det(2\boldsymbol{T}_{\check{\Delta}})=\det(2\boldsymbol{T}_c)\cdot\det\Big[\frac{x+x^*}{1-|z_\rho|^2}-\\&\qquad(b_T(\bar{z}_\rho)x-c_T(\bar{z}_\rho))^*(2\boldsymbol{T}_c)^{-1}(b_T(\bar{z}_\rho)x-c_T(\bar{z}_\rho))\Big]\\&=\det(2\boldsymbol{T}_c)\cdot\det[-(c_T(\bar{z}_\rho))^*(2\boldsymbol{T}_c)^{-1}c_T(\bar{z}_\rho)-x^*\check{K}x+x^*\check{L}^*+\check{L}x]\\&=\det(2\boldsymbol{T}_c)\cdot\det[\check{E}-(x-\check{K}^{-1}\check{L}^*)^*\check{K}(x-\check{K}^{-1}\check{L}^*)]\\&\leqslant 2^{mq}\det\boldsymbol{T}_c\cdot\det\check{E},\end{aligned}$$

with equality if and only if $x=\check{K}^{-1}\check{L}^*$. From the previously mentioned rank invariance [21, Theorem 2] it also follows that $\boldsymbol{P}_{\check{\Delta}(\frac{1}{\tau_\rho!}G_0^{(\tau_\rho)}(z_\rho))}>\boldsymbol{O}$, since

$$\operatorname{rank}\boldsymbol{P}_{\Delta'(\frac{1}{\tau_\rho!}G_0^{(\tau_\rho)}(0))}=\operatorname{rank}\boldsymbol{P}_{\check{\Delta}(\frac{1}{\tau_\rho!}(G_0^{(\tau_\rho)})(z_\rho))}=(m+1)q.$$

From [20, Section 12, the formulas (52) and (53)] we see that

$$\check{K}^{-1}\check{L}^*=(F_{z_\rho}(z_\rho))^*.$$

Thus, we have that

$$0<\max\{\det\boldsymbol{P}_{\check{\Delta}}:D_{\rho,\tau_\rho}\in\check{\mathcal{A}}_\Delta\}=2^{mq}\,|\det\textstyle\sum_{\check{\Delta}}|^2\cdot\det\boldsymbol{T}_c\cdot\det\check{E},$$

and this maximum value is attained if and only if $F(z_\rho):=F_{z_\rho}(z_\rho)$, or equivalently, if and only if

$G(z_\rho):=G_{z_\rho}(z_\rho)$, where $G\in\boldsymbol{C}_\Delta$ is connected with F by $G(z)=w(z)+A(z)F(z)$, $z\in D$. Then and only then

$$\begin{aligned}D_{\rho,\tau_\rho}&=\frac{1}{\tau_\rho!}G^{(\tau_\rho)}(z_\rho)=\frac{1}{\tau_\rho!}[W(z)+A(z)F(z)]^{(\tau_\rho)}\Big|_{z=z_\rho}\\&=\frac{1}{\tau_\rho!}[W^{(\tau_\rho)}(z_\rho)+A^{(\tau_\rho)}(z_\rho)F(z_\rho)]=\frac{1}{\tau_\rho!}[W^{(\tau_\rho)}(z_\rho)+A^{(\tau_\rho)}(z_\rho)F_{z_\rho}(z_\rho)]\end{aligned}$$

$$=\frac{1}{\tau_\rho!}[W(z)+A(z)F_{z_\rho}(z)]^{(\tau_\rho)}\Big|_{z=z_\rho}=\frac{1}{\tau_\rho!}\frac{\mathrm{d}^{\tau_\rho}}{\mathrm{d}z^{\tau_\rho}}G_{z_\rho}(z)\Big|_{z=z_\rho},$$

as needed. ■

Note that for the sake of definiteness in Theorem 12, we restrict the one-step extension of Δ with respect to the date $\check{\Delta}_\rho$ associate with z_ρ and $\boldsymbol{D}_{\rho,\tau_\rho}$. On the basis of the proof of Theorem 12, we see that the assertion formulated holds true as well when the date $\check{\Delta}_\rho$ are replaced by some $\check{\Delta}_s$ for $s=1,2,\cdots,\rho-1$. This also applies to Corollary 8 below.

The development carried out in the preceding corollary for the maximum determinant of $\boldsymbol{T}_{\check{c}}$ may be adapted to the maximum determinant of $\boldsymbol{T}_{\check{c}}$, *where* $\check{\boldsymbol{c}}$ takes the values in the set

$\check{\mathcal{B}}_\Delta:=\{\check{\boldsymbol{c}}:\check{\boldsymbol{c}}$ is the block Toeplitz vector of Problem $(\mathrm{NP})_{\check{\Delta}}$ such that $\boldsymbol{T}_{\check{c}}\geqslant\boldsymbol{O}\}$.

Corollary 8 *Let* $m=\sum_{i=0}^{\rho}\tau_i\geqslant 2$, *and let* $\mathbf{c}=(\mathbf{C}_0,\mathbf{C}_1,\cdots,\mathbf{C}_{m-1})$ *be the block Toeplitz vector of Problem* $(NP)_\Delta$ *such that* $\mathbf{T}_c=(\mathbf{C}_{j-1})_{i,j=0}^{m-1}>\mathbf{O}$. *Furthermore, let* $\mathbf{T}_{\check{c}}$ *be the information matrix of Problem* $(C)_{\check{c}}$. *Then*

$$0<\max\{\det \boldsymbol{T}_{\check{c}}:\check{c}\in\check{\mathcal{B}}_\Delta\}=2^{-q}\det \boldsymbol{T}_c\cdot\det\check{E},$$

where $\check{E}$ *is as in* (41), *and this maximum value is attained if and only if*

$$\check{\boldsymbol{c}}=\check{\boldsymbol{c}}_{\rho,z_\rho},$$

where $\check{\mathbf{c}}_{\rho,z_\rho}$ *is the block Toeplitz vector of Problem* $(NP)_{\check{\Delta}_{\rho,z_\rho}}$, *in which* $\check{\Delta}_{\rho,z_\rho}$ *is defined by* (25) *with* $s=\rho$ *and* $w=z_\rho$, *generated by the* $\mathbf{C}^{q\times q}$-*valued function* $\check{F}_{\rho,z_\rho}$ *defined by* (19″) *with* $s=\rho$ *and* $w=z_\rho$, *i. e.*,

$$\check{\boldsymbol{c}}_{\rho,z_\rho}=\left(\check{F}_{\rho,z_\rho}(0),\frac{1}{2}\check{F}'_{\rho,z_\rho}(0),\cdots,\frac{1}{2(m!)}\check{F}^{(m)}_{\rho,z_\rho}(0)\right).$$

Acknowledgements

The authors acknowledge with many thanks helpful comments and suggestions of the referee.

References

[1] N I Akhiezer. The Classical Moment Problem and Some Related Questions in

Analysis . Oliver and Boyd ,London,1965.

[2] D Z Arov,H Dym . J-Contractive Matrix Valued Functions and Related Topics. Cambridge Universify Press,New York,2008.

[3] D Z Arov,M G Krein. Problems of the search of minimum of enropy in indeterminate extension problems. Funct. Anal. Appl. ,1981,15:123—126.

[4] D Z Arov,M G Krein . On computations of entropy functionals and their minimums in indeterminate continuation problems. Acta Sci. Math. (Szeged), 1983,45:33—50(in Russian).

[5]J A Ball,I Gohberg,L Rodman. Interpolation of Rational Matrix Functions. Operator Theory:Advances and Applications ,Vol. 45,Birkhäuser,Basel,1990.

[6] V Bolotnikov, H Dym. On degenerate interpolation, entropy and extremal problem for matrix Schur functions. Integral Equations Operator Theory, 1998,32:367—435.

[7] G N Chen, Y J Hu. On the multiple Nevanlinna-Pick matrix intrepolation in the calss C_p and the Carathéodory matrix coefficient problem. Linear Aigebra Appl. 1998,283:179—203.

[8] G N Chen, Y J Hu. The Nevanlinna-Pick Interpolation and the Matrix Moment Problem. Beijing Normal University Press,Beiijing,2003(in Chinese).

[9] P Delsarte, Y Genin, Y Kamp. The Nevanlinna-Pick problem for matrix-valued functions. SIAM J. Appl. Math. 1979,36:47—61.

[10]V K Dubovoj,B Fritzsche,B Kirstein. Matricial Version of the Classical Schur Problem. Teubner-Texte Mathematik,Vol. 129,Teubner ,Leipzig,1992.

[11]H Dym. On Hermitian block Hankel matrices,matrix polynomials,the Hamburger moment problem,interpolation and maximum entropy. Integral Equations Operator Theory,1989,12:757—812.

[12] H Dym. J-Contractive matrix functions ,Reproducing Kernel Hilbert Spaces and Interpolation. CBMS Regional Conference Series in Mathmatics,Vol. 71, American Mathematical Society,Providence,RI,1989.

[13] H Dym,More on maximum entropy interpolants and maximum determinant completions of associated Pick matrices. Integral Equations Operator Theory, 1996,24:188—229.

[14] H Dym, I Gohberg. On maximum entropy interpolants and maximum determinant completions of associated Pick matrices. Integral Equations Operator Theory, 1995, 23: 61—88.

[15] B Fritzsche, S Fuchs, B Kirstein. An inverse entropy optimization problem for matrix -valued Carathéodory functions. Optimization, 1994, 29: 1—32.

[16] B Fritzsche, B Kirstein. An extension problem for nonnegative Hermitian block Toeplitz matrices. part Ⅰ: Math. Nachr. 1987, 130: 121—135; Part Ⅱ: Math. Nachr., 1987, 131: 287—297; Part Ⅲ: Math. Nachr. 1988: 135, 319—341; Part Ⅳ: Math, Nachr., 1989, 143: 329—354; Part Ⅴ: Math, Nachr., 1989, 144: 283—308.

[17]B Fritzsceh, B Kirstein. On the largest minorants associated with a matrix - valued Carathéodory function and spectral factorization. Z. Anal. Anwendungen, 1993, 12: 471—490.

[18] B Fritzsche, B Kirstein, A Lasarow. The matricial Carathéodory problem in both nondegenerate and degenerate cases. In: D Alpay, I Gohberg (Eds.), Interpolation, Schur Functions and Moment Problems, Oper. Theory Adv. Appl., Vol. 165, Birkhäuser. Basel, 2006: 251—290.

[19] B Fritzsche, B Kirstein, A Lasarow. On a class of extremal sulutions of the nondegenerate matricial Carathéodory problem. Analysis (Munich), 2007, 27: 109—164.

[20]I V Kovalishina. Analytic theory of a class of interpolation problems. Izv. Akad. Nauk SSSR Ser. Mat., 1983, 47: 455—497(in Russian).

[21] A Lasarow. On rank invariance of generalized Schwarz-Pick-Potapov block matrices of matrix-valued Carathéodory functions. Linear Algebra Appl., 2006, 413: 36—58.

[22] A Lasarow. Solution of a multiple Nevanlinna-Pick problem for Carathéodory functions via orthogonal rational functions in the matrix case. Math. Nachr., 2008, 281: 75—108.

[23] L A Sakhnovich. Interpolation Theory and its Applications. Mathematics and its Applications, Vol. 428, Kluwer, Dordrecht, 1997.

[24]E C Titchmarsh. The Theory of Functions, second ed., Oxford University Press, London, 1960.

Linear and Multilinear Algebra
2012,60(11－12):1 271－1 294.

对 Carathéodory 矩阵函数的矩阵 Carathéodory 问题与 Nevanlinna-Pick 型插值在非退化情形的极端解①

Extremal Solutions of Matricial Carathéodory Problem and Nevanlinna-Pick-Type Interpolation for Carathéodory Matrix Functions in the Nondegenerate Case

Abstract This article is devoted to the investigation of remarkable properties of extremal solutions $F^{[U]}$ and $G^{[U]}$ with a $q\times q$ unitary parameter U and their distinguished one-parameter families F_w^h and G_w^h with $|w|=1$ within the solution sets of Problem (C) (the matricial Carathéodory problem) and Problem (NP) (a certain Nevanlinna-Pick-type interpolation for Carathéodory matrix functions), respectively, in the nondegenerate case. We show connections between the poles of the functions $F^{[U]}$ and F_w^h at the unit circle of the complex plane and their Riesz-Herglotz measures, and discuss closely related matters for the extremal solutions $F^{[U]}$ and F_w^h. These results further serve as a starting

① Received 8 April 2011; final version received 13 October 2011.
Dedicated to the memory of Professor Ky Fan.
本文与胡永建合作.

point to look for corresponding properties of the functions $G^{[U]}$ and G_w^h based on the so-called modified block Toeplitz vector approach.

Keywords Nevanlinna-Pick-type interpolation; matricial Carathéodory problem; matrix-valued Carathéodory function; Riesz-Herglotz measure; extremal solution

§ 1. Introduction

In this article we study the extremal solutions $F^{[U]}(z)$ and $G^{[U]}(z)$ with a parameter $\boldsymbol{U} \in \mathfrak{U}_q$ (the set of $q \times q$ unitary matrices) and their related families $F_w^h(z)$ and $G_w^h(z)$ with a complex parameter w belonging to the unit circle T within the solution sets of two distinct interpolation problems for $q \times q$ Carathéodory functions in the nondegenerate case. The first problem is the matricial Carathéodory problem (see Problem (C) in Section 2) which, in essence, is nothing other than a certain truncated trigonometric matrix moment problem (see the problem (2.3)). The second problem is a certain matricial Nevanlinna-Pick-type interpolation possibly with zero as one of multiple nodes (see Problem (NP) in Section 3.1). In the nondegenerate case of Problem (C), i. e. in the particular situation that a specific block matrix $\boldsymbol{T}_c$ defined by (2.4) is positive Hermitian, it is well known that (see, e. g. [2,3,5~9,12,13,15~17], [14, Parts IV and V], [20,23]) Problem (C) has infinitely many (in a strong sense) solutions, which can be described in terms of a linear fractional transformation, the coefficient matrix of which is constructed with the aid of orthogonal $\mathbf{C}^{q \times q}$-valued polynomials (or in a different way). In a similar manner, the nondegenerate case of Problem (NP) means the particular situation that a specific block matrix $\boldsymbol{P}_\Delta$ defined by (3.3)~(3.4) is positive Hermitian. Then Problem (NP) also has infinitely many (in a strong sense) solutions (see, e. g. [1,3,5,6,12,13,21] and the references cited there).

Let m be the number of interpolation nodes (accounting for multiplicities) of Problem (NP). It turns out that, by using the so-called modified block Toeplitz vector approach (see [6, Chapter 9], and our brief sketch at Section 3.1), for Problem (NP) there is an m-tuple $\boldsymbol{c}=(\boldsymbol{C}_0, \boldsymbol{C}_1, \cdots, \boldsymbol{C}_{m-1})$ of $q\times q$ matrices with $\boldsymbol{C}_0=\boldsymbol{C}_0^*$, dependent only on the given data in Problem (NP), called the block Toeplitz vector of Problem (NP), and also that the solution of Problem (NP) can be reduced to what amounts to the solution of Problem (C) associated with the block Toeplitz vector $\boldsymbol{c}=(\boldsymbol{C}_0, \boldsymbol{C}_1, \cdots, \boldsymbol{C}_{m-1})$ of Problem (NP), or equivalently, the solution of the truncated trigonometric matrix moment problem (2.3) with $\boldsymbol{C}_0, \boldsymbol{C}_1, \cdots, \boldsymbol{C}_{m-1}$ as matrix moments. In particular, in the nondegenerate case if σ is the Riesz—Herglotz measure of $G^{[U]}$ and if μ is the Riesz—Herglotz measure of the corresponding $F^{[U]}$, then we have

$$\sigma(\mathrm{d}u) = A(u)\mu(\mathrm{d}u), \quad \forall u \in T, \tag{1.1}$$

where $A(z)$ is a certain scalar quasi-polynomial, defined by (3.5), such that $A(u)>0$, $\forall u\in T$.

In order to clarify the issues under consideration, in Section 2 we shall begin naturally with the first and major theme of this article: it is required to give intensive connections of the poles of the functions $F^{[U]}$ with $\boldsymbol{U}\in \mathfrak{U}_q$ and their Riesz-Herglotz measures. Here we are guided and motivated by the work of Duran and Lopez-Rodriguez [10,11] on N-extremal matrices of measures for an indeterminate matrix moment problem, and of Fritzsche et al. [16] on the extremal solutions of (nondegenerate) Problem (C).

For each $\boldsymbol{U}\in \mathfrak{U}_q$ the function $F^{[U]}$ is by definition the restriction of a certain rational matrix-valued function to the unit disk D of the complex plane among the solutions of (nondegenerate) Problem (C), or rather, for each $z\in D$,

$$F^{[U]}(z) := H_{\#}^{[U]}(z)(H^{[U]}(z))^{-1}, \tag{1.2}$$

where $H^{[U]}$ and $H_{\#}^{[U]}$ are $q\times q$ matrix polynomials of degree m, and their leading coefficient matrices are nonsingular (see (2.11)). As well, det $H^{[U]}$ and det $H_{\#}^{[U]}$ are the scalar polynomials of degree mq. This is to show that the poles of such $F^{[U]}$ are exactly the zeros of det $H^{[U]}$, all of which are located on T, and that $t_0 \in T$ is a zero of det $H^{[U]}$ of multiplicity l if and only if rank $H^{[U]}(t_0)=q-l$, and in that case, $1 \leqslant l \leqslant q$ (see (b) of Lemma 2.2). In the language of matrix polynomials with nonsingular leading coefficient matrices, the last result can be reformulated such that the $q \times q$ matrix polynomials $H^{[U]}$ of degree m have a set of l Jordan chains at $t_0 \in T$, each of which is of unit length, and thus dim Ker $H^{[U]}(t_0)=l$. Furthermore, for any $\boldsymbol{U} \in \mathfrak{U}_q$ the fact that $t_0 \in T$ is a zero of det $H^{[U]}$ of multiplicity l is also equivalent to that the rank of the matrix mass $\mu_U(\{t_0\})$ of the Riesz-Herglotz measure μ_U of $F^{[U]}$ at $t_0 \in T$ is equal to l (Theorem 2.3). In this case, for any $\boldsymbol{x}$, $\boldsymbol{y} \in$ range $\mu_U(\{t_0\})$,

$$\begin{aligned} &\boldsymbol{y}^* \mu_U(\{t_0\}) \Big(\sum_{k=0}^{m-1} (P_k(t_0))^* P_k(t_0) \Big) \boldsymbol{x} \\ &= \boldsymbol{y}^* \Big(\sum_{k=0}^{m-1} (P_k(t_0))^* P_k(t_0) \Big) \mu_U(\{t_0\}) \boldsymbol{x} = \boldsymbol{y}^* \boldsymbol{x}, \end{aligned} \tag{1.3}$$

where each P_k is a certain $\mathbf{C}^{q \times q}$-valued matrix polynomial, of exact degree k and of a nonsingular leading coefficient matrix, satisfying orthonormality with respect to the sequence $(\boldsymbol{C}_0, \boldsymbol{C}_1, \cdots, \boldsymbol{C}_{m-1})$ (see (2.5)), and thus $\sum_{k=0}^{m-1} (P_k(t_0))^* P_k(t_0)$ is positive Hermitian (see [7,8] and [9, Section 3.6] for more detail). In particular, if $\mu_U(\{t_0\}) \in \mathbf{C}^{q \times q}$ is positive Hermitian, then from (1.3) it follows that (Theorem 2.6)

$$\mu_U(\{t_0\}) = \Big(\sum_{k=0}^{m-1} (P_k(t_0))^* P_k(t_0) \Big)^{-1}.$$

Thus if $\boldsymbol{U} \in \mathfrak{U}_q$ and if $t_1, t_2, \cdots, t_r \in T$ are all pairwise different zeros of det $H^{[U]}$ with respective multiplicities $l_1, l_2, \cdots, l_r$, $\sum_{j=1}^{r} l_j = mq$, then the ranks of matrix masses $\mu_U(\{t_1\})$, $\mu_U(\{t_2\})$, $\cdots$, $\mu_U(\{t_r\})$ of the Riesz-Herglotz measure μ_U of $F^{[U]}$ at $t_1, t_2, \cdots, t_r$ are equal to $l_1, l_2, \cdots, l_r$, respectively. In that case, $m \leqslant r \leqslant mq$, and $F^{[U]}$ admits the simple fraction decomposition, for each $z \in D$,

$$F^{[U]}(z) = \sum_{j=1}^{r} \frac{t_j + z}{t_j - z} \mu_U(\{t_j\}),$$

where for $j=1, 2, \cdots, r$

$$\operatorname{rank} \mu_U(\{t_j\}) = \dim \operatorname{Ker} H^{[U]}(t_j) = l_j,$$

and (1.3) holds with t_0 replaced by any $t_j (1 \leqslant j \leqslant r)$ (Theorem 2.8). Moreover, the statement of Theorem 2.8 can be viewed as an intensified result of Proposition 9.8(a) of [16].

Among the extremal solutions $F^{[U]}$ for any $\boldsymbol{U} \in \mathfrak{U}_q$, there exists a remarkable one-parameter family F_w^h for any $w \in T$. Now let $w \in T$. Then there exists only a $\boldsymbol{U} \in \mathfrak{U}_q$ denoted by $\boldsymbol{U}_w$ such that $H^{[U_w]}(w) = \boldsymbol{O}$, the $q \times q$ null matrix. The corresponding $F^{[U_w]}$, which is as in (1.2) with $\boldsymbol{U}$ replaced by $\boldsymbol{U}_w$, then has at least w as one pole of multiplicity q, that is, $w \in T$ is a zero of det $H^{[U_w]}$ of multiplicity q such that

$$\mu_U(\{w\}) = \Big(\sum_{k=0}^{m-1} (P_k(w))^* P_k(w)\Big)^{-1}.$$

Such $F^{[U_w]}$, denoted by F_w^h with $w \in T$, consist of a distinguished subclass of the class of $F^{[U]}$ with any $\boldsymbol{U} \in \mathfrak{U}_q$. Indeed, for each $w \in T$, the function F_w^h has some interesting properties in the solution set $\mathbf{C}_q[D, \boldsymbol{c}]$ of (nondegenerate) Problem (C). We only note that its Riesz-Herglotz measure μ_w reaches uniquely at the point w the maximum matrix mass $\mu(\{w\})$ when μ varies over the Riesz-Herglotz measures of all solutions in $\mathbf{C}_q[D, \boldsymbol{c}]$, more concretely,

$$\mu(\{w\}) \leqslant \mu_w(\{w\}) = \Big(\sum_{k=0}^{m-1} (P_k(w))^* P_k(w)\Big)^{-1} \tag{1.4}$$

(Theorem 2.6), and F_w^h is the only solution of (nondegenerate) Problem (C), the Riesz-Herglotz measure μ_w of which has the maximum matrix mass at the point w (see [22, Theorem 3.2] or Theorem 2.7 of this article). In particular, for the scalar case of $q=1$ (and thus F_w^h has exactly m different simple poles), we note that the last result is an interesting and especially important assertion of (nondegenerate) Problem (C): taking any $w \in T$ there exists one and only one solution in $\mathbf{C}_1[D, \mathbf{c}]$ of which the Riesz-Herglotz measure has a concentrated mass at the point w equal to the maximum mass. This solution is F_w^h, the maximum mass at $w \in T$

is $1/\sum_{k=0}^{m-1} | P_k(w) |^2$ (see [18, Theorem 20.1] or [16, Theorem 9.18]).

In Section 3, we study the Riesz-Herglotz measures of the extremal solutions $G^{[U]}(z)$ with any $\boldsymbol{U} \in \mathfrak{U}_q$ of (nondegenerate) Problem (NP) and the subclass $G_w^h(z)$ with any $w \in T$, the second main theme of this article. Based on the modified block Toeplitz vector approach (Section 3.1), we obtain readily all the corresponding results given in Section 2 for $F^{[U]}$ and F_w^h in the case of (nondegenerate) Problem (NP). As far as we know, these observations have not appeared in the other papers on the matrix interpolation problem of Nevanlinna-Pick-type.

Throughout this article, let $\mathbf{C}$ be the set of all complex numbers, let q be a fixed positive integer and let $\mathbf{C}^{q\times q}$ denote the set of all complex $q\times q$ matrices. Let D and T denote the unit open disc of $\mathbf{C}$ and the unit circle: $D := \{z\in\mathbf{C}: |z|<1\}$, $T=\{z\in\mathbf{C}: |z|=1\}$. For each $\boldsymbol{A}\in\mathbf{C}^{q\times q}$, det $\boldsymbol{A}$ stands for the determinant of $\boldsymbol{A}$, Re $\boldsymbol{A}$ and Im $\boldsymbol{A}$ are the real part and imaginary parts of $\boldsymbol{A}$, respectively, i.e. $\mathrm{Re}\,\boldsymbol{A} := \frac{1}{2}(\boldsymbol{A}+\boldsymbol{A}^*)$ and $\mathrm{Im}\,\boldsymbol{A} := \frac{1}{2\mathrm{i}}(\boldsymbol{A}-\boldsymbol{A}^*)$. The null matrix will be denoted by $\boldsymbol{O}$, in general its size is obvious. The notation $\boldsymbol{I}_q$ stands for the identity matrix of order q. We shall write $\boldsymbol{A}\geqslant\boldsymbol{B}$ or $\boldsymbol{B}\leqslant\boldsymbol{A}$ [resp., $\boldsymbol{A}>\boldsymbol{B}$ or $\boldsymbol{B}<\boldsymbol{A}$] to indicate that $\boldsymbol{A}$ and $\boldsymbol{B}$ are Hermitian matrices of the same size such that $\boldsymbol{A}-\boldsymbol{B}$ is a nonnegative [resp., positive] Hermitian. Note that sometimes we use the notation $f(z)$ to indicate the function (or matrix-valued function) f of the variable z in $\mathbf{C}$. Moreover, we shall carry out our analysis here in terms of formalism used in [16] as far as possible.

§ 2. The Riesz-Herglotz Measures of Extremal Solutions of Nondegenerate Problem (C)

Recall that a $\mathbf{C}^{q\times q}$-valued holomorphic function Ω defined in D is called a $q\times q$ *Carathéodory function* in D if the real part $\mathrm{Re}\,\Omega(z)$ is nonnegative Hermitian for each $z \in D$. The set of all $q \times q$ Carathéodory functions in D is denoted by $\mathbf{C}_q(D)$. Each $\Omega \in \mathbf{C}_q(D)$ permits an integral representation of the form

$$\Omega(z) = \int_T \frac{u+z}{u-z} \mu(\mathrm{d}u) + \mathrm{i} \operatorname{Im} \Omega(0), \quad z \in D, \tag{2.1}$$

where $\operatorname{Im} \Omega(0)$ stands for the imaginary part of $\Omega(0)$, and μ is a nonnegative Hermitian $q \times q$ Borel measure on T, which is uniquely determined by Ω, and is called the *Riesz-Herglotz measure* of Ω.

The matricial Carathéodory problem under consideration consists of the following interpolation problem for the class $\mathbf{C}_q(D)$:

Problem (C) Let m be a given positive integer, and let $\mathbf{c} := (\boldsymbol{C}_j)_{j=0}^{m-1}$ be a sequence of $\mathbf{C}^{q \times q}$ with $\boldsymbol{C}_0 = \boldsymbol{C}_0^*$. Describe the set $\mathbf{C}_q[D, \boldsymbol{c}]$ of $q \times q$ Carathéodory functions F such that

$$F(0) = \boldsymbol{C}_0, \quad \frac{1}{k!} F^{(k)}(0) = 2\boldsymbol{C}_k, \quad k = 1, 2, \cdots, m-1 \tag{2.2}$$

hold, where $F^{(k)}(0)$ stands for the k-th derivative of F at the point $z=0$.

Problem (C), in essence, is nothing other than a certain truncated trigonometric matrix moment problem, which is required to find the nonnegative Hermitian Borel measures μ on T such that

$$\boldsymbol{C}_k = \int_T u^{-k} \mu(\mathrm{d}u), \ k = 0, 1, \cdots, m-1. \tag{2.3}$$

In the nondegenerate case which is connected with

$$\boldsymbol{T}_{\mathbf{c}} := (\boldsymbol{C}_{j-i})_{i,j=0}^{m-1} > \boldsymbol{O}, \tag{2.4}$$

where $\boldsymbol{C}_{-k} = \boldsymbol{C}_k^*$ for $k \geqslant 0$, Problem (C) (and therefore the problem (2.3)) has infinitely many (in a strong sense) solutions.

Note that for our purpose we always assume in Problem (C) that $\boldsymbol{C}_0 = \boldsymbol{C}_0^*$. Replacing $\boldsymbol{C}_0$ *by* $\operatorname{Re} \boldsymbol{C}_0$, we can reduce the general $\boldsymbol{C}_0$ to this normalized case with little effort.

The main objective of this section is to introduce and to study a class of extremal solutions of Problem (C) and its subclass in the nondegenerate case. For each $\boldsymbol{U} \in \mathfrak{U}_q$, we shall indicate the connection between the poles of such extremal solutions $F^{[U]}$ and their Riesz-Herglotz measures μ_U, and give a constructive representation of $F^{[U]}$.

§ 2.1 The Extremal Solutions to Problem (C) in the Nondegenerate Case

We begin with a description of the solution set $\mathbf{C}_q[D, \boldsymbol{c}]$ of nondegenerate Problem (C) in terms of orthogonal matrix polynomials with respect to a nonnegative Hermitian Borel measure on the unit circle T in a somewhat condensed form. For details, see, e. g. [7~9].

In the following, we always assume that the block vector $\boldsymbol{c}=(\boldsymbol{C}_0, \boldsymbol{C}_1, \cdots, \boldsymbol{C}_{m-1})$ is such that $\boldsymbol{T}_c=(\boldsymbol{C}_{j-i})_{i,j=0}^{m-1}>\boldsymbol{O}$, where $\boldsymbol{C}_{-k}=\boldsymbol{C}_k^*$ *for* $k\geqslant 0$. Starting from such $\boldsymbol{T}_c$, we derive that there exist the so-called *canonical Szegö pair* $[(P_k)_{k=0}^{m-1}, (Q_k)_{k=0}^{m-1}]$ corresponding to $\boldsymbol{c}=(\boldsymbol{C}_0, \boldsymbol{C}_1, \cdots, \boldsymbol{C}_{m-1})$ and the *dual canonical Szegö pair* $[(P_k^{\#})_{k=0}^{m-1}, (Q_k^{\#})_{k=0}^{m-1}]$ corresponding to $\boldsymbol{c}=(\boldsymbol{C}_0, \boldsymbol{C}_1, \cdots, \boldsymbol{C}_{m-1})$ such that each of the $q\times q$ matrix polynomials P_k, Q_k and $P_k^{\#}$, $Q_k^{\#}$ is of degree k and its leading coefficient matrix is nonsingular. Here $[(P_k)_{k=0}^{m-1}, (Q_k)_{k=0}^{m-1}]$ are determined by $\boldsymbol{c}=(\boldsymbol{C}_0, \boldsymbol{C}_1, \cdots, \boldsymbol{C}_{m-1})$ and form a pair of orthonormal systems corresponding to $\boldsymbol{C}_0, \boldsymbol{C}_1, \cdots, \boldsymbol{C}_{m-1}$, in view of the following orthogonal and normalized properties:

$$\int_T P_k(z)\mu(\mathrm{d}z)(P_j(z))^* = \delta_{kj}\boldsymbol{I}_q, \quad k,j=0,1,\cdots,m-1, \tag{2.5}$$

$$\int_T (Q_k(z))^*\mu(\mathrm{d}z)Q_j(z) = \delta_{kj}\boldsymbol{I}_q, \quad k,j=0,1,\cdots,m-1, \tag{2.6}$$

where μ is a solution to the problem (2.3) and δ_{kj} is the Kronecker symbol. Similar results hold as well for the dual canonical Szegö pair $[(P_k^{\#})_{k=0}^{m-1}, (Q_k^{\#})_{k=0}^{m-1}]$ corresponding to $\boldsymbol{c}=(\boldsymbol{C}_0, \boldsymbol{C}_1, \cdots, \boldsymbol{C}_{m-1})$, but with μ in (2.5) and (2.6) replaced by the reciprocal measure $\mu^{\#}$ corresponding to μ (see, e. g. [9, Section 3.6]). More explicitly, let $F(z):=\int_T \frac{u+z}{u-z}\mu(\mathrm{d}u)$ belong to $\mathbf{C}_q[D, \boldsymbol{c}]$. Then, since $\boldsymbol{C}_0=\boldsymbol{C}_0^*$ is $q\times q$ nonsingular, we derive [9, Proposition 3.6.8] that for all $z\in D$, $\det F(z)\neq 0$, and F^{-1} also belongs to $\mathbf{C}_q[D]$, which has $\mu^{\#}$, the reciprocal measure corresponding to μ, as the Riesz-Herglotz measure:

$$F^{-1}(z) = \int_T \frac{u+z}{u-z}\mu^{\#}(\mathrm{d}u), \quad z \in D.$$

We also need the notation of the *reciprocal matrix polynomial*. Let $B(z) = \sum_{i=0}^{s} \boldsymbol{B}_i z^i \ (\boldsymbol{B}_s \neq \boldsymbol{O})$, where $\boldsymbol{B}_i \in \mathbf{C}^{q\times q}$, $\forall i$. Then the reciprocal matrix polynomial $\widetilde{B}^{[s]}(z)$ of $B(z)$ with respect to T and the formal degree s is defined by

$$\widetilde{B}^{[s]}(z) = z^s\left(B\left(\frac{1}{\bar{z}}\right)\right)^* = \sum_{i=0}^{s} \boldsymbol{B}_{s-i}^* z^i.$$

Now let $\mathfrak{A}_{m-1}$ be the matrix polynomial defined, for each $z \in \mathbf{C}$, by

$$\mathfrak{A}_{m-1}(z) := \frac{1}{\sqrt{2}}\begin{bmatrix} -z\boldsymbol{Q}_{m-1}^{\#}(z) & (\widetilde{P_{m-1}^{\#}})^{[m-1]}(z) \\ -z\boldsymbol{Q}_{m-1}(z) & \widetilde{P}_{m-1}^{[m-1]}(z) \end{bmatrix}. \tag{2.7}$$

Then we have (see, e. g. [16, Lemma 6.4])

$$\begin{aligned} \mathfrak{A}_{m-1}^*(z)\boldsymbol{J}_q\mathfrak{A}_{m-1}(z) &\leqslant j_{qq}, \quad \forall z \in D, \\ \mathfrak{A}_{m-1}^*(z)\boldsymbol{J}_q\mathfrak{A}_{m-1}(z) &= j_{qq}, \quad \forall z \in T, \end{aligned} \tag{2.8}$$

where

$$\boldsymbol{J}_q := \begin{bmatrix} \boldsymbol{O} & -\boldsymbol{I}_q \\ -\boldsymbol{I}_q & \boldsymbol{O} \end{bmatrix} \quad \text{and} \quad j_{qq} := \begin{bmatrix} \boldsymbol{I}_q & \boldsymbol{O} \\ \boldsymbol{O} & -\boldsymbol{I}_q \end{bmatrix}.$$

Making use of these preliminaries, we recall a parameterization of the solutions to Problem (C) in the nondegenerate case. Here the notation $\mathbf{S}_q(D)$ stands for the class of all $\mathbf{C}^{q\times q}$-valued functions holomorphic and contractive in D.

Theorem 2.1 [14, Theorem 2.6 in Part IV] *Let* $\mathbf{c} = (\boldsymbol{C}_0, \boldsymbol{C}_1, \cdots, \boldsymbol{C}_{m-1})$ *be as in Problem* (C) *such that* $\boldsymbol{T}_{\mathbf{c}} = (\boldsymbol{C}_{j-i})_{i,j=0}^{m-1} > \boldsymbol{O}$. *Let* $[(P_k)_{k=0}^{m-1}, (Q_k)_{k=0}^{m-1}]$ *and* $[(P_k^{\#})_{k=0}^{m-1}, (Q_k^{\#})_{k=0}^{m-1}]$ *be the canonical Szegö pair corresponding to* $\mathbf{c} = (\boldsymbol{C}_0, \boldsymbol{C}_1, \cdots, \boldsymbol{C}_{m-1})$ *and the dual canonical Szegö pair corresponding to* $\mathbf{c} = (\boldsymbol{C}_0, \boldsymbol{C}_1, \cdots, \boldsymbol{C}_{m-1})$, *respectively. Then for each* $f \in \mathbf{S}_q(D)$ *and each* $z \in D$, $\widetilde{P}_{m-1}^{[m-1]}(z) + zQ_{m-1}(z)f(z)$ *is a nonsingular matrix. Moreover, the formula*

$$\begin{aligned} F(z) &= \int_T \frac{u+z}{u-z}\mu(\mathrm{d}u) \\ &= [(\widetilde{P}_{m-1}^{\#})^{[m-1]}(z) - zQ_{m-1}^{\#}(z)g(z)][\widetilde{P}_{m-1}^{[m-1]}(z) + zQ_{m-1}(z)g(z)]^{-1} \end{aligned} \tag{2.9}$$

establishes a one-to-one correspondence between $\mathbf{C}_q[D, \mathbf{c}]$[*the set* $\mathbf{M}_c$ *of solutions* μ *of the problem* (2.3), *resp.*] *and all functions* g *of the class* $\mathbf{S}_q(D)$.

The particular solutions of interest here to Problem (C) in the nondegenerate case are those F given in (2.9) for which the free parameters $g \in \mathbf{S}_q(D)$ therein are chosen to be $q \times q$ unitary matrices, that is, if $\boldsymbol{U} \in \mathfrak{U}_q$, then

$$F^{[U]}(z) := H_{\#}^{[U]}(z) H^{[U]}(z)^{-1}, \ z \in D, \tag{2.10}$$

where

$$\begin{aligned} H^{[U]}(z) &:= \widetilde{P}_{m-1}^{[m-1]}(z) + z Q_{m-1}(z) \boldsymbol{U}, \\ H_{\#}^{[U]}(z) &:= (\widetilde{P}_{m-1}^{\#})^{[m-1]}(z) - z Q_{m-1}^{\#}(z) \boldsymbol{U}. \end{aligned} \tag{2.11}$$

Note that both the $q \times q$ matrix polynomials $H^{[U]}$ and $H_{\#}^{[U]}$ for any $\boldsymbol{U} \in \mathfrak{U}_q$ are of degree m, and their leading coefficient matrices are nonsingular (since both Q_{m-1} and $Q_{m-1}^{\#}$ are of degree $m-1$, and their leading coefficient matrices are nonsingular). The class of extremal solutions to (nondegenerate) Problem (C) consists of such solutions $F^{[U]}$ with any $\boldsymbol{U} \in \mathfrak{U}_q$.

As usual, a point $t_0 \in \mathbf{C}$ is said to be a zero (sometimes, known as an eigenvalue) of $H^{[U]}$[$H_{\#}^{[U]}$, resp.] if it is a zero of the scalar polynomial $\det H^{[U]}(z)$ [$\det H_{\#}^{[U]}(z)$, resp.], and the multiplicity of t_0 as a zero of $H^{[U]}$ [$H_{\#}^{[U]}$, resp.] is the multiplicity of t_0 as a zero of $\det H^{[U]}(z)$ [$\det H_{\#}^{[U]}(z)$, resp.]. It is well known [16, Lemma 9.7] that all the zeros of $H^{[U]}$[$H_{\#}^{[U]}$, resp.] are located on T.

Lemma 2.2

(a) *If* $t_0 \in T$ *is a zero of* $H^{[U]}$ *and* $\boldsymbol{\beta} \in \mathbf{C}^q$ *is such that* $\boldsymbol{\beta}^* H^{[U]}(t_0) = \mathbf{0}$, *then*

$$H^{[U]}(t_0)(\widetilde{P}_{m-1}^{[m-1]}(t_0))^* \boldsymbol{\beta} = \mathbf{0}, \tag{2.12}$$

and

$$\begin{aligned} (H^{[U]}(t_0))'(\widetilde{P}_{m-1}^{[m-1]}(t_0))^* \boldsymbol{\beta} &= -\bar{t}_0 \Big(\sum_{k=0}^{m-1} Q_k(t_0)(Q_k(t_0))^* \Big) \boldsymbol{\beta}, \\ &= -\bar{t}_0 \Big(\sum_{k=0}^{m-1} (P_k(t_0))^* P_k(t_0) \Big) \boldsymbol{\beta}, \end{aligned} \tag{2.13}$$

where $(H^{[U]}(t_0))'$ *is the first derivative of* $H^{[U]}$ *at the point* $z=t_0$.

(b) $t_0 \in T$ *is a zero of* $H^{[U]}$ *of multiplicity* l *if and only if* $\operatorname{rank}(H^{[U]}(t_0))=q-l$, i. e. *if and only if* $\dim \operatorname{Ker}(H^{[U]}(t_0))=l$, *and in that case*, $1\leqslant l\leqslant q$.

(c) $t_0 \in T$ *is a zero of* $H_{\#}^{[U]}$ *of multiplicity* l *if and only if* $\operatorname{rank}(H_{\#}^{[U]}(t_0))=q-l$, i. e. *if and only if* $\dim \operatorname{Ker}(H_{\#}^{[U]}(t_0))=l$, *and in that case*, $1\leqslant l\leqslant q$.

Proof (a) Assume that for $t_0\in T$, $\det H^{[U]}(t_0)=0$ and $\boldsymbol{\beta}\in \mathbf{C}^q$ is such that $\boldsymbol{\beta}^* H^{[U]}(t_0)=\mathbf{0}$. Then from

$$\mathbf{0}=\boldsymbol{\beta}^* H^{[U]}(t_0)=\boldsymbol{\beta}^*[\widetilde{P}_{m-1}^{[m-1]}(t_0)+t_0 Q_{m-1}(t_0)\boldsymbol{U}]$$

we deduce that

$$\boldsymbol{\beta}^* \widetilde{P}_{m-1}^{[m-1]}(t_0)=-t_0\boldsymbol{\beta}^* Q_{m-1}(t_0)\boldsymbol{U}. \tag{2.14}$$

Thus, by using the following Christoffel-Darboux formula for $k=m-1$, $w=v=t_0$:

$$(1-w\bar{v})\sum_{j=0}^{k} Q_j(w)(Q_j(v))^* \\ =\widetilde{P}_k^{[k]}(w)(\widetilde{P}_k^{[k]}(v))^*-w\bar{v}\,Q_k(w)(Q_k(v))^*,\ \forall\, w,v\in \mathbf{C}, \tag{2.15}$$

we have

$$\begin{aligned} H^{[U]}(t_0)(\widetilde{P}_{m-1}^{[m-1]}(t_0))^*\boldsymbol{\beta} &= [\widetilde{P}_{m-1}^{[m-1]}(t_0)+t_0 Q_{m-1}(t_0)\boldsymbol{U}](\widetilde{P}_{m-1}^{[m-1]}(t_0))^*\boldsymbol{\beta} \\ &= [\widetilde{P}_{m-1}^{[m-1]}(t_0)(\widetilde{P}_{m-1}^{[m-1]}(t_0))^*-|t_0|^2 Q_{m-1}(t_0)\boldsymbol{U}\boldsymbol{U}^*(Q_{m-1}(t_0))^*]\boldsymbol{\beta} \\ &= [\widetilde{P}_{m-1}^{[m-1]}(t_0)(\widetilde{P}_{m-1}^{[m-1]}(t_0))^*-Q_{m-1}(t_0)(Q_{m-1}(t_0))^*]\boldsymbol{\beta}=\mathbf{0}. \end{aligned}$$

Furthermore, from (2.15) it follows by finding the first derivative in w that

$$-\bar{v}\sum_{j=0}^{k} Q_j(w)(Q_j(v))^*+(1-w\bar{v})\sum_{j=0}^{k}(Q_j(w))'(Q_j(v))^* \\ =(\widetilde{P}_k^{[k]}(w))'(\widetilde{P}_k^{[k]}(v))^*-\bar{v}\,Q_k(w)(Q_k(v))^*-w\bar{v}(Q_k(w))'(Q_k(v))^*.$$

Because of (2.14), in turn we have

$$\begin{aligned} &(H^{[U]}(t_0))'(\widetilde{P}_{m-1}^{[m-1]}(t_0))^*\boldsymbol{\beta} \\ &= [(\widetilde{P}_{m-1}^{[m-1]}(t_0))'+t_0(Q_{m-1}(t_0))'\boldsymbol{U}+Q_{m-1}(t_0)\boldsymbol{U}](\widetilde{P}_{m-1}^{[m-1]}(t_0))^*\boldsymbol{\beta} \\ &= (\widetilde{P}_{m-1}^{[m-1]}(t_0))'(\widetilde{P}_{m-1}^{[m-1]}(t_0))^*\boldsymbol{\beta}-|t_0|^2(Q_{m-1}(t_0))'(Q_{m-1}(t_0))^*\boldsymbol{\beta}- \\ &\quad \bar{t}_0 Q_{m-1}(t_0)(Q_{m-1}(t_0))^*\boldsymbol{\beta} \end{aligned}$$

$$=-\bar{t}_0\sum_{j=0}^{m-1}Q_j(t_0)(Q_j(t_0))^*\boldsymbol{\beta}.$$

On the other hand, because of $t_0\in T$, $Q_k(t_0)(Q_k(t_0))^*=(P_k(t_0))^*\cdot P_k(t_0)$ for $k=0,1,\cdots,m-1$ (see, e.g. [9, Lemma 3.6.16]), we infer that

$$-\bar{t}_0\Big(\sum_{j=0}^{m-1}Q_k(t_0)(Q_k(t_0))^*\Big)\boldsymbol{\beta}=-\bar{t}_0\Big(\sum_{j=0}^{m-1}(P_k(t_0))^*P_k(t_0)\Big)\boldsymbol{\beta},$$

and therefore (2.13) holds.

(b) First assume that $t_0\in T$ is a zero of $H^{[U]}$ of multiplicity l, that is, $(\det H^{[U]}(z))^{(j)}=0$, $j=0,1,\cdots,l-1$, $(\det H^{[U]}(z))^{(l)}(t_0)\neq 0$, where $(\det H^{[U]}(z))^{(j)}(t_0)$ is the j-th derivative of $\det H^{[U]}(z)$ at the point t_0. By the theory of matrix polynomials, it is well known (see, e.g. [19, Section 1.6]) that the multiplicity l of t_0 as a zero of $\det H^{[U]}$ coincides with the sum $\sum_{i=1}^{s}\mu_i$ of the lengths μ_i of Jordan chains in a canonical set corresponding to an eigenvalue t_0 of the matrix polynomial $H^{[U]}$ (i.e. $\operatorname{Ker} H^{[U]}(t_0)\neq\{\mathbf{0}\}$) of degree m with a certain nonsingular leading coefficient matrix, where s is the dimension of $\operatorname{Ker} H^{[U]}(t_0)$. We will prove that each number $\mu_i=1$. Suppose, on the contrary, that this is not true. Then, by the definition of the Jordan chain, there exist $\mathbf{0}\neq\boldsymbol{v}_0\in\mathbf{C}^q$ and $\boldsymbol{v}_1\in\mathbf{C}^q$ such that $\boldsymbol{v}_0^*H^{[U]}(t_0)=\mathbf{0}$ and $\boldsymbol{v}_1^*H^{[U]}(t_0)+\boldsymbol{v}_0^*(H^{[U]}(t_0))'=\mathbf{0}$. It follows from (2.12) and (2.13) that

$$\begin{aligned}0&=[\boldsymbol{v}_1^*H^{[U]}(t_0)+\boldsymbol{v}_0^*(H^{[U]}(t_0))'](\widetilde{P}_{m-1}^{[m-1]}(t_0))^*\boldsymbol{v}_0\\&=\boldsymbol{v}_1^*H^{[U]}(t_0)(\widetilde{P}_{m-1}^{[m-1]}(t_0))^*\boldsymbol{v}_0+\boldsymbol{v}_0^*(H^{[U]}(t_0))'(\widetilde{P}_{m-1}^{[m-1]}(t_0))^*\boldsymbol{v}_0\\&=\boldsymbol{v}_0^*(H^{[U]}(t_0))'(\widetilde{P}_{m-1}^{[m-1]}(t_0))^*\boldsymbol{v}_0\\&=-\bar{t}_0\boldsymbol{v}_0^*\sum_{k=0}^{m-1}Q_k(t_0)(Q_k(t_0))^*\boldsymbol{v}_0\neq 0,\end{aligned}$$

a contradiction. Thus, $\mu_i=1, i=1,2,\cdots,s$ and therefore $l=\mu_1+\mu_2+\cdots+\mu_s=s=\dim\operatorname{Ker} H^{[U]}(t_0)$, so that $1\leqslant l\leqslant q$ and $\operatorname{rank} H^{[U]}(t_0)=q-l$.

The converse is immediate. Indeed, from $\operatorname{rank} H^{[U]}(t_0)=q-l$ and $1\leqslant l\leqslant q$, we have that $\det H^{[U]}(t_0)=0$. Now assume that t_0 is a zero of $\det H^{[U]}(z)$ of multiplicity d. By the argument just given, $\operatorname{rank} H^{[U]}(t_0)=$

$q-d$, so that $l=d$.

(c) Observe that $[(P_k^{\#})_{k=0}^{m-1}, (Q_k^{\#})_{k=0}^{m-1}]$ may be viewed as a pair of orthonormal systems corresponding to the reciprocal measure $\mu^{\#}$ of μ, as indicated above. Thus, the statements (a) and (b) of this lemma are valid as well if $H^{[U]}$, $\widetilde{P}_{m-1}^{[m-1]}$, P_k and Q_k are replaced by $H_{\#}^{[U]}$, $(\widetilde{P}_{m-1}^{\#})^{[m-1]}$, $P_k^{\#}$ and $Q_k^{\#}$, respectively. ■

For the later analysis, we need the following three results:

$$(H^{[U]}(v))^* H_{\#}^{[U]}(v) + (H_{\#}^{[U]}(v))^* H^{[U]}(v) = \mathbf{0}, \quad \forall v \in T, \tag{2.16}$$

$$P_{m-1}(v) H_{\#}^{[U]}(v) + P_{m-1}^{\#}(v) H^{[U]}(v) = 2v^{m-1}\boldsymbol{I}_q, \quad \forall v \in \mathbf{C}, \tag{2.17}$$

$$(\widetilde{P}_{m-1}^{[m-1]}(v))^* H_{\#}^{[U]}(v) + ((\widetilde{P_{m-1}^{\#}})^{[m-1]}(v))^* H^{[U]}(v) = 2\boldsymbol{I}_q, \quad \forall v \in T. \tag{2.18}$$

Indeed, from (2.11) and the second formula of (2.8) it follows directly that (2.16) holds true for any $v \in T$. As for (2.17), it can be verified as follows:

$$\begin{aligned} & P_{m-1}(v)[(\widetilde{P_{m-1}^{\#}})^{[m-1]}(v) - v Q_{m-1}^{\#}(v)\boldsymbol{U}] + P_{m-1}^{\#}(v)[\widetilde{P}_{m-1}^{[m-1]}(v) + v Q_{m-1}(v)\boldsymbol{U}] \\ &\quad = P_{m-1}(v)(\widetilde{P_{m-1}^{\#}})^{[m-1]}(v) + P_{m-1}^{\#}(v)\widetilde{P}_{m-1}^{[m-1]}(v) - \\ &\quad\quad vP_{m-1}(v)Q_{m-1}^{\#}(v)\boldsymbol{U} + vP_{m-1}^{\#}(v)Q_{m-1}(v)\boldsymbol{U} \\ &\quad = 2v^{m-1}\boldsymbol{I}_q, \quad \forall v \in \mathbf{C}. \end{aligned}$$

Here we make use of the identities [8, Formulas (26) and (27)]

$$P_{m-1}^{\#}(v)Q_{m-1}(v) = P_{m-1}(v)Q_{m-1}^{\#}(v),$$

$$P_{m-1}(v)(\widetilde{P_{m-1}^{\#}})^{[m-1]}(v) + P_{m-1}^{\#}(v)\widetilde{P}_{m-1}^{[m-1]}(v) = 2v^{m-1}\boldsymbol{I}_q.$$

Finally, (2.18) is also easy to verify:

$$\begin{aligned} & (\widetilde{P}_{m-1}^{[m-1]}(v))^*[(\widetilde{P_{m-1}^{\#}})^{[m-1]}(v) - v Q_{m-1}^{\#}(v)\boldsymbol{U}] + \\ & ((\widetilde{P_{m-1}^{\#}})^{[m-1]}(v))^*[\widetilde{P}_{m-1}^{[m-1]}(v) + v Q_{m-1}(v)\boldsymbol{U}] \\ & = (\widetilde{P}_{m-1}^{[m-1]}(v))^*(\widetilde{P_{m-1}^{\#}})^{[m-1]}(v) + ((\widetilde{P_{m-1}^{\#}})^{[m-1]}(v))^*\widetilde{P}_{m-1}^{[m-1]}(v) - \\ & \quad v[(\widetilde{P}_{m-1}^{[m-1]}(v))^* Q_{m-1}^{\#}(v) - ((\widetilde{P_{m-1}^{\#}})^{[m-1]}(v))^* Q_{m-1}(v)]\boldsymbol{U} \\ & = 2\boldsymbol{I}_q, \quad \forall v \in T, \end{aligned}$$

with the aid of the second result given in (2.8) for $v \in T$.

For each $w \in T$, it is well known (see, e. g. [16, Lemma 6. 4]) that the matrix

$$\boldsymbol{U}_w := -\overline{w}((\widetilde{P}_{m-1}^{[m-1]}(w))^{-1}\boldsymbol{Q}_{m-1}(w))^* \tag{2.19}$$

is unitary. For $w \in T$, let

$$\begin{aligned} F_w^h(z) := &[(\widetilde{P_{m-1}^{\#}})^{[m-1]}(z) - z\boldsymbol{Q}_{m-1}^{\#}(z)\boldsymbol{U}_w] \times \\ &[\widetilde{P}_{m-1}^{[m-1]}(z) + z\boldsymbol{Q}_{m-1}(z)\boldsymbol{U}_w]^{-1}, \forall z \in D. \end{aligned} \tag{2.20}$$

Then F_w^h with any $w \in T$ belongs to $\mathbf{C}_q[D, \boldsymbol{c}]$ and is one of the extremal solutions of (nondegenerate) Problem (C). Thus, all the F_w^h with $w \in T$ form a subclass of the extremal solutions $F^{[U]}$ for $\boldsymbol{U} \in \mathfrak{U}_q$.

§ 2.2 The Riesz-Herglotz Measure of $F^{[U]}$ for any $U \in \mathfrak{U}_q$

Based on Lemma 2. 2, we are able to prove the main results of this section. First of all, we prove the following result.

Theorem 2. 3 *Let $t_0 \in T$ and let l be an integer with $0 \leqslant l \leqslant q$. Let, for $\boldsymbol{U} \in \mathfrak{U}_q$,*

$$F^{[U]}(v) = \int_T \frac{u+v}{u-v}\,\mu_U(\mathrm{d}u) = H_{\#}^{[U]}(v)(H^{[U]}(v))^{-1}, \ \forall v \in D. \tag{2.21}$$

Then

$$\dim \operatorname{Ker} H^{[U]}(t_0) = l \quad \text{if and only if} \quad \operatorname{rank}(\mu_U(\{t_0\})) = l.$$

Proof First, assume that $\dim \operatorname{Ker} H^{[U]}(t_0) = l$, and thus $\operatorname{rank} H^{[U]}(t_0) = q-l$. In the case of $l=0$, $H^{[U]}(t_0)$ is nonsingular, and thus $F^{[U]}$ is continuous at t_0 so that $\mu_U(\{t_0\}) = \boldsymbol{O}$ obviously, where μ_U is the Riesz-Herglotz measure of $F^{[U]}$ given in (2. 21). In the case of $1 \leqslant l \leqslant q$, Lemma 2. 2 (b) says that t_0 is a zero of $H^{[U]}$ of multiplicity l, so that $H^{[U]}(t_0)$ is singular with $\dim \operatorname{Ker} H^{[U]}(t_0) = l$. Then by using [10, Lemma 2. 1], we have in turn that $(\operatorname{adj}(H^{[U]}(v)))^{(r)}(t_0) = \boldsymbol{O}$, $r = 0, 1, \cdots, l-2$ and $(\operatorname{adj}(H^{[U]}(v)))^{(l-1)}(t_0) \neq \boldsymbol{O}$, so that by [4, Lemma 7. 1],

$$\begin{aligned} \mu_U(\{t_0\}) &= \frac{1}{2}\lim_{\rho \to 1^-}(1-\rho)F^{[U]}(\rho t_0) \\ &= \frac{1}{2}\lim_{\rho \to 1^-}(1-\rho)H_{\#}^{[U]}(\rho t_0)(H^{[U]}(\rho t_0))^{-1} \end{aligned}$$

$$= \frac{1}{2} \lim_{\rho \to 1^-} (1-\rho) \frac{H_{\#}^{[U]}(\rho t_0) \operatorname{adj}(H^{[U]}(\rho t_0))}{\det(H^{[U]}(\rho t_0))}$$

$$= \frac{1}{2} \lim_{\rho \to 1^-} (1-\rho) H_{\#}^{[U]}(\rho t_0) \frac{\frac{1}{(l-1)!}(\operatorname{adj}(H^{[U]}(v)))^{(l-1)}(t_0)(\rho-1)^{l-1} t_0^{l-1} + \cdots}{\frac{1}{l!}(\det(H^{[U]}(v)))^{(l)}(t_0)(\rho-1)^l t_0^l + \cdots}$$

$$= -\frac{l}{2t_0} H_{\#}^{[U]}(t_0) \frac{(\operatorname{adj}(H^{[U]}(v)))^{(l-1)}(t_0)}{(\det(H^{[U]}(v)))^{(l)}(t_0)}. \tag{2. 22}$$

In view of the fact that rank $(\operatorname{adj}(H^{[U]}(v)))^{(l-1)}(t_0)=l$ (see also [10, Lemma 2. 1]), we obtain in turn

$$\operatorname{rank} \mu_U(\{t_0\}) \leqslant \operatorname{rank} (\operatorname{adj}(H^{[U]}(v)))^{(l-1)}(t_0) = l.$$

To see that this rank is exactly equal to l, it is enough to prove that if $\boldsymbol{\beta} \in \mathbf{C}^q$ is such that $\boldsymbol{v} := (\operatorname{adj}(H^{[U]}(v)))^{(l-1)}(t_0)\boldsymbol{\beta} \neq \mathbf{0}$, then $\mu_U(\{t_0\})\boldsymbol{\beta} \neq \mathbf{0}$, or equivalently,

$$\operatorname{range}(\mu_U(\{t_0\}))^* \supseteq \operatorname{range}((\operatorname{adj}(H^{[U]}(v)))^{(l-1)}(t_0))^*.$$

Indeed, since $H^{[U]}(t_0)(\operatorname{adj}(H^{[U]}(v)))^{(l-1)}(t_0) = (\det H_{(v)}^{[U]})^{(l-1)}(t_0)\boldsymbol{I}_q = \boldsymbol{O}$, by using (2. 17) we infer

$$P_{m-1}(t_0)H_{\#}^{[U]}(t_0)\boldsymbol{v} + P_{m-1}^{\#}(t_0)H^{[U]}(t_0)\boldsymbol{v}$$
$$= P_{m-1}(t_0)H_{\#}^{[U]}(t_0)\boldsymbol{v} = 2t_0^{m-1}\boldsymbol{v}.$$

Hence from (2. 22) we deduce in turn

$$P_{m-1}(t_0)\mu_U(\{t_0\})\boldsymbol{\beta} = P_{m-1}(t_0)\left[-\frac{l}{2t_0}H_{\#}^{[U]}(t_0)\boldsymbol{v}/(\det H^{[U]}(v))^{(l)}(t_0)\right]$$
$$= -l t_0^{m-2}\boldsymbol{v}/(\det(H^{[U]}(v)))^{(l)}(t_0) \neq \mathbf{0},$$

so that $\mu_U(\{t_0\})\boldsymbol{\beta} \neq \mathbf{0}$. *Thus, we have that rank* $\mu_U(\{t_0\}) = l$.

Now we come to the other round of the proof of the theorem. Assume that rank $\mu_U(\{t_0\}) = l$. *Then for* $1 \leqslant l \leqslant q$, $H^{[U]}$ *has a zero at* t_0 *with multiplicity* s, *say. Now Lemma* 2. 2 (*b*) *concludes that rank*$(H^{[U]}(t_0)) = q - s$ *and thus dim Ker*$(H^{[U]}(t_0)) = s$. *The result that we have just proved implies that rank*$(\mu_U(\{t_0\})) = s$. *Thus*, $s = l$. *In the case of* $l = 0$, $F^{[U]}$, *defined by* (2. 21), *is continuous at* $t_0 \in T$, so that $H^{[U]}(t_0)$ is invertible, i. e. Ker $H^{[U]}(t_0) = \{\mathbf{0}\}$. This completes the proof. ■

Remark that if $t_0 \in T$ is a zero of $H^{[U]}$ of multiplicity l, then from

Theorem 2.3 and (2.22) we see that rank $H_{\#}^{[U]}(t_0)\leqslant l$. This means that $H_{\#}^{[U]}(t_0)$ is either nonsingular or $t_0\in T$ is a zero of $H_{\#}^{[U]}$ of multiplicity at most $q-l$.

By virtue of Theorem 2.3 and Lemma 2.2(b), the Proposition 9.8 of [16] can be intensified as below.

Corollary 2.4 *Let $t_0\in T$ and $\mathbf{U}\in\mathfrak{U}_q$. Let μ_U be the nonnegative Hermitian measure of $F^{[U]}$ defined by* (2.21). *Then the following are equivalent*:

(a) $H^{[U]}(t_0)=\mathbf{O}$.

(*b*) $\mu_U(\{t_0\})>\mathbf{O}$.

(*c*) t_0 is a zero of $H^{[U]}$ of multiplicity q.

(*d*) There exists a $v\in D\setminus\{0\}$ *such that* $F^{[U]}(v)=F_{t_0}^{h}(v)$, *where* $F_{t_0}^{h}$ *is as in* (2.20) *with w replaced by t_0.*

(e) $F^{[U]}=F_{t_0}^{h}$.

Proof Use Theorem 2.3 in combination with Proposition 9.8 of [16]. ■

In the next theorem, we will indicate the connection between the matrix mass $\mu_U(\{t_0\})$ of the Riesz-Herglotz measure of $F^{[U]}$ at $t_0\in T$ and $\sum_{k=0}^{m-1}(P_k(t_0))^*P_k(t_0)$. It is convenient to first prove a preliminary result.

Lemma 2.5 *Let t_0 be a zero of $H^{[U]}$ of multiplicity l, and let μ_U be the Riesz-Herglotz measure of $F^{[U]}$, where $\mathbf{U}\in\mathfrak{U}_q$. If $\boldsymbol{\beta}\in\mathbf{C}^q$ is an eigenvector of $\mu_U(\{t_0\})$ such that $\mu_U(\{t_0\})\boldsymbol{\beta}=\alpha\boldsymbol{\beta}$, where $\alpha>0$ is an eigenvalue of $\mu_U(\{t_0\})$, then*

$$\boldsymbol{\beta}^*H^{[U]}(t_0)=\mathbf{0},$$

$$-\frac{l}{t_0}(\operatorname{adj}(H^{[U]}(v)))^{(l-1)}(t_0)\boldsymbol{\beta}=\alpha(\det(H^{[U]}(v)))^{(l)}(t_0)(\widetilde{P}_{m-1}^{[m-1]}(t_0))^*\boldsymbol{\beta}. \tag{2.23}$$

Proof Observe that, by (2.22) and the fact that $\mu_U(\{t_0\})\boldsymbol{\beta}=\alpha\boldsymbol{\beta}$,

$$\begin{aligned}\alpha(\det(H^{[U]}(v)))^{(l)}(t_0)\boldsymbol{\beta}&=\mu_U(\{t_0\})(\det(H^{[U]}(v)))^{(l)}(t_0)\boldsymbol{\beta}\\&=-\frac{l}{2t_0}H_{\#}^{[U]}(t_0)(\operatorname{adj}(H^{[U]}(v)))^{(l-1)}(t_0)\boldsymbol{\beta}.\end{aligned}\tag{2.24}$$

On the other hand, from (2.16) we deduce

$$[(H_{\#}^{[U]}(t_0))^* H^{[U]}(t_0) + (H^{[U]}(t_0))^* H_{\#}^{[U]}(t_0)](\mathrm{adj}(H^{[U]}(v)))^{(l-1)}(t_0)\boldsymbol{\beta} = \boldsymbol{O}$$

and therefore, since

$$H^{[U]}(t_0)(\mathrm{adj}(H^{[U]}(v)))^{(l-1)}(t_0) = (\det(H^{[U]}(v)))^{(l-1)}(t_0)\boldsymbol{I}_q = \boldsymbol{O},$$

we have

$$(H^{[U]}(t_0))^* H_{\#}^{[U]}(t_0)(\mathrm{adj}(H^{[U]}(v)))^{(l-1)}(t_0)\boldsymbol{\beta} = \mathbf{0}.$$

Taking (2.24) into account, the last expression reduces to

$$\frac{2\alpha t_0}{l}(H^{[U]}(t_0))^* (\det(H^{[U]}(v)))^{(l)}(t_0)\boldsymbol{\beta} = \mathbf{0},$$

whence, since $(\det(H^{[U]}(v)))^{(l)}(t_0) \neq 0$ and $\alpha > 0$, $(H^{[U]}(t_0))^* \boldsymbol{\beta} = \mathbf{0}$, or equivalently, $\boldsymbol{\beta}^* (H^{[U]}(t_0)) = \mathbf{0}$.

For the second display of (2.23), it can be obtained by using of (2.24) and (2.18), together with the fact that

$$H^{[U]}(t_0)\ (\mathrm{adj}(H^{[U]}(v)))^{(l-1)}(t_0) = \boldsymbol{O}.$$

In fact,

$$\begin{aligned}
&\alpha\,(\det(H^{[U]}(v)))^{(l)}(t_0)(\widetilde{P}_{m-1}^{[m-1]}(t_0))^* \boldsymbol{\beta} \\
&= (\widetilde{P}_{m-1}^{[m-1]}(t_0))^* \left[-\frac{l}{2t_0} H_{\#}^{[U]}(t_0)(\mathrm{adj} H^{[U]}(v))^{(l-1)}(t_0)\right]\boldsymbol{\beta} \\
&= -\frac{l}{2t_0}[(\widetilde{P}_{m-1}^{[m-1]}(t_0))^* H_{\#}^{[U]}(t_0) + \\
&\quad ((\widetilde{P_{m-1}^{\#}})^{[m-1]}(t_0))^* H^{[U]}(t_0)](\mathrm{adj}(H^{[U]}(v)))^{(l-1)}(t_0)\boldsymbol{\beta} \\
&= -\frac{l}{2t_0}\cdot 2\boldsymbol{I}_q(\mathrm{adj}(H^{[U]}(v)))^{(l-1)}(t_0)\boldsymbol{\beta} \\
&= -\frac{l}{t_0}(\mathrm{adj}(H^{[U]}(v)))^{(l-1)}(t_0)\boldsymbol{\beta},
\end{aligned}$$

as claimed. ■

Theorem 2.6 *If $t_0 \in T$ and μ_U is the Riesz-Herglotz measure of $F^{[U]}$, where $\boldsymbol{U} \in \mathfrak{U}_q$, then for any $\boldsymbol{\beta}, \boldsymbol{\xi} \in$ range $\mu_U(\{t_0\})$,*

$$\boldsymbol{\beta}^* \mu_U(\{t_0\})\Big(\sum_{k=0}^{m-1}(P_k(t_0))^* P_k(t_0)\Big)\boldsymbol{\xi} = \boldsymbol{\beta}^* \Big(\sum_{k=0}^{m-1}(P_k(t_0))^* P_k(t_0)\Big)\mu_U(\{t_0\})\boldsymbol{\xi}$$

$$= \boldsymbol{\beta}^* \xi. \tag{2.25}$$

Furthermore, if $\mathbf{U} \in \mathfrak{U}_q$ *is chosen such that* $H^{[U]}(t_0) = \mathbf{O}$, *then* $\mu_U = \mu_{t_0}$, *the Riesz-Herglotz measure of* $F^{[U]} = F^h_{t_0}$, *and for each solution* $\boldsymbol{v}$ *of the nondegenerate problem* (2.3), *the inequality*

$$\boldsymbol{v}(\{t_0\}) \leqslant \mu_{t_0}(\{t_0\}) = \Big(\sum_{k=0}^{m-1} (P_k(t_0))^* P_k(t_0)\Big)^{-1} \tag{2.26}$$

holds.

Proof Assume that $\mu_U(\{t_0\}) \geqslant \mathbf{O}$ is of rank l, say, and has $\alpha_1, \alpha_2, \cdots, \alpha_l$ as all positive eigenvalues, and $\boldsymbol{\beta}_1, \boldsymbol{\beta}_2, \cdots, \boldsymbol{\beta}_l \in \mathbf{C}^q$ as the corresponding orthonormal eigenvectors with respect to the standard inner product of $\mathbf{C}^q$, which form an orthonormal basis of the space range $\mu_U(\{t_0\})$. To prove (2.25), it is sufficient to verify, for $i, j = 1, 2, \cdots, l$,

$$\boldsymbol{\beta}_i^* \mu_U(\{t_0\}) \Big(\sum_{k=0}^{m-1} (P_k(t_0))^* P_k(t_0)\Big) \boldsymbol{\beta}_j$$

$$= \boldsymbol{\beta}_i^* \Big(\sum_{k=0}^{m-1} (P_k(t_0))^* P_k(t_0)\Big) \mu_U(\{t_0\}) \boldsymbol{\beta}_j = \delta_{ij}.$$

Indeed, by using of (2.13) and (2.23), and the fact that $\mu_U(\{t_0\})\boldsymbol{\beta}_i = \alpha_i \boldsymbol{\beta}_i$, $i = 1, 2, \cdots, l$, we see that the first term of the last equation is equal to δ_{ij}, which can be verified by the following computation:

$$\alpha_i \boldsymbol{\beta}_i^* \Big(\sum_{k=0}^{m-1} (P_k(t_0))^* P_k(t_0)\Big) \boldsymbol{\beta}_j$$

$$= -t_0 \alpha_i \boldsymbol{\beta}_i^* (H^{[U]}(t_0))' (\widetilde{P}^{[m-1]}_{m-1}(t_0))^* \boldsymbol{\beta}_j$$

$$= -t_0 \frac{\alpha_i}{\alpha_j} \boldsymbol{\beta}_i^* (H^{[U]}(t_0))' \Big[-\frac{l}{t_0} \frac{(\mathrm{adj}(H^{[U]}(v)))^{(l-1)}(t_0) \boldsymbol{\beta}_j}{(\det H^{[U]}(v))^{(l)}(t_0)} \Big]$$

$$= \frac{\alpha_i}{\alpha_j} \frac{\boldsymbol{\beta}_i^*}{(\det H^{[U]}(v))^{(l)}(t_0)} [l (H^{[U]}(t_0))' (\mathrm{adj}(H^{[U]}(v)))^{(l-1)}(t_0) + H^{[U]}(t_0) (\mathrm{adj}(H^{[U]}(v)))^{(l)}(t_0)] \boldsymbol{\beta}_j$$

$$= \frac{\alpha_i}{\alpha_j} \frac{\boldsymbol{\beta}_i^*}{(\det(H^{[U]}(v)))^{(l)}(t_0)} ((H^{[U]}(v) \,\mathrm{adj}\, H^{[U]}(v))^{(l)}(t_0)) \boldsymbol{\beta}_j$$

$$= \frac{\alpha_i}{\alpha_j} \boldsymbol{\beta}_i^* \boldsymbol{\beta}_j = \frac{\alpha_i}{\alpha_j} \delta_{ij} = \delta_{ij}, \quad i, j = 1, 2, \cdots, r.$$

Furthermore, by the above argument, we have

$$\boldsymbol{\beta}_i^* \Big(\sum_{k=0}^{m-1}(P_k(t_0))^* P_k(t_0)\Big)\mu_U(\{t_0\})\boldsymbol{\beta}_j$$
$$= \boldsymbol{\beta}_i^* \Big(\sum_{k=0}^{m-1}(P_k(t_0))^* P_k(z)\Big)\alpha_j\boldsymbol{\beta}_j$$
$$= \Big(\alpha_j\boldsymbol{\beta}_j \Big(\sum_{k=0}^{m-1}(P_k(t_0))^* P_k(z)\Big)\boldsymbol{\beta}_j\Big)^*$$
$$= \delta_{ji} = \delta_{ij},\ i,j = 1,2,\cdots,r.$$

Thus (2.25) holds. Let $\boldsymbol{U}\in \mathfrak{U}_q$ be such that $H^{[U]}(t_0)=\boldsymbol{O}$. Then, by Corollary 2.4, $F^{[U]}(z)=F_{t_0}^h(z)$, $\forall z\in D$, and the matrix mass of $\mu_U = \mu_{t_0}$ at the point t_0 is of rank q: $\mu_{t_0}(\{t_0\})>\boldsymbol{O}$, so that by (2.25),

$$\mu_{t_0}(\{t_0\}) = \Big(\sum_{k=0}^{m-1}(P_k(t_0))^* P_k(t_0)\Big)^{-1}. \tag{2.27}$$

On the other hand, for any solution $\boldsymbol{v}$ of the nondegenerate problem (2.3), by (2.5) we have

$$\sum_{k=0}^{m-1}(P_k(t_0))^* P_k(t_0) = \int_T\Big(\sum_{k=0}^{m-1}(P_k(t_0))^* P_k(z)\Big)\boldsymbol{v}(\mathrm{d}z)\Big(\sum_{k=0}^{m-1}(P_k(t_0))^* P_k(z)\Big)^*$$
$$\geqslant \Big(\sum_{k=0}^{m-1}(P_k(t_0))^* P_k(t_0)\Big)\boldsymbol{v}(\{t_0\})\Big(\sum_{k=0}^{m-1}(P_k(t_0))^* P_k(t_0)\Big),$$

which implies

$$\Big(\sum_{k=0}^{m-1}(P_k(t_0))^* P_k(t_0)\Big)\Big[\Big(\sum_{k=0}^{m-1}(P_k(t_0))^* P_k(t_0)\Big)^{-1} - \boldsymbol{v}(\{t_0\})\Big]\times$$
$$\Big(\sum_{k=0}^{m-1}(P_k(t_0))^* P_k(t_0)\Big)\geqslant \boldsymbol{O}.$$

Thus

$$\Big(\sum_{k=0}^{m-1}(P_k(t_0))^* P_k(t_0)\Big)^{-1} \geqslant \boldsymbol{v}(\{t_0\}). \tag{2.28}$$

The assertion (2.26) then follows from (2.27) and (2.28). ■

Somewhat surprisingly, for the extremal problem of maximal matrix mass at $t_0\in T$ for the Riesz-Herglotz measures corresponding to $\boldsymbol{C}_q[D, \boldsymbol{c}]$ studied in Theorem 2.6, the solution $F_{t_0}^h$ is actually uniquely determined by that extremal feather. In fact, recently, Lasarow [22] gave for the first time a basis proof for the fact indicated by verifying that if the equality holds in (2.28), then $\boldsymbol{v}=\mu_{t_0}$, the Riesz-Herglotz measure of $F_{t_0}^h$.

Therefore, taking (2.26) into account, we have the following result.

Theorem 2.7 [22, Theorem 3.2] *Let $t_0 \in T$ and let $F^h_{t_0}$ be given as in (2.20) with w replaced by t_0. Then the inequality (2.28) holds for each solution $\boldsymbol{v}$ of the nondegenerate Problem (2.3) with equality if and only if $\boldsymbol{v}=\mu_{t_0}$, the Riesz-Herglotz measure of $F^h_{t_0}$.*

§2.3 The Representation of $F^{[U]}$ for any $U \in \mathfrak{U}_q$

Due to Theorems 2.3 and 2.6, (2.18), and the statement of (b) of Lemma 2.2 together with [16, Proposition 9.8], we obtain directly the representation of $F^{[U]}$ with any $U \in \mathfrak{U}_q$ as given below.

Theorem 2.8 *Let r denote the number of pairwise different zeros $t_1, t_2, \cdots, t_r$ of $H^{[U]}$. Then $m \leqslant r \leqslant mq$, $t_1, t_2, \cdots, t_r \in T$, $H^{[U]}_{\#}(t_s) \neq \boldsymbol{O}$, $1 \leqslant s \leqslant r$, and there exists a sequence $(\boldsymbol{M}_s)_{s=1}^{m}$ of nonnegative nonzero Hermitian $q \times q$ matrices such that*

$$\boldsymbol{y}^*\Big(\boldsymbol{M}_s\Big(\sum_{k=0}^{m-1}(P_k(t_s))^* P_k(t_s)\Big)\Big)\boldsymbol{x} = \boldsymbol{y}^*\Big(\Big(\sum_{k=0}^{m-1}(P_k(t_s))^* P_k(t_s)\Big)\boldsymbol{M}_s\Big)\boldsymbol{x}$$
$$= \boldsymbol{y}^*\boldsymbol{x}, \quad \forall\ \boldsymbol{x},\boldsymbol{y} \in \mathrm{range}(\boldsymbol{M}_s)$$

holds for $s=1, 2, \cdots, r$, rank $\boldsymbol{M}_s$ is equal to the multiplicity of t_s as a zero of $H^{[U]}$, $1 \leqslant s \leqslant r$, $\sum_{s=1}^{r}$ rank $\boldsymbol{M}_s =$ deg det $H^{[U]}(z) = mq$, and $F^{[U]}(z) = H^{[U]}_{\#}(z)(H^{[U]}(z))^{-1}$ admits the representation

$$F^{[U]}(z) = \sum_{s=1}^{r} \frac{t_s + z}{t_s - z}\boldsymbol{M}_s, \forall\, z \in D.$$

Furthermore, $F^{[U]}(z) = F^h_{t_j}(z)$, $\forall\, z \in D$, if $\boldsymbol{M}_j \in \mathbf{C}^{q\times q}$ is positive Hermitian, where $1 \leqslant j \leqslant r$. In particular, in the case of $r=m$, $F^{[U]}$ admits the representation

$$F^{[U]}(z) = \sum_{s=1}^{m} \frac{t_s + z}{t_s - z}\boldsymbol{M}_s, \forall\, z \in D,$$

where each $\boldsymbol{M}_s$ is certain positive Hermitian $q \times q$ matrix, and satisfies

$$\boldsymbol{M}_s = \Big(\sum_{k=0}^{m-1}(P_k(t_s))^* P_k(t_s)\Big)^{-1}, \ s = 1,2,\cdots,m.$$

Furthermore, $F^{[U]}(z) = F^h_{t_s}(z)$, $\forall\, z \in D$, $s=1, 2, \cdots, m$; in the case of $r=mq$, $F^{[U]}$ admits the representation

$$F^{[U]}(z) = \sum_{s=1}^{mq} \frac{t_s + z}{t_s - z} \mathbf{M}_s, \forall z \in D,$$

where each $\mathbf{M}_s$ *is certain nonnegative Hermitian* $q \times q$ *matrix of rank* 1.

As a direct consequence of Theorem 2. 8, in the scalar case of $q=1$, $F^{[U]}(z)$ has exactly $t_1, t_2, \cdots, t_m$ as m simple poles, which are located on T, and admits the representation

$$F^{[U]}(z) = \sum_{s=1}^{m} \frac{t_s + z}{t_s - z} \mathbf{M}_s = F^h_{t_j}(z), \ j = 1,2,\cdots,m,$$

where each $\mathbf{M}_s$ is certain positive number and fulfils (see, e. g. [18, Section 1] or [24, Theorem 2. 2. 12])

$$\mathbf{M}_s = \Big(\sum_{k=0}^{m-1} (P_k(t_s))^* P_k(t_s)\Big)^{-1}.$$

The following results are, in essence, given in [16, Corollary 9. 15] with the exception of the statements (b), (g) and (h).

Corollary 2. 9 *Let* $w \in T$, *let* F^h_w *be given as in* (2. 20) *for each* $z \in D$, *and* $\mathbf{U} := \mathbf{U}_w$ *as in* (2. 19), *and let* μ_w *be the Riesz-Herglotz measure of* F^h_w. *Then the following statements are equivalent*:

(a) *For each* $u \in T$, *the matrix mass* $\mu_w(\{u\})$ *is either a positive Hermitian* $q \times q$ *matrix or the zero matrix*.

(b) *For each* $u \in T$, *the matrix* $H^{[U]}(u)$ *is either the zero matrix or a nonsingular* $q \times q$ *matrix*.

(c) *For some positive integer* r, *some sequence* $(t_s)_{s=1}^{r}$ *of points on* T *and some sequence* $(\mathbf{M}_s)_{s=1}^{r}$ *of positive Hermitian* $q \times q$ *matrices*, F^h_w *admits the representation*

$$F^h_w(z) = \sum_{s=1}^{r} \frac{t_s + z}{t_s - z} \mathbf{M}_s, \ \forall z \in D. \tag{2. 29}$$

(d) F^h_w *satisfies* (2. 29) *for each* $z \in D$ *with* $r = m$, *some sequence* $(t_s)_{s=1}^{m}$ *of pairwise different points on* T, *and the particular choice*

$$\mathbf{M}_s = \Big(\sum_{k=0}^{m-1} (P_k(t_s))^* P_k(t_s)\Big)^{-1}, \ s = 1,2,\cdots,m.$$

(e) F^h_w *satisfies* (2. 29) *for each* $z \in D$ *with* $r = m$, *some sequence* $(t_s)_{s=1}^{m}$ *of points on* T *and some sequences* $(\mathbf{M}_s)_{s=1}^{m}$ *of nonnegative Hermitian* $q \times q$ *matrices*.

(f) *There exists an* $F \in \boldsymbol{C}_q[D, \boldsymbol{c}]$ *of the form*

$$F(z) = \sum_{s=1}^{m} \frac{t_s + z}{t_s - z} \boldsymbol{M}_s,$$

where $t_1 = w, t_2, \cdots, t_m \in T$, *and* $\boldsymbol{M}_1, \boldsymbol{M}_2, \cdots, \boldsymbol{M}_m$ *are nonnegative Hermitian* $q \times q$ *matrices.*

(g) *There exist exactly* m *pairwise different points* $t_1 = w, t_2, \cdots, t_m$ *on* T *such that* $H^{[U]}(t_s) = \boldsymbol{O}$ *for* $s = 1, 2, \cdots, m$.

(h) *There exist exactly* m *pairwise different points* $t_1 = w, t_2, \cdots, t_m$ *on* T *such that* $\mu_w(\{t_j\}) > \boldsymbol{O}$ *for* $j = 1, 2, \cdots, m$.

Proof Use Corollary 9. 15 of [16] in combination with Theorems 2. 3 and 2. 6. ■

§ 3. The Riesz-Herglotz Measures of Extremal Solutions of Nondegenerate Problem (NP)

In this section, similar results to that given in Section 2 are considered but for the extremal solutions $G^{[U]}$ of (nondegenerate) Problem (NP). Our method is based on the so-called modified block Toeplitz vector approach, which provides us with a combination of the theory of Problem (C) with Problem (NP). In the first half of this section we review some of results on this method, largely without proofs. For details, see [5,6].

§ 3.1 Problem (NP) and Its Related Problem (C)

Problem (NP) under consideration is the following multiple Nevanlinna-Pick-type interpolation for the class $\boldsymbol{C}_q(D)$:

Problem (NP) Let ρ be a fixed nonnegative integer, let $z_0, z_1, \cdots, z_\rho$ be pairwise different points in D, let $\tau_0, \tau_1, \cdots, \tau_\rho$ be positive integers and let $\boldsymbol{C}_{ik}$ be given in $\mathbf{C}^{q\times q}$ for $k = 0, \cdots, \tau_i - 1$ and $i = 0, 1, \cdots, \rho$. Denote by Δ the interpolation data:

$$\Delta := \{(z_k)_{k=0}^{\rho}; (\tau_k)_{k=0}^{\rho}; \boldsymbol{C}_{ik}, i = 0, 1, \cdots, \rho, \; k = 0, 1, \cdots, \tau_i - 1\}. \tag{3.1}$$

Describe the set $\mathbf{C}_\Delta$ of all $G\in\mathbf{C}_q(D)$ satisfying the conditions

$$\frac{1}{k!}\frac{\mathrm{d}^k}{\mathrm{d}z^k}G(z)\mid_{z=z_i}=\boldsymbol{C}_{ik},\ i=0,1,\cdots,\rho,\quad k=0,1,\cdots,\tau_i-1. \tag{3.2}$$

The so-called *generalized block Pick matrix* $\boldsymbol{P}_\Delta$ will play an important role by solving Problem (NP). Let $m:=\sum_{j=0}^{\rho}\tau_j$ and let G be a $\mathbf{C}^{q\times q}$-valued function holomorphic in D and subject to (3.2) (such a function exists always). We define

$$\boldsymbol{P}_\Delta:=(\boldsymbol{P}_{ij})_{i,j=0}^{\rho}\in\mathbf{C}^{mq\times mq}, \tag{3.3}$$

where $\boldsymbol{P}_{ij}:=(\boldsymbol{P}_{ij}^{kl})_{k=0,l=0}^{\tau_i-1,\tau_j-1}\in\mathbf{C}^{\tau_i p\times\tau_j q}$ with entries $\boldsymbol{P}_{ij}^{kl}\in\mathbf{C}^{q\times q}$ determined by

$$\boldsymbol{P}_{ij}^{kl}:=\frac{1}{k!l!}\frac{\partial^{k+l}}{\partial\lambda^k\partial\mu^l}\left(\frac{G(\lambda)+G(\bar{\mu})^*}{1-\lambda\mu}\right)\mid_{\lambda=z_i,\mu=\bar{z}_j}. \tag{3.4}$$

It is well known (see, e.g. [5, Corollary 2.7]) that Problem (NP) is solvable if and only if $\boldsymbol{P}_\Delta\geqslant\boldsymbol{O}$ (see also Corollary 3.3 below).

Throughout this article, we assume that $0\in\{z_0,z_1,\cdots,z_\rho\}$ and $z_0=0$, so that Problem (NP) takes Problem (C) as one particular case. Write $\boldsymbol{D}_i:=\boldsymbol{C}_{0i}$, $i=0,1,\cdots,\tau_0-1$ for short. Similar to [6], let

$$A(z):=\prod_{i=1}^{\rho}(z-z_i)^{\tau_i}\left(\frac{1}{z}-\bar{z}_i\right)^{\tau_i}=:\sum_{i=-n}^{n}A_iz^i, \tag{3.5}$$

where $n:=\tau_1+\tau_2+\cdots+\tau_\rho=m-\tau_0$. Then $A_{-n}=\bar{A}_n=(-1)^n\prod_{i=1}^{\rho}z_i^{\tau_i}\neq 0$, $A(z)=A(\bar{z}^{-1})^*$, $\forall z\neq 0$, and $A(z)>0$, $\forall z\in T$. It is verified readily that there exists a unique quasi-$q\times q$ matrix polynomial of the form

$$\widetilde{\boldsymbol{W}}(z):=\sum_{i=-n}^{n-1}\widetilde{\boldsymbol{W}}_iz^i,\ \widetilde{\boldsymbol{W}}_i\in\mathbf{C}^{q\times q},$$

satisfying both (3.2) with $i\neq 0$ and the conditions

$$\frac{1}{k!}\frac{\mathrm{d}^k}{\mathrm{d}z^k}G\left(\frac{1}{z}\right)\mid_{z=\bar{z}_i}=-\boldsymbol{C}_{ik}^*,\ i=1,2,\cdots,\rho,\quad k=0,1,\cdots,\tau_i-1. \tag{3.6}$$

Now let

$$W(z):=\frac{1}{2}\left(\widetilde{W}(z)-\widetilde{W}\left(\frac{1}{\bar{z}}\right)^*\right),\ \forall z\neq 0. \tag{3.7}$$

Then $W(z)$ is also subject to both (3.2) with $i\neq 0$ and (3.6). Consider the Taylor series representation of $W(z)/A(z)$ at $z=0$:

$$\frac{W(z)}{A(z)}=-\widetilde{\boldsymbol{C}}_0-2\widetilde{\boldsymbol{C}}_1 z-\cdots-2\widetilde{\boldsymbol{C}}_n z^n+\cdots(z\to 0)$$

where $\widetilde{\boldsymbol{C}}_j\in\mathbf{C}^{q\times q}$. Then we have that $\widetilde{\boldsymbol{C}}_0=\widetilde{\boldsymbol{C}}_0^*$. Let $\boldsymbol{C}_0,\boldsymbol{C}_1,\cdots,\boldsymbol{C}_{m-1}$ be defined recursively by

$$\begin{aligned}\boldsymbol{C}_i&:=\widetilde{\boldsymbol{C}}_i,\ i=0,1,\cdots,n-1,\\ \boldsymbol{C}_n&:=\frac{1}{2}A_{-n}^{-1}\boldsymbol{D}_0+\widetilde{\boldsymbol{C}}_n,\\ \boldsymbol{C}_{n+k}&:=A_{-n}^{-1}\left(\frac{1}{2}\boldsymbol{D}_k-\sum_{l=-n+1}^{n}A_l\boldsymbol{C}_{k-l}\right),\ k=1,2,\cdots,\tau_0-1,\end{aligned}\tag{3.8}$$

where $\boldsymbol{C}_{-k}=\boldsymbol{C}_k^*$ for $k\geqslant 0$. We refer to the block vector $\boldsymbol{c}:=(\boldsymbol{C}_0,\boldsymbol{C}_1,\cdots,\boldsymbol{C}_{m-1})$ as the *block Toeplitz vector* of Problem (NP), which is uniquely defined by the data Δ of Problem (NP) given in (3.1).

With such $\boldsymbol{c}=(\boldsymbol{C}_0,\boldsymbol{C}_1,\cdots,\boldsymbol{C}_{m-1})$ we can associate a certain matricial Carathéodory problem, i. e. the Problem (C) given in Section 2 but with $\boldsymbol{c}=(\boldsymbol{C}_0,\boldsymbol{C}_1,\cdots,\boldsymbol{C}_{m-1})$ replaced by the block Toeplitz vector of Problem (NP). In the following, we shall also call such Problem (C) associated with the Toeplitz vector $\boldsymbol{c}$ of Problem (NP), and use the same notation as that given in Section 2.

To indicate the congruence relation of $\boldsymbol{P}_\Delta$ with $\boldsymbol{T}_c:=(\boldsymbol{C}_{j-i})_{i,j=0}^{m-1}$, we need $\Sigma\in\mathbf{C}^{m\times m}$ defined via

$$\left[\operatorname{row}\left[\frac{1}{z^i}B(z)\right]_{i=\tau_0}^{1},\ \operatorname{row}[\operatorname{row}[B_{ik}(z)]_{i=1}^{\rho}]_{k=0}^{\tau_i-1}\right]^{\mathrm{T}}=\Sigma[\operatorname{row}[z^i]_{i=-\tau_0}^{n-1}]^{\mathrm{T}},$$

where

$$B(z)=\prod_{i=1}^{\rho}(z-z_i)^{\tau_i},\ B_{ik}(z)=\frac{B(z)}{(z-z_i)^{k+1}},1\leqslant i\leqslant\rho,\ 0\leqslant k\leqslant\tau_i-1.$$

Then $\Sigma\in\mathbf{C}^{m\times m}$ is nonsingular, since it is a transition matrix of a pair of bases of the space of all quasi-polynomials of the form $\sum\limits_{j=-\tau_0}^{n-1}b_jz^j$.

Theorem 3.1 [6, Theorem 9.6.2] *Let* $\boldsymbol{c}=(\boldsymbol{C}_0,\boldsymbol{C}_1,\cdots,\boldsymbol{C}_{m-1})$ *be the block Toeplitz vector of Problem (NP). Let* $\boldsymbol{P}_\Delta$ *be as in* (3.3)~(3.4) *and* $\boldsymbol{T}_c=(\boldsymbol{C}_{j-i})_{i,j=0}^{m-1}$. *Then*

$$\boldsymbol{P}_\Delta=2(\Sigma\otimes\boldsymbol{I}_q)\boldsymbol{T}_c(\Sigma\otimes\boldsymbol{I}_q)^*,\tag{3.9}$$

where the symbol $\otimes$ *stands for the tensor product of two matrices.*

Despite differences in formulation, Problem (NP) and Problem (C) have an intrinsic unity because their solutions $\mathbf{C}_\Delta$ and $\mathbf{C}_q[D, \boldsymbol{c}]$ are directly connected with a particular one-to-one correspondence as follows [6, Theorem 9.5.1].

Theorem 3.2 *Problem (NP) is solvable if and only if Problem (C) is solvable. In that case, if G is a solution to Problem (NP) and permits an integral representation of the form*

$$G(z) = \mathrm{i}\ \mathrm{Im}\ G(0) + \int_T \frac{u+z}{u-z}\ \sigma(\mathrm{d}u), \forall z \in D, \tag{3.10}$$

then F defined via

$$G(z) = W(z) + A(z)F(z) \tag{3.11}$$

belongs to $\mathbf{C}_q(D)$ *and is a solution to Problem (C), which has an integral representation of the form*

$$F(z) := \int_T \frac{u+z}{u-z}\ \mu(\mathrm{d}u), \tag{3.12}$$

where for a Borel subset $\mathscr{B}$ *of the unit circle T,*

$$\mu(\mathscr{B}) := \int_{\mathscr{B}} A(u)^{-1}\sigma(\mathrm{d}u).$$

Conversely, if $F \in \mathbf{C}_q(D)$ *is a solution to Problem (C) with the integral representation (3.12), then G defined by (3.11) belongs to* $\mathbf{C}_q(D)$ *and is a solution to Problem (NP), which has the integral representation of the form*

$$G(z) = \mathrm{i}\ \mathrm{Im}\ \boldsymbol{D}_0 + \int_T \frac{u+z}{u-z}\ \sigma(\mathrm{d}u), \quad \forall z \in D,$$

where $\boldsymbol{D}_0 = \boldsymbol{C}_{00}$ *and for a Borel subset* $\mathscr{B}$ *of the unit circle T,*

$$\sigma(\mathscr{B}) := \int_{\mathscr{B}} A(u)\mu(\mathrm{d}u).$$

As a consequence of Theorems 3.1 and 3.2, we have the following corollary.

Corollary 3.3 *The following statements are equivalent:*

(a) *Problem (NP) is solvable.*

(b) *Problem (C) associated with the block Toeplitz vector* $\boldsymbol{c}$ *of*

Problem (NP) *is solvable.*

(c) *The problem* (2.3) *with the block Toeplitz vector* **c** *as its matrix moments is solvable.*

(d) $\boldsymbol{P}_\Delta \geqslant \boldsymbol{O}$.

(e) $\boldsymbol{T}_c = (\boldsymbol{C}_{j-i})_{i,j=0}^{m-1} \geqslant \boldsymbol{O}$.

Moreover, $\boldsymbol{P}_\Delta > \boldsymbol{O}$ *if and only if* $\boldsymbol{T}_c > \boldsymbol{O}$, *that is*, *Problem* (NP) *is nondegenerate if and only if Problem* (C) *is.*

Furthermore, from Theorems 3.2 and 2.1 we deduce the following theorem.

Theorem 3.4 *Let* $W(z)$, $A(z)$ *and* $\boldsymbol{D}_0$ *be as before. Let* $\boldsymbol{P}_\Delta$ *defined by* (3.3)~(3.4) *be positive Hermitian*: $\boldsymbol{P}_\Delta > \boldsymbol{O}$. *Then Problem* (NP) *has infinitely many solutions, which can be parameterized, for* $z \in D$, *by*

$$\begin{aligned} G(z) &= \mathrm{i}\ \mathrm{Im}\ \boldsymbol{D}_0 + \int_T \frac{u+z}{u-z}\sigma(\mathrm{d}u) \\ &= W(z) + A(z)[(\widetilde{P_{m-1}^{\#}})^{[m-1]}(z) - zQ_{m-1}^{\#}(z)g(z)] \times \\ &\quad [\widetilde{P}_{m-1}^{[m-1]}(z) + zQ_{m-1}(z)g(z)]^{-1}, \end{aligned} \tag{3.13}$$

where g *varies over the class* $\mathbf{S}_q(D)$, *and* $\sigma(\mathscr{B}) := \int_{\mathscr{B}} A(u)\mu(\mathrm{d}u)$ *for any Borel subset* $\mathscr{B}$ *on* T.

§ 3.2 The discrete representation of the extremal solution of nondegenerate Problem (NP)

Due to Theorem 3.2, the solution of Problem (NP) can be reduced to the solution of Problem (C). Thus, much of Section 2 on Riesz-Herglotz measures μ of the extremal solutions $F^{[U]}$ given in (2.21) of (nondegenerate) Problem (C) can serve as a starting point and a model for more general constructions in (nondegenerate) Problem (NP). Now we turn our attention to the extremal solutions to Problem (NP), which means those solutions in $\mathbf{C}_\Delta$ are of the form

$$G^{[U]}(z) = \mathrm{i}\ \mathrm{Im}\ \boldsymbol{D}_0 + \int_T \frac{u+z}{u-z}\ \sigma(\mathrm{d}u),\ \boldsymbol{U} \in \mathfrak{U}_q; \tag{3.14}$$

furthermore, the $G^{[U]}$ are related with $F^{[U]}$ by

$$G^{[U]}(z) = W(z) + A(z)F^{[U]}(z) = W(z) + A(z)\int_T \frac{u+z}{u-z}\,\mu_U(du)$$
$$= W(z) + A(z)H_{\#}^{[U]}(z)H^{[U]}(z)^{-1},\ \forall z \in D, \tag{3.15}$$

where

$$\sigma_U(du) = A(u)\mu_U(du), \forall u \in T. \tag{3.16}$$

As well, we define, for $w \in T$, G_w^h by

$$G_w^h(z) = W(z) + A(z)F_w^h(z),\ \forall z \in D. \tag{3.17}$$

Let us begin with an interesting consequence of Theorem 2.3.

Theorem 3.5 *Let $t_0 \in T$ and $\mathbf{U} \in \mathfrak{U}_q$, and let l be an integer with $0 \leqslant l \leqslant q$. Let, for $\mathbf{U} \in \mathfrak{U}_q$, $G^{[U]}$ be defined and connected with $F^{[U]}$ by the formulas* (3.14) *and* (3.15), *respectively. Then*

$$\dim \operatorname{Ker} H^{[U]}(t_0) = l \text{ if and only if } \operatorname{rank} \sigma_U(t_0) = l.$$

Proof In view of (3.16) together with the fact that $A(u) > 0$, $\forall u \in T$, the assertion is an immediate consequence of Theorem 2.3. ■

We have the following corollary similar to Corollary 2.4.

Corollary 3.6 *Let $t_0 \in T$ and $\mathbf{U} \in \mathfrak{U}_q$. Let σ_U be the nonnegative Hermitian measure of $G^{[U]}$ determined by* (3.15). *Then the following statements are equivalent*:

(a) $H^{[U]}(t_0) = \mathbf{O}$.

(b) $\sigma_U(\{t_0\}) > \mathbf{O}$.

(c) t_0 is a zero of $H^{[U]}$ of multiplicity q.

(d) There exists a $v \in D \backslash \{0\}$ *such that* $G^{[U]}(v) = G_{t_0}^h(v)$, *where* $G_{t_0}^h$ *is as in* (3.17) *with w replaced by t_0*.

(e) $G^{[U]} = G_{t_0}^h$.

The following result is a slight reformulation of Theorem 2.6.

Theorem 3.7 *If $t_0 \in T$ and σ_U is the Riesz-Herglotz measure of $G^{[U]}$ defined by* (3.14), *where $\mathbf{U} \in \mathfrak{U}_q$, then for any* , $\boldsymbol{\beta}, \boldsymbol{\xi} \in$ range $\sigma_U(\{t_0\})$,

$$\boldsymbol{\beta}^* \sigma_U(\{t_0\})\Big(\sum_{k=0}^{m-1}(P_k(t_0))^* P_k(t_0)\Big)\boldsymbol{\xi}$$
$$= \boldsymbol{\beta}^* \Big(\sum_{k=0}^{m-1}(P_k(t_0))^* P_k(t_0)\Big)\sigma_U(\{t_0\})\boldsymbol{\xi}$$
$$= A(t_0)\boldsymbol{\beta}^* \boldsymbol{\xi}.$$

Furthermore, if $\boldsymbol{U}\in\mathfrak{U}_q$ *is chosen such that* $H^{[U]}(t_0)=\boldsymbol{O}$, *then* $\sigma_U=\sigma_{t_0}$, *the Riesz-Herglotz measure of* $G_{t_0}^h$, *and for the Riesz-Herglotz measure* $\boldsymbol{\sigma}$ *of each solution of the nondegenerate Problem* (NP), *the inequality*

$$\sigma(\{t_0\})\leqslant\sigma_{t_0}(\{t_0\})=A(t_0)\Big(\sum_{k=0}^{m-1}(P_k(t_0))^*P_k(t_0)\Big)^{-1}\qquad(3.18)$$

holds.

Now we state the announced result about the discrete representation of the function $G^{[U]}$, which is a consequence of Theorems 2. 8 and 2. 7 and the formula (3. 16).

Theorem 3. 8 *Let* $\boldsymbol{U}\in\mathfrak{U}_q$ *and let* r *denote the number of pairwise different zeros* $t_1, t_2,\cdots, t_r$ *of* $H^{[U]}$. *Then* $m\leqslant r\leqslant mq$, *and there exists a sequence* $(\boldsymbol{B}_s)_{s=1}^m$ *of nonnegative Hermitian* $q\times q$ *matrices unlike the zero matrix such that*

$$\begin{aligned}\boldsymbol{y}^*\boldsymbol{B}_s\Big(\sum_{k=0}^{m-1}(P_k(t_s))^*P_k(t_s)\Big)\boldsymbol{x}&=\boldsymbol{y}^*\Big(\sum_{k=0}^{m-1}(P_k(t_s))^*P_k(t_s)\Big)\boldsymbol{B}_s\boldsymbol{x}\\&=A(s)\boldsymbol{y}^*\boldsymbol{x},\ \forall\,\boldsymbol{x},\boldsymbol{y}\in\text{range }(\boldsymbol{B}_s)\end{aligned}$$

holds for $s=1, 2,\cdots, r$, rank $\boldsymbol{B}_s$ *is equal to the multiplicity of* t_s *as a zero of* $H^{[U]}$, $1\leqslant s\leqslant r$, $\sum_{s=1}^{r}\text{rank }\boldsymbol{B}_s=mq$ *and* $G^{[U]}$ *defined by* (3. 14) *admits the representation*

$$G^{[U]}(z)=\mathrm{i}\ \mathrm{Im}\ \boldsymbol{D}_0+\sum_{s=1}^{r}\frac{t_s+z}{t_s-z}\boldsymbol{B}_s,\ \forall\,z\in D.$$

Furthermore, $G^{[U]}(z)=G_{t_j}^h(z)$, $\forall\,z\in D$, *if* $\boldsymbol{B}_j\in\mathbf{C}^{q\times q}$ *is positive Hermitian, where* $1\leqslant j\leqslant r$. *In particular, in the case of* $r=m$, $G^{[U]}=G_{t_s}^h$, $s=1, 2,\cdots,m$, *and it admits the representation*

$$G^{[U]}(z)=\mathrm{i}\ \mathrm{Im}\ \boldsymbol{D}_0+\sum_{k=0}^{m-1}\frac{t_s+z}{t_s-z}A(t_s)\Big(\sum_{k=0}^{m-1}(P_k(t_s))^*P_k(t_s)\Big)^{-1},\forall\,z\in D,$$

where $A(z)$ *is as in* (3. 5), $A(t_s)>0$ *for* $s=1, 2,\cdots,m$; *in the case of* $r=mq$, $G^{[U]}$ *admits the representation*

$$G^{[U]}(z)=\mathrm{i}\ \mathrm{Im}\ \boldsymbol{D}_0+\sum_{s=1}^{mq}\frac{t_s+z}{t_s-z}\boldsymbol{B}_s,\forall\,z\in D,$$

where each $\boldsymbol{B}_s$ *is certain nonnegative Hermitian* $q\times q$ *matrix of rank* 1.

Furthermore, in general, for each $t\in T$, *there exists one and only* *one solution G of (nondegenerate) Problem (NP) such that its Riesz-Herglotz measure σ has a matrix mass at t equal to the maximum matrix mass* $A(t)\left(\sum_{k=0}^{m-1}(P_k(t))^* P_k(t)\right)^{-1}$. *This solution is* G_t^h.

As a direct consequence of Theorem 3. 8, in the scalar case of $q=1$, $G^{[U]}$ has exactly $t_1, t_2, \cdots, t_m$ as m simple poles, which are located on T, and admits the representation, for each $z \in D$,

$$G^{[U]}(z) = \mathrm{i}\ \mathrm{Im}\ D_0 + \sum_{s=1}^{m} \frac{t_s + z}{t_s - z} B_s = G_{t_j}^h(z),\ j = 1, 2, \cdots, m,$$

where each B_s is the positive number, and satisfies

$$B_s = A(t_s)\left(\sum_{k=0}^{m-1}(P_k(t_s))^* P_k(t_s)\right)^{-1}.$$

In conclusion, we consider the counterpart of Corollary 2. 9 in (nondegenerate) Problem (NP) as follows.

Corollary 3. 9 *Let* $w \in T$, *let* G_w^h *be as in* (3. 17) *for each* $z \in D$, *and* $\boldsymbol{U} := \boldsymbol{U}_w$ *as in* (2. 19), *and let* σ_w *be the Riesz-Herglotz measure of* G_w^h. *Then the following statements are equivalent*:

(a) *For each* $u \in T$, *the matrix mass* $\sigma_w(\{u\})$ *is either a positive Hermitian* $q \times q$ *matrix or the zero matrix.*

(b) *For each* $u \in T$, *the matrix* $H^{[U]}(u)$ *is either the zero matrix or a nonsingular* $q \times q$ *matrix.*

(c) *For some positive integer* r, *some sequence* $(t_s)_{s=1}^r$ *of points on* T *and some sequence* $(\boldsymbol{B}_s)_{s=1}^m$ *of positive Hermitian* $q \times q$ *matrices*, G_w^h *admits the representation*

$$G_w^h(z) = \mathrm{i}\ \mathrm{Im}\ \boldsymbol{D}_0 + \sum_{s=1}^{r} \frac{t_s + z}{t_s - z} \boldsymbol{B}_s,\ \forall z \in D. \tag{3. 19}$$

(d) G_w^h *satisfies* (3. 19) *for each* $z \in D$ *with* $r = m$, *some sequence* $(t_s)_{s=1}^m$ *of pairwise different points on* T, *and the particular choice*

$$\boldsymbol{B}_s = A(t_s)\left(\sum_{k=0}^{m-1}(P_k(t_s))^* P_k(t_s)\right)^{-1}, s = 1, 2, \cdots, m.$$

(e) G_w^h *satisfies* (3. 19) *for each* $z \in D$ *with* $r = m$, *some sequence* $(t_s)_{s=1}^m$ *of points on* T *and some sequence* $(\boldsymbol{B}_s)_{s=1}^m$ *of nonnegative Hermitian* $q \times q$ *matrices.*

(f) *There exists* $G \in \mathbf{C}_\Delta$ *of the form*

$$G(z) = \mathrm{i}\ \mathrm{Im}\ \boldsymbol{D}_0 + \sum_{s=1}^{m} \frac{t_s + z}{t_s - z} \boldsymbol{B}_s,$$

where $t_1 = w$, $t_2, \cdots, t_m \in T$, *and* $\boldsymbol{B}_1, \boldsymbol{B}_2, \cdots, \boldsymbol{B}_m$ *are nonnegative Hermitian* $q \times q$ *matrices.*

(g) *There exist exactly* m *pairwise different points* $t_1 = w$, $t_2, \cdots, t_m$ *on* T *such that* $H^{[U]}(t_j) = \boldsymbol{O}$ *for* $j = 1, 2, \cdots, m$.

(h) *There exist exactly* m *pairwise different points* $t_1 = w$, $t_2, \cdots, t_m$ *on* T *such that each* $\sigma_w(\{t_j\})$ *is positive Hermitian* $q \times q$ *matrix for* $j = 1, 2, \cdots, m$.

It is worth noting that if for each $z \in \mathbf{C}$, $A(z) = 1$ and $W(z) = \boldsymbol{O}$, Problem (C) may be considered as an extreme case of Problem (NP) when $\tau_0 = m > 0$, $z_0 = 0$ and $\rho = 0$ in the data Δ (see (3.1)), and furthermore, Theorems 3.5, 3.7 and 3.8 and Corollaries 3.6 and 3.9 may be viewed as the equivalent statements and generalizations of Theorems 2.3, 2.6 and 2.8 (together with Theorem 2.7) and Corollaries 2.4 and 2.9 given in Section 2, respectively.

Acknowledgements

The authors thank the referee for helpful comments and for drawing our attention to the article in [22] by A. Lasarow. This work was supported by the National Natural Science Foundation of China (No. 11071017) and the Program for New Century Excellent Talents in University.

References

[1] N I Akhiezer. The Classical Moment Problem and Some Related Questions in Analysis. Oliver and Boyd, London, 1965.

[2] D Z Arov and M G Kreĭn. Problems of the search of minimum of entropy in indeterminate extension problems. Funct. Anal. Appl., 1981, 15: 123—126.

[3] J A Ball, I Gohberg, and L. Rodman. Interpolation of Rational Matrix

Functions. Operator Theory: Advanced Application, Vol. 45, Birkhäuser, Basel, 1990.

[4] V Bolotnikov and H Dym. On boundary interpolation for matrix valued Schur functions. Mem. Amer. Math. Soc., 2006, 181: 856.

[5] G N Chen and Y J Hu. On the multiple Nevanlinna—Pick matrix interpolation in the class C_p and the Carathéodory matrix coefficient problem. Linear Algebra Appl., 1998, 283: 179—203.

[6] G N Chen and Y J Hu. The Nevanlinna—Pick Interpolation and the Matrix Moment Problem. Beijing Normal University Press, Beijing, 2003 (Chinese).

[7] P Delsarte, Y Genin, and Y Kamp. Orthogonal polynomial matrices on the unit circle. IEEE Trans. Circuits Syst., 1978, CAS 25: 149—160.

[8] P Delsarte, Y Genin, and Y Kamp. Schur parametrization of positive definite block-Toeplitz systems. SIAM J. Appl. Math., 1979, 36: 34—46.

[9] V K Dubovoj, B Fritzsche, and B Kirstein. Matricial Version of the Classical Schur Problem. Teubner-Texte zur Mathematik, Vol. 129, Teubner, Leipzig, 1992.

[10] A J Duran and P Lopez-Rodriguez. N-extremal matrices of measures for an indeterminate matrix moment problem. J. Funct. Anal., 2000:174, 301—321.

[11] A J Duran and P Lopez-Rodriguez. The matrix moment problem. in Margarita Mathematica EN Memoria De José Javier (Chicho) Guadalupe Hernández, Luis Y. Español and Juan L. Varona, eds., Servicio de Publicaciones, Universitat de La Rioja, Logrono, Spain, 2001:333—348.

[12] H Dym. J-contractive Matrix Functions, Reproducing Kernel Hilbert Spaces and Interpolation. Regional Conference Series in Mathematics, Vol. 71, AMS, Providence, RI, 1989.

[13] C Foias and A E Frazho. The Commutant Lifting Approach to Interpolation Problems. Operator Theory: Advanced Application, Vol. 44. Birkhäuser, Basel, 1990.

[14] B Fritzsche and B Kirstein. An extension problem for nonnegative Hermitian block Toeplitz matrices, Part I, Math. Nachr, 1987, 130: 121—135, Part II, Math. Nachr., 1987, 131: 287 — 297; Part III, Math. Nachr. 1988, 135: 319—341; Part IV, Math. Nachr., 1989, 143: 329—354; Part

V, Math. Nachr., 1989, 144:283—308.

[15] B Fritzsche, B Kirstein, and A Lasarow. The matricial Carathéodory problem in both nondegenerate and degenerate cases. in Interpolation, Schur Functions and Moment Problems, D. Alpay and I. Gohberg, eds., Operator Theory: Advanced Application, Vol. 165, Birkhäuser, Basel, 2006: 251—290.

[16] B Fritzsche, B Kirstein, and A Lasarow. On a class of extremal solutions of the nondegenerate matricial Carathéodory problem. Analysis (Munich), 2007,27:109—164.

[17] B Fritzsche, B Kirstein, and A Lasarow. On the Weyl Matrix Balls Corresponding to the Matricial Carathéodory Problem in both Nondegenerate and Degenerate Cases, Operator Theory: Advanced Application, Vol. 199, Birkhäuser, Basel, 2010: 289—332.

[18] Ja L Geronimus. Polynomials Orthogonal on a Circle and Their Applications. AMS Translations (English translation), Series 1, Vol. 3, AMS, Providence, RI, 1962: 1—78.

[19] I Gohberg, P Lancaster, and L Rodamn. Matrix Polynomials. Academic Press, New York, 1982.

[20] I V Kovalishina. Analytic theory of a class of interpolation problems. (Russian) Izv. Akad. Nauk SSSR, Ser. Mat., 1983: 47: 455—497.

[21] A Lasarow. Solution of a multiple Nevanlinna-Pick problem for Carathéodory functions via orthogonal rational functions in the matrix case. Math. Nachr., 2008,281:75—108.

[22] A Lasarow. On maximal weight solutions in a truncated trigonometric matrix moment problem. Funct. Approx. Comment. Math., 2010, 43:117—128.

[23] L A Sakhnovich. Interpolation Theory and its Applications. Mathematics and its Applications, Vol. 428, Kluwer, Dordrecht, 1997.

[24] B Simon. Orthogonal Polynomials on the Unit Circle, Part 1: Classical Theory. AMS, Providence, RI, 2004.

附录

Appendix

论文和著作目录

Bibliography of Papers and Works

论文目录

[序号] 作者. 论文题目. 杂志名称,年份,卷(期):起页一止页.

[1]陈公宁. 马绍方程及其在质谱学中的应用. 分析仪器,1974,(1):1—17.

[2] 陈公宁,方寿生. 布希涅斯克 (Boussinesq) 问题的一种推广. 北京师范大学学报(自然科学版),1978,(4): 6—24.

[3]陈公宁. 二阶矩阵概述. 中学理科教学,1979,(2):1—7;(4):1—5.

[4]王伯英,陈公宁. G-函数的行列式界限及矩阵非异性定理. 北京师范大学学报(自然科学版),1979,(3):41—49.

[5]陈公宁. 关于 Vitali 收敛定理的注记. 北京师范大学学报(自然科学版),1983,(3):17—22.

[6]Chen Gongning. On some properties of singular matrices. Linear Algebra and Its Applications,1983,49:137—151.

[7]Chen Gongning. Note on lower bounds for the rank of a matrix. Linear Algebra and Its Applications,1983,55:125—132.

[8]Chen Gongning. Some generalizations of diagonal dominance associated with G-functions. Linear Algebra and Its Applications,1983,55:169—180.

[9]陈公宁. 矩阵的正则性的若干条件. 北京师范大学学报(自然科学版),1984,(1):1—8.

[10]Chen Gongning. Iteration for holomorphic maps of the open unit

ball and the generalized upper half-plane of $\mathbf{C}^n$. Journal of Mathematical Analysis and Applications，1984，98(2)：305－313.

[11]陈公宁. 关于复矩阵秩的下界的注记. 数学学报，1985，28(1)：16－22.

[12]陈公宁. 关于 Hilbert 空间上算子解析函数的若干结果. 北京师范大学学报(自然科学版)，1985，(4)：14－22.

[13]Chen Gongning. Iteration of holomorphic maps in Hilbert spaces. Journal of Mathematical Analysis and Applications，1985，112(1)：274－278.

[14]陈公宁. Banach 代数中对谱半径的 Schwarz 引理. 北京师范大学学报(自然科学版)，1987，(1)：13－17.

[15]陈公宁. 关于 Banach 代数上的全纯函数的若干结果. 北京师范大学学报(自然科学版)，1987，(4)：9－12.

[16]Chen Gongning. On simplified proofs of theorems of von Neumann, Heinz, and Ky Fan, and their extended versions. Nonlinear and Convex Analysis (Proceedings in honor of Ky Fan，1985)，Lecture Notes in Pure and Appl. Math.，107，Dekker，New York，1987：15－21.

[17]陈公宁. 弱半正矩阵与弱对角稳定矩阵的一些特性. 数学杂志，1988，8(3)：263－268.

[18]陈公宁. 关于 von Neumann-Heinz 定理与樊墉定理的推广. 数学年刊，1990，11A(3)：314－320.

[19]陈公宁. 多项式零点定位基本定理的简化证明. 北京师范大学学报(自然科学版)，1990，26(4)：14－18.

[20]陈公宁. 关于矩阵惯性的若干基本定理. 北京师范大学学报(自然科学版)，1991，27(2)：152－157.

[21]杨淑蓉，张惠品，陈公宁. Bezout 和 Hankel 矩阵. 北京师范大学学报(自然科学版)，1991，27(增刊 2)：20－26.

[22]Chen Gongning. Bezoutian representation via Vandermonde matrices. 北京师范大学学报(自然科学版)，1991，27(3)：307－308.

[23]陈公宁. 对"关于半正定 Hermite 矩阵乘积迹的一个不等式"一文的注记. 数学学报，1992，35(5)：620－622.

[24]陈公宁,张惠品. 奇异 Hankel 矩阵(研究快报). 北京师范大学学报(自然科学版),1992,28(3):292－294.

[25]陈公宁,赵斌. 一般有理插值问题：Loewner 与 Hankel 矩阵解法(研究快报). 北京师范大学学报(自然科学版),1992,28(3):371－373.

[26] 陈公宁,张惠品. Bezout 和 Hankel 矩阵(Ⅱ)—非奇异情形. 北京师范大学学报(自然科学版),1992,28(4):438－445.

[27]陈公宁,张惠品. 一般有理插值问题的完全解. 北京师范大学学报(自然科学版),1993,29(4):465－466.

[28] Chen Gongning, Yang Zhenghong. Bozoutian representation via Vandermonde matrices. Linear Algebra and Its Applications,1993,186:37－44.

[29]张惠品,陈公宁. Hankel 矩阵的生成函数. 北京师范大学学报(自然科学版),1994,30(1):32－40.

[30]Chen Gongning, Zhang Huipin. More on Loewner matrices. Linear Algebra and Its Application,1994,203/204:265－300.

[31] Chen Gongning, Zhang Huipin. Note on products of Bezoutians and Hankel matrices. Linear Algebra and Its Applications,1995,225:23－35.

[32]胡永建,陈公宁. 有关实正定矩阵的一些性质. 北京师范大学学报(自然科学版),1996,32(1):40－46.

[33]Chen Gongning, Zhao Bin, Zhang Huipin. The general rational interpolation problem in the scalar case and its Hankel vector. Linear Algebra and Its Applications, 1996,244:165－201.

[34]陈公宁,胡永建. 广义块 Pick 型矩阵和带导数的 Nevanlinna-Pick 矩阵插值问题. 北京师范大学学报(自然科学版),1998,34(3):305－312.

[35]白庆余,胡永建,陈公宁. 一类带重点的退化 Nevanlinna-Pick 插值问题. 北京师范大学学报(自然科学版),1998,34(4):456－461.

[36]Chen Gongning. The general rational interpolation problem and its connection with the Nevanlinna-Pick interpolation and power moment problem. Linear Algebra and Its Applications,1998,273:83－117.

[37]Chen Gongning, Hu Yongjian. The truncated Hamburger matrix moment problem in the nondegenerate and degenerate cases, and matrix continued fractions. Linear Algebra and Its Applications, 1998,277(1－3):199－236.

[38]Chen Gongning, Hu Yongjian. On the multiple Nevanlinna-Pick matrix interpolation in the class $\mathbf{C}_p$ and the Carathéodory matrix coefficient problem. Linear Algebra and Its Applications, 1998, 283(1－3): 179－203.

[39]赵斌，陈公宁. 广义 Loewner 矩阵的核与因子分解. 北京师范大学学报(自然科学版),1999,35(2):147－153.

[40]胡永建,陈公宁. 退化的矩阵 SM 问题与 NP(S_p)问题. 北京师范大学学报(自然科学版),1999,35(4):431－436.

[41]Chen Gongning, Hu Yongjian. Inversion of a generalized block Loewner matrix, the minimal partial realization, and matrix rational interpolation problem. Linear Algebra and Its Applications, 1999,286(1－3): 223－259.

[42]Chen Gongning, Li Xiaoqing. The Nevanlinna-Pick interpolation problems and power moment problems for matrix-valued functions. Linear Algebra and Its Applications,1999,288(1－3):123－148.

[43] 胡永建，郝辉,陈公宁. 一类 Nevanlinna-Pick 问题的可解性准则. 北京师范大学学报(自然科学版),2000,36(2):166－171.

[44]陈公宁，李小青. 对矩阵值函数的 Nevanlinna-Pick 问题与幂矩量问题(Ⅱ). 北京师范大学学报(自然科学版),2000,36(5):588－596.

[45] Hu Yongjian, Chen Gongning. Matrix versions of interpolation problems of Akhiezer type. 北京师范大学学报(自然科学版), 2000, 36(6):722－729.

[46]Chen Gongning, Hu Yongjian. The Nevanlinna-Pick interpolation problems and power moment problems for matrix-valued functions Ⅲ. Linear Algebra and Its Applications,2000,306(1－3):59－86.

[47]Yang Zhenghong, Chen Gongning. Generalized-confluent Cauchy and Cauchy-Vandermonde matrices. Linear Algebra and Its Appli-

cations,2000，308(1—3):31—64.

[48]吴化璋，陈公宁. 关于无限广义块 Toeplitz ＋ Hankel 矩阵的求逆. 北京师范大学学报(自然科学版),2001,37(1):10—15.

[49]杨正宏，胡永建，陈公宁. 广义柯西与柯西-范德蒙块矩阵的求逆公式. 北京师范大学学报(自然科学版),2001,37(3):307—311.

[50]杨正宏,陈公宁. 带重点的多项式 Vandermonde 型矩阵位移结构与快速求逆公式. 北京师范大学学报(自然科学版),2001,37(6):726—731.

[51]Chen Gongning，Hu Yongjian. A unified treatment for the matrix Stieltjes moment problem in both nondegenerate and degenerate cases. Journal of Mathematics Analysis and Applications,2001,254(1):23—34.

[52]Chen Gongning，Hu Yongjian. Multiple Nevanlinna-Pick interpolation with both interior and boundary data and its connection with the power moment problem. Linear Algebra and Its Applications,2001,323(1—3):167—194.

[53]陈公宁. 近世代数教育家张禾瑞. 见:王淑芳,邵红英主编:师范之光. 北京师范大学出版社,2002:462—466.

[54]吴化璋，陈公宁. N[a,b] 类中 Nevanlinna-Pick 插值与 Hausdorff 矩量问题. 北京师范大学学报(自然科学版),2002,38(3):316—323.

[55]吴化璋，陈公宁. S[a,b] 类中 Nevanlinna-Pick 插值与 Hausdorff 矩量问题. 北京师范大学学报(自然科学版),2002,38(6):720—728.

[56]吴化璋，陈公宁,杨尚骏. 无限广义块 Toeplitz 和 Hankel 矩阵求逆的统一方法. 工科数学，2002,18(1):9—16.

[57]杨正宏，陈公宁,胡永建. 带重点的广义 Cauchy 矩阵及其相关问题. 数学学报，2002,45(5):1 017—1 024.

[58]胡永建,郝辉,陈公宁. 一类 Nevanlinna-Pick 问题的 Toeplitz 向量方法. 数学物理学报,2002,22A(3):310—315.

[59]Chen Gongning，Jia Lixin. Boundary Nevanlinna-Pick interpolation for meromorphic functions. Acta Mathematica Applicatae Sinica, English Series,2002,18(2)：255—262.

[60]梁燕来，杨正宏，陈公宁. 关于一般多项式基的 Bezout 矩阵的若干性质. 北京师范大学学报(自然科学版),2003,39(1):14—20.

[61] 张晓南，胡永建，陈公宁. 切矩阵多项式插值的埃尔米特公式. 北京师范大学学报(自然科学版),2003,39(1):41—46.

[62]杨正宏，胡永建，陈公宁. q-adic 范德蒙型矩阵的位移结构与逆公式. 北京师范大学学报(自然科学版),2003,39(2):153—158.

[63]赵斌,陈公宁. Hermite 型 Toeplitz 矩阵的惯性. 北京师范大学学报(自然科学版),2003,39(2):159—164.

[64]Hu Yongjian, Chen Gongning. A unified treatment for the matrix Stieltjes moment problem, Linear Algebra and Its Applications, 2004,380(2):227—239.

[65]Hu Yongjian, Yang Zhenghong, Chen Gongning. Tangential Nevanlinna-Pick interpolation and its connection with Hamburger matrix moment problem. Integral Equations and Operator Theory, 2004,49(4):445—460.

[66]Hu Yongjian, Yang Zhenghong, Chen Gongning. Generalized Hermite formula for the two-sided Lagrange-Sylvester interpolation. Journal of Computational and Applied Mathematics, 2005,173(2):345—358.

[67]秦建国,陈公宁,何红亚. Cauchy 矩阵及其相关的插值问题. 数学的实践与认识,2006,36(9):233—237.

[68]Qin Jianguo, Tong Yunzhan, Chen Gongning. Determinants of Cauchy and Loewner matrices. 工程数学学报,2006,23(4):713—716.

[69]Hu Yongjian, Boubakar K d A, Chen Gongning. On boundary Nevanlinna-Pick interpolation for Carathéodory matrix functions. Linear Algebra and Its Applications, 2007,423(2—3):209—229.

[70]Hu Yongjian, Wei Wei, Chen Gongning. Boundary Nevanlinna-Pick interpolation for Nevanlinna matrix functions and the related Hamburger matrix moment problem. Analysis (Munich),2009,29(1):41—69.

[71]Hu Yongjian, Chen Gongning. On rank variation of block matrices

generated by Nevanlinna matrix functions. Mathematische Nachrichten, 2009,282(4): 611—631.

[72]陈公宁,秦建国. 完全不确定 Hamburger 矩阵矩量问题的有限阶解. 数学物理学报,2010,30A(3):577—583.

[73]陈公宁, 岳晓鹏, 孟晓然. 关于 Sylvester 惯性律的一种变型. 数学的实践与认识, 2010,40(17):233—235.

[74] Hu Yongjian, Song Yanping, Chen Gongning. A multiple point Nevanlinna-Pick type interpolation problem with both interior and boundary data for Carathéodory matrix-valued functions. Mathematische Nachrichten, 2011,284(10):1 304—1 328.

[75] Chen Gongning, Hu Yongjian. Minimum w-entropy interpolants for matricial Carathéodory functions and maximum determinant completions of associated block Pick matrix. Linear Algebra and Its Applications,2011,434(8): 1 851—1 878.

[76]陈公宁. 樊瓛. 见:钱伟长总主编,王元数学分卷主编:20 世纪中国知名科学家学术成就概览,数学卷,第 1 分册. 科学出版社,2011, 314—325.

[77]Hao Huifeng, Hu Yongjian, Chen Gongning. Note on roots location of a symmetric polynomial with respect to the imaginary axis. Linear and Multilinear Algebra, 2012,60(4):499—510.

[78] 贺勤,徐清舟,陈公宁. 矩阵值 Carathéodory 函数广义块 Pick 矩阵的秩不变性. 数学物理学报,2012,32A(1):103—112.

[79]Hu Yongjian, Zhan Xuzhou, Chen Gongning. Note on extremal solutions of the truncated trigonometric matrix moment problem. Proceedings of the Seventh Workshop on Matrices and Operators, Harbin, China, 2012:167—170.

[80] Chen Gongning, Hu Yongjian. Extremal solutions of matricial Carathéodory problem and Nevanlinna - Pick -type interpolation for Carathéodory matrix functions in the nondegenerate case. Linear and Multilinear Algebra, 2012,60(11—12):1 271—1 294.

著作

[序号] 著者. 书名. 出版地:出版社, 出版年份

[1]陈公宁, 沈嘉骥. 计算方法导引, 第 1 版. 北京:北京师范大学出版社, 1988; 第 2 版. 2000; 第 3 版. 2009.

[2]陈公宁. 矩阵理论与应用,第 1 版. 北京:高等教育出版社, 1990; 第 2 版,现代数学基础丛书第 115 卷. 北京:科学出版社,2007.

[3]陈公宁,沈嘉骥. 计算方法. 北京:高等教育出版社,2002.

[4]陈公宁,胡永建. Nevanlinna-Pick 插值和矩量问题. 北京:北京师范大学出版社, 2003.

后记

Postscript by the Chief Editor

北京师范大学数学科学学院系统地开展几个现代数学方向的科学研究，是从王世强老师在20世纪50年代初进行数理逻辑的研究工作，发表了一批论文开始的. 50年代中期，严士健老师在华罗庚教授的指导下，进行了环上的线性群、辛群的自同构的研究，首次得到了它们的完整形式，还用自己提出的方法得到了 n 阶模群的定义关系. 刘绍学老师于1956年在莫斯科大学获得了副博士学位，他对结合环、李环、若当环和交错环做了统一的处理，获得完整的结果. 回国以后，在国内带动了环论的研究. 1958年2月，孙永生老师在莫斯科大学完成了他的副博士学位论文《关于乘子变换下的函数类利用三角多项式的最佳逼近》，结果深刻，当时受到数学界前辈陈建功的称赞. 回国后，又解决了余下的困难问题. 这4位老师在数学科学学院被戏称为“四大金刚”. 正是由于他们开创性的工作，使得数学科学学院的科学研究逐渐形成了一支具有相当学术素养的队伍；有一批确定的研究方向，形成了自己的风格和传统；获得了丰富而系统的、达到世界学科前沿的科研成果，其中有一些已经达到世界先进水平；在国内具有一定的学术地位，在国际上有一定的知名度；对国家的数学发展做出了一定的贡献. 之所以能取得这样的成绩，是经过几代人的探索和努力，遭受了诸多的困苦和磨难. 1984年5月29日，王梓坤老师被国务院任命为北京师范大学校长并到学院工作，大大加强了学院概率论学科的力量. 这5位老师均是学院的学术带头人，为学院的学科建设和人才培养，花费了毕生的精力，做出了重大的贡献. 5位老师均在1981年被批准为首批博士生导师(我校理科首批博士生导师还有5位：黄祖洽、刘若庄、陈光旭、汪堃仁和周廷儒老师)，此次批准的博士生导师的数量，提高了学院在学校中的地位，且此举对学院在全国数学界的地位奠定了重要基础，开创了近30多年来的良好局面．由于5位老师在学院学科建设中的重要地位和学术贡献，将他们的论著整理出来，作为《北京师范大学数学家文库》系列出版，是一件意义重大的事情．在北京师范大学出版社的大力支持下，该文集系列已经在2005年由北京师范大学出版社出版．

在5部数学文集出版之后，2005年12月25日，学校隆重举行了北京师范大学数学系成立90周年庆祝大会暨王世强、孙永生、严士健、王梓坤和刘绍学教授文集首发式．5部文集的出版在国内数学界产生了很好的影响．5部文集的作者按年龄排序为：王世强、孙永生、严士健、王梓坤和刘绍学，除了王世强教授在1927年出生外，其余4位均在1929年出生，广泛一点，在20世纪20年代出生．考虑学院在30年代出生的博士生导师们，按批准为博士生导师先后的顺序为：陆善镇、汪培庄、王伯英、李占柄、刘来福、陈公宁、罗里波．由于他们出生在1936～1939年，按年龄从大到小排序为：李占柄、罗里波、汪培庄、王伯英、刘来福、陈公宁和陆善镇．他们均为学院的发展和建设作出了重要贡献，出版他们的文集是学院的基本建设．因此，学院将在近几年内陆续出版他们的文集，并由院党委书记李仲来教授任主编．《文集》的结构为：照片、序、论文选、发表的论文和著作目录、后记．

《陈公宁文集》是这套文库的第14部．该文集的出版得到了北京师范大学出版社的大力支持，在此表示衷心的感谢．

华罗庚教授说："一个人最后余下的就是一本选集．"（龚升论文选集，中国科学技术大学出版社，2008）这些选集的质量反映了我们学院某一学科，或几个学科，或学科群的整体学术水平．而将北京师范大学数学科学学院著名数学家、数学教育家和科学史专家论文进行整理和选编出版，是学院学科建设的一项重要的和基础性的工作，是学院的基本建设之一．它对提高学院的知名度和凝聚力，激励后人，有着重要的示范作用．当然，这项工作还在继续做下去，搜集和积累数学科学学院各种资料的工作还在继续进行．

主编　李仲来

2013-01-16